Digital Image Processing for Remote Sensing

OTHER IEEE PRESS BOOKS

Reflector Antennas, *Edited by A. W. Love*
Phase-Locked Loops & Their Application, *Edited by W. C. Lindsey and M. K. Simon*
Digital Signal Computers and Processors, *Edited by A. C. Salazar*
Systems Engineering: Methodology and Applications, *Edited by A. P. Sage*
Modern Crystal and Mechanical Filters, *Edited by D. F. Sheahan and R. A. Johnson*
Electrical Noise: Fundamentals and Sources, *Edited by M. S. Gupta*
Computer Methods in Image Analysis, *Edited by J. K. Aggarwal, R. O. Duda, and A. Rosenfeld*
Microprocessors: Fundamentals and Applications, *Edited by W. C. Lin*
Machine Recognition of Patterns, *Edited by A. K. Agrawala*
Turning Points in American Electrical History, *Edited by J. E. Brittain*
Charge-Coupled Devices: Technology and Applications, *Edited by R. Melen and D. Buss*
Spread Spectrum Techniques, *Edited by R. C. Dixon*
Electronic Switching: Central Office Systems of the World, *Edited by A. E. Joel, Jr.*
Electromagnetic Horn Antennas, *Edited by A. W. Love*
Waveform Quantization and Coding, *Edited by N. S. Jayant*
Communication Satellite Systems: An Overview of the Technology, *Edited by R. G. Gould and Y. F. Lum*
Literature Survey of Communication Satellite Systems and Technology, *Edited by J. H. W. Unger*
Solar Cells, *Edited by C. E. Backus*
Computer Networking, *Edited by R. P. Blanc and I. W. Cotton*
Communications Channels: Characterization and Behavior, *Edited by B. Goldberg*
Large-Scale Networks: Theory and Design, *Edited by F. T. Boesch*
Optical Fiber Technology, *Edited by D. Gloge*
Selected Papers in Digital Signal Processing, II, *Edited by the Digital Signal Processing Committee*
A Guide for Better Technical Presentations, *Edited by R. M. Woelfle*
Career Management: A Guide to Combating Obsolescence, *Edited by H. G. Kaufman*
Energy and Man: Technical and Social Aspects of Energy, *Edited by M. G. Morgan*
Magnetic Bubble Technology: Integrated-Circuit Magnetics for Digital Storage and Processing, *Edited by H. Chang*
Frequency Synthesis: Techniques and Applications, *Edited by J. Gorski-Popiel*
Literature in Digital Processing: Author and Permuted Title Index (Revised and Expanded Edition), *Edited by H. D. Helms, J. F. Kaiser, and L. R. Rabiner*
Data Communications via Fading Channels, *Edited by K. Brayer*
Nonlinear Networks: Theory and Analysis, *Edited by A. N. Willson, Jr.*
Computer Communications, *Edited by P. E. Green, Jr. and R. W. Lucky*
Stability of Large Electric Power Systems, *Edited by R. T. Byerly and E. W. Kimbark*
Automatic Test Equipment: Hardware, Software, and Management, *Edited by F. Liguori*
Key Papers in the Development of Coding Theory, *Edited by E. R. Berlekamp*
Technology and Social Institutions, *Edited by K. Chen*
Key Papers in the Development of Information Theory, *Edited by D. Slepian*
Computer-Aided Filter Design, *Edited by G. Szentirmai*
Laser Devices and Applications, *Edited by I. P. Kaminow and A. E. Siegman*
Integrated Optics, *Edited by D. Marcuse*
Laser Theory, *Edited by F. S. Barnes*
Digital Signal Processing, *Edited by L. R. Rabiner and C. M. Rader*
Minicomputers: Hardware, Software, and Applications, *Edited by J. D. Schoeffler and R. H. Temple*
Semiconductor Memories, *Edited by D. A. Hodges*
Power Semiconductor Applications, Volume II: Equipment and Systems, *Edited by J. D. Harnden, Jr. and F. B. Golden*
Power Semiconductor Applications, Volume I: General Considerations, *Edited by J. D. Harnden, Jr. and F. B. Golden*
A Practical Guide to Minicomputer Applications, *Edited by F. F. Coury*
Active Inductorless Filters, *Edited by S. K. Mitra*
Clearing the Air: The Impact of the Clean Air Act on Technology, *Edited by J. C. Redmond, J. C. Cook, and A. A. J. Hoffman*

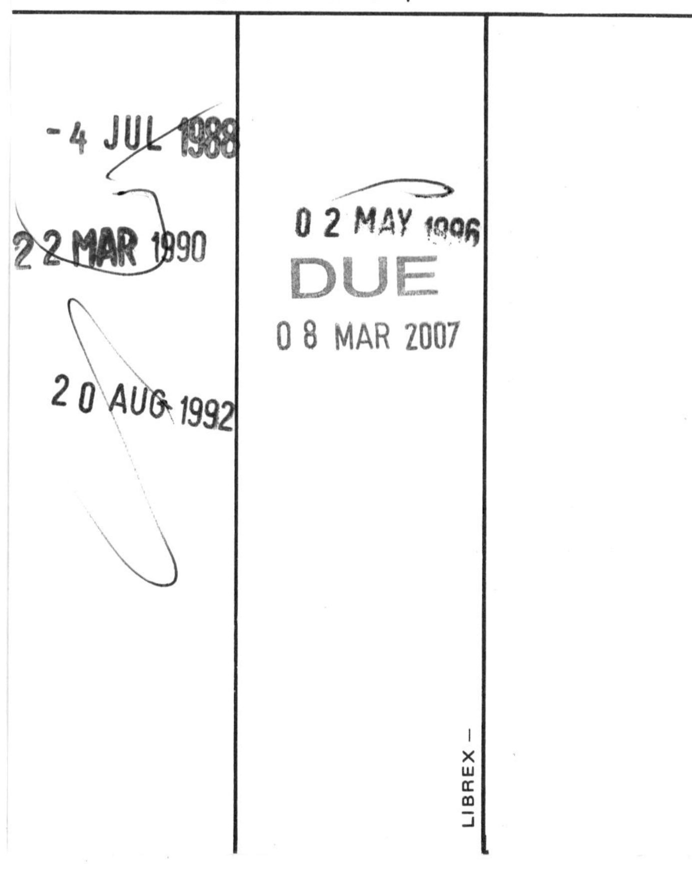
This book is to be returned on or before
the last date stamped below.
-4 JUL 1988
22 MAR 1990
20 AUG 1992
02 MAY 1996
DUE
08 MAR 2007
LIBREX

ANDERSONIAN LIBRARY
WITHDRAWN
FROM
LIBRARY
STOCK
UNIVERSITY OF STRATHCLYDE

Digital Image Processing for Remote Sensing

Edited by

Ralph Bernstein
Senior Engineer and Manager
Advanced Image Processing Analysis and Development Department
Federal Systems Division
IBM Corporation

Associate Editors
Paul E. Anuta, Associate Program Leader, Laboratory for Applications of Remote Sensing, Purdue University
Ruzena Bajcsy, Associate Professor, Computer and Information Science Department, University of Pennsylvania
Raul Hunziker, Senior Engineer, Federal Systems Division, IBM Corporation
Azriel Rosenfeld, Research Professor, Computer Science Center, University of Maryland

A volume in the IEEE PRESS Selected Reprint Series, prepared under the sponsorship of the IEEE Geoscience Electronics Society.

The Institute of Electrical and Electronics Engineers, Inc. New York

PRINTED IN THE UNITED STATES OF AMERICA

IEEE International Standards Book Numbers: Clothbound: 0-87942-105-3
Paperbound: 0-87942-106-1

Library of Congress Catalog Card Number 77-94520

Sole Worldwide Distributor (Exclusive of the IEEE):

JOHN WILEY & SONS, INC.
605 Third Ave.
New York, NY 10016

Wiley Order Numbers: Clothbound: 0-471-04939-5
Paperbound: 0-471-04938-7

Contents

Preface and Introduction

Historical Perspective

Remote sensing, as its name implies, is the acquisition of physical data relating to an object or feature in a manner which does not involve direct contact in any way. In most cases, the term refers primarily to the sensing of the electromagnetic field. In the past, due to technology limitations, remotely sensed data were derived primarily from the visible portion of the electromagnetic spectrum using eyes and then cameras for sensors. It is interesting to note that a camera was first used to view the earth from a balloon in the 1850's. Since then, there has been impressive progress in sensors, platforms for elevating the sensors to suitable altitudes, and systems for processing the remotely sensed data. Recently, new sensors have been developed, such as the Landsat Multispectral Scanner (MSS), that provide data in the visible and infrared portion of the spectrum in a digital form. The MSS uses an oscillating mirror to scan the earth and detectors to convert the radiance data into voltages which are digitized within the sensor system. Sensors of this type are increasingly being used to provide multispectral image data over a wide spectral region, and new linear array sensors are fast becoming operational.

The last two decades have seen an explosive growth in digital processing technology. This technology has brought about the beginning of a second industrial revolution—one that multiplies man's mental energy, as opposed to increasing his physical energy. This technology, when applied to scientific objectives, and in particular to image processing, has supported the achievement of remarkable scientific results that would not have been possible with earlier available technology. This has been particularly true in the earth, lunar, planetary, space physics, and astronomical observation programs. Concurrently, there has been impressive progress in image technology, transitioning from photography to multiband two-dimensional detector array sensors. It is anticipated that future imaging systems will merge the imaging sensors with the digital processing technology.

This book is a compilation of articles and papers dealing with a dynamic and powerful technology—digital processing of remotely sensed data. The concept is not new; original experiments were conducted shortly after digital computers first became available. However, the digital processing of remotely sensed data in a routine fashion has only recently become practical and common for a number of reasons: computers have become faster and less expensive, sensor technology has transitioned from film systems to solid-state detectors with digital readout, and many investigators have devised new and useful algorithms and techniques to convert data into information.

Digital processing of image data has been applied to many applications. They include biomedical, industrial, surveillance, and earth and space applications. The papers that have been compiled deal primarily, although not exclusively, with techniques that have been developed for the earth and space applications. The earth observation applications that have been extensively developed, and for which image processing has been successfully used, are provided in Table I. It is apparent from this table that the use of remotely sensed data coupled with innovative data processing and information interpretation results in many useful applications that benefit man and his environment.

Organization and Scope

We have attempted to select technical papers and articles that provide a foundation for digital image processing, and have organized the material in a manner that nearly parallels the flow of data from the remotely sensed image to the ultimate user of the data. With reference to Fig. 1, the parts in this book have been structured in the manner shown. This follows the flow of data from the raw, uncorrected image to derived information.

The first part, Image Restoration, deals with methods of compensating for effects which degrade the accuracy of the remotely sensed data by mathematically inverting some of the degrading phenomenon. Various degrading sources are identified, and restoration techniques to compensate for these sources are introduced. The elimination of noise and the compensation of the data for processes that have attenuated spatial frequency characteristics are developed in various papers. Methods to restore data corrupted by image motion blur, noise, and other degrading influences are developed.

Part II, Image Preprocessing and Correction, deals with geometric and radiometric correction of the sensor data and the conversion of the data into products that do not contain intensity errors and geometrically conform to a desired map projection or geometry. The implementation of these operations requires calibration processing and resampling of the image data (computing intensity values between given values). An increasingly important operation also involves mosaicking multiple images into one composite image useful for geological and hydrological analyses. These techniques and others are described in this part, and systems that have been used to implement them are also presented.

Image registration is addressed in Part III. This is a particularly important operation, in that more information can be extracted from remotely sensed data if various image sources are first geometrically registered with each other, that is, are in geometric conformance. Thus, image data in different spectral bands, acquired at different times and scales, from different types of sensors, can be spatially registered to provide data with improved information extraction potential. This part also addresses the technical aspects of cross correlation and the estimates of residual errors after data registration in the presence of noise.

Part IV addresses technology, applicable to the manual extraction of information from an image. Information extrac-

TABLE I
SUMMARY OF APPLICATIONS OF LANDSAT DATA IN THE VARIOUS EARTH RESOURCES DISCIPLINES[1]

Agriculture, Forestry, and Range Resources	Land Use and Mapping	Geology	Water Resources	Oceanography and Marine Resources	Environment
(1) Discrimination of vegetative types: Crop types, Timber types, Range vegetation	(1) Classification of land uses	(1) Recognition of rock types	(1) Determination of water boundaries and surface water area and volume	(1) Detection of living marine organisms	(1) Monitoring surface mining and reclamation
(2) Measurement of crop acreage by species	(2) Cartographic mapping and map updating	(2) Mapping of major geologic units	(2) Mapping of floods and flood plains	(2) Determination of turbidity patterns and circulation	(2) Mapping and monitoring of water pollution
(3) Measurement of timber acreage and volume by species	(3) Categorization of land capability	(3) Revising geologic maps	(3) Determination of areal extent of snow and snow boundaries	(3) Mapping shoreline changes	(3) Detection of air pollution and its effects
(4) Determination of range readiness and biomass	(4) Separation of urban and rural categories	(4) Delineation of unconsolidated rock and soils	(4) Measurement of glacial features	(4) Mapping of shoals and shallow areas	(4) Determination of effects of natural disasters
(5) Determination of vegetation vigor	(5) Regional planning	(5) Mapping igneous intrusions	(5) Measurement of sediment and turbidity patterns	(5) Mapping of ice for shipping	(5) Monitoring environmental effects of man's activities (lake eutrophication, defoliation, etc.)
(6) Determination of vegetation stress	(6) Mapping of transportation networks	(6) Mapping recent volcanic surface deposits	(6) Determination of water depth	(6) Study of eddies and waves	
(7) Determination of soil conditions	(7) Mapping of land-water boundaries	(7) Mapping landforms	(7) Delineation of irrigated fields		
(8) Determination of soil associations	(8) Mapping of wetlands	(8) Search for surface guides to mineralization	(8) Inventory of lakes		
(9) Assessment of grass and forest fire damage		(9) Determination of regional structures			
		(10) Mapping linears (fractures)			

[1] From *Mission to Earth: Landsat Views the World*, N. M. Short, P. D. Lowman, Jr., and S. C. Freden, NASA Scientific and Technical Office, 033-000-00659-4, 1976.

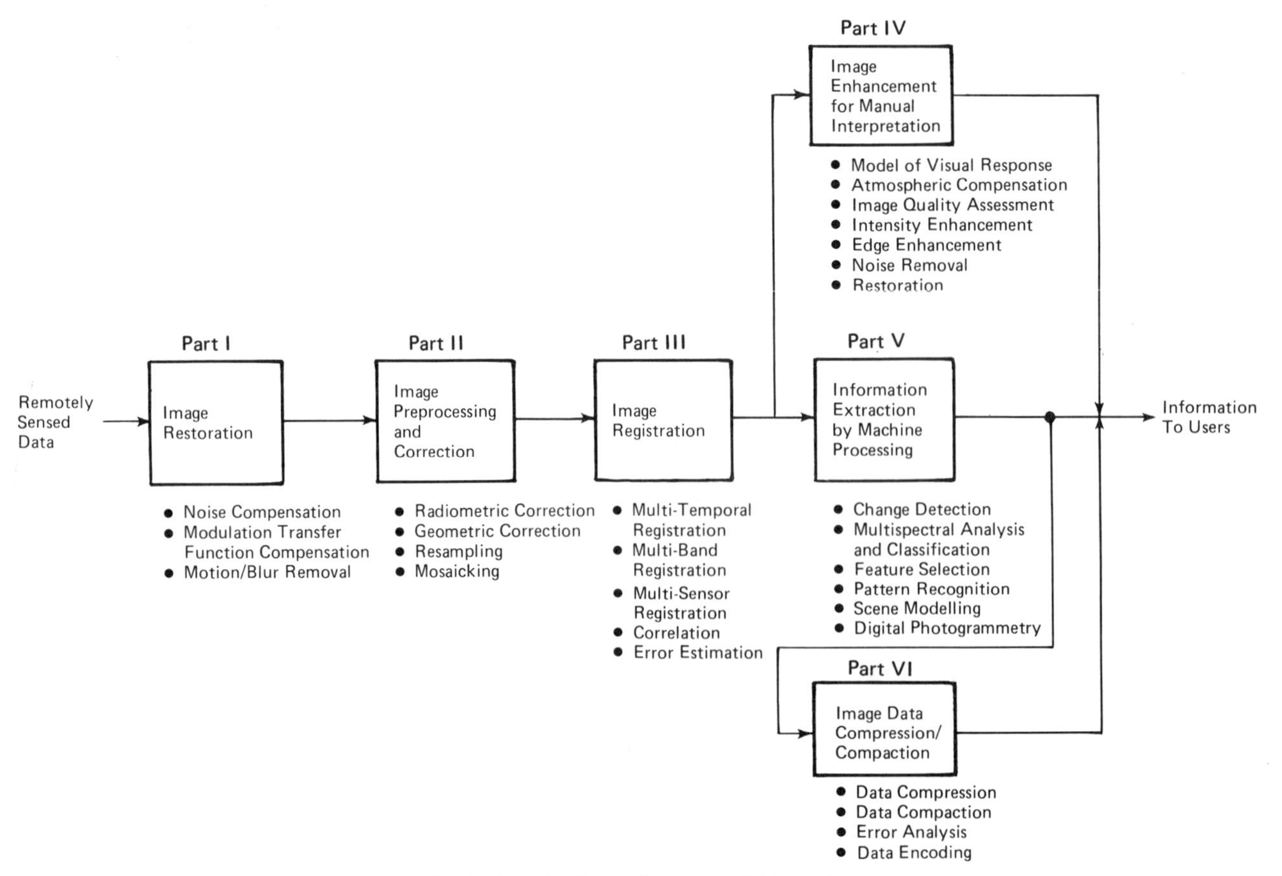

Fig. 1. Organization and content of this book.

tion from image data has been performed in the past by investigators viewing images and extracting information using their eyes and brains. Improvements have come about by developing a better understanding of the human visual response mechanism, restoring and improving the quality of the data, and enhancing the data for visual perception to separate the information from the background. The papers in Part IV develop these concepts and provide a number of illustrative results.

The very recent and important development of techniques for the extraction of information by computers, as opposed to human operators, is addressed in Part V. This development, which is still in its infancy, has resulted in a significant reduction of time and costs and an improvement in accuracy in a number of important applications. The algorithms used in some cases mimic the operation of the human; in other cases they are developed from physical laws and processes. In general, machine processing exploits the fact that computers can examine each image data sample, can discriminate subtle intensity differences, and can perform sophisticated statistical tests on anything from individual samples to large aggregates of data. Papers which discuss the latest developments in such areas as multispectral analysis and classification, digital photogrammetry, pattern recognition, and change detection are included in this part. The algorithms that exist and those that will be developed will have a profound impact on remote sensing.

Part VI encompasses the compression and/or compaction of image data. Generally, image compaction involves the reduction of the number of bits in a scene by eliminating redundant data in a reversable manner. Data compression involves eliminating data to reduce bandwidth/storage requirements and generally is an irreversable process. Various coding techniques and results are presented in this part.

Clearly, the papers that have been selected do not simply or totally fall into the categories discussed. It will be apparent that some papers cover the subjects of more than one part and could have been placed in several of the other parts. However, the papers do provide in one book information that should be useful to the scientist and engineer interested in and involved with remote sensing applications and digital image processing technology.

The book also contains an extensive bibliography of books and papers, covering remote sensing, vision and perception, resolution and image quality, sampling, digital image processing, and multispectral sensing and image analysis. Further, sources of image data are identified to aid scientists and engineers in obtaining remotely sensed data and, in particular, digital data. A glossary has been compiled that provides common definitions of frequently used terms in remote sensing and digital image processing. The Bibliography and Glossary appear at the end of the book.

Technology Directions

It is interesting to view the remote sensing and image processing activities on a historical basis and to project future technologies and applications. This provides a basis for understanding technology limitations and trends.

Fig. 2 summarizes the past, present, and future ten... celestial observation and image processing activities. The pas... and many current activities involve the use of camera systems with photo-optical processing and manual information extraction. Although this will continue to some extent into the future, the trend appears to be towards the use of more advanced multispectral sensors and sophisticated image processing operations to reduce the labor-intensive nature of manual processing. The present approach involves the use of multispectral scanners that provide data in numerous spectral bands, including many outside the spectral response of the human eye. The data, in a digital form, are transmitted to the ground, where high-speed digital processors are used to correct the data and convert them into information products for distribution. Initially, general-purpose computers were programmed to ex-

THE PAST

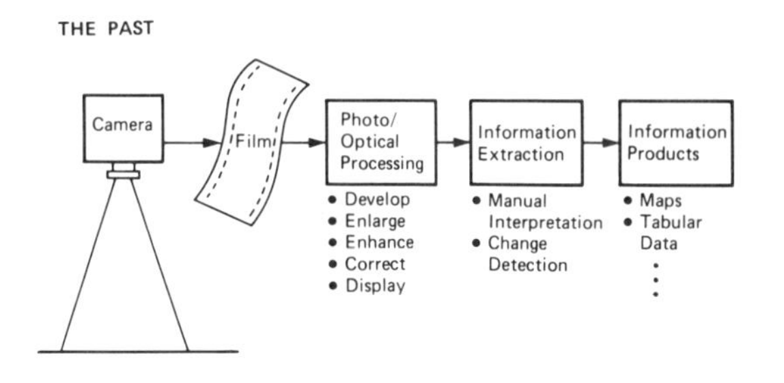

THE PRESENT

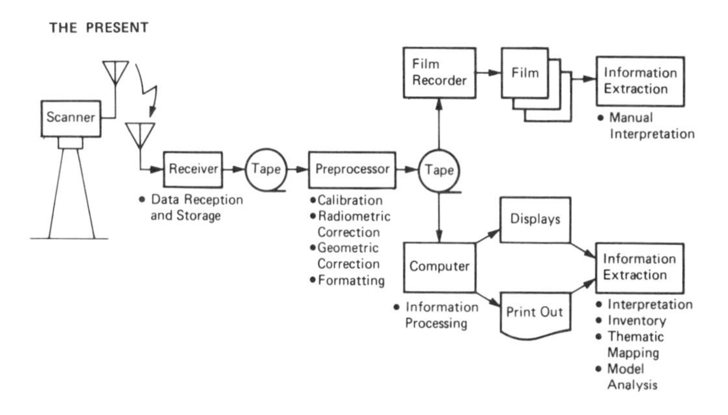

THE FUTURE

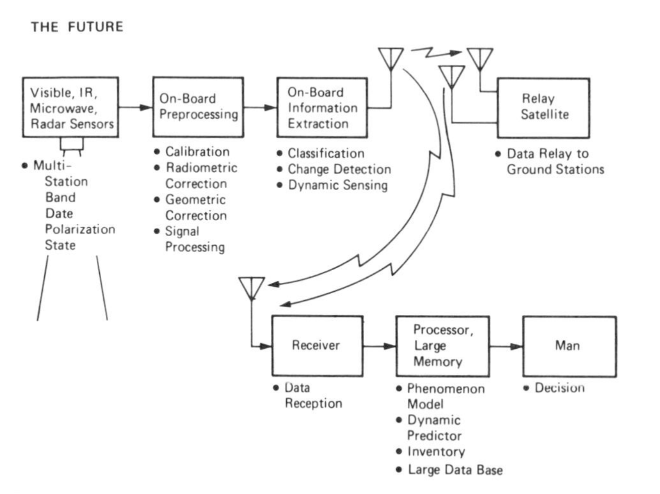

Fig. 2. Remote sensing and image processing: the past, present, and future.

perimentally implement the image processing algorithms. Increasingly, the processors are being specially designed to implement the algorithms, improving the performance and reducing the cost of image processing.

The future approach may go in the direction shown in the figure. Sensors in all spectral bands, including the ultraviolet and microwave, will be used to acquire remotely sensed data. It will be possible in the next two decades to implement selected preprocessing and information extraction algorithms with on-board digital image processing systems and relay the data and information to the ground via communications satellites. This will provide near real-time distribution of information to users so that dynamic events can be detected and monitored.

Clearly, an integrated data collection approach which will provide data acquired at various altitudes from different platforms (space, aircraft, ground) at different viewing angles and times must be and will be designed. A large data base, containing spatial, spectral, and temporal data and information acquired over all geographical areas of interest, will develop. It is anticipated that a data base for terrestrial applications alone could be global in nature and contain over a trillion bits of information. Physical phenomenon models can be structured, which, when provided with past and present data, will allow the prediction of future events more reliably. Thus, the role of man in future systems will be elevated to a higher level and will consist of an interesting interaction with a large data base and phenomenon models, as opposed to the conventional visual image processing of the past.

ACKNOWLEDGMENTS

The need for a book on this subject was first recognized by Dr. John W. Rouse, Jr. of Texas A&M, Past President of the IEEE Geoscience Electronics Professional Group. His suggestion was key to the book's preparation. The IEEE, and in particular the IEEE Press, deserve credit for supporting the publication and production of the book. In order to obtain the most representative and best papers in the field, I asked an associate and several leading investigators in the field to assume responsibility for editing the various parts in the book. The papers that have been selected are the result of their analyses and recommendations, and their contributions and help are greatly appreciated. Clearly, acknowledgment and appreciation is due to all of the authors whose papers and articles have been selected to be in this book. Finally, a greal deal of stimulus to the field of remote sensing, and in particular to digital image processing, is the result of the activities of the United States National Aeronautics and Space Administration, and this technology has greatly benefited by the motivation, applications, and support that has been provided by NASA.

Gratitude is due my employer, the Federal Systems Division of the International Business Machines Corporation. I have learned much from a number of engineers and scientists, and IBM management people have encouraged the pursuit of high technology activities and applications that benefit mankind.

This book is dedicated to my wife, Leah, and my family, who have provided me with the inspiration and time to prepare it.

Part I
Image Restoration

Organized by ~~Ralph Bernstein, Editor~~

Raul Hunziker Assoc. Editor

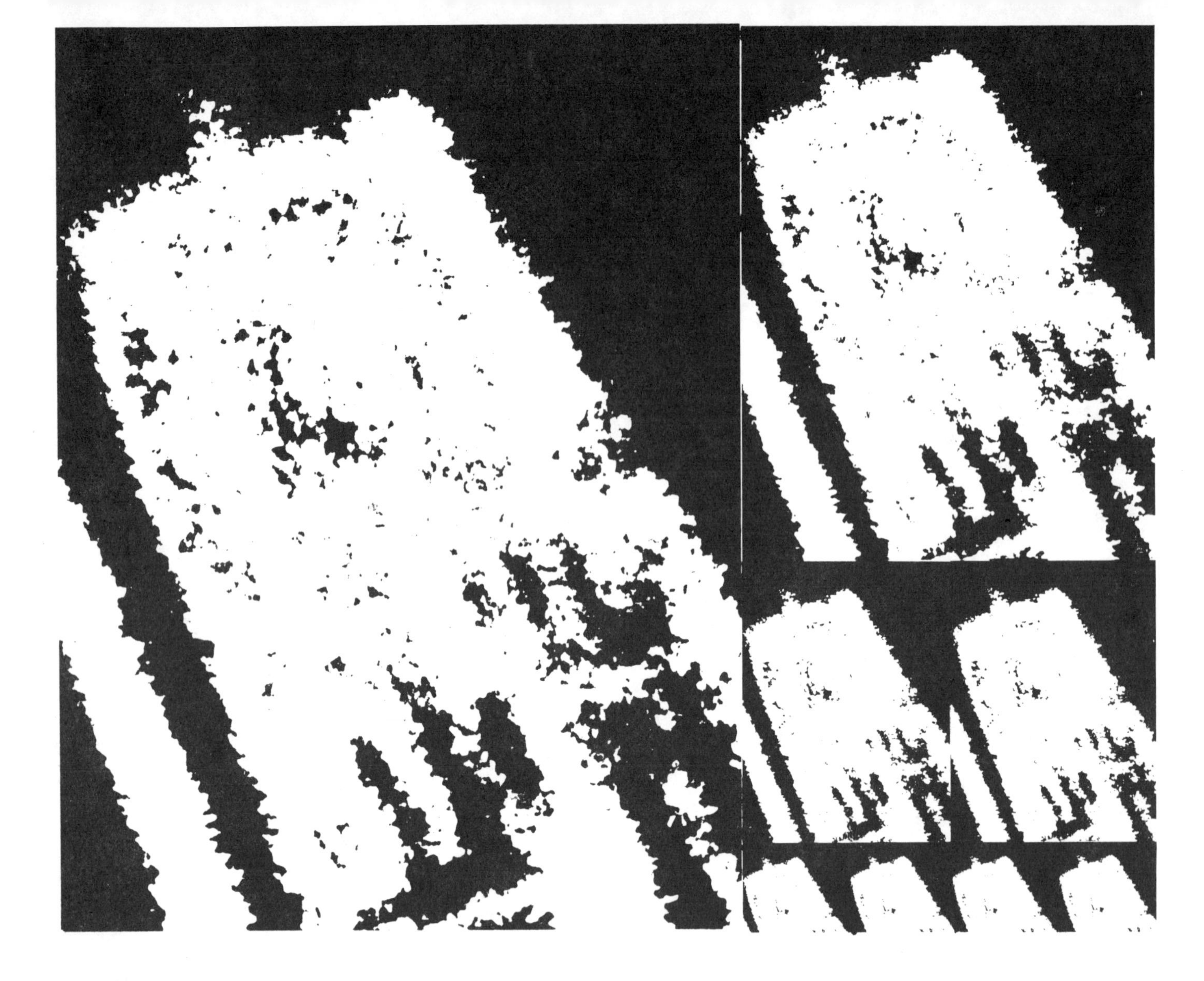

Digital Image Restoration: A Survey

Harry C. Andrews
Image Processing Institute
University of Southern California

Introduction

The state of the art in large-scale digital computers has recently opened the way for high resolution image processing by digital techniques. With the increasing availability of digital image input/output devices it is becoming quite feasible for the average computing facility to embark upon high-quality image restoration and enhancement. The motivation for such processes becomes self evident when one realizes the tremendous emphasis man puts on his visual senses for survival. Considering the relative success achieved in one-dimensional (usually time) signal processing, it is to be expected that far greater strides could be made in the visual two-dimensional realm of signal processing.

The areas of space imagery, biomedical imagery, industrial radiographs, photoreconnaissance images, television, forward looking infrared (FLIR), side looking radar (SLR), and several multispectral or other esoteric forms of mapping scenes or objects onto a two-dimensional format are all likely candidates for digital image processing. Yet many non-natural images are also subject to digital processing techniques. By non-natural images one might refer to two-dimensional formats for general data presentation for more efficient human consumption. Thus range range-rate planes, range-time planes, voiceprints, sonargrams, etc., may also find themselves subject to general two-dimensional enhancement and restoration techniques.

The concepts of digital restoration and enhancement are relatively recent in origin due to the need for usually large scale computing facilities. For the sake of semantics we will define restoration to be the reconstruction of an image toward an object (original) by inversion of some degrada-

Reprinted from *IEEE Computer*, vol. 7, pp. 36–45, May 1974.

tion phenomena. Enhancement will be the attempt to improve the appearance of an image for human viewing or subsequent machine processing. While the above definitions may seem somewhat artificial at this point, hopefully the distinction between restoration and enhancement will become more evident as we progress through the material which follows. (See reference 1 and bibliography in reference 2.)

Restoration techniques require some form of knowledge concerning the degradation phenomena if an attempt at inversion of that phenomena is to be made. This knowledge may come in the form of analytic models, statistical models, or other *a priori* information, coupled with the knowledge (or assumption) of some physical system which provided the imaging process in the first place. Thus considerable emphasis must be placed on sources and their models of degradation – the subject of the discussions that follow.

Enhancement techniques have really resulted from the power and generality provided by the general-purpose computer. Essentially any technique is fair game for enhancement if the resulting image provides additional information about the object which was not readily apparent in the original image. While such a definition in itself appears a little risky, emphasis is placed on the psychophysical aspects of the human visual system coupled with heuristic but mathematically defined operations for image manipulation.[3]

For notational convenience, consistency, and ease of understanding the following format will be established (see Figure 1):

> $f(\zeta,\eta)$ will be the object;
> $g(x,y)$ will be the image;
> $n(x,y)$ will be a two-dimensional sample from a noise process;
> $h(x,y,\zeta,\eta)$ will be known as the impulse response or point spread function (PSF) if the imaging system is linear.

Emerging from the potpourri of methods in use for digital image processing are a set of models in which attempts are made for the restoration of images by the effective inversion of degradation phenomena through which the object itself was imaged. Underlying many of these techniques is a basic assumption of linearity (questionable in itself but of sufficient value for analysis purposes) which provides for the following general model. Let $g(x,y)$ be the image of the object $f(\zeta,\eta)$ which has been degraded by the linear operator $h(x,y,\zeta,\eta)$ such that

$$g(x,y) = \iint_{-\infty}^{\infty} f(\zeta,\eta)h(x,y,\zeta,\eta)d\zeta d\eta + n(x,y) \qquad (1)$$

The system degradation, $h(\cdot)$, is known as the impulse response or point spread function and is physically likened to the output of the system when the input is a delta function or point source of light. If, as the point source explores the object plane, the form of the impulse response remains fixed except for position in the image plane, then the system is said to be spatially invariant – i.e., a spatially invariant point spread function (SIPSF) exists. If this is not the case, then a spatially variant point spread function system results (SVPSF). In this case equation (1) holds, and in the SIPSF case

$$g(x,y) = \iint_{-\infty}^{\infty} f(\zeta,\eta)h(x-\zeta, y-\eta)d\zeta d\eta + n(x,y) \qquad (2)$$

Most researchers have been satisfied with the model of equation (2), with variations such as additive noise, multiplicative noise, etc. Fourier techniques work well in attempting to obtain $f(\zeta,\eta)$ from $g(x,y)$ through the inversion of $h(\cdot)$ of equation (2) due to the Fourier-convolution relationship. Thus, in the absence of noise

$$G(u,v) = H(u,v)F(u,v) \qquad (3)$$

where G, H, and F are the Fourier transforms of g, h, f, and the determination of $F(u,v)$ simply requires the inversion of H, if it exists. If the SVPSF model is used, then Fourier techniques are no longer applicable and more general brute force inversion methods must be resorted to for object restoration.

There are three basic approaches which are often used for inversion of either of the two systems as described above. They could be referred to as a) continuous-

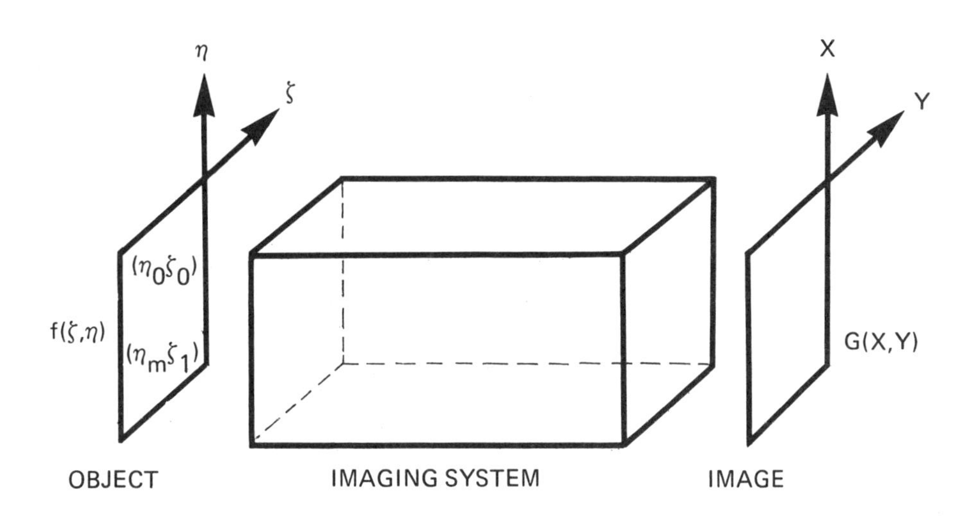

Figure 1. A Linear Imaging System Model

continuous, b) continuous-discrete, and c) discrete-discrete.

The first analysis method looks at the entire image restoration process in a continuous fashion (although ultimate implementation will necessarily be discrete). The second analysis method assumes the object is continuous but the image is sampled and therefore discrete. The third technique assumes completely discrete components and utilizes purely numerical analysis and linear algebraic principles for restoration. In equation form we would have the following imaging models for each of the three assumptions.

a) continuous-continuous:

$$g(x,y) = \iint_{-\infty}^{\infty} f(\zeta,\eta)h(x,y,\zeta,\eta)d\zeta d\eta + n(x,y) \tag{4}$$

b) continuous-discrete:

$$g_i = \iint_{-\infty}^{\infty} h_i(\zeta,\eta)f(\zeta,\eta)d\zeta d\eta + n_i \qquad i = 1,\ldots,N^2 \tag{5}$$

c) discrete-discrete:

$$\underline{g} = [H]\,\underline{f} + \underline{n} \tag{6}$$

The continuous-continuous model of equation (4) says that the image is simply the integration of the object and point spread function in an analog two-dimensional environment. When the model is put into the computer, the image, object, and point spread function will be sampled by N^2, N^2, N^4 points, repsectively.

The continuous-discrete model of equation (5) implies that the object is continuous (as it would be in the real world) but the sensor defining the image is discrete and already sampled by N^2 points. Thus there are N^2, g_i scalar values and of course N^2, $h_i(\zeta,\eta)$ different point spread functions.

Finally, the discrete-discrete system implies that the object and image are one-dimensional vectors, N^2 long, which represent the original object and image $\underline{f}$, and $\underline{g}$, respectively. The vectors can be raster-scanned versions of two-dimensional functions or any other scanning method such that all N^2 points are obtained. The four-dimensional function $h(\zeta,\eta,x,y)$ is now reduced to a two-dimensional array by the raster scan and thus is a matrix of size $N^2 \times N^2$.

Degradation Sources

There are obviously quite a few sources of degradation in imaging systems, but often they can be grouped into the following general categories and their combinations: a) point degradations, b) spatial degradations, c) temporal degradations, d) chromatic degradations, and e) combinations of the above. To handle the above degradations rigorously in a mathematical sense, and maintaining our assumptions of linearity, we would have to generalize our model in equation (1) to the following:

$$g(x,y,\tau,\omega) = \iiiint_{-\infty}^{\infty} f(\zeta,\eta,t,\lambda)h(x,\zeta,y,\eta,\tau,t,\omega,\lambda)d\zeta d\eta dt d\lambda + n(x,y,\tau,\omega) \tag{7}$$

While such a generalization may be useful for analysis purposes, we will not pursue it further, due to our major concern for spatial systems only. In passing, then, let it suffice to say that temporal and chromatic degradations imply time and color deterioration of their respective axes.

Returning now to equation (1) we see that the "perfect" imaging system would have no noise and an impulse response such that

$$g(x,y) = \iint_{-\infty}^{\infty} f(\zeta,\eta)\delta(x-\zeta, y-\eta)d\zeta d\eta \tag{8a}$$

$$= f(x,y) \tag{8b}$$

Possibly the first step from perfect imaging is that system which provides no spatial smearing [i.e., no (ζ,η) integration] but induces a point degradation. Thus we might have an impulse response such that

$$g(x,y) = \iint_{-\infty}^{\infty} f(\zeta,\eta)h(x,y)\delta(p_1(x,y)-\zeta, p_2(x,y)-\eta)d\zeta d\eta + n(x,y) \tag{9a}$$

$$= h(x,y)f(p_1,p_2) + n(x,y) \tag{9b}$$

where $p_1(x,y)$ and $p_2(x,y)$ are geometrical coordinate transformations. The above equation describes imaging systems which do not blur but which introduce a distortion due to a coordinate change. When the imaging system does not introduce a coordinate distortion – i.e.,

$$p_1(x,y) = x$$

$$p_2(x,y) = y$$

we then obtain

$$g(x,y) = \iint_{-\infty}^{\infty} f(\zeta,\eta)h(x,y)\delta(x-\zeta, y-\eta)d\zeta d\eta + n(x,y) \tag{9c}$$

$$= h(x,y)f(x,y) + n(x,y) \tag{9d}$$

This system allows for both multiplicative and additive point degradation effects. The former might be due to film grain, lens or tube shading, or other sensor defects. The latter might be due to electronic scanner effect as well as scattered light and other such phenomena. Equation (9) also provides for some point nonlinear degradations such as gamma curves for film saturation and other intensity distortions if we allow $h(x,y)$ to become a function of the object, $h(x,y,f)$. Unfortunately now, the linearity assumption of our imaging system no longer holds. However, under such generalizations when the impulse response becomes $h(x,y,\zeta,\eta,f)$ we then have a good (albeit nonlinear) model of object-dependent SVPSF phenomena, an example of which might be high energy x-ray imaging where forward scattering (and therefore blur of point sources) becomes a function of the density of the object being imaged.

When our impulse response becomes a function of the object coordinates (ζ,η), we then have some form of smearing or loss of resolution due to the integration of the imaging system over those coordinates. Examples of these so-called spatial degradations are numerous – some of the common of which are a) diffraction-limited optical systems; b) first, second, and higher-order optical system aberrations; c) atmospheric turbulence; d) object-film plane image motion blur; and e) defocused systems.

While the above is certainly not an exhaustive list of spatial degradations, it does provide the flavor of some of the problems faced in image restoration. The models associated with the various defects can be simple SIPSF convolutions or much more complicated SVPSF representations.

The next section addresses the problem of restoring an image which has experienced some form of degradation.

Restoration Techniques

"Restoration techniques" are methods which attempt the inversion of some degrading process the object experienced in being imaged onto some form of hard copy. It goes without saying that the success of the restoration attempt will depend upon how badly degraded the object is, how well one's model fits the physical degrading phenomena the object actually experienced, and how well one's computer algorithm inverts the modeled degradation. Initially it might appear that restoration may in fact be a useless endeavor, since the complexity of object-distorting mechanisms is indeed large. However, there are certain instances in which considerable knowledge is available for correct modeling and in which successful restoration is easily achieved.

In general, restoration techniques can be compartmentalized (for the sake of this discussion) into four general descriptive areas: a) *a priori* knowledge, b) *a posteriori* knowledge, c) signal-processing approaches, and d) numerical analysis approaches. The amount of *a priori* knowledge concerning an imaging system or circumstance obviously plays an important role in the inversion attempt. However, we can often learn about the degradation the object experienced by observing the image itself, and thus the *a posteriori* category. Traditional one-dimensional signal processing attempts have often been utilized successfully in two dimensions, and consequently some of these methods will be surveyed. Finally, more recent numerical analysis techniques have been brought to bear on the restoration effort and are included to provide motivation for possible direction of future methodologies.

A priori knowledge includes a variety of parameters available which might allow for image restoration beforehand. For instance, it may be known that certain geometrical equations of motion or position existed between object and image film plane during exposure, thereby introducing a specific form of motion or coordinate blur (rotation, camera pitch, linear motion blur, coma, tilt, etc.). Algorithms can then be developed for geometrical coordinate transformations for correction of such geometrical blur.[4-7] Other forms of *a priori* models might include assuming the image is formed from a Maxwell Boltzmann distribution[8] and then restoring with a maximum likelihood and maximum entropy algorithm[9,10] or assuming the image is a two-dimensional probability density function and using Bayes theorem for restoration.[11] The above models all have a common thread woven throughout their motivations, and that thread has become known as *positive restoration.* The concept of positive restoration comes from the fact that objects, images, and point spread functions are all non-negative functions.[12]

$$f(\zeta, \eta) \geqslant 0 \tag{10a}$$

$$g(x, y) \geqslant 0 \tag{10b}$$

$$h(x, y, \zeta, \eta) \geqslant 0 \tag{10c}$$

This reality stems from the fact that all optical sensing devices are energy-sensitive and therefore detect non-negative quantities. This fact, though seldom used, has considerable implications in restoration and analysis systems. For instance, the positiveness of g(x,y) forces its Fourier transform to have a deterministic upper limit provided by the Lukosz bound,[13] implying that any restoration technique had better result in a filtered image (estimated object) whose Fourier transform also lies below this bound. The model provided by homomorphic filtering[14] also fits nicely into the positive restoration framework and has been used quite effectively for inherent low frequency illumination removal for image restoration.[15] In addition to positive restoration, one might even consider positive bounded restoration thereby utilizing the *a priori* knowledge that only a finite amount of light could in fact exist at any given point in the original object. This leads to imaging system models in which the assumption of conservation of light flux (lossless imaging) results in an energy equality in image and object – i.e.,

$$\iint_{-\infty}^{\infty} f(\zeta, \eta) d\zeta d\eta = \iint_{-\infty}^{\infty} g(x, y) dx dy \tag{10d}$$

Similarly, such energy conservation assumptions imply that a point source of light should result in no loss of energy or

$$\iint_{-\infty}^{\infty} h(x, y, \zeta, \eta) dx dy = 1 \quad \forall \zeta, \eta \tag{10e}$$

In other words, no matter where the point source of light is in the object (ζ, η) plane, the resulting image (point spread function) always has the same amount of energy.

The above discussion might be the basis for utilizing *a priori* knowledge concerning the effects imposed by an imaging system. However, *a posteriori* knowledge can also play a major role in determining the degradation the object has experienced in being imaged. By *a posteriori* knowledge we mean utilization of the image g(x,y) as an aid in determination of parameters describing the degradation. Obvious examples might include point spread function determination from edges or points in the image that are known to exist in the object.[16] This is often done with test scenes such as resolution charts and point targets (stars and other point sources). Other examples of *a posteriori* image use might include obtaining estimates of the noise variance and possible power spectrum from relatively smooth regions in the image. A third example might be one in which scanner-induced noise (jitter) becomes immediately evident in the Fourier transform of the image although such noise is subtly obscured in the spatial representation of the image.[17] Finally, one might use the Fourier domain of the image to determine and correct for badly defocused imaging systems and linear motion blurred images. While these two degradations may initially seem unrelated, they both have the property that the object spectrum has been modified by functions which have zeros and negative lobes in the frequency domain. In the former case the point spread function is a circular aperture in a SIPSF system with Fourier representation as

$$H(u, v) = \frac{J_1(a\rho)}{a\rho} \tag{11a}$$

where

$$\rho = \sqrt{u^2 + v^2} \tag{11b}$$

In the latter case the SIPSF takes on the form

$$H(u, v) = \frac{\sin a\omega}{\omega} \tag{12a}$$

$$\omega = u \cos\theta + v \sin\theta \tag{12b}$$

Here θ is the direction of motion of the object with respect to the image film plane. Figure 2 represents the concepts. By measuring the zeroes (dark bands) in the Fourier planes it is possible to quantitatively determine the amount of defocus or motion blur and consequently to then remove the effect by traditional filtering techniques.

While the above examples of *a posteriori* use of the image for restoration parameter determination are not exhaustive, hopefully they provide the flavor of such techniques. Naturally parameter determination is important in defining the restoration filter for many of the signal processing approaches which have been commonly employed in image restoration. Traditionally the signal processing approach makes use of the SIPSF model described by equation (2) and repeated here in the absence of noise.

$$g(x, y) = \iint_{-\infty}^{\infty} f(\zeta, \eta)h(x-\zeta, y-\eta)d\zeta d\eta \tag{2}$$

We mentioned earlier the convolutional nature of the above equation and its Fourier equivalent representation in equation (3)

$$G(u, v) = H(u, v)F(u, v) \tag{3}$$

The most immediate filter that comes to mind is the inverse filter which multiplies G(u,v) by $H(u,v)^{-1}$, and then the object $f(\zeta,\eta)$ is obtained by inverse Fourier transforming the result. The restoration filter then becomes:

Inverse Filter

$$R(u, v) = H(u, v)^{-1} \tag{13}$$

Unfortunately the inverse filter may not exist due to zeroes in H(u,v) (see Figure 2) and we have ignored the

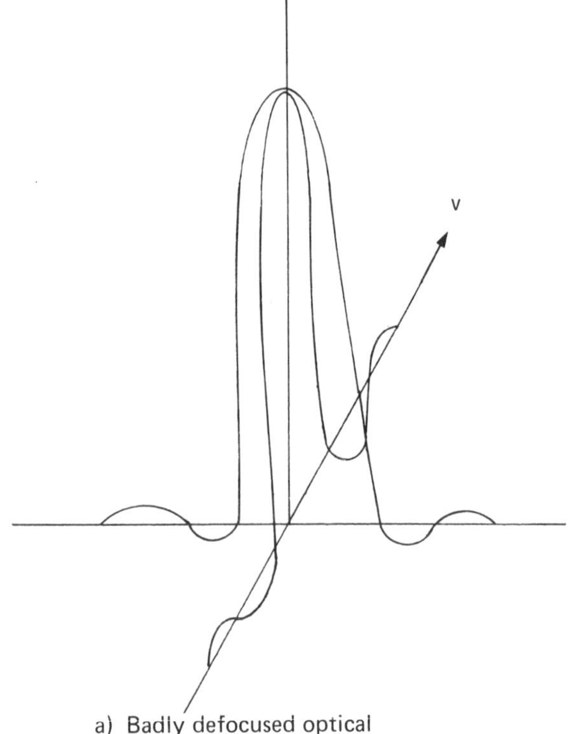

a) Badly defocused optical transfer function, H(u, v)

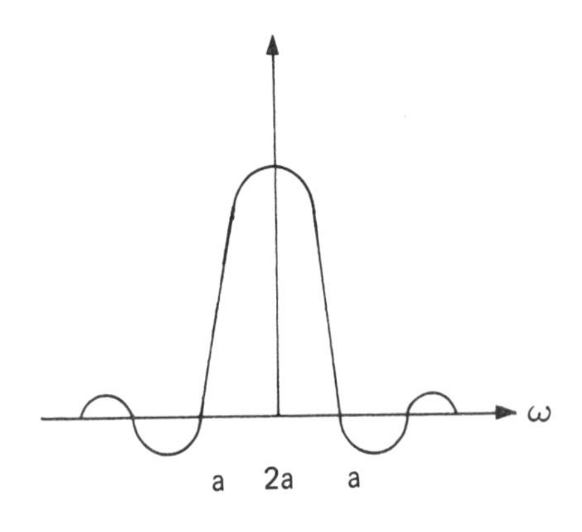

b) Linear motion blur, H(u, v)

c) Bell dummy g(x, y) (moving horizontally)

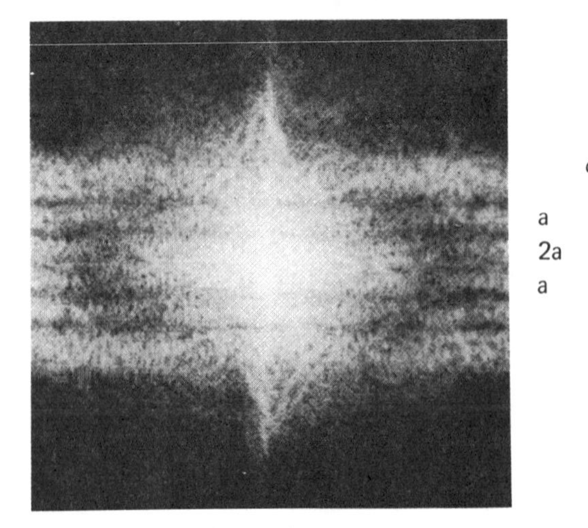

d) Fourier transform of image

Figure 2. Zeros and Negative Lobe Modulation

noise in equation (2). The best filter in a mean square error sense is known as the Wiener filter and takes the form:[18]

Wiener Filter

$$R(u,v) = \frac{H^*(u,v)}{|H(u,v)|^2 + \frac{\varphi_n(u,v)}{\varphi_f(u,v)}} \tag{14}$$

where φ_n and φ_f are the noise and object power spectra, respectively. The fidelity criterion of mean square error is suspect in image restoration, and variations on the Wiener filter have been proposed. Hunt[19] has proposed a constrained least-squares filter which, for SIPSF systems, becomes

Constrained Least-Squares Filter

$$R(u,v) = \frac{H^*(u,v)}{|H(u,v)|^2 + \gamma|C(u,v)|^2} \tag{15}$$

and by judicious choice of C(u,v) and the constant γ one can minimize higher order derivatives, eye models, or even achieve the Wiener filter. An additional restoration filter has been suggested by Stockham and Cole,[20] which is a geometrical mean filter between the inverse filter and Wiener filter:

Geometrical Mean Filter

$$R(u,v) = \left(\frac{1}{H(u,v)}\right)^s \left(\frac{H^*(u,v)}{|H(u,v)|^2 + \frac{\varphi_n(u,v)}{\varphi_f(u,v)}}\right)^{1-s} \tag{16a}$$

where $0 \leqslant s \leqslant 1$. For a symmetric phaseless filter ($H(u,v)^* = H^*(u,v)$) with $s = ½$ we have:

Stockham-Cole Filter

$$R(u,v) = \left(|H(u,v)|^2 + \frac{\varphi_n(u,v)}{\varphi_f(u,v)}\right)^{-½} \tag{16b}$$

The above filter has been implemented using *a posteriori* image parameter determination with very successful results. While the list of possible filters can be extended ad nauseum, it should be emphasized that implementation of any of these filters need not always be in the Fourier domain. Convolutional, recursive, and Kalman-Bucy filter implementations are also possibilities and have been used successfully.[1]

The fourth area of discussion on restoration techniques falls under the heading of numerical analysis approaches and is still in its infancy with respect to digital image restoration applications. Here the underlying model is either the continuous-discrete or discrete-discrete representations of equations (5) and (6) respectively.

$$g_i = \iint_{-\infty}^{\infty} h_i(\zeta,\eta) f(\zeta,\eta) d\zeta d\eta + n_i \tag{5}$$

$$\underline{g} = [H]\,\underline{f} + \underline{n} \tag{6}$$

Because most of the numerical analysis approaches in image processing to date have used the discrete-discrete representation, this will be the form we will concentrate on here. If the system is spatially invariant, then [H] of equation (6) becomes block circulant when circular convolution is used to replace normal convolution. Under these circumstances Fourier series expansions are appropriate and traditional Fourier signal processing techniques become applicable. If the imaging system is spatially variant, then the form of [H] becomes more complex. Practically speaking [H] will be singular due to its enormous size ($N^2 \times N^2$) and due to the fact that most imaging systems irreversibly remove certain aspects of the original object. Thus even in the absence of noise we cannot form $\underline{f}$ from $[H]^{-1}$. This suggests the use of pseudoinverse techniques[21] as a means of getting better estimates of $\hat{\underline{f}}$ even when [H] is singular. Thus

$$\hat{\underline{f}}_k = \sum_{i=1}^{k} (\lambda_i^{-½} u_i^t g)\, v_i \tag{17a}$$

where

$$[H] = \sum_{i=1}^{N^2} \lambda_i^{½} u_i v_i^t \tag{17b}$$

and the u and v vectors are the singular valued decompositions[22] of the space-variant impulse response matrix [H]. Another approach at matrix inversion is given by the power series method where

$$\hat{\underline{f}}_k = \sum_{i=0}^{k} [I-H]^i \underline{g} \tag{18a}$$

or

$$\hat{\underline{f}}_k = \underline{g} + [I-H]\, \hat{\underline{f}}_{k-1} \tag{18b}$$

We see from both equations (17) and (18) that numerically we have a technique (at least in theory) to iteratively move an image back through the imaging system as a function of k and allow a human to participate interactively in the stepwise inversion process before singularity is reached. The two above techniques have not utilized the positive restoration criterion and thus appropriate modifications would be necessary and have been implemented in one dimension in a modified Van Cittert method.[23] Often the optimization criteria for numerical techniques are referred to as unconstrained or constrained least-squares where $\|\underline{g}-[H]\,\hat{\underline{f}}\|^2$ is minimized. Simply minimizing the above without constraints results in

$$\hat{\underline{f}} = [H^tH]^{-1} [H]^t \underline{g} \tag{19a}$$

which we recognize as a generalized inverse filter. If we constrain the technique to minimize $\|[C]\underline{f}\|^2$ subject to $\|\underline{g}-[H]\hat{\underline{f}}\|^2 = \|\underline{n}\|^2$ we obtain[19] the objective function $\|[C]\underline{f}\|^2 + \gamma\|\underline{g}-[H]\underline{f}\|^2 - \gamma\|n\|^2$ for optimization. This solution results in the generalized constrained least-squares filter of equation (15):

$$\hat{\underline{f}} = [H^tH + \gamma C^tC]^{-1} [H]^t \underline{g} \tag{19b}$$

Similarly we will obtain the generalized Wiener filter by setting $C^tC = 1/\gamma\, [\varphi_n]\, [\varphi_f]^{-1}$ where the $[\varphi]$ matrices are noise and signal covariance matrices respectively. By introducing the inequality constraint of positive restoration (i.e., $\hat{\underline{f}} \geqslant 0$ component-wise) we then must resort to semi-infinite linear and quadratic programming techniques utilizing gradient projection or conjugate gradient algorithms.[24] Finally for the continuous-discrete model it is possible to formulate the objective function for optimization as

$$\|\underline{g}-\hat{\underline{g}}\|^2 + \gamma \iint_{-\infty}^{\infty} (\nabla^2 f(\zeta,\eta))^2 d\zeta d\eta.$$

Here the second derivative of the continuous object is being minimized in the estimate of that object. D. Ferguson has suggested the use of two variable spline functions as a suitable basis system for solution of this continuous-discrete representation.

Enhancement Techniques

Enhancement techniques are inherently quite different from restoration methods because the fidelity criterion of attempting better object representation no longer governs the motivation. In fact if one were to be required to define "enhancement," he would be particularly hard-pressed because one man's enhancement may be another man's. noise. As a rule enhancement broadly refers to the manipulation of imagery to present to the viewer (or subsequent machine) additional information or insight into some factor concerning the pre-enhanced image. Broad categories of enhancement techniques might be as follows: a) intensity mappings, b) eye modeling, c) edge sharpening, and d) pseudocolor.

Intensity mappings refer to usually nonlinear operations on a point-by-point basis to map one gray scale into another. Figure 3 presents some typical operations that have been

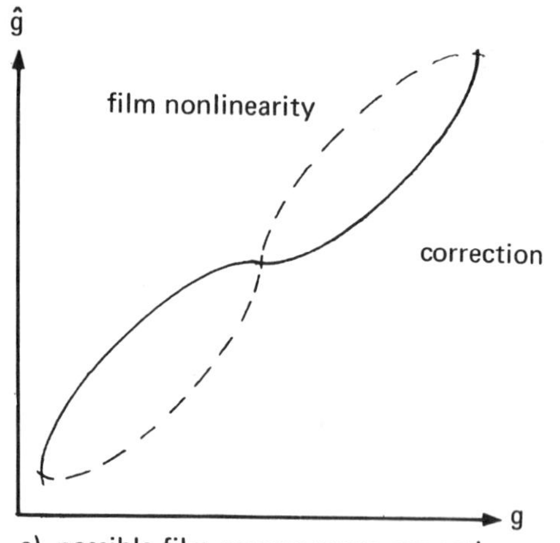

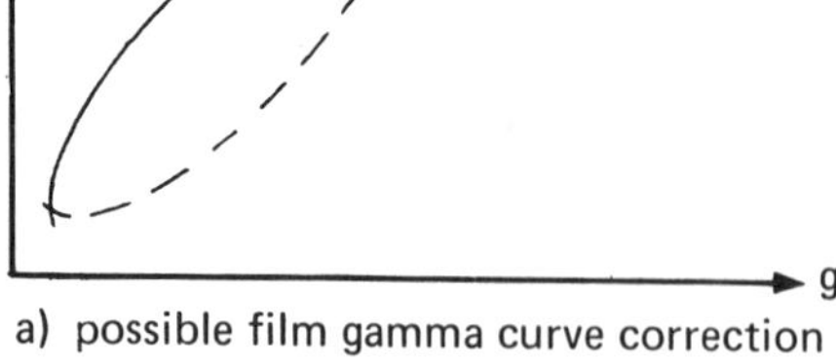

a) possible film gamma curve correction

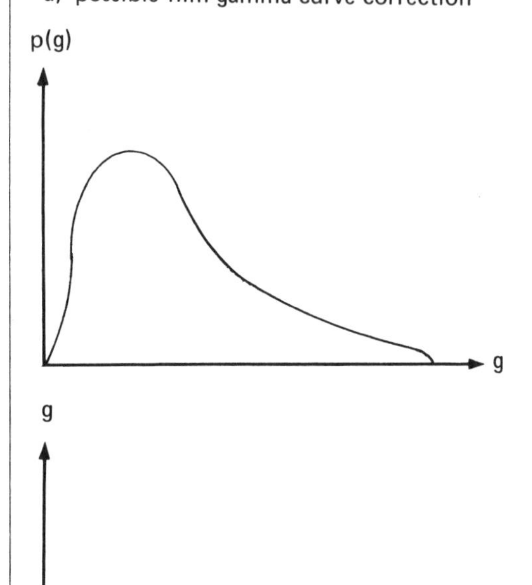

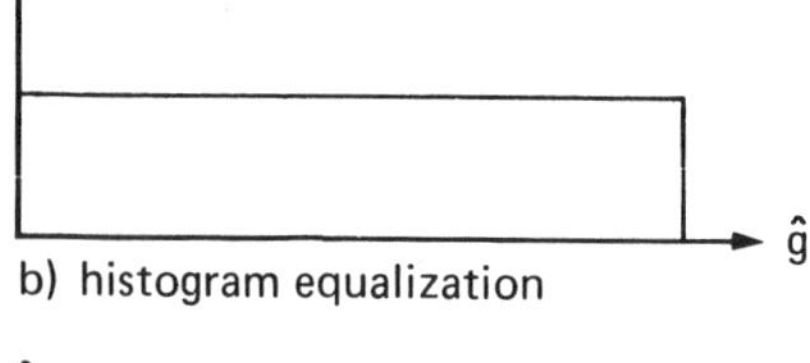

b) histogram equalization

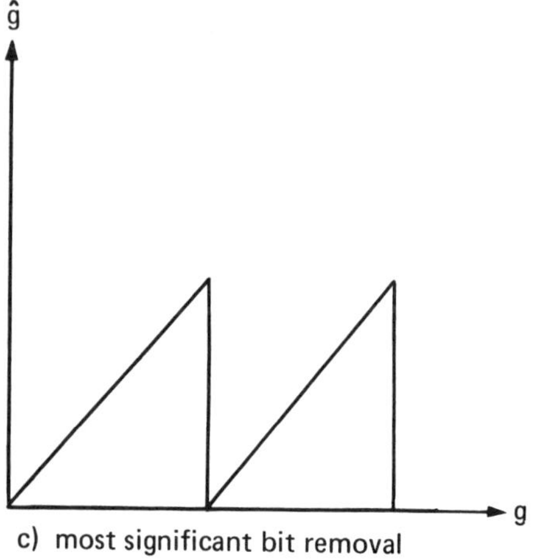

c) most significant bit removal

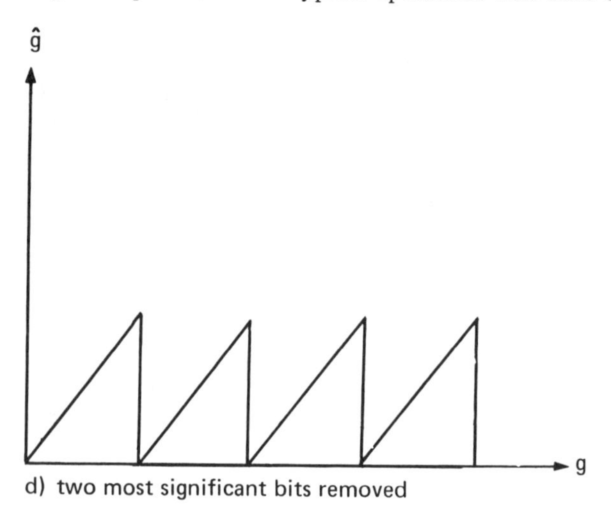

d) two most significant bits removed

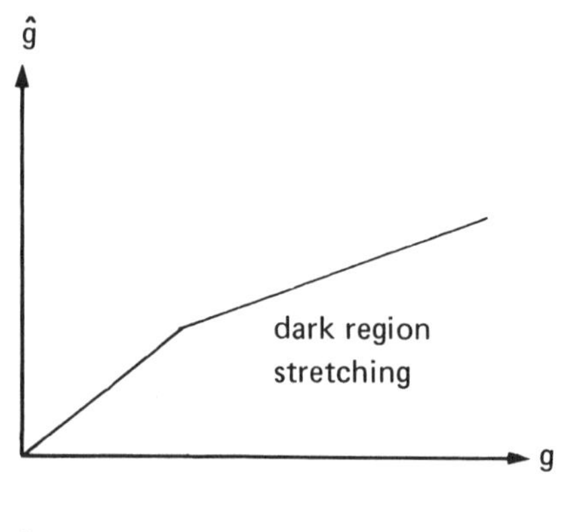

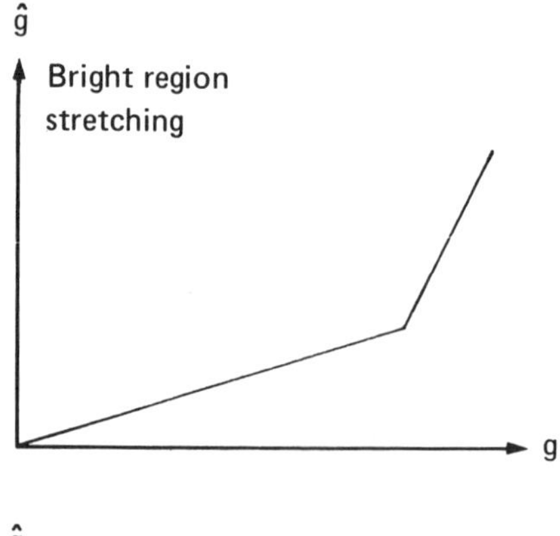

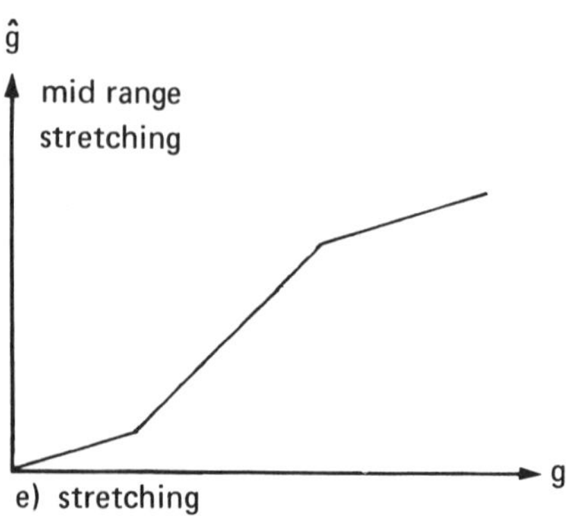

e) stretching

Figure 3. Intensity Mapping Enhancements

utilized by various image processing facilities. If the input image is g(x,y) then the intensity mapping enhanced image becomes

$$\hat{g}(x, y) = I(g(x, y)) \tag{20}$$

where I(•) is a nonlinear mapping of g(•) independent of the position (x,y) in the image. Often the first operation attempted is correction for film and display nonlinearities, as illustrated in Figure 3a. The next mapping is known as histogram equalization (Figure 3b[3]) and is achieved by redistributing the gray levels of the image g(x,y) such that $\hat{g}$(x,y) has as uniform a histogram as is possible. Figures 3c and 3d represent the mappings for the single most and two most significant bit removal and are often useful for discovering underlying contours in the image, representing structural information. Finally, Figure 3e represents various possible stretching algorithms enhancing the dark, bright, and mid ranges of gray respectively. Figure 4 illustrates an example of the point intensity mapping technique applied to actual imagery. It is evident from *a priori* information that the original object was binary (i.e., the Air Force Resolution Chart had only two gray levels). Therefore an appropriate intensity mapping might be one which essentially binarizes the extreme blacks and whites while leaving the mid gray region linear. Thus a <b in Figure 4a. (Note that for a = b the enhanced image $\hat{g}$(x,y) would also be binary.) The image and enhanced version are presented in Figure 4b and 4c, and it is evident that the interference patterns present in the image due to scanning are eliminated in the enhanced version.

The next category of enhancement listed above is that of eye modeling. Here the objective is to understand the mechanism of the psychophysics of human perception well enough to then allow the computer to enhance an image for visual consumption by precompensating for the visual system.[15] Such models would take into account mach banding phenomena, intensity response, spectral (chromatic) response, temporal response, etc. One particularly important visual system response is that of spatial frequency, for the eye has a tendency to differentiate at the lower spatial frequencies while emphasizing the higher spatial frequencies. Thus edge sharpening becomes a particularly attractive enhancement mechanism because of the desire of the human to see sharp edges. There have been various models suggested for the response of the eye; the interested reader is referred to Cornsweet.[27]

Edge sharpening is often implemented in the Fourier domain of an image with ramp or other monotonically

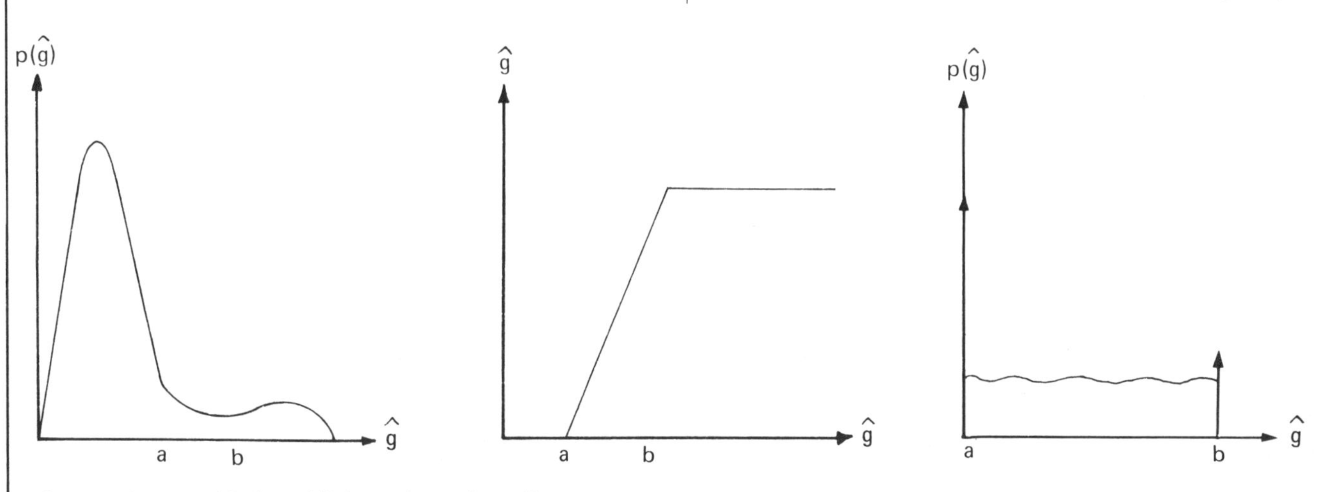

a) intensity map (dark and light regions clipped)

b) image g(x, y)

c) "enhanced" image $\hat{g}$(x, y)

Figure 4. Noise Clipping by Intensity Mapping

increasing functions in the spatial frequency plane. Unsharp masking, the technique of subtracting out the low-frequency portion of an image, thereby leaving an enhanced version, is a well-established technique which has found considerable success in the wirephoto industry.[28] Certain nonlinear operations which result in effective edge enhancement have been referred to as α (alpha) processing, in which the Fourier coefficients are nonlinearly amplified as[3]

$$\hat{G}(u, v) = |G(u, v)|^{\alpha} e^{j\varphi(u, v)} \qquad (21)$$

$$0 \leqslant \alpha \leqslant 1$$

For $\alpha = 0$ we have a phase-only image which, surprisingly enough, retains a high degree of original structure and which has heavily emphasized the edge information. This simply illustrates the importance of phase information in the Fourier representation of an image, a fact which is well established by kinoforms and phase-only (bleached) holograms. The nonlinear α process has been generalized to other unitary transforms but will not be further discussed here.[29]

The final topic for enhancement discussion is the use of color on originally monochrome imagery, or pseudocolor. Because the original object was monochrome, the addition of color results in pseudocolor effects as opposed to the use of *falsecolor* to make viewable electromagnetic spectral energy imaged in the invisible region in multiband cameras (i.e., usually infrared or ultra violet). The underlying motivation in the pseudocolor applications is to increase the effective viewing dynamic range of the original gray scale by appealing to the human's visual response in color. Examples of pseudocoloring techniques have been presented by both intensity mapping techniques, and by spatial frequency mapping methods.[3] The former method simply maps a particular gray shade to a given intensity, hue, and saturation defining a color shade. The latter method maps a particular spatial frequency range to a particular color shade. Quantitatively it is difficult to evaluate the "optimum" color map for a particular objective because of both the subjectivity of enhancement as well as color viewing. As an illustration, Figure 5 shows a pseudocolored portion of the image in Figure 4. In the color version the noise is made readily apparent and the wing tip definition might be enhanced. Also, the scanner-induced jitter becomes apparent. The author leaves it to the reader to decide whether the addition of color is an aid in this example.

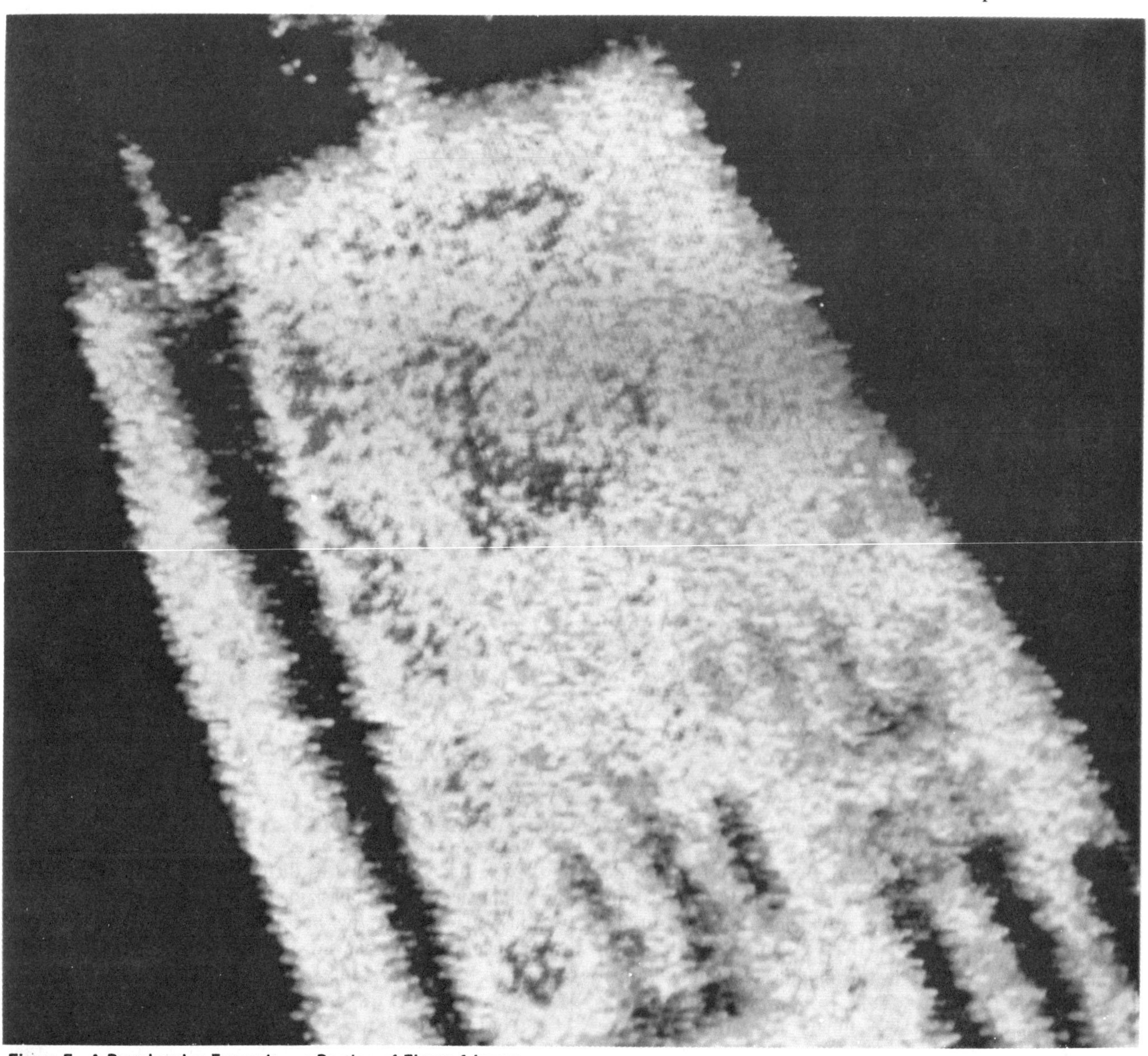

Figure 5. A Pseudocolor Example — a Portion of Figure 4 Image
(*Note:* The reader is referred to the original publication for color illustrations.)

Summary

It should be emphasized that restoration and enhancement are two different subjects. Restoration depends upon the model used for the imaging system and the degradations thereby imposed. The space variant point spread functions (SVPSF) and space invariant point spread functions (SIPSF) were defined and then models developed around continuous-continuous, continuous-discrete, and discrete-discrete notation. Then degradation sources were described where point, spatial, temporal, chromatic, and combinations thereof were mentioned. Particular emphasis was placed on point and spatial deteriorations as these seem to be the ones most easily corrected. Following the sources of degradation, four classes of restoration techniques were presented, covering use of *a priori* knowledge, *a posteriori* knowledge, signal processing techniques, and numerical analysis methods. Finally a few enhancement methodologies were discussed, the heavy emphasis there being in intensity mappings, color mappings, eye modeling, and edge sharpening. Examples illustrating a few of the techniques are included.

Acknowledgements

The author wishes to acknowledge the support provided by the University of Southern California and particularly the director of the Image Processing Institute at USC. The major support for research in this field is provided by the Advanced Research Projects Agency of the Department of Defense, and was monitored by the Air Force Eastern Test Range under contract number F08606-72-C-0008. Finally the author wishes to thank A. G. Tescher for permission to use some illustrations from his doctoral dissertation. The author is indebted to a reviewer for his critical comments concerning certain sections of this paper.

References

1. *Proceedings of the IEEE,* "Special issue on Digital Picture Processing," Vol. 60, No. 7, p. 763 (July 1972).

2. H. C. Andrews, "N Topics in Search of an Editorial: Heuristics Superresolution, and Bibliography," *Proceedings of the IEEE,* Vol. 60, No. 7, pp. 891-894 (July 1972).

3. H. C. Andrews, A. G. Tescher, and R. P. Kruger, "Image Processing by Digital Computer," *IEEE Spectrum,* Vol. 9, No. 7, p. 20-32, (July 1972).

4. G. M. Robbins, and T. S. Huang, "Inverse Filtering for Linear Shift-Varient Imaging Systems," *Proceedings of the IEEE,* Vol. 60, No. 7, pp. 862-872 (July 1972).

5. A. A. Sawchuk, "Space-Variant Image Motion Degradation and Restoration," *Proceedings of the IEEE,* Vol. 60, No. 7, pp. 854-861 (July 1972).

6. A. A. Sawchuk, "Space-Variant System Analysis of Image Motion," *JOSA,* Vol. 63, No. 9, pp. 1052-1063 (September 1973).

7. A. A. Sawchuk, "Space-Variant Image Restoration by Coordinate Transformations," *JOSA,* Vol. 64, No. 2 (February 1974).

8. R. S. Hershel, "Unified Approach to Restoring Degraded Images in the Presence of Noise," Optical Sciences Center, University of Arizona Technical Report No. 72 (December 1971).

9. B. R. Freiden, "Restoring with Maximum Likelihood and Maximum Entropy," *JOSA,* Vol. 62, No. 4, pp. 511-518 (April 1972).

10. B. R. Freiden, and J. J. Burke, "Restoring with Maximum Entropy II: Superresolution of Photographs of Diffraction-Blurred Images," *JOSA,* Vol. 62, No. 10, pp. 1207-1210 (October 1972).

11. W. H. Richardson, "Bayesian-Based Iterative Method of Image Restoration," *JOSA,* Vol. 62, No. 1, pp. 55-59 (January 1972).

12. H. C. Andrews, "Positive Digital Image Restoration Techniques: A Survey," ATR-73 (8139)-2, Aerospace Corporation Technical Report (February 1973).

13. W. Lukosz, "Transfer of Nonnegative Signals Through Linear Filters," *Optical Acta,* Vol. 9, pp. 335-364 (1962).

14. A. V. Oppenheim, R. W. Schafer, and T. G. Stockham, Jr., "Nonlinear Filtering of Multiplied and Convolved Signals," *Proceedings of the IEEE,* Vol. 56, pp. 1264-1291 (August 1968).

15. T. G. Stockham, Jr., "Image Processing in the Context of a Visual Model," *Proceedings of the IEEE,* Vol. 60, No. 7, pp. 828-842 (July 1972).

16. A. G. Tescher, and H. C. Andrews, "Data Compression and Enhancement of Sampled Images," *Applied Optics,* Vol. 11, No. 4, pp. 919-925, (April 1972).

17. H. C. Andrews, "Digital Fourier Transforms as a Means for Scanner Evaluation," *Applied Optics,* Vol. 13, No. 1 (January 1974).

18. C. W. Helstrom, "Image Restoration by the Method of Least Squares," *JOSA,* Vol. 57, No. 3, pp. 297-303, (March 1967).

19. B. R. Hunt, "The Application of Constrained Least Squares Estimation to Image Restoration by Digital Computer," *IEEE Transaction on Computers,* Vol. C-22, No. 9, pp. 805-812 (September 1973).

20. E. R. Cole, "The Removal of Unknown Image Blurs by Homomorphic Filtering," Ph.D Dissertation, Dept. of Electrical Engineering, Univ. of Utah, Salt Lake City (June 1973).

21. A. Albert, *Regression and the Moore-Penrose Pseudoinverse,* Academic Press (1972).

22. G. H. Golub, and C. Reinsch, "Singular Value Decomposition and Least Squares Solutions," *Numer. Math.,* Vol. 14, pp. 403-420 (1970).

23. P. A. Jansson, R. H. Hunt, and E. K. Plyler, "Resolution Enhancement of Spectra," *JOSA,* Vol. 60, No. 5, pp. 596-599 (May 1970).

24. J. Philip, "Reconstruction from Measurements of Positive Quantities by the Maximum-Likelihood Method," *J. Math. Analysis and Appl.,* Vol. 7, No. 3, pp. 327-347 (December 1963).

25. D. Ferguson, Private Communications, Aerospace Corporation,

26. T. N. E. Greville, *Theory and Applications of Spline Functions,* Academic Press (1969).

27. T. N. Cornsweet, *Visual Perception,* Academic Press (1970).

28. W. F. Schrieber, "Wirephoto Quality Improvement by Unsharp Masking," *Pattern Recognition,* Vol. 2, pp. 117-123 (May 1970).

29. A. G. Tescher, "The Role of Phase in Adaptive Image Coding," Ph.D Dissertation, Dept. of Electrical Engineering, University of Southern California, Los Angeles, Calif., (January 1974).

Transfer Function Compensation of Sampled Imagery

ROGER J. ARGUELLO, HARVEY R. SELLNER,
AND JOHN A. STULLER

Abstract—With the availability of the computer for two-dimensional picture processing, digital restoration of deterministically degraded sampled images is a practical reality. This correspondence presents a mathematical formulation of discrete modulation transfer compensation of sampled imagery. The analysis makes no assumption regarding the band limitedness of the imaging system and accounts explicitly for the effects of spectrum foldover.

A series of simulation experiments were performed to determine the minimum-size processing array that would result in scene quality subjectively equivalent to that obtained using an array that is large enough to be unnoticeably affected by truncation. When the optimum truncation function is applied to the optimum processing array, virtually complete compensation is achieved utilizing a processing array of the same relative extent as the uncompensated intensity-point spread function.

Index Terms—Aliasing, image restoration, modulation function compensation, point spread function, processing array, sampled image, transfer function compensation, truncation.

Manuscript received November 6, 1971; revised March 3, 1972. A preliminary version of this correspondence was presented at the IEEE, UMC, Two-Dimensional Digital Signal Processing Conference, Columbia, Mo., October 6-8, 1971. This special correspondence contains valuable expository and original information and was published in full at the Guest Editors' request.

R. J. Arguello and H. R. Sellner are with the Perkin-Elmer Corporation, Danbury, Conn. 06810.

J. A. Stuller is with the Department of Electrical Engineering, University of New Brunswick, Fredericton, N. B., Canada.

I. Introduction

The quality of a reconstructed sampled image can be improved by processing the digital scene samples prior to recording on hard copy to remove certain degrading effects in the system where the functional nature of the degrading process is known [1]–[3]. A type of digital data processing that is presented in this correspondence is transfer function compensation (TFC). This correspondence provides a mathematical formulation of discrete TFC of sampled imagery, describes the effects of TFC, gives a basis for choosing processing array coefficients, and shows the effects of truncation on the processing array.

TFC is the process of boosting high-spatial frequency content in imagery to compensate for the previous degrading effects of the optics, motion of the image on the sensor or film, and the (spatial) frequency-response limitations of the recording medium. These degradations often act as a series of low-pass filters (shown in Fig. 1) that can be considered as being cascaded. In such cases, the net modulation transfer function (MTF) is the product of the individual MTF's. If these are known, a filter can be constructed that is the inverse of the net MTF. When imagery is applied to such a filter, its high-frequency content (detail) can be restored as illustrated in Fig. 1. When applied to two-dimensional picture processing, discrete compensation presents a viable technique. The scene is sampled such that the intensity at each point in a two-dimensional array is represented by a scene value. Scene values in the array form are then convolved with values of a two-dimensional processing filter array to arrive at a new array —the inverse filtered-image values. These values may then be re-

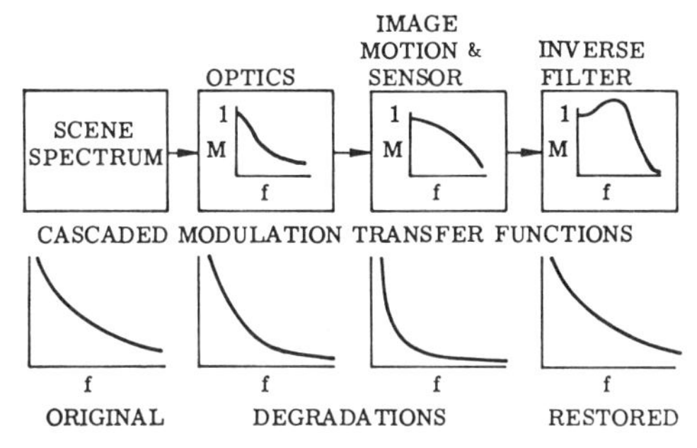

Fig. 1. Transfer function compensation.

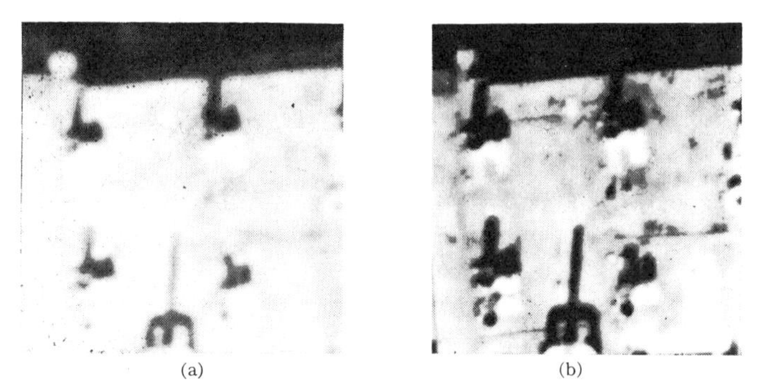

Fig. 2. Digital image restoration. (a) Before. (b) After.

corded on film as the final copy. Fig. 2(a) shows a sampled image that is degraded because of the effects of optics and image motion. Fig. 2(b) is the restored image after application of discrete TFC.

Assumptions found necessary to provide analytical tractability sometimes conflict with accurate modeling of the subjective trade-offs involved. The problem is compounded because the proper merit functions for picture quality are not as yet identified. In view of these difficulties, a series of simulation experiments were conducted depicting the effects of TFC. The approach taken in this correspondence is to generate sampled-image simulations using a facsimile-type scanner and associated minicomputer [4].

For notational simplicity, the analyses are performed in one dimension wherever no loss of generality results. Extension to two dimensions is straightforward.

II. Analytical Representation of Transfer Function Compensated System

For a sampled incoherent optical-imaging system with an intensity-point spread function $b(x)$ (that may include contributions from optics, image motion, and sensor), the imaged and sampled scene,

$$[b(x) * s_{\text{in}}(x)] \sum_n \delta(x - n\Delta x)$$

has a periodic spectrum

$$\frac{1}{\Delta x} \sum_n S_{\text{in}}(f - nf_s) B(f - nf_s)$$

where Δx is the sample spacing (reciprocal of sampling frequency f_s) and $B(f)$, $S_{\text{in}}(f)$ are Fourier transforms of $b(x)$ and the input scene $s_{\text{in}}(x)$, respectively [5, ch. 1]. It is assumed that the sampling array is infinite in extent $(-\infty < n < \infty)$. The transmission function $B(f)$ may be modified by performing a digital convolution on the imaged- and sampled-scene elements with a discrete processing array $[p_n]$. By representing $[p_n]$ as an array of impulses, one can express the result of the digital convolution as the product

$$\left[\frac{1}{\Delta x} \sum_n S_{\text{in}}(f - nf_s) B(f - nf_s)\right] Q(f)$$

where $Q(f)$ is the Fourier transform of the processing impulse array.

$$Q(f) = \sum_n p_n e^{-j2\pi f n \Delta x}. \tag{1}$$

It can be seen from (1) that $Q(f)$ is a periodic function in f with period f_s.

$$Q(f - mf_s) = Q(f), \qquad m = 0, \pm 1, \pm 2, \cdots. \tag{2}$$

Using (2) the spectrum of the processed-scene data sequence may be written

$$\left[\frac{1}{\Delta x} \sum_n S_{\text{in}}(f - nf_s) B(f - nf_s)\right] Q(f) = \frac{1}{\Delta x} \sum_n S_{\text{in}}(f - nf_s) B'(f - nf_s)$$

$$= \frac{1}{\Delta x} \sum_n S_{\text{proc}}(f - nf_s) \tag{3}$$

where

$$B'(f) = Q(f) B(f) \tag{4}$$

$$S_{\text{proc}}(f) = S_{\text{in}}(f) B'(f). \tag{5}$$

Equation (4) gives the modified transmission function of the sampled-imaging system. It is seen that sampling has constrained the form of the possible modifications of $B(f)$; $Q(f)$ must be periodic.

By taking the inverse transform of (4), one obtains the modified system intensity-point spread function:

$$b'(x) = b(x) * \sum_n p_n \delta(x - n\Delta x) = \sum_n p_n b(x - n\Delta x). \tag{6}$$

Finally, if the sampled system uses an output device whose point spread and MTF's are denoted by $m(x)$ and $M(f)$, then the output signal $s_{\text{out}}(x)$ and its spectrum $S_{\text{out}}(f)$ will be

$$s_{\text{out}}(x) = \left\{\sum_n s_{\text{proc}}(n\Delta x) \delta(x - n\Delta x)\right\} * m(x)$$

$$= \sum_n s_{\text{proc}}(n\Delta x) m(x - n\Delta x) \tag{7}$$

and

$$S_{\text{out}}(f) = M(f) \frac{1}{\Delta x} \sum_n S_{\text{proc}}(f - nf_s)$$

$$= M(f) \frac{1}{\Delta x} \sum_n B'(f - nf_s) S_{\text{in}}(f - nf_s). \tag{8}$$

Figs. 3 and 4 illustrate possible forms for $B(f)$, $Q(f)$, $B'(f)$, and their transforms in a sampled-imaging system. In this context, (4) and (6) are fundamental to an understanding of TFC in sampled-imaging systems. It should be noted that these equations, as well as the other results obtained, have made no assumptions regarding the bandlimitedness of the original system.

Note that if the original transmission function $B(f)$ had a cutoff frequency less than $f_s/2$, then no foldover terms would exist in the output spectrum in the region $|f| < f_s/2$. It is also noteworthy for this case that the periodicity of $Q(f)$ would have no effect upon the product $B(f)\ Q(f)$. Therefore, if the system samples occur at the Nyquist rate, no loss of freedom in choosing $B'(f)$ results from sampling.

It has been determined experimentally that certain scene features such as edges become sharper when the cutoff of $B(f)$ is extended beyond the critical frequency. An improvement in subjectively judged scene quality results by accepting some aliasing artifacts for an increase in apparent edge sharpness.

A. *Specification of the Processing Array*

We now derive the processing array coefficients $[p_n]$ for a system which is not band limited at the critical frequency.

Since $Q(f)$ is periodic, it may be expanded into a Fourier series. Such a series will be unique and is, in fact, given by (1). It follows that the coefficient p_n is given by

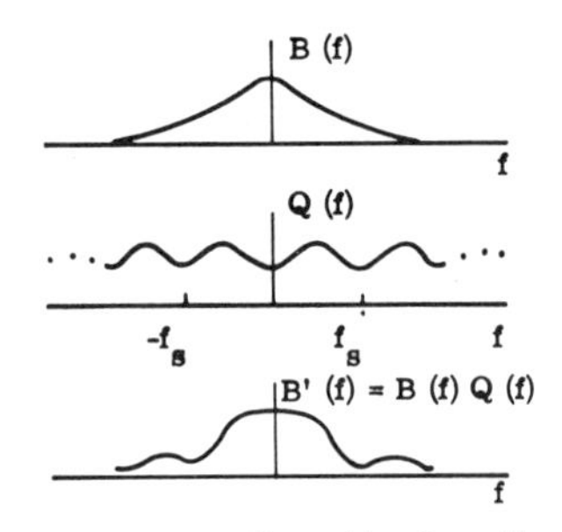

Fig. 3. Modification of $B(f)$ resulting from discrete transfer function compensation of sampled imagery.

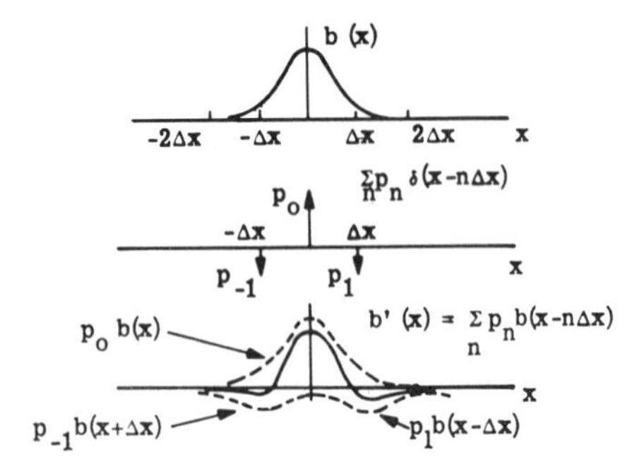

Fig. 4. Modification of $b(x)$ resulting from discrete transfer function compensation of sampled imagery.

$$p_n = \frac{1}{f_s}\int_{-f_s/2}^{+f_s/2} Q(f)e^{-j2\pi f n\Delta x}\,df. \tag{9}$$

$Q(f)$ may be specified by its values within a period $(-f_s/2 \le f \le +f_s/2)$ and p_n subsequently found from (9). Alternatively, $Q(f)$ may be specified in terms of an arbitrary function $P(f)$ (not necessarily band limited) by setting

$$Q(f) = \sum_n P(f - nf_s). \tag{10}$$

Note that the inverse Fourier transform of $P(f)$, $p(x)$, has samples $p(n\Delta x) = p_n$:

$$F^{-1}\left[\frac{1}{\Delta x}\sum_n P(f - nf_s)\right] = \sum_n p_n\delta(x - n\Delta x). \tag{11}$$

Hence,

$$p_n = \int_{-\infty}^{+\infty} P(f)e^{-j2\pi f n\Delta_x}\,df. \tag{12}$$

The array $[p_n]$ is obtained by computing (12) rather than (10). $P(f)$ is specified from the division

$$P(f) = \frac{D(f)}{B(f)} \tag{13}$$

where $D(f)$ is a desired transmission function. Substituting (13) and (10) into (4) and rearranging gives

$$B'(f) = D(f) + B(f)\sum_{n\neq 0} P(f - nf_s). \tag{14}$$

For a circularly symmetric $D(f)$, the terms not centered at the origin of the spatial frequency domain are attenuated by $B(f)$ and become zero beyond the cutoff of $B(f)$.

B. Processing Array Truncation

Fig. 4 illustrates how a system blur spot $b(x)$ covering essentially three detector elements may be sharpened by a processing array $[p_n]$ containing three elements. A problem with designing a system purely from such spatial domain considerations is that this approach does not readily permit control over the boost that may result from frequency components for which aliasing has occurred. Such a description is found by considering the shape of $P(f)$ [or $Q(f)$].

As described previously, a degree of control on the compensated system frequency response can be obtained by deriving the processing array $[p_n]$ from a specification of the desired frequency response $D(f)$.

For an arbitrary $D(f)$, (13) and (12), in general, will yield an infinite number of nonzero coefficients p_n. The minimum number of coefficients required to approach or match the subjectively adjudged quality attained by a scene that has been transfer function compensated with an arbitrarily large processing array must ultimately be determined by viewing experiments. The optimum method for choosing the coefficients for a small array must also be dependent upon an ultimate subjective determination. Hence, standard analytical truncation criteria, such as minimization of out-of-band ripple, minimization of mean-square-error measures, and others, will not necessarily lead to the optimum subjective tradeoffs between resolution, aliasing, and overshoot.

However, it seems clear that the elements of a small processing array should be chosen in such a way that the resulting system transfer function approximates in some sense the transfer function $B'(f)$ found to be optimum for a very large array. [Hence the number of coefficients p_n required will be dependent upon $D(f)$.]

The effect of processing spot truncation on $Q(f)$, denoted $\hat{Q}(f)$, can be expressed as a convolution of $Q(f)$ and a truncation window $T(f)$:

$$\hat{Q}(f) = Q(f) * T(f). \tag{15}$$

The effects of processing spot truncation on $P(f)$ may also be expressed as a convolution of $P(f)$ with the truncation window $T(f)$:

$$\hat{P}(f) = P(f) * T(f). \tag{16}$$

Hanning and rectangular truncation windows were studied to determine the impact on output image quality and are discussed in the TFC experiments described in Section III of this correspondence.

C. Noise Considerations

In a scanner-transmission system, noise in introduced by the sensor in the process of scanning. Noise statistics may vary, depending on the particular sensor type and parameters. In this correspondence the sensor noise model chosen is that of a zero mean additive process having a Gaussian probability density function and white power spectral density.

An often noticeable property of the output noise on a transfer function compensated scene is that it appears to have spatial dependence from one scene element to another. That is, it appears to form a series of random patterns made up of line segments and contours. For systems using a linear interpolation reconstruction point spread function, the essential properties of these noise patterns may be conveniently described in terms of the patterns formed by the discrete noise samples $[n_{ij}]$, where n_{ij} is the noise on the ijth sample lattice point on the output scene.

Consider any two sample lattice points (say the ijth and lmth) on the transfer function compensated scene. The noise values at these points are given by

$$n_{ij} = \sum_{q,v} w_{qv}p_{i-q,j-v} \tag{17}$$

and

$$n_{lm} = \sum_{r,s} w_{rs}p_{l-r,m-s} \tag{18}$$

where w_{kh} is the noise value at the khth-sampled scene element and $p_{k,h}$ is the khth element of the two-dimensional processing array. Since the detector noise values $\{w_{kh}\}$ are zero mean and jointly Gaussian, it follows that the output noise elements n_{ij} and n_{lm} are zero mean and jointly Gaussian. The sensor noise values are uncorrelated (in fact, orthogonal):

$$E\{w_{qv}w_{rs}\} = \begin{cases} 0, & \text{for } (q, v) \neq (r, s) \\ \sigma_w^2, & \text{for } (q, v) = (r, s). \end{cases} \tag{19}$$

It follows from (17)–(19) that

$$E\{n_{ij}\,n_{lm}\} = \sum_{q,v}\sum_{r,s} E\{w_{qv}\,w_{rs}\}\,p_{i-q,j-v}\,p_{l-r,m-s}$$
$$= \sigma_w^2 \sum_{q,v} p_{i-q,j-v}\,p_{l-q,m-v}$$
$$= \sigma_w^2 \sum_{k,h} p_{k,h}\,p_{k+I,h+J} \quad (20)$$

where

$$I \equiv l - \quad (21)$$
$$J \equiv m - j. \quad (22)$$

Hence $E\{n_{ij}\,n_{lm}\}$ is a function only of $(l-i)$ and $(m-j)$ so that the output Gaussian noise array is stationary along its rows and columns.

Note that the power (variance) of the ijth-output noise sample is

$$\sigma_n^2 = E\{n_{ij}^2\} = \sigma_w^2 \sum_{k,h} p_{k,h}^2. \quad (23)$$

Finally, it follows from (20) and (23) that the correlation coefficient between n_{ij} and n_{lm} is

$$\rho_{ij,lm} = \frac{E\{n_{ij}\,n_{lm}\}}{\sigma_n^2} = \frac{\sum_{k,h} p_{k,h}\,p_{k+I,h+J}}{\sum_{k,h} p_{k,h}^2} = \rho_{I,J} \quad (24)$$

where I and J are given by (21) and (22).

Correlation coefficient matrices may be computed from (24) and a determination can be made of the correlated spatial noise extent as a function of the processing array size. A comparison of the noise extents for several processing arrays is given in Section III, which is concerned with noise experiments.

III. TFC Experiments

The approach taken in accomplishing the TFC objectives consisted of three separate parts. Part one consisted of a theoretical and experimental determination of the kind of system response that is desirable as a design goal using a large processing array (13×13). The second part of the effort consisted of a series of truncation experiments designed to determine the minimum-sized processing array that would result in scene quality subjectively equivalent to that attained by the best 13×13 array. The third part consisted in an experimental determination of the effects of additive noise on the simulated imagery.

A. System Response Design Goal

On the basis of a preliminary study, three candidate desired transfer functions were chosen for initial comparison. They are the: 1) "ideal MTF"—$D_1(f_r)$ [circular symmetry]; unity response to desired rolloff frequency, cosine rolloff to cutoff frequency, zero elsewhere; 2) MTF of a circular diffraction limited aperture—$D_2(f_r)$; and 3) "quarter-wave zero-order Bessel MTF"—$D_3(f_r)$ [circular symmetry].

Each of the above three transfer functions are shown in Fig. 5 and given analytically in Table I. Each represents a system property that is regarded to be in some sense fundamental. $D_1(f)$ passes a certain band of scene frequency components without distortion. $D_2(f)$ yields a monotonically increasing edge response. The point spread function (PSF) corresponding to $D_3(f)$ attains the minimum possible circular second moment for any circularly symmetric band-limited function [5, p. 218]. While an infinite variety of other candidate design MTF's exist, experience gained from the use of the above three provided some insight into the types of compensation possible and the degree of scene quality differences that result from MTF's having basically different forms.

A series of 13×13 processing arrays was derived for each candidate MTF in order to permit qualitative study of the resulting scenes. The primary purpose of these experiments was to obtain an understanding of the effects of the various transfer functions on the scene component of a given image. Therefore, emphasis was placed upon processing the highest signal-to-noise case of the assigned set. The

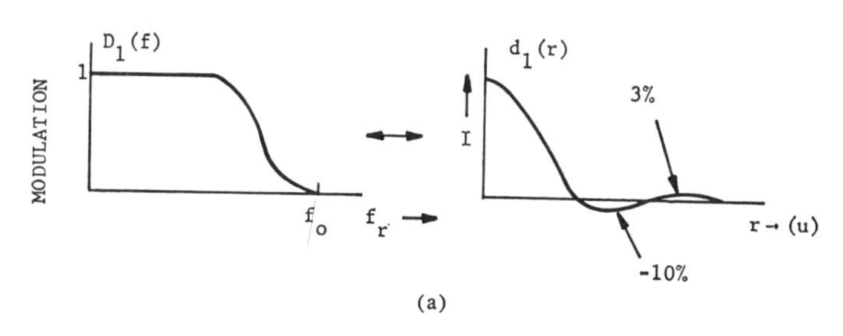

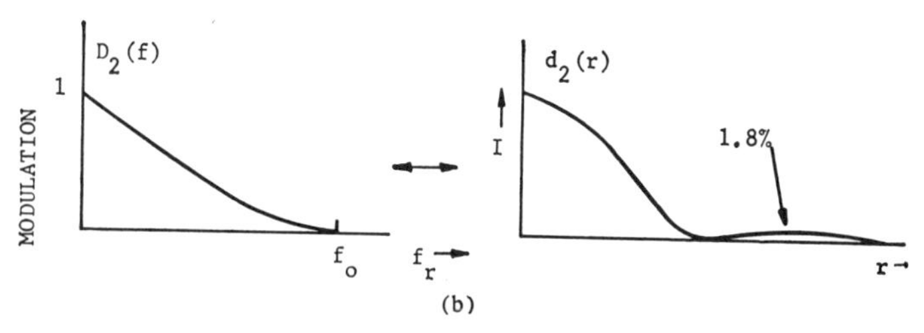

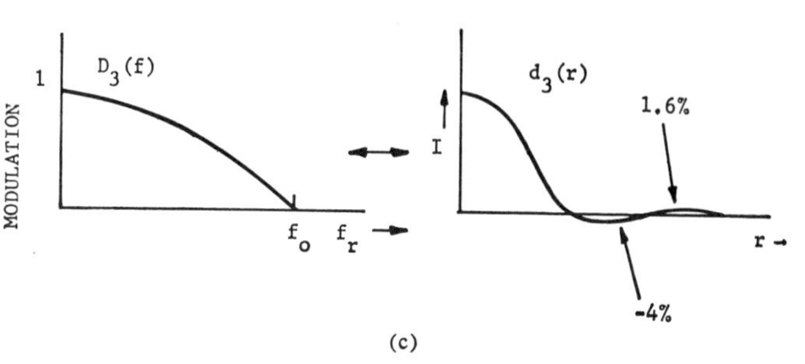

Fig. 5. Candidate "desired" transfer functions and their transforms. (a) "Ideal" MTF. (b) Circular diffraction-limited MTF. (c) Quarter-wave Bessel MTF.

TABLE I

Analytical Expressions for Desired Transfer Functions

Type	Analytical Expression
"Ideal MTF"	$D_1(f_r) = \begin{cases} 1 \text{ for } 0 \le f_r \le f_o - f_w \\ \cos\{\pi(f_r - f_o + f_w)/f_w\} \text{ for } f_o - f_w \le f_r < f_o \\ 0 \text{ for } f_o \le f_r \end{cases}$ f_w = width of cosine function rolloff
Circular Diffraction Limited MTF	$D_2(f_r) \begin{cases} \frac{2}{\pi}\cos^{-1}(f_r/f_o) - \frac{2f_r}{\pi f_o}\sqrt{1-(f_r/f_o)^2} \\ \text{for } 0 \le f_r < f_o \\ 0 \text{ for } f_o \le f_r \end{cases}$
Quarter-Wave Bessel MTF	$D_3(f_r) = \begin{cases} J_o(2\pi\alpha f_r) & \text{for } 0 \le f_r < f_o \\ 0 & f_o \le f_r \end{cases}$ $2\pi\alpha f_o = 2.405$

f_r = radial spatial frequency $= \sqrt{f_x^2 + f_y^2}$

f_o = cutoff spatial frequency

processed scenes in each case displayed characteristics that clearly conformed with the theoretical properties of the design MTF. Thus, the scenes processed for the ideal MTF case displayed high resolution accompanied by visible overshoot artifacting. The diffraction-limited circular aperture desired MTF resulted in apparent monotonic edge response (no overshoot) but lower resolution. Design transfer func-

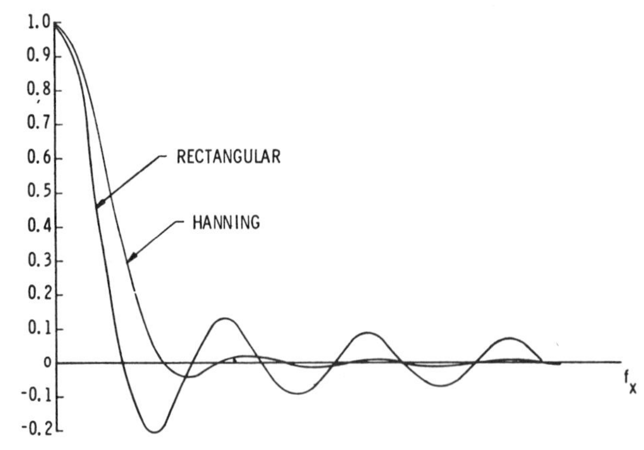

Fig. 6. Fourier transforms of rectangular and Hanning truncation functions.

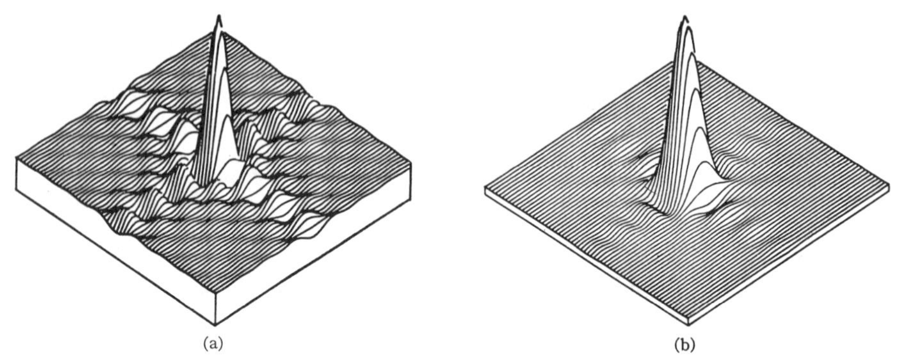

Fig. 7. Isometric views of truncation function transforms. (a) Rectangular function. (b) Hanning function.

tion $D_3(f)$ resulted in edge sharpness similar to that of $D_1(f)$ but with decreased overshoot artifacting.

On the basis of subjective comparison of imagery, $D_3(f)$ was chosen as giving the best overall quality image.

B. Processing Array Truncation

The second part of the study consisted of a series of truncation experiments. The optimum processing spot was truncated to 11×11, 9×9, 7×7, 5×5, and 3×3 arrays by Hanning and rectangular truncation windows t_H and t_R. The Hanning window is a raised cosine function evaluated between $-\pi/2$ and $\pi/2$. The rectangular window is a rectangular function, unity within the processing area, and zero elsewhere. The spatial frequency domain plot of these windows (Fig. 6) shows that an improved frequency response with less overshoot may be obtained with a Hanning truncation window. The larger amount of ripple created by the rectangular window is further evidenced by comparing the three-dimensional plots in Fig. 7.

These experiments showed that truncation was found to cause no significant difference in the output print until spot sizes 7×7 for rectangular truncation and 5×5 for Hanning truncation had been reached It was also observed that Hanning truncation causes a very gradual and predictable lowering of the net system MTF and, therefore, the loss in scene quality could be recovered by suitably boosting the processing MTF before truncating the processing spot. For these reasons, the Hanning truncation window was used in the remainder of experiments described in this section.

Fig. 8 illustrates the drop in compensated system MTF, $\hat{D}(f_x, 0)$, resulting from Hanning truncation of the processing array. The carat denotes approximation to the desired MTF, $D_3(f_x, 0)$. The system MTF was then recovered by preboosting the processing MTF prior to spot truncation. The processing function $P(f_x, f_y)$ was divided by $\text{sinc}^n (f_x/f_s)\ \text{sinc}^n (f_y/f_s)$ before truncation of the processing array as a method of implementing the preboost. The parameter n was adjusted to compensate for distortion introduced by the truncation operation. Fig. 9 demonstrates the recovery in net system MTF obtained for the 3×3, 5×5, and 7×7 element array sizes. The uncompensated MTF, $B(f)$ is also shown for comparison in Fig. 9.

Fig. 10(a) is a sampled image of a concrete yard that has undergone degradations created by the combined optics, image motion, and sensor MTF, $B(f)$. Fig. 10(b) and (c) show the effects of digital restoration with the use of 3×3 and 7×7 element processing spots, respectively. Note that the restoration employing the 3×3 processing spot results in imagery subjectively equivalent to that obtained by the larger array.

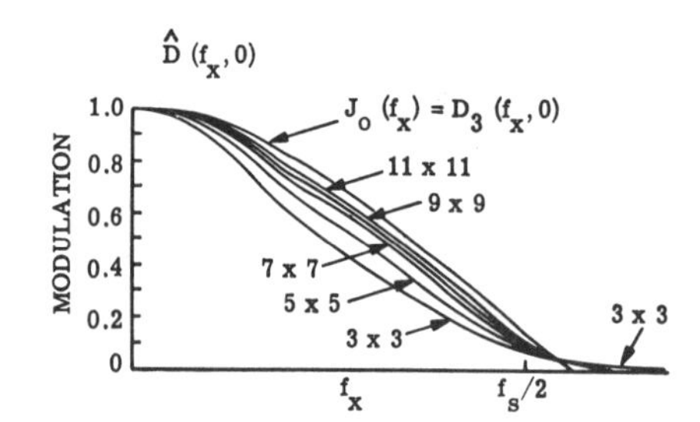

Fig. 8. Impact of Hanning truncation on desired MTF.

The optimized 3×3, 5×5, and 7×7 element processing arrays used in these simulation experiments are given in Table II. Note that the algebraic sum of the elements in each array is unity so that the corresponding modulation transfer function is maintained at unity at zero spatial frequency. This normalization insures that the processing array does not alter the mean value of the scene.

C. Compensation of Imagery Degraded by Noise

In order to quantitatively assess the effect of compensation on imagery that has been degraded by additive Gaussian noise, a series of processing-array noise-correlation coefficient matrices were computed. These matrices, whose elements are in percent correlation ($100\ \rho_{I,J}$), are listed in Table III for the three processing-array cases considered. Elements of the correlation matrix, whose magnitudes are a fraction of one percent, are truncated to zero.

The entry in the first row first column of the correlation array gives the correlation between a chosen noise element n_{ij} and itself and is accordingly always 100 percent. The entry to the immediate right of this element gives the correlation between the chosen noise element

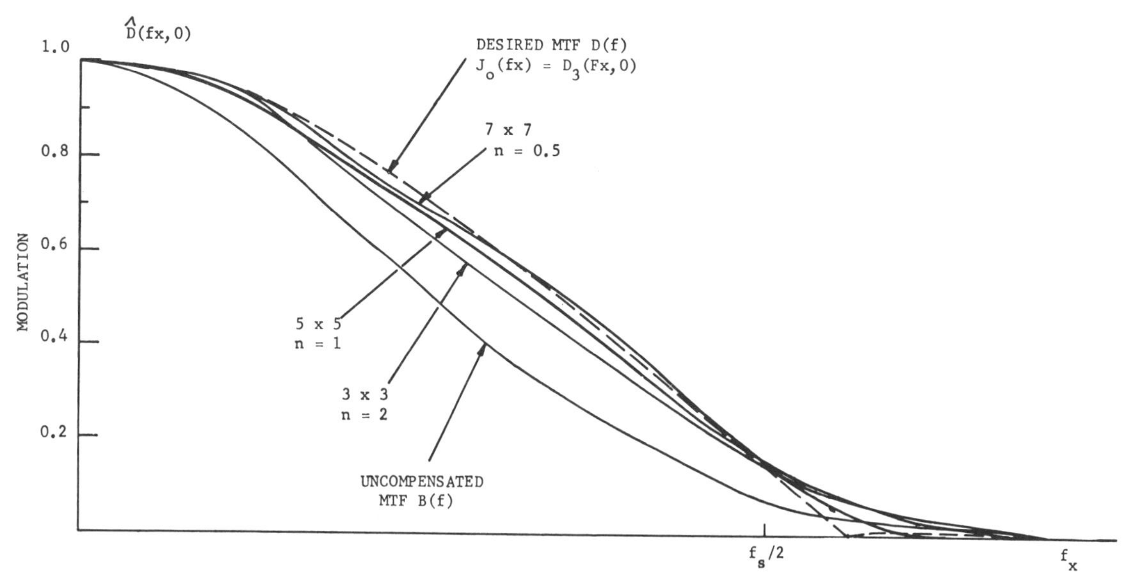

Fig. 9. Effect of preboost on truncated MTF's.

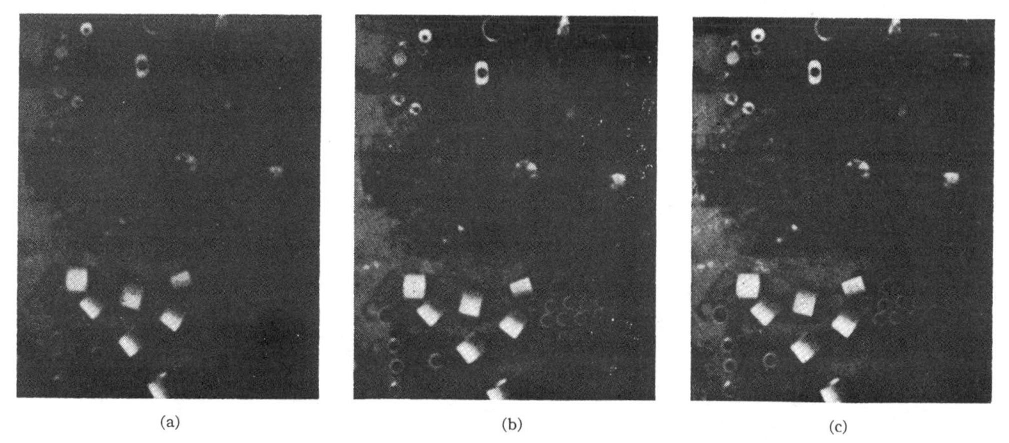

Fig. 10. Concrete yard. (a) Before TFC. (b) After TFC 3×3 processing array. (c) After TFC 7×7 processing array.

TABLE II

PROCESSING ARRAYS FOR TFC EXPERIMENTS

-0.157	0.006	-0.157
0.030	1.558	0.030
-0.157	0.006	-0.157

(a) 3 x 3 Array

0.012	0.013	-0.017	0.013	0.012
0.009	-0.168	0.059	-0.168	0.009
0.023	0.080	1.337	0.080	-0.023
0.009	-0.168	0.059	-0.168	0.009
0.012	0.013	-0.017	0.013	0.012

(b) 5 x 5 Array

0.001	-0.005	-0.003	0.003	-0.003	-0.005	0.001
-0.005	0.021	0.013	-0.033	0.013	0.021	-0.005
-0.001	0.008	-0.174	0.090	-0.174	0.008	-0.001
0.004	-0.039	0.111	1.309	0.111	-0.039	0.004
-0.001	0.008	-0.174	0.090	-0.174	0.008	-0.001
-0.005	0.021	0.013	-0.033	0.013	0.021	-0.005
0.001	-0.005	-0.003	0.003	-0.003	-0.005	0.001

(c) 7 x 7 Array

TABLE III

CORRELATION COEFFICIENT MATRICES

100	4	2
0	-19	0
2	0	1

(a) 3 x 3 Processing Array

100	9	0	0	0
5	-22	0	0	0
1	1	3	0	0
-1	0	0	0	0
0	0	0	0	0

(b) 5 x 5 Processing Array

100	11	-1	0	0	0	0
8	-22	-2	1	0	0	0
-1	0	2	-1	0	0	0
0	0	-1	0	0	0	0
0	0	0	0	0	0	0
0	0	0	0	0	0	0
0	0	0	0	0	0	0

(c) 7 x 7 Processing Array

n_{ij} and the noise element to the immediate right of it on the scene. The other array elements are treated accordingly.

The correlation between a noise element n_{ij} and an element corresponding to a point outside the limits of these arrays is zero. An examination of these matrices verifies the expectation that the smaller processing arrays tend to have smaller spatial regions of noise dependence than the larger arrays.

Fig. 11(a) shows a sampled-image ship scene that has been defocused to display the effects of optics, sensor, and image motion MTF degradation. Computer-generated noise with a Gaussian prob-

(a) (b)

Fig. 11. Ships scene signal-to-noise ratio = 46 dB. (a) Before TFC. (b) After TFC 13×13 processing array.

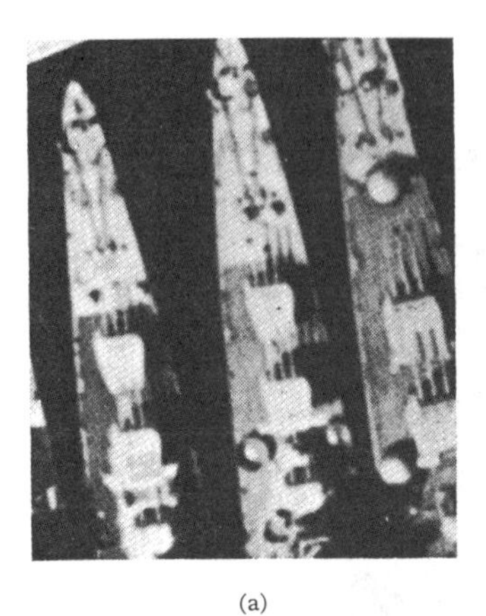
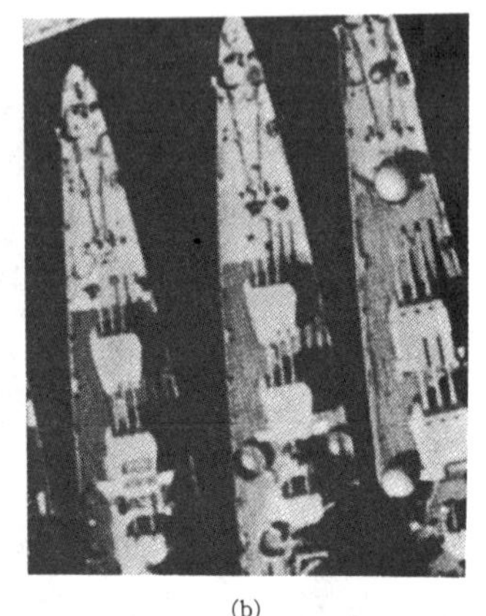

(a) (b)

Fig. 12. Ships scene signal-to-noise ratio = 40 dB. (a) Before TFC. (b) After TFC 13×13 processing array.

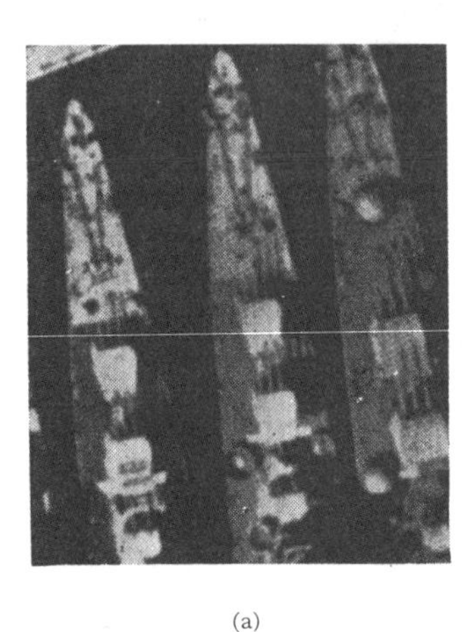
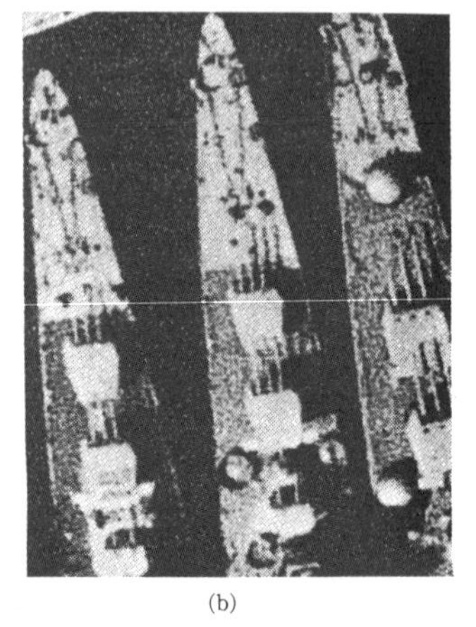

(a) (b)

Fig. 13. Ships scene signal-to-noise ratio = 26 dB. (a) Before TFC. (b) After TFC 13×13 processing array.

ability density function was added to this scene to simulate the effect of scanner noise appearing in a real scanner/transmission system having a 46 dB peak-to-peak signal-to-rms-noise ratio. Fig. 11(b) shows the effect of digital restoration by use of the MTF compensation algorithm. A 13×13 element processing array was used in the ship scene images. Note that even as the signal-to-noise ratio deteriorates from 46 to 40 dB and finally to 26 dB, as shown in the sequence of images (Figs 12 and 13), that the imagery is crispened to make fine detail more accessible to the viewer. However, the noise appears to be more structured as the signal-to-noise ratio decreases and the effectiveness of digital restoration is reduced.

IV. Simulation System

The instrument employed for the above simulations is the line scan image generator [4]. It is a drum-type photographic scanner that transforms a photographic transparency into a precisely controlled digitized line scan image via a 12-bit analog-to-digital converter for use in computer processing. In addition, the line scan image generator performs the inverse operations, transforming 8-bit digital line scans (or computer tape) into a photographic transparency. The two-dimensional convolutions were performed on the digitized scene values using full 12-bit arithmetic. All scenes were prepared from positive transparencies, which were photographically processed to provide a linear relationship between transmittance and scene reflectance.

V. Discussion

Processing arrays found to be optimum with respect to high signal-to-noise ratio scenes are also optimum (or very nearly optimum) for a wide range of signal-to-noise ratios. This result is attributed to the fact that the spatial patterns formed by the sensor noise are typically very different from those of the scene "signal" objects. Therefore, a viewer will prefer to have the sharpest possible scene signal (i.e., that obtained by processing with the array found best for high signal-to-noise ratio) since he can easily distinguish noise from the signal and can see around or through it. However, there appears to be a threshold condition where processed noise can be so great that it obliterates substantial areas of the signal. With noise above this threshold, less scene sharpening is desirable. The nature of this noise and signal interaction clearly depends heavily upon the spatial structure of the signal. The processing array should not tend to cause the output noise to have a structure similar to the signal.

This correspondence has shown both analytically and experimentally that digital processing can be applied to reduce the net system blur function (in the absence of aliasing) when applied to sampled imagery. Further, the extent of the processing array need be no larger than the extent of the blur function. In many systems where the MTF falloff is not too rapid and near-Nyquist sampling is employed, either a 3×3 or 5×5 element processing array can be used effectively. A significant implication of this conclusion is that this type of processing lends itself well to minicomputer operation, since the number of required operations per scene element are few and storage requirements are only three to five lines of scene information. Use of a small processing array implies that efficient compensation may be carried out by digital convolution rather than operations in the frequency domain. Since processing time per scene element via convolution is proportioned to n^2 where $n \times n$ is the array size, even more incentive is given to keep the array sizes small. Thus the 7×7 array required a computation time approximately five times longer than the 3×3 array to process per scene element. For most applications, the nine multiplications and additions required for a 3×3 array are possible in "real time," while the images are being either transmitted or received. If the extent of image degradation requires an array size that precludes real-time minicomputer convolution, one may utilize either a hardwired processor or perform compensation operations in the spatial frequency domain using fast Fourier transform techniques.

Acknowledgment

The authors are indebted to M. I. Crockett for performing the line scan image generation of all scans shown in this correspondence.

References

[1] D. P. MacAdam, "Digital image restoration by constrained deconvolution," *J. Opt. Soc. Amer.*, vol. 60, pp. 1617–1627, Dec. 1970.
[2] T. S. Huang, W. F. Schreiber, and O. J. Tretiak, "Image processing," *Proc. IEEE*, vol. 59, pp. 1586–1609, Nov. 1971.
[3] J. L. Harris, Sr., "Image evaluation and restoration," *J. Opt. Soc. Amer.*, vol. 56, pp. 569–574, May 1966.
[4] F. Scott, "A line scan image generator," *Photographic Sci. Eng.*, vol. 11, pp. 348–351, Sept.–Oct. 1967.
[5] A. Papoulis, *Systems and Transforms with Application in Optics.* New York: McGraw-Hill, 1968, ch. 4.

Space-Variant Image Motion Degradation and Restoration

ALEXANDER A. SAWCHUK, MEMBER, IEEE

***Abstract*—A description of motion degradation in linear incoherent optical systems is presented. Given a mechanical description of the motion, an equivalent linear space-variant system containing all the motion effects is derived, and detailed examples of common types of variant and invariant motion are included. Following a review of restoration techniques for motion blur, a method for image restoration applicable to a large class of space-variant systems is presented. This method is based on the decomposition of the degradation into geometrical coordinate distortions and a space-invariant operation. A computer simulation of space-variant restoration is included.**

I. Introduction

IN THE PHOTOGRAPHY of the Earth, Moon, and planets by aerial vehicles such as aircraft and spacecraft, degradations due to motion blurring and geometrical distortion are often the factors limiting resolution in the recorded image [1]. Motion effects have received little attention in the past because optical system degradations from diffraction, aberrations, and recording medium response [2] were more severe. With improvements in these areas and longer exposures necessary to photograph the faintly radiating far planets of the solar system, the study of motion has become more important. In this paper we analyze motion effects in two-dimensional incoherent imaging systems by modeling the degradation as a linear operation. Previous work in this field has been limited almost exclusively to space-invariant imaging in which the same degradation is applied to each point of the object intensity function. With knowledge of imaging system motion and orientation, we extend the linear system model to the space-variant case in which the blurring of the object varies with position. Aerial imaging is often space-variant, and we discuss the motion functions and linear system operations for this case in detail.

Once the motion degradation has been described mathematically, a more difficult task is to process the recorded image by *a posteriori* methods in the hope of restoring the image to the original object intensity function. With the availability of large high-speed digital computers, a number of techniques are known for the restoration of space-invariant motion blur. Unfortunately, restoration of space-variant degradations has been limited by the enormous computational effort required for a general solution. By examining motion blur from a system viewpoint, we show that many space-variant degradations can be decomposed into space-invariant systems whose variables are related to object and image by geometrical coordinate transformations. When the decomposition is possible, space-variant restoration reduces to reversing the geometrical transformations and using space-invariant restoration methods. We illustrate many types of degradation and restoration by examples, including a digital simulation of space-variant processing.

Manuscript received December 17, 1971; revised April 20, 1972. This work was performed in part at Stanford University, Stanford, Calif., with the support of the Office of Naval Research, and in part at the University of Southern California, Los Angeles, with the support of the Advanced Research Projects Agency of the Department of Defense, and monitored by the Air Force Eastern Test Range under Contract F08606-72-C-0008.

The author is with the Department of Electrical Engineering, University of Southern California, Los Angeles, Calif. 90007.

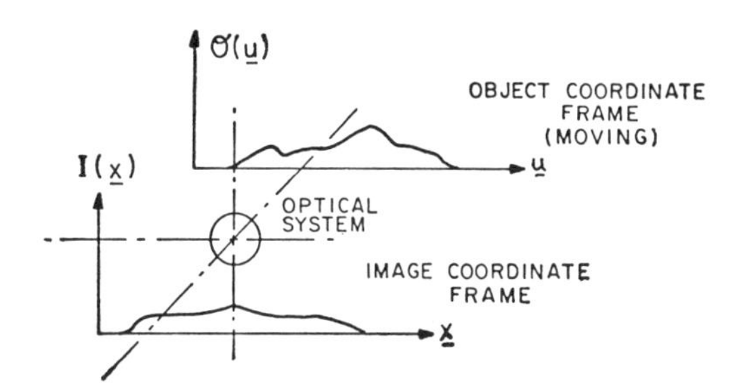

Fig. 1. Coordinate frames.

II. Image Motion Degradation

A. General Expressions

When motion is the major source of degradation in incoherent optical imaging systems, an equivalent linear system model can be derived which includes all the motion effects. We assume geometrical optics approximations are valid and derive the degradation for a system with two spatial dimensions.

In our analysis of general motion, the two coordinate frames shown in Fig. 1 are used. Given a mechanical description of the motion, a general linear model for describing the degradation from object to image is the superposition integral

$$\mathcal{I}(\boldsymbol{x}) = \int_{-\infty}^{\infty} h(\boldsymbol{x}, \boldsymbol{u})\mathcal{O}(\boldsymbol{u})d\boldsymbol{u} \tag{1}$$

where $\mathcal{O}(\boldsymbol{u})$ is the original object intensity function, $\mathcal{I}(\boldsymbol{x})$ is the image intensity recorded by the system, and $h(\boldsymbol{x}, \boldsymbol{u})$ is the response in the image coordinates $\boldsymbol{x} = (x_1, x_2)$ to a unit impulse at $\boldsymbol{u} = (u_1, u_2)$ in the object coordinates [3]. Generally, the response $h(\boldsymbol{x}, \boldsymbol{u})$ in the image space varies with the position $\boldsymbol{u}$ of the input impulse and is called a space-variant point-spread function (SVPSF) in an optical context. If $h(\boldsymbol{x}, \boldsymbol{u})$ is a function only of the difference between $\boldsymbol{x}$ and $\boldsymbol{u}$, then $h(\boldsymbol{x}, \boldsymbol{u})$ is a space-invariant point-spread function (SIPSF) and (1) simplifies to a convolution.

Assume now that a mechanical description of the motion is available parametrically in the form

$$u_1 = g_1(x_1, x_2; t) = g_1(\boldsymbol{x}; t) \tag{2a}$$

$$u_2 = g_2(x_1, x_2; t) = g_2(\boldsymbol{x}; t). \tag{2b}$$

These functions uniquely relate any object point $\boldsymbol{u}$ to the location $\boldsymbol{x}$ of its image in the fixed frame as a function of time for the exposure interval $[0, T]$. Denoting an element of area in $\boldsymbol{x}$ by $d\boldsymbol{x} = dx_1dx_2$, and in $\boldsymbol{u}$ by $d\boldsymbol{u} = du_1du_2$, we adopt the simple model that the power collected by the imaging system which is radiated from a particular spatial region of the object is conserved through the imaging process at any time instant

Reprinted from *Proc. IEEE*, vol. 60, pp. 854–861, July 1972.

during movement. Thus the power from a small region du around object point $\boldsymbol{u}$ is measured as

$$\mathcal{I}(\boldsymbol{x}, t)d\boldsymbol{x} = \mathcal{O}(\boldsymbol{u})d\boldsymbol{u} \tag{3}$$

in the image space at time instant t. Using the motion relations (2), we substitute into (3) to find

$$\mathcal{I}(\boldsymbol{x}, t) = \mathcal{O}(g_1(\boldsymbol{x}; t), g_2(\boldsymbol{x}; t))J_g(\boldsymbol{x}, t) \tag{4}$$

where $J_g(\boldsymbol{x}, t)$ is the Jacobian function

$$J_g(\boldsymbol{x}, t) = \begin{vmatrix} \dfrac{\partial g_1(\boldsymbol{x}; t)}{\partial x_1} & \dfrac{\partial g_1(\boldsymbol{x}; t)}{\partial x_2} \\ \dfrac{\partial g_2(\boldsymbol{x}; t)}{\partial x_1} & \dfrac{\partial g_2(\boldsymbol{x}; t)}{\partial x_2} \end{vmatrix}. \tag{5}$$

We now sum $\mathcal{I}(\boldsymbol{x}, t)$ over the exposure interval $[0, T]$ to get the recorded image

$$\mathcal{I}(\boldsymbol{x}) = \int_0^T \mathcal{O}(g_1(\boldsymbol{x}; t), g_2(\boldsymbol{x}; t))J_g(\boldsymbol{x}, t)dt. \tag{6}$$

Recognizing that (6) holds for all $\boldsymbol{x}$ in the image coordinates, we eliminate the integration over time by substituting

$$t = k_1(u_1; \boldsymbol{x}) = k_2(u_2; \boldsymbol{x}) \tag{7}$$

for t in (6). The expression (7) is obtained by rewriting the motion functions to show the dependence of t on $\boldsymbol{u}$ for fixed $\boldsymbol{x}$. The relation (7) is the path of integration followed in the $\boldsymbol{u}$ plane to obtain the image intensity at point $\boldsymbol{x}$, and may be multiple-valued when the motion retraces itself. Denoting by ds_u the path variable in object coordinates, we have

$$ds_u = \left[\left(\frac{\partial g_1}{\partial t}\right)^2 + \left(\frac{\partial g_2}{\partial t}\right)^2\right]^{1/2} dt \tag{8}$$

and substitution into (6) gives

$$\mathcal{I}(\boldsymbol{x}) = \int_{\substack{u_1=g_1(\mathbf{x};0)\\ u_2=g_2(\mathbf{x};0)\\ k_1(u_1;\mathbf{x})=k_2(u_2;\mathbf{x})}}^{\substack{u_1=g_1(\mathbf{x};T)\\ u_2=g_2(\mathbf{x};T)}} \frac{\mathcal{O}(\boldsymbol{u}) \left| J_g(\boldsymbol{x}, k_1(u_1; \boldsymbol{x}))\right| ds_u}{\left[\left(\dfrac{\partial g_1}{\partial t}\right)^2 + \left(\dfrac{\partial g_2}{\partial t}\right)^2\right]^{1/2}_{\substack{t=k_1(u_1;\mathbf{x})\\ =k_2(u_2;\mathbf{x})}}} \tag{9}$$

as the general representation for motion degradation. By inspection of (9), we can identify $h(\boldsymbol{x}, \boldsymbol{u})$ in the linear superposition integral form (1).

Note that the absolute value brackets and positive square root in (9) ensure that the point-spread function (PSF) is always ≥ 0 in this incoherent optical system. This rather complicated expression is generally space-variant, and there are two effects which combine to determine the amplitude of the point response. The Jacobian function arises from the change of coordinates in maintaining a constant power flux, and the denominator is proportional to the speed of movement of an object point. Thus as an object point moves faster, its image is spread out over a greater extent with reduced intensity. This fact was pointed out by Shack [4] in his analysis of space-invariant systems by Fourier techniques.

B. Space-Invariant Motion Degradation

The general expression (9) reduces to some well-known space-invariant motion degradation expressions when the

(a)

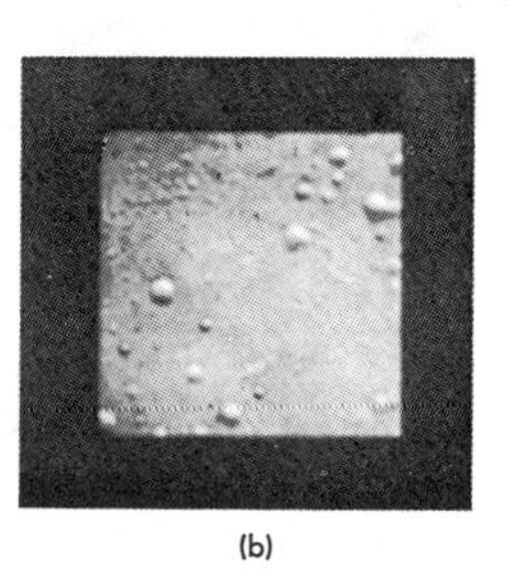
(b)

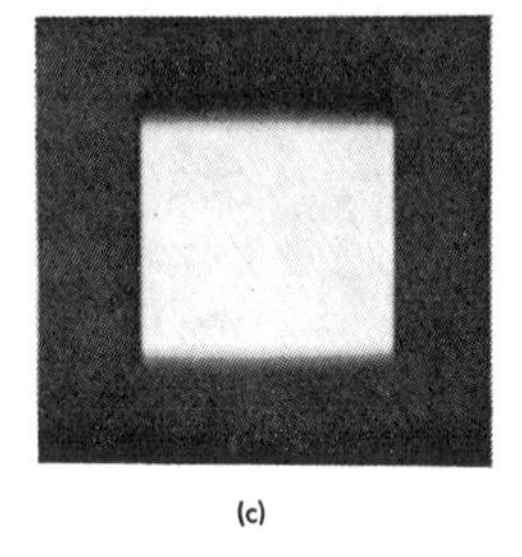
(c)

(d)

Fig. 2. Uniform motion degradation. (a) Constant intensity object. (b) Lunar object. (c) Constant intensity object blurred by uniform motion. (d) Lunar object blurred by uniform motion.

object and image intensity function planes are parallel and translate as a function of time during exposure. In this case, there is no coordinate distortion of the object as recorded by the imaging system, and the motion functions (2) take the form

$$\begin{aligned} u_1 &= g_1(\boldsymbol{x}; t) = x_1 - m_1(t) \\ u_2 &= g_2(\boldsymbol{x}; t) = x_2 - m_2(t) \end{aligned} \tag{10}$$

where only the coordinate difference $\boldsymbol{x}-\boldsymbol{u}$ is a time function. The Jacobian $J_g(\boldsymbol{x}, t)$ of (5) is unity for this type of motion and the degradation (9) reduces to

$$\mathcal{I}(\boldsymbol{x}) = \int_{\substack{u_1=x_1-m(0)\\ u_2=x_2-m(0)\\ m_1^{-1}(x_1-u_1)=m_2^{-1}(x_2-u_2)}}^{\substack{u_1=x_1-m_1(T)\\ u_2=x_2-m_2(T)}} \frac{\mathcal{O}(\boldsymbol{u})ds_u}{\left[(\dot{m}_1(t))^2 + (\dot{m}_2(t))^2\right]^{1/2}_{\substack{t=m_1^{-1}(x_1-u_1)\\ =m_2^{-1}(x_2-u_2)}}} \tag{11}$$

which is a space-invariant expression.

The most elementary example of space-invariant motion blur occurs when the object translates at constant velocity V along a straight line during the exposure interval. The motion function is simply $u=x-Vt$ and the equivalent PSF is

$$h(x-u) = \begin{cases} \dfrac{1}{V}, & 0 \leq x-u \leq VT \\ 0, & \text{elsewhere} \end{cases} \tag{12}$$

which is just an integration over a small finite region of the object [4]–[8]. Fig. 2 shows a computer simulation of a constant intensity object and a lunar scene blurred from top to bottom by uniform image motion. The degradation is severe and much of the information in the higher spatial frequencies is lost.

For constant linear acceleration between parallel planes, the motion function is

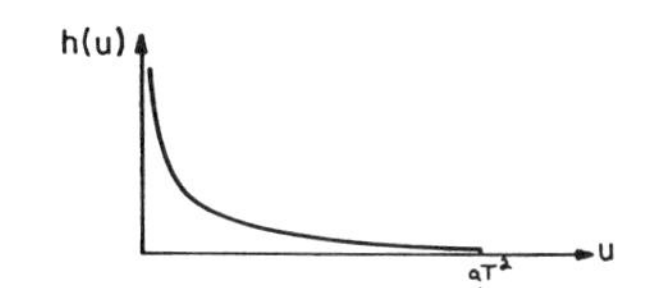

Fig. 3. Constant acceleration PSF.

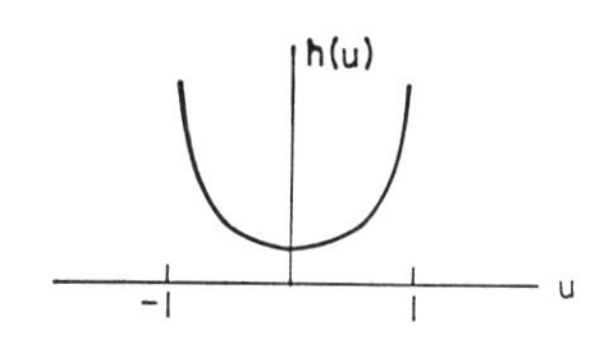

Fig. 4. Oscillatory motion PSF.

$$x - u = m(t) = at^2, \qquad a > 0, \quad 0 \le t \le T \tag{13}$$

with equivalent PSF [8]

$$h(x, u) = h(x-u) = \begin{cases} \dfrac{1}{2a^{1/2}(x-u)^{1/2}}, & x - aT^2 < u < x \\ 0, & \text{elsewhere} \end{cases} \tag{14}$$

and this equation is shown in Fig. 3 plotted as a function of the single variable u.

For oscillatory motion of one complete cycle during exposure the motion function is

$$x - u = m(t) = \sin\frac{2\pi t}{T}, \qquad 0 \le t \le T \tag{15}$$

with equivalent PSF [9]

$$h(x, u) = h(x-u) = \begin{cases} \dfrac{T}{\pi(1-(x-u)^2)^{1/2}}, & -1 \le x-u \le 1 \\ 0, & \text{elsewhere} \end{cases} \tag{16}$$

shown in Fig. 4. When evaluating the denominator of (11) to find (16), special care is taken when substituting for t to sum the absolute values of the PSF's due to each part of the path retraced by the oscillation.

When parallel planes translate with simultaneous constant velocity and acceleration along orthogonal coordinate axes, the motion functions are

$$x_1 - u_1 = Vt = m_1(t), \qquad V, a > 0 \tag{17a}$$

$$x_2 - u_2 = at^2 = m_2(t), \qquad 0 \le t < T \tag{17b}$$

and the PSF is

$$h(\mathbf{x} - \mathbf{u}) = \begin{cases} \dfrac{1}{[V^2 + 4a(x_2 - u_2)]^{1/2}} = \dfrac{V}{[V^4 + 4a^2(x_1 - u_1)^2]^{1/2}}, & \left(\dfrac{x_1 - u_1}{V}\right)^2 = \left(\dfrac{x_2 - u_2}{a}\right) \\ & 0 \leqq x_1 - u_1 \leqq VT \\ & 0 \leqq x_2 - u_2 \leqq aT^2 \\ 0, & \text{elsewhere} \end{cases} \tag{18}$$

shown in Fig. 5 as a ribbon-like function in the image plane.

C. Space-Variant Motion Degradation

When the object and image coordinates are not translating parallel planes, the movement of each object point varies with position during exposure, and the equivalent PSF is space-variant. If the motion functions (2) are known for any general moving system, we can evaluate (9) to find the PSF.

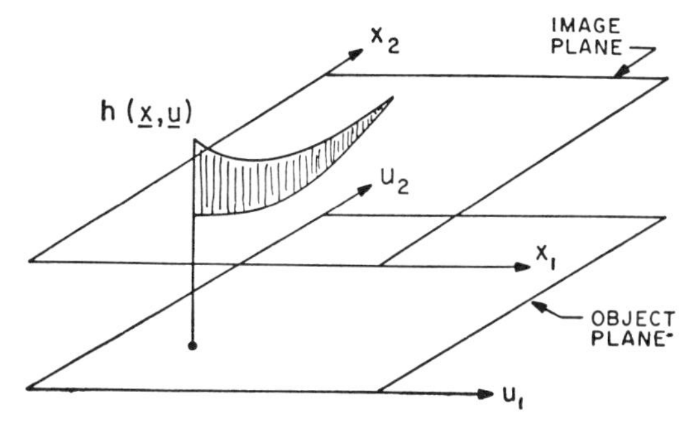

Fig. 5. Orthogonal acceleration and velocity PSF.

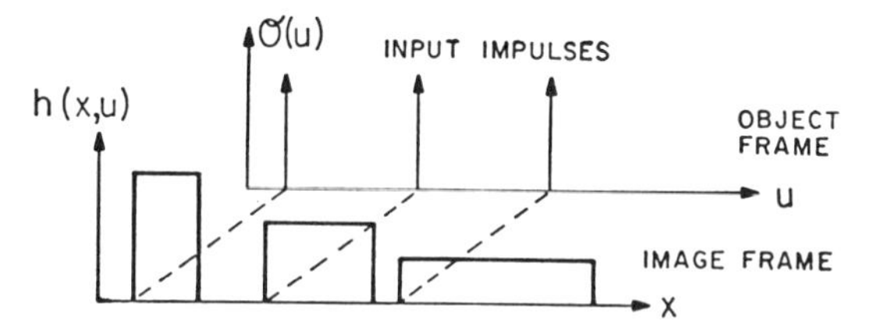

Fig. 6. Space-variant point-spread function.

As an example, consider the one-dimensional motion

$$u = g(x; t) = \frac{ax - \alpha t}{t + a}, \qquad a, \alpha, u > 0, \quad t \in [0, T] \tag{19}$$

with equivalent PSF

$$h(x, u) = \begin{cases} \dfrac{a}{u + \alpha}, & \dfrac{ax}{T+a} - \dfrac{\alpha T}{T + a} \le u \le x \\ 0, & \text{elsewhere} \end{cases} \tag{20}$$

shown in Fig. 6. Here, there is a stretching of coordinates in addition to the motion, giving a position-dependent object velocity and an SVPSF whose duration and amplitude changes with the location of the input impulse.

When object and image planes are parallel and there is relative rotation between them, the equivalent PSF is space-variant. The motion functions

$$u_1 = x_1 \cos \omega t + x_2 \sin \omega t, \qquad t \in [0, T] \tag{21a}$$

$$u_2 = -x_1 \sin \omega t + x_2 \cos \omega t, \qquad \omega > 0 \tag{21b}$$

describe a relative rotation of object and image planes at constant angular velocity ω during exposure. Assuming an exposure time T equal to $\pi/3\omega$, there is a total rotation of $\pi/3$ radians during the imaging process, and for this case, we use (9) to obtain the SVPSF

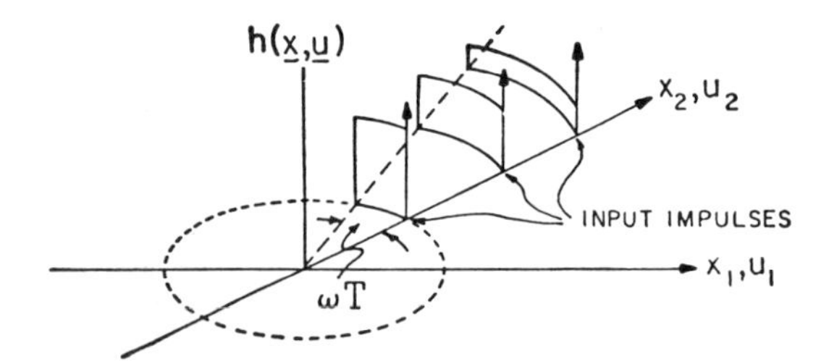

Fig. 7. SVPSF for constant velocity rotation.

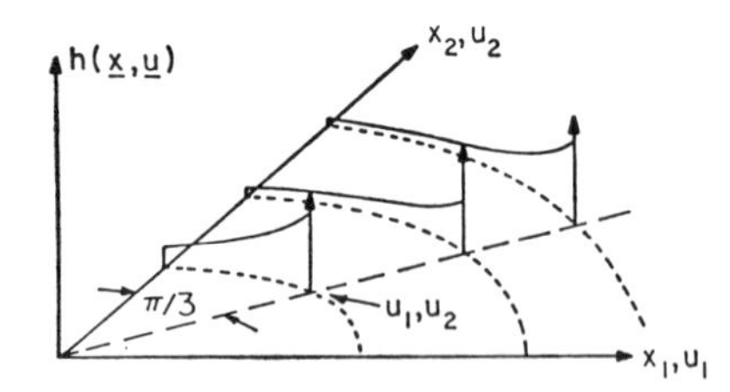

Fig. 8. SVPSF for accelerating rotation.

$$h(x, u) = \begin{cases} \dfrac{1}{\omega[u_1^2 + u_2^2]^{1/2}}, & x_1^2 + x_2^2 = u_1^2 + u_2^2 \\ & \dfrac{\sqrt{3}}{2} - \dfrac{u_2}{2} \le x_1 \le u_1 \\ & u_2 \le x_2 \le \dfrac{u_1}{2} + \dfrac{\sqrt{3}}{2} u_2 \\ 0, & \text{elsewhere.} \end{cases} \tag{22}$$

This PSF is space-variant because the image response falls off inversely with the object distance from the origin. Fig. 7 shows this function plotted in the image plane for input impulses at various distances from the origin.

For the related case of rotation with constant angular acceleration through $\pi/3$ radians during exposure, Fig. 8 shows the equivalent PSF for object impulses at various distances from the origin. The analytic expression for $h(\boldsymbol{x}, \boldsymbol{u})$ is rather complicated in this case [10].

D. *Motion Degradation with Coordinate Transformations*

When the motion of the object is combined with a coordinate transformation or geometrical distortion from object to image, the overall motion degradation is generally space-variant. Aerial imaging by aircraft and spacecraft are important cases of this type of motion blur, and such degradations are often the factor limiting resolution [1].

Earlier work in motion effects on resolution [4]–[6], [8], [9], [11]–[15] was directed primarily at finding velocities of point images and computing spatial frequency response both experimentally and analytically. In many types of aerial imaging, some means are provided to move the film, lens, or both as *a priori* compensation for motion during exposure. This image motion compensation (IMC) is useful for partial reduction of gross motion blur, but is almost always imperfect because translation in one direction cannot compensate for coordinate stretching.

With no movement, the geometrical coordinate transformation C uniquely relates a point location $\boldsymbol{x}$ in the image to the location $\boldsymbol{u}$ of its conjugate point in the object space. The transformation is a one-to-one mapping given by

$$\boldsymbol{u} = c(\boldsymbol{x}) \tag{23}$$

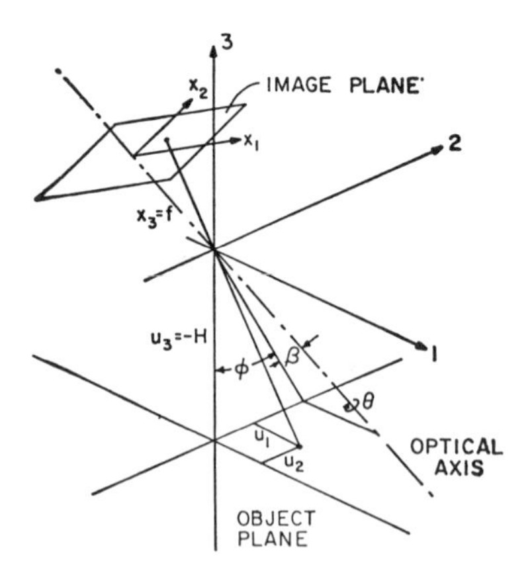

Fig. 9. Aerial system geometry.

and in component form by

$$u_1 = c_1(x_1, x_2) \tag{24a}$$

$$u_2 = c_2(x_1, x_2) \tag{24b}$$

and may be a nonlinear function of the coordinate variables. With these properties, C may be inverted to give

$$\boldsymbol{x} = c^{-1}(\boldsymbol{u}) \tag{25}$$

with corresponding component form.

The aerial imaging geometry is shown in Fig. 9 and is similar to a version of the transformation used in photogrammetric work [10], [15]. The camera orientation is specified by the angles ϕ, β, and θ, and the center of the orientation axis translates with known motion over the object plane at constant altitude H. With the assumption that the distance of the object plane from the origin is large compared to the focal length f, the components of (25) are given by an orthographic projection

$$x_1 = c_1^{-1}(u_1, u_2) = f \frac{m_{11}u_1 + m_{12}u_2 - m_{13}H}{m_{31}u_1 + m_{32}u_2 - m_{33}H} \tag{26a}$$

$$x_2 = c_2^{-1}(u_1, u_2) = f \frac{m_{21}u_1 + m_{22}u_2 - m_{23}H}{m_{31}u_1 + m_{32}u_2 - m_{33}H} \tag{26b}$$

where the coefficients m_{ij} for $i, j = 1, 2, 3$ are constants. The m_{ij} are the components of an orientation matrix determined by a sequence of rotations through ϕ, β, and θ to the fixed orientation.

After inverting (26) to the form (24), the motion functions (2) for fixed altitude and orientation aerial translation can be expressed in the form

$$u_1 = c_1(x_1, x_2) - m_1(t) \tag{27a}$$

$$u_2 = c_2(x_1, x_2) - m_2(t) \tag{27b}$$

where the time functions $m_i(t)$ describe a translation of the distorting system over the object plane which is similar to (10) of Section II-B. If the image $\mathcal{g}(\boldsymbol{x}, t)$ is found from (4) assuming no motion during exposure ($m_1(t) = m_2(t) = 0$) there is a geometrical distortion from object to image given by

$$\mathcal{g}(\boldsymbol{x}) = \mathcal{O}(c_1(\boldsymbol{x}), c_2(\boldsymbol{x})) \, | c'(\boldsymbol{x}) | \tag{28}$$

where $|c'(\boldsymbol{x})|$ is the time-independent Jacobian

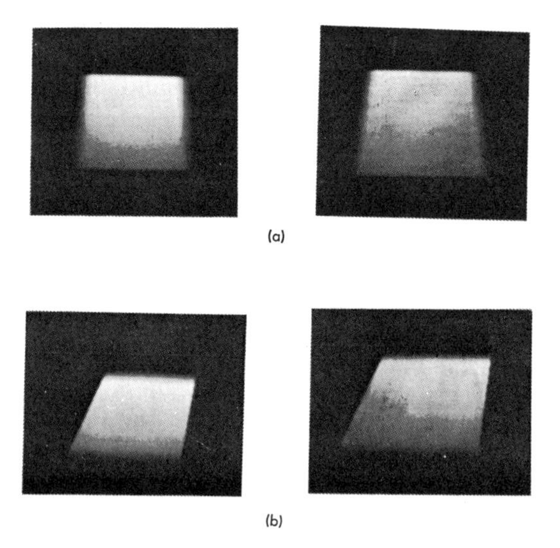

Fig. 10. Space-variant motion degradation. (a) Objects degraded by moving side oblique system. (b) Objects degraded by moving forward oblique system.

$$| c'(x) | = \begin{vmatrix} \frac{\partial u_1}{\partial x_1} & \frac{\partial u_1}{\partial x_2} \\ \frac{\partial u_2}{\partial x_1} & \frac{\partial u_2}{\partial x_2} \end{vmatrix} \tag{29}$$

obtained by evaluating (5). Substituting (27) into the general expression (9) gives

$$\mathscr{g}(x) = \int_{\text{path in object plane}} \frac{\mathcal{O}(u) \mid c'(x) \mid}{[(\dot{m}_1(t))^2 + (\dot{m}_2(t))^2]^{1/2}_{\substack{t=m_1^{-1}(c_1(x)-u_1) \\ =m_2^{-1}(c_2(x)-u_2)}}} ds_u, \tag{30}$$

as the linear space-variant expression for aerial motion degradation. For simplicity, the complicated limits of integration in (30) are replaced with the notation that integration is over a path in the object plane.

A common example of aerial imaging is the side oblique view—when the camera moves at constant velocity V along the 2 axis of Fig. 9 while tilted sideways at an angle β. The geometrical transformation (26) is given in this case by

$$x_1 = f \frac{-u_1 \cos \beta + H \sin \beta}{u_1 \sin \beta + H \cos \beta} \tag{31a}$$

$$x_2 = \frac{-fu_2}{u_1 \sin \beta + H \cos \beta} \tag{31b}$$

and the Jacobian is

$$| c'(x) | = \left| \frac{H^2 f}{(x_1 \sin \beta + f \cos \beta)^3 V} \right| . \tag{32}$$

To demonstrate space-variant motion blur, this side oblique system degradation was simulated by computer for $\beta = 20°$ and constant velocity blur using the constant intensity and lunar objects. The results are shown in Fig. 10. The imaging operation smears and distorts the objects, and the visible blurring along the edges of the images varies with position as the extent of the integration decreases with distance from the origin. There is no blur in the x_1 direction—just geometrical distortion, although the Jacobian factor modifies the intensity over the field. A commonly observed example of this type of imaging occurs when a nearly planar landscape is photographed from a moving car.

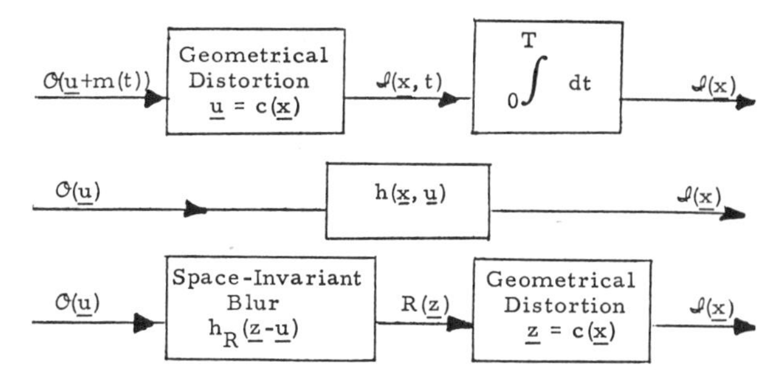

Fig. 11. Equivalence of motion blur—planar translation.

An example similar to the side oblique case is the forward oblique system. Here the camera translates with velocity V along the 2 axis while tilted forwards at an angle ϕ. The transformations in this case are

$$x_1 = \frac{-fu_1}{u_2 \sin \phi + H \cos \phi} \tag{33a}$$

$$x_2 = f \frac{-u_2 \cos \phi + H \sin \phi}{u_2 \sin \phi + H \cos \phi} \tag{33b}$$

and the Jacobian is

$$| c'(x) | = \left| \frac{H^2 f}{(x_2 \sin \phi + f \cos \phi)^3 V} \right| . \tag{34}$$

Fig. 10 also shows the degradations of this system for $\phi = 20°$, where the object is located in the quadrant ($u_1 \leq 0$, $u_2 \geq 0$) of the u plane. The forward oblique system is also space-variant since the blurring length and direction changes with position. In contrast to the previous example there is blur in both the x_1 and x_2 directions, even though the movement is only along the u_2 axis.

Returning to the general degradation expression (30), the Jacobian $|c'(x)|$ can be factored out from the integration because it is not dependent on u. This occurs only because the Jacobian is time independent for fixed orientation planar translation. Comparing (30) with the geometrical distortion (28) and referring to (11) describing space-invariant blur, the overall space-variant degradation takes the form of a space-invariant operation followed by a fixed geometrical distortion. This equivalence is shown by the block diagram of Fig. 11. The top line is the physical process of forming a moving image followed by time integration, the second line is the SVPSF, and the third line shows the decomposition into space-invariant and distortion components. We can consider the object to be degraded first by the SIPSF $h_R(z-u)$ given by (11) with intermediate variable z replacing x and with the $m_i(t)$ of (27). This produces the intermediate function $R(z)$, which is then geometrically distorted to produce the same $\mathscr{g}(x)$. Although $R(z)$ does not exist physically it is useful when we consider restoration.

When the camera altitude changes with time or position in aerial imaging, or when the orientation of the camera changes during exposure, it is still possible to obtain the

Fig. 12. Inverse filter restoration. (a) Restored constant intensity object. (b) Restored lunar object.

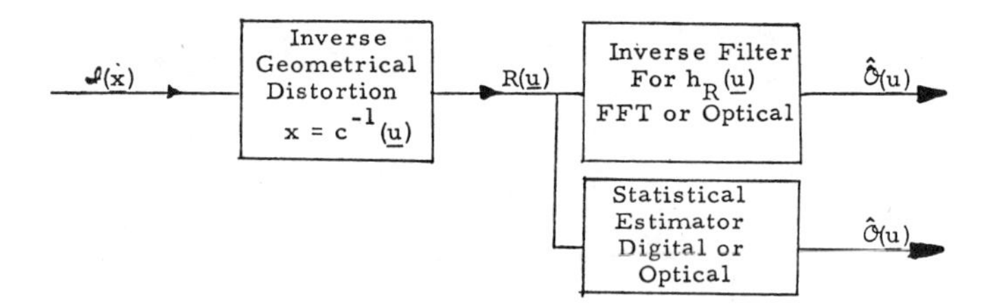

Fig. 13. Coordinate transformation restoration.

motion function (2) and equivalent SVPSF (9). Unfortunately, these expressions are complicated [10] and decompositions as in Fig. 11 are not generally possible because the Jacobian $J_g(x, t)$ cannot be factored out. In the future, more general coordinate distortions may be useful in solving this problem and allowing decomposition.

III. Image Motion Restoration

A. General Restoration

Given a record of a degraded image $\mathcal{I}(x)$ and a description of the degradation process, the goal of image restoration and enhancement is to produce an estimate $\hat{O}(u)$ of the original object intensity function $O(u)$. In imaging by space vehicles, it is usually easier to provide measurements of spacecraft motion and orientation with respect to the object than to stabilize the spacecraft or recording system for *a priori* motion compensation. Indeed, perfect *a priori* compensation is usually impossible in aerial imaging, and *a posteriori* restoration of motion blur is of great value in improving the quality of mapping and exploratory photography in space.

A wide variety of digital and optical methods have been used to restore images degraded by space-invariant operations, and a recent article by Huang, Schreiber, and Tretiak [16] summarizes these methods.

When the motion degradation is space-invariant with no noise present during recording, and when the recorded image extends over an infinite or semi-infinite interval in space, an estimate of the Fourier transform of the object function can be obtained by dividing the Fourier transform of the SIPSF into the transform of the recorded image. The infinite interval conditions hold when a small moving object is photographed on a background of known intensity, or when the recording area is much larger than the extent of the blur. This technique, known as inverse filtering, has been described in some detail [5], [7], [10], [16]–[20] and is applicable to all of the space-invariant degradations of Section II-B.

Fig. 12 shows the result of applying ideal inverse filters to the images of Fig. 2 blurred by constant velocity space-invariant motion. The filter is implemented by a fast Fourier transform (FFT) discrete inverse technique, and a great deal of computer time is saved by realizing that the restoration can be applied on a line-by-line basis to each vertical line in $\mathcal{I}(x)$. The restored $O(u)$ is almost exactly the same as the original object function, although if recording noise were present, we would expect $\hat{O}(u)$ to be an imperfect noisy estimate of $O(u)$.

When the Fourier transform of the blur is zero at some spatial frequency, dividing it into the transform of the image may amplify any noise present and swamp out the restored object. Slepian [7] suggested that the difficulties with zeros were due to the constraint that the filter must reproduce the background over an infinite region. He and Cutrona and Hall [19] effectively removed the zeros by finding a modified inverse filter which gives $O(u)$ only over the region where its image $\mathcal{I}(x)$ was recorded, with arbitrary output elsewhere. In the presence of noise, linear minimum mean-square error (MMSE) filters are effective for restoration and estimation [20]. These filters exist for discrete and continuous space models, and optical [21] and digital [16], [18]–[20] implementation has been reported.

When the degrading PSF is space-variant, image restoration becomes immensely more complex. Fourier transforms cannot be used to solve the degradation equation (1) for inverse filters, and estimation techniques are much more difficult to derive and implement. One technique for space-variant restoration involves breaking up $h(x, u)$ into regions which have different space-invariant PSF's [22], [23]. Robbins and Huang show [23] that a closed form solution to the inversion is possible in some cases. Another general approach is to convert all continuous space functions to discrete form, either by sampling or expansion in terms of some set of functions. Theoretically, discrete inversion could be used in the noise-free case, and statistical estimation could be used in the noisy case to find $\hat{O}(u)$. However, these brute force methods require enormous computational capacity and are not practically useful at present.

Motivated by these facts, it is evident that space-variant restoration could be easily accomplished if the inversion or estimation operation could be converted to an invariant equivalent.

B. Space-Variant Restoration of Constant Altitude Translation

The constant altitude translation PSF discussed in Section II-D can be decomposed into a space-invariant blur and a geometrical distortion. This fact immediately reveals a method for transformation to a space-invariant system for restoration. Because the distortion $z = c(x)$ in Fig. 11 is one-to-one and not a function of time, it is invertible by digital or optical means even though it is nonlinear in the spatial coordinates. If an inverse distortion $x = c^{-1}(u)$ given by (25) is applied to $\mathcal{I}(x)$, the result is ideally the intermediate object function $R(u)$ in Fig. 11. In the block diagram of Fig. 13, this operation is the first stage of coordinate transformation restoration.

In Fig. 11, $R(u)$ is produced by a space-invariant system $h_R(u)$ operating on $O(u)$, so the space-invariant restoration and estimation techniques of the previous section are directly applicable to the problem of finding $\hat{O}(u)$ from $R(u)$. This is the second operation in Fig. 13 and can also be accomplished by digital or optical means. It is generally possible to use Fourier techniques to find an inverse filter for any $h_R(u)$ re-

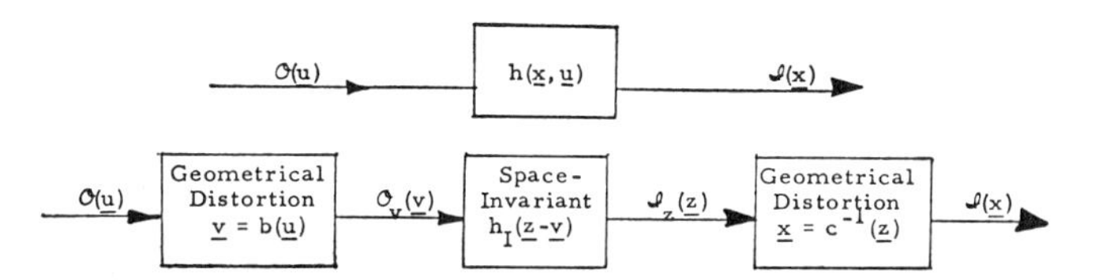

Fig. 14. General space-variant decomposition.

sulting from translational imaging system motion, and if the zeros of the transform of $h_R(\mathbf{u})$ are a problem, modifications can be made using the previous results for inverse filtering over some part of the object field.

Given the space-variant blurred images of Fig. 10, the appropriate inverse geometrical distortion was applied using this method to get an intermediate image $R(\mathbf{u})$. This intermediate image for each object is identical to the uniform motion blurred objects of Fig. 2 for both types of oblique systems, because the translation over the object plane is the same. Using the inverse FFT filter for space-invariant uniform blur gives the restored images of Fig. 12. The restoration is almost exact for these two cases of noise-free space-variant imaging.

If the recording noise is nonzero, the transformation restoration method is still useful for estimating $\hat{\mathcal{O}}(\mathbf{u})$, but a statistical estimator is required because of the noise. A difficulty that arises here is that many MMSE estimators assume stationary object and noise statistics. In general, this stationarity is lost when the object and noise pass through the inverse geometrical distortion of Fig. 13 because of the nonlinear spatial coordinate shift. Although little is known about the exact alteration of the covariance function after passage through a nonlinear spatial coordinate shift, good results might be obtained by assuming stationary object and noise statistics, although the estimate would not minimize the mean-square error.

The coordinate transformation restoration has the advantage of greatly reducing the system dimensionality. In the two-dimensional discrete case, image and object values are matrices, and the space-variant filters have components which are functions of four space variables. If the movement in the space-invariant filter $h_R(\mathbf{u})$ follows a linear path, the reduction from space-variant form means that processing could be done on a line-by-line basis with filter coefficients which are functions only of the coordinate difference.

C. General Coordinate Transformation Restoration

Fig. 14 shows how the decomposition technique of Section II-D can be generalized to allow for another invertible coordinate transformation $\mathbf{v}=b(\mathbf{u})$ preceding the space-invariant system. By an obvious generalization of (27), a space-variant degradation $h(\mathbf{x}, \mathbf{u})$ due to motion may be decomposed into the system of Fig. 14 when the system motion function (2) can be expressed as

$$b_1(u_1, u_2) = c_1(x_1, x_2) - m_1(t) \tag{35a}$$

$$b_2(u_1, u_2) = c_2(x_1, x_2) - m_2(t) \tag{35b}$$

where the $m_i(t)$ express a translation between transformed coordinate planes [10].

Given a record of $\mathcal{G}(\mathbf{x})$ produced by a decomposable degradation operating on $\mathcal{O}(\mathbf{u})$, we can follow the outline of the previous section to find an estimate $\hat{\mathcal{O}}(\mathbf{u})$. The technique is summarized in the block diagram of Fig. 15. The coordinate transformation $\mathbf{x}=c^{-1}(\mathbf{z})$ is inverted first to produce an intermediate image $\mathcal{G}_z(\mathbf{z})$. Following a space-invariant inverse filter or statistical estimator which produces the intermediate object $\hat{\mathcal{O}}_v(\mathbf{v})$, a second inverse distortion $\mathbf{u}=b^{-1}(\mathbf{v})$ is applied to obtain the estimate $\hat{\mathcal{O}}(\mathbf{u})$. For this general type of restoration, the comments of the previous section on changes in noise statistics in estimation and reduction of system dimensionality are valid.

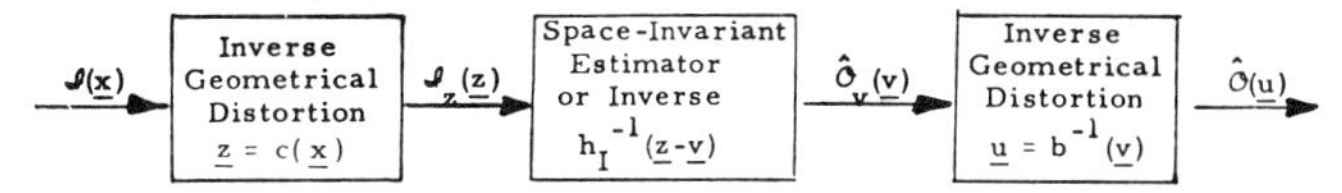

Fig. 15. General coordinate transformation restoration.

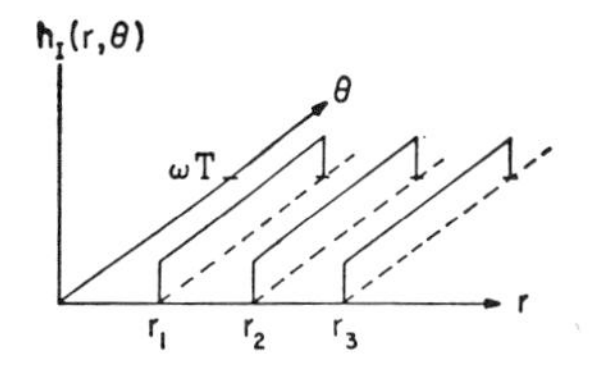

Fig. 16. Polar transformation of rotational blur.

The general coordinate transformation technique may be used to restore images blurred by the rotating systems of (22) and Figs. 7 and 8. Writing z_1, z_2, v_1, and v_2 as r_z, θ_z, r_v, and θ_r, respectively, for clarity, the space-variant operation may be decomposed by the identical polar coordinate transformations

$$x_1 = r_z \cos \theta_z \tag{36a}$$

$$x_2 = r_z \sin \theta_z \tag{36b}$$

$$u_1 = r_v \cos \theta_v \tag{36c}$$

$$u_2 = r_v \sin \theta_v \tag{36d}$$

into a space-invariant $h_I(\mathbf{z}-\mathbf{v})$ in which transformed object points follow identical paths in the transformed image space. For the constant ω rotation example, we have

$$h_I(r, \theta_z - \theta_v) = \begin{cases} \frac{1}{\omega}, & 0 \leqq \theta_z - \theta_v \leqq \omega T \\ 0, & \text{elsewhere} \end{cases} \tag{37}$$

where we let $r=r_z=r_v$ because there is no blur in the radial direction. If the three arcs of Fig. 7 are at radii r_1, r_2, and r_3, then Fig. 16 is a plot of the space-invariant (37), which is just a uniform motion blur in the θ direction. The polar mappings (36) and their inverses are applied so that $h_I(\cdot)$ is defined only for $0 \leq \theta < 2\pi$ while $\mathcal{G}_z(\mathbf{z})$ and $\mathcal{O}_v(\mathbf{v})$ have values for all $\theta_v \geq 0$, $\theta_z \geq 0$, and are periodic with period 2π. In this form, the space-invariant inverse filter used in Section III-A is used for intermediate restoration, although a total rotation of 2π during exposure makes restoration impossible. In this case the image along a circle of constant r is

$$A(r) = \int_0^{2\pi} \frac{1}{\omega} \mathcal{O}_v(r, \theta_v) d\theta_v \tag{38}$$

and knowledge of the origin of rotation is lost [7]. When the total rotation is a multiple $2\pi n$ plus some fraction $\omega T'$ of a full rotation, the n rotations add a function depending on r to the recorded image. With knowledge of this function, restoration is still possible. Due to the polar mapping, recovery is not uniquely possible at the origin, although it is theoretically possible at an arbitrarily close distance.

The polar coordinate method can also be applied to the accelerating rotation of Fig. 8, giving a transformed $h_I(\cdot)$ similar to the acceleration PSF of Fig. 3. In general, any circular rotation is decomposable by these methods, and the geometrical distortions can be accomplished electronically by nonlinear scanning and display.

Images degraded by the space-variant moving coordinate stretch the example given in Section II-C by (20) and shown in Fig. 6 can be restored using the coordinate transformations

$$z = c(x) = \ln (x + \alpha) \tag{39a}$$

$$v = b(u) = \ln (u + \alpha) \tag{39b}$$

$$\mathcal{G}_z(z) = \mathcal{G}(e^z - \alpha) \tag{39c}$$

$$\mathcal{O}_v(v) = a\mathcal{O}(e^v - \alpha). \tag{39d}$$

The space-invariant equivalent blur for this example is found to be

$$h_I(z - v) = \begin{cases} 1, & 0 \leqq z - v \leqq \ln \left(\dfrac{T + a}{a} \right) \\ 0, & \text{elsewhere} \end{cases} \tag{40}$$

which is just a uniform blur. This might be expected because the degradation integrates over variable regions of the object as a function of position.

The space-invariant function $h_I(\cdot)$ is not limited to the ribbon-like PSF's which result from motion. If a decomposition can be performed on any $h(\boldsymbol{x}, \boldsymbol{u})$ giving an $h_I(\cdot)$, then the coordinate transformation technique is useful in analyzing systems more general than incoherent systems with motion degradation. Huang *et al.* [16] and Robbins and Huang [23] have also discussed a coordinate transformation of this form for decomposition in the restoration of optical aberrations and tilt in cylindrical lens systems. It can be shown that the Mellin transform [24] used for some of these restorations is a special case of the general coordinate transformation method [25]. Coordinate transformation restoration is also useful in analyzing coherent optical systems. Although linear in field amplitude, these systems generally have space-variant impulse responses which may be decomposed into a space-invariant equivalent.

IV. Conclusions

In this paper, we model motion degradation processes in linear coherent optical systems by equivalent space-variant linear operations. Given the mechanical system motion, the space-variant system describes degradations which vary over the object. A number of examples of motion blur have been discussed, including the space-variant operations of aerial imaging. Although many techniques are known for the restoration of space-invariant degradations, including motion blur, little has been done to restore images blurred by space-variant operations. For a large class of moving imaging systems, we show that the space-variance can be decomposed into the cascade of geometrical distortion operations with space-invariant systems. Once in this form, space-variant degradations are restored by inverting the distortions and applying simple space-invariant estimation or inverse filtering to recover the object function. This method of inversion is applied to restore space-variant rotational and translational motion, and can be applied to certain optical aberrations.

Acknowledgment

The author wishes to thank Prof. J. W. Goodman of Stanford University for his advice during the course of this work, and the Image Processing Laboratory of the Jet Propulsion Laboratory for the use of the lunar photographs in the computer simulations.

References

[1] N. Jensen, *Optical and Photographic Reconnaissance Systems.* New York: Wiley, 1968, p. 102.
[2] J. W. Goodman, *Introduction to Fourier Optics.* New York: McGraw-Hill, 1968.
[3] A. W. Lohmann and D. P. Paris, "Space-variant image formation," *J. Opt. Soc. Amer.*, vol. 55, pp. 1007–1013, 1965.
[4] R. V. Shack, "The influence of image motion and shutter operation on the photographic transfer function," *Appl. Opt.*, vol. 3, pp. 1171–1181, 1964.
[5] E. L. O'Neill, *Introduction to Statistical Optics.* Reading, Mass.: Addison-Wesley, 1963.
[6] D. P. Paris, "Influence of image motion on the resolution of a photographic system—I," *Photogr. Sci. Eng.*, vol. 6, pp. 55–59, 1962.
[7] D. Slepian, "Restoration of photographs blurred by image motion," *Bell Syst. Tech. J.*, vol. 46, pp. 2353–2362, 1967.
[8] S. C. Som, "Analysis of the effect of linear smear on photographic images," *J. Opt. Soc. Amer.*, vol. 61, pp. 859–864, 1971.
[9] L. Levi, "Motion blurring with decaying detector response," *Appl. Opt.*, vol. 10, pp. 38–41, 1971.
[10] A. A. Sawchuk, "Space-variant image motion degradation and restoration," Ph.D. dissertation, Dep. Elec. Eng., Stanford Univ., Stanford, Calif. 1972.
[11] T. Trott, "The effects of motion on resolution," *Photogrammetric Eng.*, vol. 26, pp. 819–827, 1960.
[12] L. O. Hendeberg and W. E. Welander, "Experimental transfer characteristics of image motion and air conditions in aerial photography," *Appl. Opt.*, vol. 2, pp. 379–386, 1963.
[13] D. A. Kawachi, "Image motion and its compensation for the oblique frame camera," *Photogrammetric Eng.*, vol. 31, pp. 154–165, 1965.
[14] M. D. Rosenau, Jr., "Parabolic image motion," *Photogrammetric Eng.*, vol. 27, pp. 421–427, 1961.
[15] E. B. Brown, "V/H image motion in aerial cameras," *Photogrammetric Eng.*, vol. 31, pp. 308–323, 1965.
[16] T. S. Huang, W. F. Schreiber and O. J. Tretiak, "Image processing," *Proc. IEEE*, vol. 59, pp. 1586–1609, Nov. 1971.
[17] J. L. Harris, Sr., "Image evaluation and restoration," *J. Opt. Soc. Amer.*, vol. 56, pp. 569–574, 1966.
[18] ——, "Potential and limitations of techniques for processing linear motion-degraded imagery," in *Eval. of Mot. Deg. Images*, M. Nagel, Ed. NASA Tech. Rep. SP-193, Washington, D. C., 1968, pp. 131–138.
[19] L. J. Cutrona and W. D. Hall, "Some considerations in post-facto blur removal," in *Eval. of Mot. Deg. Images*, M. Nagel, Ed. NASA Tech. Rep. SP-193, Washington, D. C., 1968, pp. 139–148.
[20] C. W. Helstrom, "Image restoration by the method of least squares," *J. Opt. Soc. Amer.*, vol. 57, pp. 297–303, 1967.
[21] J. L. Horner, "Optical spatial filtering with the least mean-square-error filter," *J. Opt. Soc. Amer.*, vol. 59, pp. 553–558, 1969.
[22] E. M. Granger, "Restoration of images degraded by spatially varying smear," in *Eval. of Mot. Deg. Images*, M. Nagel, Ed. NASA Tech. Rep. SP-193, Washington, D. C., 1968, pp. 161–165.
[23] G. M. Robbins and T. S. Huang, "Inverse filtering for linear shift-variant imaging systems," this issue, pp. 862–872.
[24] R. Bracewell, *The Fourier Transform and Its Applications.* New York: McGraw-Hill, 1965, pp. 254–257.
[25] G. M. Robbins, "Image restoration for a class of linear spatially variant degradations," *Pattern Recogn.*, vol. 2, pp. 91–103, 1970.

Image restoration by singular value decomposition

T. S. Huang and P. M. Narendra

We demonstrate by a computer simulation example that singular value decomposition is a powerful tool for restoring noisy linearly degraded images. We also discuss a way of reducing the computation time requirement.

I. Introduction

Singular value decomposition (SVD) of matrices has recently found many applications in image processing. It has been used in designing two-dimensional recursive filters[1,2] in character recognition[3] and in image restoration.[4,10]

In this paper we demonstrate by a computer simulation example that SVD could be a powerful tool in restoring noisy linearly degraded images.

II. Restoration by Matrix Pseudoinverse

Adopting a discrete model, we can represent the linear degradation of an image by a matrix equation,

$$g = [H]f + n, \tag{1}$$

where **f** and **g** are column matrices containing the samples from the original object and the degraded image, respectively. The numbers of elements in **g** and **f** need not be equal. The rectangular matrix $[H]$ is derived from the impulse response (generally spatially varying) of the degradating system. And **n** is a column matrix containing noise samples. The noise may, for example, be due to the detector.

To clarify our notation, it might be mentioned that if the samples of the original two-dimensional image is represented by a matrix $[a_{ij}]$, $i = 1,2,\ldots, M$, and $j = 1,2,\ldots, N$, the elements of the column matrix

$$f = \begin{bmatrix} f_1 \\ f_2 \\ \vdots \\ f_p \end{bmatrix},$$

where $p = MN$ are related to a_{ij} by $a_{ij} = f_{(i-1)N+j}$, similarly for **g** and **n**.

The problem is: given **g** and $[H]$, estimate **f**.

A good estimate is

$$\hat{f} = [H]^{+}g, \tag{2}$$

where $[H]^+$ is the Moore-Penrose pseudo-inverse of $[H]$.[5,6] The nice thing about the pseudo-inverse is that it always exists so that we do not have to worry about whether the set of linear equations represented by Eq. (1) has a solution or whether the solution is unique. In fact, $\hat{\mathbf{f}}$ is the minimum-norm least-square solution to Eq. (1) when **n** = 0.

In the presence of noise, we have

$$\hat{f} = [H]^{+}[H]f + [H]^{+}n, \tag{3}$$

where the first term on the right-hand side of the equation is the minimum-norm least-square estimate in the absence of noise, and the second term represents the contribution due to noise. Unfortunately, in many cases, the noise effect dominates, and the signal part of Eq. (3) may be totally obscured.

III. Use of SVD to Combat Noise

One way of dealing with the noise is to use SVD to calculate the pseudoinverse. It is shown in Ref. 7 that any matrix $[H]$ can be decomposed thus

$$[H] = \sum_{i=1}^{R} (\lambda_i)^{1/2} U_i V_i^{t}, \tag{4}$$

where R is the rank of $[H]$, $\mathbf{U}_i$ and $\mathbf{V}_i$ are eigenvectors (considered as column matrices) of $[H][H]^t$ and $[H]^t[H]$, respectively, and λ_i the eigenvalues of either. We have used a superscript t to denote matrix transposition. Furthermore, the pseudoinverse of $[H]$ is given by

$$[H]^{+} = \sum_{i=1}^{R} \frac{1}{(\lambda_i)^{1/2}} V_i U_i^{t}. \tag{5}$$

The effectiveness of SVD in the presence of noise can be demonstrated by noting that we can trade off between the amount of noise and the signal quality by choosing the number of terms we use in the SVD of the pseudoinverse. Using the SVD of $[H]^+$, Eq. (3) becomes

The authors are with the School of Electrical Engineering, Purdue University, West Lafayette, Indiana 47907.

Received 13 January 1975.

Reprinted with permission from *Appl. Optics*, vol. 14, pp. 2213-2216, Sept. 1975.

(a)

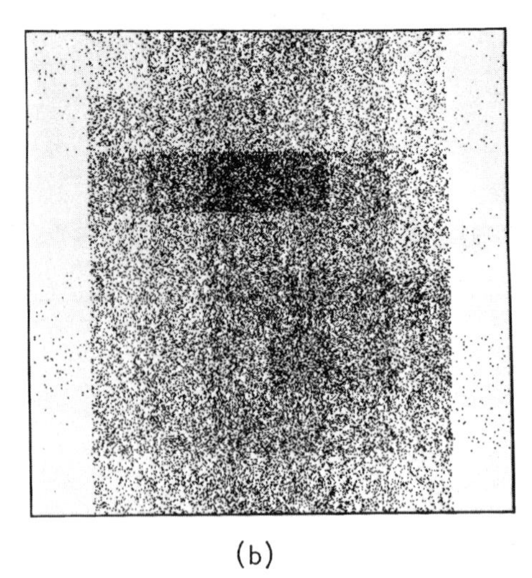

(b)

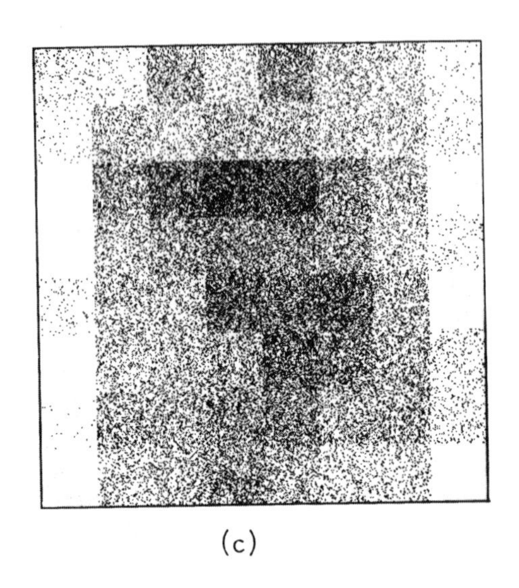

(c)

Fig. 1. (a) Original; (b) smeared image with additive Gaussian noise (mean = 0, standard deviation = 0.1); (c) smeared image with additive Gaussian noise (mean = 0, standard deviation = 0.5).

$$\mathrm{f} = \sum_{i=1}^{R} \lambda_i^{-1/2} \mathrm{V}_i \mathrm{U}_i^t \{[H]\mathrm{f}\} + \sum_{i=1}^{R} \lambda^{-1/2} \mathrm{V}_i \mathrm{U}_i^t \mathrm{n}. \tag{6}$$

Each term in the first summation has more or less comparable magnitudes, while the magnitudes of the terms in the second summation increase as $1/(\lambda_i)^{1/2}(\lambda_i$ are in the order of decreasing magnitudes). When we use more and more terms in the summations in Eq. (6), the first summation becomes closer and closer to the original object, but the S-NR (the ratio of the first summation to the second summation) becomes smaller and smaller. What we would like to do is to achieve a reasonable balance between the two effects. One possibility is to stop at the term where the noise magnitude becomes comparable to the signal magnitude. A better alternative is to look at the result after adding in each new term and stop at the visually best restoration.

IV. Computer Simulation Results

A computer-simulation example is shown in Figs. 1–3. The original is a character **5,** sampled with 8 × 8 points. The 8 × 8 matrix representing the digitized original is shown in Fig. 1(a). Each point inside the character was given a value 7; each point outside, a value 0. We blurred this picture by replacing each point by the average of nine points located in the 3 × 3 block centered around the point in question. Then zero-mean Gaussian random noise was added to it. Two degraded images are shown in Figs. 1(b) and 1(c), the noise variances in these images being 0.1 and 0.5, respectively.

The restoration was done using the equation

$$\hat{\mathrm{f}} = \sum_{i=1}^{P} \lambda_i^{-1/2} \mathrm{V}_i \mathrm{U}_i^t \mathrm{g}. \tag{7}$$

For each degraded image, we tried $P = 1,2,\ldots, 64$ and looked at all the 64 restorations. Some of the selected restored images are shown in Figs. 2 and 3. These images were plotted on an electrostatic plotter using dot density modulation. Each point of the 8 × 8 point image is represented by a square block. An estimation of the sixty-four eigenvalues of $[H][H]^t$ re-

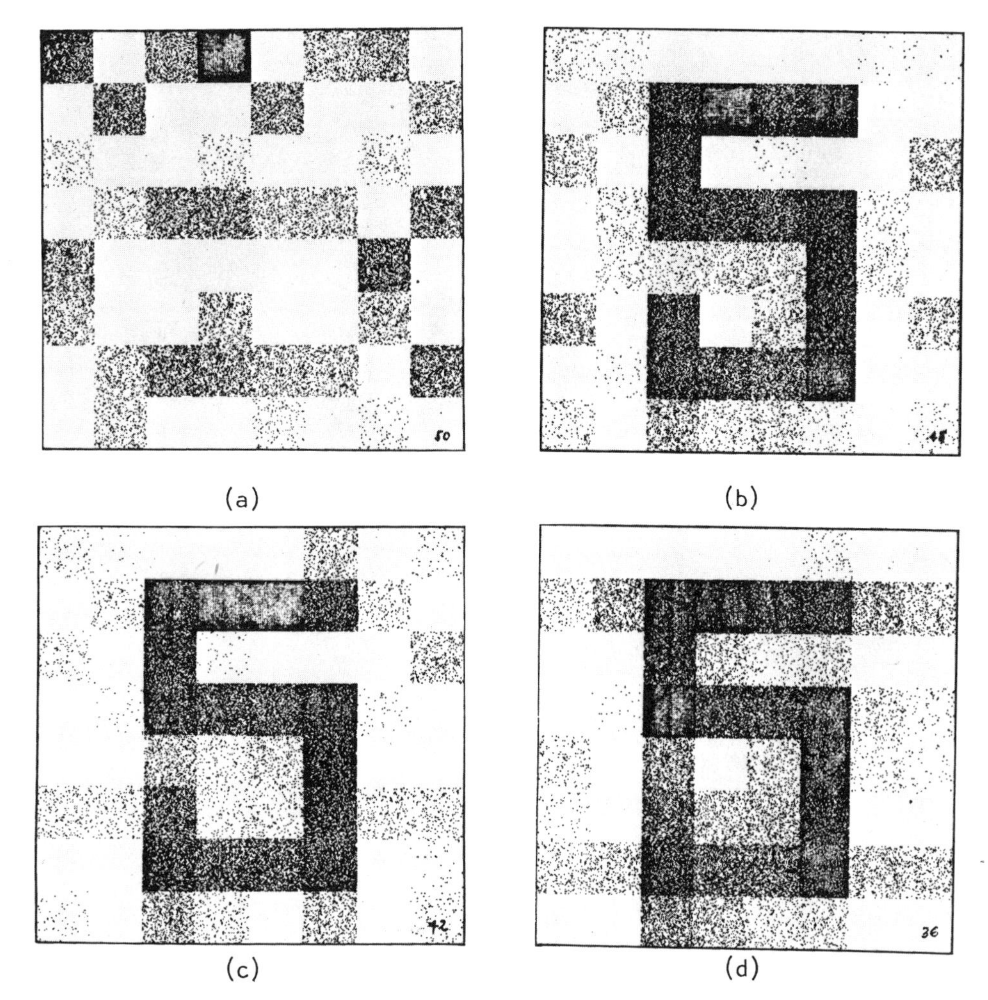

Fig. 2. Restored image from Fig. 1(b) using SVD. The number of terms used are (a) 50; (b) 48; (c) 42; (d) 36.

vealed that fifteen of them are practically zero. (They are much smaller than the others.) Therefore, one obviously should not use more than forty-nine terms in Eq. (7). This is demonstrated in Fig. 2(a). It is important to note that when the noise is large, one may want to use even fewer terms. Thus, for the degraded image in Fig. 1(c), thirty-eight terms seemed to give the best (visually) restoration (see Fig. 3).

V. Discussions

We have seen that the SVD approach of calculating the pseudoinverse is quite suitable for restoring noisy linearly degraded images. However, there is one major drawback: viz., even for moderately sized images, we have to find the eignevectors and the eigenvalues of very large matrices. For example, for a 100×100 point image, the matrix will be $10{,}000 \times 10{,}000$. As discussed in Ref. 8, if the degrading impulse response is separable, we can simplify the problem considerably. But what can we do in the nonseparable case? One possibility is suggested in Ref. 8: to approximate the nonseparable impulse response by a sum of separable ones. An alternative approach is the following. Usually, the spatial extent of the degrading impulse response is much smaller than that of the picture. Therefore, we can reduce the matrix size by dividing the degraded image into smaller subpictures (the size of each one is still much larger than that of the degrading impulse response) and restore each one separately. The problem one will encounter is how to treat the border effect. The points in each subpicture that are near the border are dependent on points in the neighboring subpictures;

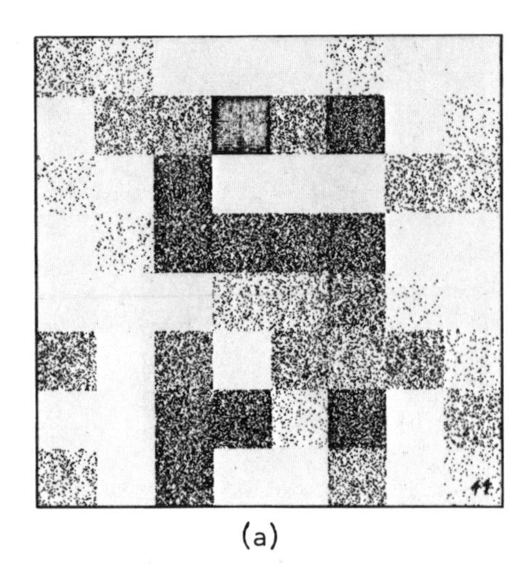

(a)

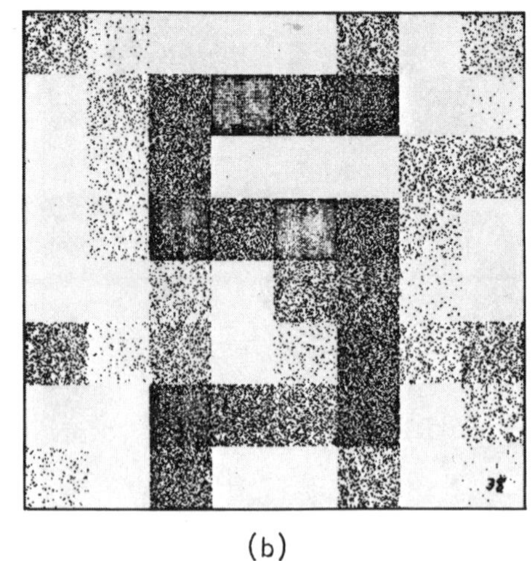

(b)

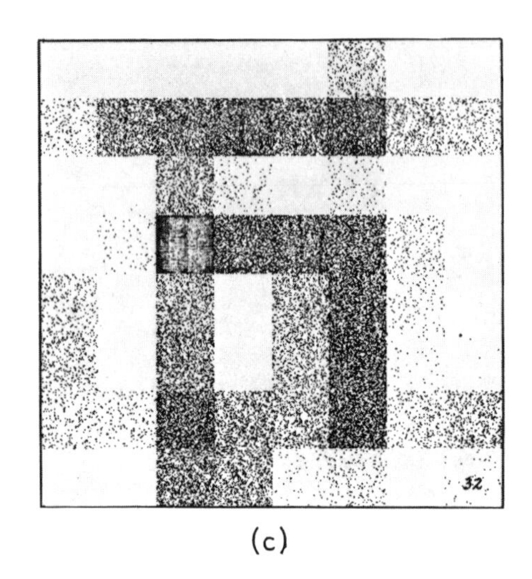

(c)

Fig. 3. Restored image from Fig. 1(c) using SVD. The number of terms used are (a) 44; (b) 38; (c) 32.

therefore, in theory, we cannot treat each subpicture independent of the others. A mathematical solution to this problem is not available at this time. However, some related work[9] indicates that it might involve Wiener-Hopf techniques. From a practical point of view, we can use this subdivision method by choosing overlapping subpictures, doing the restoration of each separately, and throwing away the borders.

This work was supported by ARPA under contract MDA 903-74-C-0098.

References

1. S. Treitel, and J. L. Shanks, IEEE Trans. Geosci. Electron. **GE-9,** 10 (1971).
2. T. S. Huang, W. F. Schreiber, and O. J. Tretiak, Proc. IEEE **59,** 1586 (1971).
3. G. Sherman, IEEE Trans. Computers, in press.
4. H. Andrews, Computer **7,** (5), 36 (May 1974).
5. A. Albert, *Regression and the Moore-Penrose Pseudoinverse* (Academic, New York, 1972).
6. W. Pratt, Semi-annual Reports under an ARPA Research Contract, Department of Electrical Engineering, University of Southern California, Los Angeles (1974).
7. C. Lanczos, *Linear Differential Operators* (Van Nostrand, New York, 1961), Chap. 3.
8. H. Andrews, "Two-Dimensional Transforms," in *Digital Picture Processing,* T. S. Huang, Ed. (Springer-Verlag, Berlin, 1975), Chap. 2.
9. G. M. Robbins and T. S. Huang, "Inverse Filtering for Linearly Shift-Variant Imaging Systems," in *Proc. Symp. Bildverarbeitung und Interaktive Systeme,* DLR-MITT, 73-11, DFVLR-Forschungszentrum, Oberpfaffenhofen, West Germany (Dec. 1971), pp. 40–49.
10. H. C. Andrews, and C. L. Patterson, Amer. Math. Monthly, **82,** 1 (Jan. 1975).

Restored Pictures of Ganymede, Moon of Jupiter

Digital restoration of two space pictures of Ganymede has revealed some interesting surface features.

B. Roy Frieden and William Swindell

Ganymede is the largest moon of Jupiter, having a diameter of about 5000 km. Because earth-based telescopes can barely resolve it, the details of Ganymede's surface are largely unknown. Other, nonvisual evidence has led to the belief that its surface is very rough, largely composed of rocky or metallic material embedded in ice (*1*). The detailed pictures presented here provide a body of visual information on the surface makeup of Ganymede.

During its mission to Jupiter, the Pioneer 10 spacecraft acquired two pictures of Ganymede (*2*), which provided a much improved view of its surface. The pictures were obtained with two different color filters, one in red (5950 to 7200 Å) and one in blue (3900 to 5000 Å). Unfortunately, these pictures are quite blurred because of the small scale of details on Ganymede relative to the size of the image blur spot (the total instrument response function).

We report here the results of an attempt to restore the pictures—that is, to remove the blur due to the instrument response function. Such removal is at least theoretically possible, because the instrument response function is deterministic, and largely known.

The authors are professors of optical sciences at the University of Arizona, Tucson 85721.

Let $s(x,y)$ represent the instrument response function, with x,y the usual space coordinates. Mathematically, the restoration problem consists in inverting the imaging equation

$$i(x_m,y_n) = \iint_{\text{scene}} dx'dy' O(x',y') \times s(x_m - x', y_n - y') \qquad (1)$$
$$m,n = 1,2,\ldots,M$$

for the unknown $O(x',y')$, the "restoration." The irradiance image data $i(x_m,y_n)$ and response function $s(x,y)$ are assumed known, from measurements, and hence contain noise. Such noise is the chief impediment to estimating $O(x',y')$.

Three factors aided in making such restoration practicable. First, the irradiance image is a linear function of the image data; hence, there are no problems of estimating the irradiance image such as occur when the image is photographic.

Second, and most important, the image was sampled at a sufficiently fine subdivision to allow some degree of enhancement. There were about 28 sampled image values within the central core of the two-dimensional instrument response function.

Third, the instrument response function is very nearly separable. That is, if $s(x,y)$ represents the general response function, with x,y the usual space coordinates, in our case

$$s(x,y) \simeq s_1(x)s_2(y) \qquad (2)$$

Functions s_1 and s_2 are the x- and y-component marginal distributions of s. Although Eq. 2 is an approximation, the maximum discrepancy between the left- and right-hand sides is about 2 percent of the central maximum in s. Figure 1 shows the marginal imaging kernels s_1 and s_2.

Separability is important because it permits use of a restoration procedure—the maximum entropy algorithm—whose output is constrained to be positive (or zero) everywhere (*3*). The general two-dimensional case would otherwise require too much computer time. Because of separability, the two-dimensional image may be restored as a sequence of one-dimensional, or line, restorations. These may be implemented with enough speed to permit the positive constraint to be enforced on the moderate-sized Ganymede pictures discussed below.

One negative aspect of the problem was the occasional existence of artifacts in the image data. Even worse, the artifacts were systematic—that is, highly correlated—and hence indistinguishable from true detail. We discuss below the steps we took to minimize this problem.

The images were restored in two different ways: by conventional linear filtering and by the maximum entropy algorithm cited above. To the best of our knowledge, the latter is the first published use of this kind of algorithm on real (nonsimulated), moderately extended image data.

The linear restoring algorithm was of the type used by Nathan (*4*)—inverse filtering, with a maximum permitted boost in amplitude specified by the user. Phase was always fully corrected. All operations on the image data were in direct (compared to frequency) space. Hence, the image was restored by convolution with a function whose Fourier transform is the upper-bounded, inverse filter. We tried maximum boosts of 2, 4, 5, and 10 before settling on 2 as the most reliable.

Image Geometry and Sampling Rates

The images were two arrays of data, each 33 points across (y direction) by 23 points down (x direction). The data spacings were 0.33 mrad in x and 0.195 mrad in y.

The restored images are constructed on a finer mesh. For the maximum entropy outputs (*5*), these are 65 points across by 45 points down, with point spacings of 0.165 mrad in x and 0.0975 mrad in y. The corresponding array size for the linear restorations was 65 by 67. The existence of a finer mesh potentially permits higher resolution in the output. All restoring methods were carried through by first restoring in x—that is, down each column—and then in y, across each row.

For effective data processing, it is necessary that the sampling rate in each coordinate direction be sufficiently high. As a rule of thumb, five sampling points per direction, within the central core of the instrument response function, are required. In our case, the situation is as in Fig. 1, where sampling positions of the Ganymede data are marked along each instrument response curve. Because there are four sampling points in x within the instrument response function s_1 and seven in y within s_2, we see that sampling was somewhat deficient in x but more than adequate in y. We can therefore expect in the restorations more accuracy in the y direction (horizontal) than in the x direction.

Reliability of Data

When the Pioneer 10 optics were tested before the Jupiter mission, inaccuracies in the image data were small. For the level of brightness in Ganymede, maximum image errors of about 7 percent were found, with a root-mean-square error of about 2 percent. This is an adequate accuracy for image restoration.

However, once the spacecraft was in flight, it became apparent that the red channel occasionally records substantial artifact information. When present, this appears as additive noise in the form of wavy, parallel lines (somewhat resembling a fingerprint). A particular red image may suffer this problem over all, or part, of it. Pictures of Jupiter taken immediately before and after those of Ganymede showed the defect, which implies that it is present in the Ganymede photos.

Fortunately, the blue channel suffers little, if any, error of this kind, and presumably still provides preflight performance. This gives a check on the red information and ultimately a measure of reliability for it, as follows.

Ganymede is most often modeled as having little atmosphere and consisting mostly of ice and rock. Such features would have little coloration. Hence, where the red channel lacks artifacts, the information it provides should be nearly proportional to the blue channel information—that is, the two sets of data should have a very high correlation coefficient. Conversely, when red artifacts are present, the two data sets should not correlate well.

This allowed us to establish a measure of reliability for different portions of the red Ganymede image. We first computed the cross-correlation coefficient between the entire red and blue images. This was .73, which indicates the presence of some artifacts (*6*). Next, we computed the correlation coefficients for corresponding quadrants of the image, with the intersection point at the center of the disk.

The results were most informative. As shown in Fig. 2, the upper left quadrant has a red-blue correlation of .35, the lower left .56, the lower-right .73, and the upper right .92. We use these as measures of our confidence in the red data over those regions.

Restoring procedures cannot distinguish true image details from systematic artifacts (the type present) and will equally enhance both. Therefore, given the above correlation figures, we should, for example, be skeptical of restorations of the upper left quadrant of the red Ganymede image, assuming the hypothesis of little coloration to be correct. As mentioned before (*6*), Pioneer 11 data seem to corroborate this hypothesis.

Image Data

In the images shown here, the local central meridian is approximately 103° and runs approximately from the upper left to the lower right. The subspacecraft latitude is −18°. The terminator is on the right-hand side of the pictures and the solar phase angle is 37°.

Figure 3a shows the red channel image of Ganymede and Fig. 3b the blue channel image. The minimum resolvable length in these images is about 390 km (*2*) (on a scale where disk diameter is 5000 km).

To enhance the visual appearance of these pictures, the data were stretched out geometrically by a factor of about 8 to 1 and linearly interpolated. Also, the otherwise weak internal features were accentuated by use of a photographic gamma exceeding unity and of high exposure.

With these visual aids, a few features of interest (see pointers, Fig. 3a), which are common to the two images, become apparent. First, there is a dark, caplike region on the terminator in the upper right quadrant of the disk, with a complementary bright region in the lower left quadrant. There is a small bright feature within the lower right quadrant, and a large dark area in the upper left quadrant. These gross features became more detailed, and interesting, in the restored pictures below.

The blue image in Fig. 3b is of further interest in that it has some more pronounced circular features (see pointers). In view of the appreciable diameter of Ganymede (about 5000 km) these are quite large. That these are maria or ice fields seems a plausible working hypothesis.

Restoring Algorithms

All data processing was done with a Control Data Corp. 6400 computer. Computer time for each maximum entropy picture was about 30 seconds. To the best of our knowledge, this is a much shorter computer time for an image the size of Ganymede than has been required in any previous use of the maximum entropy algorithm. For example, Wernecke (*7*) reported a time requirement on the order of an hour for a 21 by 21 array of data.

All maximum entropy restorations were formed with a sharpness factor (*3*) $\rho = 5$, which yielded a modest level of enhancement. In this situation, a negligible level of artifact details is created by the restoring technique itself. The exception is at the periphery of the disk, where the familiar Gibbs oscillation phenomenon arises because of the abrupt change in intensity. These oscillations are, however, easily recognized and suppressed. At higher test values of ρ than 5, some artifacts occur, somewhat resembling orthogonal sets of lines at 45° and 135°. However, these values of ρ were not used in the restorations reported below.

All linear restorations were of the type previously described, and were formed using a boost factor of 2. With this choice of boost, there was also a negligible level of restoration-induced artifact. Artifacts occurred for higher levels of boost, and these also took the form described above. We used the linear restoration method (i) to obtain an independent check of the reliability of the maximum entropy algorithm, and (ii) to see which restoring technique gives a higher-quality output when both are done at the same (negligible) artifact level. Previous tests (*3, 7*) have shown maximum entropy to yield better results. However, these were for pointlike objects, such as star fields, which significantly differ from an extended object such as Ganymede.

Restorations

Figure 4 shows the red channel image, its restoration by maximum entropy, and its restoration by linear filtering. The restorations have been geometrically stretched to coincide in size with the image data.

Let us now examine the maximum entropy restoration, Fig. 4b. Visually, this has a much higher level of detail than Fig. 4a. The finest resolution length in Fig. 4b is about 190 km, compared to 390 km in Fig. 4a.

It is informative to observe the enhanced versions of the image features previously described (see pointers, Figs. 3a and 4b). First, the caplike region at the upper right-hand edge in Figs. 3a and 4a is more sharply delineated in Fig. 4b. This is consistent with the way a crater, or other large hole, would appear on the dark limb side. Beneath it in Fig. 4b is an elliptical, darkened region that resembles a large, shallow crater, as one would appear within a generally darkened limb region and in an oblique view. This has a faint counterpart in the image, Fig. 4a.

Nearby in Fig. 4b are features with a craterlike appearance. Finally, the small bright feature within the lower right quadrant of Fig. 4a is more sharply delineated and smaller in Fig. 4b, where it now appears as the center of a large, scallop-shaped bright arc. Notice that this arc was previously seen in the blue image, Fig. 3b. Thus, the red restoration and the blue image confirm one another, and this is important because the two channels are independent sources of information. The bright arc must be a real feature. Regarding overall confirmation of results, we may note that all features discussed so far lie in the upper right and lower right quadrants, regions of high data reliability according to Fig. 2.

Continuing the comparison of Fig. 4a and Fig. 4b, the generally dark region of the upper left-hand quadrant of Fig. 4a is restored in Fig. 4b as a series of bands, running approximately from upper left to lower right. Figure 2 leads us to suspect that this feature is merely an enhancement of the artifacts in the red data. But there are features within the bands that are probably not artifacts, because these will be seen to correlate well from blue to red channels. However, the bands themselves will not, which lessens their credibility.

We next examine the linear restoration, Fig. 4c. Comparing it with Fig. 4b, we observe a resemblance between grosser details of the two. That is, except for a lower state of resolution in Fig. 4c, every feature in it is also present in Fig. 4b. However, the reverse is not true. Figure 4b provides a great deal more visual information than does Fig. 4c, in the form of (i) sharper edge gradients for the details they share in common and (ii) details of finer structure lying between these. This comparison, we believe, shows the improved resolution obtained with the maximum entropy technique compared to linear techniques.

A detail of further interest in Fig. 4, b and c, is the conspicuous bright band that appears to encircle the dark cap at the top (see pointer). According to Fig. 2, this is probably a real feature. As a working hypothesis, we suggest that the feature is a circular ridge, perhaps of ice. Large, circular, bright features seem to be common on Ganymede (see previous discussions of Figs. 3b and 4b).

The blue channel image data, maximum entropy restoration, and linear restoration are shown, respectively, in Fig. 5, a through c. Because not much coloration effect is expected for Ganymede, our original interest in the blue results was in the extent to which they agree with and verify the red results. There is good agreement in the following features of Figs. 4b and 5b, or 4c and 5c: the dark cap; the small white feature and its bright, encircling arc; and the white region at the lower left edge.

However, Fig. 5, b and c, contain something new, not found in Fig. 4. In the upper left-hand quadrant of Fig. 5, b and c, there are three round details (pointers) that very much resemble maria. If so, the maria are very large. Figure 5b also shows some structure within the topmost round feature, which makes it reminiscent of certain lunar maria.

In retrospect, these apparent maria can also be seen, albeit very faintly and greatly

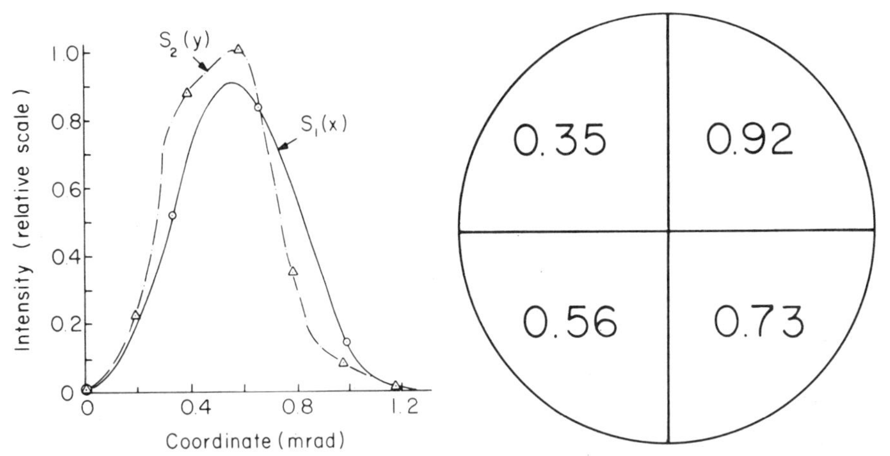

Fig. 1 (left). Marginal instrument response functions $s_1(x)$, $s_2(y)$ for the overall imaging system of Pioneer 10. Experimental data establishing these curves are spaced at 0.05 mrad. Points indicated on each curve show the sampling spacing, in each coordinate direction, for the Ganymede image data. Fig. 2 (right). Red-blue correlation coefficient for each quadrant of the input data arrays. The extent to which these data correlate offers a measure of confidence for the red data.

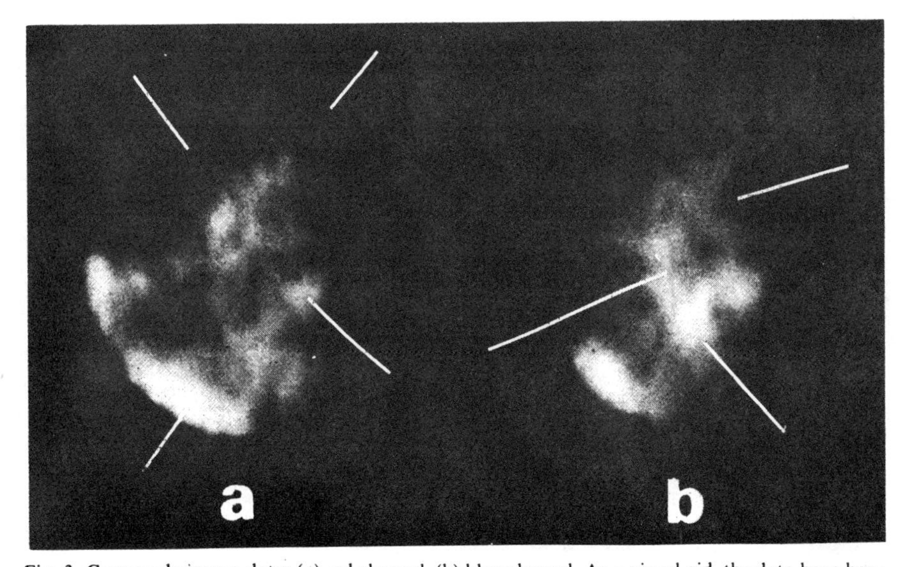

Fig. 3. Ganymede image data: (a) red channel, (b) blue channel. As a visual aid, the data have been developed with high photographic gamma, geometrically stretched, and linearly interpolated. The white pointers indicate key features that will show some interesting structure in the restorations to follow.

blurred, in the red and blue images (Figs. 3a, 3b, 4a, and 5a). Comparisons of these images with Fig. 5b illustrates the power of restoring methods to render marginally visible details strongly visible, and with additional detail. In view of the clarity with which Fig. 5b shows the three apparent maria, it is rather surprising that others are not seen elsewhere in the disk. Perhaps this implies that such maria are rare on Ganymede.

The linear restoration, Fig. 5c, also shows the maria, but with less resolution. Furthermore, it does not restore the internal detail of the topmost one, shown in Fig. 5b. This again illustrates the resolution advantage enjoyed by the maximum entropy technique over linear methods, when all operate at the same (negligible) level of artifact detail.

Color Composites

Color pictures of Ganymede may be formed by superimposing corresponding red and blue images, or corresponding restorations. However, for a proper visual color effect three primary colors are needed. In our case the third color must be added artificially, somehow based on the two available colors (red and blue). The method chosen was arbitrary, but works well when applied to Jupiter. We created an artificial green intensity at each pixel by forming a linear combination of the red and blue values there. By this method a pure green could not be formed, nor could a pure purple. But, for Jupiter at least, these hues do not exist in significant amounts, so the method works well.

The applicability of this method for Ganymede is, however, more speculative. It is not known a priori what distribution of hues Ganymede contains, so that green or purple regions cannot be ruled out. However, its overall hue may be observed from the earth, and this information may be used as input to the coloration scheme. The coefficients that weight the red and blue contributions may be chosen so that the acquired color image has the same overall hue as in an earth-based view. In this way, the top image on the cover was formed from the red and blue image data of Fig. 3. (Note that this has an overly enhanced green component due to an oversight in darkroom procedures.)

There is another benefit, aside from obtaining a color photograph, to be gained by such a superposition. It may also give us a means of increasing the signal-to-noise ra-

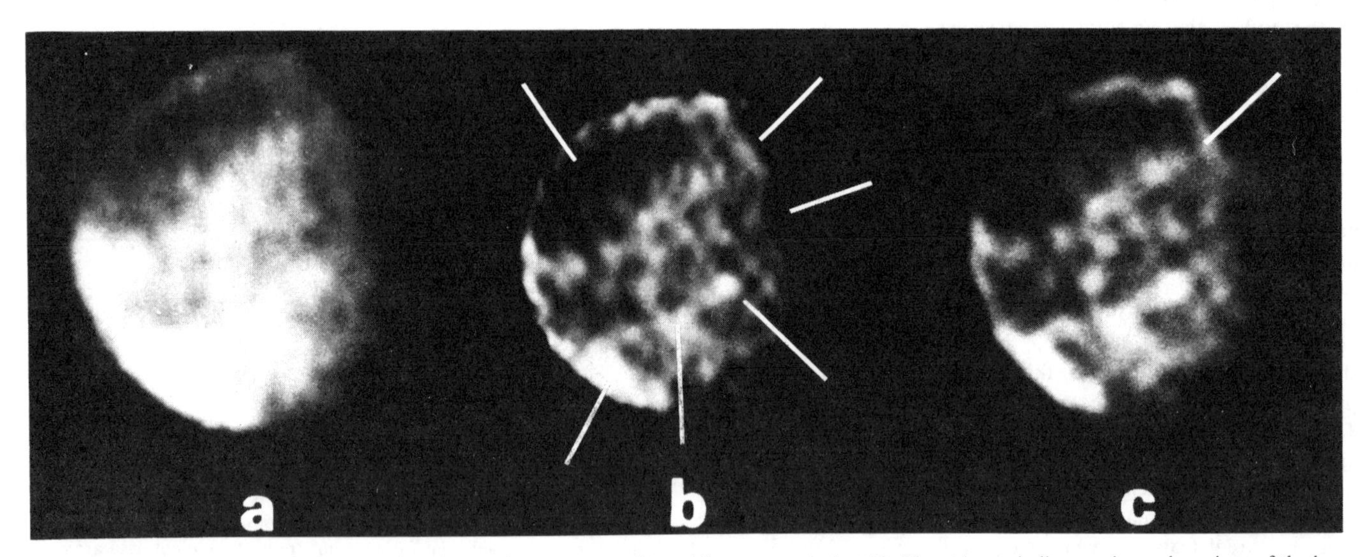

Fig. 4. Red channel image (a) and restorations by maximum entropy (b) and linear convolution (c). The pointers indicate enhanced versions of the key features pointed out in Fig. 3a.

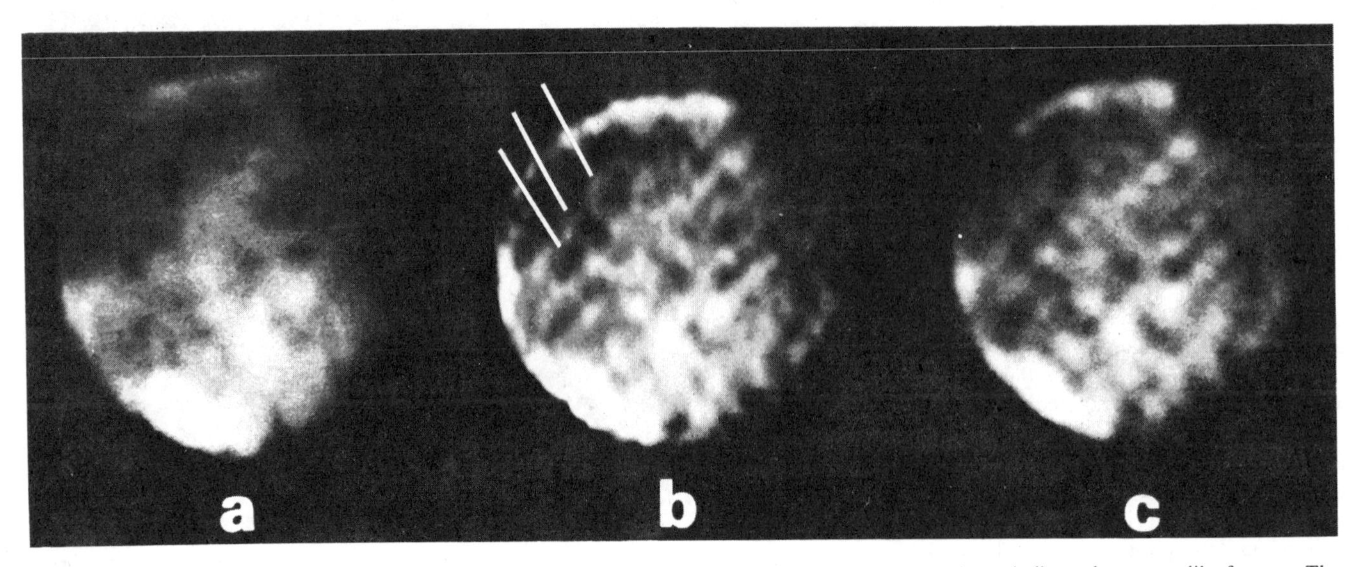

Fig. 5. Blue channel image (a) and restorations by maximum entropy (b) and linear convolution (c). The pointers indicate three mare-like features. The top one actually appears to show some internal structure. The bottom mare seems to intrude into the middle one, which is reminiscent of certain lunar maria.

tio across the picture. Since Ganymede is expected to have weak coloration (*1*) the red and blue images (or corresponding restorations in these colors) should strongly correlate spatially. So, therefore, should the artificial green with both the red and the blue. Now, in a composite, correlated features come through strongly, while uncorrelated or weakly correlated features do not. By the hypothesis of weak color effect, the latter should tend to be artifacts anyhow. The upshot, then, is that the ratio of true to false details in the Ganymede pictures should be enhanced by the color superposition. A corollary, however, is that any strong color (departure from the gray range) is suspect.

Color versions of the restorations were produced in a similar manner. At each pixel, an amount of green was generated equal to the arithmetic average of the red and blue intensities there. Before this step, the red and blue pictures were equalized in total light intensity. The effect of the green addition, then, is to produce an equal-energy white color where red and blue are equal. The philosophy here is that, if red and blue are equal, and nothing is known about green, the simplest assumption to make about green is that it equals both the red and the blue value, thereby producing a net color that is a shade of gray (the most unbiased or conservative choice of "color" in these circumstances).

In this manner, the linear restorations in Figs. 4c and 5c were used to produce the color output in the middle image on the cover; and the maximum entropy restorations in Figs. 4b and 5b were used to produce the color output in the bottom image. As before, the maximum entropy picture exhibits more detail than does the linear one.

Summary

Restored pictures of Ganymede have been produced that have some identifiably reliable features and some identifiable artifacts. The latter arise from artifacts in parts of the red image data. Among the presumably reliable features are some mare-like objects (perhaps with some internal structure), and a few rather large, bright rings. Whether the latter are ice, or arise from near-specular reflection from smooth surface features, is left for future investigation.

One of the restoring methods used, maximum entropy, has been shown to be applicable to moderately extended images. In view of its short time requirements (30 seconds per picture), the method should be applicable to moderately larger images, for example with twice the given number of data points.

References and Notes

1. R. M. Goldstein and G. A. Morris, *Science* **188**, 1211 (1975).
2. T. Gehrels, in *Planetary Satellites*, J. A. Burns, Ed. (Univ. of Arizona Press, Tucson, in press); see also R. O. Fimmel, W. Swindell, E. Burgess, *NASA Spec. Publ. 349* (1974).
3. The importance of the positive constraint and examples of its incorporation into various restoring methods, such as maximum entropy, are discussed by B. R. Frieden, in *Tropics in Applied Physics*, T. S. Huang, Ed. (Springer-Verlag, New York, 1975), vol. 6, pp. 219–245; see also (*5*).
4. R. Nathan, in *Pictorial Pattern Recognition*, G. C. Cheng, Ed. (Thompson, Washington, D.C., 1968).
5. If the output restoration is denoted $\hat{o}(x,y)$, its entropy H is defined as

$$H = -\int\int \hat{o}(x,y) \ln \hat{o}(x,y)\,dx\,dy$$

This H is maximized, subject to the input image data as constraints.
6. An identical calculation for Pioneer 11 data on Ganymede gave a correlation coefficient of .91. By comparison, Pioneer 11 data suffer little, if any, artifact detail. (However, they were sampled more coarsely than the Pioneer 10 data, and so are less appropriate to restore.)
7. S. Wernecke, abstract of paper presented at the meeting on image processing for 2-D and 3-D Reconstruction from Projections, Stanford University, Stanford, California, 4 to 7 August 1975.
8. The authors gratefully acknowledge the assistance provided by members of the Pioneer Imagery Photo-Polarimeter team of the University of Arizona, and personnel of the Pioneer Project Office of NASA-Ames Research Center In addition, we thank D. Wells of Kitt Peak National Observatory for his strong encouragement and for permitting access to the Comtal display equipment at his facility. This work was supported in part by NASA contract NAS2-6265 and by Air Force contract F33659-72-C-0605.

Part II
Image Processing and Correction

Organized by ~~Raul Hunziker, Associate Editor~~

Ralph Bernstein Editor

DIGITAL MAPPING AND ASSOCIATED DIGITAL IMAGE PROCESSING

Lawrence A. Gambino
Director, Computer Sciences Laboratory
and
Michael A. Crombie
Mathematician, Advanced Technology Division
Computer Sciences Laboratory
US Army Engineer Topographic Laboratories
Fort Belvoir, Virginia 22060

BIOGRAPHICAL SKETCH

Mr. Gambino is responsible for several programs in Exploratory and Advanced research under the sponsorship of the Defense Mapping Agency (DMA) and the Office, Chief of Engineers (OCE). These programs are primarily associated with digital image processing and digital mapping as applied toward problems of interest to the mapping and charting community. Special emphasis is being given to digital processors and their applications in these efforts.

Mr. Crombie is responsible for carrying out the Computer Sciences Laboratory's efforts in digital mapping, which is broken down into three related efforts: namely, digital mensuration, digital stereocompilation, and digital line-of-sight. Mr. Crombie has extensive experience in the areas related to analytical photogrammetry, and he has written numerous documents for both the US Army Engineer Topographic Laboratories (USAETL) and for what is now the Defense Mapping Agency/Topographic Center. His expertise in mathematics and statistics as applied toward the new and challenging area of digital processing for mapping and charting purposes, will greatly facilitate these programs at USAETL.

ABSTRACT

Digital Mapping, which is defined later, and Digital Image Processing are discussed in fairly general terms along with discussions of related software and hardware. Some information is given concerning the capability of a digital, associative array processor as it might be applied to correlation as well as to numerous other applications. Preliminary examples of derived parallax data from digitized grey shade data are presented. Our Digital Image Manipulation and Enhancement System (DIMES) software is discussed indicating its potential as a flexible R&D tool for conducting studies of the application of digital image processing to mapping and charting problems.

INTRODUCTION

We will briefly describe our efforts in two related areas: namely, digital photogrammetry and digital image processing. The similarity between the two is that both involve the manipulation of digitized, grey shade data. They are separate problems in that they differ in their objectives; that is, digital photogrammetry implies working with grey shade data from stereo pairs of photographs to extract three-dimensional information. On the other nand, digital image processing implies working with a single photograph for enhancement or, the inverse, degradation, for whatever purpose. Degradation, for example, has been used in our simulation of Plan Position Indicator (PPI) radar scenes beginning with an aerial photograph.

Both of these efforts may not be mutually exclusive for it may be that we wish to perform image processing type operations on a stereo pair of photographs prior to correlation. For example, we may wish to normalize the contrast on both photos using digital image processing techniques. Also, digital image processing implies correlation operations but for other purposes than stereocompilation; that is, we may wish to correlate dissimilar records such as a photograph with a map for map revision purposes. Here there would be much digital image processing performed on the photograph prior to correlating these records with each other. This is another facet of our work.

Additional clarification is necessary when discussing digital photogrammetry as opposed to digital mapping. Our definition of digital photogrammetry includes only that portion of the mapping operation which involves the stereocompilation process to produce topographic data. These data may be used for other purposes such as the production of ortho-photos, topographic products, or terrain profile data. Further, a subset of our digital photogrammetric effort includes digital mensuration. In addition, our work in the digital area should not be confused with digital line rectification operations.

We will also indicate the potential of digital array processors in these areas of endeavor, and emphasize the fact that we are working in a "general purpose" environment as opposed to developing "special purpose" hardware. We feel that the freedom of investigation afforded by the "general purpose" environment is necessary in the early stages of digital processing for whatever purpose and that very fundamental questions can be better answered in this type environment.

DIGITAL IMAGE PROCESSING

DICOMED SYSTEM - The DICOMED image processing system was procured as a relatively low cost means of entering into the scientific discipline of digital image processing. This hardware system is commercially available and it is flexible, reliable, and is of sufficient accuracy and resolution to accomplish much research in software techniques for handling and manipulating digitized, grey shade data. The hardware is a means of gathering and displaying results while the problem solving aspects are relegated to the development of unique, efficient and oftentimes complex mathematical processes. These two ingredients, the hardware and the software, are currently separate items, but they will soon be brought together into a very versatile, interactive system.

Currently, the hardware is off-line; that is, the digitized photographic data are put on magnetic tape and carried to a computer for subsequent processing. However, the system comes equipped with programmable hardware which will allow us to implement an interactive capability as soon as we take delivery of our control computer system. This will allow an analyst additional flexibility in performing various sophisticated mathematical operations on selected portions of a photograph.

The system consists of five basic components: namely, an image digitizer, image display storage tube, 7 and 9 track tape units, black and white and color printing unit and a coordinate entry unit which allows selective scanning, including a single spot. Therefore, the system provides both raster scans and random point scanning. Each unit has approximately 17 different commands which can be activated once the system is interfaced with a computer.

To afford the reader an idea of hardware capability, we will describe, in general, three of the five basic components: namely, the digitizer, tape units, and display. Information on all components of the DICOMED system is readily available from the company itself.

The digitizer offers a choice of scanning speeds and scanning resolutions. For example, a maximum of 2048 by 2048 points can be scanned in 2.5 minutes or in a maximum of 1.5 hours where the integrate time for spot is 1280 microseconds. The higher quality afforded by the slower scan capability is visible on the display screen. At the other extreme, scanning at a resolution of 256 by 256 points is accomplished in 4.3 seconds and in 87 seconds at the slow speed. The output from all scanning variations can be at either 64 (6 bits) or 256 (8 bits) grey levels, and the effective spot size at the film plane is approximately 25 micrometers. There are numerous other functional versatilities associated with this piece of equipment which will not be discussed here.

The Image Display unit is a directly viewed image display which constructs visual images from digital information. The dark-trace storage-type cathode ray tube used eliminates the need for periodic refreshing and results in a completely stable presentation. It will construct either single or multiple images and uses either a raster scan or random position format. A command structure is provided which allows the image display to be operated either manually or under program control. It constructs images with a resolution of up to 2048 points per axis where each point assumes one of 64 possible intensity levels.

The tape unit portion of the system is capable of operating in one of five modes: Bypass, Write, Read, Read-after-Write, or Computer. The Bypass mode allows direct operation between Digitizer and Display thereby bypassing the tape transport. The Read mode allows reading from tape onto the display while the Write mode allows recording on tape from the digitizer. In the Read-after-Write mode, data are recorded on tape from the digitizer and 25 milliseconds later the data are read from the tape and transmitted to the Image Display. The tape unit, as with the other components, can be interfaced with a computer thereby allowing approximately 17 different commands to be activated between computer and tape unit.

This brief summary of only three of the DICOMED components was presented primarily to reflect the versatility of the system. When the system is interfaced with our computer, which in turn will be capable of tasking our digital image processing software system, we feel that we will have a versatile tool for conducting research in digital image processing with applications in photo and radar interpretation, perspective scene generation, line-of-sight problems, creating a learning process for pattern recognition, simulating various types of hardware, etc.

SOFTWARE-DIGITAL IMAGE PROCESSING - The development of a flexible software system for digital image processing is in operation at USAETL. The system is called "Digital Image Manipulation and Enhancement System" (DIMES) and was developed for USAETL by the Computer Sciences Corporation (CSC). Application routines are modular and are called for execution through a free-form, user-oriented source language similar to image processing languages in use at the Jet Propulsion Laboratory, Goddard Space Flight Center, and Rome Air Development Center on other computers. It cannot be stressed too strongly that the system is user oriented. Users are not burdened by data handling problems and complex mathematical developments.

The extreme flexibility of the digital method in image processing makes a wide variety of linear and non-linear processes possible. Some of these uses are as follows:

1. Picture generation
2. Intensity manipulation
3. Geometric manipulation
4. Spatial frequency operations
5. Analysis
6. Multipicture analysis
7. Emphasize details
8. Sharpen the picture
9. Modify tonal range
10. Aid picture interpretation
11. Remove anomalies
12. Detect differences between pictures

Two scenarios of typical applications are given here. For example, one problem might involve improving the visibility of features in a photographic image. This might include the following steps:

1. Read the image into direct access storage from the tape produced by the image scanner.
2. Generate a histogram of grey-level distributions.
3. Stretch the contrast for the range of grey levels which contained most of the image information.
4. Use Fourier transform techniques to generate a digital filter, appropriate to the optical transfer and noise characteristics of the camera system, for reducing image blur.
5. Carry out the digital filtering by convolution.
6. Add a descriptive alphanumeric text to the image.
7. Write the image back to tape in a format acceptable by the image display or recording device.

A second problem might involve the registration of an image with a base map or reference image. The steps for achieving this include:

1. Select prominent features which can be identified in both the problem image and the reference image.
2. Determine the precise position of each feature in the problem image relative to the reference image by means of digital correlation over the neighborhood of the feature.

3. Compute coefficients for a suitable linear or non-linear geometric transformation which will bring the selected features in the problem image into registration with the reference image.

4. Carry out the geometric transformation.

5. If a composite is being constructed, average together the overlapping image areas.

To execute problems similar to those outlined above, the analyst simply writes a group of command verbs, each followed by fields and subfields, which provide more specific information about the image. Several samples of these commands are as follows:

RESERVE.....Allocate space for a new image on tape or disk,

FIND........Locate an existing image on tape or disk,

PARAMS......Name a parameter set for use by multiple tasks, etc.

The basic manipulation and enhancement routines required to carry out the sequence of events outlined above may consist of user oriented verbs such as

HISTO.......Compute a histogram of intensity values for a specified image segment and display it graphically on the line printer.

CONTRAST....Carry out a position-independent contrast conversion using a user-specified linear, piecewise-linear, or nonlinear relation between the old intensity values and the new.

MOSAIC......Form a mosaic by inserting pieces from two or more input images into the output image.

FFT.........Perform a one or two-dimensional fast Fourier transform or its inverse on a data set in floating point format.

FILTGEN.....Compute a frequency - space or convolution (image-space) filter corresponding to a user-specified modulation transfer function.

etc.

As examples of other applications, pattern recognition and feature recognition may be facilitated by contrast manipulation to accentuate density thresholds, convolution filtering to detect lines and edges, and generation of phase, amplitude and logarithmic amplitudal maps for the two-dimensional frequency spectrum of the image. In addition, comparison of images taken in different spectral bands or at different times may be facilitated by computing intensity differences, and photometric corrections for nonuniform response of video-type cameras may be applied using position-dependent contrast correction [1].

With this brief and nontechnical description of DIMES, it is intuitively evident that a great deal of mathematics and data handling is being performed through the use of simple verbs specified by the user. It is this aspect of DIMES which should lead to expanded applications of digital image processing for experimental purposes and which should eventually find its way into production environments.

As stated earlier, we will have DIMES available through our DICOMED system which will be interfaced with a control computer. This computer in turn will communicate with the CDC 6600 computer at 50,000 bits per second transfer rate.

DIGITAL PHOTOGRAMMETRY

Our efforts in digital photogrammetry are primarily concerned with the development of mathematical strategies associated with the problem of correlating digitized grey shade data obtained from the common area of two overlapping photographs. The key words are "development of mathematical strategies," thereby implying more than simply adopting one of several well known equations, such as the linear correlation coefficient equation, to determine when correlation of conjugate imagery occurs. Our feeling is that the strategies to be developed for "total" automation are still futuristic and that they have a better chance of evolving in a somewhat "general purpose" environment as opposed to outright development of "special purpose" hardware. That this is essentially the case, we point out that numerous questions still arise with current, specialized stereocompilation equipment even though it has been in production for a number of years.

The data we have been working with thus far comes from an unrectified, stereo pair of photographs taken over the Arizona and Fort Sill areas. Approximately thirty different scenes have been digitized using the USAETL microdensitometer. The photographs are at the scale of 1:47,000, and the thirty scenes fall roughly into eight categories, as follows:

1. Urban area
2. Small furrows
3. Dark fields
4. Drainage patterns
5. Mountainous regions
6. Flat terrain
7. Orchards
8. Forests

SOFTWARE - All of our software is being written for the CDC 6600 computer. We are using the digital data management capabilities of our DIMES package and all routines written thus far will be converted to modules of DIMES. This will give us added flexibility in manipulating the digital data.

Software is being written in-house to test essentially one concept but enbodying several correlation measures. The concept is called the "Infiltration Process" [2] and it in turn uses a "Mixed Coordinate" scheme. These will be defined shortly. It is made clear that testing a concept does not imply that we would eventually recommend it as the best way to do digital photogrammetry. We realize immediately that a large computer memory may be necessary to implement this concept and that it may not be cost effective until such time that high speed, mass meories become commercially available. Possibly, with greater consideration given to data organization, this may not be a stringent requirement. This becomes part of our systems analysis. On the other hand, it does offer the possibility of "total" automation and we find that numerous mathematical strategies evolve which would be tested regardless of the concept.

INFILTRATION AND MIXED COORDINATE SCHEME - Usually the independent coordinates of a stereo pair are selected on one image in a prescribed way, say along raster lines, and the dependent coordinates on the stereo mate are then calculated. The mixed coordinate scheme defines a coordinate on each image as the independent pair and the two remaining coordinates, one on each image, are then calculated. This approach allows for a significant decrease in interpolation for grey levels over the usual approach.

The matching process generates a matched point, say point P, from a neighboring matched point, say point Ø. Let JXAØ, YAØ, XBØ and IYBØ be the image coordinates of point Ø on images A and B. JXAØ and IYBØ are the independent coordinates and YAØ and XBØ are the derived dependent coordinates. The independent coordinates of point P are defined to be:

$$JXBP = JXA\emptyset + \Delta X$$

$$IYBP = IYB\emptyset + \Delta Y$$

where ΔX and ΔY are defined incremental changes. The essential idea in the infiltration process is that ΔX and ΔY are chosen so that the matching proceeds along paths that show the greatest promise for successful correlation. If ΔX and ΔY are sufficiently small then the dependent coordinates of P are estimated from the following prediction equations.

$$YAP' = YA\emptyset + R\Delta X + S\Delta Y$$

$$XBP' = XB\emptyset + T\Delta X + U\Delta Y$$

R and S are the partial derivatives of YAP with respect to the independent X-coordinate and with respect to the independent Y-coordinate respectively. T and U are similarly defined. The partial derivatives are estimated from previously matched points at and around point Ø. The prediction equations are derived by Taylor expansion after assuming the existence of a functional relationship between the dependent coordinates and the independent coordinates. The prediction equations can easily be enlarged to include second or higher order terms.

The predicted coordinates are refined by generating a correlation function at and around (YAP',XBP') and then defining the coordinates associated with the peak correlation as (YAP,XBP). The value of the peak correlation and the shape of the correlation function are used to evaluate the quality of the match.

RESULTS - The results listed below pertain to an urban scene taken from the Phoenix model. Input included the exterior orientation of each photograph. The spot spacing is 12.5 μm and the spot diameter is 25.0 μm. In the matching process, 7 arrays were generated in the direction of major parallax, i.e., the array centers were selected along the computed epipolar line. The array sizes were (7 x 7), the spacing within arrays was 2 spots, and the central array was located at the predicted match point. Seven arrays were also generated in the Y-direction. Note that the central array of the major parallax set is identical to the central array of the Y-set. Two correlation functions were generated by correlating the central array of each set with the 7 arrays of the other

PATH HISTORY

TRIAL	JXAØ	IYBØ	JXBP	IYBP	CP	CQ	CØNF	PK	QK	PX	PY
1762	-46	-10	-48	-10	.156	-.659	-.131	-1.58	-.01	3.48	.14
1763	-32	-48	-34	-48	-.559	.042	-.025	.85	.01	2.25	.04
1764	-34	-46	-34	-48	-.392	-.098	.389	.02	.01	2.61	.10
1765	-36	-44	-36	-46	-.254	-.031	.215	- .19	-.01	2.87	-.05
1766	-38	-42	-38	-44	-.194	-.034	.172	- .04	-.01	2.06	-.07
1767	-40	-40	-40	-42	-.214	-.091	.147	.291	-.01	- .87	-.24
1768	-40	-40	-40	-42	-.470	-.042	.319	.439	.01	-1.64	-.20

CORRELATION HISTORY - MAJOR PARALLAX DIRECTION

TRIAL	-3	-2	-1	0	1	2	3
1762	.523	.380	.392	.505	.339	.195	.170
1763	.226	.186	.333	.682	.876	.512	-.175
1764	.496	.514	.718	.922	.733	.142	-.451
1765	.607	.738	.886	.964	.789	.295	-.168
1766	.307	.503	.882	.971	.865	.509	.003
1767	-.031	.378	.567	.737	.692	.563	.425
1768	-.152	.274	.463	.904	.876	.681	.295

set. The correlation measure is the well-known linear correlation coefficient.

The process began at the center of the scene (JXAØ = IYBØ = 0) where 4 matched points were generated by a TOHOLD routine. The infiltration technique proceeds from the 4 points and progresses through the scene by attempting to match those points with even independent coordinates. That is $|\Delta X|$ and $|\Delta Y|$ are 0 or 25 μm depending on the infiltration path. The tabulated results pick up after the process has matched or attempted to match 1761 points. Therefore, elevations at approximately 1000 points in this scene can be computed using these data where the spacing of the points are 25 micrometers, or less; that is, every 4 feet at the scale of the photography.

CORRELATION HISTORY - Y-DIRECTION

TRIAL	-3	-2	-1	0	1	2	3
1762	-.112	.025	.048	.505	.312	.058	-.330
1763	.726	.676	.668	.682	.626	.528	.563
1764	.846	.830	.877	.922	.868	.768	.808
1765	.913	.925	.949	.964	.949	.912	.917
1766	.904	.921	.953	.971	.954	.917	.907
1767	.134	.334	.616	.737	.755	.682	.647
1768	.781	.846	.893	.904	.873	.853	.829

CP and CQ are the values of the second derivatives of the correlation functions with respect to X(major parallax direction) and Y respectively. CØNF is a function of CP, CQ, and the peak correlation in the X-direction. If CØNF > 0 the match is regarded as successful; if CØNF $\leq$ 0 the match is regarded as a failure. PK and QK are the computed shifts along the major parallax and Y directions, respectively. Since the orientation parameters are known, the Y shift is constrained to $\pm$ .01 where the sign is determined from the Y correlation function. PX and PY are parallax values. They are the differences between the independent coordinates and the derived dependent coordinates. The column headings of the correlation values pertain to the shift of the corresponding array center from the central array. In this particular example a shift of one unit pertains to 25 μm.

The infiltration process employs a restart feature whenever the process bogs down. In the restart mode the process leaves the troublesome area and begins anew in an area determined by the process. For example, if the infiltration process produces 21 failures in a row, the process goes into the restart mode. It turns out that trial 1762 was the 21st failure in a row. (Trial 1742 was the 1st in this particular sequence.) Note that trial 1763 takes place a good distance from trial 1762. Unfortunately trial 1763 produced another failure (CØNF = -.025); however, trial 1764 was a successful match (CØNF = .389). The next three trials were successes, but note that point Ø and P of the last trial are the same as trial 1767. This is because the infiltration process will attempt a new match at the same point with refined values of R, S, T and U whenever |PK| is greater than a given test value. In this example, the test value was 0.25; the confidence improved (.147 to .317) but the shift got larger. This indicates that the process is running into a sharp elevation change. Note that the peak correlation improved.

With these results, we see that the digital photogrammetric approach gives us information on the quality of a match as a byproduct of the operation. This kind of information will be available to operators of future digital systems and should be of tremendous value in producing quality products for which information on the degree of "goodness" is available at each step. By the same token, at the input end of the operation, operators will be able to specify various parameters designating the quality required as output. These input parameters will be developed as part of our effort, also.

HARDWARE - ASSOCIATIVE ARRAY PROCESSOR - Special efforts have been made to investigate current digital technology and its development for processing data in parallel. Probably the most well known of the parallel processors is the ILLIAC IV computer which is now installed and operating at NASA's Ames Research Center, Moffet Field, California. However, during the development of this computer, industry began implementation of this type digital technology into much more special purpose parallel processors thereby reducing costs significantly. As time passes, we read more and more about these special, digital parallel processors being used in place of so-called analog systems [3]. One reason for their popularity is that they offer a degree of flexibility not afforded by other type systems in that they are "programmable." We believe that the degree of "programmability" is directly related to their cost. In any event, among numerous other reasons, which are beyond the scope of this paper, the "programmable" aspect of these processors offers the opportunity to change strategies without significantly changing hardware. This was a recognized fact by personnel at the Goodyear Aerospace Corporation (GAC), Akron, Ohio, a number of years ago which in turn led to their development of GAC's Associative Array Processor (AAP) [4], called STARAN, and is now commercially available.

Let this brief and very general discussion of parallel processors indicate our interest in their potential for processing photogrammetric, digitized, grey shade data. As a result, we developed an idealistic scenario which could be supplied to several manufacturers so that they could provide timing estimates based on exactly the same ground rules. This scenario is as follows:

1. 20-micrometer spot size
2. 40-square-inches of stereo area
3. Assume digitized, grey shade data is readily available, i.e., idealistic high speed mass memory system (no data transfer rates involved).
4. Arrays are 11 x 11
5. Move the array throughout the 40-square-inch area
6. Assume a match at each move
7. There are approximately 65 x 10^6 moves or matches.
8. No computational short cuts
9. Evaluate timing estimates for the absolute difference algorithm between two arrays (Algorithm #2 in Table I).

TABLE I - ETL CORRELATION ALGORITHM EXAMPLES

ALGORITHM NO.	ALGORITHM
1	$\sigma_1 = \sum_{i=1}^{n} \sum_{j=1}^{n} X_{ij}Y_{ij}$
2	$\sigma_2 = \sum_{i=1}^{n} \sum_{j=1}^{n} \mathrm{ABS}\left(X_{ij}-Y_{ij}\right)$
3	$\sigma_3 = \sum_{i=1}^{n} \sum_{j=1}^{n} \left(X_{ij}-Y_{ij}\right)^2$
4	$\sigma_4 = \dfrac{\sum_{i=1}^{n} \sum_{j=1}^{n} X_{ij}Y_{ij} - \left(\sum_{i=1}^{n} \sum_{j=1}^{n} X_{ij}\right)\left(\sum_{i=1}^{n} \sum_{j=1}^{n} Y_{ij}\right)}{n^2 - 1}$
5	$\sigma_5 = \dfrac{\sum_{i=1}^{n} \sum_{j=1}^{n} X_{ij}Y_{ij}}{\left[\left(\sum_{i=1}^{n} \sum_{j=1}^{n} X_{ij}^2 \quad \sum_{i=1}^{n} \sum_{j=1}^{n} Y_{ij}^2\right)\right]^{1/2}}$
6	$\sigma_6 = \dfrac{\sigma_{xy}}{\left[\left(\sigma_{xx}\right)\left(\sigma_{yy}\right)\right]^{1/2}}$ where: $\sigma_{xy} = \dfrac{\sum_{i=1}^{n} \sum_{j=1}^{n} X_{ij}Y_{ij} - \left(\sum_{i=1}^{n} \sum_{j=1}^{n} X_{ij}\right)\left(\sum_{i=1}^{n} \sum_{j=1}^{n} Y_{ij}\right)}{n^2 - 1}$ $\sigma_{xx} = \dfrac{\sum_{i=1}^{n} \sum_{j=1}^{n} X_{ij}^2 - \left(\sum_{i=1}^{n} \sum_{j=1}^{n} X_{ij}\right)^2}{n^2 - 1}$ $\sigma_{yy} = \sum_{i=1}^{n} \sum_{j=1}^{n} Y_{ij}^2 - \left(\sum_{i=1}^{n} \sum_{j=1}^{n} Y_{ij}\right)^2$
NOTE:	n=11

Personnel in the Computer Sciences Laboratory, USAETL, developed the timing estimates for the CDC 6600 computer using the minor cycle times of this machine, i.e., 200 nanoseconds for an add. The results of this timing game are given in Table II, comparing the CDC 6600 with GAC's S-1000, STARAN, which is a 4-array processor. Not only is the aforementioned algorithm evaluated, but several others, also.

TABLE II - COMPARISON OF STARAN S-1000 AND CDC 6600 COMPUTATION TIMES ESTIMATES

ALGORITHM	STARAN S-1000	CDC 6600	TIME RATIO
#1	0.488 hrs.	4.04 hrs.	8.3
#2	0.257 hrs.	4.66 hrs.	18.1
#3	0.519 hrs.	12.0 hrs.	23.2
#4	0.892 hrs.	5.9 hrs.	6.6
#5	1.47 hrs.	12.0 hrs.	8.1
#6	1.87 hrs.	15.84 hrs.	8.5
			12.1 AVG.

As stated earlier, this scenario was evaluated by other parties using their special purpose signal processors. Let it suffice to say that there were slight differences in the estimates among these special processors but large differences when compared to the CDC 6600, as is the case between the CDC 6600 and STARAN. It is very difficult to get a "normalized" set of values between various parallel processors because they can be configured to the problem.

CONCLUSIONS

We have presented some aspects of our current research in two broad areas: namely, Digital Image Processing and Digital Photogrammetry. Some results of our digital photogrammetric effort was presented primarily to indicate the degree of flexibility we have in operating in a somewhat "general purpose" environment without being strapped with special purpose hardware. This is especially evident when one considers that we are currently independent of a scanning system. Many specific features of our software for correlating digitized grey shade data have been omitted simply because we have the capability to add, change, and delete strategies. We have availed ourselves of this flexibility many times and we expect that we will continue to do so. However, we see already that the information obtained from our digital approach will prove invaluable in making judgments concerning the quality of the end product in future systems.

Our efforts in Digital Image Processing have proven to be worthwhile from the point of view that we have already produced simulated PPI radar scenes starting with an aerial photograph. In generating these scenes, extensive use was made of the DIMES software, especially the digital filtering capabilities. Other applications thus far include taking a photograph, digitizing it, and then generating various perspective views of the scene. Views can be made from inside or outside of the data base while simultaneously handling the hidden terrain problem; that is, there are no false representations of terrain after manipulating the data base. We expect that this capability will be useful in radar scene generation where a proper perspective of objects will give a radar analyst a better chance of assigning proper reflectance values. Also, the process can be used for placement of various perspectives of a portion of the terrain in the border of special map products. The basic philosophy in these efforts is that the photograph itself is the data base, or mass storage device, so-to-speak, and we must devise efficient means of exploiting this fact.

Not overlooking the fact we must quickly and efficiently handle digitized grey shade data, we have investigated recent developments in digital signal processors, array processors, associative array processors, or parallel processors, however closely these processors resemble each other in digital logic, as the means of performing tasks for which they are best suited as opposed to the serial computer. In general, we are interested in maintaining flexibility by being able to program these special processors simply because requirements change.

Finally, we expect that we will soon have a Digital Photogrammetric Simulation System (DPSS) in which we will be independent of scanners or any other special purpose equipment. This will give us an unprecedented opportunity to develop parameters which control the quality of the final product, and this will be as variable as one wishes it to become.

REFERENCES

[1] Steffel, Jerry, Unpublished Correspondence, Computer Sciences Corporation, Falls Church, Virginia, January 1973.

[2] Erickson, K.E., Rosenberg, P., Digital Mapping System Study, Keuffel and Esser Co., Morristown, New Jersey, October 22, 1971, USAETL Contract No. DAAK02-71-C-0079.

[3] Klass, P.J., USAF Pushes Digital Avionics for Aircraft, Aviation Week & Space Technology, June 11, 1973.

[4] Rudolph, J.A., A Production Implementation of an Associative Array Processor - STARAN, Goodyear Aerospace Corporation, December 5-8, 1972, Fall Joint Computer Conference, Los Angeles, California.

R. Bernstein

Digital Image Processing of Earth Observation Sensor Data

Abstract: This paper describes digital image processing techniques that were developed to precisely correct Landsat multispectral Earth observation data and gives illustrations of the results achieved, e.g., geometric corrections with an error of less than one picture element, a relative error of one-fourth picture element, and no radiometric error effect. Techniques for enhancing the sensor data, digitally mosaicking multiple scenes, and extracting information are also illustrated.

Introduction

• *Landsat*

In July 1972 the U.S. National Aeronautics and Space Administration (NASA) launched the first Earth Resources Technology Satellite (ERTS-1), recently renamed "Landsat-1," and a second satellite was launched in January 1975. A number of significant scientific discoveries and practical benefits have already resulted [1–8]; see Table 1 for the major disciplines and applications of the Landsat program.

Figure 1 Return Beam Vidicon (RBV) camera and its field of view.

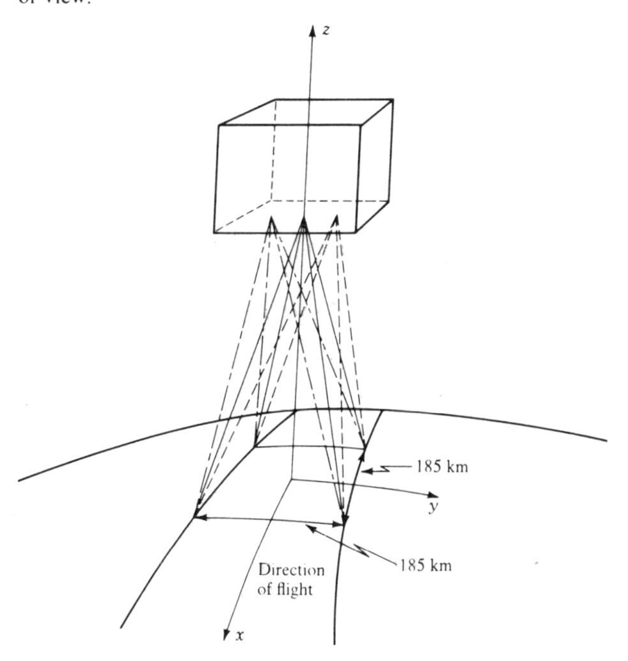

Each unmanned Landsat makes 14 orbits per day, viewing 185-km-wide strips of the Earth [9]. Each satellite provides global coverage once every 18 days, so that the two satellites provide such coverage every nine days. All equatorial crossings occur at approximately the same *local* time each day, 9:30 a.m.; this holds for all parts of the world. The satellite orbit plane precesses at about one degree per day and the sun-synchronous orbit has a constant orbit-plane angle relative to the sun-Earth line, which minimizes misinterpretation of satellite-sensor data due to shadow effects from the sun. The physical parameters of the Landsat mission are summarized in Table 2.

Landsat carries two kinds of sensors to detect and record sunlight reflected from the Earth's surface in particular spectral bands [9]:

The *Return Beam Vidicon* (*RBV*) is a three-band, three-camera television system (Fig. 1) using conventional lenses and shutters, and vidicons for image scanning and storing prior to transmission of the image data to ground stations. The cameras are sensitive to scene radiance in wavelengths from 0.48 to 0.83 μm. Their field of view is an area 185 km × 185 km.

The *Multispectral Scanner* (*MSS*) consists of an oscillating mirror and an optical system which reflect and direct scene radiance into a solid-state detector array that is sensitive to wavelengths in four spectral bands between 0.5 and 1.1 μm (not including thermally generated radiation); see Fig. 2. Six scan lines are simultaneously swept in each spectral band with one mirror oscillation. The detectors subtend an instantaneous field of

Table 1 Landsat (ERTS) data and applications.

Agriculture, Forestry, Range	Crop census Crop yield Identification of vegetation disease Land use inventory
Oceanography, Marine resources	Fish production Ship routing Sea state and ice conditions
Hydrology	Water resources inventory Fresh-water source identification Flood monitoring Health monitoring of lakes Pollution monitoring
Geology	Tectonic feature identification Geologic and physiographic mapping Mineral and field exploration Earthquake-area studies Temporal studies (glaciers, volcanos, shoreline erosion)
Geography	Thematic maps of land use Physical geography (to improve land use)

view on the ground 79 m on a side; their outputs are in digital form for transmission to ground stations. Characteristics of the Multispectral Scanner are listed in Table 3.

• *Digital image processing concepts*

Earth observation data acquired by on-board spacecraft sensors are affected by a number of electronic, geometric, mechanical, and radiometric distortions that, if left uncorrected, would diminish the accuracy of the information extracted and thereby reduce the utility of the data.

Previous methods of correction, using electro-optical processing techniques, have had some limitations [9]. Recent investigations [10–12] have shown the superiority of a digital approach over that of electro-optics as a consequence of the former's processing flexibility, fewer required data conversions, and improved accuracy and quality of the information developed.

To correct sensor data, internal and external errors must be determined—they must be either predictable or measurable. Internal errors are due to sensor effects; they are systematic or stationary, i.e., constant (for all practical purposes), and can be determined from prelaunch calibration measurements. External errors are due to platform perturbations and scene characteristics, which are variable in nature but can be determined from ground control and tracking data. Thus, the information required for correcting data distortion can be obtained (within certain limits of precision). Figure 3 presents a simplified illustration of the sequence of sensor-data processing from acquisition to application.

Table 2 Landsat (ERTS) mission parameters.

Apogee	917 km
Perigee	898 km
Semi-major axis	7285.7 km
Inclination	99.0 degrees
Anomalistic period	103 minutes
Eccentricity	0.0012
Local time at descending node (equatorial crossing)	9:30
Coverage cycle duration	18 days
Distance between adjacent ground tracks	159.38 km

Table 3 Multispectral Scanner (MSS) characteristics.

Instantaneous field of view	0.086 mrad
Earth area subtended	6240 m^2
Mirror oscillation range	±2.89 degrees
Mirror oscillation frequency	13.62 Hz
Scan lines per oscillation	6
Cross track field of view	11.56 degrees
Cross track scan	185 km
Spectral band range: 4	0.5 to 0.6 μm
5	0.6 to 0.7 μm
6	0.7 to 0.8 μm
7	0.8 to 1.1 μm
Number of detectors	24
Sampling interval of detector output	9.95 μs
Sample word length	6 bits
Samples per line	3240
Lines per bands	2340
Information per band	7.6 × 10^6 bytes
Information per scene	30.4 × 10^6 bytes

Figure 2 Multispectral Scanner (MSS) and its field of view.

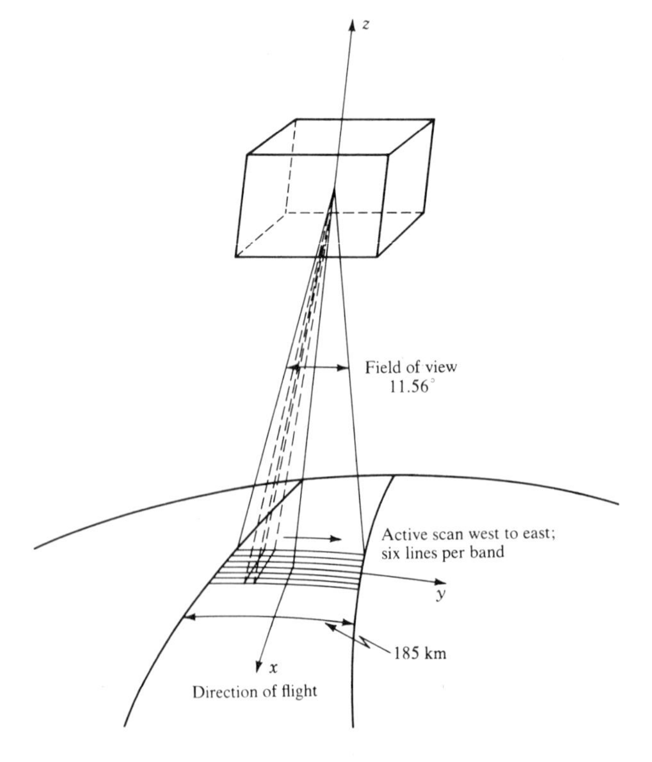

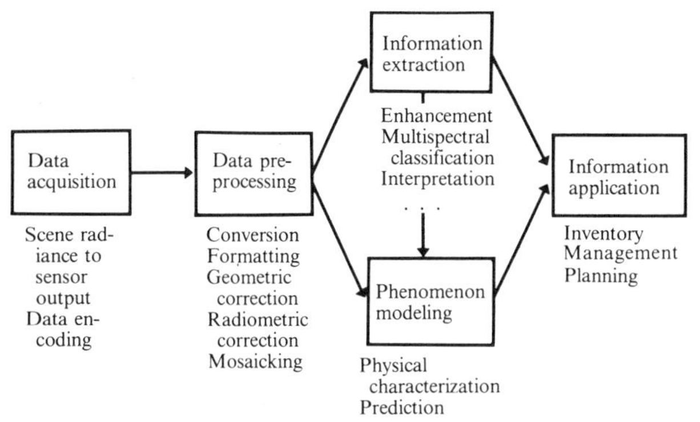

Figure 3 Simplified scheme of sensor-data processing stages.

The principal error sources (for the Multispectral Scanner) and the maximum compensation for the errors, as determined by statistical analysis, are the following:

Platform effects

Altitude Departures of the spacecraft from nominal altitude produce scale distortions in the sensor data. For the MSS, this distortion is along-scan only and varies with time; the magnitude of correction is 1.5 km.

Attitude Nominally, the sensor axis system is maintained with one axis normal to the Earth's surface and another parallel to the spacecraft velocity vector. As the sensor departs from this attitude, geometric distortion results. For the MSS, the complete attitude time-history must be known to compensate for the distortion; magnitude of correction: pitch, 12 km; roll, 12 km; yaw, 2.46 km; pitch rate, 0.93 km; roll rate, 0.54 km; and yaw rate, 0.040 km.

Scan-skew During the time required for the MSS mirror to complete an active scan, the spacecraft moves along the ground track. Thus, the ground swath scanned is not normal to the ground track but is slightly skewed, which produces cross-scan geometric distortion; the magnitude of correction is 0.082 km.

Velocity If the spacecraft velocity vector departs from nominal values, the ground track covered by a given number of successive mirror sweeps changes, producing along-track scale distortion; the magnitude of correction is 1.5 km.

Scene effects

Earth rotation As the MSS mirror completes successive scans, the Earth rotates beneath the sensor. Thus, there is a gradual westward shift of the ground swath being scanned. This causes along-scan distortion; the magnitude of correction is 13.3 km.

Map projection For Earth resources use, image data are usually required in a specific map projection. Although map projection does not constitute a geometric error, it does require a geometric transformation of the input data, and this can be accomplished by the same operations that compensate for distortion in the data; the magnitude of correction is 3.7 km along scan and along track (for the continental U.S.).

Sensor effect

Mirror sweep The MSS mirror scanning rate varies nonlinearly across a scan because of imperfections in the electromechanical driving mechanism. Since data samples are taken at regular intervals of time, the varying scan rate produces along-scan distortion; the magnitude of correction is 0.37 km.

Scene and sensor effects

Panorama The imaged ground area is proportional to the tangent of the scan angle rather than to the angle itself and, since data samples are taken at regular intervals, this produces along-scan distortion; the magnitude of correction is 0.12 km.

Perspective For most Earth resources applications, the desired Landsat images represent the projection of points on the Earth on a plane tangent to the Earth at the nadir, with all projection lines normal to the plane. The sensor data, however, represent perspective projections, i.e., projections whose lines meet at a point above the tangent plane. For the MSS, this produces only along-scan distortion; the magnitude of correction is 0.08 km.

Atmospheric effects

The scene radiance is dispersed and attenuated by the atmosphere between the sensor and the ground. Compensation for these effects is difficult but, if it can be done, increases the accuracy of information extraction still further. For example, see the discussion of sensor requirements by Kidd and Wolfe [13].

Elements of digital correction

The procedure developed for correction of satellite imagery is based on the computation of geometric coefficients from models that describe the distortion in the image. After obtaining these coefficients, a mapping function can be synthesized for geometric and radiometric correction of the sensor data. In this section we discuss the digital techniques that were developed to preprocess (correct) Earth observation data. These techniques have been applied to a variety of problems in addition to those used here as examples.

• *Ground Control Points (GCPs)*
Because tracking and spacecraft attitude data are not known precisely, Ground Control Points are used to obtain external reference information. A GCP is a physical feature detectable in a scene, whose location and elevation are known precisely. Typical GCPs are airports, highway intersections, land-water interfaces, geological and field patterns, etc. A registration operation is used to match a small image (subimage) area containing the GCP with a scene to be corrected, acquired possibly at a later time but containing the same feature.

Computationally efficient techniques have been developed to locate GCPs in digital data arrays. These are based on implementations [14, 15] of Sequential Similarity Detection Algorithms (SSDA), originally developed by Barnea and Silverman [14].

Sequential Similarity Detection Algorithms [14]
In this technique, also known as template matching, it is necessary to determine the similarity, or distance (in our case the difference in intensity), between two elements. A class of functions called metrics, which is a conventional abstraction of the notion of Euclidian distance, is used.

The principle of operation involves differencing an intensity-normalized random sample of corresponding points in the GCP and search areas and summing the absolute values of the differences. When the sum exceeds a selected threshold value, the subimages are considered to be dissimilar, the GCP area is displaced in relation to the search area, and another comparison is made. This process is repeated until a minimum sum function is found, which corresponds to the coordinates x, y of a best match between the GCP and the search area.

Consider two images, the search area S and the "window" (GCP subimage) W, illustrated in Fig. 4. The search area is an $L \times L$ array of digital picture elements which can assume one of G gray (intensity) levels. The window is an $M \times M$ $(M < L)$ array of digital picture elements having the same gray-scale range. By superimposing the window on the search area, and by constraining the translation of the window so that at all times it is contained entirely within the search area, subimages $S_M^{i,j}$ are defined, which are $M \times M$ arrays of digital picture elements.

Each subimage is identified by the coordinates i, j of its upper left corner. Because of the constraint on the translation of the window (within S only), the domain of definition of the subimages—called the allowed range of reference points—is smaller than the search area.

The search for a point with coordinates i^*, j^*, in this domain, that indicate the position of the subimage which is most "similar" to the given window is called translational registration. For each point i, j there are M^2 points of the subimage to be compared with the M^2 corresponding points of the window. Since there are $(L - M + 1)^2$ points in the domain of definition, conventional correlation methods require comparison of $M^2(M - L + 1)^2$ pairs of points. Barnea and Silverman used a Constant Threshold Algorithm, in which the pairs of points to be compared are selected in random order from a non-repeating sequence of integers, $1, 2, \cdots, M^2$. The advantage of this technique relative to conventional correlation methods or the fast Fourier transform method is that few computations are performed when no similarity exists, which results in a decrease in registration time of about two orders of magnitude [14, 15].

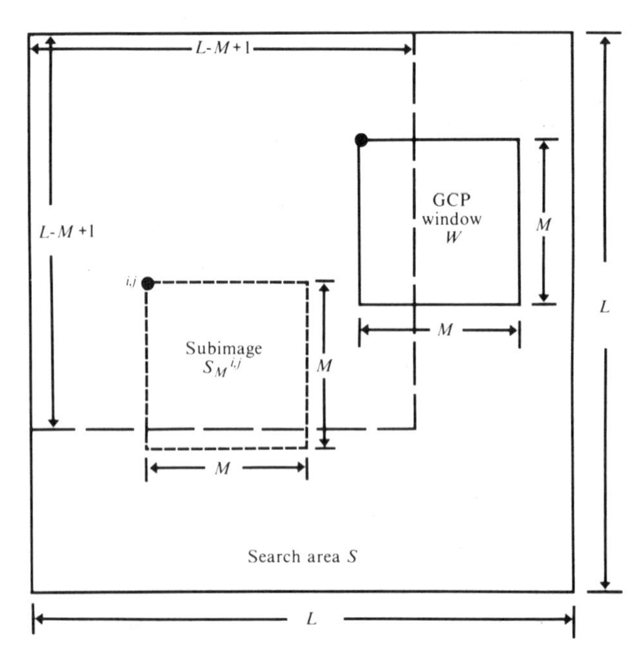

Figure 4 Illustration of the search area, allowed range of reference points, and the associated window.

The Sequential Similarity Detection Algorithms are usually used to get to the neighborhood of the best match. Then a 5×5 grid of correlation coefficients, centered at the best integral picture element (pixel) match, as determined by the SSDA, is computed. Next, a smooth surface is fitted to the grid and the Fletcher-Powell method [16] is used to attain sub-pixel GCP registration accuracy.

We define a *success ratio* as the number of correct registrations divided by the total number of registration attempts. The primary factor affecting the success ratio is the temporal and seasonal variation between the GCP and the scene to be processed. Table 4 shows that the registration success ratio is reasonably high, even for data separated in time by one year [12]. Another factor involves the spectral characteristics of the GCP feature.

For example, macadam airfields can be detected and registered more successfully in the infrared bands (6 and 7) than in the visible spectral bands (4 and 5). Generally, to increase reliability, redundant GCPs are processed.

• *Mapping operation*

Spacecraft roll, pitch, and yaw data are not provided with sufficient accuracy by the satellite attitude determination system, and the ephemeris data do not provide either altitude or ground position with sufficient accuracy; these parameters must be calculated from knowledge of the GCP locations.

Differences between actual and observed GCP locations are used to evaluate the coefficients of cubic polynomial time functions of roll, pitch, and yaw, and of a linear time function of altitude deviation [12]. The GCPs are first located in the input image (sensor data) and then mapped into the Earth tangent plane using models based on all those errors that can be predicted or determined from tracking data. The tangent plane projection contains a Cartesian coordinate system with its origin at the center of the image and the x-y plane tangent to the Earth ellipsoid at this origin. The positive x axis is in the direction of the nominal spacecraft ground track; the z axis is oriented away from the center of the Earth; and the y axis completes a right-handed coordinate system. The functional steps that generate the mapping function are summarized in Fig. 5.

• *Geometric correction function*

The input image is an array of digital data which represents a geometrically distorted, one-dimensional perspective projection of some portion of the Earth's surface. The output image is a geometrically corrected map projection of the same ground area.

A network of grid points spanning the output image area is mapped into the input image using a pair of bivariate polynomials; the following mapping functions [12, 17] provide this transformation:

$$v = v(x, y) = \sum_{p=0}^{N} \sum_{q=0}^{N-p} a_{pq} x^p y^q;$$

$$u = u(x, y) = \sum_{p=0}^{N} \sum_{q=0}^{N-p} b_{pq} x^p y^q.$$

Rather than apply the mapping functions to all points of the output image, an interpolation grid is established on the output image. The grid is constructed so that if the four corner points of any grid mesh are mapped with the aid of the mapping function polynomials, all points internal to the mesh can be located in the input image, with sufficient accuracy, by bilinear interpolation from the corner points.

After determining the position of an output picture element on the input image, several methods can be used to calculate the intensity value of the output element. For example:

Table 4 Registration success ratios as a function of the temporal separation between the acquistion of search area and window area data (Chesapeake Bay vicinity).

		Success ratio			
Surface feature	*Separation (days)*	*Band 4*	*Band 5*	*Band 6*	*Band 7*
Large land-water interfaces	18	0.40	0.33	0.93	1.00
	108	0.36	0.36	0.57	0.57
	288	0.79	0.64	1.00	0.93
	378	0.64	0.71	1.00	1.00
Small land-water interfaces	18	0.71	0.57	1.00	1.00
	108	0.86	0.71	0.71	0.43
	288	0.67	0.67	0.33	0.33
	378	0.86	0.71	0.86	0.71
Interstate-grade highways	18	1.00	1.00	0.94	0.94
	108	1.00	1.00	0.63	0.38
	288	0.88	1.00	0.63	0.44
	378	0.88	0.88	0.75	0.88
Airfields (macadam)	18	0.00	0.00	1.00	1.00
	108	1.00	1.00	1.00	1.00
	288	1.00	1.00	1.00	1.00
	378	1.00	1.00	1.00	1.00
Airfields (concrete)	18	1.00	1.00	1.00	1.00
	108	0.00	1.00	1.00	0.00
	288	0.00	1.00	0.00	0.00
	378	1.00	1.00	1.00	1.00

The *nearest-neighbor* method, which selects the intensity of the closest input element and assigns that value to the output element:

$$I(x, y) = I(u, v).$$

The *bilinear interpolation* method, which uses four neighboring input values to compute the output intensity by two-dimensional interpolation:

$$I(x, y) = a_1 I(u, v+1) + a_2 I(u, v) + a_3 I(u+1, v+1) + a_4 I(u+1, v).$$

The *cubic convolution* method [11, 12, 18], originally suggested by Rifman and McKinnon, which uses 16 neighboring values to compute the output intensity (see the Appendix):

$$I(x, y) = \sum_{m,n} a_{m,n} I(u+m, v+n), -1 \leq m, n \leq 2.$$

Resampling of MSS image data can be used to eliminate spatial discontinuities due to nonsimultaneous detector sampling and geometric image correction operations.

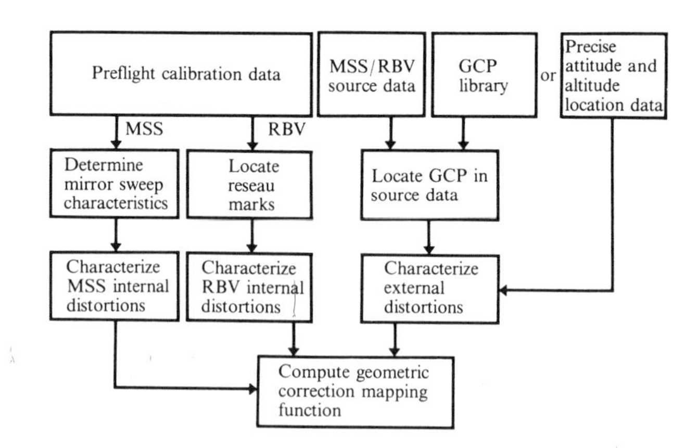

Figure 5 Functional steps for the generation of the mapping functions.

Figure 6 shows an MSS subimage, extracted from a central California scene (Fig. 8), before and after use of the three resampling algorithms [12]. Discontinuities in the input data have been eliminated by the bilinear interpolation and cubic convolution methods. Some high-frequency loss, due to low-pass filtering of the image data, can be noted in the corrected image for which bilinear interpolation was used.

Figure 6 MSS subimage area (band 5) before and after application of resampling algorithms: (a) original data; (b) nearest-neighbor method; (c) bilinear interpolation method; and (d) cubic convolution method.

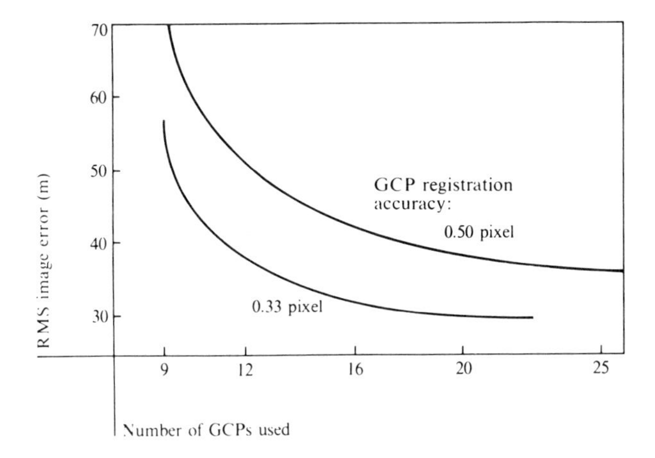

Figure 7 Root-mean-square (RMS) distance error as a function of the number and registration accuracy of the GCPs used.

Table 5 Truncated radiometric correction table (for three-bit data).

Detector output V	*Corrected value R for function*[a] *1*	*2*	*3*	*4*
000	000	000	000	000
001	000	001	000	000
010	000	011	000	001
100	010	110	010	100
011	001	100	010	100
101	011	111	100	110
110	100	111	110	111
111	101	111	111	111

[a]Functions: 1) $R = V - 2$; 2) $R = 1.5V$; 3) $R = 2(V - 3)$; and 4) $R = 0$ if $V \leq 1$, $R = 1.5(V - 1)$ if $1 < V < 6$, and $R = 7$ if $V \geq 6$.

Table 6 Geometric MSS error analysis (ninety percent of the GCPs have errors less than that indicated).

Scene	*Number of GCPs*	*Relative geometric accuracy* (m)
Chesapeake Bay[a]	16	72
Monterey Bay[b]	14	80
New Jersey[c]	15	80
New Jersey[c]	16	79

[a]Area in Fig. 11.
[b]Area in Figs. 8 and 15.
[c]Areas in Fig. 11.

Figure 8 (facing page) Digitally processed images of a central California area: (a) MSS image (band 7) before and (b) after geometric correction (see the data in Table 6); (c) MSS image (band 7) before radiometric enhancement, (d) subimage after radiometric road network enhancement, and (e) subimage after radiometric water pattern enhancement; (f) RBV image (band 2) before and (g) after radiometric correction (see the data in Table 7).

• *Radiometric correction*

Multispectral Scanner

The MSS has 24 solid-state detectors, six for each band. Both bias and gain errors can exist with uncalibrated detector data. Each detector voltage output is digitized into 64 values, or counts. An internal MSS calibration lamp is scanned by the detectors and the data are used to provide absolute (in terms of the lamp strength) and relative (in terms of each detector) response information. The corrections thus generated are implemented by simple table-lookup operations; see Table 5 for an example of radiometric corrections. The detector output (V) is used as an address or pointer to the correct value in the table, and the table value (R) is used as the corrected image-element radiance response. In Table 5 the response of Function 1 shows bias error compensation; Function 2, gain error compensation with rounding and truncation; Function 3, bias and gain compensation; and Function 4, nonlinear compensation. Highly nonlinear compensation and enhancement transformations can be implemented in the same manner.

Return Beam Vidicon

The RBV computer-compatible tapes contain image data that have been sampled and digitized with six-bit quantization. Thus, there are 64 possible input values. If a table which specifies the correct output intensity for each of the 64 input intensities can be defined, radiometric correction of the RBV images can also be accomplished by table-lookup operations. This is essentially the technique used, but it is complicated by the fact that RBV radiometric errors vary across the image; this results in the need to use multiple correction tables.

The RBV data suffer from significant shading (nonuniform response) effects. Since the objective of RBV

Table 7 RBV error analysis and radiometric correction zones.[a]

Band	*Number of GCPs*	*RMS error* (m)	*Maximum error* (m)	*Number of horizontal fields*	*Number of vertical fields*	*Total zones*
1	9	37.1	64.9	191	138	26 358
2	9	36.3	53.2	184	119	21 896
3	8	42.1	68.0	181	106	19 186

[a]Data relate to Figs. 8(f) and (g).

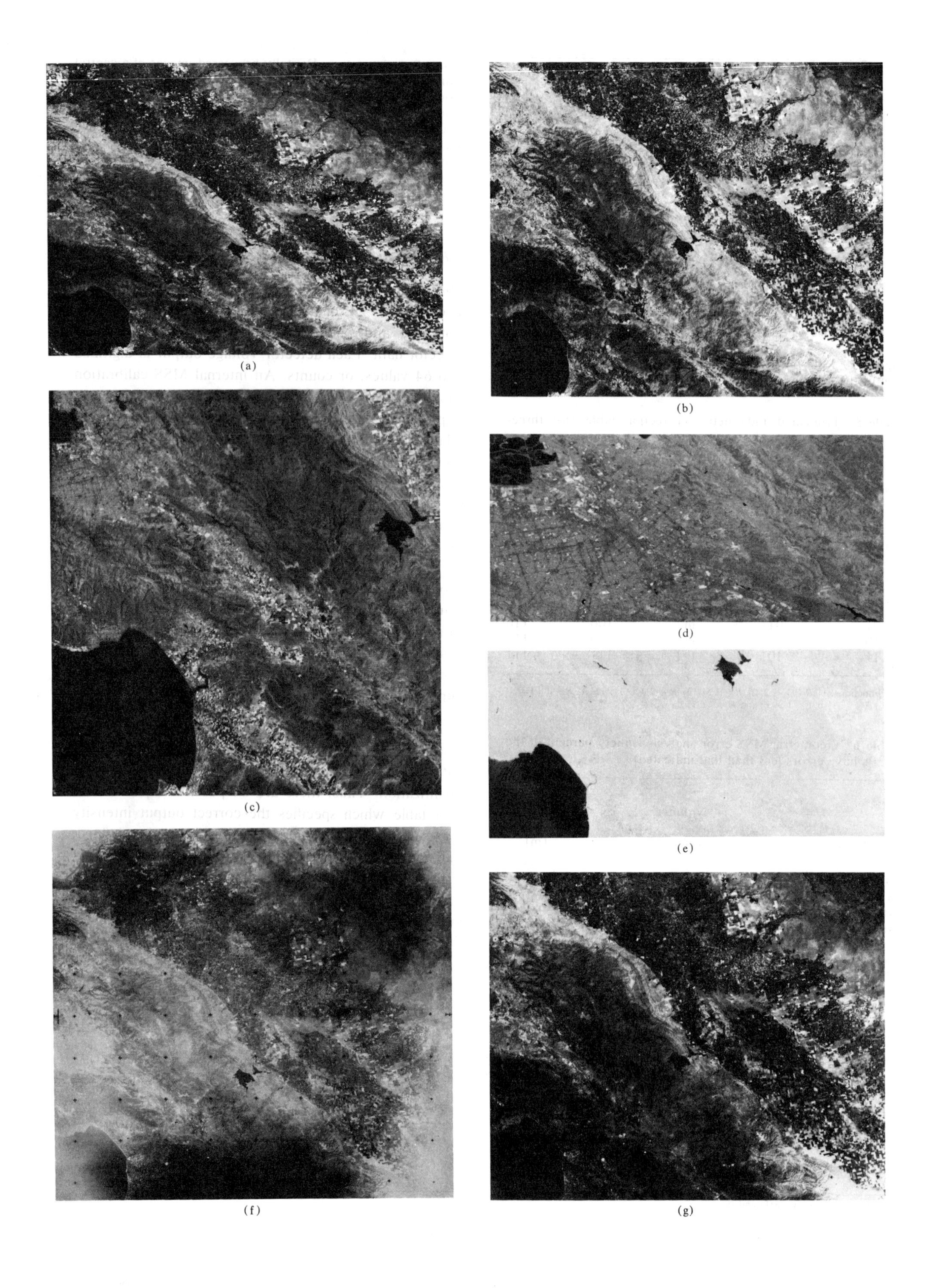

(a) (b) (c) (d) (e) (f) (g)

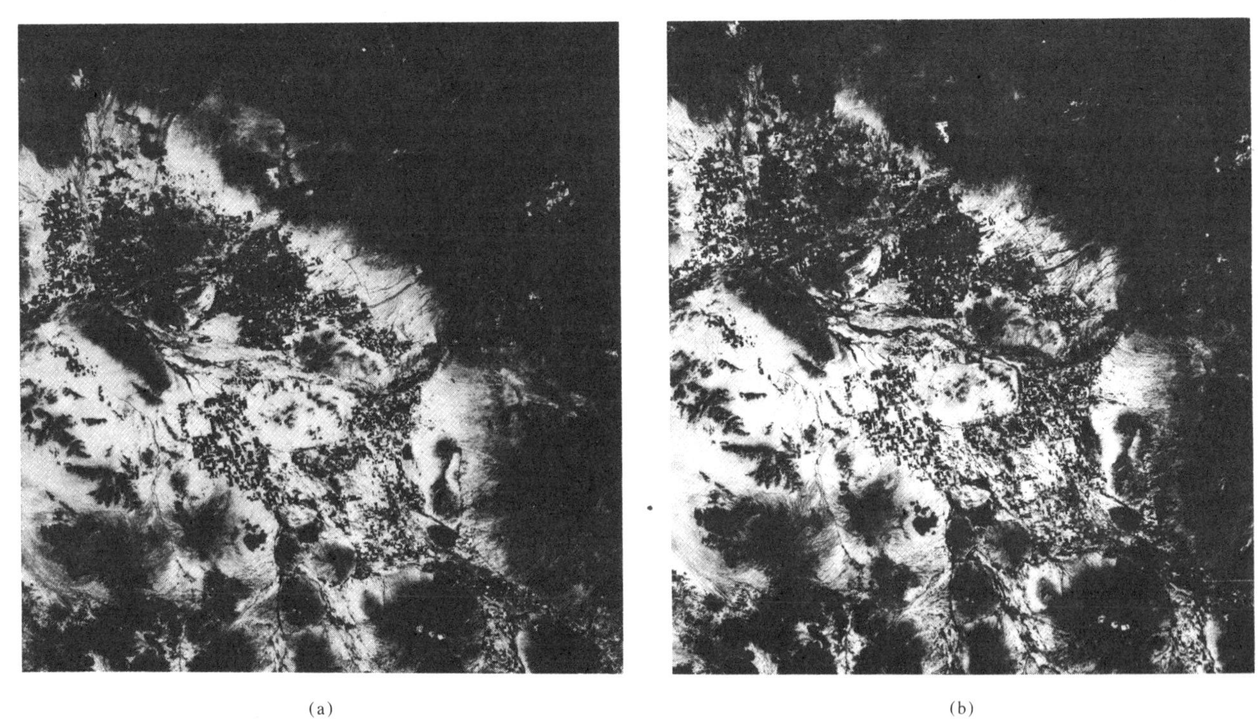

(a) (b)

Figure 9 MSS scenes of the Phoenix, Arizona area geometrically corrected to be in registration; (a) acquired on October 16, 1972 and (b) acquired on November 21, 1972.

Figure 10 Subimage of a digital mosaic of two MSS scenes showing a circular geologic feature.

correction is to modify the image data intensity to compensate for the spatially variable response characteristics, a method has been developed that mathematically structures the RBV image into correction zones. (A sufficient number of such zones must be established to have a nearly continuous radiometric response function.) Each zone has a unique radiometric correction table to be used for compensation of the RBV errors. In effect, RBV radiometric correction is performed in two stages: generation of the correction tables (an off-line operation) and then application of the correction.

Pre-flight calibration readings at several intensity levels from a uniform light source provide an 18×18 array of points in the RBV image. These readings, in terms of voltage ranging from 0.32 to 1.10 V, are then scaled to the digital range 0 to 63. From these uniform input values, 18×18 arrays of gain (G) and bias (B) are computed for the correction equation, $V_{out} = G(V_{in} + B)$.

These input readings are distributed uniformly throughout the image but do not include the edges. Various extrapolation techniques, with polynomial functions up to third order, were tested for use in estimating the

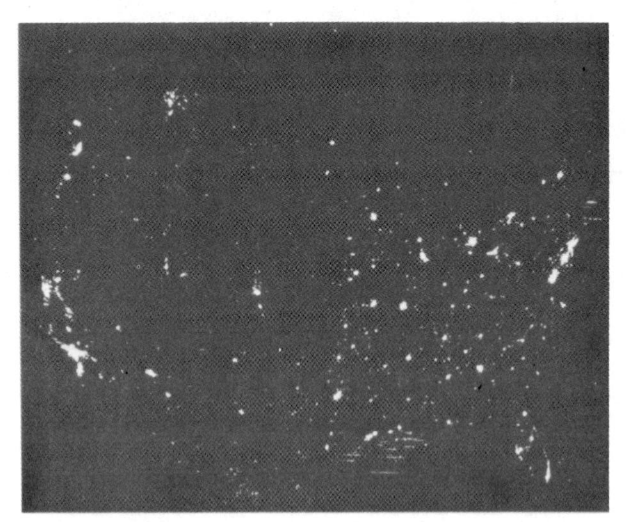

Figure 12 Digital mosaic of the United States produced from visible and infrared radiation recorded at night.

Figure 11 (at left) Digital mosaic color composite (bands 4, 5, and 7) of a portion of the eastern seaboard of the United States, centered on the city of Philadelphia.

Figure 13 Enlargement of the New York City subimage area in Fig. 11.

(*Note:* The reader is referred to the original publication for color illustrations.)

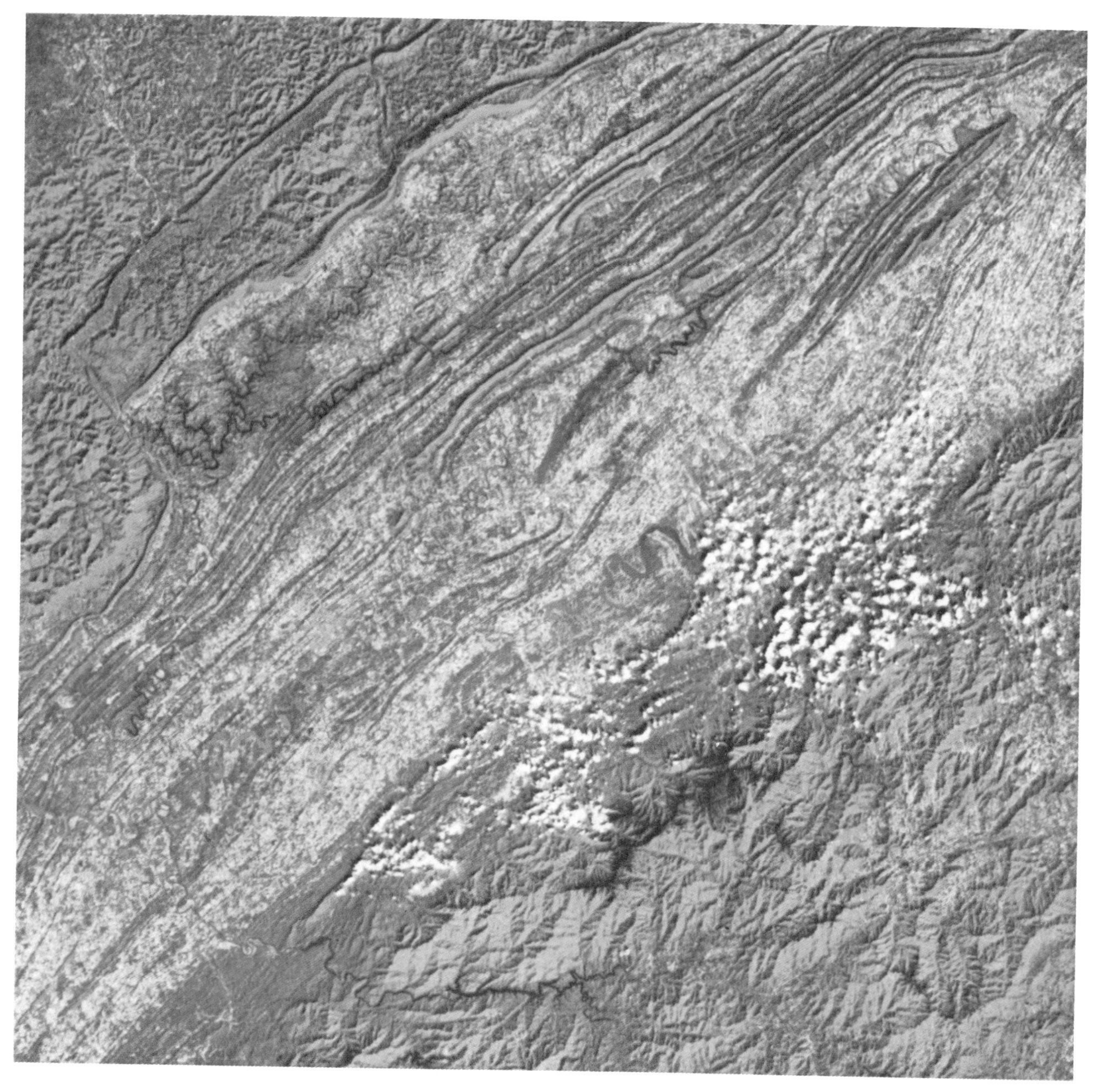

Figure 14 Color composite (bands 4, 5, and 7) of the Tennessee valley region around Knoxville.

edge data but, because the radiometric distortion is extreme near the edges, zero-order extrapolation proved to yield the best results [12].

With the inclusion of edge values, these computations finally produce 20 × 20 arrays of gain and bias values which completely span the image. The values can be fit in a least-square-error sense with spatially dependent functions $G(x, y)$ and $B(x, y)$. For this, it is computationally efficient to divide the image into zones, within which constant values of G and B can be used with an acceptably small error.

Digital image processing results

A number of Multispectral Scanner and Return Beam Vidicon scenes have been processed by these correction elements. In this section we present some of the results in both numerical and pictorial form.

• *Geometric correction*

It was found that for MSS data the number, distribution, and registration accuracy of GCPs influence the accuracy of the output image data. Figure 7 shows the root-mean-square distance error as a function of the number of

(*Note:* The reader is referred to the original publication for color illustrations.)

Figure 15 Color composite (bands 4, 5, and 7) of the central California area also shown in Fig. 8; Monterey Bay is at the lower left.

Figure 16 Representational-color composite subimage of the San Joaquin valley showing agricultural features: red – alfalfa and other healthy green crops; orange – safflower; light yellow – barley; blue-green – fallow fields; blue – water; and black – barley fields recently harvested (and burned over).

(*Note:* The reader is referred to the original publication for color illustrations.)

GCPs, from a well distributed set, used to establish the distortion characteristics of a scene; 16 seems to be a reasonable number of GCPs to use for an acceptable degree of accuracy.

The relative geometric accuracy of the processed image data is determined by comparing the positions of picture elements in the processed GCP scene with U.S. Geological Survey maps, with translation errors removed, through the use of metric functions to determine similarity. Table 6 is a summary of the typical accuracy obtained; for 90 percent of the GCPs, a relative geometric accuracy of about one picture element (79 m) can be achieved. (The use of the 90-percent figure is consistent with a National Map Accuracy Standard for error measurement.)

Figures 8(a) and (b) are MSS images of a central California area before and after correction. The "before" image is a rectangle because of the overlapping of the along-scan sample values (1.4:1 oversampling) and because no Earth rotation correction has yet been made. The "after" image has been fully corrected and can be used as a map product at a scale of about 1:250000.

A similar geometric error analysis was made on corrected RBV data. Since each band corresponds to an independent sensor, an error value was determined for each band. Relatively high accuracy can be obtained with few GCPs because the sensor data are essentially independent of the satellite attitude rate (an RBV scene is imaged in 4 to 16 ms).

• *Radiometric enhancement*

Figure 8(c) shows an MSS image of the central California area with normal radiometric processing. Figures 8(d) and (e) show subimages after radiometric enhancement. Generally, in the infrared band, water appears black because of the low radiance response of water in the 0.8 to 1.1 μm wavelength range. A radiometric "stretching" operation was done on the water data, Fig. 8(e), to determine whether any additional information could be extracted for the Monterey Bay region. In addition, the area of San Jose north of Monterey Bay was processed to enhance the road network, Fig. 8(d). It is apparent that the radiometrically modified images exhibit information that would not have been normally detected. These examples demonstrate that digital sensor data have a wider dynamic range than photographic film can provide. We estimate that twice the film range exists in the sensor data and that the additional information can be extracted by simple radiometric processing.

Three RBV bands were radiometrically corrected and the results, in terms of the number of zones, i.e., individual correction tables for each band, are given in Table 7.

Figures 8(f) and (g) show the RBV image of the same central California area before and after radiometric correction. Most of the shading (non-uniform RBV response characteristic) has been removed, with the recovery of a significant amount of data.

• *Temporal registration experiment*

In some applications it is useful to have two or more images in geometric conformance, i.e., in registration with each other. This is particularly useful in change detection applications where a difference in ground features is of interest, such as shoreline erosion. In agricultural feature extraction applications, the use of multiple scenes of the same area over a crop maturation period improves crop classification accuracy. For these types of applications, extremely precise geometric correction of the various scenes must be made so that corresponding ground features of both scenes are assigned the same geographic location and are thus in conformance.

An experiment was conducted to determine the geometric similarity of two digitally processed scenes. A scene that included Phoenix, Arizona was processed (using 20 GCPs) to correct all geometric errors; this served as a reference image. Then a scene, acquired 36 days later, was processed using the earlier scene as a geometric ground control reference. Both processed scenes were recorded on film and are shown in Fig. 9.

To evaluate relative error, 69 features were located in each scene and the differences in corresponding line and sample coordinates were computed. The results showed temporal registration errors less than 0.24 picture element in sample location and less than 0.11 picture element in line location for 90 percent of the features. When the scene transparencies are overlaid, no geometric difference can be discerned. This experiment demonstrates that multiple scenes can be corrected to be in conformance within about one-fourth picture element.

• *Mosaicking*

Techniques have been developed to digitally combine two or more scenes [19]. The object of mosaicking is to allow a larger area to be viewed; e.g., some geologic faults or lineaments and some hydrologic features are thousands of kilometers in length. By combining multiple scenes with precise correction, information extraction can be improved. Figure 10 shows a circular geologic feature that was recorded midway between two scenes; its unusual shape was not evident before the scenes were mosaicked.

Figure 11 is a digital mosaic of a portion of the east coast of the United States. This mosaic was produced by merging three contiguous MSS scenes and geometrically correcting the swath. No geometric discontinuity is apparent in the processed scene. Bands 4, 5, and 7 have been combined into a representational-color composite "photograph" in which, for example, vegetation is shown in red.

Reference 20 contains a digital mosaic of a Montana-Wyoming region, produced from eight MSS scenes. Those scenes were acquired 24 hours apart and were digitally mosaicked to produce one composite image. First, the scenes of the first swath were merged and geometrically corrected. Next, the scenes of the second swath were merged and corrected, so that common GCPs were in coincidence. Finally, the overlapped region was removed from one swath and the two swaths were combined.

Figure 12 is a digital mosaic of the entire United States, produced from infrared and visible light sensor data (0.4 to 1.0 μm) accumulated on three nighttime south-to-north passes of a meteorological satellite at an altitude of 835 km. Each sweep provided ground coverage of a 3000-km wide swath but, because of the wide (112°) field of view, the sensor data contained severe panoramic distortion. Geometric correction programs were used to eliminate this distortion and to convert the data to a Lambert conformal conic projection. The data from the three swaths were then adjusted for intensity variations and mosaicked into the scene shown. Los Angeles, San Francisco, Seattle, and other cities outline the west coast; Miami, Washington, Philadelphia, New York and other cities can be identified on the east coast (clouds obscured Boston and other northern cities); and most medium-to-large cities in the interior of the United States can also be identified.

- *Enlargement*

An image can be enlarged by a number of digital techniques. One involves simple picture element and line repeating, which enlarges an image by an integral factor. If the enlargement factor is too great, however, the image may have the appearance of discrete blocks. Resampling of the data can enlarge an image without this effect. Figure 13 shows a subimage area, centered on New York City in Fig. 11, which has been digitally enlarged using a cubic convolution resampling function. Note that great detail can be resolved despite the fact that the sensors were at an altitude of 500 nautical miles (915 km).

- *Information extraction*

Figure 14 is a digitally processed color composite scene of the southwest-northeast valley region of Tennessee between the Cumberland and the Great Smoky Mountains. Numerous intricate land-water interfaces are visible, e.g., the serpentine Douglas Lake near the center, which was formed by the dam at its western end, and Cherokee Lake, north of Douglas Lake, also formed by a dam at its western end. Digitally processed images such as this are used to monitor the condition of lakes, predict and assess flood damage, anticipate irrigation requirements, and develop hydrologic models for water management.

In Fig. 14, as well as in the other color composite illustrations in this paper, although water is represented by a medium-to-dark blue color, which appears "natural" to the human eye, this color is actually a representational one based on integration of digitally recorded information from two wavelength bands in the visible range (bands 4 and 5) and one band (7) in the infrared region, which is not perceived by the eye. In this system, urbanized areas such as Kingsport, Tennessee, at the upper right and Knoxville, west of Douglas Lake, are generally represented by a light blue color (which in part indicates the infrared reflectivity of asphalt roads and roofs).

Figure 15 is a color composite of the central California scene in Fig. 8. This image was also produced by photographic combination of digitally processed bands 4, 5, and 7. Of significance are the agricultural, geologic, hydrologic, and urban features that are readily discerned. For example, the San Andreas fault, with a northwest-southeast orientation, can be seen at the lower left, and the Diablo Mountain ranges are easily identifiable. Crops in the San Joaquin valley, including alfalfa, barley, rice, and safflower, can be identified visually by color (see the agricultural subimage in Fig. 16) and can be accurately and rapidly classified by computer processing. Such computer-programmed spectral classification has achieved crop identification and area determination accuracy as high as 98 percent. Also visible in this scene are reservoirs (the San Luis reservoir is in the center) and irrigation canals and aqueducts leading from the San Luis reservoir. Various urban areas in central California such as San Jose, Merced, and Salinas are quite apparent, and this kind of data is often useful to land-use planners.

An MSS scene of the northwestern part of Pakistan was used in a mineral exploration experiment [8] by U.S. scientists in cooperation with Pakistani government agencies to seek previously undiscovered mineral deposits using multispectral classification techniques. Digital image data for a known copper-bearing region were used in a pattern-recognition program to identify similar rock and surficial materials generally associated with subsurface copper ore, and five previously unknown mineral-rich areas were discovered. The Pakistan scene is reproduced in Ref. 20.

Computer configurations and analysis

Our image processing program development and experimentation were performed at the Gaithersburg image processing facility. This facility consists of a general

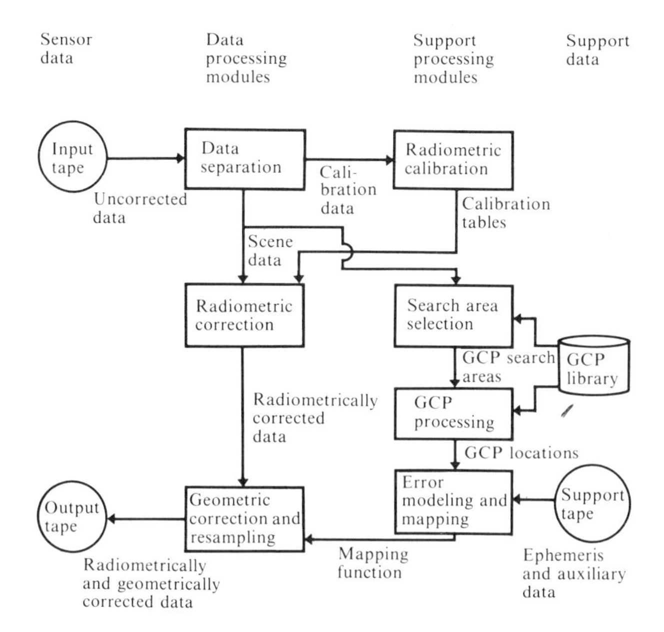

Figure 17 Image processing operations and associated data flow.

purpose computer, an interactive image display sy a variety of application software, and image reco [20].

An image processing configuration analysis addressed the requirements of an operational environment, i.e., the processing of multiple Landsat scenes on a production basis. Figure 17 is a schematic of the image processing operations and the associated data flow.

Several general purpose computer configurations and one special purpose configuration were analyzed in an implementation of the image processing algorithms. Several system configurations were considered and their performance evaluation was predicted through the use of analytical models [12]. For each, the following processing steps were considered:

Step 1

Input image data are read from magnetic tape and transformed to a pixel-interleaved (by band) format.

Supporting data (e.g., ephemeris) are read.

Image data are radiometrically corrected by table-lookup.

Table 8 Configuration definitions.

Configuration	*Input tape type*	*Reformatting and radiometric correction processor*	*Geometric correction processor*	*Geometric correction function*	*Output tape type*
A	800-BPI[a] CCT[b]	One GPP[c]		Individual scene	6250-BPI CCT
B	HDDT[d]	One GPP		Individual scene	6250-BPI CCT
C	HDDT	One micropro-grammed SPP[e]	One GPP	Individual scene	6250-BPI CCT
D	HDDT	One SPP	One GPP	Entire pass for MSS; individual scene for RBV	HDDT
E	HDDT	One SPP	One SPP supported by one GPP	Entire pass for MSS; individual scene for RBV	HDDT

[a]BPI: Bits(s)-per-inch
[b]CCT: Computer-compatible tape
[c]GPP: General purpose processor
[d]HDDT: High density digital tape
[e]SPP: Special purpose processor

Table 9 Scene throughput summary (in scenes processed per 12-hour day).

	IBM System/370 Model														
	135			*145*			*155*			*158*			*168*		
Configuration	*MSS*	*RBV*	*Both*	*MSS*	*RBV*	*Both*	*MSS*	*RBV*	*Both*	*MSS*	*RBV*	*Both*	*MSS*	*RBV*	*Both*
A	25	14	9	43	23	15	86	48	31	107	59	38	141	74	48
B	25	14	9	43	23	15	94	48	32	153	69	48	400	180	124
C	44	34	19	68	54	30	144	93	56	211	118	75	400	200	133
D	64	36	23	94	57	35	211	110	72	313	146	100	720	300	211
E[a]	194	40	33	313	64	53	514	126	101	654	175	138	1200	450	327

[a]With *and* without laser beam recorder.

Ground Control Point (GCP) and reseau search areas (as required) are extracted from the input stream and stored for subsequent detection operations.

Step 2

GCPs are located in the image data.

Reseau (for RBV images only) are located in the image data.

Geometric correction functions are generated.

Step 3

Geometric corrections are applied using different techniques.

Corrected, annotated data are written on magnetic tape in picture-element-interleaved format.

Step 4

User-requested scenes are selected from the master output tape and copied onto user tapes.

The five computer configurations considered are defined in Table 8, which summarizes the assumed processing requirements and environment. Two alternative I/O media were assumed—computer-compatible tape and high density digital tape. For the highest-throughput configuration (E), a variation including direct film output to a laser beam recorder was also evaluated. The detailed configurations and analysis can be found in Ref. 12.

A summary of the results of the configuration analysis is given in Table 9. Throughput figures are in scenes per day (a scene consists of all spectral band images of a 185-km × 185-km area). Many throughput figures are much larger than the daily processing requirement anticipated but are listed as a measure of the computing capacity available for other tasks, for example, production of user tapes. The smaller general purpose processors are CPU-bound, but the use of an attached, microprogrammed, special purpose processor and the use of high density digital tape were found to significantly increase system throughput.

Conclusions

Several conclusions can be drawn from the work that has been performed:

Utility Because of the digital nature of advanced sensors, and information extraction techniques that use digital computers, it is advantageous to correct sensor data using digital rather than electro-optical techniques. Corrected sensor data can be generated with no data degradation, resulting in improved information extraction.

Accuracy Digital correction of digital sensor data results in geometric errors of less than one picture element and in full preservation of sensor radiometry. No radiometric error or loss is introduced (as occurs in electro-optical processing with its data conversions and multiple photographic generations).

Throughput Production image processing systems can be configured to allow high speed processing of Earth observation data. Microprogrammed signal processors can be used to improve the cost/performance ratio relative to conventional general purpose computer configurations, where high throughput systems are required.

Flexibility Digital processing provides a significant degree of processing and operational flexibility. Correction, enhancement, mosaicking, and information extraction can all be done with the same hardware by sequential selection of software routines.

Feasible technology In the past, the use of digital technology and, in particular, digital computers for image processing, was considered impractical because the parallel nature of the data conflicted with the serial nature of digital processors. However, advanced sensors provide serial digital data, and advanced digital hardware provides parallel processing capability. This inversion of organization, combined with high speed circuitry and efficient algorithms, has made a major impact on image processing technology. It is likely that most future ground systems for processing Earth observation data will use digital technology and techniques.

Acknowledgments

The significant contributions of M. L. Cain, C. P. Colby, R. D. Depew, D. G. Ferneyhough, T. Gaidelis, S. W. Murphrey, C. W. Niblack, J. J. Przybocki, and H. Silverman to the research, analysis, programming, and film processing described in this paper are gratefully acknowledged. The basic concepts of reseau detection and image processing were suggested by R. Bakis, M. A. Wesley, and P. M. Will. The support and encouragement of the National Aeronautics and Space Administration was instrumental in implementing this work, which was partially funded by NASA contract NAS5-21716.

Appendix: Derivation of cubic convolution resampling filter

It can be shown that a band-limited sampled signal can be reconstructed (resampled) without information loss by the use of a filter which, in one dimension, has the form

$$I(x_j) = \sum_k I(x_k) f(x_j - x_k), \qquad \text{(A1)}$$

where $I(x_k)$ is the input signal being reconstructed, x_k is the picture element location of the input signal, and $f(x)$ is the reconstruction filter.

Ideally, a reconstruction filter of the form $f(x) = \sin x/x$ should be used. This function, however, requires an infinite amount of data. The use of an equivalent digital filter operator requires an infinite number of input signal terms, due to the continuous nature of the operator.

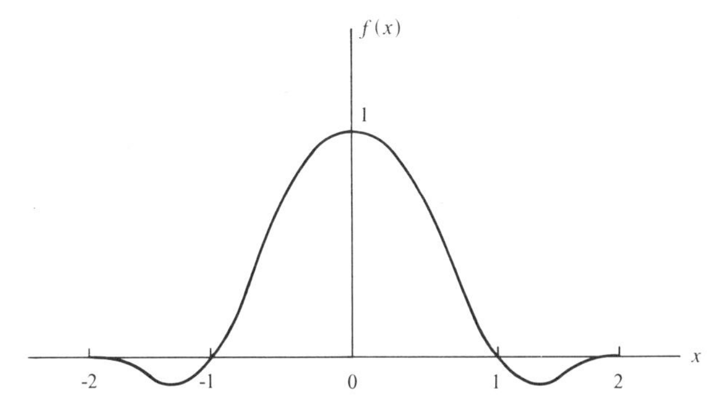

Figure A1 Approximation of sin x/x resampling function used in cubic convolution.

An approximation of $f(x)$, shown in Fig. A1, is desirable. No term beyond $x = 2$ exists. This form reduces the input data requirement to four values (one dimension) or sixteen (two dimensions).

Such a filter can be developed as a cubic spline curve where

$$f(x) = \begin{cases} f_1(x) = a_1|x|^3 + b_1|x|^2 + c_1|x| + d_1, \ 0 \le |x| < 1; \\ f_2(x) = a_2|x|^3 + b_2|x|^2 + c_2|x| + d_2, \ 1 \le |x| < 2; \\ f_3(x) = 0, \ 2 \le |x|. \end{cases} \tag{A2}$$

The conditions used [18] to approximate sin x/x, cubic convolution, are slope and value continuity, symmetry about $x = 0$, and a cutoff value. Further, the resampling function should have a value of one at the center and zero at all other points separated from the center by an integral multiple of the sampling interval.

The functions obtained under these constraints are

$$f_1(x) = (a_2 + 2)|x|^3 - (a_2 + 3)|x|^2 + 1;$$

$$f_2(x) = a_2|x|^3 - 5a_2|x|^2 + 8a_2|x| - 4a_2. \tag{A3}$$

Bounds on the value of a_2 can be obtained by requiring that f_2 be concave upward at $|x| = 1$, and f_1 be concave downward at $x = 0$. Thus

$$\frac{d^2}{dx^2} f_2(x)|_{x=1} = -4a_2 > 0 \rightarrow a_2 < 0; \tag{A4}$$

$$\frac{d^2}{dx^2} f_1(x)|_{x=0} = -2(a_2 + 3) < 0 \rightarrow a_2 > -3. \tag{A5}$$

With $a_2 = -1$, as in [18], the coefficients determined above produce the cubic convolution polynomial

$$f(x) = \begin{cases} f_1(x) = 1 - 2|x|^2 + |x|^2 + |x|^3, \ 0 \le |x| < 1; \\ f_2(x) = 4 - 8|x| + 5|x|^2 - |x|^3, \ 1 \le |x| < 2; \\ f_3(x) = 0, \ 2 \le |x|. \end{cases} \tag{A6}$$

This function is the one plotted in Fig. A1.

References

1. E. P. Mercanti, "ERTS-1, Teaching Us a New Way to See," *Astronautics & Aeronautics*, Vol. 11, p. 36, September 1973.
2. J. C. Fletcher, "ERTS-1, Toward Global Monitoring," *Astronautics & Aeronautics*, Vol. 11, p. 32, September 1973.
3. *Proceedings of the Symposium on Significant Results Obtained from the Earth Resources Technology Satellite-1*, March 1973, Vol. I, Section B, NASA SP-327, compiled and edited by S. C. Freden, E. P. Mercanti, and M. A. Becker, NASA Goddard Space Flight Center, Greenbelt, Maryland.
4. *Proceedings of the Third Earth Resources Technology Satellite-1 Symposium*, December 1973, NASA SP-351, compiled and edited by S. C. Freden, E. P. Mercanti, and M. A. Becker, NASA Goddard Space Flight Center, Greenbelt, Maryland.
5. G. Bylinsky, "ERTS Puts the Whole Earth Under a Microscope," *Fortune Magazine*, Vol. XCI, p. 117, February 1975.
6. Abstracts of the NASA Earth Resources Survey Symposium, June 1975, NASA Johnson Space Center, Houston, Texas.
7. R. B. MacDonald, F. G. Hall, and R. B. Erb, "The Use of LANDSAT Data in a Large Area Crop Inventory Experiment (LACIE)," *Proceedings of the NASA Earth Resources Survey Symposium*, June 1975, NASA Johnson Space Center, Houston, Texas.
8. R. G. Schmidt, B. B. Clark, and R. Bernstein, "A Search for Sulfide-Bearing Areas Using LANDSAT-1 Data and Digital Image-Processing Techniques," *Proceedings of the NASA Earth Resources Survey Symposium*, June 1975, NASA Johnson Space Center, Houston, Texas.
9. *ERTS Data Users Handbook*, NASA 715D4249, NASA Goddard Space Flight Center, Greenbelt, Maryland.
10. R. Bernstein, "Results of Precision Processing (Scene Correction) of ERTS-1 Images Using Digital Image Processing Techniques," *Proceedings of the Symposium on Significant Results Obtained from the Earth Resources Technology Satellite-1*, March 1973, Vol. I, Section B, NASA SP-327, NASA Goddard Space Flight Center, Greenbelt, Maryland.
11. R. Bernstein, "Scene Correction (Precision Processing) of ERTS Sensor Data Using Digital Image Processing Techniques," *Proceedings of the Third Earth Resources Technology Satellite-1 Symposium*, December 1973, Vol. I, Section B, NASA SP-351, NASA Goddard Space Flight Center, Greenbelt, Maryland.
12. R. Bernstein, "All-Digital Precision Processing of ERTS Images," IBM Final Report to NASA, Contract NAS5-21716, April 1975, NASA Goddard Space Flight Center, Greenbelt, Maryland.
13. R. H. Kidd and R. H. Wolfe, "Performance Modeling of Earth Resources Remote Sensors," *IBM J. Res. Develop.* **20**, 29 (1976, this issue).
14. D. T. Barnea and H. Silverman, "A Class of Algorithms for Fast Digital Image Registration," *IEEE Trans. Computers* **C-21**, 179 (1972).
15. R. Bernstein and H. Silverman, "Digital Techniques for Earth Resources Image Data Processing," *Proceedings of the American Institute of Aeronautics and Astronautics, 8th Annual Meeting and Technology Display*, Washington, D.C., October 1971, Vol. C21, No. 2, AIAA 71-978, New York, N.Y.
16. R. Fletcher and M. I. D. Powell, "Rapidly Convergent Descent Methods for Minimizing," *Computer J.* **6**, 163 (1963).
17. H. Markarian, R. Bernstein, D. G. Ferneyhough, L. E. Gregg, and F. S. Sharp, "Digital Correction for High-Resolution Images," *Photogrammetric Engineering and Remote Sensing (J. Am. Soc. Photogrammetry)* **39**, 1311 (1973).

18. S. S. Rifman and D. M. McKinnon, "Evaluation of Digital Correction Techniques for ERTS Images," TRW Corporation Final Report, *TRW 20634-6003-TU-00*, NASA Goddard Space Flight Center, Greenbelt, Maryland, March 1974.
19. "Feasibility of Generating Mosaics Directly from ERTS-1 Digital Data," IBM Final Report to U.S. Department of the Interior, Contract 08550-CT3-12, *TR FSC72-0140*, IBM Federal Systems Division, Gaithersburg, Maryland, April 30, 1974.
20. R. Bernstein and D. G. Ferneyhough, "Digital Image Processing System," *Photogrammetric Engineering and Remote Sensing* (*J. Am. Soc. Photogrammetry*) **41**, 1465 (1975).

Received April 16, 1975; revised September 19, 1975

The author is with the IBM Federal Systems Division, 18100 Frederick Pike, Gaithersburg, Maryland 20760.

RALPH BERNSTEIN
DALLAM G. FERNEYHOUGH, JR.
International Business Machines Corporation
Gaithersburg, MD 20760

Digital Image Processing

The IBM Image Processing Facility is described and the mosaicking of LANDSAT imagery and multispectral classification of mineral deposits are discussed.

ABSTRACT: *The use of digital sensors in earth resources applications appears to be well-established. The signals sent to the ground from the* LANDSAT *(previously known as the Earth Resources Technology Satellite,* ERTS*) Multispectral Scanner* (MSS) *are digitized prior to transmission.*[1,2] *For future earth-observation programs, both the sensor outputs and the ground processing will be digital.*[3] *If such sensors are to serve a useful role in the surveying and management of the earth's resources, efficient methods for correcting and extracting information from the sensor outputs must be developed. The Federal Systems Division of IBM has developed an image processing facility to experimentally process, view, and record digital image data. This facility has been used to support* LANDSAT *digital image processing investigations and advanced image processing research and development. A brief description of the facility is presented, some techniques that have been developed to correct the image data are discussed, and some results obtained by users of the facility are described.*

IBM IMAGE PROCESSING FACILITY

AN INTERACTIVE Image Processing Facility (IPF) has been developed to process earth-observation image data. The capability of the system allows a user to perform geometric and radiometric correction of image data, enhance the data, determine image characteristics and statistics, perform information extraction operations, and view the image products. The objectives of the facility are to provide an investigator with the means to process data in a research and development manner in order to develop and evaluate advanced techniques and programs.

SYSTEM DESCRIPTION

Imagery can be provided to the IPF on computer compatible tape (CCT), high-resolution photographic film, or hard-copy form (maps, charts, or line graphic prints). The processing may be performed by digital, analog, or standard photographic processes or by combinations of these techniques. Figure 1 shows the system configuration of the IPF. Off-line equipment digitizes and records image data. The processing of the data is performed on-line in an operator interactive manner by the use of displays and terminals. Table 1 summarizes the performance characteristics of the IPF peripheral devices. Most forms of human readable image data can be digitized, displayed, and recorded.

IPF SOFTWARE

The IPF operates under the Time Sharing Option (TSO) of the IBM Operating System. The IPF software structure is shown in Figure 2. The IPF Interactive Programming System provides a user with a set of analysis tools for digital image manipulation and processing. It employs a user-oriented language which includes a set of commands and associated operands that selects the image to be processed, implements the processing func-

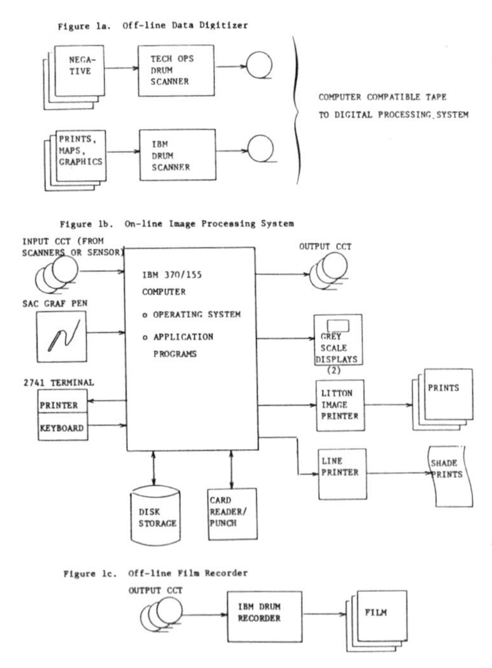

FIG. 1. IBM Image Processing Facility.

tions to be invoked, and outputs the processed image. Table 2 summarizes the applications programs that can be selected by the operator.

IMAGE PROCESSING TECHNIQUES

A significant amount of effort has been expended in the development of digital techniques for radiometric correction, geometric correction, and mosaicking of LANDSAT MSS scenes.[4-7] Each of these areas is discussed in the following sections.

RADIOMETRIC CORRECTION

This section discusses both the nominal radiometric calibration/correction of the MSS data and the supplemental calibration which can be applied if the nominal methods fail fully to compensate for differences in the detectors.

Nominal Calibration/Correction. The nominal method for calibrating the outputs of the 24 MSS detectors is adequately described elsewhere in the literature.[2] Basically, an on-board incandescant lamp and a variable neutral density filter are used to establish points on the input/output curve of each detector. Occasional solar observations are used to correct for changes in the ouput of the lamp. In this way, gradual changes in the output of each detector can be detected and measured.

The MSS detector outputs are digitized prior to transmission to the ground. Since six-bit quantization is used, each detector can produce only 64 discrete values. The calibration data can be used to construct for each detector a table which specifies the "correct" radiometric intensity for each value output by the detector. Radiometric correction can thus be reduced to a simple table look-up operation in which a value output by a given detector is used to extract the "correct" value from the correction table. Although a unique table is required for each of the 24 detectors, the storage to accommodate these tables is quite small by present computer standards.

Supplemental Calibration. These nominal techniques do not always fully compensate for differences in the outputs of the various MSS detectors. When this occurs, the effects are sometimes sufficiently large to produce visible horizontal "stripes" or "banding" in the images. Supplemental calibration can often be used to reduce these effects below the level of visual detectability.

There are several possible approaches to supplemental calibration. One is to note the response of each detector in areas of uniform radiance at different intensity levels. Another approach is to compile histograms of the responses of the detectors over a large number of data samples. In all cases, the object is to measure the differences in the "corrected" response curves of the detectors in each spectral band. These measurements then can be used to modify the nominal radiometric correction tables to produce identical corrected response curves for the detectors in each band.

It should be emphasized that supplemental calibration provides only a cosmetic correction. Unlike the nominal calibration techniques, supplemental calibration does not attempt to make the detector outputs correct. It only attempts to make them equal.

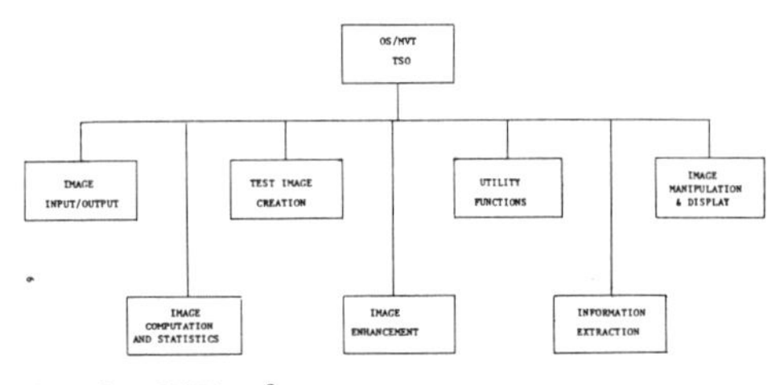

FIG. 2. IPF software structure.

GEOMETRIC CORRECTION

This section discusses the geometric errors present in the MSS data, methods of measuring those errors, formation and application of a geometric correction function, and resam-

TABLE 1. CHARACTERISTICS OF IPF PERIPHERALS AND TERMINALS.

	IBM Drum Scanner	Tech-ops Film Scanner	IBM Film Plotter	Litton Imagery Printer	PEP GSGT Grey Scale Display	PEP 801 Grey Scale Display
Input Form	Opaque documents, prints, maps	Photo film	9 track tape (800/1600 bpi)	On-line to computer	On-line to computer	On-line to computer
Output Form	9 track tape (800/1600 bpi)	9 track tape (800 bpi)	Photo film (Negative or positive)	Photo print	CRT display	CRT display
Largest I/O document or display (mm)	610 by 762	200 by 250	610 by 762	216 by 211	254 by 254	254 by 254
Largest Raster	30,000/24,000	9598/7998	30,000/24,000	1536/1536	1024/1024	1024/1024
Resolution (spot size, μm)	25, 50, or 100	25, 50, or 100	25, 50, or 100	130	horizontally 4 pixels/mm	horizontally 4 pixels/mm
Number of Grey Scale Levels	256	256	256	256	32	32
Density Range (D) or screen brightness (L)	0.15 to 2.5	0.1 to 3.0	0.1 to 2.5	0.15 to 1.8	50-0 ft. Lamberts	50-0 ft. Lambert
Relationship of Grey Levels	Linear	Linear or Log.	Linear or Log.	Linear	Linear	Linear

TABLE 2. IMAGE PROCESSING APPLICATION PROGRAMS.

Image Input:

KINGIN—loads an image from an IBM Drum Scanner tape into an image data set on disk.
TECHIN—loads an image from a TECHOPS scanner tape into an image data set on disk.
TAPEIN—loads an image from a standard tape into an image data set on disk.
REFORMAT—reformats input data into selected processing format.
GRAF—accepts line graphic input from Graf Pen.

Test Image Creation:

GRID—generates a test image on disk of a geometric grid.
BAR—generates a test image on disk of a resolution bar chart.
STAR—generates a test image on disk of a radial bar chart.
LINWEDGE—generates a linear gray-scale step wedge (successive steps change by an equal increment).
SQ2WEDGE—generates a non-linear gray-scale step wedge (successive steps change by $\sqrt{2}$).

Image Geometry Modification:

MAGNIFY—performs image magnification.
REDUCE—performs image reduction.
EXPAND—performs expansion of image by repeating pixel values.
GEOM—performs image geometry modification by use of n'th order mapping function.

Image Enhancement:

DIRECT—computes directional derivative of an image.
LAPLACE—computes Laplacian of an image.
SPATIAL—applies a spatial filter algorithm to an image.
ADJUST—converts intensity values of an image using a replacement table.
COMBINE—combines two or more images into a composite image.

Information Extraction:

CLASSIFY—performs multispectral classification on registered multispectral images.
ROSE—generates a rose diagram useful for geological fault analysis.

Image Computations & Statistics:

AREA—measures area within a specified intensity range included in a specified polygon.
AREARECT—measures the area within a specified intensity range included in a specified rectangle.
HISTOG—outputs a histogram of an image on terminal or printer.
SHADE—outputs a shade print of an image on terminal or printer.
COEFF—computes mapping function coefficients for GEOM command.
ACCUR—computes accuracy of mapping function.

Image Output:

KINGOUT—outputs an image to tape in format compatible with IBM Film Plotter.
TAPEOUT—outputs an image to tape in standard image processing format.

Image Display:

IMPOUT—directs image data to the on-line Imagery Printer
GSGERASE—erases on-line Gray Scale Display screen.
GSGOUT—directs image data to the on-line Gray Scale Display.
GCTERASE—erases on-line graphic computer terminal screen.
GCTOUT—directs image data to the on-line GCT display.

Utility Functions:

LOGON—initiates a terminal session allocating resources required for the session.
LOGOUT—terminates a terminal session.
SAVE—permanently saves a temporary image under a user specified name.
TABLE—creates or updates a data table for input to the ADJUST, AREA, CLASSIFY, SHADE or ROSE commands.
SELECT—selects an image area for further processing by creating a new disk image dataset.
CATALOG—lists images and tables saved by users.
WHOUSER—lists the IDs of the active users.
REMOVE—removes a specified dataset from the system.
CHANGE—changes the name of an existing dataset.
RESERVE—allocates the initializes disk space for an image.
SIZE—lists the size of an image dataset in terms of pixels and lines.
CLEAR—frees all system files.
TIME—displays CPU, execution, and session time used during terminal session.

pling of the input data to obtain output intensity values.

Geometric Errors. The principal geometric errors associated with the data received from the MSS are illustrated in Figure 3. Brief explanations of these errors are—

- Earth Rotation—As the MSS completes successive scans, the earth rotates beneath the sensor. Thus the ground swaths scanned by each mirror sweep gradually migrate westward. This effect causes along-scan distortion.
- S/C Velocity—If the spacecraft velocity departs from nominal, the ground track covered by a fixed number of successive mirror sweeps changes. This produces a cross-scan scale distortion.
- Altitude—Departures of the spacecraft altitude from nominal produce scale distortions in the sensor data. For the MSS, the distortion is along-scan only and varies with time.
- Attitude—Nominally the sensor axis system is maintained so that one axis is normal to the earth's surface and another is aligned with the spacecraft velocity vector. As the sensor departs from this attitude, geometric distortions result. For the MSS, the full attitude time history contributes to the distortion.
- Perspective Projection—For some applications it is desired that LANDSAT images represent the projection of points on the earth upon a plane tangent to the earth, with all projection lines normal to the plane. The sensor data represent perspective projections, i.e., projections whose lines all meet at a point above the tangent plane. For the MSS, this produces only along-scan distortion.
- Scan Skew—During the time that the MSS mirror completes one active scan, the spacecraft moves along the ground track. Thus the ground swath scanned is not normal to the ground track but is slightly skewed. This produces cross-scan geometric distortion.
- Panoramic Distortion—Nominally, data samples are taken at regular intervals of time (and, hence, nominally at regular spatial intervals on the ground). In reality, the ground area imaged is proportional to the tangent of the scan angle rather than to the angle itself. This effect produces along-scan geometric distortion.
- Mirror Velocity—The scanning mirror of the MSS nominally moves at a constant angular rate. In reality, the mirror rate varies across a scan. Since data samples are taken at regular intervals of time, the varying mirror rate produces along-scan geometric distortion.
- Map Projection—For some applications, production of output products in a specific map projection is desired. Although map projection does not constitute an actual geometric error, it does require a geometric transformation of the input data and can be accomplished in the same operation that compensates for the distortions present in the data.

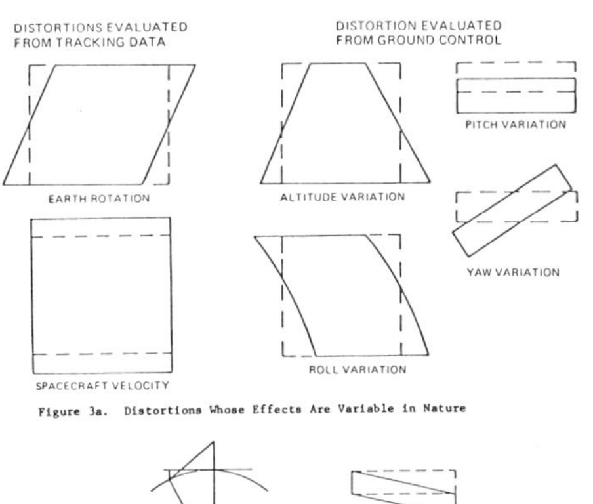

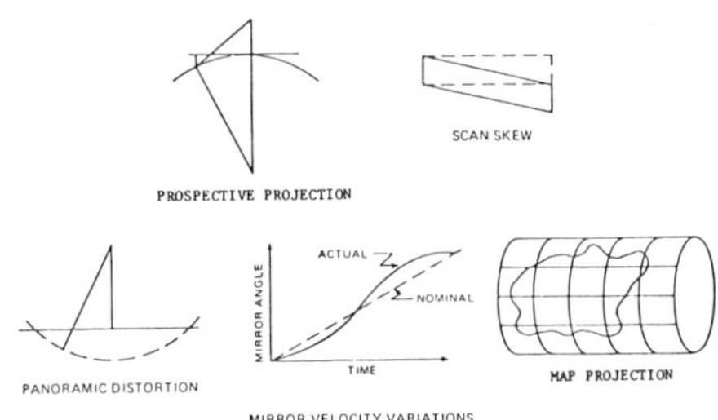

FIG. 3. MSS geometric distortions.

Determination of Error Magnitudes. If such errors are to be corrected, they must be either predictable or measurable. Errors due to MSS mirror velocity, panoramic distortion, scan skew, and perspective projection are systematic and stationary. That is, the effects are constant (for all practical purposes) and can be predicted in advance. Errors due to spacecraft velocity are a known function of that velocity, which can be obtained from tracking data. Errors due to earth rotation are a function of spacecraft latitude and orbit and thus also can be predicted from tracking data.

Attitude and altitude errors are neither systematic nor stationary. If they are to be corrected, their effects must be measured for each image. The measurement technique used here involves apparent displacements of ground control points (GCP's), detectable and recognizable geographic features whose geographic positions are known. The image locations of the GCP's are determined by application of a control location algorithm to appropriate areas of sensor data. For the MSS, differences between the actual and observed GCP locations are used to evaluate the coefficients of cubic time functions of roll, pitch and yaw, and a linear time function of altitude.

Geometric Correction Function. The image spaces and transformations used in the geometric correction of MSS data are shown in Figure 4. The input image is an array of digital data which represents a geometrically distorted one-dimensional perspective projection of some portion of the earth's surface. The output image is a geometrically correct map projection of the same ground area.

GCP's are located in the input image and are mapped into the tangent plane by using models based on all those errors which can be predicted or determined from tracking data. The nominal GCP locations are mapped from the map space to the tangent plane through the equations that relate points in map or tangent plane space to points on the earth's surface. The nominal and observed GCP locations in the tangent plane are then used to evaluate the coefficients of the attitude and altitude models. The error models and the map projection equations together provide the correction functions needed to relate points in the output space to points in the input space.

Rather than apply the correction functions to all points of the output image, an interpolation grid is established on the output image. This grid is constructed so that, if the four corners points of any grid mesh are mapped with the correction functions, all points interior to the mesh can be located in the input image with sufficient accuracy by bilinear interpolation on the corner points. (See Figure 5.)

Resampling. If the input data values are considered to lie at points on a regular lattice, the situation shown in Figure 6 occurs. The input space has been sampled at the points represented by the data values. When an output image point is mapped into the image space, its location does not generally coincide with any of the input sample points. In order to establish a data value for the output point, the input space must be resampled at the output point location.

There are at present three different resampling techniques being advocated for LANDSAT data. The simplest of these is nearest-neighbor assignment, in which the value of the closest input sample (point 11 in the example shown in Figure 6) is assigned to the output point. The second technique is two-dimensional linear (hence, bilinear) interpolation over the four surrounding input values (points 6, 7, 10, and 11 in Figure 6). Bilinear interpolation takes approximately ten times as long as nearest-neighbor assignment when implemented on a general-purpose computer. The third resampling technique is cubic convolution.[9] Cubic convolution uses the sixteen input values closest to the output point in question and provides a higher order approximation to a sin x/x interpolator (theoretical resampling function). It runs approximately 20 times as long as nearest-neighbor assignment when implemented on a general-purpose computer. Figure 7 presents examples of LANDSAT MSS data resampled by IBM using these three techniques.

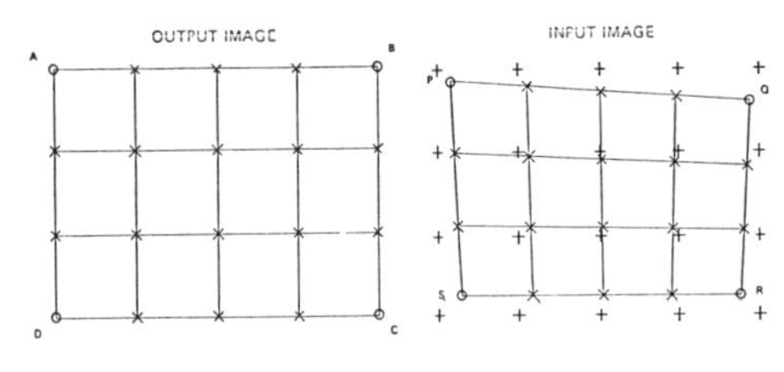

FIG. 5. User of linear interpolation in the mapping operation.

TYPICAL RESULTS

The IBM Image Processing Facility has

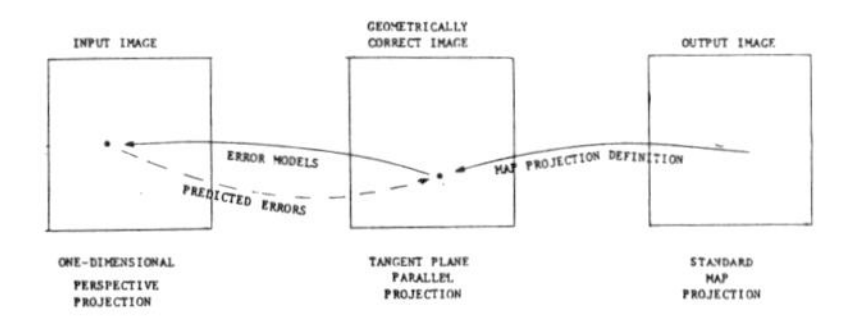

FIG. 4. MSS image spaces and transformations.

+ - INPUT IMAGE DATA VALUES
o - MAPPED OUTPUT IMAGE POINT

NEAREST NEIGHBOR ASSIGNMENT USES POINT 11
BILINEAR INTERPOLATION USES POINTS 6, 7, 10, AND 11
CUBIC CONVOLUTION USES POINTS 1 THROUGH 16

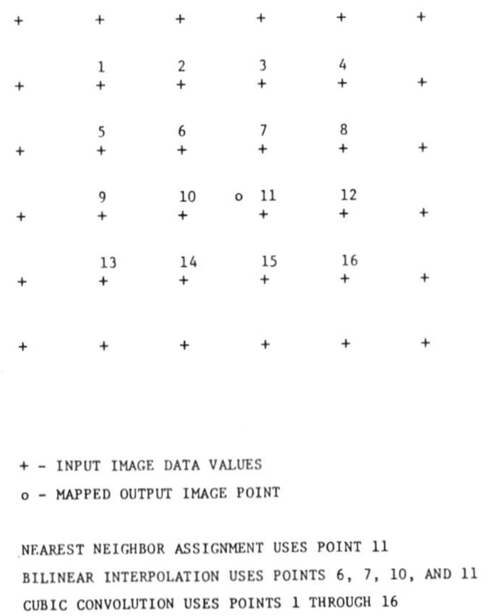

FIG. 6. Resampling geometry.

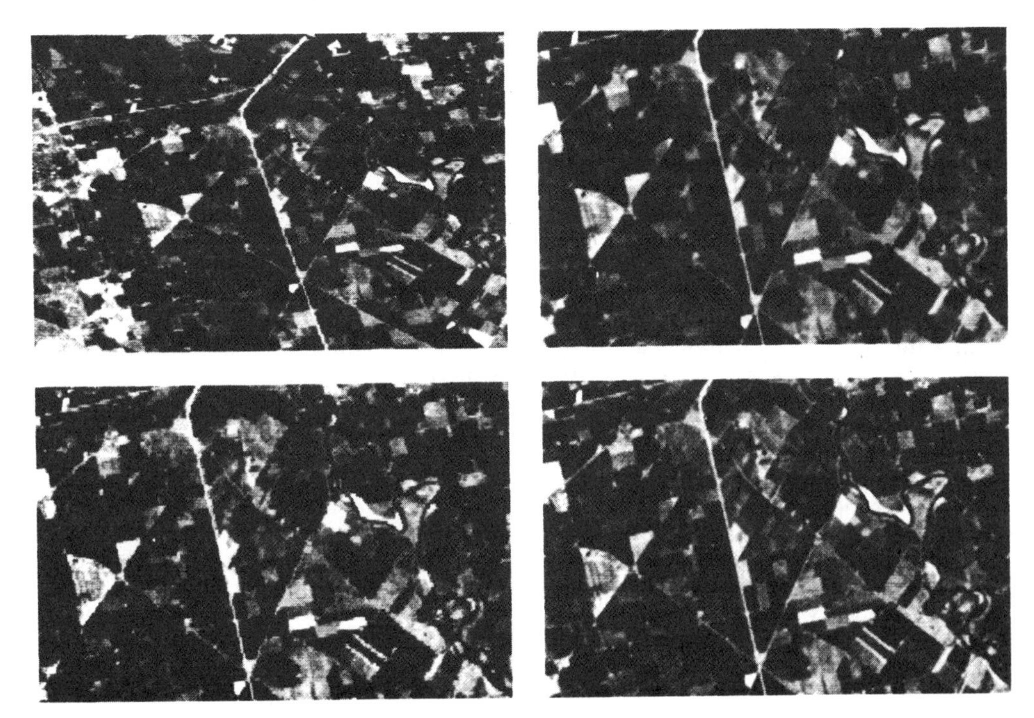

FIG. 7. Resampling results.
Upper left—original data.
Lower left—nearest-neighbor assignment.
Upper right—bilinear interpolation
Lower right—cubic convolution.

been used to support a variety of investigations. Two typical experiments, mosaicking and multispectral classification of mineral deposits, are discussed.

MOSAICKING

A growing number of applications require the combination or mosaicking of several LANDSAT scenes. Under contract to the Department of Interior Bureau of Land Management, IBM completed an experiment to demonstrate the feasibility of forming such mosaics digitally.[7] Eight MSS scenes, whose relative geometry is shown in Figure 8, were chosen for the experiment. Since the four scenes from each pass were originally continuous strips of data, the first step in the processing was to reformat the data as two continuous strips, eliminating the along-track overlap.

Geodetic coordinates for 75 GCP's, whose approximate locations are shown in Figure 8, were measured by the BLM from 1:250,000 or 1:24,000 scale maps and provided to IBM. The image coordinates of these GCP's were determined by using computer-generated shade prints. For each strip, a computer pro-

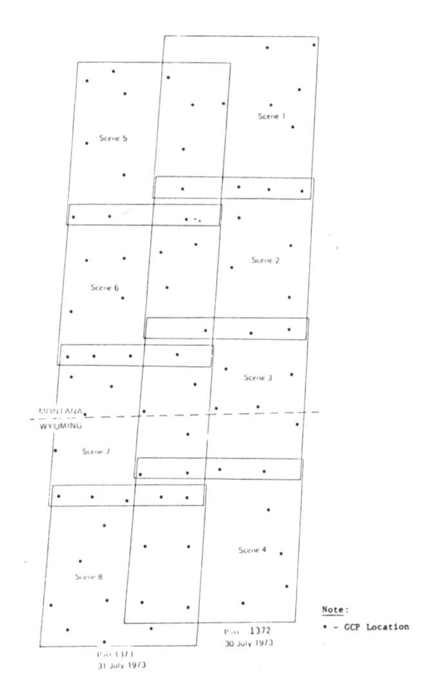

FIG. 8. Mosaicked scenes.

gram was used to compute correction functions which transformed the GCP locations so that they were located in their proper positions in a UTM projection. For the pass 1372 strip, a second function also was computed. This function transformed the GCP's in the overlap region only to bring them into coincidence with those of the pass 1373 strip. The two correction functions for the pass 1372 strip were then combined in a single composite transformation. The two strips of data were then corrected geometrically by using the composite transformation. Registration of the two corrected strips was achieved by examination of shade prints.

The remaining problem concerned the elimination of the duplicate data in the overlap region. This was accomplished by a program which accepted as input a boundary specified as a sequence of straight line segments. Data to the left of the boundary were taken from the left-hand strip; data to the right of the boundary were taken from the right-hand strip. The mosaicked array was trimmed down to fit the capacity of IBM's drum film recorder, and a border including annotation and geodetic tick marks was added. The composite array was recorded on film at 1:1,000,000 scale using a 50 μm square spot.

The processed mosaic is shown in Plate 1. Excellent geometric fitting has been accomplished and scene discontinuity results only from cloud patterns and radiance differences due to time separation between the data.

MULTISPECTRAL CLASSIFICATION OF MINERAL DEPOSITS

Under a contract with the U. S. Geological Survey, a LANDSAT MSS scene of the western Chagai District, Pakistan was processed, and computer-aided information-extraction experiments were conducted to identify potential sulfide ore-bearing localities.[8] The experimental approach is summarized in Figure 9. Shown there are the source data used, the digital processing applied to the source data, the products generated, the analysis conducted, and the final products. By a combination of digital image processing and information extraction, and manual analysis and evaluation, three processing operations were performed: digital image generation, support data generation and analysis, and multispectral classification.

Digital image generation. The uncorrected LANDSAT MSS data was reformatted into 185 km × 185 km areas, and each band was radiometrically (intensity) adjusted and systematically geometrically corrected. The resulting computer-compatible tapes (CTT) were then recorded on film from which black-and-white and color prints were made. These prints were used as aids in the selection of the field prospecting sites during the evaluation of the classification results and also during the field checking. A color composite of the processed scene is reproduced on the cover.

Support data generation and analysis. The formatted but uncorrected CCT's were used for analysis prior to the multispectral classification operation. Shade prints (computer printouts providing the reflectance sensed in each spectral band) for selected areas were prepared and used as maps for precise location of individual data rectangles (pixels) relative to known ground features and known rock types. Numeric data for the 4 MSS bands

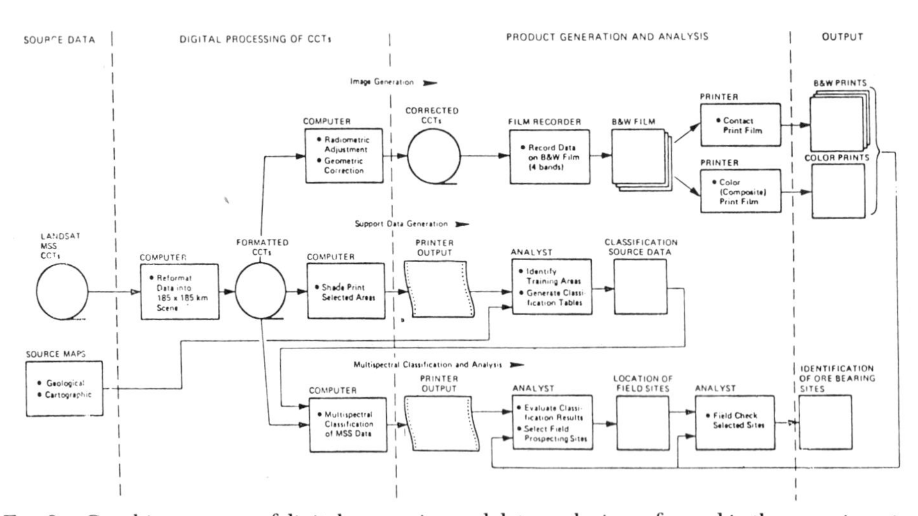

FIG. 9. Graphic summary of digital processing and data analysis performed in the experiment.

PLATE 1. Digitally mosaicked LANDSAT MSS scenes.

(*Note:* The reader is referred to the original publication for color illustrations.)

were extracted for each pixel in the known areas, and maximum and minimum sensed-reflectance limits were chosen for each rock type. A known copper sulfide-bearing deposit at Saindak was the source of data used to prepare the classification tables. Five revisions were made and tested, and one alternate classification table was tried. These tables were then used on an interactive basis to classify a nearby region within the same LANDSAT scene in which copper sulfide-bearing areas were suspected but in which no deposits were known (application area).

Multispectral classification and analysis. A spectral-intensity discrimination program was used for multispectral classification on the application area using the tables prepared for the Saindak deposit. The program tested the reflectance of each picture element within the application area against the maximum and minimum reflectance limits in the table and determined into which surface class (rock type) the picture element belonged. The symbol for that class was printed on a computer listing as part of a classification map. When the observed values fit more than one class (when classes were set up with overlapping limiting values), a pixel was placed in the class that was considered first in the search sequence.

The classification table resulting from five revisions was used to evaluate an adjacent area of 2100 km² considered to have good potential for porphyry copper deposits in the western Chagai Hills. The results were printed-out in 13 computer-generated vertical strip maps. These maps were examined for groups of pixels classified as mineralized quartz diorite and pyritic rock, and about 50 groups or concentrations were identified. Each was then evaluated for probability of correct classification, relationship to concentrations of other classes, and comparison with known rock types and occurrences of hydrothermal mineralization. From this examination, 30 localities most deserving reconnaissance checking in the field were chosen. The locations of these targets were marked on an enlarged (1:250,000) digitally enhanced image of MSS band 5 in order to simplify location on aerial photographs and in the field.

As part of the field check, all anomalous areas were first examined on stereoscopic pairs of 1:40,000-scale aerial photographs; at this point, it was possible to reject seven areas as related to windblown sand. Nineteen sites were examined in the field, and four desirable sites were not reached in the field checks. Five sites were found to be extensive outcrops of hydrothermally altered sulfide-rich rock. Two additional sites contain altered rock with some sulfide but seem less attractive for prospecting at this time.

Conclusions

Digital image processing techniques and systems will play a progressively larger role in future imaging applications. This is due in part to the emerging digital sensors that have demonstrated their spatial and spectral utility on the LANDSAT program, the flexibility and performance that can be achieved with general and special purpose digital hardware, and the rapid advance of digital information extraction programs such as multispectral classification.

Acknowledgments

A number of people have contributed to the technology and system discussed in this paper. These people include M. Cain, R. Cannizzaro, B. Clark, C. Colby, R. Depew, S. Murphrey, W. Niblack, N. Rossi, and S. Shapiro. The mineral deposit experiment was performed cooperatively with R. G. Schmidt of the U.S. Geological Survey, and his significant effort is acknowledged. The mosaicking experiment was sponsored by G. Torbert of the U. S. Department of Interior, Bureau of Land Management; his contribution and support are sincerely appreciated. The image correction work described in this paper was partially supported under NASA Contract NAS5-21716.

References

1. Mercanti, E. "The ERTS-1 Experiments Teaching Us a New Way to See," *Astronautics and Aeronautics*, Sept. 1973, Vol. 11, No. 9.
2. *ERTS Data Users Handbook*, NASA—Goddard Space Flight Center, GE Document #71SD4249.
3. *Specifications for EOS System Definition Studies*, Earth Observatory Satellite (EOS) Project, Goddard Space Flight Center, Document No. EOS-410-02, September 13, 1973.
4. R. Bernstein, "Results of Precision Processing (Scene Correction) of ERTS-1 Images Using Digital Image Processing Techniques", *Symposium on Significant Results Obtained from the Earth Resources Technology Satellite-1*, Vol. II, NASA Document #SP-327, March 5-9, 1973.
5. R. Bernstein, "Scene Correction (Precision Processing) of ERTS Sensor Data Using Digital Image Processing Techniques", *Third ERTS Symposium*, Vol. 1, Section A, NASA SP-351, December 10-14, 1973.
6. R. Bernstein "All-Digital Precision Processing

of ERTS Images", Final Report, NASA Contract NAS5-21716, April 1975.

7. Bernstein, R., D. G. Ferneyhough and S. W. Murphrey, *Final Report—Feasibility of Generating Mosaics Directly from ERTS-1 Digital Data,* IBM Report No. FSC 74-0140, April 30, 1974.
8. Schmidt, R. G., B. B. Clark, and R. Bernstein, "A Search for Sulfide-Bearing Areas Using LANDSAT-1 Data and Digital Image Processing Techniques," presented at the NASA Earth Resources Survey Symposium, Houston, Texas, June 1975.
9. Rifman, Samuel S., "Evaluation of Digital Correction Techniques for ERTS Images—Final Report," TRW Systems Group Report 20634-6003-TU-00, March 1974.

DIGITAL IMAGE RECONSTRUCTION AND RESAMPLING FOR GEOMETRIC MANIPULATION

K. W. Simon

TRW Systems Group, Redondo Beach, Calif.

I. ABSTRACT

The problems of digital image registration and geometric correction can be subdivided into two parts: 1) determination of the warping function which will transform the geometry of the scene to the desired geometric coordinate system; and 2) processing of the digital image intensity samples, given the warping function, to produce image samples on the desired coordinate grid. The latter process, called "resampling", is a subset of the problems of image reconstruction, i.e., determination of the continuous (analog) image from a set of samples of the image, and is the subject of this paper.

This paper defines the process of image resampling in more detail in terms of general imager system models, the requirements of digital image geometric manipulation and constraints of available digital processing systems. The problem is then formulated as a constrained linear estimation problem with suitable image models and optimization criteria. The resulting reconstruction filters are compared to more heuristic approaches, such as nearest neighbor, bilinear interpolation, Lagrange interpolation, and cubic convolution (cubic and quartic spline interpolators). Finally, the various resampling techniques are compared against theoretical image models, synthetically generated imagery, and actual ERTS MSS data. Nearest neighbor, bilinear, and Lagrange interpolation resamplers are shown to give significantly poorer reconstruction accuracy than TRW Cubic Convolution and the optimal constrained linear estimator.

II. PROBLEM DEFINITION

For purposes of definition of the resampling or reconstruction process, consider the imaging system shown functionally in Figure 1. The scene f(x) is assumed to be a random process of the two-dimensional spatial parameter x, observed through an aperture a(x) as an image g(x), which is sampled by the sampler s(x). The resulting samples $\underline{g}$ are available to the digital processor. In general, the imager system contains geometric error sources which preclude specification of the ideal sampler phase at the time of imaging. Thus, image samples are not available at required locations, e.g., a given map projection grid system or at the same locations sampled on an earlier imaging pass. Assuming a function is available which describes the actually-sampled locations in terms of the desired grid locations (the distortion, or warp, function), an estimator, or reconstruction filter w(x) can be derived which will estimate the continuous image g(x) prior to the sampler. This continuous image estimate can then be effectively evaluated at the desired grid locations, hence, the terminology "resampling". (Alternatively, the estimator could be derived to estimate the original scene f(x) prior to the imaging aperture, resulting in "aperture correction," or "image restoration," as contrasted to image reconstruction. Aperture correction suffers from noise sensitivity and is not always appropriate to the processing discussed here.)

Given an infinite number of sufficiently closely-spaced, uncorrupted samples of a band-limited image, it is well-known that the original unsampled (continuous) image can be reconstructed without error by using a two-dimensional sinc function for the interpolation kernel. However, an infinite number of contributions to each interpolated value requires an infinite amount of time to process. In reality, computation time and storage limitations restrict the estimate $\hat{r}(x)$ to be a function of at most $N^2 \ll \infty$ image samples. In addition, imagery is seldom perfectly band-limited to an extent compatible with realizable sampling rates. The specific problem of interest here can be stated as follows: "Given an image with specified spectral density and N^2_{TOTAL} samples of that image, perhaps corrupted by measurement noise, find the appropriate interpolation function which uses $N^2 (\leq N^2_{TOTAL})$ subscene samples to estimate the image value at each point in the image with zero mean error and minimum error variance."

Heretofore, because of the processing time limitations of general purpose digital computers, image resampling has generally been accomplished by "nearest neighbor" resampling for which N=1 (each point is a function of only one sample), or by bilinear interpolation. Nearest neighbor (so-called because the intensity of the sample nearest the desired location is ascribed to the desired location) is extremely fast to compute, but causes deletion or replication of image samples and position errors of up to $\pm$ 1/2 pixel (sample spacing), significantly degrading change detection performance and giving a blocked appearance to images with large warp functions. Bilinear interpolation of the four samples surrounding the desired location resolves difficulties of nearest neighbor (at increase in number of computer operations required), but causes noticeable resolution degradation in resampled

Reprinted from *Proc. IEEE Symp. on Machine Processing of Remotely Sensed Data*, June 3-5, 1975, pp. 3A-1-3A-11.

images due to straight-line truncation of intensity peaks in the image.

Hard-wired algorithm approaches to image resampling recently have made feasible interpolators with larger N, i.e., 4 or larger. Interpolators for N=4 have been studied extensively and results are reported here and elsewhere. (Rifman, 1974; Rifman, 1975; Taber, 1973; Caron, 1974.)

III. OPTIMAL LINEAR RECONSTRUCTION ESTIMATOR

Consider a signal $g(x)$ in one-dimension with a specified autocovariance, $C_g(x)$. (Extrapolation to two-dimensions is straightforward and avoided here for clarity.) The signal mean is assumed unknown. A number of equally-spaced samples are available:

$$\underline{g}^T = [g^*(x_0),\ g^*(x_1),\ \ldots,\ g^*(x_{N-1})]$$

where the measurements are corrupted by an uncorrelated zero-mean white noise sequence $\{v_k\}$ with variance σ_v^2

$$g^*(x_k) = g(x_k) + v_k$$

A linear unbiased estimator of $g(x)$ is desired such that the estimate error variance, $J(x) \equiv E[(g(x) - \hat{g}(x))^2]$, is minimized at all x. The form of the estimator is

$$\hat{g}(x) = \underline{W}^T(x)\underline{g} + u(x)$$

A constraint is added to the minimization problem requiring a constant input to the estimator to result in the same constant estimate, i.e., $g^*(x_{k+1}) = g^*(x_k)$ all $k \Rightarrow \hat{g}(x) = g^*(x_k)$. This is equivalent to requiring that

$$\underline{W}^T(x)\ \underline{1} + u(x) = 1$$

where $\underline{1}$ is an N-vector of all ones. Using the error variance as a cost functional to which we append the constraint with a Lagrange multiplier,

$$J(x) = \underline{W}^T\hat{R}\underline{W} + R_g(o) + \sigma_v^2 - 2\underline{W}^T\underline{G} + 2M_g(\underline{W}^T\underline{1}-1)\,u + u^2 + \lambda(\underline{W}^T\underline{1}+u-1)$$

With algebraic manipulation:

$$J(x) = \underline{W}^T\hat{C}\underline{W} + C_g(o) + \sigma_v^2 - 2\underline{W}^T\underline{H} + \lambda(\underline{W}^T\underline{1}-1 + u) + [u+M_g(\underline{W}^T\underline{1}-1)]^2$$

where $R_g(x)=C_g(x)+M_g^2$ and M_g is the mean of g. Also $\hat{C}$ is the autocovariance of g and $\hat{R}$ is the autocorrelation of g. But $u(x)=W^T(x)\underline{1}+1$ from the constraint, so

$$J = \underline{W}^T\hat{C}\underline{W} + C_g(o) + \sigma_v^2 - 2\underline{W}^T\underline{H} + \lambda(\underline{W}^T\underline{1}\underline{1} + u) + (1-M_g)^2 u^2$$

Note that the last additive term in J is the only term involving M_g and is non-negative. Since M_g is unspecified, J must be minimized over all M_g, implying that $u=0$. In this event, J becomes

$$J = \underline{W}^T\hat{C}\underline{W}+C_g(o)+\sigma_v^2 - 2\underline{W}^T\underline{H}+\lambda(\underline{W}^T\underline{1}-1)$$

and the constraint becomes

$$\underline{W}^T(x)\underline{1} = 1$$

Minimizing J with respect to $\underline{W}(x)$ yields

$$\underline{W}^\circ(x) = \hat{C}^{-1}[\underline{H}(x) - \frac{1}{2}\lambda(x)\ \underline{1}]$$

Substituting this into the constraint equation yields

$$\lambda = \frac{2(\underline{1}^T\hat{C}^{-1}\underline{H}-1)}{\underline{1}^T\hat{C}^{-1}\underline{1}}$$

The corresponding value of J is

$$J^\circ(x) = C_g(o)+\sigma_v^2-\underline{W}^{\circ T}(x)\ \underline{H}(x) - \frac{1}{2}\lambda(x)$$

or:

$$J^\circ(x) = C_g(o)+\sigma_v^2-\underline{H}^T(x)\hat{C}^{-1}\underline{H}(x) + \frac{(\underline{1}^T\hat{C}^{-1}\underline{H}-1)^2}{\underline{1}^T\hat{C}^{-1}\underline{1}}$$

If a suboptimal estimator $\underline{W}'(x)$, still subject to the constraint, were used, the estimate error variance would be

$$J(x) = \underline{W}'^T(x)\hat{C}\underline{W}'(x) + C_g(o)+\sigma_v^2-2\underline{W}'^T(x)\underline{H}(x)$$

From earlier, if the mean M_g is known, then

$$\underline{W}(x) = \hat{C}^{-1}\underline{H}(x)$$

$$u(x) = -M_g(\underline{1}^T\hat{C}^{-1}\underline{H}-1)$$

and

$$J^\circ(x)= C_g(o)+\sigma_v^2- \underline{H}^T(x)\hat{C}^{-1}\underline{H}(x)$$

However, if M_g is erroneously estimated as $\hat{M}_g$, then

$$\hat{J}(x) = C_g(o)+\sigma_v^2-\underline{H}^T\hat{C}^{-1}\underline{H}+(M_g-\hat{M}_g)^2(\underline{1}^T\hat{C}^{-1}\underline{H}(x)-1)^2$$

If the mean were estimated by

$$\hat{M}_g = \frac{\underline{1}^T\hat{C}^{-1}\underline{g}}{\underline{1}^T\hat{C}^{-1}\underline{1}}$$

then both the estimators and the error variances for the two approaches would coincide.

In summary, the desired estimator is

$$\hat{\underline{g}}(x) = \underline{W}^T(x)\,\underline{g} \qquad \text{estimate}$$

$$\underline{W}(x) = \hat{C}^{-1}[\underline{H}(x) - \frac{1}{2}\lambda(x)\underline{1}] \qquad \text{filter}$$

$$\lambda(x) = \frac{2(\underline{1}^T\hat{C}^{-1}\underline{H}(x) - 1)}{\underline{1}^T\hat{C}^{-1}\underline{1}} \qquad \text{Lagrange multiplier}$$

$$J^{\circ}(x) = C_g(0) + \sigma_v^2 - \underline{H}^T(x)\hat{C}^{-1}\underline{H}(x) + \frac{(\underline{1}^T\hat{C}^{-1}\underline{H}-1)^2}{\underline{1}^T\hat{C}^{-1}\underline{1}} \qquad \text{error variance}$$

where $\hat{C}$ is the autocovariance of $\underline{g}$, and $\underline{H}(x)$ is the crossvariance of $\underline{g}$ and $g(x)$. (The problem of aperture correction can be handled similarly by replacing $\underline{H}(x)$ with the crossvariance of the original scene $f(x)$ and the samples $\underline{g}$.) The problem remaining is the determination of the signal and measurement covariance $\underline{H}(x)$ and the measurement autocovariance $\hat{C}$.

Utilization of theoretical autocovariances or those derived from test images with much greater resolution than the subject imager and convolved with theoretical sensor and electronics apertures generally result in filters with noticeable image resolution degradation. (For N=4 and the image model of Reference 2 (PoPP, 1972), the optimum estimator is very nearly linear interpolation.) The cause of this is the relatively low spectral power at high frequencies in images relative to low spatial frequencies, i.e., high frequencies, are sparse in images in spite of their importance to visual information content. Consequently, minimum rms filters for this type of image spectrum sacrifice accuracy at the high frequencies for slight improvements at low frequencies.

In order to give suitable emphasis to the higher frequencies in the image, an error criterion weighted by an appropriate function of image spatial frequency content at each point is required. Alternatively, the image can be prewhitened for derivation of the filter, thus resulting in a filter which emp-asizes all spatial frequencies equally (up to the Nyquist rate). In the latter case, the filter is designed for an autocovariance:

$$C_g(x) = \sigma_g^2 \operatorname{sinc}(\pi a x)$$

For N=4, σ_v=0, and a=1, the resulting reconstruction filter is shown in Figure 2A.

In practice, the filter is used to estimate only points between the central two samples of the N samples, with other points being estimated from other appropriate sets of N samples.

IV. COMPARISON WITH HEURISTIC APPROACHES

Several heuristic approaches to image interpolation suggest themselves. As an example, for N=4-point reconstruction filters, the 4-point Lagrange interpolator is well-known (passes a cubic polynomial through the four points) and is shown in Figure 2B.

A more popular approach, developed at TRW Systems and called cubic convolution, utilizes a 4-section cubic spline function as the N=4-point interpolator kernel. The spline is chosen to satisfy the following boundary conditions:

$$w(0) = 1$$
$$w(\pm 1) = 0$$
$$w(x) = 0, \ |x| \geq 2$$
$$w(x) = w(-x)$$
$$w(x)+w(1+x)+w(1-x)+w(2-x) = 1$$

thus guaranteeing exact interpolation of constant intensity areas. The first derivative of w(x) is further constrained to be continuous, guaranteeing continuity of the first derivative of the interpolated signal. The resulting interpolator has one remaining degree of freedom, a. If the parameter a is chosen for exact constant slope interpolation, i.e., $-w(1+x)+w(1-x)+2W(1-x) = x$, then the resulting interpolator is as shown in Figure 2C. For continuity of second derivative of w(x) at $|x| = 1$, the resulting interpolator is as shown in Figure 2D. For the derivative of w(x) at x=1 to be the same as that of $\operatorname{sinc}\pi x$, the interpolator is as shown in Figure 2E. Alternatively, a quartic spline can be defined to satisfy the above boundary conditions plus the additional constraint of continuity of second derivative of w(x). This interpolation is shown in Figure 2F.

Reconstruction error for a Gaussian test function $e^{-x^2/2}$ was calculated for each of the above filters and several others for several sample phasings. Some of these errors are plotted in Figure 3 for two sampler phasings. The comparison using an error function as test signal gave similar results.

A second comparison was made using ERTS MSS data. The data samples were resampled on a grid shifted from the input sampled grid by 1/2 pixel along-scan using a high-order sinc interpolator (N=30, or 900 samples per output point). The same was then done for nearest neighbor (N=1), bilinear interpolation (N=2), cubic convolution (N=4), and a truncated sinc interpolator with N=10. The resampled images were differenced pixel-by-pixel with the 900-point sinc interpolation. Difference images and corresponding histograms are shown in Figure 4. Note that the 16-point cubic convolution yields lower error than the 100-point truncated sinc interpolator.

A third comparison was made by taking a high resolution digital image (3.4m sample spacing) and convolving it with an aperture similar to the EOS thematic mapper, i.e., a scanning square detector shape with an integrate-and-dump sampler. Samples were then extracted every 20m along-scan and every 28m across-scan and used to reconstruct the continuous image (as convolved with the aperture) using several of the above techniques. Difference images and error histograms are shown in Figure 5. The resolution degradation inherent in 2-point interpolators is apparent. Remaining errors in the 4-point cubic convolution are due primarily to insufficient sampling rate (aliasing).

A fourth comparison involved registration of two successive scenes of the same area (Baltimore, Md.) taken by ERTS MSS using nearest neighbor resampling and cubic convolution. The registered images were differenced and the difference images are shown in Figure 6. The errors inherent in low-order resampling are apparent.

V. REFERENCES

1. R.H. Caron, "Application of Advanced Signal Processing Techniques to the Rectification and Registration of Spaceborne Imagery," Proc. of First Houston Tech. Transfer Conference, Sept. 1974, pp 245-255.

2. D.J. Popp, D.S. McCormack, and J.L. Sedwich, "Imagery Correlation and Sampling Study," Report MDC A1740, McDonnell Aircraft Co., St. Louis, Mo., June 1972.

3. S.S. Rifman, "Digital Rectification of ERTS Multispectral Imagery," Symp. on Significant Results Obtained from ERTS-1, (NASA SP-327), Vol. I, Section B, pp. 1131-1142, Mar. 5-9, 1973.

4. S.S. Rifman and D.M. McKinnon, "Evaluation of Digital Correction Techniques for ERTS Images - Final Report," Report No. 20634-6003-TU-00, TRW Systems, Redondo Beach, Calif., July 1974.

5. S.S. Rifman, W.B. Allendoerfer, D.M. McKinnon, and K.W. Simon, "Experimental Study of Digital Image Processing Techniques for ERTS Data - Task I Final Report," Report No. 26232-6001-RU-01, TRW Systems, Redondo Beach, Cal., Jan. 1975.

6. J.E. Taber, "Evaluation of Digitally Corrected ERTS Images," Third ERTS-1 Symposium, Vol. I, Tech. Presentations, (NASA SP-351), Dec. 1973, pp. 1837-1843.

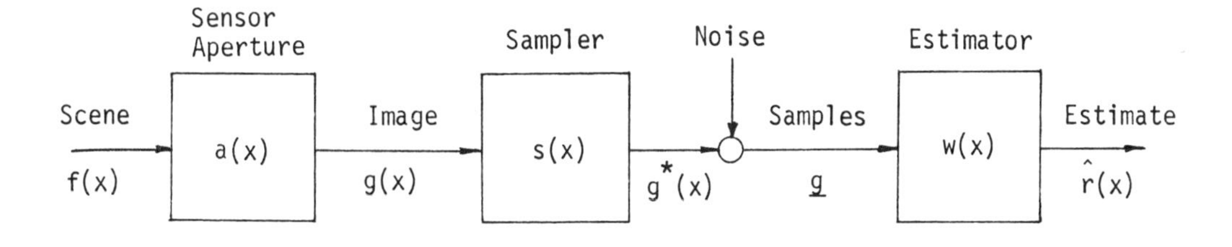

FIGURE 1. IMAGER MODEL

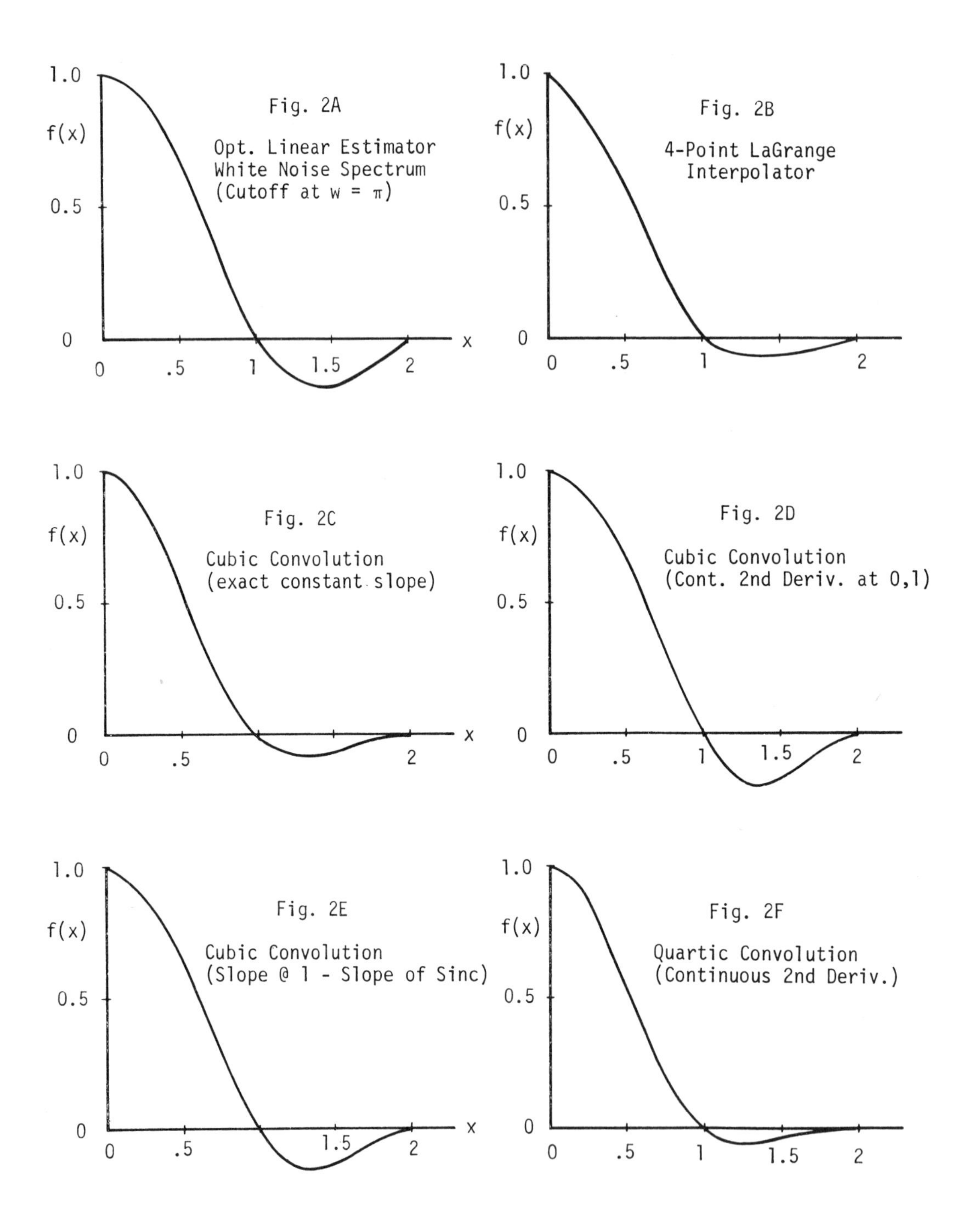

FIGURE 2. INTERPOLATION KERNELS

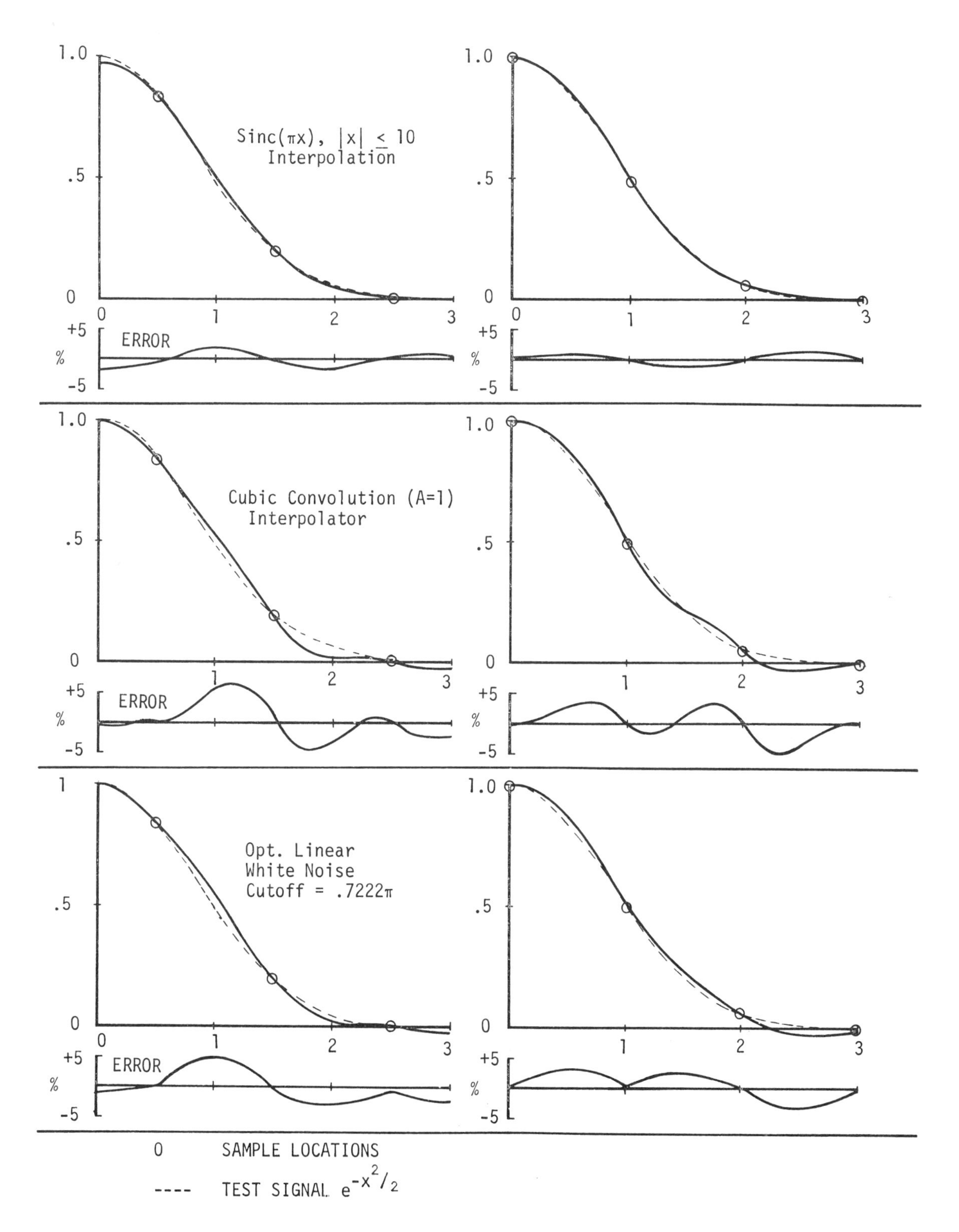

FIGURE 3. INTERPOLATION ERRORS

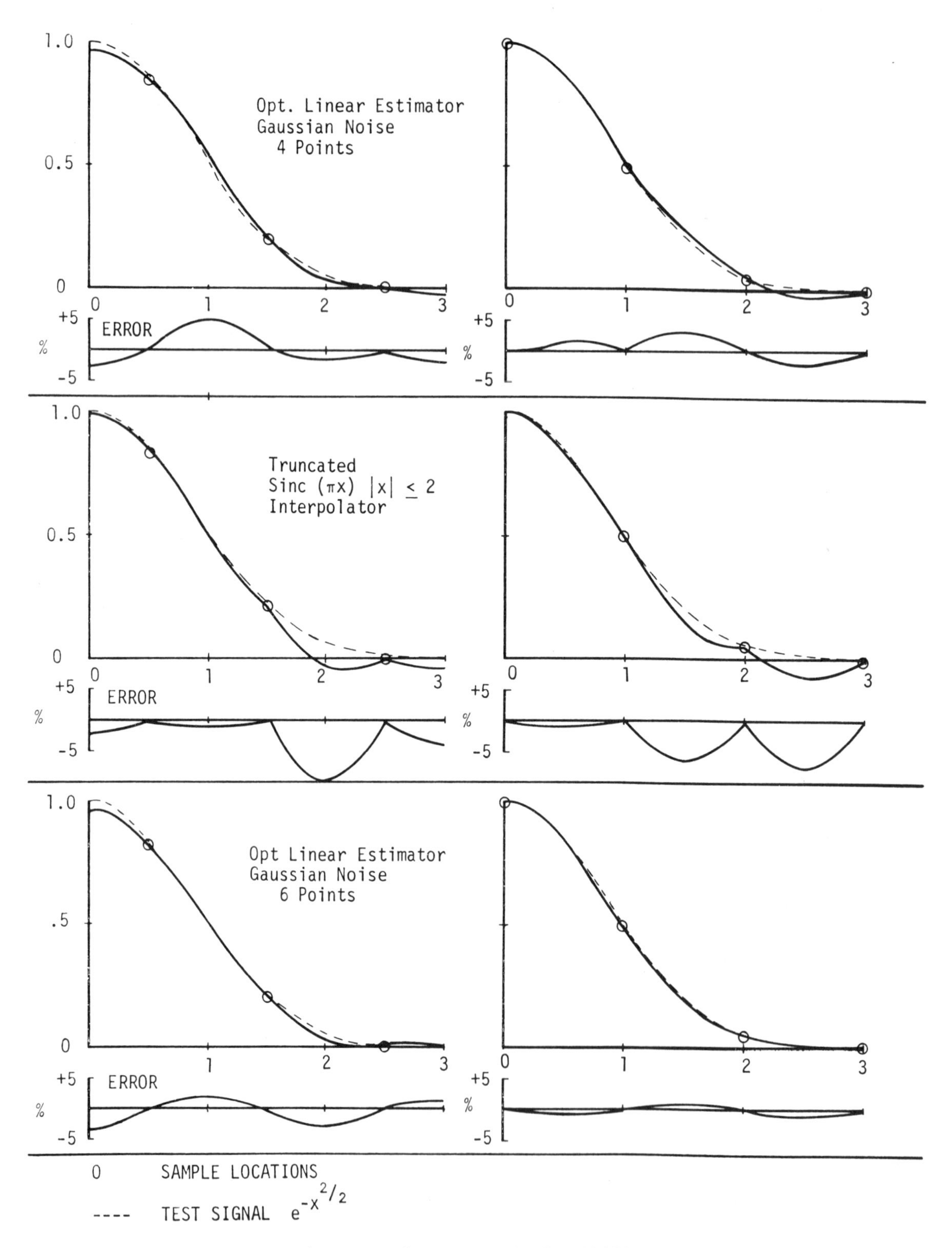

FIGURE 3 (CONTINUED) INTERPOLATION ERRORS

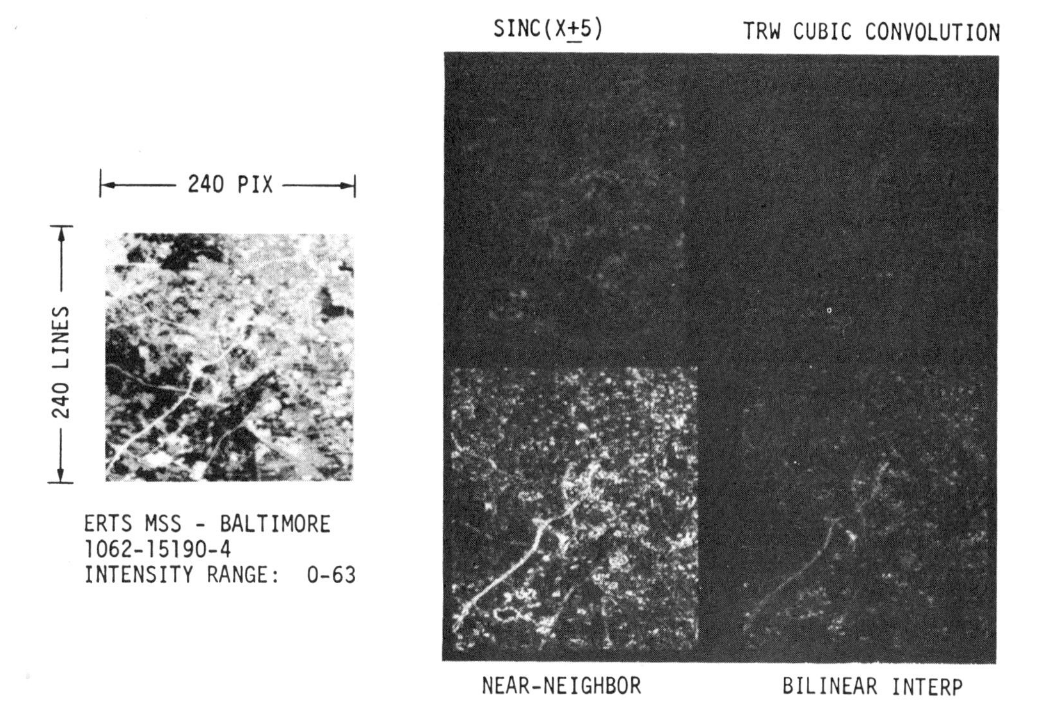

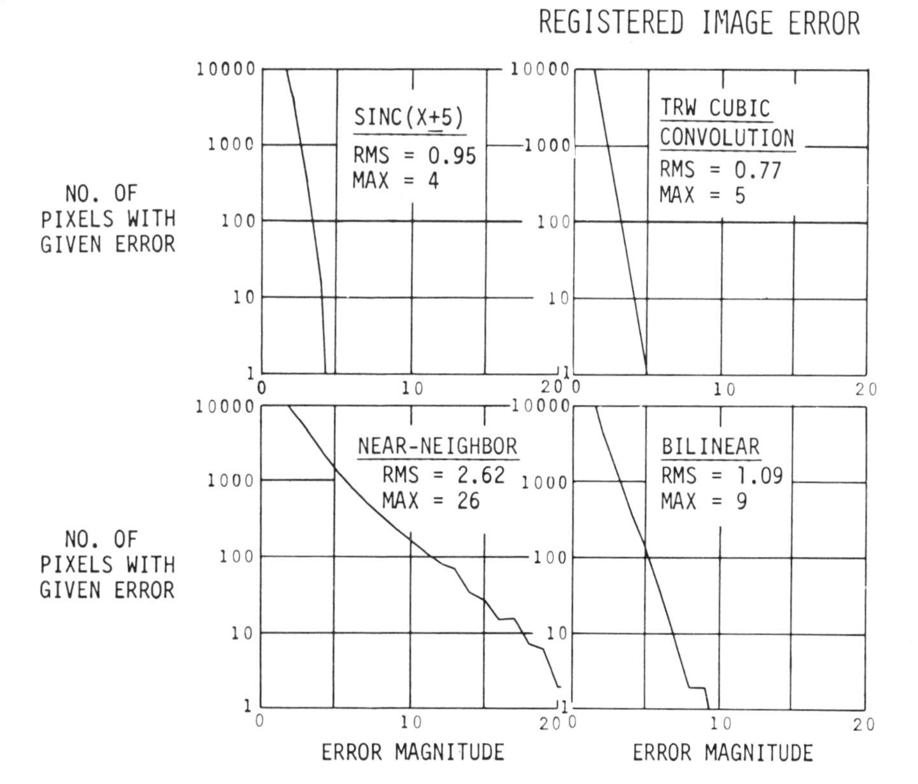

FIGURE 4 ERROR BETWEEN REGISTERED IMAGES DUE TO RESAMPLING

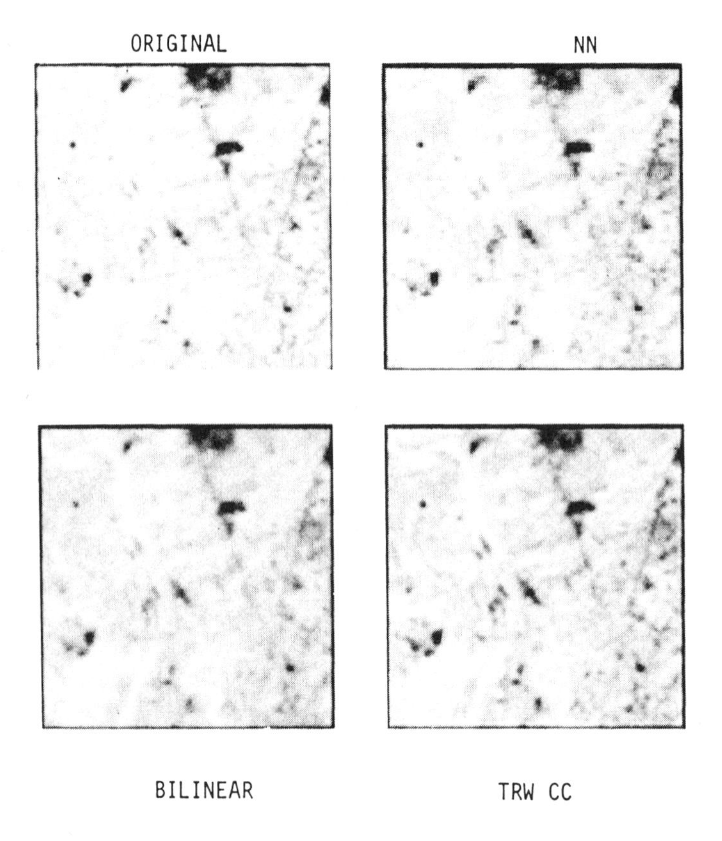

FIGURE 5A - 30M RESOLUTION IMAGE RECONSTRUCTION (64 x 78 PIXELS)

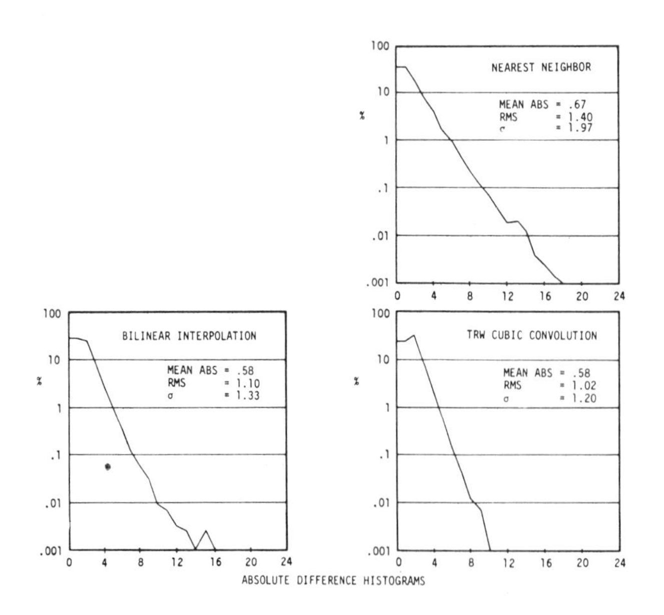

FIGURE 5B - RECONSTRUCTION DIFFERENCE IMAGES

FIGURE 6C - DIFFERENCE IMAGE USING TRW CUBIC CONVOLUTION FOR WARPING
(GRAY = NO ERROR)

FIGURE 6D - DIFFERENCE IMAGE USING NEAREST NEIGHBOR RESAMPLING

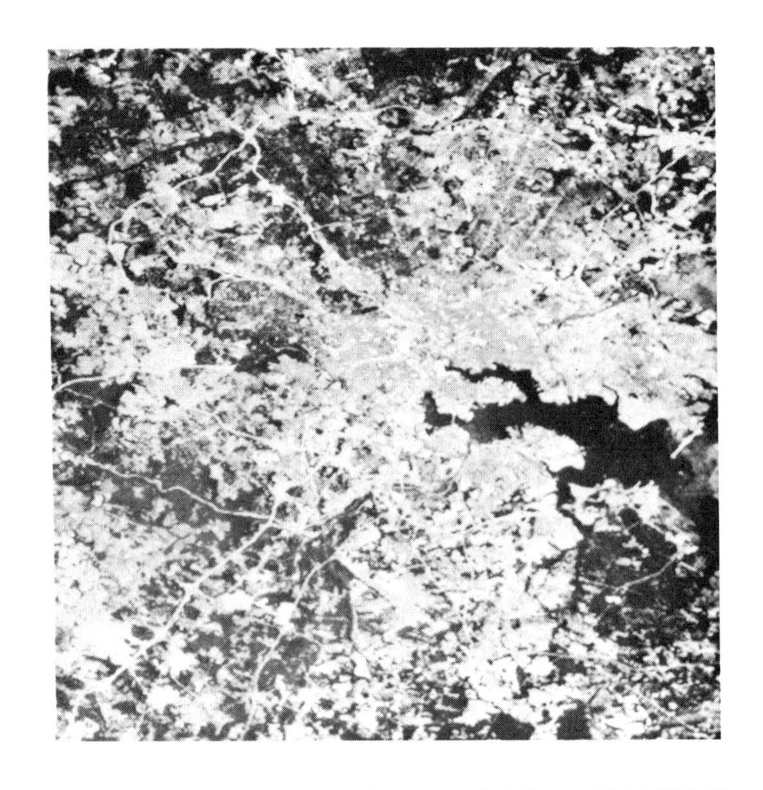

FIGURE 6A - ERTS BALTIMORE (SEPT '72) (512 x 512 PIXELS)

FIGURE 6B - ERTS BALTIMORE (OCT '72) REGISTERED TO PREVIOUS SCENE
(512 x 512 PIXELS)

Stephen W. Murphrey
Rex D. Depew
Ralph Bernstein
International Business Machines Corp.
Gaithersburg, MD 20760

Digital Processing of Conical Scanner Data

Skylab S-192 conical scanner data, digitally processed to remove systematic errors, were recorded on film and provided imagery of good quality.

Introduction

In the past, the correction of earth observation image data has been done primarily with electro-optical image processing techniques[1]. Extensive research in digital processing has been conducted [2,3,4]. Digital processing of earth observation sensor data is now accepted as the best technology for geometric and radiometric correction[5].

Most of the work that has been performed has dealt with linear scanners[2,3,4]. A class of instruments that scan the earth along a curved path (as opposed to a linear path) is that of the conical scanners. They have unique geometric properties that require a different approach to correction[8]. This paper presents the techniques and results of a recent investigation in which Skylab S-192 conical data was geometrically corrected, by using digital image processing methods, in order to eliminate systematic errors.

Abstract: *An experimental software system to remove systematic errors and to geometrically correct S-192 conical scanner data by using digital techniques has been developed. The digital image processing programs were implemented on an IBM 370/168 computer and were used on a September 15, 1973 S-192 image of Lake Havasu, Arizona. The resulting digital image was recorded on film and was of good quality. The experiment described demonstrates that digital image processing techniques can be used to correct conical scanner data.*

Statement of Problem

The principal objective of the experiment described in this paper is to show that conical scanner data can be geometrically corrected by using digital techniques. A second objective is to determine ways for minimizing the computer resources required to geometrically correct conical scanner data.

Background

The S-192 multispectral scanner is an optical-mechanical scanner, together with a spectral dispersion system. It uses a rotating mirror to perform a conical scanning of the earth. The cone angle is 5° 32' about the instrument axis. Data are collected during the front 116°15' of the 360° scanning cycle. There are 13 detectors which cover spectral regions with wavelengths be-

Reprinted with permission from *Photogrammetric Engineering and Remote Sensing*, vol. 43, pp. 155–167, Feb. 1977.

tween 0.41 and 12.5 micrometers[6]. The scan geometry of the conical scanner is shown in Figure 1, where

γ is the cone angle,
η is the scan angle,
V is the spacecraft position, and
$\overline{VP}$ is the vector along which the sensor points.

The rectangular coordinate axes represent the local sensor coordinate system (see Appendix A for definitions of all coordinate systems used in this paper). As the point P moves around a circle which is perpendicular to the Zs-axis, the line VP generates a cone (hence, the term "conical scanner").

The input data used in this experiment consisted of type 51-2 computer-compatible tapes[7]. These processed data tapes contain GMT-correlated calculated aperture radiances, as well as some ancillary data. That is, the data have been radiometrically corrected but not geometrically corrected.

Outline of Experiment

The presence of the S-192 conical scanner on Skylab provided a unique opportunity to use real conical data in this experiment. The processing was done in the five steps shown in Figure 2. These same steps were used in demonstrating all-digital processing of LANDSAT-1 MSS data[4]. Each step will be described.

Data Reformatting

There were two reasons for reformatting the input data. First, the type 51-2 computer-compatible tape format is not usable by the geometric correction program. It was much simpler to write a reformatting program than it would have been to modify the geometric correction program. Second, it was necessary to strip the ancillary data from the type 51-2 tape and to save it on a direct access data set for use by the error modeling program. Consequently, a computer program that reformats the data was written. Each record contains data from all spectral bands in line-interleaved format.

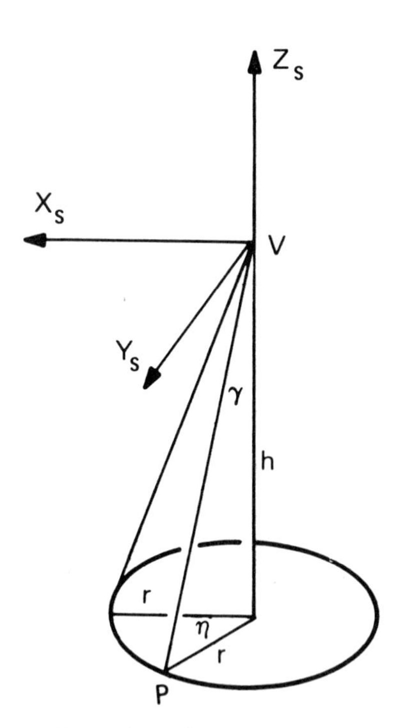

Fig 1. Conical scan geometry.

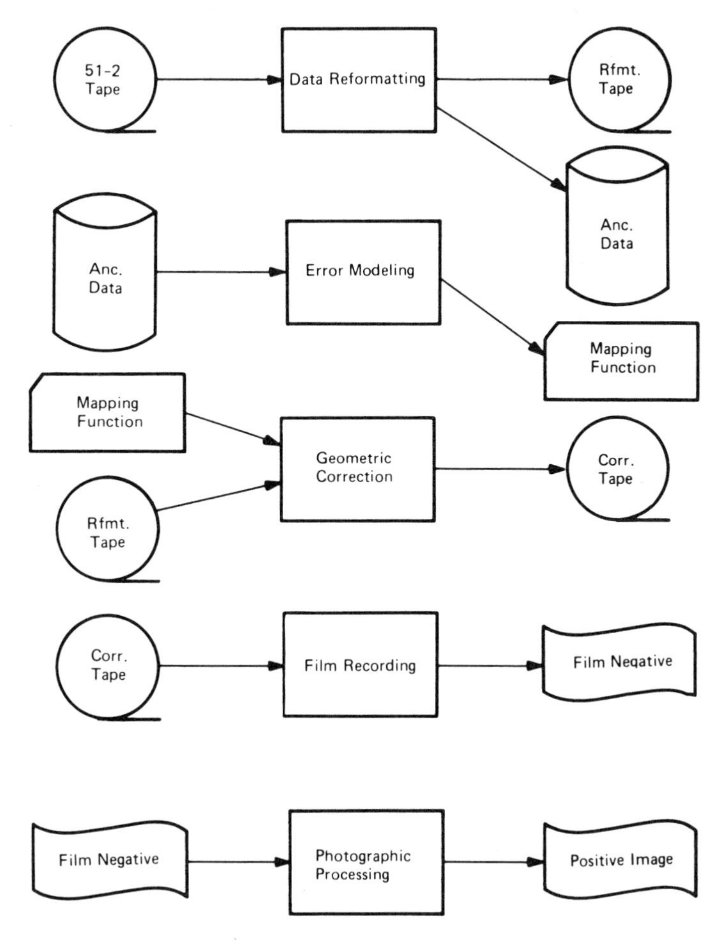

FIG 2. S-192 processing steps and data flow.

ERROR MODELING

The output image is defined to be a rectangular array of square pixels. Each output-image pixel represents a square of 50.8 metres in length and width. The particular pixel size was chosen so that the resulting film products will be at a scale of 1:500,000 when recorded on an IBM Drun Scanner/Plotter. The output image represents a Universal Transverse Mercator (UTM) projection. The input image is a rectangular array of pixels taken from a type 51-2 tape.

The function of the error modeling program is to create an output-image to input-image mapping function as indicated in Figure 3. The domain of the mapping function consists of

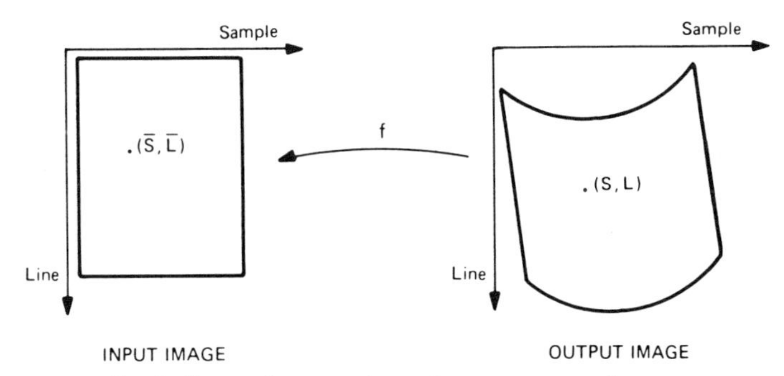

FIG 3. Output-image to input-image mapping function.

locations (i.e., line and sample coordinates) in the output image, and the range of the mapping function consists of locations in the input image. An output-image is physically created by using the mapping function to aid in computing the intensity (radiance) of each output-image pixel. The function is evaluated at an output-image pixel location (S, L). The intensities of the pixels near the resulting input-image location $(\bar{S}, \bar{L})$ are used in an appropriate resampling algorithm to compute the intensity of the out-put image pixel (S, L). The computer program which performed the error modeling for the experiment was written in FORTRAN and was compiled with a FORTRAN IV H EXTENDED compiler. Essentially all computations are done with 8-byte floating-point arithmetic. The main transformations which make up the error modeling program are all done in the subroutines described below.

INITIALIZATION

The initialization routines, in addition to reading various input parameters, define an array of points in the input-image space. These points, called anchor points, are regularly spaced 100 pixels apart as illustrated in Figure 4. In the actual image that was processed, there were 15 columns and 20 rows of anchor points (a total of 300 anchor points). Pixel coordinates (sample number and line number) are used for the input-image anchor points.

INPUT IMAGE TO SCAN ANGLE AND ANCILLARY PARAMETERS

The input-image sample coordinate of each anchor point is used to calculate the scan angle of that point:

$$\eta = \left(\bar{S} - \frac{N' + 1}{2}\right)\left(\frac{W\pi}{180\,(N' - 1)}\right) \tag{1}$$

where

η is the scan angle measured counter clockwise from the satellite velocity vector (as shown in Figure 1) in radians
$\bar{S}$ is the sample coordinate in pixels,
N' equals 1240 which is equal to the number of samples in one input scan line, and
W equals 116.25 which is equal to the total scanning arc width in degrees.

All calculations in Equation 1 are done in 8-byte floating-point arithmetic.

The input-image line coordinate of each anchor point is used to obtain certain ancillary parameters that were originally on the type 51-2 input data tape. The following parameters are used by the error modeling programs:

- Satellite roll
- Satellite pitch
- Satellite yaw
- Nadir latitude
- Nadir longitude
- Spacecraft altitude
- Orbital inclination angle

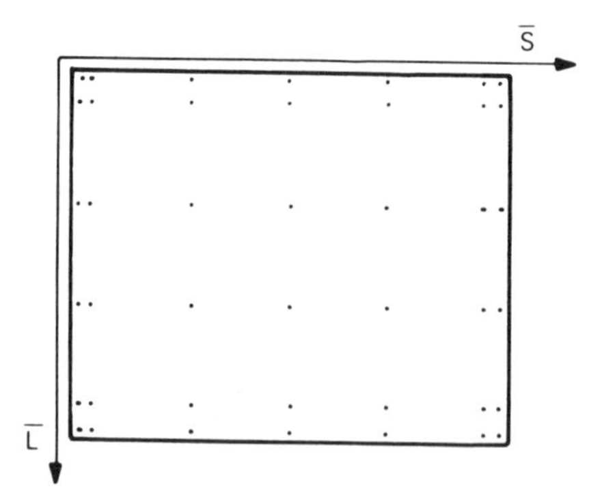

FIG 4. Input-image anchor points.

These seven values correspond to the time at which the first sample of the given line was observed. There is a set of ancillary parameters for each scan line.

SCAN ANGLE AND ANCILLARY PARAMETERS TO GEODETIC LATITUDE AND LONGITUDE

This routine computes the geodetic latitude and longitude of each anchor point. The treatment used essentially is that of Eppes[8]. In Figure 1, the S-192 instrument is assumed to be pointing along the line *VP*. The coordinates of the intersection of the line *VP* with the earth ellipsoid are computed. From these coordinates, the geodetic latitude and longitude are calculated.

In order to compute the intersection of the line *VP* and the earth ellipsoid, it is necessary to have the equations of both the line and the ellipsoid with respect to the same coordinate system. These are obtained by first finding the equation of the line in local sensor coordinates and then using several coordinate transformations to get the equation of the line in local earth-centered coordinates. (See Appendix A for definitions of coordinate systems.)

The equation of the line *VP* has the form

$$\frac{Xs}{A'} = \frac{Ys}{B'} = \frac{Zs}{C'} \tag{2}$$

in local sensor coordinates. The coordinates of the point P are $(r \cos \eta, r \sin \eta, -h)$. Hence, the direction cosines in Equation 2 are

$$A' = \frac{(1,0,0)\cdot\overline{VP}}{|\overline{VP}|} = \frac{r \cos \eta}{h^2 + r^2} = \sin \gamma \; \cos \eta$$

$$B' = \frac{(0,1,0)\cdot\overline{VP}}{|\overline{VP}|} = \frac{r \sin \eta}{h^2 + r^2} = \sin \gamma \; \sin \eta$$

$$C' = \frac{(0,0,1)\cdot\overline{VP}}{|\overline{VP}|} = \frac{-h}{h^2 + r^2} = -\cos \gamma$$

The equation of the line *VP* can also be written in the parametric form:

$$\begin{aligned} Xs &= p \sin \gamma \cos \eta = pA' \\ Ys &= p \sin \gamma \sin \eta = pB' \\ Zs &= -p \cos \gamma \qquad = pC' \end{aligned} \tag{3}$$

The equation of the line *VP* in local earth-perpendicular coordinates will now be found. The attitude transformation matrix can be written in the form

$$\mathbf{A} = \begin{bmatrix} \cos AP & 0 & -\sin AP \\ 0 & 1 & 0 \\ \sin AP & 0 & \cos AP \end{bmatrix} \begin{bmatrix} 1 & 0 & 0 \\ 0 & \cos AR & \sin AR \\ 0 & -\sin AR & \cos AR \end{bmatrix} \begin{bmatrix} \cos AY & \sin AY & 0 \\ -\sin AY & \cos AY & 0 \\ 0 & 0 & 1 \end{bmatrix}$$

$$= (a_{ij}).$$

where

AP is the pitch angle (right-handed rotation about the Yp-axis),
AR is the roll angle (right-handed rotation about the Xp-axis), and
AY is the yaw angle (right-handed rotation about the Zp-axis).

Then the coordinate transformation equation is

$$\begin{bmatrix} Xs \\ Ys \\ Zs \end{bmatrix} = \mathbf{A} \begin{bmatrix} Xp \\ Yp \\ Zp \end{bmatrix} \tag{4}$$

The Cramer matrices **A**1, **A**2, **A**3, are defined:

$$\mathbf{A}1 = \begin{bmatrix} A' & a_{12} & a_{13} \\ B' & a_{22} & a_{23} \\ C' & a_{32} & a_{33} \end{bmatrix}$$

$$\mathbf{A}2 = \begin{bmatrix} a_{11} & A' & a_{13} \\ a_{21} & B' & a_{23} \\ a_{31} & C' & a_{33} \end{bmatrix}$$

$$\mathbf{A}3 = \begin{bmatrix} a_{11} & a_{12} & A' \\ a_{21} & a_{22} & B' \\ a_{31} & a_{32} & C' \end{bmatrix}$$

Then Equation 4 has the solution:

$$\begin{aligned} Xp &= p(|\mathrm{A}1|/|\mathrm{A}|) = p|\mathrm{A}1| \\ Yp &= p(|\mathrm{A}2|/|\mathrm{A}|) = p|\mathrm{A}2| \\ Zp &= p(|\mathrm{A}3|/|\mathrm{A}|) = p|\mathrm{A}3| \end{aligned} \tag{5}$$

since $|A| = 1$. After algebraic manipulation, Equation 5 can be written in the form

$$C_1Xp = C_2Yp = C_3Zp \tag{6}$$

where

$$\begin{aligned} C_1 &= |\mathrm{A}2|\ |\mathrm{A}3| \\ C_2 &= |\mathrm{A}1|\ |\mathrm{A}3| \\ C_3 &= |\mathrm{A}1|\ |\mathrm{A}2| \end{aligned}$$

The equation of the line *VP* in local earth-tangent coordinates is derived from the transformation:

$$\begin{bmatrix} Xp \\ Yp \\ Zp \end{bmatrix} = \begin{bmatrix} Xt \\ Yt \\ Zt - H \end{bmatrix} \tag{7}$$

where *H* is the spacecraft altitude measured along a perpendicular to the earth ellipsoid. Substituting Equation 7 into Equation 6 gives the resulting equation of the line VP in local earth-tangent coordinates:

$$C_1Xt = C_2Yt = C_3(Zt - H). \tag{8}$$

The final transformation of the equation of the line *VP* will be to local earth-centered coordinates. The transformation consists of a translation followed by two rotations. The azimuth rotation matrix

$$\mathbf{AZM} = \begin{bmatrix} \cos AZS & -\sin AZS & 0 \\ \sin AZS & \cos AZS & 0 \\ 0 & 0 & 1 \end{bmatrix}$$

represents a counter-clockwise rotation through the angle *AZS* around the *Zc*-axis. The geodetic-latitude rotation matrix

$$\mathbf{LDM} = \begin{bmatrix} \sin LDN & 0 & -\cos LDN \\ 0 & 1 & 0 \\ \cos LDN & 0 & \sin LDN \end{bmatrix}$$

represents a clockwise rotation through the angle ($\pi/2 - LDN$) around the Yc-axis. Then the transformation from local earth-centered coordinates to local earth-tangent coordinates is given by

$$\begin{bmatrix} Xt \\ Yt \\ Zt \end{bmatrix} = (\boldsymbol{AZM})\,(\boldsymbol{LDM}) \left\{ \begin{bmatrix} Xc \\ Yc \\ Zc \end{bmatrix} - \begin{bmatrix} Xcn \\ 0 \\ Zcn \end{bmatrix} \right\} \tag{9}$$

where

$AZS = \arcsin\left(\dfrac{\cos OIE}{\cos LCN}\right)$ = instantaneous heading or azimuth measured clockwise from south

OIE = inclination angle of orbital plane from equator

LCN = geocentric latitude of the spacecraft nadir

LDN = geodetic latitude of the spacecraft nadir

$Xcn = E \cos LDN$, $Zcn = E(1-e^2)\sin LDN$ } earth-centered coordinates of spacecraft nadir

$E = \dfrac{a}{\sqrt{1 - e^2 \sin^2 LDN}}$

$e = \dfrac{\sqrt{a^2 - b^2}}{a}$ = eccentricity of the earth ellipsoid

a = semi-major (equatorial) axis of the earth ellipsoid

b = semi-minor (polar) axis of the earth ellipsoid.

If the matrix product $(\boldsymbol{AZM})(\boldsymbol{LDM}) = (b_{ij})$, then substituting Equation 9 into Equation 8 gives

$$\begin{aligned} &C_1[b_{11}(Xc - Xcn) + b_{12}Yc + b_{13}(Zc - Zcn)] \\ &= C_2[b_{21}(Xc - Xcn) + b_{22}Yc + b_{23}(Zc - Zcn)] \\ &= C_3[b_{31}(Xc - Xcn) + b_{32}Yc + b_{22}(Zc - Zcn) - H]\,. \end{aligned} \tag{10}$$

Equations 10 can be written in parametric form with Zc as the parameter:

$$\begin{aligned} R_1Xc + R_2Yc &= -R_3Zc + R_1Xcn + R_3Zcn \\ S_1Xc + S_2Yc &= -S_3Zc + S_1Xcn + S_3Zcn - C_3H \end{aligned} \tag{11}$$

where

$$\begin{aligned} R_1 &= C_1b_{11} - C_2b_{21} \\ R_2 &= C_1b_{12} - C_2b_{22} \\ R_3 &= C_1b_{13} - C_2b_{23} \\ S_1 &= C_1b_{11} - C_3b_{31} \\ S_2 &= C_1b_{12} - C_2b_{32} \\ S_3 &= C_1b_{13} - C_3b_{33}\,. \end{aligned}$$

By using Cramer's rule, Equations 11 can be solved for Xc and Yc in terms of Zc (Zc is regarded as a parameter):

$$\begin{aligned} Xc &= \bar{A}Zc + \bar{B} \\ Yc &= \bar{C}Zc + \bar{D} \end{aligned} \tag{12}$$

where

$$\bar{A} = \frac{R_2S_3 - R_3S_2}{R_1S_2 - R_2S_1}$$

$$\bar{B} = \frac{(R_1S_2 - R_2S_1)Xcn + (R_3S_2 - R_2S_3)Zcn + C_3R_2H}{R_1S_2 - R_2S_1}$$

$$\bar{C} = \frac{R_3S_1 - R_1S_3}{R_1S_2 - R_2S_1}$$

$$\bar{D} = \frac{(R_1S_3 - R_3S_1)Zcn - C_3R_1H}{R_1S_2 - R_2S_1}.$$

Equations 12 can be substituted into the equation for the earth ellipsoid,

$$\frac{Xc^2 + Yc^2}{a^2} + \frac{Zc^2}{b^2} = 1,$$

to get the following, after collecting terms:

$$(a^2 + b^2\bar{A}^2 + b^2\bar{C}^2)Zc^2 + 2b^2(\bar{A}\bar{B} + \bar{C}\bar{D})Zc - b^2(a^2 - \bar{B}^2 - \bar{D}^2) = 0.$$

By using the quadratic formula:

$$d = b^2(\bar{A}\bar{B} + \bar{C}\bar{D})^2 + (a^2 - \bar{B}^2 - \bar{D}^2)\left[a^2 + b^2(\bar{A}^2 + \bar{C}^2)\right] \qquad (13)$$

$$Zc = \frac{-b^2(\bar{A}\bar{B} + \bar{C}\bar{D}) \pm b\sqrt{d}}{a^2 + b^2(\bar{A}^2 + \bar{C}^2)}.$$

the Xc and Yc coordinates of the ray-ellipsoid intersection are now calculated by substituting Equation 13 into Equation 12. The positive sign is used in Equation 13 for a northern-hemisphere intersection, and the negative sign is used for a southern-hemisphere intersection.

The local earth-centered coordinates of the intersection of the line VP with the earth ellipsoid are converted to geodetic latitude and longitude by the well-known equations:

$$Ld = \arcsin \sqrt{\frac{Xc^2 + Yc^2 - a^2}{(Xc^2 + Yc^2)e^2 - a^2}}$$

$$Ln = Lnn + \arctan\left(\frac{Yc}{Xc}\right)$$

where
- Lnn is the longitude of the spacecraft nadir,
- Ld is the geodetic latitude of an anchor point, and
- Ln is the longitude of an anchor point.

GEODETIC LATITUDE AND LONGITUDE TO UTM

The geodetic latitude and longitude coordinates of the anchor points are converted to UTM coordinates by means of a modified version of a program obtained from the United States Geological Survey.

UTM TO OUTPUT IMAGE

The array of anchor point UTM coordinates is converted to output-image pixel coordinates by the equation:

$$\begin{bmatrix} S \\ L \end{bmatrix} = \begin{bmatrix} -\sin\alpha & \cos\alpha \\ \cos\alpha & \sin\alpha \end{bmatrix} \begin{bmatrix} \dfrac{E - E_c}{S_s} \\ \dfrac{N - N_c}{S_l} \end{bmatrix} + \begin{bmatrix} \dfrac{W_s}{2} \\ \dfrac{W_l}{2} \end{bmatrix}$$

where

- α = angle measured clockwise from the positive E-axis to the positive L-axis
- E = UTM easting of an anchor point
- N = UTM northing of an anchor point
- E_c = UTM easting of format center of output image
- N_c = UTM northing of format center of output image
- S_s = width of one output-image pixel = 50.8 metres
- S_l = length of one output-image pixel = 50.8 metres
- W_s = number of samples in one output-image line
- W_l = number of lines in the output image
- S = output-image sample coordinate
- L = output-image line coordinate

MAPPING FUNCTION

The routines described compute the output-image coordinates (S, L) of each of 300 input-image anchor points $(\bar{S}, \bar{L})$. The mapping function will be two least-squares polynomials, P_s and P_l, of degree 5 (21 terms each) that map each point (S, L) onto the corresponding input-image anchor point $(\bar{S}, \bar{L})$. That is,

$$\bar{S} = P_s(S, L)$$
$$\bar{L} = P_l(S, L)$$

or, in the notation of Figure 3,

$$f(S, L) = (P_s(S, L), P_l(S, L)).$$

Geometric Correction

Geometric correction was performed by using the nearest-neighbor assignment algorithm. The technique is adequately described elsewhere[2,3] and only will be summarized here. Nearest-neighbor assignment means that, for each output-image pixel, the intensity (radiance) of the particular input-image pixel whose location is closest to the location in the input image of that output-image pixel is used. That is, for each output-image pixel location (S, L), the corresponding input-image location $(\bar{S}, \bar{L})$ is found. Note that S and L are integers, but $\bar{S}$ and $\bar{L}$ usually are not integers. The input-image location $(\bar{S}', \bar{L}')$, which is the closest point to $(\bar{S}, \bar{L})$ having integral coordinates, is found. Then the intensity of the Point $(\bar{S}', \bar{L}')$ is used as the intensity of the point (S, L).

Theoretically, the mapping function should be used on every point in the output image. Doing this would be quite expensive computationally. Instead of mapping every point, a rectangular array of points was mapped to the input image. This array of grid points contained 35 rows and 31 columns, totalling 1085 points. The spacing between the rows was 100 pixels, and between the columns it was 50 pixels. The remaining 5,098,915 points of the output image were mapped to the input image by using successive linear interpolations on the grid points. This method is considerably cheaper than the mapping of all 5,100,000 points through the two polynomials would be.

Photo Processing

The geometrically corrected image was then recorded on photographic film by using an IBM Drum Scanner/Plotter[9]. These negatives were developed, and positive false-color image products were made from the negatives.

Results and Conclusions

The experiment described consisted of systematic geometric correction of Skylab S-192 conical data. The error modeling program ran in 4.77 seconds on an IBM 370/168 computer. The running time for each anchor point is about the same as that of IBM's LANDSAT error modeling programs. However, since the conical scanner has larger distortions than a linear scanner (due to the curvature of the scan lines), it requires more anchor points than a linear scanner. It is estimated that a conical scanner may require up to twice as much Central Processing Unit (CPU) time as a comparable linear scanner in order to perform the error modeling step. This difference between conical and linear scanners is likely to be insignificant relative to the total correction times.

The geometric correction of the S-192 data confirmed that conical scanner data can be geometrically corrected by the same methods that have been used successfully for linear scanner data. Either type of sensor requires the same amount of computer time per pixel. However, it is necessary to have all of the input-image data needed to construct one ouput-image line in core at one time. For linear scanner data, whose distortions are on the order of only a few input-image lines, this restriction is negligible. A buffer of 10 or 20 input-image data lines is all that is required. The curvature of a conical scan line dictates that an output-image line be constructed from data from hundreds of input-image scan lines. In the case of Skylab S-192 data, the large buffer required was not prohibitive because there were only 1240 samples per line. Future conical scanners will have many more samples per line, and the size of the input buffer will assume major importance.

The quality of the photographic products produced by this experiment were good (see Plate 1). They appeared to be about equal to LANDSAT photographs produced on the same equipment and in a similar manner.

This experiment has demonstrated that a conical scanner can be a viable choice for an earth observation sensor. Ground processing of conical data does require that large amounts of data be stored on fast access devices such as disks or computer memory. The choice of sensors, whether linear or conical, should be based primarily on sensor performance and cost.

Unanswered Questions

This experiment did not address several important topics pertaining to ground processing of conical scanner data:

- Terrain relief,
- Ground control point processing, and
- Large input data buffer,

Terrain relief is much more of a problem with a conical scanner than it is with a linear scanner due to the large scanning angle. It is probably necessary to correct for terrain relief in order to produce images that have precise geometric accuracy. It is also necessary to use ground truth information in order to obtain high geometric accuracy. Neither of these aspects has been considered in this investigation. The large buffer needed for conical data may be reduced in size by using a clever, efficient disk data management system. As yet, no such scheme has been tested.

Acknowledgments

The authors would like to thank several people from Honeywell Radiation Center in Lexington, Massachusetts for their assistance during the course of the S-192 experiment. They are D. A. Koso, M. F. Harris, H. W. Robinson, Jr., and O. E. Toler.

References

(1) NASA/Goddard Space Flight Center, *ERTS Data User's Handbook*, GE Document No. 71SD4249.

(2) Bernstein, R. "Results of Precision Processing (Scene Correction) of ERTS-1 Images Using Digital Image Processing Techniques", *Symposium* on Significant Results Obtained from the Earth Resources Technology Satellite-1, Vol. II, NASA Document #SP-327, March 5-9, 1973.

(3) Bernstein, R. "Scene Correction (Precision Processing) of ERTS Sensor Data Using Digital Image Processing Techniques", Third ERTS *Symposium*, Vol. 1, Section A, NASA SP-351, December 10-14, 1973.

(4) Bernstein, R. *All-Digital Precision Processing of ERTS Images*, Final Report, NASA Contract NAS5-21716, April 1975.

(5) NASA/Goddard Space Flight Center, *Master Data Processing (MDP) Systems*, RFP No. 5-70301-156, June 4, 1975.

(6) NASA/JSC, *Earth Resources Production Processing Requirements for EREP Electronic Sensors*, PHO-TR524 Rev. B, June 1975.

(7) NASA/Manned Spacecraft Center, *Earth Resources Data Format Control Book*, PHO-TR543, March 1973.

Plate 1. Systematically processed S-192 image of Lake Havasu area.

(8) Forrest, R. B., T. A. Eppes, and R. J. Ouellette, *EOS Mapping Accuracy Study,* Final Report for contract NAS-5-21727, March 1973.

(9) Friar, M. E., R. D. Hogan, T. J. Min, J. V. Sharp, and D. R. Thompson, *System and Design Study for an Advanced Drum Plotter,* Final Technical Report, USAETL contract DAAK-02-69-C-0015, April 1970.

Appendix A

DEFINITIONS OF COORDINATE SYSTEMS (RIGHT-HANDED)

Notation	Terminology	Definition
Xs, Ys, Zs	Local sensor coordinates	Origin at spacecraft; X, Y, Z-axes determined by roll, pitch, and yaw relative to local earth-perpendicular axes.
Xp, Yp, Zp	Local earth-perpendicular coordinates	Origin at spacecraft; X-direction of spacecraft orbital vector; Z-axis normal to earth ellipsoid.
Xt, Yt, Zt	Local earth-tangent coordinates	Translate of local earth-perpendicular coordinates to spacecraft nadir point.
Xc, Yc, Zc	Local earth-centered coordinates	Origin at geocenter; X-axis in meridional plane of nadir; Z-axis is polar axis.

DIGITAL TECHNIQUES FOR EARTH RESOURCE IMAGE DATA PROCESSING

R. Bernstein
IBM Corporation, Federal Systems Division
Gaithersburg, Maryland
H. Silverman
IBM Corporation, T. J. Watson Research Center
Yorktown Heights, New York

Abstract

The growing availability of faster and more versatile digital hardware, along with the development of more efficient processing algorithms has brought about an improvement in the ability of digital systems to process high resolution imagery. In this paper, techniques for implementing image corrections are described and the results of image processing experiments are presented. The techniques include methods for automatically locating reseau marks and reference ground-control points, and for computing and applying both geometric and radiometric image corrections. A fast method of correlation (based on a class of sequential-similarity-detection algorithms) which solves the problem in digital image processing of correlating the working image to a reference image is also presented.

I. Introduction

The National Aeronautics and Space Administration (NASA), supported by other government agencies, has a continuing earth observation program whose objective is the acquisition of data for environmental research applicable to earth resource inventory and management.(1,2,3) These programs include:

a. Earth Resources Technology Satellite (ERTS)
b. Aircraft Program
c. Earth Observatory Satellite (EOS)
d. Skylab
e. Space Station

Sensors carried on the ERTS platform(4) include multispectral imaging systems such as Return Beam Vidicon (RBV) camera systems which form consecutive central perspective images, and Multispectral Scanner (MSS) systems which cover a fixed swath width by transverse scan lines. Future earth observation programs will include microwave radiometers, multispectral imaging radiometers, and a scanning spectrophotometer.(5) The U.S. Army and other government agencies use aerial photos extensively for mapping and surveillance.

The image data received from these sensors will contain both geometric and radiometric distortions because of their many sources.(6,7,8) This paper presents promising experimental digital techniques that allow for both determination of the image error characteristics and methods which will correct these errors.

II. Sensor Error Characteristics

The sensor errors can be broadly categorized as geometric distortions and radiometric distortions.

Geometric Distortions

Sensor errors in this category are characterized by a geometric or spatial distortion of the image from a particular cartographic projection, such as Universal Transverse Mercator or orthophoto projection. The error sources are due to internal errors (sensor-related) and external errors (for example, platform attitude, scene characteristics). The effects of these errors are shown in Figure 1.

Radiometric Distortions

Radiometric distortions are characterized by an incorrect intensity distribution, spatial frequency filtering of the scene data, blemishes in the imagery, banding of the image data, etc. These distortions are caused by camera or scanner shading effects, detector gain variations, atmospheric and sensor induced filtering, sensor imperfections, sensor detector gain errors, etc. Examples of sensor intensity errors are shown in Figure 2. The only radiometric distortions considered here are intensity distortions. Digital techniques to reduce the effects of other radiometric distortion effects are discussed elsewhere.(9,10)

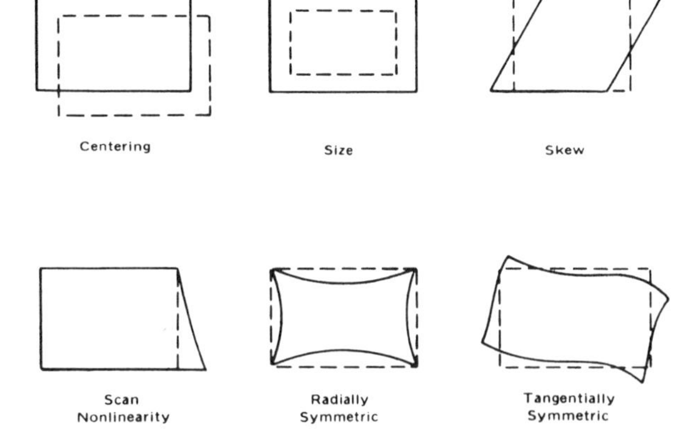

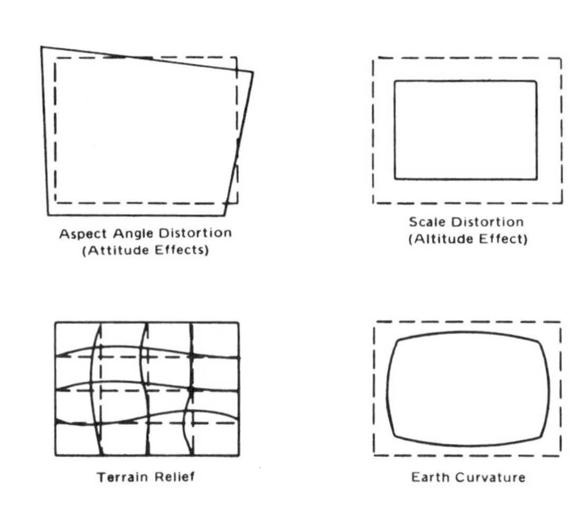

Figure 1. Characteristic Geometric Image Distortions

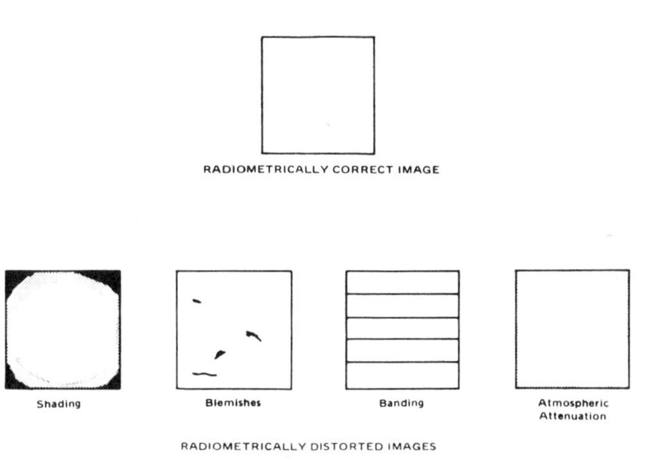

Figure 2. Characteristic Radiometric Image Distortions

Reprinted with permission from *Proc. Amer. Inst. Aeronautics and Astronautics 8th Annu. Meeting*, vol. C21, AIAA paper no. 71-978, Oct. 1971, pp. 1-1-1-14.

III. Sources of Image Correction Data

Information is generally available to determine the geometric and radiometric distortion of sensor data.(6,7) This information includes reseau marks, platform attitude and altitude data, ground control point information, and intensity calibration data. The characteristics of the data will be briefly discussed in this section; the manner in which this information is used to correct the sensor data will be discussed in later sections.

Reseau Marks

One common method of obtaining measurements of internal distortions in camera-type sensors is to inscribe a network of fiducial or reseau marks on the faceplate of the sensor. As an example, consider the ERTS RBV. A reseau pattern composed of an ordered 9 x 9 array of opaque cruciform marks is inscribed on the RBV faceplate. In addition, along the four sides, within the image format are four anchor marks with a particular pattern for identification and orientation as shown in Figure 3.(6) Since the locations of the reseau marks can be measured quite accurately prior to use of the sensor, measured differences between the apparent locations of the reseau marks in the RBV images and their known locations will give a map of the internal distortions present.

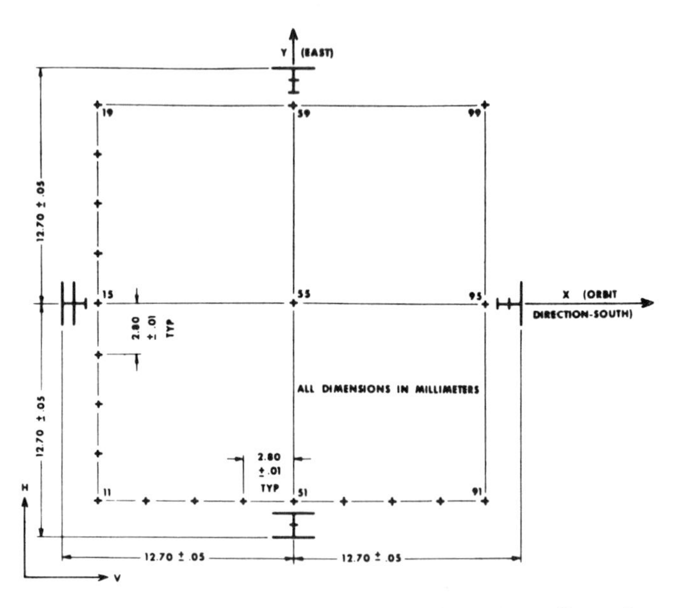

Figure 3. RBV Reseau Marks Pattern—Used to Obtain Sensor (Internal) Geometric Errors

Ground Control Point and Attitude Data

Errors external to the sensor will cause additional geometric distortion in the sensor's images. A suitable number (9 to 12)(7,11) of identifiable natural or cultural ground control points can be used to determine the sensor attitude errors (roll, pitch, yaw), earth curvature and rotation, and can be combined to mathematically characterize the external errors. In addition, geographic ground reference is obtained. Alternatively, sensor attitude errors can be independently determined at the moment of camera exposure or scan by precision attitude determination equipment (such as star trackers, horizon scanner, and attitude and rate gyros) on-board the platform. This data can be combined with known earth curvature and terrain effects to characterize external distortions. Precise orbit ephemeris data would be required for geographic ground referencing.

Intensity Data

Preflight sensor calibration data (see Figure 5) provides RBV shading information, and MSS detector gain errors.(6,7) Inflight sensor data using calibrated light sources or taken over uniform intensity scene areas provides differential intensity data that can be used to determine changes in the preflight shading characteristics of the RBV and differential gains of the MSS detectors.

IV. Method of Image Correction

Overall Description

Figure 4 shows the overall method of the geometric mapping function characterization. The reseau marks on the source image are located, and their deviations from their correct positions are used to generate a global polynomial function that characterizes internal errors. Alternatively, or additionally, preflight (or inflight) calibration data can be used to characterize the sensor related distortions. To determine the external errors, a number of ground control points are detected and located, and their known geographic positions are compared with the source image positions. This information is used to generate an external distortion function, which, in combination with the internal distortion function, completely characterizes the mapping function which allows transformation of the source image to a corrected output image. If platform attitude, altitude and orbit ephemerides data were known precisely, the external errors could be directly characterized and there would be no need for ground control point correlation. Generally, this information is not precisely known, and ground control point correlation provides an accurate and reliable method of determining the external errors.

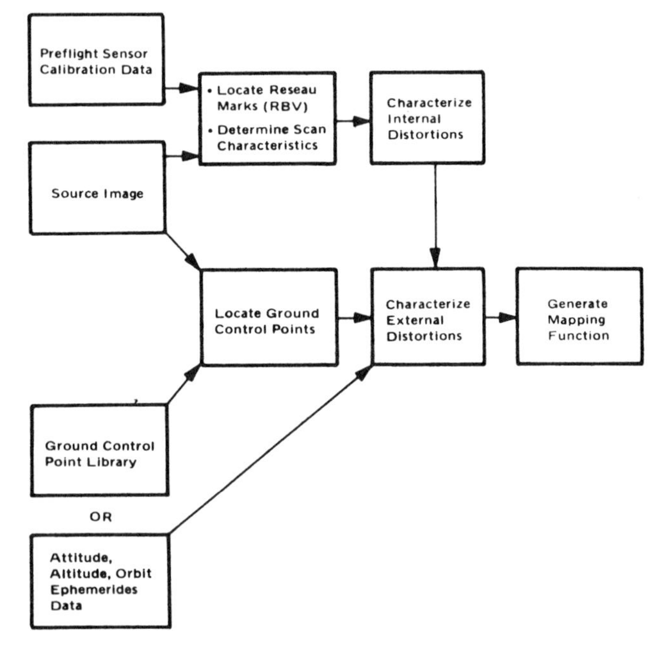

Figure 4. Mapping Function Generation Summary

Figure 5 shows the approach to radiometric correction determination. Preflight and inflight intensity calibration data, combined with source image data is used to characterize the radiometric (intensity) errors, which are then converted to a table which relates correct intensities to observed values. Subsequent intensity corrections of each picture element (pixel) are accomplished by a simple table look-up operation.

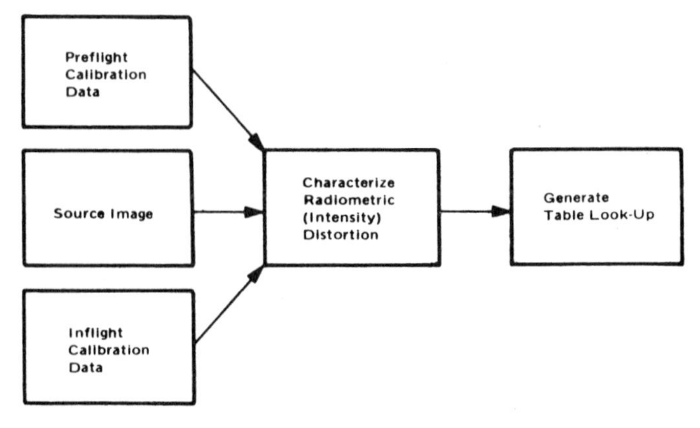

Figure 5. Intensity Correction Summary

Reseau Mark Detection and Location

The ERTS RBV reseau mark is shown in Figure 6. Since the exact position of each of the 81 reseau marks is not known, a search area sufficiently large must be defined about each reseau mark, and a method developed to detect and locate them. The marks are cross-shaped with an arm width of 25 ± 5 μm and an overall length of 200 ± 12.5 μm. This corresponds to approximately 4 and 32 pixels, respectively. A geometric error variation of ± 1 percent (typical RBV performance) creates an uncertainty of ± 41 pixels in the X and Y directions about the known location of the reseau mark. Given that the reseau mark can be inscribed inside a square with side dimension of 32 pixels, a search area of 128 x 128 pixels allows sufficient coverage for a ± 1 percent geometric error.

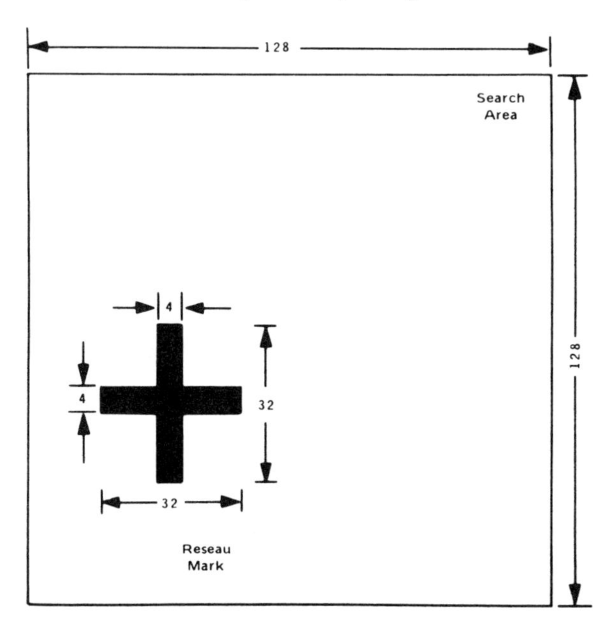

Figure 6. Search Area and Reseau Mark Shapes and Dimensions in Pixel Units

The reseau detection routine developed by IBM is based on the following operational sequence:

a. The last known locations of reseau marks are inputs to the program.
b. Within blocks of 128 x 128 elements, each centered around a previous reseau mark location, individual row and column sums of pixel gray levels are computed. This operation is called "shadow casting." Thus, along the nth column, the sum would be

$$S_n = \sum_{m=1}^{128} g_{m,n} \tag{1}$$

where $g_{m,n}$ is the gray level of the pixel located at the mth row and the nth column of the 128 x 128 block.
c. The reseau mark contained within a block is detected by the application of the detection algorithm to the row and column sequences $\{S_m\}$ and $\{S_n\}$. The algorithm described with reference to Figure 7 is based on moving a quadratic along the column sequence $\{S_n\}$, by fitting it to the sums S_{n-5}, S_{n+5}, and detecting the presence of sums at the locations n + 1, n + 2, n + 3, n + 4 whose values exceed the quadratic function at those locations by a computed dynamic threshold. That is, each point at these four locations is tested according to the condition

$$S_{n+p} - \hat{S}_{n+p} > \delta_n$$

where (2)

p = 1, 2, 3, 4

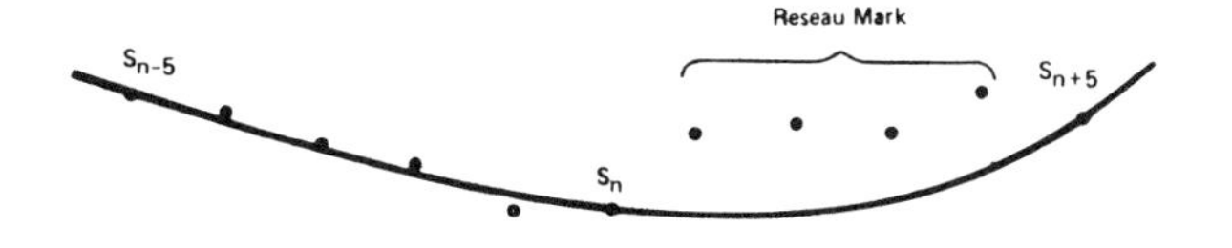

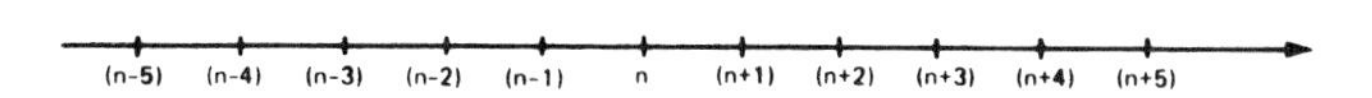

Figure 7. Reseau Detection Algorithm Geometry

S_{n+p} = Actual sum at the n + p location

$\hat{S}_{n+p}$ = Estimated value of the sum at the n + p location computed from the quadratic

$$\hat{S}_{n+p} = \left\{\frac{S_{n+5} + S_{n-p} - 2S_n}{50}\right\} p^2 + \left\{\frac{S_{n+5} - S_{n-5}}{10}\right\} p + S_n \tag{3}$$

δ_n = Dynamic threshold

$= 1000 - 0.123\, S_n$

A similar procedure is used for the row sequence $\{S_m\}$.

Additional detail regarding the technique and development of the method is available.(12,13)

Ground Control Point Correlation

The Translational Registration Problem

Registration is an inherently basic requirement for any image processing system. The Sequential Similarity Detection Algorithms (SSDAs) introduced in this paper are efficient digital implementation for determining points of correspondence between two images. If the images do not differ in magnification and rotation, the "best" translational fit will yield the required registration. (The problems arising from magnification and rotation will not be considered here. However, the methods are applicable when proper modifications are introduced.)

Let two images, S, the search area, and W, the window be defined as shown in Figure 8. S is taken as an LxL array of digital picture elements which may assume one of J gray levels; that is,

$$0 \leq S(i,j) \leq J-1$$

$$1 \leq i,j \leq L$$

W is considered to be an MxM, M smaller than L, array of digital picture elements having the same gray/scale range; that is,

$$0 \leq W(\ell,m) \leq J-1$$

$$1 \leq \ell,m \leq M$$

It will be convenient to introduce a notation for MxM wholly contained subimages

$$S_M^{i,j}(\ell,m) \equiv S(i+\ell-1, j+m-1)$$

$$\begin{cases} 1 \leq \ell,m \leq M \\ 1 \leq i,j \leq L-M+1 \end{cases} \tag{4}$$

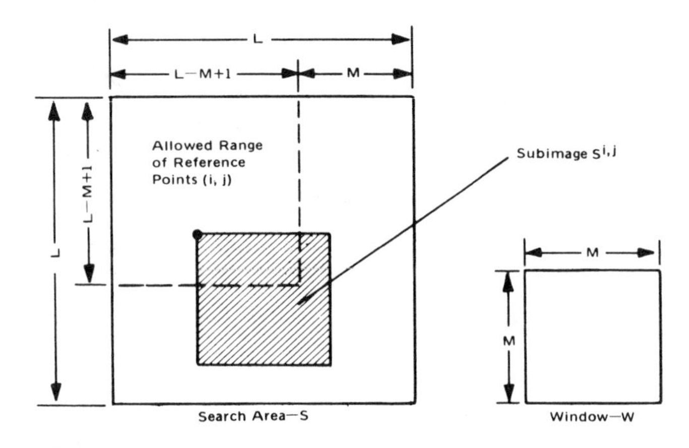

Figure 8. Illustration of Search Space

Each MxM subimage of S can be uniquely referenced by the specification of its upper left corner's coordinates (i,j). These will be used to define reference points. It will be assumed that enough a priori information is known about the dislocation between the window and search area so that the parameters L and M may be selected with the virtual guarantee that, at registration, a complete subimage is contained in the search area as shown in Figure 8.

Translational registration, therefore, is a search over some subset of the allowed range of reference points to find a point (i^*,j^*) which indicates a subimage that is "most similar" to the given window.

Computational Efficiency of Correlation Methods

The method most widely used for the automatic determination of translation is correlation.[15,16] The elements of the unnormalized cross-correlation surface R(i,j) are defined to be

$$R(i,j) \equiv \sum_{\ell=1}^{M} \sum_{m=1}^{M} W(\ell,m)\, S_M^{i,j}(\ell,m) \qquad (5)$$

$$1 \leqslant i,j \leqslant L\text{-}M+1$$

In the correlation scheme, a representative output surface such as R(i,j) is searched for a maximum, $(\hat{i},\hat{j})$. The procedure is successful if $(\hat{i},\hat{j})$ and (i^*,j^*) are equivalent. As a counterexample, however, consider the unnormalized cross-correlation of equation (5) even in the ideal case where W exactly matches some subimage; that is,

$$W = S_M^{i^*,j^*} .$$

Then

$$R(i^*,j^*) = \sum_{\ell=1}^{M} \sum_{m=1}^{M} W^2(\ell,m) \qquad (6)$$

Also, for this ideal case, consider the nonmatching point, $(\hat{i},\hat{j})$, where

$$S_M^{\hat{i},\hat{j}}(\ell,m) \equiv \max_{\ell,m} W(\ell,m) = W_M, \text{ for all } (\ell,m) \qquad (7)$$

Clearly,

$$R(\hat{i},\hat{j}) = W_M \sum_{\ell=1}^{M} \sum_{m=1}^{M} W(\ell,m) \geqslant R(i^*,j^*) \qquad (8)$$

Therefore, even in the ideal case a search for a maximum over R(i,j) does not necessarily yield the registration point. Normalization is therefore necessary in even the simplest of cases. For completeness, the usual normalized correlation surface is defined in equation (9).

$$R_N^2(i,j) = \frac{\left(\sum_{\ell=1}^{M} \sum_{m=1}^{M} W(\ell,m)\, S_M^{i,j}(\ell,m)\right)^2}{\left[\sum_{\ell=1}^{M} \sum_{m=1}^{M} W^2(\ell,m)\right]\left[\sum_{\ell=1}^{M} \sum_{m=1}^{M} {S_M^{i,j}}^2(\ell,m)\right]} \qquad (9)$$

$$1 \leqslant i,j \leqslant L\text{-}M+1$$

The choice of a similarity detection algorithm should be justifiable by its probability for error and its computational complexity, rather than by tradition or expediency. Perhaps the two reasons which are generally given for using the correlation method are:

a. Correlation appears to be a natural solution for the mean-square error criteria.[17]
b. Analog/optical methods implement correlation easily.[18]

However, there is no guarantee for any method that a solution is correct or unique. There seems to be, therefore, no adequate justification for the use of correlation to solve all digital registration problems. Algorithms, such as those presented in this paper, which have selectable distance measure properties and lower computational complexity, appear to be a more fitting choice.

The Cost of Correlation

The normalized correlation surface may be calculated by direct means or by FFT methods.[19] In each case the amount of computation required for normalization is the same. The numerator of equation (9) is all that may be treated by FFT.

In Table 1 the approximate numbers of calculations for each procedure are shown. One should note that although the FFT method requires fewer operations for L and M large, this method also requires a memory capacity of $2L^2$ real words which may be unfeasible for L larger than 256.

The Basic Concept

For a particular reference point (i,j), there are M^2 points of the subimage $S_M^{i,j}$ which may be compared with the M^2 corresponding points in W. (Each set of points for comparison (for example, $\langle S_M^{i,j}(\ell,m), W(\ell,m) \rangle$) will be called a windowing pair.) In correlation the maximum number of windowing pairs or $M^2(L\text{-}M+1)^2$ are compared. Each reference point, regardless of content, is therefore processed with very high precision. However, accuracy is required only for those relatively few points near surface maxima. Hence, there is considerable waste in performing high accuracy calculations at a vast majority of points.

SSDA reduces this redundancy by performing a sequential search which may be terminated before all M^2 windowing pairs for a particular reference point are tested. Furthermore, the algorithms do not implicitly contain any fixed error measure or measure evaluation method.

Constant Threshold Algorithm

A simple, but important SSDA has been extensively studied and will be used as a vehicle to introduce the concept. Here, a search over each of the $(L\text{-}M+1)^2$ reference points is performed, as in correlation. However, the criteria for "similarity" at each reference point is significantly different from correlation.

In the constant threshold algorithm, windowing pairs are selected for comparison in a random order so that, in general, a great deal of "new"

Table 1. The Cost of Normalized Correlation

Direct Correlation	FFT Correlation	Ordinary Normalization	"Fast" Normalization
$M^2(L-M+1)$–Mults.	$6L^2(\log_2 L)$–Complex Mult–Adds	L^2+M^2–Mults.	L^2+M^2–Mults.
$M^2(L-M+1)^2$–Adds	FFT of S,W and IFFT of Product	$M^2(L-M+1)^2$–Adds	$4(L-M+1)^2$–Adds
$(L-M+1)^2$–Squarings	L^2–Complex Mults.	$(L-M+1)^2$–Mults. and Divides	$(L-M+1)^2$–Mults. and Divides
	$(L-M+1)^2$–Squarings		

information is considered in each test; that is, a random nonrepeating sequence of the integers $1,2,\ldots,M^2$ is generated and used to yield the random, nonrepeating sequence of coordinates (ℓ_n,m_n), $n = 1,2,\ldots,M^2$. Thus the windowing pairs $\langle S_M^{i,j}(\ell_n,m_n), W(\ell_n,m_n)\rangle$ are compared in random order as n increases.

Non-normalized or normalized measures for evaluating the error between windowing pairs may be defined respectively as:

$$\epsilon'(i,j,\ell_n,m_n) \equiv |S_M^{i,j}(\ell_n,m_n) - W(\ell_n,m_n)| \qquad (10)$$

$$\epsilon(i,j,\ell_n,m_n) \equiv |S_M^{i,j}(\ell_n,m_n) - \hat{S}(i,j) - W(\ell_n,m_n) + \hat{W}| \qquad (11)$$

where:

$$\hat{W} \equiv \frac{1}{M^2}\sum_{\ell=1}^{M}\sum_{m=1}^{M} W(\ell,m) \qquad (12)$$

and

$$\hat{S}(i,j) \equiv \frac{1}{M^2}\sum_{\ell=1}^{M}\sum_{m=1}^{M} S_M^{i,j}(\ell,m) \qquad (13)$$

Unlike the correlation methods cited previously, in the ideal case, where $W = S_M^{i^*,j^*}$, a minimum of zero is guaranteed for the non-normalized case; that is, for

$$\| E(i,j) \| \equiv \sum_{\ell=1}^{M}\sum_{m=1}^{M} | S_M^{i,j}(\ell,m) - W(\ell,m) | \qquad (14)$$

$$0 = \| E(i^*,j^*) \| \leqslant \| E(i,j) \| \qquad (15)$$

Thus, in this ideal case, no normalization is necessary and obviously a comparison of very few points will yield the answer. The error measure, based upon the L_1 norm between two images, is also computationally simpler than the multiplicative measure of correlation.

In this SSDA implementation, a constant threshold T is introduced. As the error for randomly selected windowing pairs is accumulated, a test is made against T. When the accumulated error exceeds T at test N, operations cease for reference point (i,j) and the value N is recorded. The SSDA surface I(i,j) is therefore defined as:

$$I(i,j) = \left\{ r \,\Big|\, \min_{1 \leqslant r \leqslant M^2} \left\{ \sum_{n=1}^{r} \epsilon(i,j,\ell,m_n) \geqslant T \right\} \right\} \qquad (16)$$

Reference points where I(i,j) is large—reference points which require many windowing pair tests to exceed T—are considered points of similarity.

It is clear that if a suitable value for T is selected, many fewer than M^2 tests will be required for reference points which rapidly accumulate error. It is this property which significantly reduces the computational complexity of an SSDA. This fact is made clear by Figure 9. Curves A, B and C depict the cumulative error for three different reference points as a function of test. A and B accumulate error rapidly and operations for their reference points terminate early with I(i,j) obtaining values 7 and 11, respectively. Curve C, however, accumulates error more slowly. It is, therefore, much more likely to be a candidate for registration, and will accordingly have a value of 44 assigned to I(i,j).

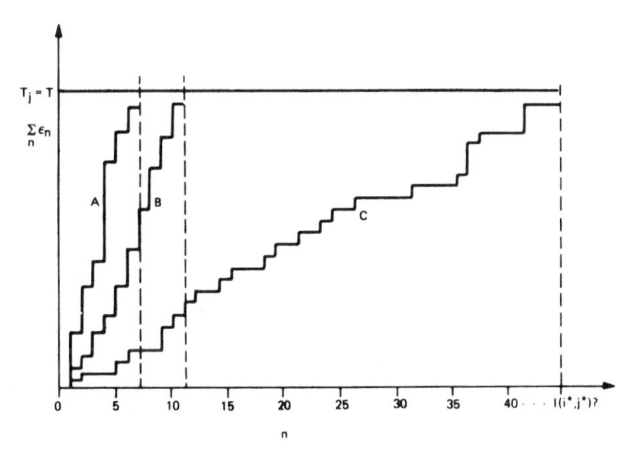

Figure 9. Growth Curves for Algorithm A with T = Constant

Geometric Correction Concept

The theory underlying the geometric correction techniques described here has been detailed in a variety of documents.[9,10,12,13] A summary of this theory is shown as Figure 10.

The sampled and digitized RBV image (the geometrically distorted "input image") is considered to be a uniform, two-dimensional array of

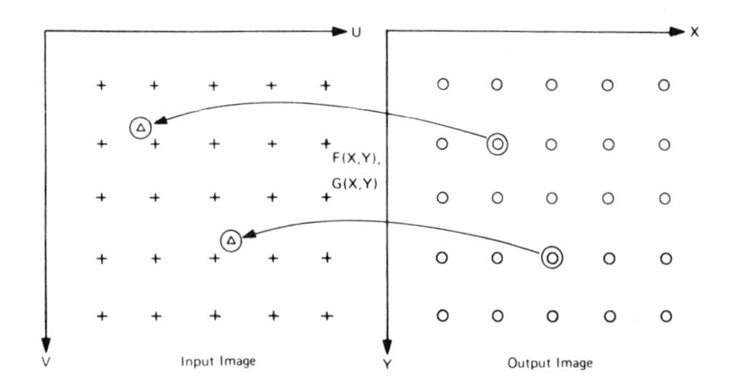

Figure 10. Geometric Correction Concepts

picture elements (pixels) each of which has a specific gray level. Since the array is uniform (that is, in a rectangular coordinate frame with axes U and V, the input pixels lie only at integer values of U and V), only the gray levels are stored in the computer; that is, no coordinate data need be stored. The (U, V) coordinates of each input pixel are the column and row indices of its gray level in the input data array.

In order to take advantage of both this "implied position" property of data defined on a uniform array and the accuracy of film recording devices (for example, laser beam and drum recorders) which record uniform rasters of image points, the geometrically correct output image is also defined as a uniform, two-dimensional array of pixels.

A pair of global, bivariate "mapping polynomials" of the form

$$U = F(X, Y) = \sum_{p=0}^{N} \sum_{q=0}^{N-p} a_{pq} X^p Y^q$$

$$V = G(X, Y) = \sum_{p=0}^{N} \sum_{q=0}^{N-p} b_{pq} X^p Y^q \qquad (17)$$

are used to determine the (U, V) input image coordinates of the pixel at coordinates (X, Y) in the output image. These polynomials account for the low-frequency sensor-associated distortions (centering, size, skew, pincushion, barrel, and S-term), as well as for distortions caused by earth curvature and camera attitude and altitude deviations.

Initial values for the polynomial coefficients are obtained by performing a least-squared-error fit to the vector differences between the observed and nominal reseau locations. A least-squared-error fit of standard photogrammetric resection equations to the vector differences between the observed and nominal geodetic control point locations is used to determine the attitude and altitude of the camera. This information is used to modify the polynomial coefficients to account for the effects of earth curvature and camera attitude and altitude.

These global mapping polynomials will not account for distortions which are high-frequency and/or local in nature. It is anticipated that there will be no such errors in the sensors. The only high-frequency external error source is terrain relief, and it has been estimated that this will cause no significant error in the general ERTS case.[8] Therefore, use of techniques which correct only low-frequency errors appears justified.

As Figure 10 shows, the (U, V) coordinates of an output pixel mapped onto the input image plane do not generally coincide with the coordinates of any input pixel [that is, the values of U = F(X, Y) and V = G(X, Y) are not generally integers]. The mapped output pixel generally lies somewhere within a square defined by the four surrounding input pixels, and some form of interpolation on the gray levels of the input pixels must be used to determine a gray level for the output pixel. Therefore, the general geometric correction procedure involves mapping each pixel of the output image into the input image plane and interpolating on the gray levels of the surrounding input pixels to determine the output gray level.

Point Shift Algorithm

The point shift algorithm is a mapping procedure that is based on the recognition that a point mapped on the input image has a location which is at most one-half pixel spacing removed in each axis from some input image pixel location. Therefore, in all cases where the error budget allows a one-half pixel location error in this phase of image processing, it is possible to assign to each pixel in the output image the gray level of the nearest pixel in the input image. The consequence of this extends beyond the mere resolution of the interpolation problem for the gray level assignment. It simplifies the entire mapping procedure, and as a result significantly decreases the execution time of the geometric correction process.

The procedure is illustrated in Figure 11, where the large rectangle subdivided into square blocks represents a portion of an input image. Each square block constitutes the half-pixel neighborhood of the input pixel location marked with an "x" at its center. The slanting line represents the mildly curved map of some horizontal line in the output plane. The points marked "0" on this line are the mapped locations of the points that make up the horizontal line in the output image. The short arrows indicate the "nearest neighbor" assignment of levels. The assignment rule is that if the input pixel with coordinates (m, n) (that is, m^{th} row and n^{th} column) is the pixel closest to the actual mapped location of the $(i, j)^{th}$ output pixel on the input image, the gray level $g_{m,n}$ is assigned to the $(i, j)^{th}$ output pixel.

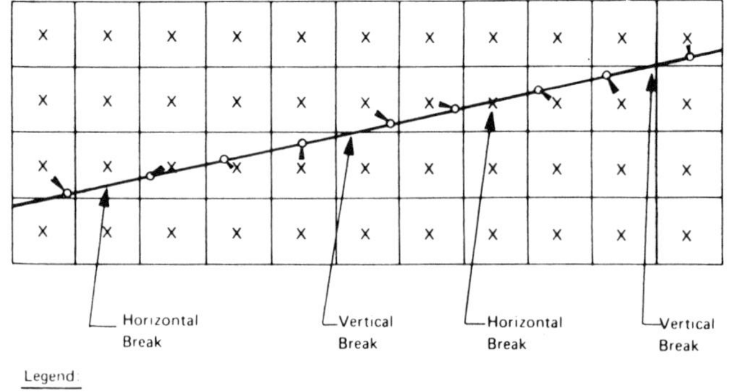

Figure 11. Nearest Neighbor Assignment

Nearest neighbor assignment simplifies the entire mapping problem because the expected total geometric distortion is small. Thus, if a line segment is mapped from the output plane onto the input plane, the first pixel of the output line segment with coordinates (i, j) is assigned the gray level of the pixel (m, n) in the input image, $g_{i,j} = g_{m,n}$, and this relationship is maintained for a number of successive points on the line segment. The break in this pattern occurs for the (i, j + k) output pixel if it maps somewhere with a closest input image neighbor at the location.

(m, n + k ± 1), a horizontal break

or

(m ± 1, n + k), a vertical break

instead of mapping at (m, n + k).

Figure 11 illustrates both types of breaks. It is clear that if, on the average, the breaks occur P pixels apart, input gray levels can be transferred to the output image in strings on the average P pixels long. If use is made of instructions which manipulate strings of data as single units (for example, the Move Character instruction of IBM System/360 and System/370), the computer operations required to transfer the input gray levels to the output array will be reduced by a factor of P relative to the operations required to transfer the data one pixel at a time.

In the implementation of the point shift algorithm, an irregularly spaced, rectangular grid of lines is established on the output image plane (as depicted in Figure 12). The intersections of the grid lines define a lattice of "anchor points" in the output image. The separation (horizontal and vertical) of the anchor points is chosen such that, when they are mapped from the output to the input image and connected through straight lines into a distorted grid, these lines at no place deviate more than a small fraction of the interpixel spacing (for instance, 0.1) from the curved lines that would have been obtained if every point on the output grid had been mapped.

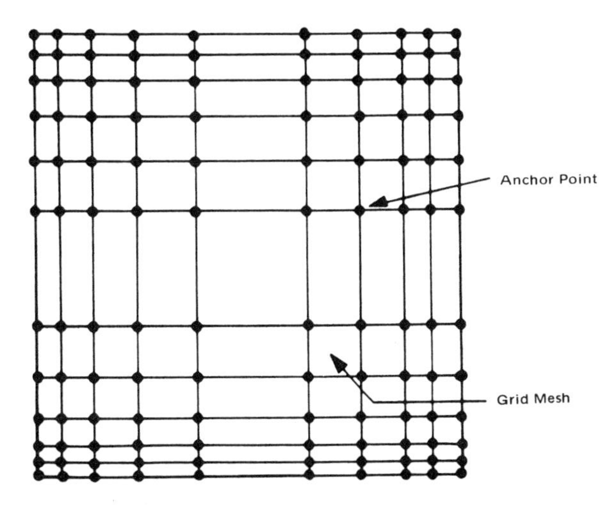

Figure 12. Typical Anchor Point Grid

As the geometric error on an RBV image increases with distance from the center of the image, the grid resulting from computation of anchor points (that is, grid intersections) is one with a large mesh at the center of the image and increasingly smaller meshes moving towards the periphery. The advantage of this is that in comparison to a uniformly spaced grid, the nonuniform grid requires fewer meshes for the same allowable error. In one example based on typical ERTS RBV errors, a variable grid set for a maximum of 0.1 pixel interpolation error required only a 23 x 23 mesh (that is, required mapping of 24 x 24 = 576 anchor points). A uniformly spaced grid for the same error would have required a 50 x 50 mesh (that is, 51 x 51 = 2601 anchor points).

The purpose of mapping anchor points and, hence, the grid they represent from the output image onto the input image in the manner described is that the four anchor points that represent a single grid mesh are sufficient to determine, through geometric interpolation, the location of every point in the mesh. Therefore, by this means, the problem of the mapping (through equation 17) of 1.7×10^7 pixels of the RBV has been reduced to the precise mapping of a few hundred points (namely 576 in the example cited).

To complete the description of the algorithm, it is sufficient to describe the operation on the boundary and inside a single mesh, since the procedure repeats for each grid mesh in the image.

The operation inside and on the boundary of a mesh consists of the following computations:

a. With reference to Figure 13, the location of every pixel on the leading vertical edge of the input grid mesh is computed by interpolation between the coordinates of the pixels a and c,

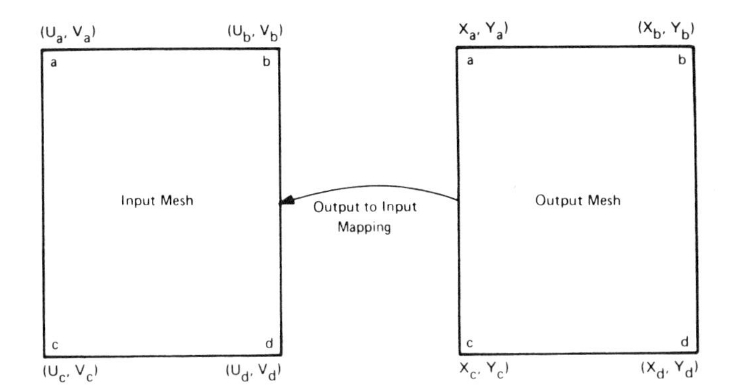

Figure 13. Typical Grid Meshes

(U_a,V_a) and (U_c,V_c), respectively. When the entire image is considered, for the cited example this amounts to 24 x 4096 interpolations.

b. On the leading vertical edge of each mesh the partials

$$A = \frac{\partial[F(X,Y) - X]}{\partial X} = \frac{\partial F}{\partial X} - 1$$

and,

$$B = \frac{\partial G}{\partial X}$$

are computed by interpolation between their values at the points a and c (Figure 13), respectively. The values of A and B are assumed to be constant on horizontal lines of the mesh.

c. For each horizontal line in the output image, the break points are computed, one line at a time, in the input image. To do this, note that the partials A and B indicate the relative motion of the pixels mapped from output into input with respect to the pixel locations of the input image. Thus, for example, in the case of horizontal spacings, $A > 0$ indicates that the horizontal spacing between two consecutive mapped pixels, ΔU, is larger than the regular spacing of unity between the input pixel locations. Continued mapping of additional pixels would indicate the mapped locations gaining on the input pixel locations. Still as an example, if $A = 0.1$, it means that the spacing ΔU between two consecutive mapped pixels is

$$\Delta U = 1 + 0.1 - 1.1.$$

Therefore, after $\frac{1}{A} = \frac{1}{0.1} = 10$ pixels, the total distance covered by the mapped pixels is 10 x 1.1 = 11 units instead of 10 units, and the nearest neighbor relationship for this string of pixels has broken down. In this case, while the 10th mapped pixel derives its gray level from the 10th pixel of the input image, the 11th mapped pixel is assigned the gray level of the 12th input pixel. The 11th input pixel is skipped in this case.

If the value of the partial A is negative, the mapped pixels migrate leftward relative to the pixel locations in the input image. In this case, at a break point the last input pixel gray level has to be used twice. For vertical break points, which involve the crossing of horizontal lines of data of the input image, the same reasoning applies. The partials B are used to compute the break points in this case.

d. For strings of pixels terminated by horizontal and vertical break points, video values are assigned according to the nearest neighbor rule. Once the next break point on a line has been determined as P pixels ahead, computer instructions (for example, the Move Character instruction) can be used to move the P pixel string of data points to the output image in one operation.

To transfer gray level data from the input to the output array at high speed, the data should be resident in computer memory. Since a single output image line will map across a swath of input image lines, the entire swath of input data must be available to the correction process. For the ±4 percent combined worst case errors of ERTS RBV images, this could require more than 1.3 megabytes of data storage. However, if the output image is processed one line at a time, the terms of the mapping polynomials which are constant across an output image line can be evaluated and the problem can be reduced to that of accommodating ±1 percent random error, which requires less than 340 kilobytes of data storage.

Radiometric (Intensity) Correction

High speed radiometric corrections for nonuniform sensor sensitivity can be implemented by table look-up techniques. In Figure 14, it can be seen that a unique correct sensor response can be determined from and related to the actual response. In general, the digitized image data will be in the form of "bytes" n bits in length. (For the ERTS RBV and MSS sensors, the bytes will be six bits long; for future sensors they will be longer). Since a byte n bits long can represent at most 2^n different data values, any relationship between incorrect input data and correct output data can be expressed in a table of 2^n values. This principle is illustrated in Figure 15, which shows the input-to-output conversion tables resulting from four different "correction functions" for the case where n = 3.

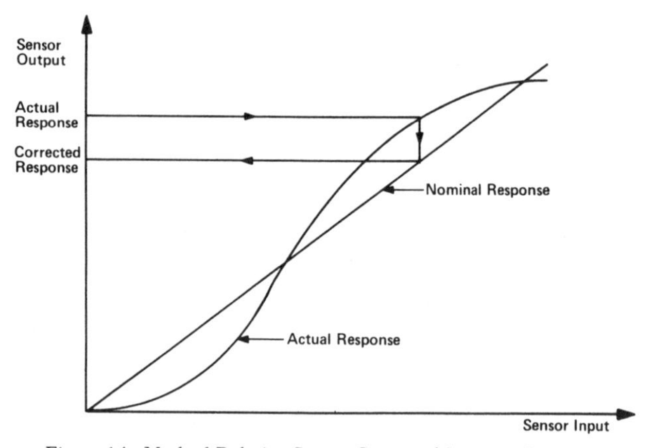

Figure 14. Method Relating Sensor Corrected Intensity Response to Actual Response

CORRECTION FUNCTION # 1 (bias correction)

OUTPUT = INPUT −2

CORRECTION FUNCTION # 2 (scale correction)

OUTPUT = 1.5 x INPUT

CORRECTION FUNCTION # 3 (scale & bias correction)

OUTPUT = 2 x (INPUT −3)

CORRECTION FUNCTION # 4 (video stretching)

OUTPUT = 0 IF INPUT ⩽ 1

OUTPUT = 1.5 x (INPUT −1) IF 1 < INPUT < 6

OUTPUT = 7 IF 6 ⩽ INPUT

INPUT VALUES	OUTPUT VALUES			
	Function # 1	Function # 2	Function # 3	Function # 4
000	000	000	000	000
001	000	001	000	000
010	000	011	000	001
011	001	100	000	011
100	010	110	010	100
101	011	111	100	110
110	100	111	110	111
111	101	111	111	111

Figure 15. Sample Radiometric Correction Tables for 3-Bit Data

The table look-up method is independent of the complexity of the input-to-output conversion function that is to be applied to the input image. Whatever the given correction function, the correction table can be generated by applying the function to each of the 2^n possible input values. Each input data value can then be used as an index to extract the proper output value from the correction table, thereby eliminating the need for further evaluation of the correction function itself.

For scanning sensors such as the ERTS MSS, one correction table for each detector would be used. These tables could, if required, be updated from scan to scan. For imaging sensors such as the ERTS RBV, if required, the image could be divided into subimage areas and a different correction table generated for each area. In either case, the inner loop of the table look-up radiometric correction routine can be implemented with a set of four instructions.[9]

V. Experimental Results

Various experiments have been conducted to evaluate the techniques described in this paper. The results obtained to date are explained in the paragraphs which follow in this subsection.

Ground Control Point Correlation

Data from the NOAA weather satellite ITOS-1 has been used to test the constant threshold algorithm. Figure 16 is an example of a typical data set. Images a and b are segments of uncorrected vidicon output taken over Baja California on August 9 and 11, 1970, respectively. In this application, the interest is in registering land masses. Therefore, in addition to the noise due to optics and scanners, there is a great deal of intense noise due to clouds and cloud shadow, as well as from the fiducial mark (the x) in picture 2. Figure 16c shows the window function W taken from the upper left corner of picture 2. In d, the answer is displayed. The window of 16c is put into the search area (16a) at the position indicated by the algorithm.

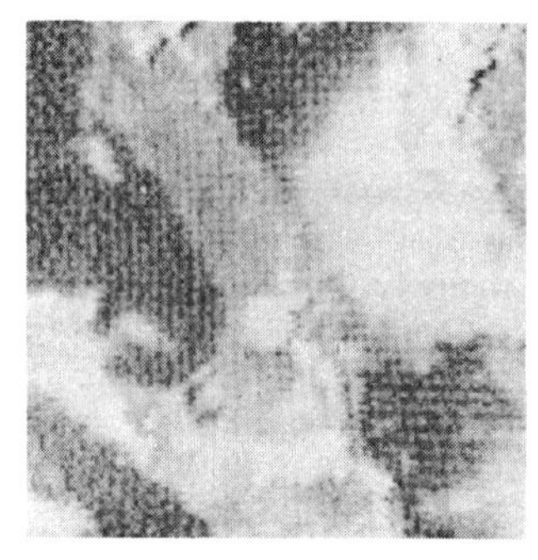

a. Search Area

b. Picture Number 2

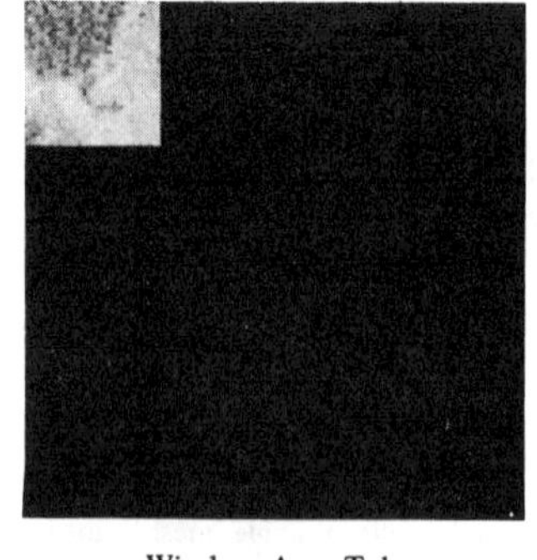

c. Window Area Taken from Number 2

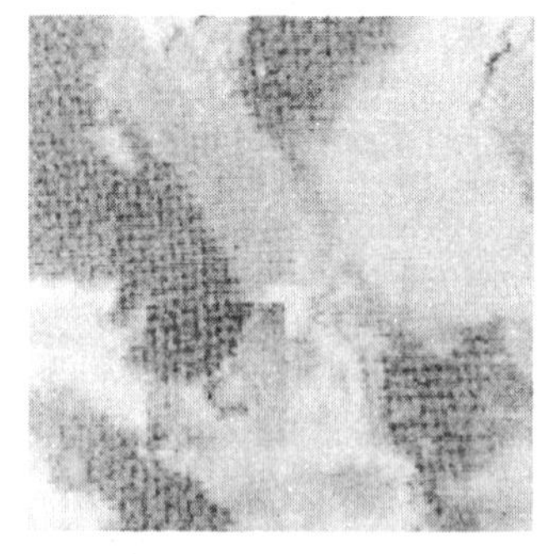

d. Window inserted into Search Area at (i*, j*)

Figure 16. An Example of Registered Images Taken From NOAA Data

The fit is visually accurate, but as real, rather than contrived, data was tested, there is virtually no way to absolutely check this fit. However, the same data was run in an FFT correlation-registration scheme, for the purposes of a relative comparison. As expected, there were some complications involved in the application of this more traditional method. Only after clouds had been detected, and their bright surface values replaced by random noise would correlation yield a meaningful peak. (No corrections whatsoever were necessary for SSDA.) For this data, correlation gave the registration point as one picture element higher than did this SSDA. As correct registration might easily lie between two points, this difference is not significant.

The data in Figure 16 has the following attributes:

a. $L = 128$
b. $M = 32$
c. $0 \leq S(i, j) \leq 225$
d. $E[S(i, j)] \approx 75$ (Expected Value)

For this data set, tests were made for various values of T. The important results from these tests are given in Figure 17. The average value of a point in the surface I(i, j)–and hence the number of calculations required–grows linearly with threshold. The maximum value in the surface I(i, j)–an indicator of the "accuracy" of the method–also grows linearly with threshold value, although at a somewhat higher rate. The ratio of these two items can be considered as a cost/performance measure. For the case of a constant threshold, this measure is shown to be relatively constant. This implies that "accuracy" may be achieved only by an associated increase in computation.

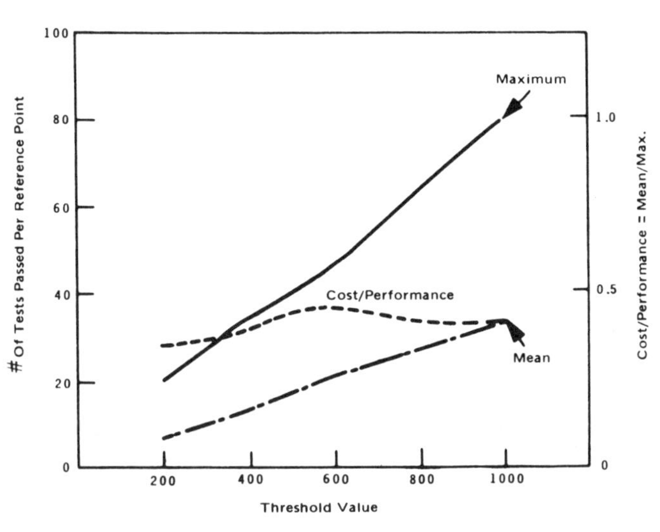

Figure 17. Statistical Results for Constant Threshold Algorithm

Monotonic Increasing Threshold Sequence Algorithm

The growth curve for a particular reference point, (three of which are shown in Figure 9), is a monotonically increasing function. It is the average slope of the growth curve that is important in the determination of a threshold crossing. It therefore seems reasonable that the replacement of the constant threshold T by a threshold, T_j, increasing monotonically with test, would improve performance when the following criteria are considered:

a. The sequence T_j should have "shape" approximately that of the growth curve for I(i*, j*), but should bound this growth curve from above for arbitrarily large n.
b. The T_j sequence should have initial values high enough so that a trend might be established even for reference points far from registration.

An example of a monotonically increasing threshold sequence is shown in Figure 18. Growth curves A and B, for reference points far from registration, are eliminated earlier than in 9, at 4 and 7, respectively. As most reference points do exhibit rapid growth, the total number of tests will be diminished significantly. Growth curve C, however, which appears to be a strong candidate for registration, will undergo a larger number of tests than before. Therefore a high degree of accuracy will be

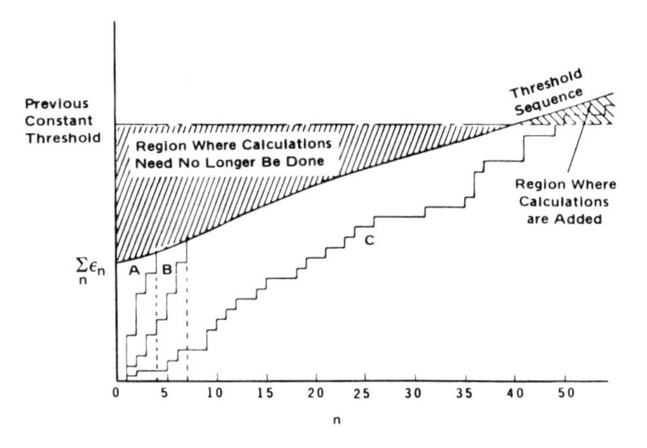

Figure 18. Growth Curve for Algorithm Q with T_j a Monotonically Increasing Function

achieved for those few points which have growth curves with low average slope.

A complete discussion of some design criteria is available.[20],[21] In general, a much more efficient and accurate algorithm results by the application of monotonic increasing threshold sequences. This is illustrated in Figure 19. One should note the presence of a minimum in cost/performance for these algorithms.

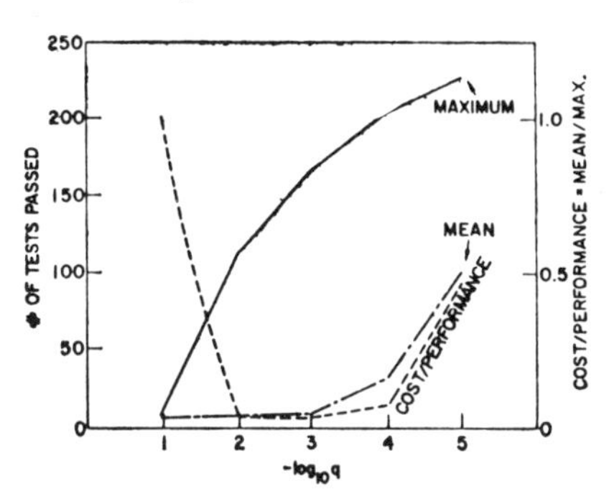

Figure 19. Statistical Results for Constant q Algorithm

Computational Aspects

The efficiency of SSDA has been demonstrated in the given examples. The saving which results can be illustrated by relating numbers of operations required for various implementations. Time ratios relating arithmetic operations for the IBM System/360 Model 65 will be used, namely:

a. $\frac{\text{Real Multiply Time}}{\text{Real Add Time}} = 3$

b. $\frac{\text{Real Add Time}}{\text{Integer Add Time}} = 2$

c. $\frac{\text{Integer Multiply Time}}{\text{Integer Add Time}} = 3.5$

d. Compare Time = Integer Add Time

e. Complex Add = 2 Read Adds

f. Complex Multiply = 3 Real Multiplies + 5 Real Adds

From Table 2, the direct cross correlation method, not including normalization, reduces to:

$$\#\,(\text{Direct}) = 4.5M^2\,(L\text{-}M+1)^2 \text{ equivalent integer-adds,} \tag{18}$$

and the FFT cross correlation method, not including normalization, reduces to

$$\#\,(\text{FFT}) = 200\,L^2 \log_2 L \text{ equivalent integer-adds.} \tag{19}$$

Table 2. Number of Equivalent Integer Adds for Various Algorithms for Several Values of L and M.

L	M	Direct Method $4.5M^2 (L-M+1)^2$	FFT Correlation $200\ L^2 \log_2 L$	Algorithm A $4\ (1+10\,(M/32)^{1/2})\,(L-M+1)^2$
128	32	4.4×10^7	2.25×10^7	4.2×10^5
256	32	2.57×10^8	1×10^8	2.2×10^6
512	32	1.1×10^9	4.6×10^8*	1.05×10^7
1024	32	4.5×10^9	2×10^9*	4.35×10^7
2048	32	1.85×10^{10}	8.8×10^9*	1.75×10^8
128	64	8.15×10^7	2.25×10^7	2.5×10^5
256	64	6.9×10^8	1×10^8	2.2×10^6
512	64	3.7×10^9	4.6×10^8*	1.2×10^7
1024	64	1.7×10^{10}	2×10^9*	5.5×10^7
2048	64	7.4×10^{10}	8.8×10^9*	2.4×10^8
256	128	1.15×10^9	1×10^8	1.37×10^6
512	128	1.1×10^{10}	4.6×10^8*	1.25×10^7
1024	128	5.8×10^{10}	2×10^9*	6.7×10^7
2048	128	2.5×10^{11}	8.8×10^9*	2.9×10^8
512	256	2×10^{10}	4.6×10^8*	7.5×10^6
1024	256	1.8×10^{11}	2×10^9*	7×10^7
2048	256	1×10^{12}	8.8×10^9*	3.7×10^8
1024	512	2.7×10^{11}	2×10^9*	4.1×10^7
2048	512	2.6×10^{12}	8.8×10^9*	4×10^8
2048	1024	4.5×10^{12}	8.8×10^9*	5.7×10^8

*FFTs not possible on most machines without much disc accessing.

The Monotone Increasing Threshold SSDA, using the normalized distance measure, requires:

a. 4 adds/reference point to obtain normalization
b. 2 adds + 1 compare/windowing point to get the error measure
c. 1 compare/windowing point
d. An average of n windowing tests per reference point.

Thus,

$$\#\,(\text{SSDA}) = (4 + 4n)\,(L\text{-}M+1)^2 \text{ equivalent integer-adds.} \tag{20}$$

For M of 32, n should be about 10-15. However, as M increases, more points should be checked to establish conclusively that misregistration has occurred. A figure of $n = 10(M/32)^{1/2}$ is reasonable. Thus,

$$\#\,(\text{SSDA}) = 4(1 + 10(M/32)^{1/2})\,(L\text{-}M+1)^2 \text{ equivalent integer-adds.} \tag{21}$$

Table 2 lists corresponding calculated costs for the several methods for various value of M and L. Values which are a power of two are selected, so that best figures for the FFT method might be achieved. Note, however, that no other algorithm requires L and M to be a power of two. (It is recognized that this condition might be relaxed somewhat for the FFT, but almost all available routines maintain this requirement.)

Table 2 shows that the given example of an SSDA is about a factor of 50 faster than the FFT correlation method. (This is a quite conservative estimate.) The order of magnitude saving by use of SSDAs is very important to production digital image processing, as this now allows digital ground control point correction to be practically implemented in general purpose digital computers.

The General Concept of SSDA

A structure for the class of SSDAs may now be stated in a general form so that the versatility of the method is clear. There are many implementations other than those of the examples, each of which contains the following four basic elements:

a. An Ordering Algorithm 0_1: 0_1 orders the $(L\text{-}M+1)^2$ reference points. The window is successively compared to subimages which are selected by 0_1. This ordering need not be fixed in advance, and may vary in a way that depends on events occurring during the execution. In these cases, the total number of reference points which are processed is less than $(L\text{-}M+1)^2$.
b. An Ordering Algorithm 0_2: 0_2 orders the M^2 (or fewer) windowing pairs to be compared at each reference point. 0_2 may be fixed—the same for all reference points—or it may adapt in a data dependent fashion.
c. A Distance Measure (norm) $\|x\|$: $\|x\|$ is used as a measure of error when windowing pairs are compared.
d. A Sequential Measuring Algorithm Q: Q is a mapping from a subset of the M^2 possible distance measures, for a particular reference point (i, j), into an element in the inspection surface I(i, j).

I(i, j) is directly analogous to the correlation surface described in Section IV. Q operates upon distances $\|x\|$ for windowing pairs in the sequential manner specified by 0_2. This sequence of operations continues until an event (like the passing of a threshold) occurs. At this instant I(i, j) is evaluated on the basis of the measurements taken.

Each of the four basic properties may be tuned to fit a specific similarity detection problem so that a proper balance between accuracy and efficiency is realized. There are many variations which are far different from the examples presented. However, each implementation displays the basic structure cited above.

Geometric Correction and Reseau Mark Detection Results

If the nearest neighbor assignment technique intrinsic to the point shift algorithm is applied to an image of a grid of lines whose spacing is some small number of pixels, the horizontal and vertical break points will produce visible staircase or herringbone patterns in the processed image. We believed that the characteristics of the ERTS RBV images would be such as to exhibit no such objectionable cosmetic effects after a correction process using nearest neighbor assignment. In order to test this hypothesis, to gain accurate information on the execution time of the point shift algorithm, and to investigate the efficacy of the shadow casting reseau detection technique, the techniques described above were experimentally reduced to practice and quantitative results were obtained.[12]

Experimental Method

The steps of this experiment are shown in Figure 20. Two simulated ERTS RBV images were generated by scanning and digitizing two Gemini photographs on an IBM drum scanner/recorder. A 2-mil square spot was used in scanning and digitizing the 7.5-inch x 7.5-inch images, resulting in 3,750 lines of 3,750 samples for each image. Samples were quantized to 6 bits, giving 64 distinct gray levels. In order to more closely approximate the RBV image size, a uniform border was added to the data to expand each "image" to 4,096 lines of 4,096 samples.

Fifth-order mapping polynomials representative of worst case RBV errors (1 percent each for centering, skew, size, pincushion, and S-term; 0.2 percent for keystone) were assumed. An APL program was used to compute the locations of the anchor point grid lines, using the technique described. This resulted in a 25 x 25 mesh (as opposed to the 50 x 50 mesh that would have resulted if a regular grid were used). The mapping polynomials were used to compute the positions into which 81 reseau marks were inserted in the input image data. These marks were positioned so as to appear as a regular 9 x 9 array in the output image.

Experimental Results – Image Processing

A program implementing the point shift algorithm was written and executed on an IBM System/360 Model 65. The total CPU time required for the geometric correction process was 80 seconds for each image, and 450 kilobytes of core memory were used. The processed images were recorded on computer tape and were then recorded on film by an IBM drum scanner/recorder. The input and output versions of both experimental images are shown in Figure 21. No cosmetic defects are visible in the processed images.

In a separate operation, the shadow casting reseau detection algorithm was applied to both experimental images. In the first image, 71 of the 81 marks were detected and located (that is, the 4 pixels of each arm width were detected unambiguously for both the horizontal and vertical arms). Of the 81 marks in the second image, 73 were detected. There were no cases of false identification in either image.

Aerial Photo Correction Results

Under contract to the U.S. Army Engineer Topographic Laboratory (ETL), an orthophoto was generated from an aerial photo using the point shift algorithm.[14] The original aerial photo had aircraft attitude, scale, and terrain relief errors (approximately 6 percent total geometric distortion). The ETL aerial photograph was digitized and recorded using the IBM Drum Scanner/Recorder. A central image area of 8.192 inches was scanned with a 2 mil aperture and a 6-bit quantization level. A total of 4096 lines (records) of data were scanned, each line sampled 4096 times providing a digital image of 16,777,216 picture elements or bytes of data. This data was recorded on one reel of 9-track digital tape at 1600 bits per inch.

Displacement data for a total of 71 ground control points was provided by ETL and was used to define fifth order global mapping functions that characterized the photo distortions in a least-squared error sense.

The results of the geometric correction experiment are shown in Figure 22. Image 1 is the digitized aerial photo and Image 2 the geometrically corrected image. The mapping function was also applied to a line grid test image and the results of the mapping operation are shown in Image 3. Obviously, severe distortion is present in the lower left regions of Images 2 and 3. The reason for this distortion can be seen in Image 4.

As mentioned above, the global mapping functions were developed by performing a least-squared error fit to the control point displacement data

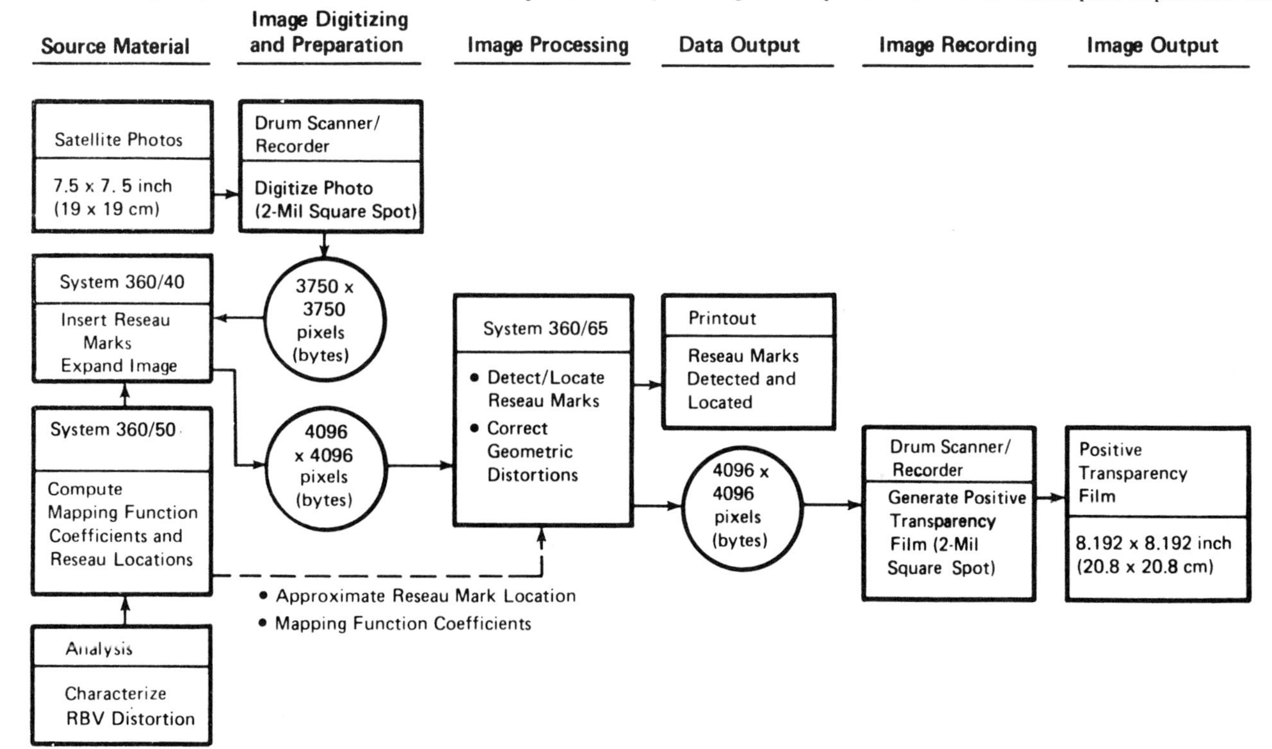

Figure 20. Summary of Experiment Steps and Equipment Used

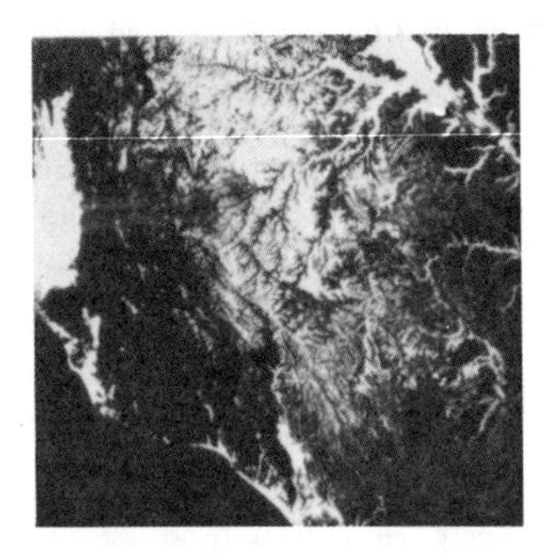
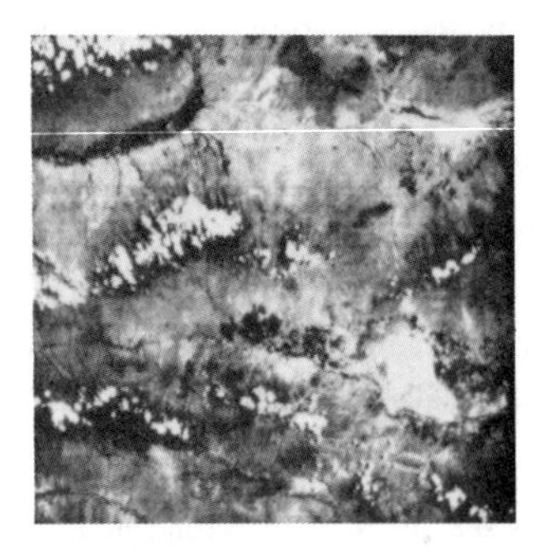

Scanned and Digitized Images

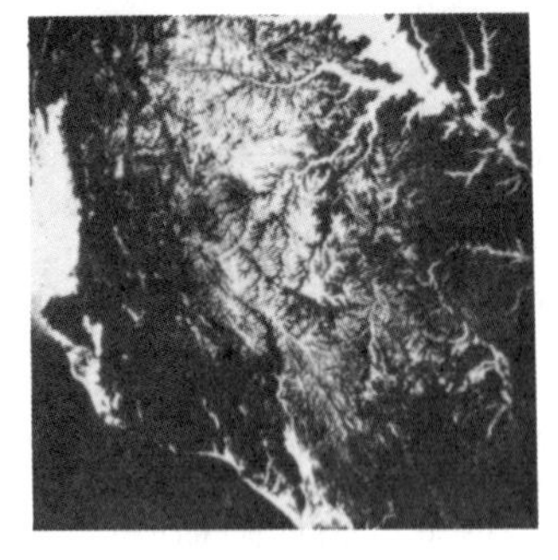
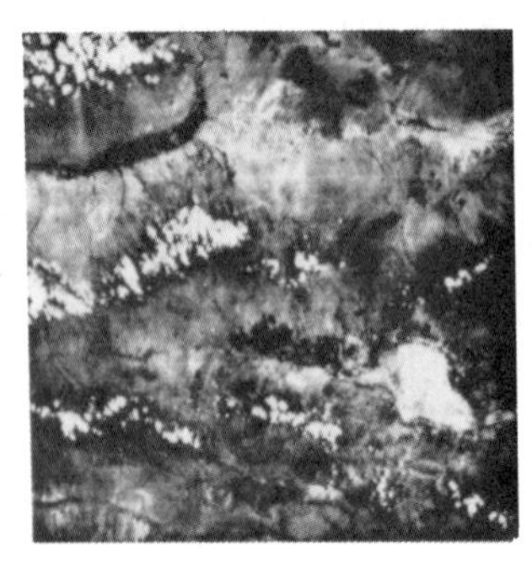

Images After Processing

Figure 21. Simulated ERTS RBV Images

supplied by ETL. Such a fit is valid only for regions spanned by the fitted data. As Image 4 shows, the control points used were almost all in the right half of the image. In the region of control point data, the original photo was corrected with a mean error of 3.6 pixels (7.2 mils or 0.185 mm) which is within map standards for the image scale. In the region not covered by control point data, the fit was uncontrolled and severe distortion resulted.

Note that the "tearing" apparent in the lower left regions of Images 2 and 3 of Figure 22 is not a consequence of the point shift algorithm, but resulted from the inability of the particular program used to deal with distortions of the magnitude of those dictated by the mapping functions in the uncontrolled region.

VI. Conclusions

These experiments show that digital techniques are a viable candidate for correction of high resolution imagery. The processed images show none of the cosmetic defects which may result from the use of nearest neighbor assignment.

The CPU time of 80 seconds per image on a System/360 Model 65 is much lower than any previous estimate. No attempt was made to achieve efficient I/O operation in these experiments, but it has been estimated that a 4096 x 4096 image can be read from or recorded on standard 800-bits-per-inch computer tape in 135 seconds. Since this I/O completely overlaps the processing, a single RBV image can be corrected in approximately five minutes. If 1600-bits-per-inch tape is used, the total correction time for each image decreases to 1.7 minutes.

The reseau detection technique discussed here is totally adequate for support of the correction of RBV images. In each of the two images tested, more than 70 of the 81 reseau marks were found and no false detections were made. Since the mapping polynomials used require the detection of a minimum of 21 reseau marks in each image, this level of performance is more than sufficient.

A general class of sequential algorithms for similarity detection has been introduced, referencing the specific problem of translational image registration. Experimental and analytic results have been presented to show orders of magnitude improvement in efficiency. Several ideas for further improvement of experimentally implemented algorithms have also been presented.

The structure of the new algorithms is ideally suited for digital similarity detections. There is a time saving of at least a factor of 50 for typical problems on a representative medium size computer—a prediction substantiated by experiment.

VII. Acknowledgements

The work reported in this paper has been supported by many individuals. The technical analysis provided by H. Markarian and D.G. Ferneyhough and the programming by L. Gregg are particularly noteworthy. The basic concepts of reseau detection and point shift image processing were suggested by P.M. Will, R. Bakis, and M.A. Wesley of the IBM T.J. Watson Research Center. Many fruitful discussions were held with H. Kobayashi, J. Mommens, D. Grossman, and P. Franaszek relating to the image registration work. Support and encouragement was provided by S. Shapiro and H.F. Branning.

VIII. References

1. Park, A.B., "NASA Flight Programs" Seventh International Symposium on Remote Sensing of the Environment, University of Michigan, May 17-21, 1971.

2. Jaffe, L., and Summers, R.A., "The Earth Resources Survey Program Jells," Astronautics and Aeronautics, April 1971.

3. George, T.A., "ERTS A and B – The Engineering System" Astronautics and Aeronautics, April 1971.

4. National Aeronautics and Space Administration, Design Study Specifications for the Earth Resources Technology Satellite ERTS A and B, document No. S-701-P-3, National Aeronautics and Space Administration, Goddard Space Flight Center, Greenbelt, Maryland, released April 1969, revised October 1969.

5 National Aeronautics and Space Administration, Earth Observatory Satellite (EOS) Definition Phase Report, Vol. 1, NASA, GSFC, August 1971.

6. McEwen, R.B., "Photogrammetric Evaluation for Use of RBV Images from ERTS," ASP-ACSM Fall Convention, Sept. 7-11, 1971.

7. Kratky, V., "Precision Processing of ERTS Imagery," ASP-ACSM Fall Convention, Sept. 7-11, 1971.

8. Colovocoresses, Alden P., "ERTS-A Satellite Imagery," Photogrammetric Engineering, Vol. XXXVI, No. 6, June 1970, pp 555-560.

9. Will, P.M., Bakis, R., Wesley, M.A., International Business Machines Corporation, ERTS Program: Final Report, On an All-Digital Approach to Image Processing for ERTS, IBM Thomas J. Watson Research Center, Yorktown Heights, New York, March 6, 1970.

10. Will, P.M., Bakis, R., Wesley, M.A., Bernstein, R., Markarian, H., International Business Machines Corporation, "Digital Image Processing for the Earth Resources Technology Satellite Data," ASP Meeting, Washington, D.C., March 7-12, 1971.

11. TRW Systems Group, Earth Resources Technology Satellite Final Report, Volume 2: ERTS System Studies, April 17, 1970.

12. Bernstein, R., Ferneyhough, D.G., Gregg, L., Higley, R., Markarian, H., Miklos, J., Mooney, P.P., Sharp, F., Experimental ERTS Image Processing, IBM Report CESC-70-0465, May 18, 1970.

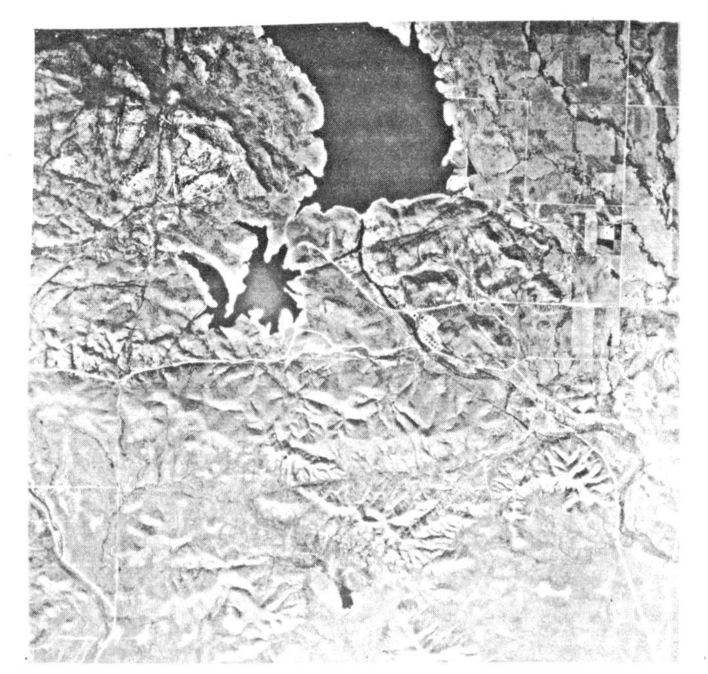

Image 1 Digitized Aerial Photo

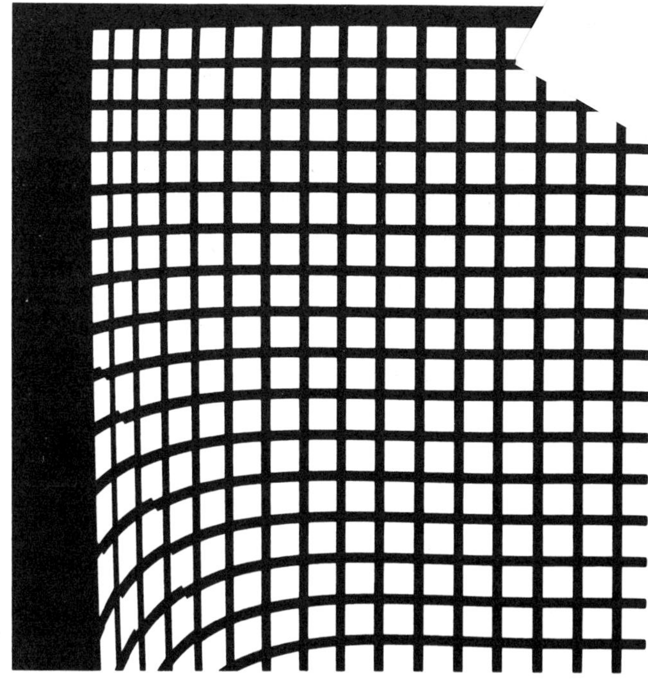

Image 3 Mapping Function Applied to Test Grid

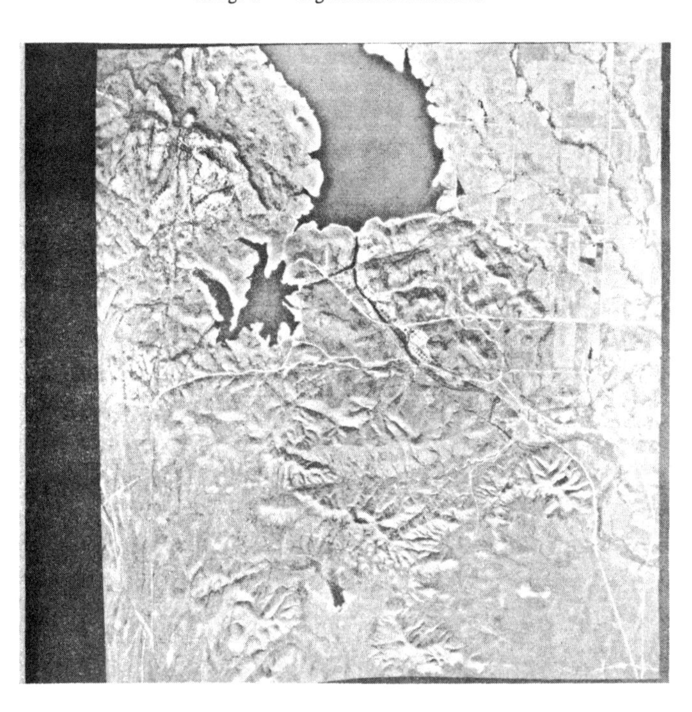

Image 2 Geometrically Corrected Aerial Photo

Note: Left portion of photo did not contain ground control points

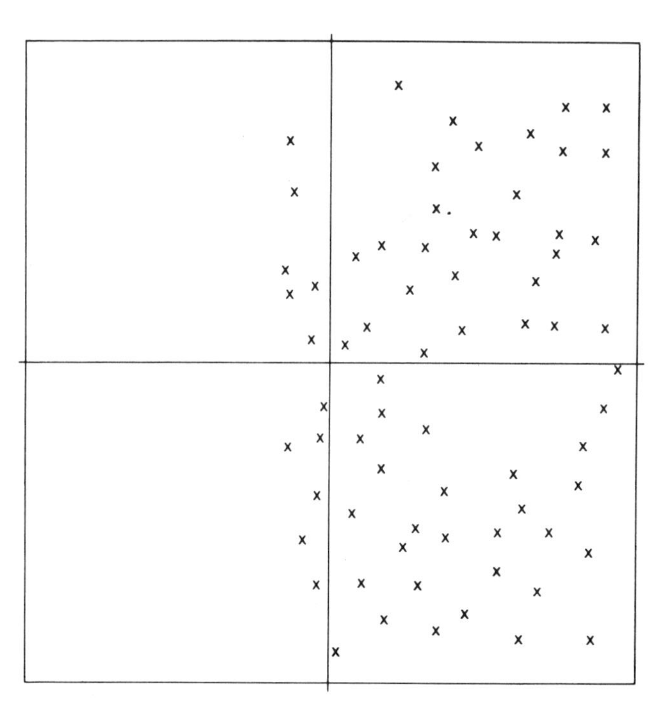

Image 4 Control Point Locations

Figure 22. Orthophoto Generation Using Digital Image Processing

13. Markarian, H., Bernstein, R., Ferneyhough, D.G., Gregg, L.E., Sharp, F.S., "Implementation of Digital Techniques for Correcting High Resolution Images," IBM FSC 71-6012. Presented at ASP-ACSM Fall Convention, Sept. 7-11, 1971.

14. Bernstein, R., Branning, H.F., Ferneyhough, D.G., Photo Geometric Correction, Final Report to U.S. Army Engineer Topographic Laboratory, Ft. Belvoir, Virginia, 22060. IBM FSC 71-0063, 2 March 1971.

15. Anuta, Paul E., "Spatial Registration of Multispectral and Multitemporal Digital Imagery Using Fast Fourier Transform Techniques," IEEE Transactions on Geoscience Electronics, Volume GE-8, No. 4, October 1970, pp. 353-368.

16 Leese, John A., Novak, Charles S., and Clark, Bruce B., "An Automated Technique for Obtaining Cloud Motion from Geosynchronous Satellite Data Using Cross-Correlation," Journal of Applied Meteorology, Vol. 10, No. 1, February 1971, pp. 110-132.

17. Papoulis, A., Probability, Random Variables and Stochastic Processes, McGraw-Hill, New York, 1965.

18. Goodman, J.W., Introduction to Fourier Optics, McGraw-Hill, San Francisco, 1968.

19. Cooley, J.W., Lewis, P.A.W., and Welch, P.D., "Application of Fast Fourier Transform to Computation of Fourier Integrals, Fourier Series, and Convolution Integrals," IEEE Transactions on Audio and Electroacoustics, Vol. AU-15, No. 2, June 1967, pp. 79-84.

20. Barnea, D.I., and Silverman, H.F., "The Class of Sequential Similarity Detection Algorithms (SSDAs) for Fast Digital Image Registration," IBM Research Report RC-3356, May 10, 1971.

21. Barnea, D. I. and Silverman, H. F., "A Class of Algorithms for Fast Digital Image Registration, IEEE Transactions on Computers, Feb. 1972.

Part III
Image Registration

Organized by ~~Ralph Bernstein, Editor~~

Paul E. Anuta Assoc. Editor

Spatial Registration of Multispectral and Multitemporal Digital Imagery Using Fast Fourier Transform Techniques

PAUL E. ANUTA, MEMBER, IEEE

Abstract—**A system for spatial registration of digitized multispectral and multitemporal imagery is described. Multispectral imagery can be obtained from sources such as multilens cameras, multichannel optical-mechanical line scanners, or multiple vidicon systems which employ filters or other spectral separation techniques to sense selected portions of the spectrum. Spatial registration is required so that multidimensional analysis can be performed on contextually similar image elements from different wavelength bands and at different times. The general registration problem is discussed first; then the fast Fourier transform (FFT) technique for cross correlation of misregistered imagery to determine spatial distances is discussed in detail. A method of achieving translational, rotational, and scaling corrections between images is described. Results of correlation analysis of multispectral scanner imagery and digitized satellite photography is presented. Use of the system for registration of multispectral airborne line-scanner imagery and space photography is described. Application of the techniques to preprocessing of earth resources satellite imagery from systems such as the earth-resources technology satellite (ERTS) scanner and vidicon system is discussed in conclusion.**

I. INTRODUCTION

REMOTE-SENSING technology is rapidly expanding into new forms of measurement. The data analysis and interpretation task is being automated to cope with the huge quantities of data generated by these advances. Photo interpretation has advanced from human analysis of panchromatic photography through similar manual analysis of color and color infrared photography to automatic computer analysis of multiband imagery. Human interpretation of multispectral photography which contains the spatial dimension and two or more spectral bands is cumbersome due to the difficulty of simultaneously observing the characteristics of multiple representations of a scene. Color and color infrared film greatly increases the information content of a single picture by introducing the spectral dimension into a single pictorial representation; however, the human speed and judgement factor is still a limitation in this form of analysis. The magnitude of the analysis task and the limited range of spectral response of film have led to the development of automatic data collection and analysis systems which can sense and process large quantities of data from many measurement dimensions. The key requirement of an automatic multidimensional analysis system is the availability of a set of congruent measurements for each resolution element in the image. Multiple measurements from each image resolution element offer a means of improving the accuracy of recognition of the properties of the surface of the scene over that attainable using one dimension. Measurements of reflectance and radiance from microwave, thermal, and reflective infrared, through the visible wavelengths and into the ultraviolet region, can be utilized for analysis of each image point if congruence of these measurements can be achieved.

Remote-sensing measurement and analysis techniques are commonly classified into spectral, spatial, and temporal methods. Polarization measurement is often considered to be a fourth method. The spectral class was discussed previously. Use of the temporal or time-varying dimension also requires that image congruence be achieved so that automatic analysis can be carried out using this dimension. The time-varying properties of spectral and spatial features in a scene are included in the temporal dimension, and image registration is required in either case. Misregistration results from the inability of the sensing system to produce congruenced data due to design characteristics or the fact that the sensors are separated in space and time such that spatial alignment of the sensor is impractical or impossible. Geometric distortion, scale differences, and look angle effects can all combine to produce misregistration.

A digital system has been developed at the Laboratory for Applications of Remote Sensing (LARS), Purdue University West LaFayette, Ind., for spatial registration of multispectral-multitemporal imagery so that research into the usefulness of various measurement dimensions could be performed. The sources of multi-imagery considered by LARS are airborne multispectral optical-mechanical line scanners and multiband photography. However, the results of the work described are applicable to imagery from any source which produces multiple images of the same scene in different wavelength bands or at different times and in different wavelength bands. LARS is concerned with the development and application of remote-sensing technology to all aspects of earth-resources survey. Our work is

Manuscript received May 1, 1970. This work was supported by the U.S. Department of Agriculture under Contracts 12-14-100-9549 (20) and 12-14-100-10292 (20). This paper was presented at the 1970 IEEE International Geoscience Electronics Symposium, Washington, D. C., April 14–17.

The author is with the Laboratory for Applications of Remote Sensing, Purdue University, West Lafayette, Ind. 47906.

Reprinted from *IEEE Trans. Geosci. Electron.*, vol. GE-8, pp. 353–368, Oct. 1970.

concentrated on development of digital multispectral and multitemporal statistical pattern-recognition methods which can identify agricultural, geological, and other features accurately and rapidly [1]. The work of Steiner [2] is an example of the use of the time dimension for pattern recognition. The digital approach was chosen by LARS to enable maximum flexibility in algorithm-development and data-analysis activities. LARS in an interdisciplinary project which includes engineers and scientists from the fields of electrical engineering, computer science, agronomy, soil science, geology, civil engineering, forestry, natural resource conservation, and others. Creative interaction between persons in these diverse areas has proven to be most easily achieved through user-oriented computer programs and a high-speed general-purpose digital computer. This form of interface allows persons with little computer or engineering knowledge to utilize the algorithms and data produced by the engineering technologists with very little training of the users. Also, testing and evaluation of data-handling and analysis schemes is easily carried out through software development rather than iterative hardware design and implementation. The decision to study image registration using sequential digital techniques was based on the success experienced using similar techniques in other areas of data-handling and analysis research at LARS.

The general multi-image registration problem is discussed first; then correlation techniques and problems are discussed. The use of the fast Fourier transform (FFT) for high-speed correlation of image arrays is discussed in detail; then the results of correlation analysis of test imagery from agricultural areas are presented. The use of enhancement techniques to improve the correlation accuracy of relatively uncorrelated wavelength bands is discussed next, and a description of the current LARS imagery registration system is presented. Application of these techniques and results to registration of aircraft-scanner imagery and Apollo 9 multispectral photography is discussed in conclusion.

II. Imagery Registration Definitions

Multiple images of the same scene are in registration if contextually coincident resolution elements in different images are uniquely addressable by one coordinate pair and an image pointer. A digital image is a two-dimensional array of numbers which represents the characteristics of a real image at discrete points. Digital image registration considerations include the sampling rate, sample point geometry, quantization effects, and other factors, as well as the geometric alignment of multiple images. Fig. 1 depicts the registration requirement for a three-member digital image set. The shaded resolution element represents an image point from the same scene point, and the coordinates (i, j) locate that element in each image. Spatial misregistration of multiple images can take the form of translational, rotational, and scale differences between image pairs.

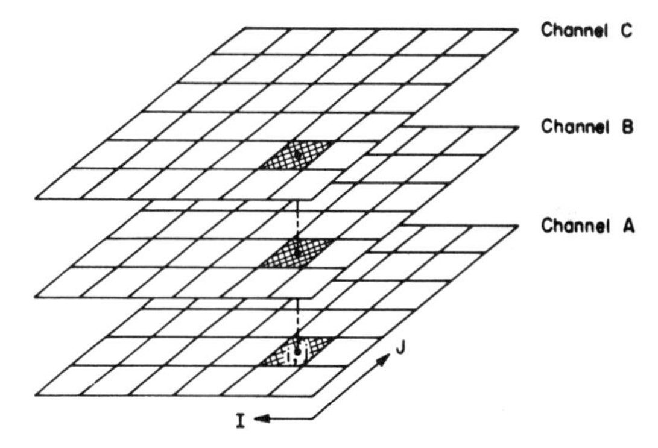

Fig. 1. Multi-image registration requirement. Resolution elements in registered multi-image are addressable by one coordinate pair and a channel pointer.

Multispectral and multitemporal images of the same scene can have markedly different characteristics. Although the context is the same in each of the images, the reflectance or emissivity of individual scene points may be totally different. Fig. 2 is an example of multispectral scanner imagery in 14 wavelength bands from 0.32 μm to 14 μm reproduced in image form on a computer line printer. The differences in spectral response can be seen by examining a particular scene area in all the bands. Multitemporal images may depict a scene which has undergone climatic or cultural changes such as seasonal changes, plowing, harvesting, or cloud-cover intrusion. A form of rotational misregistration in the multitemporal case can be seen in Fig. 3 which contains a computer printout of multispectral scanner imagery taken in June and August, 1969. The scanning aircraft flew with a large yaw angle in June, due to crosswinds, which caused the unstabilized scanner to produce the skewed imagery seen in the left photo. The August data from the same area is nearly geometrically correct. This data must be registered if it is to be used by a multitemporal pattern-recognition system. The general registration problem is thus one of determining the location of matching context points in multiple images and alteration of the geometric relationships of the images such that the registration of each context point is achieved.

The registration process is divided into three phases which aid in studying the problem: enhancement, correlation, and overlay. Enhancement refers to the preprocessing necessary to improve the accuracy of registration; correlation is the process of determining the location of matching context points; and overlay is the geometric transformation process which produces the registered imagery. A digital multi-image is defined as a set of $M \times N$-point digital images or pictures P^k, where $k=1, \cdots, K$,

$$P^k = \{x_{ij}^k\} \quad ,i = 1, \cdots, M; \; j = 1, \cdots, N.$$

One of the K multi-images q is selected as a reference or context image, and the other $K-1$ members of the set are registered with respect to P^q. The crux of the registration process is determining the spatial distance d_{ij}^k between each point in each of the $K-1$ images and the corresponding point in the context image. The distance

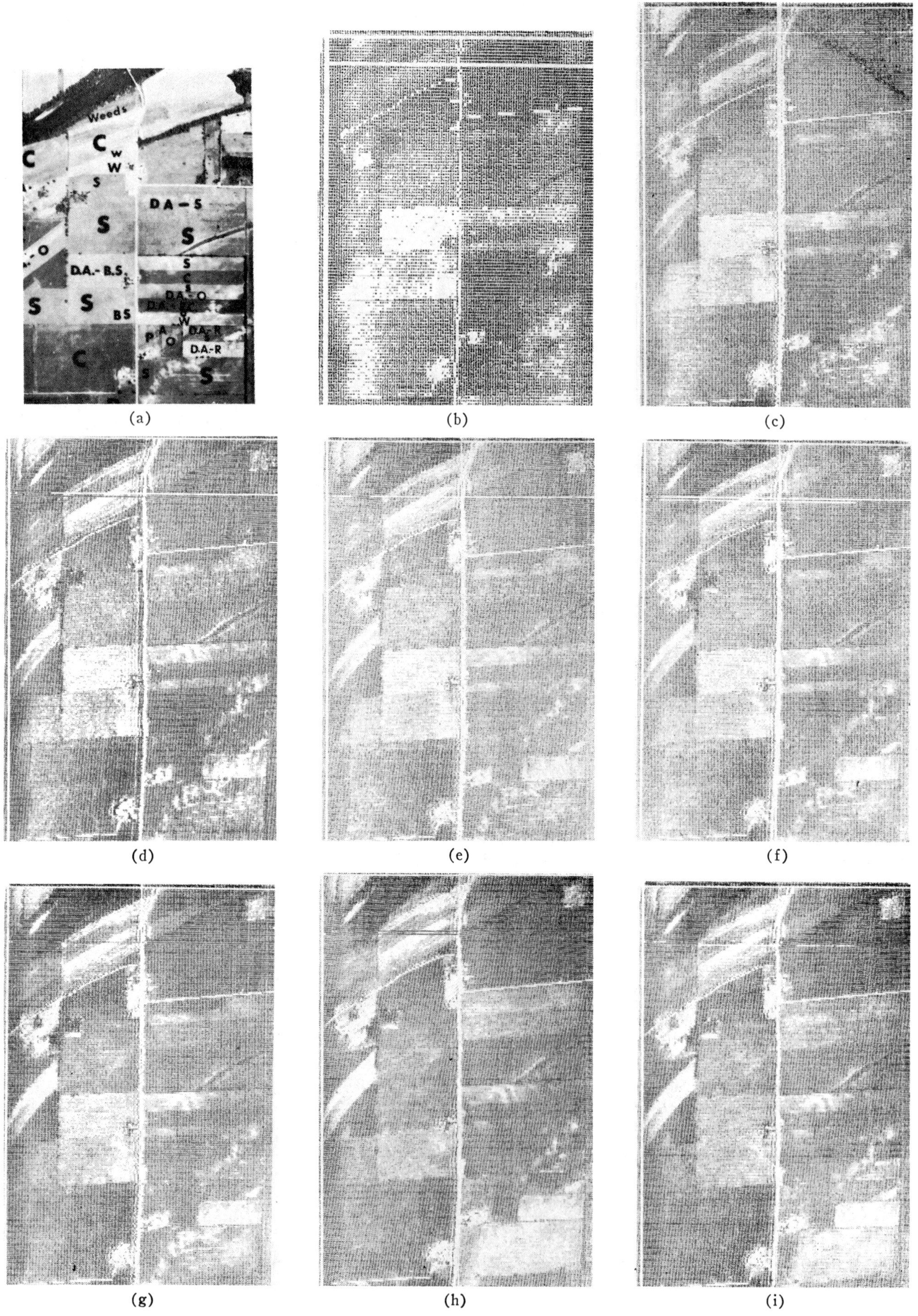

Fig. 2. Example of scanner imagery from 14 bands. (a) Aerial photo. (b) 0.32–0.38 μm. (c) 0.40–0.44 μm. (d) 0.44–0.46 μm. (e) 0.46–0.48 μm. (f) 0.48–0.50 μm. (g) 0.50–0.52 μm. (h) 0.52–0.55 μm. (i) 0.55–0.58 μm.

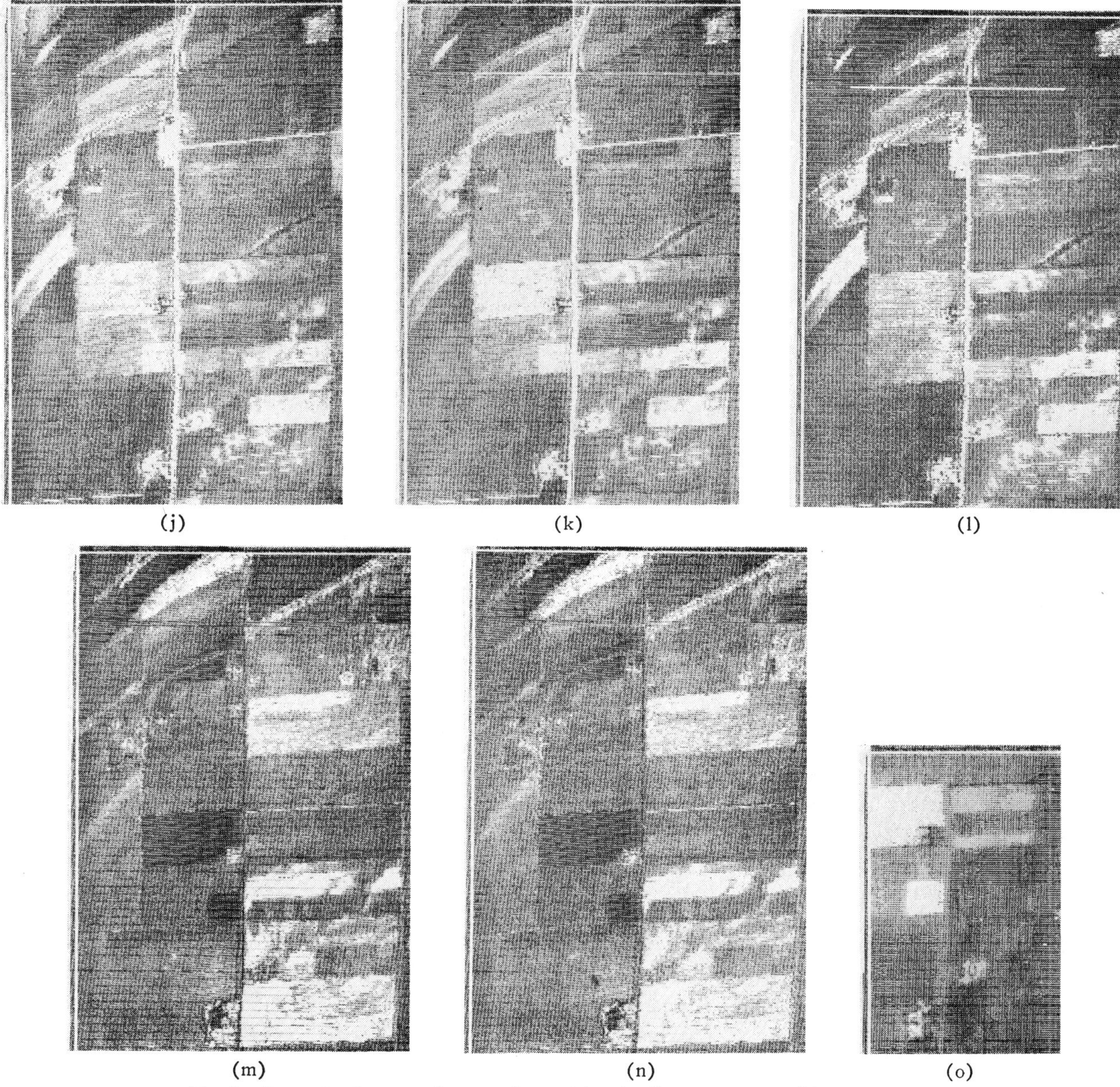

Fig. 2. Example of scanner imagery from 14 bands. (j) 0.58–0.62 μm. (k) 0.62–0.66 μm. (l) 0.66–0.72 μm. (m) 0.72–0.80 μm. (n) 0.80–1.0 μm. (o) 8–14 μm.

is defined in terms of translation, rotation, and scale differences. The general spatial distance for the image k-tuple on a point basis consists of three quantities.

1) Translational distance: $d_{T_{ij}}{}^k = (\Delta I_{ij}{}^k, \ \Delta J_{ij}{}^k)$ is an ordered pair of real numbers giving the translation on two orthogonal axes which point $x_{ij}{}^q$ would undergo to be aligned with the corresponding point in image P^k.

2) Rotational distance: $d_{R_{ij}}{}^k$ is the angular rotation a neighborhood in image P^k would have to under go to have the same angular orientation as the corresponding point in the context picture. Note that this definition is ambiguous for a single picture point and is assumed to refer to a small image neighborhood around the point $x_{ij}{}^k$.

3) Scale distance: $d_{s_{ij}}{}^k$ is the scale factor of the P^k picture with respect to P^q at point $p_{ij}{}^k$. It is an ordered pair giving the scale factor on two orthogonal axes. The scale factor is defined as the ratio of the incremental image distance represented by a set of points in P^q to the distance represented by the same size set in P^k.

An additional problem encountered in digital imagery is the situation in which a picture point may not exist at the same context point in all images due to differences in sampling rate and sample reference points. This condition requires either resampling of the original image or interpolation between available image points to obtain a sample at the desired position. Also, differences in the size, shape, and values of the image elements in different bands constitute a registration error which must be corrected before the multi-image can be considered to be in accurate registration.

III. Image Correlation

The core of the LARS registration system is an array correlator which computes the correlation coefficient of two image arrays for a set of juxtapositions in two dimensions. The correlation is carried out by numerical operations on a sequential digital computer. This approach is in contrast to the more common method using electrooptical techniques in which analog voltages from an optical sensor are correlated. Since a digital picture is composed of discrete brightness points each having a specific numerical value, the correlation function is represented by a double summation over the points being correlated. The two dimensional discrete correlation function [3] for two pictures X and Y is

$$\phi(D) = \phi(k, l) = \frac{\sum_{i=1}^{N} \sum_{j=1}^{N} x(i + k + s, j + l + s)\, y(i, j)}{\sum_{i=1}^{N} \sum_{j=1}^{N} x^2(i + k, j + l) \sum_{i=1}^{N} \sum_{j=1}^{N} y^2(i, j)} \tag{1}$$

where

x_{ij}, y_{ij} points from pictures X and Y corrected to have zero mean. Picture X is of size M by M, Y is N by N, $N < M$.

k, l shift variables, $k,\ l = -s, \cdots, s$, with $s = (M - N)/2$.

$D = (k, l)$ translational distance vector from the reference point.

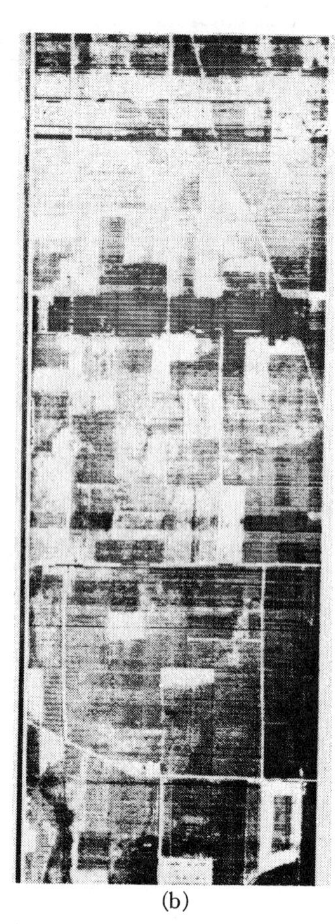

(a) (b)

Fig. 3. Example of rotational distortion in multispectral airborne scanner imagery. (a) June, 1969; 11.7° yaw distortion. (b) August, 1969; 0° yaw distortion. Flightline: PF24; altitude: 5000 feet; band: 0.66–0.72 μm.

This formula expresses the correlation on a scale from -1 to 1 where $\phi = 1$ indicates perfect positive correlation and $\phi = -1$ indicates perfect negative correlation. The expression assumes that the two pictures X and Y have the same scale, rotational alignment, and distortions. This is not the case in general, and these other degrees of freedom must be included in the correlation process. Rotational and scaling search processes must be carried out in addition to translational search if the angular and scale differences are severe. This costly search procedure can be simplified if initial checkpoints can be obtained before correlation. These checkpoints can be used to determine the preprocessing scale factor and rotation so that when correlation is carried out a reliable image match can be found. The scale factor and rotation are then updated from the translational correlation function results.

This approach was implemented at LARS early in the registration study. The search mode was used for each correlation operation, and for large search rectangles on the order of 10 to 20 points square the computation time was excessive. The track mode was unfeasible at the time since the correlation was not carried out on adjacent or nearby points and the variation of lock-on points over 100- to 200-image-point intervals tended to be large. Instead of improving the search-track method of correlation to improve processing speed, the use of the recently developed FFT for correlation was investigated.

The convolution theorem of Fourier analysis states that convolution in the time or space domain is equivalent to multiplication in the frequency domain. Since correlation is a form of convolution, an alternate method of computing the correlation function thus exists. The classical numerical integration method of computing the Fourier transform is as time consuming as direct evaluation, and the method alone offers no great advantage. With the publication of an algorithm [4] for computing the Fourier transform which is orders of magnitude faster than the conventional method, a fast means of computing the correlation function became available. The time required to compute an n-point transform using the FFT varies as $n \log n$ instead of approximately n^2 for conventional numerical integration evaluation. Therefore, the transform method was implemented in the imagery registration system.

Certain problems unique to the use of the finite Fourier transform exist, and the solutions to them are outlined here. Convolution in one dimension is expressed mathematically as

$$C(k) = \int_{-\infty}^{\infty} x(t) y(k - t) \quad dt = FT^{-1}[X(f) Y(f)] \tag{2}$$

where

$x, y(t)$ two time functions

$C(k)$ the convolution between two time functions shifted by k units with respect to each other

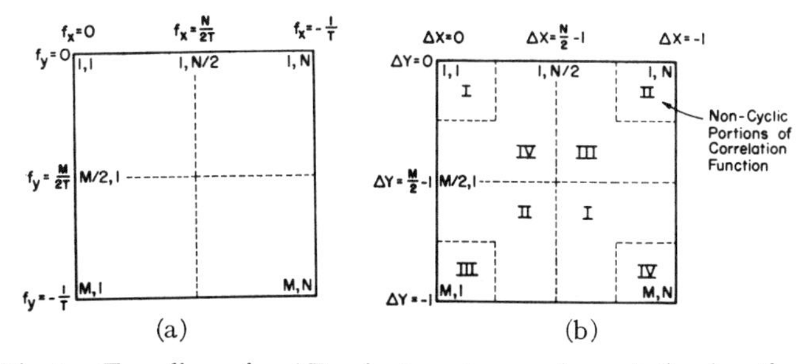

Fig. 4. Two-dimensional Fourier transform and correlation function structure. (a) Location of harmonic components of a 2-dimensional discrete Fourier transform. (b) Location of components of 2-dimensional correlation function computed by discrete Fourier transform.

X, $Y(f)$ the Fourier transforms of the two time functions; the transform is a function of the frequency variable f

FT^{-1} signifies the inverse Fourier transform operation.

Cross correlation is defined in the same way as convolution except that one of the two functions is not reversed but is simply shifted on its axis. The correlation function is expressed as

$$\phi(k) = \int_{-\infty}^{\infty} x(t)y(t-k) \quad dt = FT^{-1}[X(f)\,Y^*(f)] \quad (3)$$

where

$\phi(k)$ the correlation function of the shift variable k

$Y^*(f)$ the complex conjugate of the Fourier transform of $y(t)$.

Direct application of the preceding expression to computation of the correlation function of discrete data using the discrete Fourier transform leads to problems which may cause erroneous results. The Fourier transform algorithm computes the discrete N-term Fourier series of an N-point function. The location of the harmonic components and convolution values in the transform array is shown in Fig. 4. Inherent in the operation of the transform algorithm is the assumption that the function being transformed is periodic. The resulting N-point transform is also periodic. The result of the application of the preceding expression in the discrete case is a cyclical convolution function which is not the desired result. An example of the desired result compared to the cyclic result is shown in Fig. 5. This problem can be alleviated by increasing the size of the transformed array and including zero values for the range of shift desired. For $x(k)$ defined at M discrete points (M even) $k=0, 1, \cdots, M-1$ and $y(k)$ defined at $N<M$ discrete points (N even) centered in an array of size M, let $y(k)=0$ for $k=0, 1, \cdots, [(M-N)/2]-1$ and $k=(M+N)/2, \cdots, M-1$. Then executing the correlation process using the transform will give the correct result for a shift of $(N-M)/2$ points in each direction.

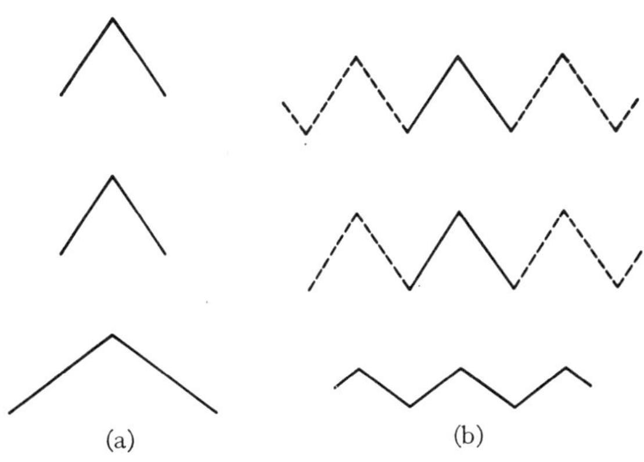

Fig. 5. Example of cyclical correlation problem. (a) Desired correlation function. (b) Cyclical correlation resulting from direct application of convolution theorem using the discrete Fourier transform.

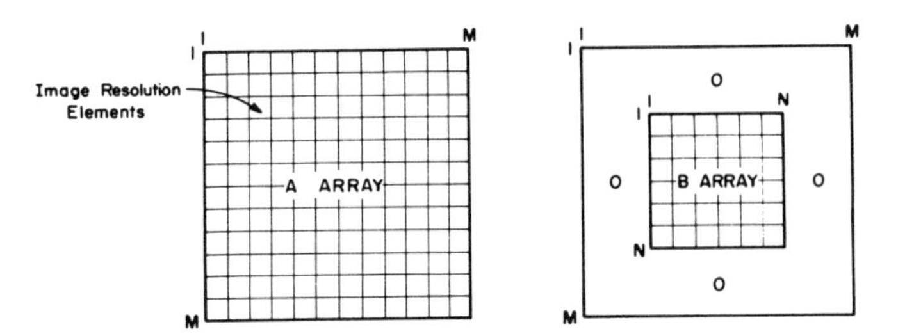

Fig. 6. Array structure for avoidance of cyclic correlation.

The correlation function

$$\phi(k) = \sum_{i=0}^{M-1} x(i)y(i+k), \quad k=0, \pm 1, \cdots, \pm \frac{M-N}{2} \quad (4)$$

is thus computed using the FFT by multiplying the M-point transforms of the x array and the y array constructed with $M-N$ 0's added as follows

$$\phi(k) = FT^{-1}[X(f)\,Y^*(f)], \quad k = 0, 1, \cdots, M-1. \quad (5)$$

The $k=0$ point is the correlation for no shift, $k=1$ is for a 1-point shift in the positive direction and so on up to $(M-N)/2$ points of shift in one direction. The $(M+N)/2$ point is the correlation for maximum shift in the opposite direction and the $k=M-1$ value is a 1-point shift in the negative direction. This split is due to the cyclic property of the transform, and behavior of this type must be accounted for in a system which uses the transform technique. The values of $\phi(k)$ for $k=[(M-N)/2]+1$ up to $k=[(M+N)/2]-1$ are invalid and are not used. They represent the correlation of y shifted such that the values of $y(i)$ for $i>k$ are wrapped around the end of x and are being correlated with $x(0)$, $x(1)$, etc., which is meaningless is most cases. This picture is changed if the 0's are included in the y array at different points. It can be stated in general that the only valid correlation function points are those with the same index values as the 0 points in the $y(i)$ function.

The cyclic convolution elimination problem for the two-dimensional case is solved in the same manner as for the one-dimensional case. The y points must be surrounded by 0's on four sides to fill it out to the size of the larger x array, as shown in Fig. 6. The valid correlation points are identified in the same manner as for the one-dimensional case except that the four quandrants of the valid correlation function are at the four corners of the total cyclical correlation function. Specifically, let the x array be of size M by M $[x(i, j); i=0, \cdots, M-1; j=0, 1, \cdots, N-1]$ and y' be N by N $[y'(i, j); i=0, \cdots, N-1; j=0, \cdots, M-1]$ with $N<M$ (M and N even). The N by N points are assumed to be in the center of an M by M-size array, and this array is padded out with 0's such that

$$y(i, j) = 0, \quad i, j = 0, 1, \cdots, \frac{M-N}{2} \quad \text{and} \quad i, j = \frac{M+N}{2}, \cdots, M-1$$

$$y(i, j) = y'(i-s, j-s), \quad i, j = \frac{M-N}{2}+1, \cdots, \frac{M+N}{2}-1; \quad s = \frac{M-N}{2} \tag{6}$$

where $y(i, j)$ is an M by M array.

This is the same basic format as for the one-dimensional case. Similarly, the valid correlation function points lie in the first $(M-N)/2$ points in each corner of the two-dimensional square result array.

In order to prevent data sets with large average values from "swamping" the correlation by effectively introducing a large square pulse into the data, the average value of the y data points is removed before padding in the 0's. The average value of x is also removed to minimize the magnitude of the correlation function. The root sum of the squares for the arrays is also computed and used to normalize the correlation function. A step-by-step account of the operations necessary for two dimensional correlation is now described.

1) Select the size in image points of the area to be covered by the correlation integral. Assuming it is square, let this value be N (N even). (The rectangular case is a trivial extension of the square case.) Next select the maximum shift of one array with respect to the other. Let this be Δ. The correlation function will then have $2\Delta+1$ points in each direction, plus and minus Δ and 1 for 0-shift.

2) Step 1) defines the necessary size for the base or x array. It is $N+2\Delta$ square, and this is called M. The average of the M by M x array is removed, and the average of the N by N y array is removed. The root sum of squares is computed for x and divided into each point of the array. The same is done for y. The y set is placed in the center of the M by M array which is padded out with 0's as defined previously.

3) The M by M Fourier transform of x and y is computed using the FFT algorithm HARM (SHARE program SDA 3425 or similar versions). The complex conjugate of the Y transform is taken; the X and Y transforms are multiplied and the inverse transform is a taken which produces the total M by M cyclical correlation function with the valid points at the corners of the array. Mathematically these operations are expressed as follows. The two-dimensional transformation is

$$X(f, g) = \frac{1}{M^2} \sum_{i=0}^{M-1} \sum_{j=0}^{M-1} x(i, j) W_M^{-fi} W_M^{-gj},$$

$$W_M = e^{2\pi\sqrt{-1}/M}. \tag{7}$$

$Y(f, g)$ is computed in the same manner.

Then

$$\phi(k, l) = \sum_{f=0}^{M-1} \sum_{g=0}^{M-1} [X(f, g) Y^*(f, g)] W_M^{kf} W_M^{lg} \tag{8}$$

where * indicates the complex conjugate.

4) The resulting total correlation function $\phi(k, l)$ is partitioned and the quadrants are interchanged as shown in the right of Fig. 6 to place the 0-shift point in the center of the $2\Delta+1$ by $2\Delta+1$ two-dimensional correlation function as follows

$$\begin{aligned}
\psi(k+\Delta, l+\Delta) &= \phi(k, l), \quad k, l = 0, 1, \cdots, \Delta; \quad \left(\Delta = \frac{M-N}{2}\right) \\
\psi(k+\Delta, l-M+\Delta) &= \phi(k, l), \quad k = 1, \cdots, \Delta; \quad l = M-\Delta-1, \cdots, M-1 \\
\psi(k-M+\Delta+1, l+\Delta) &= \phi(k, l), \quad k = M-\Delta-1, \cdots, M-1; \quad l = 0, \cdots, \Delta \\
\psi(k-M+\Delta+1, l-M+\Delta+1) &= \phi(k, l), \quad k, l = M-\Delta-1, \cdots, M-1.
\end{aligned} \tag{9}$$

The function ψ is a $2\Delta+1$ by $2\Delta+1$ array of correlation values with the 0-shift point in the center. This array is the output of the correlation routine. An example of a typical correlation function for highly correlated imagery is presented in Fig. 7.

A core storage saving scheme can be used when employing the FFT routine for transforming real data [5]. The transform algorithm is written to be able to process complex data; thus each input data point is a double computer word. To perform correlation, two separate arrays must be transformed, i.e., the x and y data sets. The total number of data storage words required for the M by M transform is thus $2 \cdot 2 \cdot M^2$. The core saving method is based on the fact that the real part of the Fourier transform of real data is even about the 0 frequency point and the imaginary part is odd. This fact is implicit in the following development for the one-dimensional case. We wish to compute X and $Y(k)$ from $x(j)$ and $y(j)$. This is expressed by the inverse transforms

$$x(j) = \sum_{k=0}^{M-1} X(k) W_M{}^{jk} \tag{10}$$

$$y(j) = \sum_{k=0}^{M-1} Y(k) W_M{}^{jk}, \quad W_M = e^{2\pi\sqrt{-1}/M}. \tag{11}$$

One real data set (x) is placed in the real part of the input data array and the other (y) is placed in the imaginary part so that a complex array is formed

$$\xi(j) = x(j) + iy(j). \tag{12}$$

The ξ array is then transformed to the complex frequency domain forming $Z(k)$

$$\xi(j) = \sum_{k=0}^{M-1} Z(k) W_M{}^{jk}. \tag{13}$$

To get the X and Y transform from the transform Z, the following development is used: multiply the y-transform expression (11) by $i = \sqrt{-1}$ and then add to and subtract it from the x-transform expression (10) which produces

$$x(j) \pm iy(j) = \sum_{k=0}^{M-1} (X(k) \pm iY(k)) W_M{}^{jk}. \tag{14}$$

The complex conjugate of (13) can be written in terms of an inverse transform by setting $k' = M - k$ as follows:

$$\xi^*(j) = \sum_{k=0}^{M-1} Z^*(k) W_M{}^{-jk} = \sum_{k'=0}^{M-1} Z^*(M - k') W_M{}^{jk'}. \tag{15}$$

Equating coefficients of (13) with those of (14) having the plus sign (+) gives

$$Z(k) = X(k) + iY(k) \tag{16}$$

and equating coefficients of (15) with those of (14) having a minus sign (−) gives

$$Z^*(M - k) = X(k) - iY(k). \tag{17}$$

Solving these two expressions for X and Y gives

$$X(k) = \tfrac{1}{2}(Z(k) + Z^*(M - k)) \tag{18}$$

$$Y(k) = \tfrac{1}{2}(Z(k) - Z^*(M - k)). \tag{19}$$

Thus the X and Y transforms can be resolved from the transform of the complex combination by applying the preceding expressions. This is implemented in the correlation program thereby cutting the core requirement for complex array storage in half. For the 32 by 32-point array being used this is a saving of $2 \cdot 32 \cdot 32 = 2048$ words or 8192 bytes of storage.

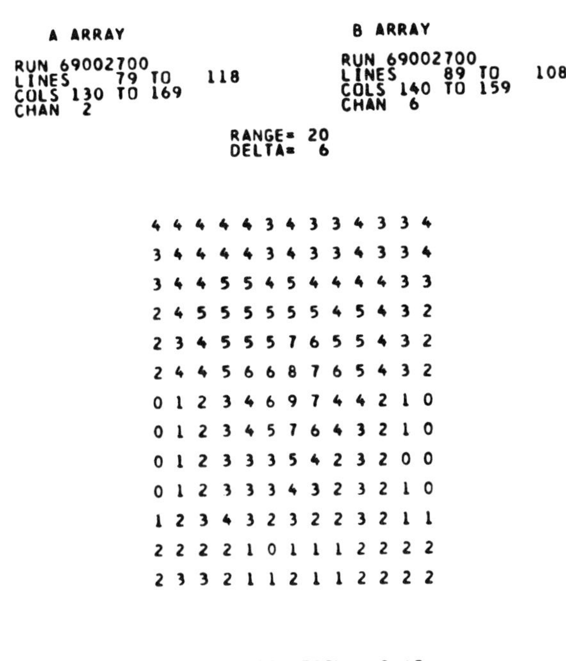

Fig. 7. Example of computer printout of 2-dimensional correlation function.

The FFT method described previously significantly reduces the time required to compute the correlation function compared to the numerical integration approach. Table I presents a comparison of the time required to compute the correlation function by the two methods. The numerical integration time refers to the conventional method of computing the correlation function. The FFT time shown is the time to compute the averages and sum of squares of the two data arrays, normalize, set up the complex array, take a forward and a reverse two-dimensional Fourier transform, unscramble the two transforms as required by the core saving method discussed previously, and to extract the valid correlation function points from the total correlation function. The time savings using the FFT averages about an order of magnitude and this savings has a great impact on the usefulness of digital registration methods.

IV. Enhancement Techniques

Imagery enhancement algorithms were experimented with in an attempt to improve registration accuracy and possibly in some cases to enable registration in areas where correlation of the unaltered imagery would be impossible. Two enhancement processes are presently under study. One implements the spatial gradient edge

TABLE I

TIME COMPARISON OF NUMERICAL INTEGRATION AND FFT METHODS OF COMPUTING CORRELATION FUNCTION

Correlation	Maximum Shift in Both Directions							
	±5 points		±10 points		±15 points		±20 points	
Area (points square)	Numerical Integration (seconds)	FFT (seconds)	Numerical Integration (seconds)	FFT (seconds)	Numerical Integration (seconds)	FFT (seconds)	Numerical Integration (seconds)	FFT (seconds)
4	1.5	0.70	5.4	3.2	11.7	13.9	20.4	13.6
8	5.7	3.4	20.4	3.2	44.5	13.9	77.8	13.6
12	12.6	3.4	45.6	3.2	99.2	13.9	173.3	13.6
16	22.4	3.4	80.8	14.4	175.6	13.9	306.9	13.6
20	34.9	3.4	125.9	14.4	273.8	13.9	478.4	13.6
24	50.1	15.1	181.1	15.5	393.7	13.9	688.1	13.6

enhancement operation which amplifies the borders of features in the imagery. The other uses clustering techniques to identify homogeneous regions in imagery. It defines borders as the edges of these regions. Both schemes are based on the assumption that borders will be invariant features in imagery from different wavelength bands, whereas the reflectance and emissivity of the material within the boundaries generally varies widely across the spectrum. Also, for imagery gathered at different times the nature of the surface features may vary; however, the borders of the features tend to remain the same.

The spatial gradient is a form of the two-dimensional first derivative of a surface [6]. For continuous functions

$$|\text{grad } F| = |\Delta F| = \left[\left(\frac{\partial F}{\partial x}\right)^2 + \left(\frac{\partial F}{\partial y}\right)^2\right]^{1/2}. \quad (20)$$

In the discrete domain the first derivative is analogous to the first difference and the gradient is

$$|\Delta F_{ij}| = [(F_{i+1,j} - F_{ij})^2 + (F_{i,j+1} - F_{ij})^2]^{1/2}. \quad (21)$$

A modified form of the gradient is used in the enhancement processor which computes the absolute value of the difference rather than the square. The results are similar, and computation time is saved. The multichannel gradient can also be computed by the processor. If several channels are known to be in registration the border representation can be summed to further improve the boundary estimation. The final gradient representation is thresholded to produce a binary border map which is correlated with the border map for the second image. An example of thresholded spatial gradient enhancement is shown in Fig. 8(b) for typical agricultural land. This form of enhancement is inherently "noisy" since the differentiation is in effect a high-pass filtering of the imagery. The method is also very sensitive to the border threshold used.

The clustering technique [7] is more stable but also more costly enhancement method. The technique can be applied to a single or multichannel image. An example of cluster-derived boundaries for a section of an agricultural test site is shown in Fig. 8(c). Comparison of the cluster-derived boundaries to the gradient bound-

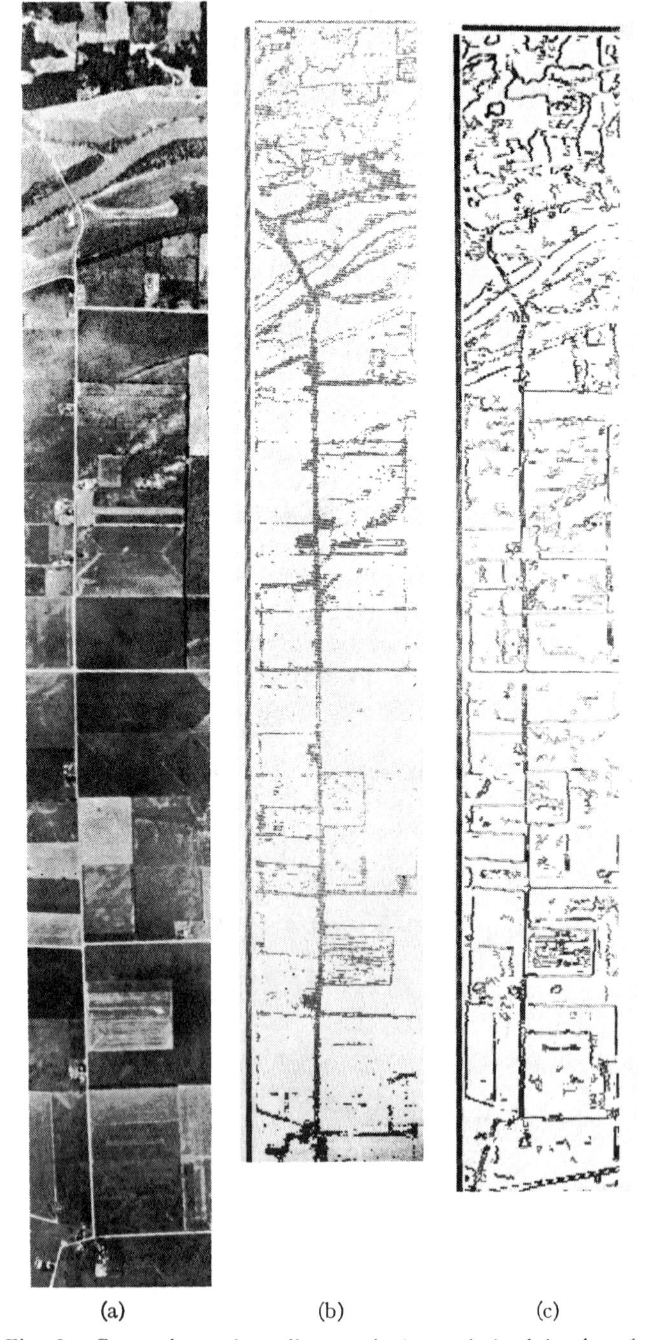

(a) (b) (c)

Fig. 8. Comparison of gradient and cluster-derived borders for agricultural land. (a) Aerial photo. (b) Gradient borders. (c) Cluster borders.

aries shows improvement in uniformity and detail using the clustering technique. Either method of enhancement can be executed for the arrays being correlated in the registration system. Some results of correlation of border enhanced imagery are presented in Section VI.

V. Registration System

The enhancement and correlation algorithms form the core of the imagery registration system. The overlay or image transformation function, control, and input-output functions complete the software system which operates via Disk Programming System 44 on an IBM System 360 Model 44 computer. A general block diagram of the system is presented in Fig. 9. The imagery to be registered is referred to in terms of apertures (derived from the concept of a segment or window of the electromagnetic spectrum). A set of one or more images in registration is called an aperture. There can be three input apertures denoted, A, B, and C. The A aperture is called the context or reference aperture. The B and C apertures are called overlay apertures and are correlated and shifted with respect to the context aperture. The segments of the system are discussed as follows.

1) Input monitor: this routine selects imagery lines from the appropriate input file and extracts the desired channels and samples from the line. A fully flexible input definition is implemented since the three apertures could exist in many combinations, i.e., context on tape A, overlay B on tape B, overlay C on tape C, or context on tape A, overlay B on tape A, overlay C on tape B, etc. This feature allows a context aperture to be used for a geometric reference, and all new imagery of the same area can be overlayed such that the same geometric dimensions are achieved for all the imagery processed. Two image sets (overlay B and C) can be overlayed with respect to the context image (A) in one pass.

2) Scaling and rotation: scale factors are applied to the imagery in two dimensions. The input monitor and scaling operation are interactive since the line required by the B or C overlay aperture depends on the scale factor for that aperture. Global rotation is implemented by reading the desired line segments from tape and placing them in the buffer in the required position.

3) Aperture buffers: three large cyclic or precessing buffers are used to store rectangular imagery arrays for correlation and overlay. This approach allows addressing of imagery lines up to several lines ahead of and behind the current reference line without tape input. Rectangular arrays for correlation can be extracted from any area in the span of available lines, and final rotation and scaling are implemented by addressing the appropriate image cell in the overlay aperture buffers.

4) Enhancement processor: the spatial gradient and cluster border-finding algorithm are presently implemented for enhancement. Also, an image complexity estimator is included with the gradient processor which

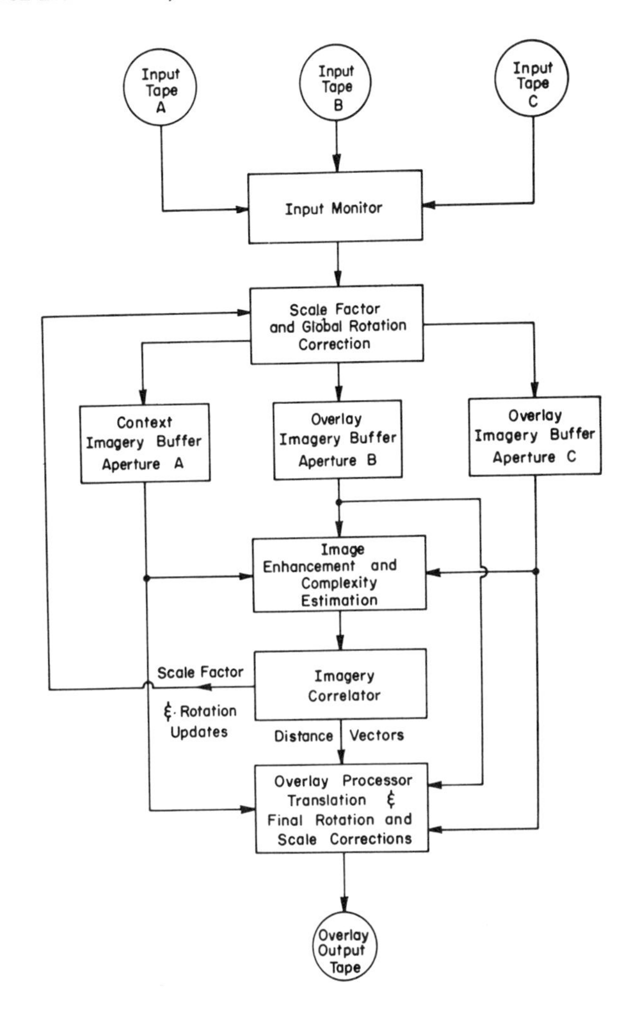

Fig. 9. Digital imagery registration system.

supplies an index of the correlatability of two image arrays [8]. If the index is too low or too high, correlation is not attempted since lock-on failure is highly likely in this case. The complexity index can also be used to vary the correlation integration range; however, the FFT correlation speed is not greatly improved as was the case for numerical integration method discussed in [8] since the total correlation time is largely input–output time. Other methods of enhancement are being studied such as frequency-domain filtering to achieve border enhancement, and work is continuing on improvement of the clustering algorithm for border finding.

5) Correlator: the FFT correlation process is implemented for square arrays up to 32 by 32 points. The correlation array size minus the integration range is twice the maximum shift in both dimensions. An integration range of 20 and an array size of 32 is most commonly used, and thus the maximum shift is ± 6 points. A numerical integration correlator is also implemented which can search over a larger range to find an initial lock-on point. The FFT correlator can then determine the point of maximum correlation for changes of up to 6 points between points of correlation. The point of maximum positive and the point of maximum negative

correlation is found, and the distance coordinates and euclidian distance are computed for each.

Correlation is carried out at the intersections of a grid as pictured in Fig. 10. The line and column interval between correlation points depends on the amount of variation in the misregistration between two apertures. In some cases, such as aircraft scanner data in which random distortion exists, correlation intervals of 25 lines and columns have been used. For scanned aerial and satellite photographs taken from the same position, the variation in misregistration is relatively small, and intervals of 100 points are used. Also, an averaging of misregistration estimations is achieved by closely spaced correlations as long as the computation time does not become excessive. The system can operate in an adaptive mode by employing either of two correlation estimators. The enhancement processor computes the average number of border points per unit area, and previous work has shown that if this estimator is above or below certain limits correlation is likely to fail. Secondly, if the absolute value of the maximum or minimum correlation is below a threshold it is also very likely that the lock-on point is badly in error. Thus the correlation output is discarded if the estimator indicates erroneous results are likely.

A correlation function interpolator is included which uses the value of the peak and its eight or 24 neighbors to estimate the true point of maximum or minimum correlation to better than one resolution element.

6) Overlay: the line and column registration errors from the correlator are used in a linear least-squares procedure which determines a two-dimensional shift function defined over the entire space between line and column correlation nodes shown in Fig. 10. The shift function is of the form

$$\begin{aligned} \Delta L &= a_{L1}L + a_{L2}C + a_{L3} \\ \Delta C &= a_{C1}L + a_{C2}C + a_{C3} \end{aligned} \tag{22}$$

where

- ΔL line shift value
- ΔC column shift value
- L line
- C column
- $a_{L,C}$ coefficients determined by a least-squares fit to correlation distances.

This function implements translation in two dimensions, small rotations and scaling, depending on the values of the coefficients. For example, if the first and second coefficients are 0, the function is a translation of a_{L3} lines and a_{C3} columns. If a_{L2} and a_{C1} are 0, then the result is scaling and translation in two dimensions. If $a_{L1} = 1 - \cos\theta$, $a_{L2} = -\sin\theta$, $a_{C1} = \sin\theta$, and $a_{C2} = 1 - \cos\theta$, the function is a rotation through an angle θ about the point $L = 0$, $C = 0$. The linear shift function is updated each line where correlation takes place. Also, the scale factor and global rotation is determined at each update

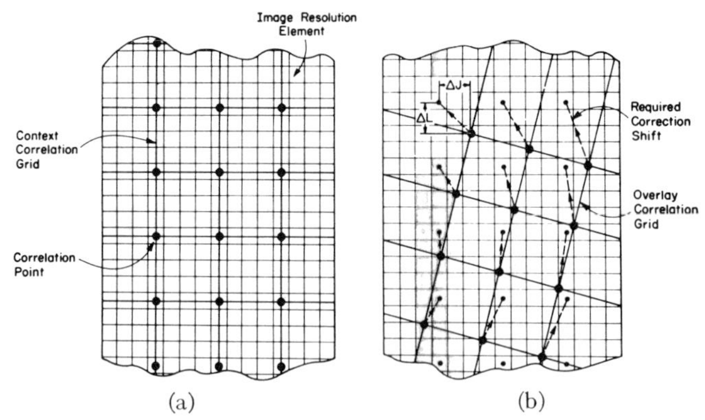

Fig. 10. Context and overlay aperture correlation and correction grids. (a) Context aperture. (b) Overlay aperture.

and is fed back to the input monitor. Translation corrections are also fed back so that the amount of correction that the shift function must handle is minimal.

Output lines are assembled in a core buffer by addressing appropriate areas in the imagery stored in core. The overlay tape format is one line per block with a maximum number of 30 channels. The total record byte length is the product of the number of channels and the samples per line and is limited only by the core buffer size available to read and write the line. All imagery is represented in byte form (8-bit words) on tape and is also processed in the computer in byte form. The processing rate of the present system is relatively slow due to nonoptimum input-output and computational procedures. The correlation time averages 3.5 seconds, gradient enhancement of two 32 by 32 arrays requires 2 seconds, and cluster border enhancement of the same arrays requires 10 seconds. The input, overlay, and output rate exclusive of enhancement averages about 1 second per line of 2800 samples. The processing rate has not been one of the primary considerations during development of the system. Once a satisfactory set of algorithms has been defined, program refinement can be carried out to optimize the processing speed.

VI. Correlation Analysis

The two dimensional FFT array correlator portion of the system was used to analyze real imagery to determine the relationship between the correlation coefficient for imagery from different bands and the spatial registration accuracy. Imagery known to be in registration from 17 bands listed in Table II was correlated over a test area to obtain basic trends and nominal values for the accuracy that the registration system can achieve for a specific example.

Correlation ranges from four to over 20 image points square were used in tests to determine the required range for consistent lock-on of the correlator. A correlation range of 20 points square tended to give satisfactory performance for the rectangular agricultural field type of ground cover encountered in the data analyzed [8].

TABLE II
SPECTRAL BANDS STUDIED

Channel Number	Aircraft Scanner Wavelength Bands: Channel Wavelength Limits (μm)	Description
1	0.40–0.44	violet
2	0.46–0.48	blue
3	0.52–0.55	green
4	0.55–0.58	green
5	0.58–0.62	yellow
6	0.62–0.66	red–orange
7	0.66–0.72	red
8	0.72–0.80	reflective infrared
9	0.8–1.0	reflective infrared
10	1.0–1.4	reflective infrared
11	1.5–1.8	reflective infrared
12	2.0–2.6	reflective infrared
13	4.5–5.5	thermal infrared
14	8–13.5	thermal infrared
	Apollo 9 Multispectral Photography Bands	
1	0.47–0.61	blue–green
2	0.59–0.715	red
3	0.68–0.89	reflective infrared

The correlation coefficient as a function of wavelength for several individual cover types and for a large area (global case) is presented in Fig. 11(a) and (b). The blue wavelength band (0.46 to 0.48 μm) was used as reference, and data in this band was correlated with all other bands available. The cover types, corn, soybeans, wheat, and bare soil, chosen from June, 1969, aircraft scanner data were used. The June data is agriculturally diverse in that many different crops are present in various stages of growth. Three dips in the correlation curve are seen at about 0.56, 0.9, and above 5 μm with a reversal centered at 0.9 μm for two cover types. The general shape of the curves is similar for the individual cover types and for the global or random sample case (Fig. 11(b)). It was hypothesized that the higher the correlation the higher will be the registration accuracy. Also, it seemed that for strongly negatively correlated regions, the negative peak of the cross correlation function should be used as the point of lock-on.

The distance of the cross correlation function peak from the origin (Euclidian distance) was then plotted and is shown in Fig. 11(c). It can be observed that the error generally follows the expected trend in which high negative or positive correlation reduces the error. Increasing distance in wavelength between the two channels being correlated consistently increases the error so the effects due to correlation error are superimposed on the general increase from 0.47 to 11 μm. A similar analysis was run for three band-digitized Apollo 9 multispectral photography from Imperial Valley, Calif. The results are presented in Fig. 12(a) and (b). The inverse relationship of correlation to registration accuracy is clearly seen here. The error magnitudes are considered to be reasonable, but in many applications even an average error magnitude of one or two resolution elements is not acceptable. The results indicate that the farther apart the bands are the greater the error and that even adjacent channels do not achieve perfect registration. This result may be pessimistic since the sample was small (30 correlations over 111 000 image points) and for agricultural land only. Further correla-

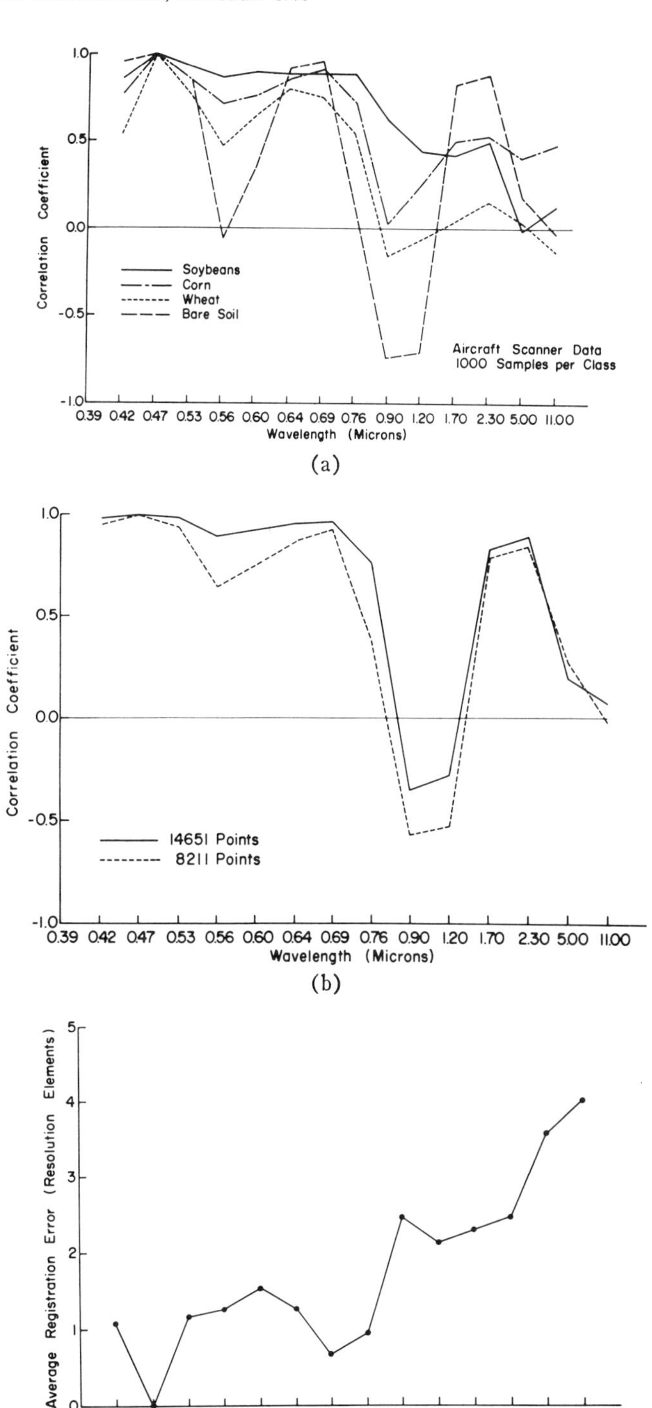

Fig. 11. (a) Correlation of blue (0.46–0.48 μm) channel with all other channels for agricultural features in June, 1969 (test site PF24). (b) Global correlation of blue (0.46–0.48 μm) channel with all other channels for agricultural test area PF24 in June, 1969. (c) Average registration error for correlation of blue (0.46–0.48 μm) channel with all other channels for agricultural land in June, 1969 (test site PF24).

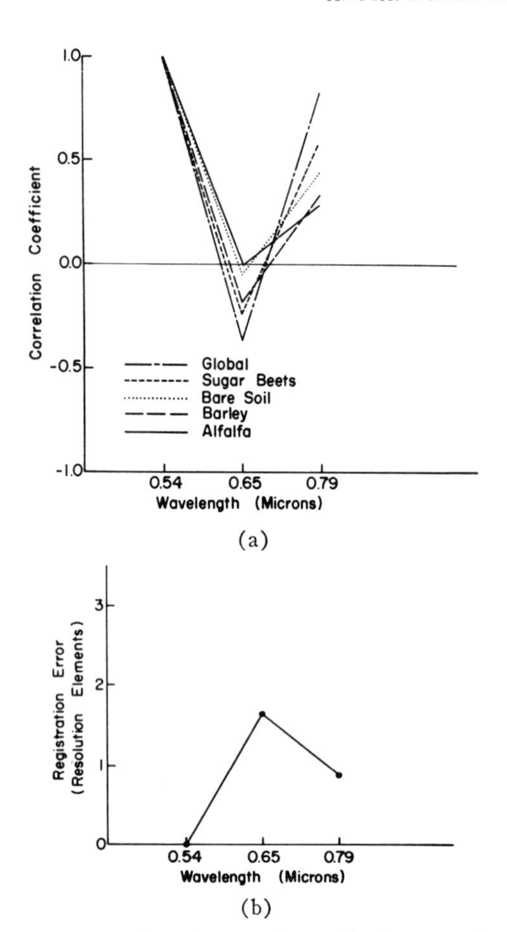

Fig. 12. (a) Correlation of green channel (0.47–0.61 μm) with red and infrared for Apollo 9 multispectral photography (SO65 test site 15A). (b) Average registration error for correlation of green band (0.47–0.61 μm) with red and infrared from Apollo 9 photography.

tion tests were run to see if an improvement could be achieved using enhanced imagery.

Test imagery enhanced by the spatial gradient method was correlated for the test area, and the error curve is plotted in Fig. 13 along with the unenhanced error curve from Fig. 11(c). The performance for this test was not significantly better than for the unaltered data case. The clustering technique was then employed for boundary enhancement, and the correlation tests were rerun using the results. The plotted error curve is also shown in Fig. 13. It can be seen that the performance is about the same as for the unenhanced case with some improvement at 0.53 and 0.60 μm and substantial improvement above 2.3 μm.

The correlation analysis presented here is intended to illustrate the worst situation in which the shortest wavelength channel is used as reference and is correlated with other channels as distant as 8 to 14 μm. In general, the optimum approach is to use the two most highly correlated channels from the misregistered image sets as input to the correlator. Often the same wavelength band may be available in the two images to be registered, and in this case highly accurate registration may be possible. The results here indicate that correlation of unaltered imagery is about as good an approach

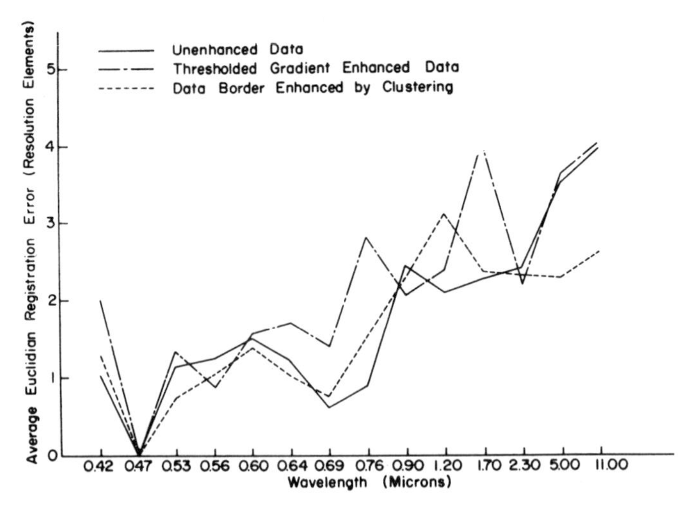

Fig. 13. Average registration error for correlation of blue (0.46–0.48 μm) channel with all other channels for agricultural test site PF24 in June, 1969.

as using costly enhancement before correlation, and in some cases it is better. If accurate correlation of widely separated bands is required, the cluster boundary enhancement approach appears promising.

VII. Applications

Multispectral and multitemporal imagery can be used as input to statistical pattern-recognition systems for automatic classification of scene features [1]. In such a process several spectral measurements from one image resolution element are used to classify that element into one of a set of scene feature classes. This process requires that all the spectral measurements come from the same resolution element, and it is the purpose of the registration system to achieve this state. Once registered imagery is obtained, the pattern-classification algorithms can be applied to any recognition task. Two examples of multispectral pattern classification made possible by image registration are shown in Figs. 14 and 15. In example 1, aircraft scanner data from 1.0–1.4-, 1.5–1.8-, 2.0–2.6-μm bands were registered with multiband data from the visible and near-reflective infrared portion of the spectrum from 0.4 to 1.0 μm. A feature selection algorithm [9] was employed to determine the most distinctive four bands for separation of the pattern classes considered, corn, soybeans, wheat, oats, pasture, grass, trees, and water, from data collected in June, 1969. The four bands chosen by the feature selection algorithm were 0.4–0.44, 0.66–0.72, 1.0–1.4, and 1.5–1.8 μm, and 290 290 points were classified for agricultural test site PF24 using the LARS multispectral pattern-recognition system. In Fig. 14 the classifier output is coded to denote the classifier decisions. The average classification accuracy based on 87 679 test points was 80 percent. Classification accuracy using only the best four bands available before registration was 60 percent. Use of the two infrared channels above 1.0 μm was made possible by the registration process.

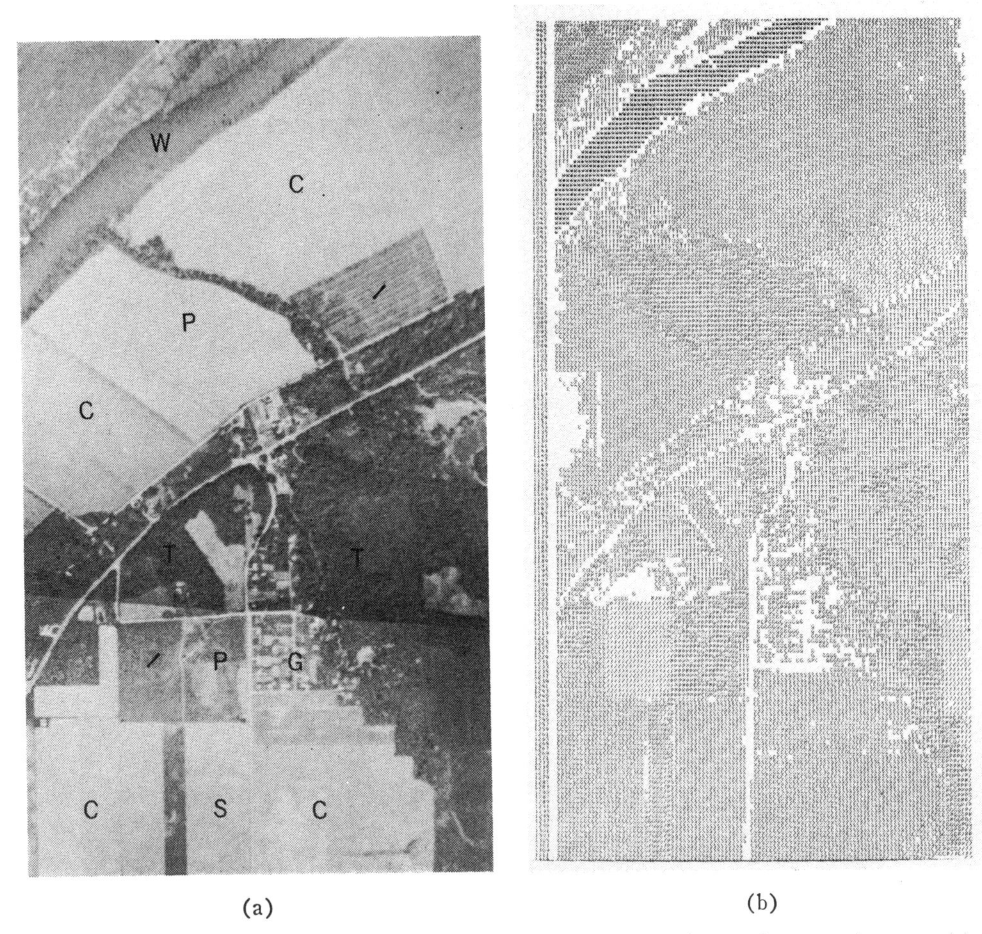

(a) (b)

Fig. 14. Computer classification of agricultural test site PF24 using registered multispectral scanner imagery. (a) Aerial photograph. (b) Classification printout. Key: C, corn; S, soybeans; /, wheat; O, oats; P, pasture; G, grass; T, trees; W, water.

In the second experiment the digitized multiband photography from Apollo 9 taken over Imperial Valley, Calif., was registered, and statistical pattern classification was attempted. The available wavelength bands were listed in Table II. The basic pattern classes, green vegetation, bare soil, salt flat, and water, were defined, and classification was carried out on 66 000 samples in the NASA S065 test site near Brawley, Calif. Fig. 15 is a printout of the results which proved to be quite accurate. The average performance was 95 percent for 880 test samples. The accuracy using only one channel was only 80 percent. Here again the multispectral registration system enabled use of automatic statistical pattern-recognition techniques for classification of imagery.

Multispectral classification has been carried out at LARS on registered imagery from agricultural, geological, and hydrological test sites. Imagery obtained at different times offers a valuable dimension, and registration of multitemporal imagery is in progress. Analysis will begin when data from several time points are available for a test site. The LARS imagery registration system is a key part of the remote-sensing research project and it is making valuable new dimensions available to the researcher.

Multi-images to be used for multidimensional analysis should ideally be obtained in registered form so that costly and, at best, inexact registration processing is not necessary. At the present time perfectly registered multispectral and multitemporal imagery is not available to all who require it. The software system described here is one approach to achieving registration where the need exists. It is suitable for a research environment where a small quantity of imagery is needed for design of algorithms and data-handling systems. Specialized high-speed image-registration and data-handling systems can then be built based on results of the research using this slower general-purpose implementation. The LARS registration software system can be operated on any general-purpose digital computer having FORTRAN IV capability and sufficient memory capacity. The largest phase of the current system requires 128 000 bytes of storage.

VIII. Conclusions

A software system for spatial registration of multiple images has been described, and some analysis results were presented to indicate likely system performance. The use of the FFT for improving the speed of calcula-

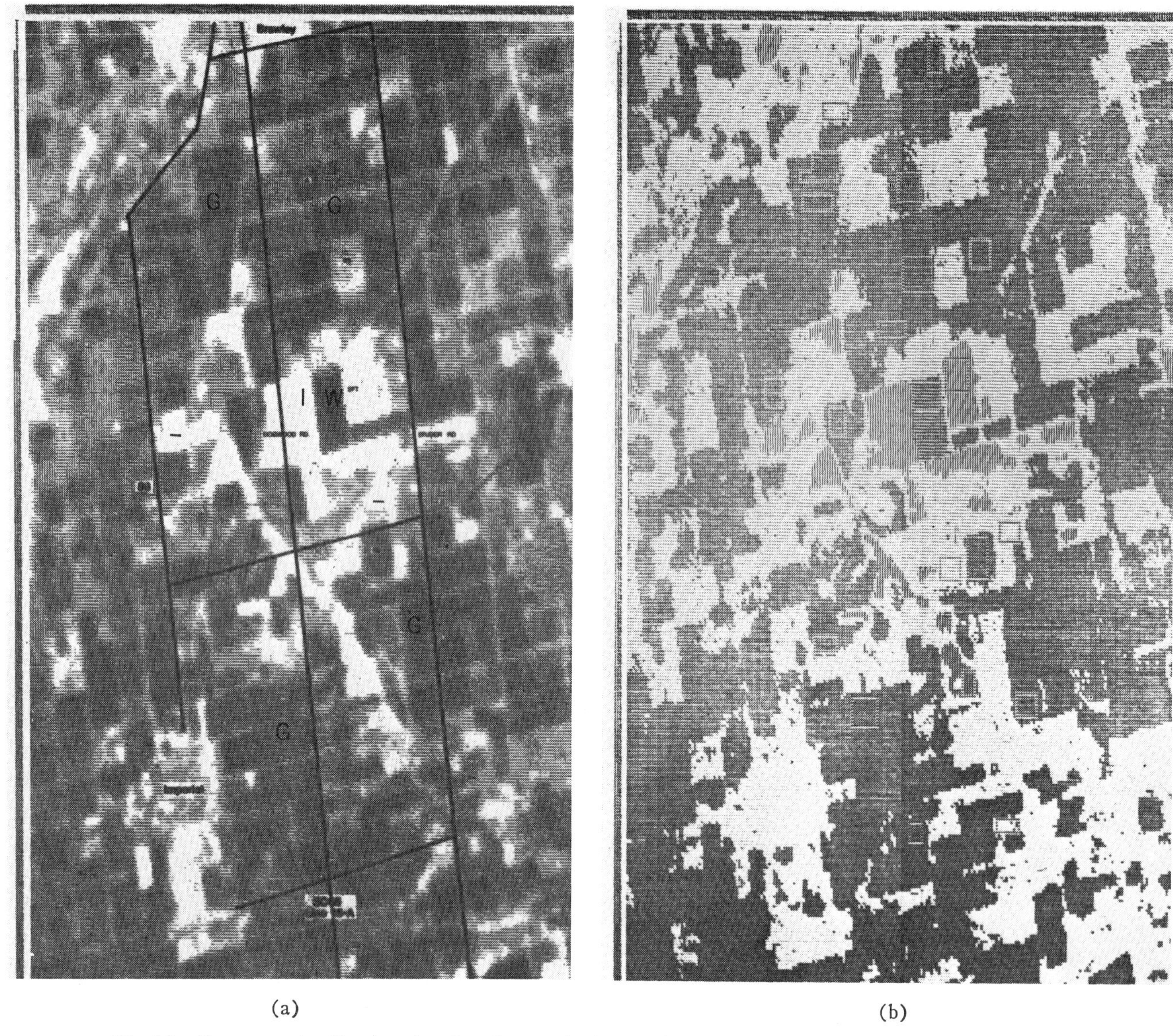

(a) (b)

Fig. 15. Computer classification of surface features in Apollo 9 photography of S-065 test site. (a) Gray-scale printout. (b) Classification printout. Key: G, green vegetation; —, soil; I, salt flats; W, water.

tion of the image correlation function was discussed in detail. Two methods of imagery enhancement were described, and test results were presented. The structure of the system was outlined, and utilization of registered imagery produced by the system was described.

The goal of the registration system development project at LARS is to achieve precise registration of digital multi-image data so that research could be performed by automatic multivariate analysis techniques for imaging remote-sensor data. The current system is producing registered imagery from an aircraft multispectral scanner and from digitized aerial and space photography. This imagery is being used at LARS for automatic classification research of earth-surface features in the categories of agriculture, geology, hydrology, transportation, forestry, and detection of pollution. Imagery collected over an area at different times is being registered to enable study of the time-varying properties of surface features. The registration system is thus enabling advanced research to be conducted into the automation of earth-resources survey processes vital to the preservation and improvement of life on the earth.

ERTS will produce imagery over large areas of the earth's surface in three bands in photographic and digital form. Each image frame will cover approximately a 100 mi^2 and will contain several million picture points in each band. Digitization if necessary and registration of this imagery over areas of interest on a frequent time base will enable detailed study of the time-varying properties of these surface features. It is predicted that ERTS will photograph every point on the earth every 20 days; thus a time history of over 18 samples per year will be available for each point. Also, registration of imagery from other sensors in other parts of the spectrum will enable multidimensional analysis similar to that made possible by wideband line scanners. Other applications for registered multi-images will arise as research progresses. The LARS system described here is designed to handle a wide range of requirements for research on limited quantities of data.

Acknowledgment

The airborne multispectral scanner data used in the analysis were gathered by the Institute of Science and Technology, University of Michigan, Ann Arbor, under contract to NASA using equipment supplied by the U. S. Army Electronics Command. The Apollo 9 satellite photography was obtained from the Earth Observations Division, Manned Spaceflight Center, NASA, Houston, Tex., and was digitized by Optronics International, Chelmsford, Mass.

References

[1] K. S. Fu, D. A. Landgrebe, and T. L. Phillips, "Information processing or remotely sensed agricultural data," *Proc. IEEE*, vol. 57, pp. 639–653, April 1969.
[2] D. Steiner, "Time dimension for crop surveys from space," *Photogrammetr. Eng.*, vol. 36, pp. 187–194, February 1970.
[3] B. Ostle, *Statistics in Research.* Ames, Iowa: Iowa State University Press, 1963.
[4] J. W. Cooley and J. W. Tukey, "An algorithm for the machine calculation of complex Fourier series," *Math. Comput.*, vol. 19, pp. 297–301, April 1965.
[5] J. W. Cooley, P. A. Lewis, and P. D. Welch, "The fast Fourier transform algorithm and its applications," IBM Watson Research Center, Yorktown Heights, N. Y., IBM Res. Rep. RC1743, February 9, 1965.
[6] A. Rosenfeld, *Picture Processing by Computer.* New York: Academic Press, 1969.
[7] A. G. Wacker, "A cluster approach to finding spatial boundaries in multispectral imagery," Laboratory for Applications of Remote Sensing, Purdue University, West Lafayette, Ind., LARS Information Note 122969.
[8] P. E. Anuta, "Digital registration of multispectral video imagery," SPIE J., vol. 7, pp. 168–175, September 1969.
[9] P. J. Min and K. S. Fu, "On feature selection in multiclass pattern recognition," School of Elec. Eng., Purdue University, Lafayette, Ind., Tech. Rep. TR-EE68-17, July 1968.

A Class of Algorithms for Fast Digital Image Registration

DANIEL I. BARNEA, MEMBER, IEEE, AND HARVEY F. SILVERMAN, MEMBER, IEEE

Abstract—The automatic determination of local similarity between two structured data sets is fundamental to the disciplines of pattern recognition and image processing. A class of algorithms, which may be used to determine similarity in a far more efficient manner than methods currently in use, is introduced in this paper. There may be a saving of computation time of two orders of magnitude or more by adopting this new approach.

The problem of translational image registration, used for an example throughout, is discussed and the problems with the most widely used method-correlation explained. Simple implementations of the new algorithms are introduced to motivate the basic idea of their structure. Real data from ITOS-1 satellites are presented to give meaningful empirical justification for theoretical predictions.

Index Terms—Registration efficiency, sequential similarity detection algorithms, spatial cross correlation, spatial registration of digital images.

I. INTRODUCTION

THE AUTOMATIC determination of local similarity between two structured data sets is fundamental to the disciplines of pattern recognition and image processing. A class of algorithms, which may be used to determine similarity in a far more efficient manner than methods currently in use, is introduced in this paper. By adopting this new approach there may be a saving of computation time of two orders of magnitude or more.

The method most widely used for similarity detection is correlation. In fact the similarity detection problem itself is generally called "correlation." In this paper, however, the distinction between the mathematical method, correlation, and the class of similarity detection problems will be maintained; for it will be shown that procedures other than those currently in practice will yield efficient and accurate results.

The class of sequential similarity detection algorithms (SSDAs) introduced here may be applied to the entire spectrum of similarity detection problems. A broad theory relating general properties of these algorithms can be developed. In this paper, however, the utility of SSDAs will be illustrated by their application to a specific problem, translational registration, which is basic to image processing.

In Section II, the translational registration problem will be introduced. Section III will describe the computational efficiency of correlation methods. In Section IV some examples of SSDAs are presented in the order they were chronologically investigated. It is intended that this section impart an intuitive feeling about SSDAs to the reader. Section V generalizes this class of algorithms and presents, in light of this general concept, some ideas for further development of the given examples.

Manuscript received May 13, 1971; revised August 6, 1971.

D. I. Barnea was with the IBM T. J. Watson Research Center, Yorktown Heights, N. Y. 10598. He is now with Eljim, Holon, Israel.

H. F. Silverman is with the IBM T. J. Watson Research Center, Yorktown Heights, N. Y. 10598.

II. THE TRANSLATIONAL REGISTRATION PROBLEM

Registration is inherently basic to any image processing system. When it is desired to detect changes or perform a mapping of two similar images, it is necessary for meaningful results to have the images registered. If the pictures do not differ in magnification and rotation, then the best translational fit will yield the required registration. (The problems arising from magnification and rotation will not be considered here for the sake of brevity. However, the methods are applicable when proper modifications are introduced.)

Let two images, S the *search area* and W the *window* be defined as shown in Fig. 1. S is taken as an $L \times L$ array of digital picture elements which may assume one of K grey levels; i.e.,

$$0 \leq S(i, j) \leq K - 1$$
$$1 \leq i, j \leq L.$$

W is considered to be an $M \times M$, M smaller than L array of digital picture elements having the same grey scale range; i.e.,

$$0 \leq W(l, m) \leq K - 1$$
$$1 \leq l, m \leq M.$$

It will be convenient to introduce a notation for $M \times M$ wholly contained subimages.

$$S_M^{i,j}(l, m) \equiv S(i + l - 1, j + m - 1),$$
$$\begin{cases} 1 \leq l, m \leq M \\ 1 \leq i, j \leq L - M + 1. \end{cases} \quad (1)$$

Each $M \times M$ subimage of S can be uniquely referenced by the specification of its upper left corner's coordinates (i, j). These will be used to define *reference points*. It will be assumed that enough *a priori* information is known about the dislocation between the window and search area so that the parameters L and M may be selected with the virtual guarantee that, at registration, a complete subimage is contained in the search area as shown in Fig. 1.

Translational registration, therefore, is a search over some subset of the allowed range of reference points to find a point (i^*, j^*) which indicates a subimage that is most similar to the given window.

Reprinted from *IEEE Trans. Comput.*, vol. C-21, pp. 179–186, Feb. 1972.

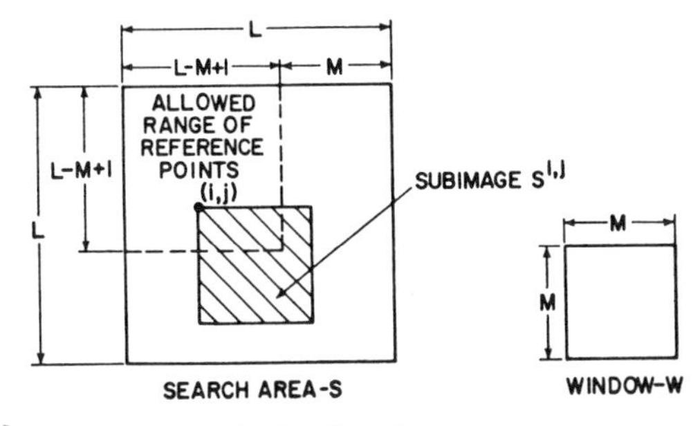

Fig. 1. Search space.

III. Computational Efficiency of Correlation Methods

A. Correlation Method

The method most widely used for the automatic determination of translation is correlation [1], [2]. The elements of the unnormalized cross-correlation surface $R(i,j)$ are defined to be

$$R(i,j) \equiv \sum_{l=1}^{M} \sum_{m=1}^{M} W(l,m) S_M{}^{i,j}(l,m), \quad 1 \leq i,j \leq L-M+1. \quad (2)$$

In the correlation scheme a representative output surface such as $R(i,j)$ is searched for a maximum $(\hat{\imath},\hat{\jmath})$. The procedure is successful if $(\hat{\imath},\hat{\jmath})$ and (i^*,j^*) are equivalent. As a counterexample, however, consider the unnormalized cross correlation of (2) even in the *ideal case* where W exactly matches some subimage; i.e., $W = S_M{}^{i^*,j^*}$. Then

$$R(i^*,j^*) = \sum_{l=1}^{M} \sum_{m=1}^{M} W^2(l,m). \quad (3)$$

Also, for this ideal case, consider the nonmatching point $(\hat{\imath},\hat{\jmath})$ where

$$S_M{}^{\hat{\imath},\hat{\jmath}}(l,m) = \max_{l,m} W(l,m) = W_M, \quad \text{for all } (l,m). \quad (4)$$

Clearly

$$R(\hat{\imath},\hat{\jmath}) = W_M \sum_{l=1}^{M} \sum_{m=1}^{M}, \quad W(l,m) \geq R(i^*,j^*). \quad (5)$$

Therefore even in the ideal case, a search for a maximum over $R(i,j)$ does not necessarily yield the registration point. Normalization is therefore necessary in even the simplest of cases! For completeness, the usual normalized correlation surface is defined in (6).

$$R_N^2(i,j) = \frac{\left(\sum_{l=1}^{M} \sum_{m=1}^{M} W(l,m) S_M{}^{i,j}(l,m) \right)^2}{\left[\sum_{l=1}^{M} \sum_{m=1}^{M} W^2(l,m) \right] \left[\sum_{l=1}^{M} \sum_{m=1}^{M} S_M^2{}^{i,j}(l,m) \right]}, \quad 1 \leq i,j \leq L-M+1. \quad (6)$$

The choice of a similarity detection algorithm should be justified by its probability for error and its computational complexity rather than by tradition or expediency. Perhaps the two reasons which are generally given for using the correlation method are that: 1) correlation appears to be a natural solution for the mean-square-error criteria [3]; and 2) that analog–optical methods implement correlation easily [4].

However, there is no guarantee for *any* method that a solution is correct or unique. There seems to be, therefore, no adequate justification for the use of correlation to solve all digital registration problems. Algorithms, such as those presented in this paper, which have selectable distance measure properties and lower computational complexity, appear to be a more fitting choice.

B. The Cost of Correlation

The normalized correlation surface may be calculated by direct means or by fast Fourier transform (FFT) methods [5]. In each case the amount of computation required for normalization is the same. The numerator of (6) is all that may be treated by FFT.

In Table I the approximate number of calculations for each procedure are shown. One should note that although the FFT method requires fewer operations for L and M large, this method also requires a memory capacity of $2L^2$ real words which may be infeasible for an L larger than 256.

IV. Some Examples of SSDA Implementation

A. The Basic Concept

For a particular reference point (i,j), there are M^2 points of the subimage $S_M{}^{i,j}$ which may be compared with the M^2 corresponding points in W. (Each set of points for comparison (e.g., $\langle S_M{}^{i,j}(l,m), W(l,m)\rangle$) will be called a *windowing pair*.) In correlation the maximum number of windowing pairs or $M^2(L-M+1)^2$ are compared. Each reference point, regardless of content, is therefore processed with very high precision. However, accuracy is required only for those relatively few points near surface maxima. Hence, there is considerable waste in performing high-accuracy calculations at a vast majority of points.

SSDA reduces this redundancy by performing a sequential search which may be terminated before all M^2 windowing pairs for a particular reference point are tested. Furthermore, the algorithms do not implicitly contain any fixed error measure or measure evaluation method.

B. Constant Threshold Algorithm

A simple but important SSDA has been extensively studied and will be used as a vehicle to introduce the concept. Here, a search over each of the $(L-M+1)^2$ reference points is performed, as in correlation. However, the criteria for similarity at each reference point is significantly different from correlation.

In the constant threshold algorithm, windowing pairs are selected for comparison in a random order so that, in general, a great deal of new information is considered in each test; i.e., a random nonrepeating sequence of the integers 1, 2, $\cdots$, M^2 is generated and used to yield the random, nonrepeating sequence of coordinates (l_n, m_n),

TABLE I

COST OF NORMALIZED CORRELATION

Direct Correlation	FFT Correlation	Ordinary Normalization	"Fast" Normalization
$M^2(L-M+1)$-Mults.	$6L^2(\log_2 L)$-Complex Mult-Adds	L^2+M^2-Mults.	L^2+M^2-Mults.
$M^2(L-M+1)^2$-Adds	FFT of S,W and IFFT of Product	$M^2(L-M+1)^2$-Adds	$4(L-M+1)^2$-Adds
$(L-M+1)^2$-Squarings	L^2-Complex Mults.	$(L-M+1)^2$-Mults & Divides	$(L-M+1)^2$-Mults & Divides
	$(L-M+1)^2$-Squarings		

$n=1, 2, \cdots, M^2$. Thus the windowing pairs $\langle S_M^{i,j}(l_n, m_n), W(l_n, m_n)\rangle$ are compared in random order as n increases.

Unnormalized or normalized measures for evaluating the error between windowing pairs may be defined, respectively, as

$$\epsilon'(i, j, l_n, m_n) \equiv |S_M^{i,j}(l_n, m_n) - W(l_n, m_n)| \quad (7)$$

$$\epsilon(i, j, l_n, m_n) \equiv |S_M^{i,j}(l_n, m_n) - \hat{S}(i, j) - W(l_n, m_n) + \hat{W}| \quad (8)$$

where

$$\hat{W} \equiv \frac{1}{M^2} \sum_{l=1}^{M} \sum_{m=1}^{M} W(l, m) \quad (9)$$

and

$$\hat{S}(i, j) \equiv \frac{1}{M^2} \sum_{l=1}^{M} \sum_{m=1}^{M} S_M^{i,j}(l, m). \quad (10)$$

Unlike the correlation methods cited previously, in the *ideal case*, where $W = S_M^{i^*,j^*}$, a minimum of zero is guaranteed for the nonnormalized case, i.e., for

$$\|E(i, j)\| \equiv \sum_{l=1}^{M} \sum_{m=1}^{M} |S_M^{i,j}(l, m) - W(l, m)| \quad (11)$$

$$0 = \|E(i^*, j^*)\| \leq \|E(i, j)\|. \quad (12)$$

Thus, in this ideal case no normalization is necessary and, obviously, a comparison of very few points will yield the answer. The error measure, based upon the L_1 norm between two images, is also computationally simpler than the multiplicative measure of correlation.

In this SSDA implementation, a constant threshold T is introduced. As the error for randomly selected windowing pairs is accumulated, a test is made against T. When the accumulated error exceeds T at test N, operations cease for reference point (i, j) and the value N is recorded. The SSDA surface $I(i, j)$ is therefore defined as

$$I(i, j) = \left\{ r \,\middle|\, \min_{1 \leq r \leq M^2} \left\{ \sum_{n=1}^{r} \epsilon(i, j, l_n, m_n) \geq T \right\} \right\}. \quad (13)$$

Reference points where $I(i, j)$ is large (those which require many windowing pair tests to exceed T) are considered points of similarity.

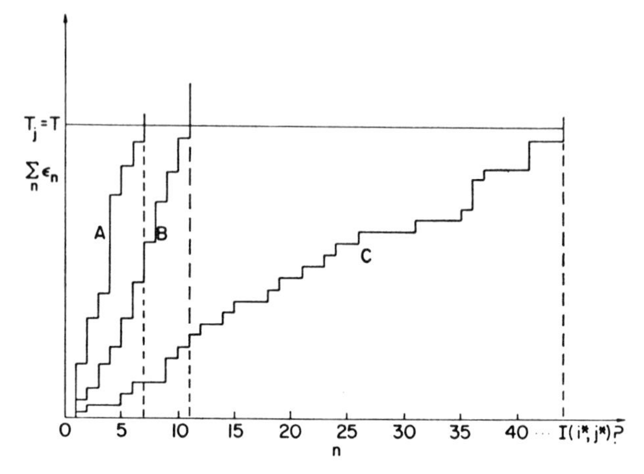

Fig. 2. Growth curves with T=const.

It is clear that if a suitable value for T is selected, many fewer than M^2 tests will be required for reference points which rapidly accumulate error. It is this property which significantly reduces the computational complexity of an SSDA! This fact is made clear by Fig. 2. Curves A, B, and C depict the cumulative error for three different reference points as a function of test. A and B accumulate error rapidly and operations for their reference points terminate early with $I(i, j)$ obtaining values of 7 and 11, respectively. Curve C, however, accumulates error more slowly. It is, therefore, much more likely to be a candidate for registration, and will accordingly have a value of 44 assigned to $I(i, j)$.

Data from the NOAA weather satellite ITOS-1 have been used to test this algorithm. Fig. 3 is an example of a typical data set. Images (a) and (b) are segments of uncorrected vidicon output taken over Baja California on August 9, and 11, 1970, respectively. In this application, one is interested in registering land masses. Therefore, in addition to the noise due to optics and scanners, there is a great deal of intense noise due to clouds and cloud shadow, as well as from the fiducial mark (the ×) in (b). Fig. 3(c) shows the window function W taken from the upper left corner of (b). In Fig. 3(d), the answer is displayed. The window of Fig. 3(c) is put into the search area [Fig. 3(a)] at the position indicated by the algorithm.

The fit is visually accurate, but as real rather than contrived data were tested, there is virtually no way to absolutely check this fit. However, the same data were run in an FFT

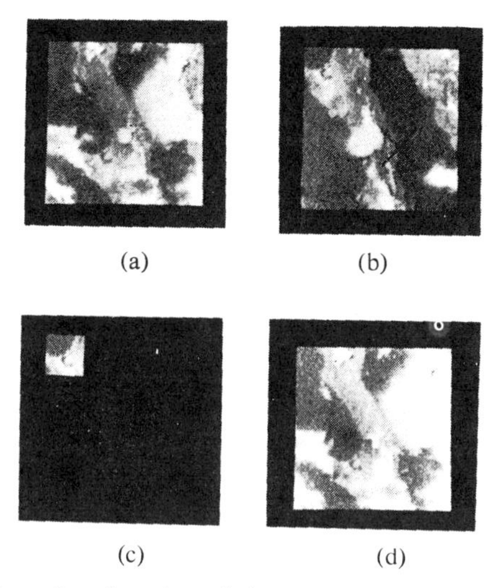

Fig. 3. Example of registered images taken from NOAA data. (a) Search area. (b) Picture 2. (c) Window area taken from (b). (d) Window inserted into search area at (i^*, j^*).

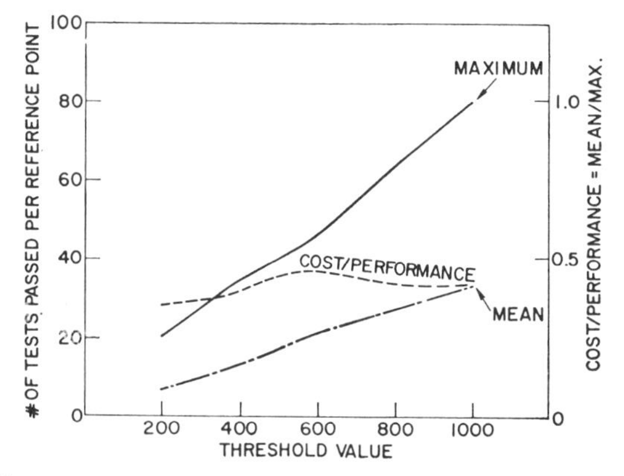

Fig. 4. Statistical results for constant threshold algorithm.

correlation-registration scheme, for the purposes of a relative comparison. As expected, there were some complications involved in the application of this more traditional method. Only after clouds had been detected and their bright surface values replaced by random noise would correlation yield a meaningful peak. (No corrections whatsoever were necessary for SSDA!) For this data, correlation gave the registration point as one picture element higher than did this SSDA. As correct registration might easily lie between two points, there is little to argue!

The data in Fig. 3 have the following attributes:

1) $L=128$;
2) $M=32$;
3) $0 \leq S(i, j) \leq 255$;
4) $E[S(i, j)] \approx 75$ (expected value).

For this data set, tests were made for various values of T. The important results from these tests are given in Fig. 4. The average value of a point in the surface $I(i, j)$—and hence the number of calculations required—grows linearly with threshold. The maximum value in the surface $I(i, j)$, an indicator of the accuracy of the method, also grows linearly with threshold value, although at a somewhat higher rate. The ratio of these two items can be considered as a cost/performance measure. For the case of a constant threshold, this measure is shown to be relatively constant. This implies that accuracy may be achieved only by an associated increase in computation.

C. *Monotonic-Increasing Threshold Sequence Algorithm—(Algorithm A)*

The growth curve for a particular reference point (three of which are shown in Fig. 2), is a monotonically increasing function. It is the *average slope of the growth curve* that is important in the determination of a threshold crossing. It therefore seems reasonable that the replacement of the constant threshold T by a threshold T_j, increasing monotonically with test, would improve performance when the following criteria are considered.

1) The sequence T_j should have "shape" approximating that of the growth curve for $I(i^*, j^*)$, but should bound this growth curve from above for an arbitrarily large n.

2) The T_j sequence should have initial values high enough so that a trend might be established even for reference points far from registration.

An example of a monotonically increasing threshold sequence is shown in Fig. 5. Growth curves A and B, for reference points far from registration, are eliminated earlier than in Fig. 2, at 4 and 7, respectively. As most reference points do exhibit rapid growth, the total number of tests will be diminished significantly. Growth curve C, however, which appears to be a strong candidate for registration, will undergo a larger number of tests than before. Therefore, a high degree of accuracy will be achieved for those few points which have growth curves with low average slope.

Monotonically increasing threshold curves were generated by several methods. A stochastic approach yielded excellent analytic and experimental results. The distances at registration $\epsilon(i^*, j^*, l, m)$ may be considered a random variable with an exponential distribution

$$f_\alpha(x) = \begin{cases} \dfrac{1}{\lambda} e^{-x/\lambda}, & 0 \leq x \leq \infty \\ 0, & \text{otherwise.} \end{cases} \tag{14}$$

At the point of registration, the sum of "ϵ terms" is a cumulative measure of the noise between two different pictures. At any other reference point $\epsilon(i, j, l, m)$ must be considered as the sum of the noise due to the fact that the pictures are different, *image noise*, and that the pictures are out of registration, *registration noise*. One assumes, in order to say a correct answer has been achieved, that the total noise is minimal at the point of registration.

Optimally, one desires to find a threshold sequence which does the following.

1) It minimizes the probability that a point other than the registration point will remain beneath the threshold as long as the "true" registration point.

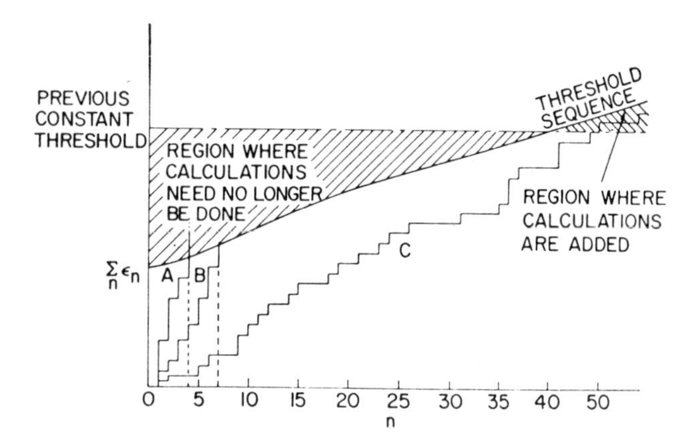

Fig. 5. Growth curve for algorithm Q with T_j a monotonically increasing function.

2) It maximizes the probability that the registration point remains below threshold.

3) It allows the total number of operations to be small.

Some analytic difficulty developed when this total optimization problem was considered. However, suboptimal threshold sequences may be found based upon the image noise alone which appear to satisfy the two hypotheses at the beginning of this section. Empirical results also tend to justify their contention for near optimality.

Fig. 6 shows a set of curves derived simply from the mean and variance of the image noise's statistics. If one has a measure of the mean of this noise λ and a bound to the deviation from the mean that he is willing to allow, he may select a threshold sequence. This method of design, however, leads to results which are somewhat less satisfying than those obtained by more careful derivation.

The curves shown in Fig. 7 are drawn from a computer solution of an analytic system of recursive equations. Here, one considers the probability P_k, which is the probability for the cumulative error at the registration point $I(i^*, j^*)$ to exceed T_k at test k, given that the cumulative error has stayed beneath the threshold for all tests $n=1, 2, \cdots, k-1$. The useful set of design curves of Fig. 7 may be derived by restricting P_k to be a constant q for all k.

While the curves in Figs. 6 and 7 appear to have the same shape, the equiprobable curves, which take into account the results from past as well as present tests, are closer to the optimal. Any particular one of these curves may be generated by an exact set of equations or by an approximate model. Once an estimate to the mean of the image noise λ has been obtained, a reasonable design follows.

Several tests were conducted on the data set shown in Fig. 8. Fig. 8(a) and (b) are each taken from the edge of a large, unnormalized NOAA vidicon output. As before the images were taken two days apart each of the Gulf coast of Texas. The data have many distortions due to sun angle, spacecraft attitude, and optical edge effects. There are also very few prominent features in evidence. The insertion in Fig. 8(d) appears to be correct upon visual inspection.

For this set of data, *a priori* knowledge of the approximate registration point allowed one to find the parameter λ

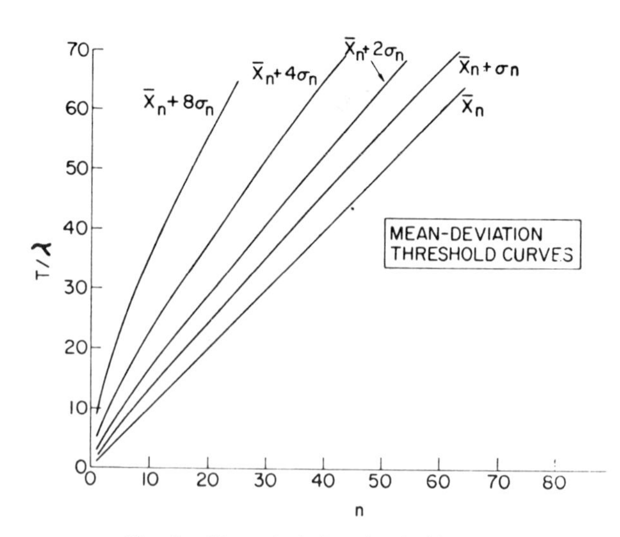

Fig. 6. Mean deviation threshold curves.

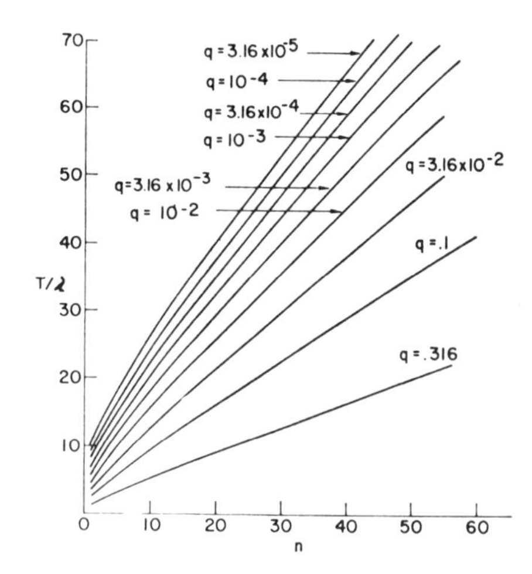

Fig. 7. Calculated equiprobable threshold curves.

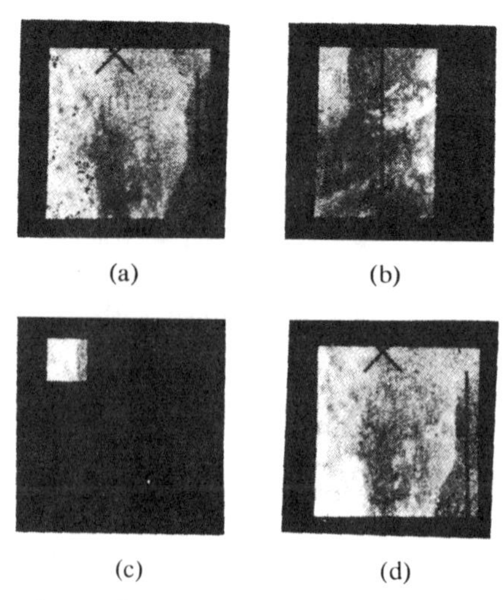

Fig. 8. Second example of registered images taken from NOAA data. (a) Search area. (b) Picture 2. (c) Window area taken from (b). (d) Window inserted into search area at (i^*, j^*).

empirically. Fig. 9 shows the sample distribution for the data of Fig. 8 for the absolute value of the image noise. The exponential distribution assumed in (14) appears to have some justification, with a value of 9.089 for λ. Based upon this data, quantized versions of threshold functions for several values of q were constructed. These are shown in Fig. 10. A quantized approximation was used so that thresholds might be stored in a lookup table and the speed of calculation increased. It was found that a coarse quantization yielded accurate results.

Curves based upon the results of tests on the data of Fig. 8, using the constant q design criteria, are given in Fig. 11. One immediately can see the benefit of using the monotone-increasing threshold sequence by looking at the cost/performance ratio. As one would expect, there is a region where this ratio has minimal—implying optimal—value(s). In the tests conducted, the optimal value(s) for q lie in the range $10^{-2} \leq q \leq 10^{-3}$. Hence, knowledge of the parameter λ and an optimal value for q can lead to a highly efficient design.

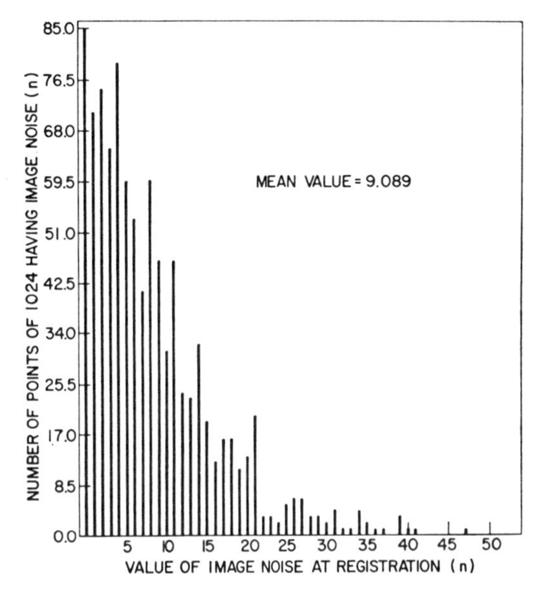

Fig. 9. Sample distribution for image noise for data.

D. Computational Aspects

The efficiency of SSDA has been demonstrated in the given examples. The saving which results can be illustrated by relating numbers of operations required for various implementations. Time ratios relating arithmetic operations for the IBM 360/65 will be used. Namely,

1) real multiply time/read add time = 3;
2) real add time/integer add time = 2;
3) integer multiply time/integer add time = 3.5;
4) compare time = integer add time;
5) complex add = 2 real adds;
6) complex multiply = 3 real multiplies + 5 real adds.

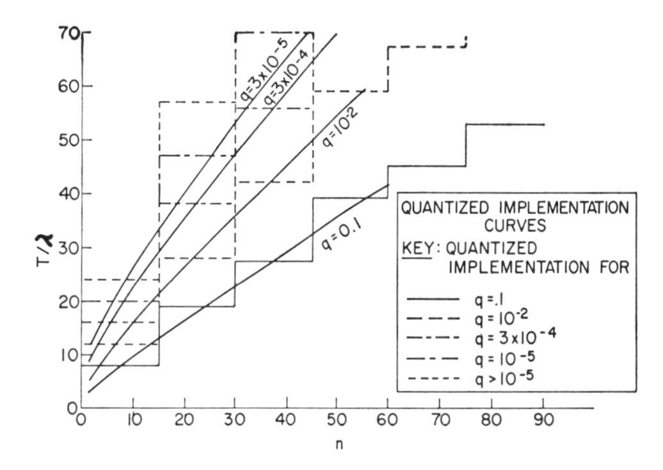

Fig. 10. Quantized implementation of equiprobable curves, where q is the probability for the growth curve at (i^*, j^*) to exceed T_k at any k given that the growth curve is below $T_{k'}$ for all $k' = 1, 2, \cdots, k-1$.

From Table I the direct cross-correlation method, *not including normalization*, reduces to

number (direct)

$$= 4.5M^2(L - M + 1)^2 \text{ equivalent integer adds} \quad (15)$$

and the FFT cross-correlation method, *not including normalization*, reduces to

number (FFT)

$$= 200L^2 \log_2 L \text{ equivalent integer adds.} \quad (16)$$

The monotone increasing threshold SSDA, *using the normalized distance measure*, requires:

1) 4 adds/reference point to obtain normalization;
2) 2 adds + 1 compare/windowing point to get the error measure;
3) 1 compare/windowing point;
4) An average of $\bar{n}$ windowing tests/reference point.

Thus

number (SSDA)

$$= (4 + 4\bar{n})(L - M + 1)^2 \text{ equivalent integer adds.} \quad (17)$$

For M of 32, $\bar{n}$ should be about 10–15. However, as M in-

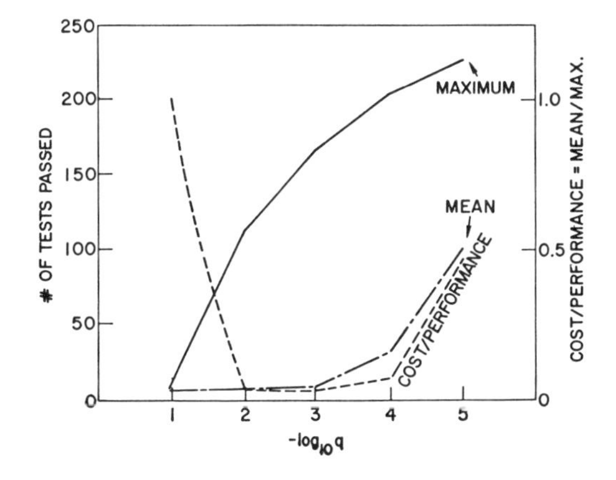

Fig. 11. Statistical results for constant q algorithm.

creases, more points should be checked to establish conclusively that misregistration has occurred. A figure of $\bar{n} = 10(M/32)^{1/2}$ is reasonable. Thus

$$\text{number (SSDA)} = 4(1 + 10(M/32)^{1/2}) \cdot (L - M + 1)^2 \text{ equivalent integer adds.} \quad (18)$$

Table II lists corresponding calculational costs for the several methods for various values of M and L. Values which are power of two are selected so that best figures for the FFT method might be achieved. One should note, however, that no other algorithm requires L and M to be a power of two. (It is recognized that this condition might be relaxed somewhat for the FFT, but most all available routines maintain this requirement.)

Table II shows that the given example of an SSDA is about a factor of 50 faster than the FFT correlation method. (This is a quite conservative estimate.) The order of magnitude saving by use of SSDAs is very important to production digital image processing.

V. Generalization and Conclusion

A. The General Concept of SSDA

A structure for the class of SSDAs may now be stated in a general form so that the versatility of the method is clear. There are many implementations other than those of the examples, each of which contains the following basic elements.

1) Ordering Algorithm O_1: O_1 orders the $(L-M+1)^2$ reference points. The window is successively compared to subimages which are selected by O_1. This ordering need not be fixed in advance, and may vary in a way that depends upon events occurring during the execution. In these cases, the total number of reference points which are processed is less than $(L-M+1)^2$.

2) Ordering Algorithm O_2: O_2 orders the M^2 (or fewer) windowing pairs to be compared at each reference point. O_2 may be fixed—the same for all reference points—or it may adapt in a data-dependent fashion.

3) Distance Measure (Norm) $\|x\|$: $\|x\|$ is used as a measure of error when windowing pairs are compared.

4) Sequential Measuring Algorithm Q: Q is a mapping from a subset of the M^2 possible distance measures for a particular reference point (i, j) into an element in the inspection surface $I(i, j)$.

$I(i, j)$ is directly analogous to the correlation surface described in Section III. Q operates upon distances $\|x\|$ for windowing pairs in the sequential manner specified by O_2. This sequence of operations continues until an event (like the passing of a threshold) occurs. At that instant $I(i, j)$ is evaluated on the basis of the measurements taken.

Each of the four basic properties may be tuned to fit a specific similarity detection problem so that a proper balance between accuracy and efficiency is realized. There are many variations that are far different from the examples presented. In the next few paragraphs, reasonable extensions to the monotonic-increasing threshold algorithm are given to show just how the general formulation can be used.

The reference points, in the examples above, were exhaustively searched. The following list presents some alternative approaches for O_1, if it is desired to sacrifice some accuracy for computation speed.

1) Apply a two-step coarse–fine uniform search, i.e., a first search on a coarse grid at every *m*th point and then do a full search in local regions about the maxima. (This procedure was implemented on several data sets, and a tradeoff, in fact, does exist. The degradation in accuracy is highly dependent upon the smoothness of the fully sampled $I(i, j)$ surface.)

2) Apply other multistep open-loop search criteria such as a coarse random search.

3) Apply, assuming that the surface $I(i, j)$ is monotone in some sense and has a global maximum, closed-loop hill climbing or gradient techniques.

The order for the windowing pairs O_2 might be easily changed to yield rapid elimination of misregistration points. Rather than selecting a completely random sequence, one might, for example, select a pseudorandom sequence based upon prominent features of W. That is, the first points in the sequence would be as random as possible while representing the key points for registration. There is a host of methods for the legitimate selection of feature points.

Each data structure has its own problem of normalization. SSDA may adapt to any necessary condition by the selection of suitable $\|x\|$. Most measures and process normalizations are acceptable. However, complex measures will somewhat degrade the ratio of saving shown in Table II.

It is apparent that there is a wide area yet unexplored for determining Q. Other mappings might be tried. It is possible, for example, to set a fixed threshold and test each of the windowing points against it independently. The surface value $I(i, j)$ might then be the number of comparisons prior to the occurrence of a certain number of threshold crossings.

Perhaps of great value to the monotonic-increasing threshold algorithm would be threshold adaptation. A conservative guess to the optimal threshold sequence might be used at first. On the basis of measurements made at succeeding reference points, the threshold sequence might be adaptively lowered. This procedure could allow an efficient algorithm to develop with very little *a priori* design on the part of the user. The algorithm would also gain a certain amount of independence from data.

B. Conclusion

A general class of sequential algorithms for similarity detection has been introduced, referencing the specific problem of translational image registration. Experimental and analytic results have been presented to show orders of magnitude improvement in efficiency. Several ideas for further improvement of experimentally implemented algorithms have also been presented.

The structure of the new algorithms is ideally suited for digital similarity detections. There is a time saving of at least 50 for typical problems on a representative medium-size computer—a prediction substantiated by experiment.

TABLE II

EQUIVALENT INTEGER ADDS FOR VARIOUS ALGORITHMS FOR SEVERAL VALUES OF L AND M

		Direct Method	FFT Correlation	Algorithm A
L	M	$4.5M^2(L-M+1)^2$	$200\ L^2 \log_2 L$	$4(1+10(M/32)^{1/2}(L-M+1)^2$
128	32	4.4×10^7	2.25×10^7	4.2×10^5
256	32	2.57×10^8	1×10^8	2.2×10^6
512	32	1.1×10^9	4.6×10^8*	1.05×10^7
1024	32	4.5×10^9	2×10^9*	4.35×10^7
2048	32	1.85×10^{10}	8.8×10^9*	1.75×10^8
128	64	8.15×10^7	2.25×10^7	2.5×10^5
256	64	6.9×10^8	1×10^8	2.2×10^6
512	64	3.7×10^9	4.6×10^8*	1.2×10^7
1024	64	1.7×10^{10}	2×10^9*	5.5×10^7
2048	64	7.4×10^{10}	8.8×10^9*	2.4×10^8
256	128	1.15×10^9	1×10^8	1.37×10^6
512	128	1.1×10^{10}	4.6×10^8*	1.25×10^7
1024	128	5.8×10^{10}	2×10^9*	6.7×10^7
2048	128	2.5×10^{11}	8.8×10^9*	2.9×10^8
512	256	2×10^{10}	4.6×10^8*	7.5×10^6
1024	256	1.8×10^{11}	2×10^9*	7×10^7
2048	256	1×10^{12}	8.8×10^9*	3.7×10^8
1024	512	2.7×10^{11}	2×10^9*	4.1×10^7
2048	512	2.6×10^{12}	8.8×10^9*	4×10^8
2048	1024	4.5×10^{12}	8.8×10^9*	5.7×10^8

* FFT's not possible on most machines without much disk accessing.

ACKNOWLEDGMENT

The authors wish to thank H. Kobayashi, J. Mommens, D. Grossman, and P. Franaszek for their help in many fruitful discussions.

REFERENCES

[1] P. E. Anuta, "Spatial registration of multispectral and multitemporal digital imagery using fast Fourier transform techniques," *IEEE Trans. Geosci. Electron.*, vol. GE-8, pp. 353–368, Oct. 1970.

[2] J. A. Leese, C. S. Novak, and B. B. Clark, "An automated technique for obtaining cloud motion from geosynchronous satellite data using cross-correlation," *J. Appl. Meteorol.*, vol. 10, pp. 110–132, Feb. 1971.

[3] A. Papoulis, *Probability, Random Variables and Stochastic Processes.* New York: McGraw-Hill, 1965.

[4] J. W. Goodman, *Introduction to Fourier Optics.* San Francisco: McGraw-Hill, 1968.

[5] J. W. Cooley, P. A. W. Lewis, and P. D. Welch, "Application of fast Fourier transform to computation of Fourier integrals, Fourier series, and convolution integrals," *IEEE Trans. Audio Electroacoust.*, vol. AU-15, pp. 79–84, June 1967.

TECHNIQUES FOR IMAGE REGISTRATION

Dr. W. F. Webber

McDonnell Douglas Astronautics Company
5301 Bolsa Avenue
Huntington Beach, California 92647

I. ABSTRACT

Techniques are developed for determining spatial or geometric distortions between two images of the same scene. The first procedure is iterative linearized least squares estimation (LLSE) for determining small geometric distortions between images. Error variances for these estimators are derived which are interpreted as noise-to-signal ratios for translational and rotational registration. The natural measure of the signal strength of an image for translational registration obtained from these variances is used to establish threshold settings in a new algorithm for fast translational registration. This algorithm belongs to the class of sequential similarity detection algorithms (SSDA's) recently developed for translational registration. Finally, an implementation of an image registration system incorporating all these techniques is described.

II. INTRODUCTION

Image registration is a procedure to determine the spatial best fit between two images that overlap the same scene. Registration is basic to image processing systems since two images of the same scene cannot be meaningfully compared (to determine temporal changes, for example) without having the images in registration. Automatic analysis of remotely sensed imagery will necessitate accurate registration.

Several digital techniques have been used for registering imagery. Principal among these are cross-correlation, normalized cross-correlation and minimum distance criteria. Fast algorithms for determining translational differences between images have been developed in recent years for all these techniques. Efficient algorithms for determining other spatial or geometric distortions such as horizontal scale or rotational differences have not been extensively developed. The primary mechanisms and error types associated with such distortions have been investigated(Bernstein and Silverman, 1971). In some applications determining these distortions will be critical to the success of further analyses on the images.

In this paper a technique, iterative linearized least squares estimation, is derived for efficiently estimating all distortions between two images of the same scene. Also techniques are suggested for improving fast translational registration procedures. A particular implementation of an image registration system is discussed which incorporates all these procedures.

III. ITERATIVE LINEARIZED LEAST SQUARES ESTIMATORS

Two images W and Z of the same scene are to be registered. The image W, referred to as the reference image, covers only a portion of the total scene described by the search image Z. W is assumed to be of good quality, i.e., no clouds are present, contrast is good and geometrical distortions are negligible. This reference image is just one of a large set of such small image available for registering new images as they become available. The search image, on the other hand, is of relatively unknown quality. Some cloud cover may be present along with geometrical

Reprinted from *Proc. IEEE Conf. on Machine Processing of Remotely Sensed Data*, Oct. 16-18, 1973, pp. 1B-1-1B-7.

distortions. Techniques are developed in this section for estimating the remaining geometrical distortions present in Z after an initial translational registration has been completed.

Denote the reference image by $W = \{w(j,k)\}$ where $j = 1,2,\ldots,M_W$, $k=1,2,\ldots,N_W$ and $w(j,k)$ is the gray level of the pixel (picture element) at the array location (j,k). Similarly the search image to be registered is $Z = \{z(j,k)\}$ where $j = 1,2,\ldots,M_Z$ and $k=1,2,\ldots,N_Z$. Since the scene covered by W is assumed to be contained in that of Z, necessarily $M_Z>M_W$ and $N_Z>N_W$. The best match sub-image of Z compared with W obtained by an initial translational registration is $Z_o = \{z(j_o+j,k_o+k)\}$ where $j = 0,1,\ldots,M_W-1$ and $k = 0,1,\ldots,N_W-1$.

It is assumed that the relationship between points in W and Z_o is of the form

$$\alpha z(x,y) = w(x',y') + n(x,y). \tag{1}$$

The coordinate systems for both Z_o and W centered at the origins of these arrays with axes parallel to the array rows and columns are denoted (x,y). The coordinate system (x',y') referred to in (1) is assumed to be related to the (x,y) system of W by the linear transformation

$$\begin{aligned} x' &= (1+c_{11})x+c_{12}y+c_{13} \\ y' &= c_{21}x+(1+c_{22})y+c_{23} \end{aligned} \tag{2}$$

where the c's are all small unknown constants. Such a transformation corresponds to small two dimensional translational, horizontal scaling and rotational errors. Anuta (Anuta, 1971) assumed this transformation to explain the total distortion over an image. At the least, it should explain local distortion in an image,which is its purpose in this paper. The unknown scale factor in (1) approximately describes the differences in brightness levels between W and Z_o. The additive noise surface $n(x,y)$ in (1) describes the remaining differences between W and Z_o.

The registration problem in this formulation is to estimate the unknown c's. Linearized least squares estimators (LLSE's) of these terms will be discussed here. To obtain such estimates write the Taylor series expansion of $w(x',y')$ in terms of $w(x,y)$ as

$$w(x',y') = w(x,y) + (x'-x)w_x(x,y) + (y'-y)w_y(x,y) + 0\,(\Delta^2) \tag{3}$$

where $w_x = \partial w/\partial x$, $w_y = \partial w/\partial y$ and $0\,(\Delta^2)$ refers to the remaining terms which contain factors of the form $(x'-x)^p(y'-y)^q$ such that $p+q\geq 2 (p,q\geq o)$. Since the c's are all small, these terms will be assumed negligible. From (1) and (3) the difference

$$\alpha z(x,y) - w(x,y) = (x'-x)w_x(x,y) + (y'-y)w_y(x,y) + n(x,y) + 0\,(\Delta^2) \tag{4}$$

is approximately linear in the unknowns, i.e., using (2) and (4)

$$\alpha z(x,y) - w(x,y) = (c_{11}x+c_{12}y+c_{13})w_x(x,y) + (c_{21}x+c_{22}y+c_{23})w_y(x,y) + n(x,y) + 0\,(\Delta^2) \tag{5}$$

This suggests obtaining LLSE's for α and the c's by minimizing the quadratic

$$\phi(\alpha,c) = \sum_x \sum_y (\alpha z-w-(c_{11}x+c_{12}y+c_{13})w_x -(c_{21}x+c_{22}y+c_{23})w_y)^2. \tag{6}$$

The double summation is over all the points (x,y) of the sub-images W and Z_o. Also the arguments of all quantities such as $w(x,y)$ have been suppressed. These conventions will be used in the remainder of this section. A generalized least squares procedure taking into account the covariance structure of the noise surface will not be considered here because it is believed to be of limited importance in applications.

The minimization of (6) is readily accomplished by setting all partial derivatives of ϕ in respect to the unknowns to zero and solving these equations for the unknowns.

The estimation of only translational and rotational errors will be examined in detail in the remainder for simplicity of presentation. In this case, to the same order of approximation as above

$$\begin{aligned} x' &= (x+x_o)\cos\theta_o + (y+y_o)\sin\theta_o = x+x_o + y\theta_o + 0\,(\Delta^2), \\ y' &= (y+y_o)\cos\theta_o - (x+x_o)\sin\theta_o = y+y_o - x\theta_o + 0\,(\Delta^2) \end{aligned} \tag{7}$$

so that the quadratic form to be minimized is

$$\phi(x_o,y_o,\theta_o) = \sum_x \sum_y (u - x_o w_x - y_o w_y - \theta_o (y w_x - x w_y))^2, \tag{8}$$

where $u(x,y) = z(x,y) - w(x,y)$. The normal equations for the LLSE's are found to be

$$\begin{aligned}
\sum_x \sum_y w_x u &= \hat{x}_o \sum_x \sum_y w_x^2 + \hat{y}_o \sum_x \sum_y w_x w_y + \hat{\theta}_o \sum_x \sum_y (y w_x - x w_y) w_x \\
\sum_x \sum_y w_y u &= \hat{x}_o \sum_x \sum_y w_x w_y + \hat{y}_o \sum_x \sum_y w_y^2 + \hat{\theta}_o \sum_x \sum_y (y w_x - x w_y) w_y \\
\sum_x \sum_y (y w_x - x w_y) u &= \hat{x}_o \sum_x \sum_y (y w_x - x w_y) w_x + \hat{y}_o \sum_x \sum_y (y w_x - x w_y) w_y + \hat{\theta}_o \sum_x \sum_y (y w_x - x w_y)^2
\end{aligned} \tag{9}$$

If it is further assumed that only translational errors are present, the estimators for x_o and y_o are given by

$$\begin{aligned}
\hat{x}_o &= \Big(\sum_x \sum_y w_y^2 \sum_x \sum_y w_x u - \sum_x \sum_y w_x w_y \sum_x \sum_y w_y u\Big)/D \\
\hat{y}_o &= \Big(\sum_x \sum_y w_x^2 \sum_x \sum_y w_y u - \sum_x \sum_y w_x w_y \sum_x \sum_y w_x u\Big)/D
\end{aligned} \tag{10}$$

where $D = \sum_x \sum_y w_x^2 \sum_x \sum_y w_y^2 - (\sum_x \sum_y w_x w_y)^2$. Similarly if only θ_o is to be estimated the LLSE is given by

$$\hat{\theta}_o = \sum_x \sum_y (y w_x - x w_y) u / \sum_x \sum_y (y w_x - x w_y)^2 \tag{11}$$

A. ERROR ANALYSIS

The variances of the last three estimators can be evaluated in the important case that $\hat{x}_o = \hat{y}_o = \hat{\theta}_o = 0$ and that the noise surface $\{n(x,y)\}$ is white with zero mean, i.e., $En(x,y) = 0$ and $En(x,y)n(x',y') = \sigma_N^2 \delta_{x-x',y-y'}$, where $\delta_{x,y}$ is the Kronecker delta. In this situation $u(x,y) = n(x,y)$ so that $E\hat{x}_o = E\hat{y}_o = E\hat{\theta}_o = 0$ and

$$\begin{aligned}
\mathrm{Var}(\hat{x}_o) &= \sigma_N^2 \sum_x \sum_y w_y^2 / D \\
\mathrm{Var}(\hat{y}_o) &= \sigma_N^2 \sum_x \sum_y w_x^2 / D \\
\mathrm{Var}(\hat{\theta}_o) &= \sigma_N^2 \sum_x \sum_y (y w_x - x w_y)^2
\end{aligned} \tag{12}$$

where D was defined above. The mean square radial error corresponding to the translational error estimates is given by

$$E(\hat{x}_o^2 + \hat{y}_o^2) = \sigma_n^2 / (D / \sum_x \sum_y (w_x^2 + w_y^2)). \tag{13}$$

This noise-to-signal type ratio is useful in summarizing the translational accuracy of a spatial registration system. The positive quantity $D / \sum_x \sum_y (w_x^2 + w_y^2)$ can be interpreted as a measure of the two dimensional signal strength of the reference image W for purpose of translational registration. This quantity will be of primary importance in the next section for establishing threshold settings in sequential similarity detection algorithms used for translational registration. The quantity $\sum_x \sum_y (y w_x - x w_y)^2$, similarly, is to be interpreted as a measure of the two dimensional signal strength of W for rotational registration.

B. NON-ZERO MEAN NOISE

In applications there is no assurance that the noise is zero mean. The usual approach to correct for non-zero mean noise would be to remove the sample means from both W and Z_o before evaluating the error estimates. It is recommended, however, that least squares planes be removed from both of these arrays to protect against two dimensional trends as well as non-zero means in the noise surface. Details for this procedure are given in the next section.

C. PARTIAL DERIVATIVE ESTIMATES

The LLSE's for the spatial errors have been derived assuming that the partial derivatives $w_x(x,y), w_y(x,y)$ are known at all lattice points for which $w(x,y)$ is defined.

In general these derivatives will not be available requiring that estimates be used in their place. The simplest such estimates are of the form

$$\hat{w}_x(x+1/2,y+1/2)=(w(x+1,y+1 \quad w(x,y))/\sqrt{2}$$

$$\hat{w}_y(x+1/2,y+1/2)=(w(x,y+1)-w(x+1,y))/\sqrt{2}$$

assuming the spatial distance between lattice points to be unity. These estimates actually correspond to a reference system translated by 1/2 grids units in x and y and rotated counter-clockwise through $\pi/4$ radians from the original reference system for W. The difference terms u(x,y) should also be adjusted to the new reference system, i.e., use

$$\hat{u}(x+1/2,y+1/2)=(u(x,y)+u(x+1,y)+u(x,y+1)+u(x+1,y+1))/4.$$

The estimates $\hat{x}_o,\hat{y}_o,\hat{\theta}_o$ obtained with these approximations can readily be transformed to the original system.

Computational requirements could be reduced if pre-computed estimates of the partial derivatives of W of the form

$$\hat{w}_x(x,y)=\sum_{k=1}^{m} a_k(w(x+k,y)-w(x-k,y))$$

$$\hat{w}_y(x,y)=\sum_{k=1}^{m} a_k(w(x,y+k)-w(x,y-k))$$

were to be used since no correction of the u(x,y) need be made in this case. Only the first estimates for w_x and w_y have been used to evaluate the LLSE technique.

D. ITERATION OF THE ESTIMATES

The many approximations and assumptions required to obtain the LLSE's make the actual worth of these estimators, at best, uncertain.

If the estimation procedure is useful,then iteration on the solution will generally improve the estimates by reducing the error due to the linearization assumption of (5).

The first step in the iteration procedure is to obtain the estimates $\hat{x}_o,\hat{y}_o,\hat{\theta}_o$ using the reference image W and the best match sub-image Z_o obtained in the initial translational registration. The best match sub-image, Z_1, corresponding to these error estimates is interpolated (using nearest neighbor four point linear interpolation for example) from the search image Z. The LLSE's, $\hat{x}_1,\hat{y}_1$ and $\hat{\theta}_1$, of the remaining spatial errors are then calculated using W and Z_1. Similarly the sub-image Z_n is interpolated from Z corresponding to the previous best estimates of the spatial errors $\hat{x}_{n-1}$, $\hat{y}_{n-1}$ and $\hat{\theta}_{n-1}$. The next LLSE's $\hat{x}_n,\hat{y}_n,\hat{\theta}_n$ are obtained from W and Z_n. This procedure can be continued until the estimates converge satisfactorily.

This iterative procedure has been evaluated using pseudo-random signal and noise surfaces. The reference image W (a 16x16 array) was obtained by linear interpolation from the search image to simulate offset and rotational errors. The rotation used to form W was 10 degrees and the offset between the sampling lattices of W and Z was 1/2 the sampling interval in both x and y. White noise, noise with the same spectrum as the search image and multiplicative noise to simulate white cloud covering were used to evaluate the sensitivity of the procedure. The multiplicative noise was simulated by first generating a correlated Gaussian noise surface and then setting the surface to zero if the original surface were below a threshold value and to one if it were above. The threshold value was selected to give a specified percent of cloud cover over the scene. In all cases very similar results were obtained. It was found that whenever the amount of noise was sufficiently low to allow accurate translational registration that the initial LLSE's reduced the spatial errors. However, these estimates were always smaller than the remaining errors. The iteration procedure converged to values that were roughly consistent with the variation predicted by the error analysis previously given.

The number of iterations required to reduce the translational error to within half a sampling interval in both x and y and the rotational error to within a degree of their final values was greater than fifteen in some cases. Several ad hoc methods were evaluated to increase the rate of convergence. One of these procedures resulted in convergence (as defined above) in only three or four iterations. This particular procedure is to use twice the horizontal error estimates and four times the rotational error estimates until the rotational error estimate changes sign. The actual error estimates are used in subsequent iteration steps.

These results indicate that the iterative LLSE approach to precision image registration would be useful for production processing of remotely sensed imagery. Techniques for increasing the rate of convergence of the interactive procedure should be investigated further to assure a technique is employed that is insensitive to the types of error sources expected in a given application.

IV. SEQUENTIAL SIMILARITY DETECTION ALGORITHMS

A very efficient class of algorithms for translational registration has recently been suggested (Barnea and Silverman, 1972). These sequential similarity detection algorithms (SSDA's) are reported to be one to two orders of magnitude faster than previously discussed algorithms such as fast cross-correlation (Anuta, 1971).

As an example of algorithms of this type consider that a reference image W and a search image Z, as described in the last section, are to be registered. One method useful for this purpose is the minimum p-distance criterion, which is to find the rectangular subset Z_0 of Z which minimizes

$$d_p^p(j',k')=\sum_{j=1}^{M_w}\sum_{k=1}^{N_w}|Z(j'+j,k'+k)-w(j,k)|^p \tag{14}$$

where $p>1$. (This criterion with $p = 1$ or 2 is commonly used. The $p = 2$ case is referred to as the Euclidean distance criterion. Computational considerations usually determine the choice of p .) Now consider a random sequence related to the distance measure. Let (j_s,k_s) for $s = 1,2\ldots, M_wN_w$ be a sequence of random samples without replacement from the lattice of points on which W is defined. Define the random sum

$$S_p(r;j',k')=\sum_{s=1}^{r}|Z(j'+j_s,k'+k_s)-w(j_s,k_s)|^2 \tag{15}$$

for $r = 1,2,\ldots,M_wN_w$ which increases monotonically to $d^p(j',k')$ as r increases to M_wN_w. The essential feature of all SSDA's is that the rate of increase of $S_p(r\,;j',k')$ is used in determining the similarity between the sub-image of Z and the reference image. If this quantity increases rapidly then the two images are dissimilar. Barnea and Silverman suggested using a threshold to measure an average rate of increase. In the simplest case a constant threshold is used and the number of steps required for each sub-image to reach this threshold is the measure of the rate of increase of S_p. The importance of this procedure is that the calculation of S_p terminates whenever the threshold is crossed, resulting in the substantial reduction in computational time quoted above. The primary difficulty with the technique is establishing a useful threshold level. Too high a threshold value reduces the technique to the original p-distance criterion, whereas too low a level can give rise to gross registration errors. Barnea and Silverman suggested adaptive threshold setting procedures to reduce the difficulties associated with a constant threshold.

Another approach is to base the threshold setting on the signal strength of W and the allowable noise level associated with W for accurate registration, i.e., a signal-to-noise criterion could be used to set the threshold. In the last section it was shown that a natural signal-to-noise ratio for translational registration assuming additive white noise with variance $\sigma_N{}^2$ is (using (13))

$$\sigma_S{}^2/\sigma_N{}^2=((\sum_x\sum_y w_x{}^2\sum_x\sum_y w_y{}^2-(\sum_x\sum_y w_x w_y)^2)/\sum_x\sum_y(w_x{}^2+w_y{}^2))\ \sigma_N{}^2 \tag{16}$$

defining the signal strength measure $\sigma_s{}^2$. Suppose that $z(x,y)=w(x,y)+n(x,y)$ for some sub-image Z_0 of the search image Z, i.e., assume that there is no geometric distortion. In this case the Euclidean distance criterion $d_2{}^2\ (j',k')$ will reduce to

$$d_2^2(j_o,k_o)=\sum_j\sum_k\ n^2(j,k)$$

for (j_o,k_o) corresponding to Z_o. The mean of this quantity is

$$Ed_2^2\ (j_o,k_o) = M_wN_w\sigma_N{}^2$$

so that at low to moderate noise levels the minimum value of the discrete function $d_2^2\ (j',k')$ is roughly proportional to $\sigma_N{}^2$. In order to assure a given level of registration accuracy the threshold should be set proportional to σ_s^2 as indicated by (16). The threshold must also be set sufficiently low that values of $d_2^2(j',k')$ away from the match point are always greater than the threshold even in the case of no noise.

A. TECHNIQUES FOR STABILIZING IMAGE SIMILARITY CALCULATIONS

The primary difficulty in using σ_s^2 to establish the threshold setting is that the minimum and shape of $d_2^2(j,k)$ are sensitive to low wave-number error sources. In general, signal energy content at low wave-numbers does not contribute to the signal strength of an image for registration (the quantity σ_s^2 is a function only of the partial derivatives of W) so that perhaps the best procedure to alleviate this sensitivity is to apply a digital high-pass filter to the reference image and the search image before the SSDA calculation. However, the computational advantages of the SSDA algorithm would be essentially lost if this approach were taken. Removing the mean from the reference image and each subset of the search image before the similarity calculations helps to reduce this sensitivity, but a better approach requiring only a moderate computational increase over mean removal is to remove the least squares plane from each sub-image.

Consider a sub-image $Z_o = \{z(x,y)\}$ where $x=j_o,j_o+1,\ldots,j_o+M_w-1$ and $y=k_o,k_o+1,\ldots,k_o+N_w-1$. The spatial center of this data set is $(x_oy_o)=(j_o+(M_w-1)/2,k_o+(N_w-1)/2)$. A plane defined on this lattice can be written as

$$P(x,y)=a(x-x_o)+b(y-y_o)+c$$

so that $P(x_o,y_o)=c$. The least squares best fit plane of this form to the sub-image Z_o minimizes

$$\phi(a,b,c) = \sum_x \sum_y (z-a(x-x_o)-b(y-y_o)-c)^2.$$

The solution is

$$\begin{aligned}\hat{a} &= \frac{12}{M_wN_w(M_w^2-1)} \sum_x \sum_y (x-x_o)z,\\ \hat{b} &= \frac{12}{M_wN_w(N_w^2-1)} \sum_x \sum_y (y-y_o)z,\\ \hat{c} &= \frac{1}{M_wN_w} \sum_x \sum_y z.\end{aligned} \tag{17}$$

B. SIMULATION RESULTS

Simulation studies have been used to evaluate the technique for setting the SSDA threshold level using the signal strength measure σ_s^2 and the least squares plane removal procedure. It was found that setting the threshold at three to four times σ_s^2 resulted in the true minimum being found in all cases at low to moderate noise levels. At high noise levels the original Barnea and Silverman constant threshold approach can be used. It is not necessary to know the noise level a priori in any case. An added advantage to using the method suggested here for setting thresholds is that whenever a sum is found that does not reach the threshold the search procedure can be stopped since the region of the registration position has been found. This early termination results in considerable time savings in many cases.

V. REGISTRATION SYSTEM IMPLEMENTATION

The implementation of a production image registration system has to be based on many considerations not covered in this paper. In particular the expected data quality and registration accuracy requirements will strongly affect design parameters. One implementation that has been extensively studied will be described here.

The initial translation registration is accomplished in two stages. The first stage is a coarse translation registration. The SSDA threshold is initially established using the signal strength measure σ_s^2 as discussed in the last section. A spiral search procedure starting with the a priori most probable sub-image of the search area is employed to locate the initial translation registration estimate. Let (x_o,y_o) be the center position of the sub-image at the center of the search image Z. The coordinates of the center points of subsequent sub-images in the spiral search procedure are (x_o+r,y_o),(x_o+r,y_o+r),(x_o,y_o+r),(x_o-r,y_o+r),(x_o-r,y_o), (x_o-r,y_o-r),(x_o,y_o-r),(x_o+r,y_o-r),(x_o+2r,y_o-r), etc. Taking $r = 2$ in this search pattern will significantly reduce the required number of calculations without compromising system performance in many applications. The parameters of the least squares planes of the sub-images can be efficiently updated for this search pattern and the resultant arithmetic error accumulation is minimized.

After the coarse registration is completed the residue at the best match point is used as the new threshold. A fine search procedure is then started about this best match point taking every sub-image (r=1). This procedure is continued until a local minimum is obtained.

At the completion of the fine search procedure the iterative least squares estimates of the remaining registration errors are computed. The actual parameters estimated in this procedure will depend on the types of geometrical distortions expected in the data. The number of iterations can be fixed or, if high precision is required, determined by the rate of convergence observed with the actual data set.

It should be noted that the user generally can select the set of reference images employed for registration. These should be chosen to have large signal strength, σ_s^2. The final registration estimates may not be accurate. This can roughly be determined by computing the sample variance of the difference between W and Z_n obtained in the final iteration of the LLSE's. This quantity can then be compared with σ_s^2 to assess the registration accuracy.

VI. ACKNOWLEDGEMENT

The research reported here was partially supported by the Air Force Office of Scientific Research (Air Force Systems Command) under Contract F44620-71-C-0052.

REFERENCES

P. E. Anuta, Spatial Registration of Multispectral and Multidimensional Digital Imagery Using Fast Fourier Transform Techniques, IEEE Transactions Geosci, Electron, GE-8, October 1970.

R. Bernstein and H. Silverman, Digital Techniques for Earth Resource Image Data Processing, American Institute of Aeronautics and Astronautics (AIAA) Paper No. 71-978, October 1971.

D. I. Barnea and H. F. Silverman, A Class of Algorithms for Fast Digital Image Registration, IEEE Transactions on Computers, C-21, pp. 179-186, February 1972.

MULTITEMPORAL GEOMETRIC DISTORTION CORRECTION UTILIZING THE AFFINE TRANSFORMATION

R. A. Emmert and C. D. McGillem

Lawrence Livermore Laboratory, Livermore, California;
Laboratory for Applications of Remote Sensing, Purdue University, West Lafayette, Indiana

I. ABSTRACT

In the analysis of multitemporal remotely sensed imagery, it is necessary to place these data into registration. To implement this operation the data are divided into subimages, and the misregistration between the data subsets is modeled by an affine transformation. The properties of the Fourier transform of a two-dimensional function under the affine transformation are given, and examples of these relations between the spatial and spatial frequency domains are shown. Techniques for the estimation of the coefficients of the distortion model using the spatial frequency information are developed, and an example of the use of this method for the correction of line scanner imagery is given.

II. INTRODUCTION

In analyzing imagery obtained by remote sensing devices, it is frequently necessary to compare data taken at different times on a point by point basis. In order to carry out such a comparison, it is necessary to register, or overlay, one set of data on the other. Thus, it is required to process one set of data such that its image under an appropriate transformation is in proper geometrical registration with the other data.

Before implementing a registration operation, a model characterizing the misregistration must first be introduced, and the choice of the size of the data set for which the model is to be valid determines its complexity. If a large data set is chosen, the resulting model is most often complex. However, if the data are divided into smaller subimages, or regions, the resultant model for each of these regions is simple.

In this study the misregistration between data regions is modeled by an affine transformation, and the properties of the Fourier transform of a two-dimensional function under this transformation are given. The coefficients of the geometrical distortion component of this model are readily estimated in the spatial frequency domain, and an example of the use of spatial frequency information for the correction of distortion in line scanner imagery is given.

III. A REGIONAL MISREGISTRATION MODEL

A general model for characterizing misregistration between two sets of remotely sensed data is the two-dimensional polynomial

$$y_i = \sum_{j=0}^{N-1} \sum_{k=0}^{N-1} {}^{i}a_{jk} \; x_1^j \; x_2^k , \qquad i = 1,2 \tag{1}$$

where $\underline{x}$ and $\underline{y}$ are respectively the coordinate systems of the arbitrarily chosen reference and background data sets. This model is useful because the linearity of the equations as a function of the coefficients permits least-squares procedures to be used for determining these coefficients. Table

Reprinted from *Proc. IEEE Conf. on Machine Processing of Remotely Sensed Data*, Oct. 16-18, 1973, pp. 1B-24-1B-32.

1 lists several polynomials obtained from this model and includes the number of coefficients that would be required to utilize the corresponding model.

For many data sets regional misregistration can be represented as having the following four components: (1) scale, (2) rotation, (3) skew, and (4) displacement. Such misregistration can be completely characterized by means of the affine transformation given by

$$\underline{y} = \underline{H}\,\underline{x} = \underline{A}\,\underline{x} + \underline{t}$$

where

$$\underline{y} = \begin{bmatrix} y_1 \\ y_2 \end{bmatrix}, \quad \underline{x} = \begin{bmatrix} x_1 \\ x_2 \end{bmatrix}, \quad \underline{t} = \begin{bmatrix} t_1 \\ t_2 \end{bmatrix} \tag{2}$$

and

$$\underline{A} = \begin{bmatrix} a_{11} & a_{12} \\ a_{21} & a_{22} \end{bmatrix}$$

The non-singular matrix $\underline{A}$ characterizes the geometrical components of the misregistration. The quantities are illustrated in Fig. 1.

The usefulness of the affine model results from the fact that the misregistration can be interpreted as consisting of two components: the first is the characterization of the geometrical distortion by a linear model, and the second is the displacement of the coordinate systems. Straightforward search techniques are available for estimating the distortion matrix coefficients by determining the differences between the moduli of the Fourier transforms of these data.

IV. TWO-DIMENSIONAL FOURIER TRANSFORM UNDER AN AFFINE TRANSFORMATION

The two-dimensional Fourier Transform of the function $f(\underline{x})$ is defined as

$$F(\underline{u}) = \int_{-\infty}^{\infty}\!\!\int f(\underline{x}) \exp\{-j2\pi(\underline{u},\underline{x})\}\, d\underline{x} \tag{3}$$

and the inverse is

$$f(\underline{x}) = \int_{-\infty}^{\infty}\!\!\int F(\underline{u}) \exp\{j2\pi(\underline{u},\underline{x})\}\, d\underline{u} \tag{4}$$

where

$$\underline{u} = \begin{bmatrix} u_1 \\ u_2 \end{bmatrix}$$

and the notations $(\underline{u},\underline{x})$ denotes the inner product of the vectors $\underline{u}$ and $\underline{x}$. The following properties of the transform of the function $f(\underline{x})$ under an affine transformation $\underline{H}$ are readily established.

(1) Similarity

If

$$f(\underline{x}) \xrightarrow{\underline{H}} f(\underline{H}\,\underline{x}) = g(\underline{y}), \tag{5}$$

then

$$G(\underline{v}) = \mathcal{F}\{f(\underline{H}\,\underline{x})\} = \frac{1}{|J|} \exp\{j2\pi(\underline{u},\underline{A}^{-1}\underline{t})\}\; F\{(\underline{A}^{-1})^T\underline{u}\} \tag{6}$$

where

$$J = \frac{\partial(y_1,y_2)}{\partial(x_1,x_2)}$$

is the Jacobian of the transformation.

(2) Energy Spectrum and Modulus Independent of $\underline{t}$

The energy density spectrum of f $(\underline{x})$ is

$$S(\underline{u}) = | F(\underline{u}) |^2, \tag{7}$$

and the linear phase term cancels. The modulus is

$$M(\underline{u}) = \{ | F(\underline{u}) |^2 \}^{1/2} \tag{8}$$

and the result is immediate.

(3) Symmetry of the Modulus

For real signals it readily follows from (3) that

$$F(-\underline{u}) = F^*(\underline{u}) \tag{9}$$

where the asterisk denotes the complex conjugate.

(4) Relation of Coordinate Systems

From (6) it follows that the relation between coordinate systems $\underline{u}$ and $\underline{v}$ of the transform moduli is

$$\underline{v} = (\underline{A}^{-1})^T \underline{u} \tag{10}$$

V. ESTIMATION OF DISTORTION MATRIX COEFFICIENTS

In a number of special cases the distortion matrix A reduces to a simpler form, and it is illustrative to examine the effects of various geometrical distortions in these domains. For scaling changes the distortion matrix is

$$\underline{A} = \begin{bmatrix} a_{11} & 0 \\ 0 & a_{22} \end{bmatrix} \tag{11}$$

where the scaling is assumed to be along the coordinate axes. The relationship between the spatial frequency domain coordinates is then

$$\underline{v} = (\underline{A}^{-1})^T \underline{u} = \begin{bmatrix} \frac{1}{a_{11}} & 0 \\ 0 & \frac{1}{a_{22}} \end{bmatrix} \underline{u} \tag{12}$$

This relation is illustrated in Fig. 2 where the dimensions of the rectangular blocks in the spatial domain have a width which is two times the height.

For rotation the distortion matrix is

$$\underline{A} = \begin{bmatrix} \cos\theta & \sin\theta \\ -\sin\theta & \cos\theta \end{bmatrix} \tag{13}$$

which is an orthogonal matrix. Thus

$$\underline{v} = (\underline{A}^{-1})^T \underline{u} = \underline{A}\,\underline{u} \tag{14}$$

and the transform of the distorted data is also rotated by the angle θ; this distortion is illustrated by the modulus of the transform of the ideal data shown in Fig. 3.

For skew distortion the geometric distortion matrix becomes

$$\underline{A} = \begin{bmatrix} 1 & 0 \\ a_{21} & 1 \end{bmatrix} \tag{15}$$

and in the spatial frequency domain the coordinates are related by

$$\underline{v} = \begin{bmatrix} 1 & -a_{21} \\ 0 & 1 \end{bmatrix} \underline{u} \tag{16}$$

This distortion and its associated modulus of the Fourier transform are shown in Fig. 4.

The overall distortion matrix $\underline{A}$ is some combination of these component distortions.

The differences between the moduli of the transforms of two corresponding regions of data provide all the information required for determining values for the coefficients of the linear distortion matrix. The zero spatial frequency component is mapped into the origin of the transform domain and higher frequency components are mapped into locations proportional to both the value of their spatial frequency and in a direction from the origin characteristic of the orientation of the component in the spatial domain. The moduli of the two-dimensional Fourier transforms of corresponding regions of data are invariant under the coordinate shift $\underline{t}$; thus the coordinate systems can be chosen arbitrarily. It is further assumed that for data of a reasonably homogenous composition, small shifts Δd of the data aperture will yield moduli which can be assumed to be unchanged for purposes of this study.

The class of agricultural imagery, which is of principal interest in this study, typically consists of a collection of rectangular fields with each of the fields having essentially a homogeneous ground cover. It has been observed that the modulus of the two-dimensional Fourier transform of a data set from this class of imagery typically exhibits a simple structure with the property that a majority of the energy in the spatial frequency domain is concentrated along linear loci or rays perpendicular to the field boundaries.

The rotative and skew components of the misregistration are obtained by determining the directions of the loci of energy maxima in each of the two transform moduli. From these angles it is then possible to compute the parameters of the transformation corresponding to the geometrical distortion in the images.

Referring to Fig. 5 the geometrical distortion matrix for skew and rotation only is given by

$$\underline{A} = \begin{bmatrix} \cos\phi_1 & \pm \sin\phi_1 \\ a_{21}\cos\phi_1 \mp \sin\phi_1, & \pm a_{21}\sin\phi_1 + \cos\phi_1 \end{bmatrix} \tag{17}$$

where

$$a_{21} = \cot(\theta_2 - \phi_1).$$

VI. APPLICATION TO MEASURED DATA

As an example of the application of this technique, images obtained from an airborne multispectral scanner system on two different days have been analyzed and the geometrical distortion components determined. The area for which the correction was computed and the corresponding Fourier transform moduli are shown by Fig. 6. The fast Fourier transform algorithm was used for computing the transform of these data. In each of these pictures the transformed data have a logarithmic amplitude scale, and the gray scale employed in the display system is linear. The size of the data set being transformed is 128 x 128 picture elements.

A regression line representing the best least-squares fit of the maximi of the modulus along a radial line was used to obtain the angle between each loci and its coordinate axis. The angular quantities given by this algorithm, measured with respect to the u_1 axis, are as follows

data set A	-7.18×10^{-2} radian
	1.57 "
data set B	-3.30×10^{-2} "
	1.52 "

With a hypothetical transform coincident with the coordinate axes used as a reference data transform, the resultant angular quantities defined in Fig. 5 are

data set A $\quad \phi_1 = -4.1°$

$\phi_2 = 0°$

data set B $\quad \phi_1 = -1.9°$

$\phi_2 = -3.3°$

The two-dimensional polynomial was used to implement the correction. Rather than mapping one data set into the other, it was chosen to rectify each data set. Examining the data sets in both the spatial and spatial frequency domain, the angular quantities chosen for the rectification differed slightly from those above. The numerical values used for the rectification are

data set A $\quad \phi_1 = -3.7°$

$\phi_2 = 0°$

data set B $\quad \phi_1 = -1.7°$

$\phi_2 = -3.3°$

with the reference transform angular quantities having the values

$$\theta_1 = 0°$$

$$\theta_2 = 90°$$

for each of the data sets.

The distortion matrices are then

$$\underline{A}_1 = \begin{bmatrix} 0.9979 & -0.0645 \\ 0.0 & 1.002 \end{bmatrix} \tag{18}$$

$$\underline{A}_2 = \begin{bmatrix} 0.9999 & -0.0297 \\ 0.0576 & 0.9987 \end{bmatrix} \tag{19}$$

The results of this geometrical correction on the spatial data are shown in Fig. 7. It is emphasized that the correction is applicable only in the center of each of these pictures as the distortion introduced by the aircraft motion is a dynamic quantity.

VII. REFERENCES

Goodman, J. W., "Introduction to Fourier Optics," McGraw-Hill, New York, 1968.

Hadley, G., "Linear Algebra," Addison Wesley, Reading, 1961.

Lendaris, G. and G. Stanley, "Diffraction-Pattern Sampling for Automatic Pattern Recognition," Proc. IEEE, vol. 58, no. 1, pp. 198-216, February, 1970.

VIII. TABLES AND ILLUSTRATIONS

Table 1. Two-Dimensional Polynomial Misregistration Models

Common Name	Degree N-1	No. Parameters To Be Determined
DISPLACEMENT	0	2
LINEAR	1	4
AFFINE	1	6
PROJECTION	2	8
QUADRATIC	2	12
CUBIC	3	20

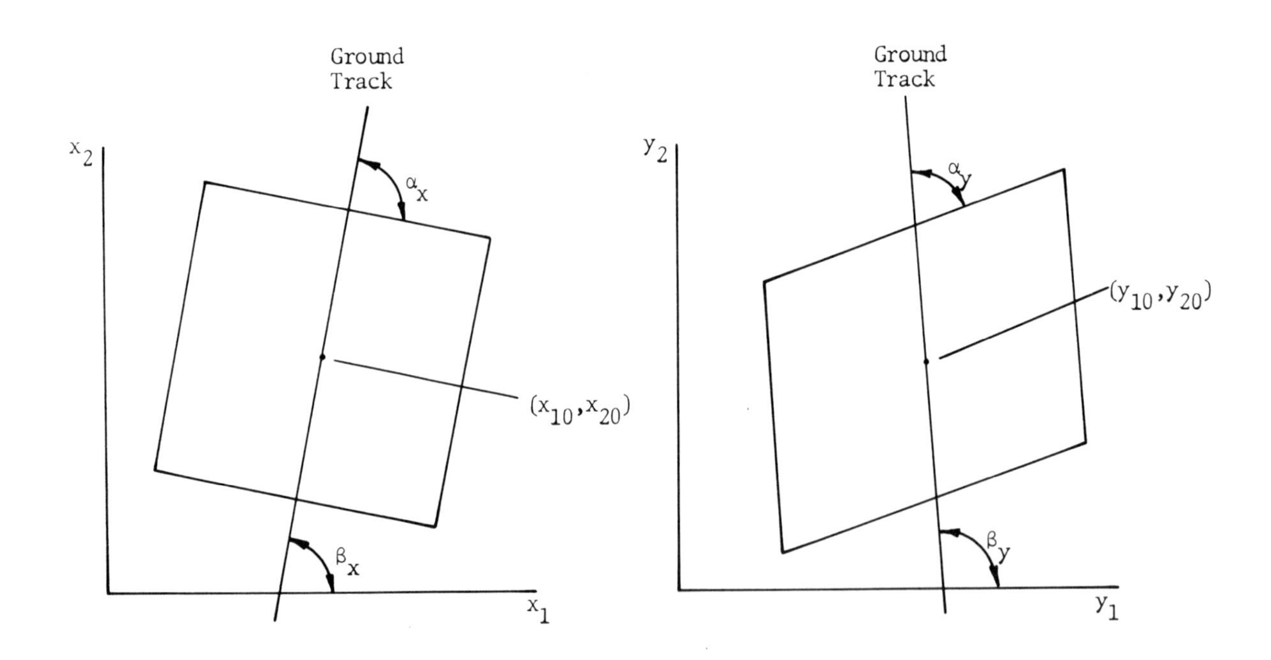

Figure 1. Misregistration Between Data Regions

Block Pattern

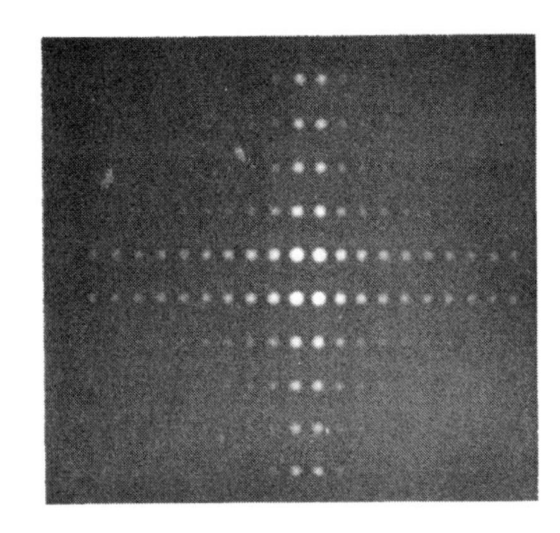

Modulus of Fourier Transform

Figure 2. Modulus of Fourier Transform of Block Pattern

Rotated Block Pattern

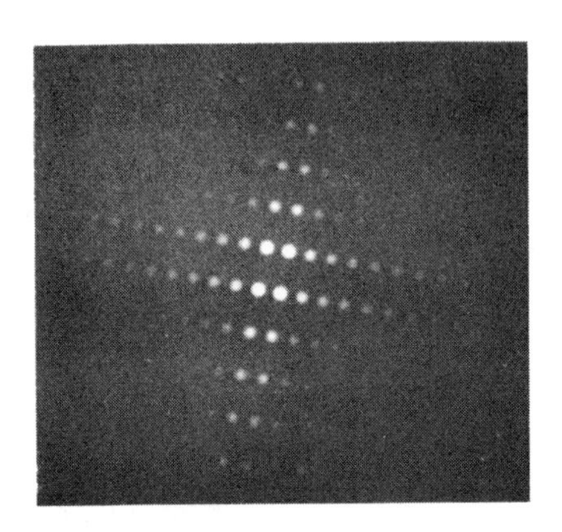

Modulus of Fourier Transform

Figure 3. Modulus of Fourier Transform of Rotated Block Pattern

Block Pattern with Skew

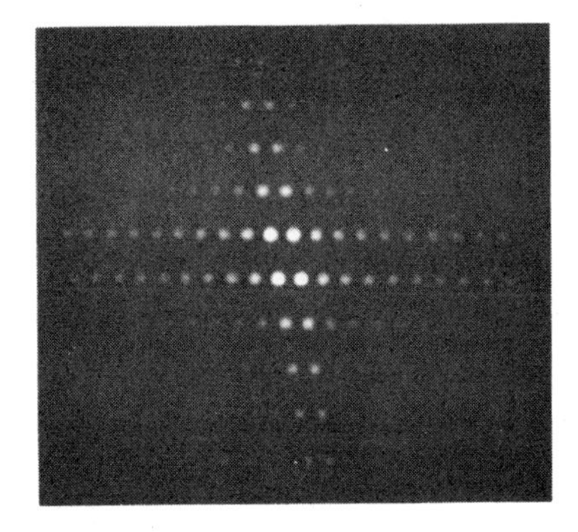

Modulus of Fourier Transform

Figure 4. Modulus of Fourier Transform of Skewed Block Pattern

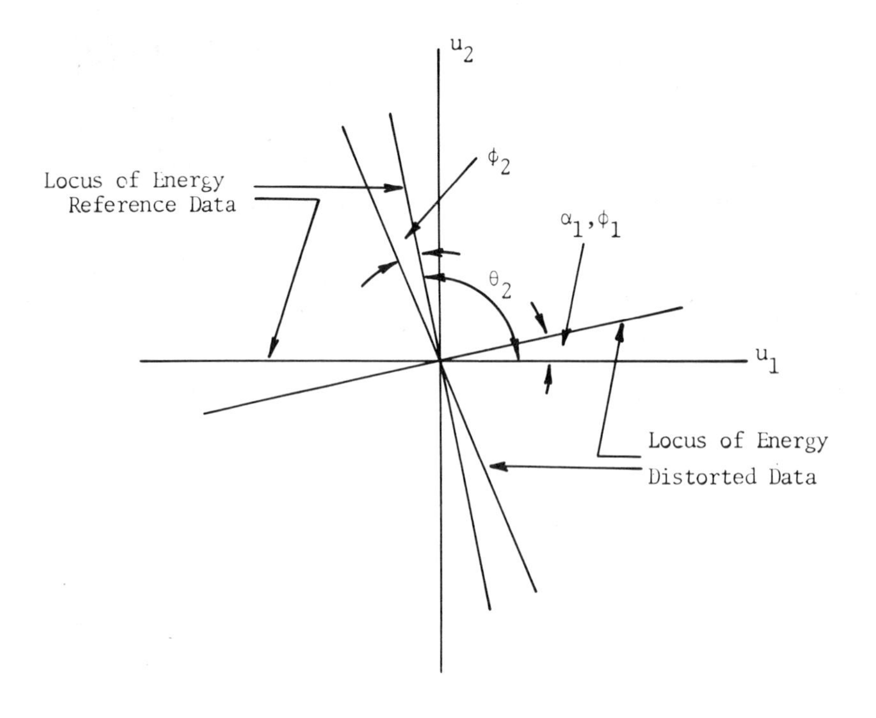

Figure 5. Location of Loci of Energy in the Modulus of the Two-Dimensional Fourier Transform, Rotational and Skew Distortion

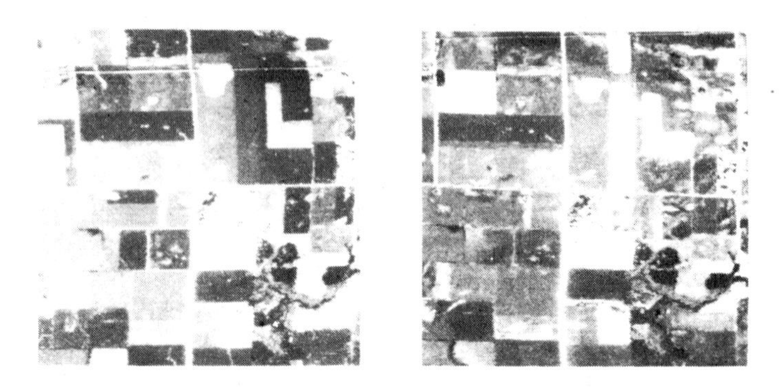

Data Set A   Data Set B

Figure 6. Multitemporal Imagery

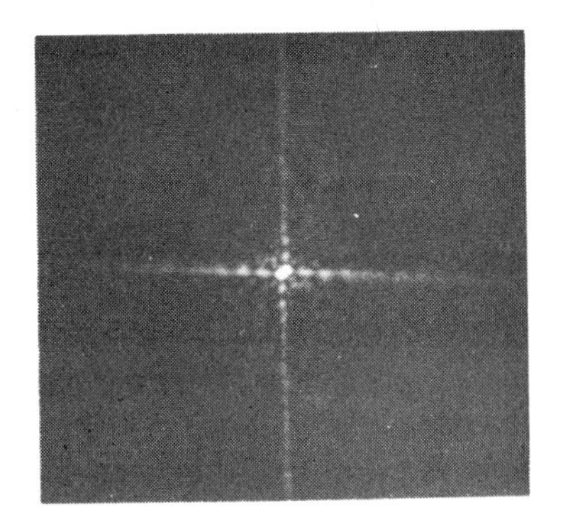

Data Set A

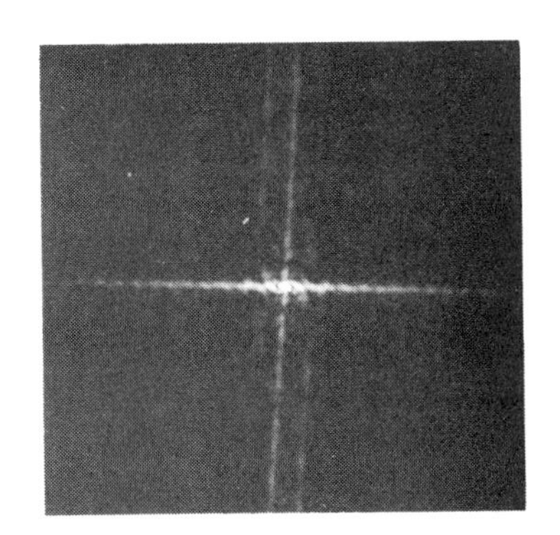

Data Set B

Data Set A

Data Set B

Figure 7. Modulus of Fourier Transform of Distorted Data and Geometrically Corrected Imagery

Correlation Techniques of Image Registration

WILLIAM K. PRATT
Image Processing Institute
University of Southern California
Los Angeles, Calif. 90007

Abstract

An extension to the basic concept of correlation detection as a means of image registration is developed. The technique involves linear spatial preprocessing of the images to be registered prior to the application of a correlation measure. This preprocessing operation utilizes the spatial correlation within each image and greatly improves the detectability of image misregistration. An analysis of the computational aspects of the algorithm is given. Also, results of a computer simulation to evaluate the technique are given.

Manuscript received May 17, 1973.

This work was supported in part by the Advanced Research Projects Agency of the Department of Defense and monitored by the Air Force Eastern Test Range under Contract F08606-72-C-0008.

I. Introduction

In many image processing applications it is necessary to form a pixel-by-pixel comparison of two images of the same object field obtained from different sensors, or of two images of an object field taken from the same sensor at different times. To form this comparison it is necessary to spatially register the images and thereby correct for relative translational shifts, magnification differences, and rotational shifts, as well as goemetrical and intensity distortions of each image. Often it is possible to eliminate or minimize many of these sources of misregistration by proper static calibration and compensation of the image sensor; in some applications misregistration detection and subsequent correction must be performed dynamically for each pair of images.

Consideration is given here to the single problem of registering images subject to translational differences. The results can be applied to the detection of rotational and magnification differences by increasing the dimensionality of the problem, or by a proper transformation of coordinates (e.g., a rotational shift is equivalent to a translational shift in polar coordinates).

II. Basic Correlation Measure

A classical technique for registering a pair of functions is to form a correlation measure between the functions and determine the location of the maximum correlation [1]. In applying this technique to two dimensions, let $f_1(j, k)$ and $f_2(j, k)$ represent two discrete images to be registered. In its simplest form, the correlation measure is defined as

$$R(u, v) = [\Sigma_{j=1}^{J} \Sigma_{k=1}^{K} f_1(j, k) \cdot f_2(j-u, k-v)] / \{ [\Sigma_{j=1}^{J} \Sigma_{k=1}^{K} f_1^{2}(j, k)]^{1/2} \cdot [\Sigma_{j=1}^{J} \Sigma_{k=1}^{K} f_2^{2}(j-u, k-v)]^{1/2} \} \quad (1)$$

where (j, k) are indices in a $J \times K$ point window area, W, that is located within an $M \times N$ point search area, S. Fig. 1

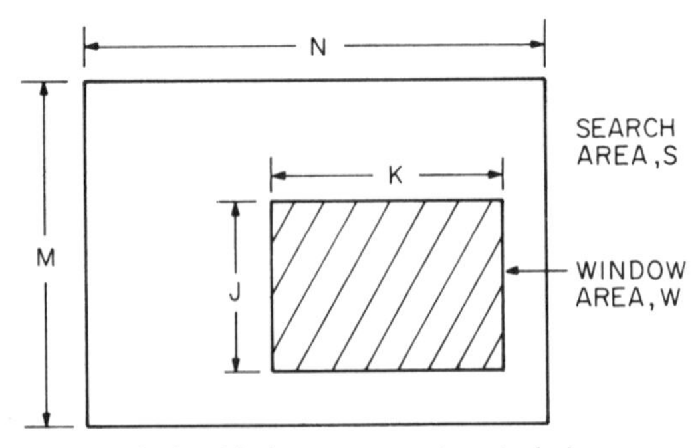

Fig. 1. Relationship between search and window areas.

illustrates the relationship between the search and window areas. In general, the correlation function $R(u, v)$ must be computed for all $(M-J+1)$ $(N-K+1)$ possible

Reprinted from *IEEE Trans. Aerosp. Electron. Syst.*, vol. AES-10, pp. 353–358, May 1974.

translations of the window area within the search area to determine its maximum value and obtain a misregistration estimate.

There are two basic problems with this simple correlation measure. First, the correlation function may be rather broad, making detection of the peak difficult. It should be noted that the simple correlation measure ignores the spatial relationship of points within each image. Second, image noise may mask the peak correlation. Both problems can be alleviated by extending the correlation measure to consider the statistical properties of each image function $f_1(j, k)$ and $f_2(j, k)$.

III. Statistical Correlation Measure

The statistical correlation measure is defined as

$$R_S(u, v) = \cdot[\Sigma_{j=1}^{J} \Sigma_{k=1}^{K} g_1(j, k) g_2(j-u, k-v)] / \{[\Sigma_{j=1}^{J} \Sigma_{k=1}^{K} g_1^2 \cdot (j, k)]^{1/2} [\Sigma_{j=1}^{M} \Sigma_{k=1}^{N} g_2^2 (j-u, k-v)]^{1/2}\} \tag{2}$$

where the $g_i(j, k)$ are obtained by spatially convolving the sampled images $f_i(j, k)$ with spatial filter functions $D_i(j, k)$. Thus,

$$g_i(j, k) = f_i(j, k) \circledast D_i(j, k). \tag{3}$$

The spatial filter function is chosen to maximize the correlation peak ratio

$$C_P = [R_S(\delta_x, \delta_y)] / [R_S(u, v)] \quad \text{for all } u \neq \delta_x \text{ and } v \neq \delta_y. \tag{4}$$

Determination of the optimum spatial filter function is facilitated by a vector space representation of each image. Let the column vector **Q** represent the image function $f_1(j,k)$ when the image is scanned in a vertical raster fashion. Thus,

$$\mathbf{Q} = \begin{bmatrix} Q(1) \\ Q(2) \\ \vdots \\ Q(J) \\ Q(J+1) \\ \vdots \\ Q(J \cdot K) \end{bmatrix} = \begin{bmatrix} f_1(1, 1) \\ f_1(2, 1) \\ \vdots \\ f_1(J, 1) \\ f_1(1, 2) \\ \vdots \\ f_1(J, K) \end{bmatrix}. \tag{5}$$

Similarly, let $\mathbf{P}_{u,v}$ represent the column scanned image $f_2(j+u, k+v)$ as given by

$$\mathbf{P}_{u,v} = \begin{bmatrix} P_{u,v}(1) \\ P_{u,v}(2) \\ \vdots \\ P_{u,v}(j) \\ P_{u,v}(J+1) \\ \vdots \\ P_{u,v}(J \cdot K) \end{bmatrix} = \begin{bmatrix} f_2(u+1, v+1) \\ f_2(u+2, v+1) \\ \vdots \\ f_2(u+J, v+1) \\ f_2(u+J+1, v+2) \\ \vdots \\ f_2(u+J, v+K) \end{bmatrix}. \tag{6}$$

The elements of **Q** and $\mathbf{P}_{u,v}$ will be highly correlated spatially since $f_1(j,k)$ and $f_2(j,k)$ are each spatially correlated to a significant extent for natural imagery. The first step in the spatial filter design process is to decorrelate or "whiten" each image vector by whitening filter matrixes $\mathbf{H}_Q$ and $\mathbf{H}_P$. Thus, let

$$\mathbf{A} = [\mathbf{H}_Q]^{-1} \mathbf{Q} \tag{7a}$$

$$\mathbf{B}_{u,v} = [\mathbf{H}_P]^{-1} \mathbf{P}_{u,v} \tag{7b}$$

where $\mathbf{H}_Q$ and $\mathbf{H}_P$ are obtained by a factorization of the image covariance matrixes

$$\mathbf{K}_Q = \mathbf{H}_Q \mathbf{H}_Q^T \tag{8a}$$

$$\mathbf{K}_P = \mathbf{H}_P \mathbf{H}_P^T. \tag{8b}$$

The factorization matrixes $\mathbf{H}_Q$ and $\mathbf{H}_P$ may be found in terms of the eigenvectors and eigenvalues of $\mathbf{K}_Q$ and $\mathbf{K}_P$. That is,

$$\mathbf{H}_Q = \mathbf{E}_Q \mathbf{\Lambda}_Q^{1/2} \tag{9a}$$

$$\mathbf{H}_P = \mathbf{E}_P \mathbf{\Lambda}_P^{1/2} \tag{9b}$$

where $\mathbf{\Lambda}_Q$ and $\mathbf{\Lambda}_P$ are diagonal matrixes containing eigenvalues along the diagonal, and $\mathbf{E}_Q$ and $\mathbf{E}_P$ are composed of the eignviectors arranged in column form.

The basic correlation operation is now performed on the whitened vectors **A** and **B**, yielding the statistical correlation measure

$$R_S(u, v) = (\mathbf{A}^T \mathbf{B}_{u,v}) / [(\mathbf{A}^T \mathbf{A})^{1/2} (\mathbf{B}_{u,v}^T \mathbf{B}_{u,v})^{1/2}] \tag{10a}$$

which can be written as

$$R_S(u, v) = [(\mathbf{K}^T)^{-1} \mathbf{Q}^T \mathbf{P}_{u,v}] / \{[(\mathbf{K}^T)^{-1} \mathbf{Q}]^T [(\mathbf{K}^T)^{-1} \mathbf{Q}] \mathbf{P}_{u,v}^T \mathbf{P}_{u,v}\}^{1/2} \tag{10b}$$

where

$$\mathbf{K} = \mathbf{H}_P \mathbf{H}_Q^T.$$

Fig. 2 illustrates two alternate implementations of the statistical correlation operation. In the first method, specified by (10a), each image is whitened before the correlation operation, while in the second method, one of the images is multiplied by a single filter matrix $\mathbf{G} = (\mathbf{K}^T)^{-1}$ before the correlation operation. The second implementation is clearly preferred, since the matrix multiplication need only be performed a single time for all u, v.

IV. Computation of Convolution Matrix

Computation of the convolution matrix $\mathbf{G}$ requires computation of two sets of eigenvectors and eigenvalues of the covariance matrixes of the two images to be registered over the window area. If the window area contains $J \times K$ pixels, the covariance matrixes $\mathbf{K}_Q$ and $\mathbf{K}_P$ will each be $(JK) \times (JK)$ matrixes. For example, if $J = K = 16$, the covariance matrixes $\mathbf{K}_Q$ and $\mathbf{K}_P$ are each of dimension 256×256. Computation of the eigenvectors and eigenvalues for such matrixes is numerically difficult for all but the largest of computers. However, the situation is not hopeless. By making a few simplifying assumptions about the statistical structure of the images to be registered, it is possible to cast the computational problem of generating the filter matrix $\mathbf{G}$ into tractable form, and yet maintain useful registration performance. The assumptions are as follows.

Assumption 1: Row/Column Separability: It is assumed that the image cross-covariance matrix K can be written in direct product form as

$$\mathbf{K} = \mathbf{K}_R \times \mathbf{K}_C \tag{11}$$

where $\mathbf{K}_R$ and $\mathbf{K}_C$ represent row and column cross-covariance matrixes.

Assumption 2: Additive White Sensor Noise: The sensor noise is assumed to be white and additive. Hence, the covariance matrixes of the row and column image data may be written as

$$\mathbf{K}_{QR} = \mathbf{K}_{S1R} + \sigma_{N1}^2\mathbf{I} \tag{12a}$$

$$\mathbf{K}_{QC} = \mathbf{K}_{S1C} + \sigma_{N1}^2\mathbf{I} \tag{12b}$$

$$\mathbf{K}_{PR} = \mathbf{K}_{S2R} + \sigma_{N2}^2\mathbf{I} \tag{12c}$$

$$\mathbf{K}_{PC} = \mathbf{K}_{S2C} + \sigma_{N2}^2\mathbf{I} \tag{12d}$$

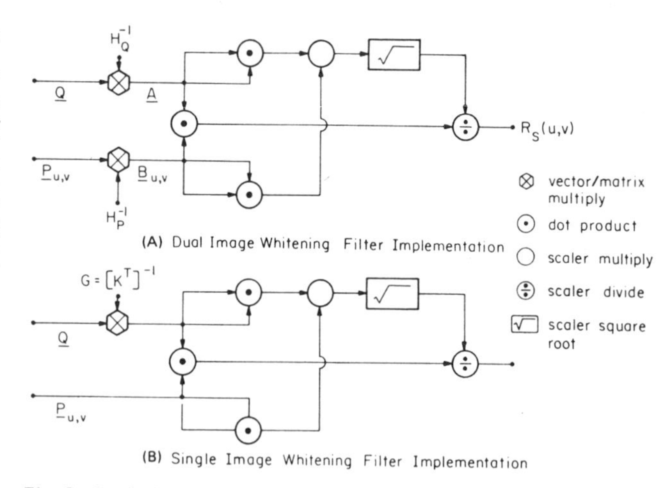

Fig. 2. Statistical image correlation implementation techniques. Symbols with an underbar correspond to boldface symbols in the text.

where $\mathbf{K}_{S1R}$ is the covariance matrix of row elements of $f_1(j,k)$ and σ_{N1}^2 is the average noise power added to the image, etc.

Assumption 3: Markov Process Image: The row and column image elements are assumed to be samples of a Markov process. Hence,

$$\mathbf{K}_{S1R} = [\rho_{S1R}^{|i-j|}], \qquad \mathbf{K}_{S1C} = [\rho_{S1C}^{|i-j|}]$$
$$\mathbf{K}_{S2R} = [\rho_{S2R}^{|i-j|}], \qquad \mathbf{K}_{S2C} = [\rho_{S2C}^{|i-j|}]$$

where ρ_{S1R} is the correlation between row elements of $f_1(i,j)$, etc.

Assumption 1 permits sequential row and column computations in the evaluation of the statistical correlation measure. Assumptions 2 and 3 enable generation of the required eigenvalues and eigenvectors recursively with a relatively simple algorithm [2].

As an example of the generation of a registration filter, consider the case when there is no sensor noise and the covariance matrixes of the two images are both of Markov form:

$$\mathbf{K}_{S1R} = \mathbf{K}_{S1C} = \mathbf{K}_{S2R} = \mathbf{K}_{S2C} = [\rho^{|i-j|}]. \tag{13}$$

In this particular example, the registration filter becomes [3]

$$\mathbf{G} = \mathbf{K}^{-1} = 1/(1-\rho^2) \begin{bmatrix} -1 & -\rho\Sigma^{-1} & 0 & 0 & \cdots & 0 \\ -\rho\Sigma^{-1} & (1+\rho^2)\Sigma^{-1} & -\rho\Sigma^{-1} & 0 & \cdots & 0 \\ 0 & -\rho\Sigma^{-1} & (1+\rho^2)\Sigma^{-1} & -\rho\Sigma^{-1} & \cdots & 0 \\ \vdots & & & & & \vdots \\ 0 & & \cdots & 0 & -\rho\Sigma^{-1} & 1 \end{bmatrix} \tag{14}$$

where

$$\Sigma^{-1} = 1/(1-\rho^2) \begin{bmatrix} 1 & -\rho & 0 & 0 & \cdots & 0 \\ -\rho & (1+\rho^2) & -\rho & 0 & \cdots & 0 \\ 0 & -\rho & (1+\rho^2) & -\rho & \cdots & 0 \\ \vdots & & & & & \vdots \\ 0 & & \cdots & & 0\ -\rho & 1 \end{bmatrix}. \quad (15)$$

Multiplication of the image vector $\mathbf{Q}$ by the registration filter $\mathbf{G}$ is equivalent to convolving the image $f_1(j, k)$ with the two-dimensional function [3]

$$D_{jk} = \begin{bmatrix} 0 & 0 & 0 & 0 & 0 \\ 0 & \rho^2 & -\rho(1+\rho^2) & \rho^2 & 0 \\ 0 & -\rho(1+\rho^2) & (1+\rho^2)^2 & -\rho(1+\rho^2) & 0 \\ 0 & \rho^2 & -\rho(1+\rho^2) & \rho^2 & 0 \\ 0 & 0 & 0 & 0 & 0 \end{bmatrix}, \quad (16)$$

as indicated by (3). If the images are completely spatially unrelated, then $\rho = 0$. The convolution operator then becomes

$$D_{jk} = \begin{bmatrix} 0 & 0 & 0 \\ 0 & 1 & 0 \\ 0 & 0 & 0 \end{bmatrix}. \quad (17)$$

Hence, the statistical correlation measure reduces to the simple correlation measure. At the other extreme, if the correlation factor $\rho = 1$, then

$$D_{jk} = \begin{bmatrix} 0 & 0 & 0 & 0 & 0 \\ 0 & 1 & -2 & 1 & 0 \\ 0 & -2 & 4 & -2 & 0 \\ 0 & 1 & -2 & 1 & 0 \\ 0 & 0 & 0 & 0 & 0 \end{bmatrix}. \quad (18)$$

This operator is a form of a spatial discrete differentiation operator. Thus, when the images are highly correlated, the statistical correlation measure concentrates on the edge outline comparison between the two scenes.

In summary, computation of the registration filter matrix G is a feasible task under the three assumptions previously stated. Preliminary simulation studies, described in a later section, indicate that the registration system still performs quite well. It should be further noted that the filter matrix G is not a function of the images directly, but rather is only dependent upon their second-order statistics. Therefore, it is possible to precompute and store a variety of filter matrixes G for various expected operating conditions involving different amounts of sensor noise and different classes of imagery.

V. Computational Requirements

The computational requirements of the statistical correlation measure, as computed by the system of Fig. 2 and using a registration filter as developed in the previous section, will now be considered. As an example, assume that the window contains $U = J \times K$ points and there are $V = (M - J + 1) \times (N - K + 1)$ possible test points of registration. The computation requirements of the statistical correlation measure $R_S(u, v)$ for a precomputed registration filter are:

ONE — $U \times U$ matrix by $U \times 1$ vector multiply

$1 + 2V$ — $U \times 1$ dot product

V — scalar multiply

V — square root

V — divide.

The major computational operations are the matrix/vector multiply and the V dot product operations. In a physical system it may be worthwhile to implement these operations by functional circuits. However, if the matrix/vector multiply and the dot product operations are implemented by a general-purpose computer, the number of computations are given by

$U^2/2 + (1 + 2V)U + V$ — scalar multiple

$U^2 + (1 + 2V)U$ — scalar add

V — square root

V — divide.

Fig. 3 contains a plot of the scalar multiple operations as a function of the sizes of the window and search areas.

The determination of the expected amount of image misregistration is obtained by searching the array $R_S(u, v)$ for its peak value and denoting the coordinates u, v at the peak. This search procedure can be performed on a general-purpose computer using conventional sorting techniques. Alternatively, the computed values of $R_S(u, v)$ can be stored in an associative memory, and the peak determined by special-purpose circuitry.

VI. Performance Evaluation

It is desireable to develop a performance measure for the statistical correlation measure in order to assess the probability of achieving registration and the expected amount of misregistration. In order to evaluate the registration probability for two images, it will be necessary to determine the joint probability density of each image. This joint density is rarely, if ever, known in most image

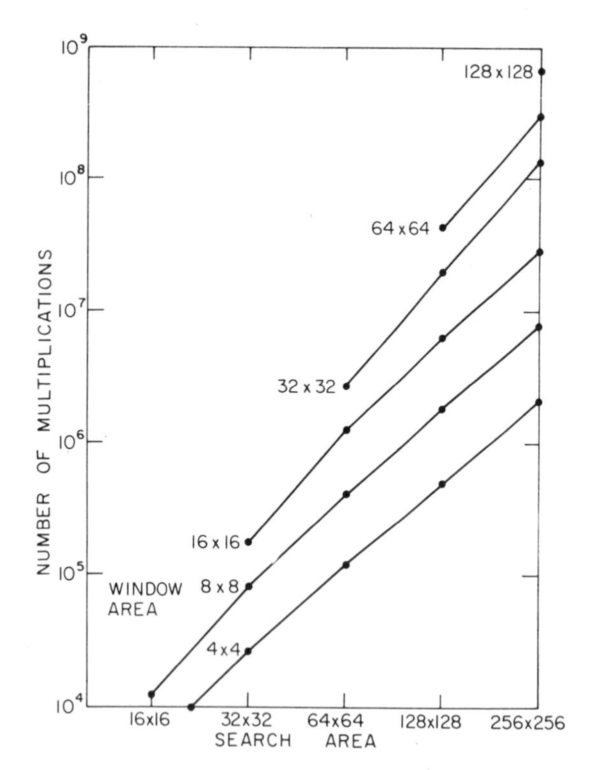

Fig. 3. Number of multiplications for statistical correlation measure.

Fig. 4. Tank image.

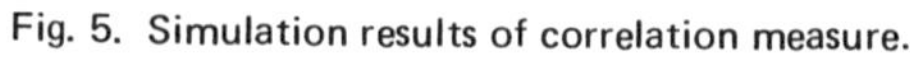

Fig. 5. Simulation results of correlation measure.

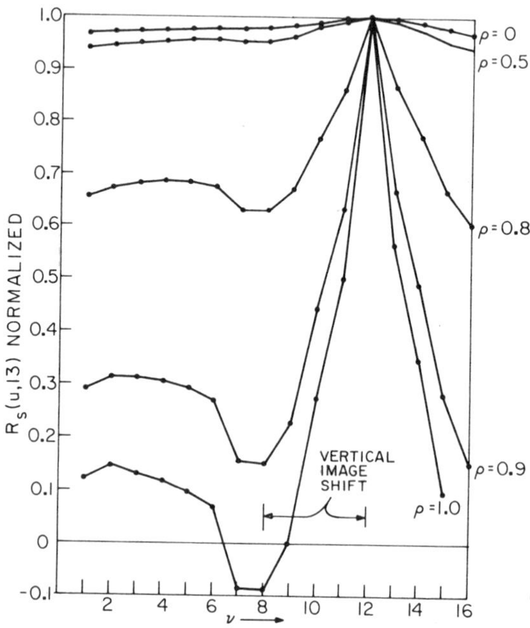

processing systems. Alternatively, one might attempt to determine a signal-to-noise ratio (SNR) for the registration system. The signal-to-noise ratio is defined as

$$S/N = \frac{(E^2\{R_S(u, v)\} \mid \text{in registration})}{(E\{R_S^2(u, v)\} \mid \text{out of registration})}$$

where, from (10a),

$$R_S(u, v) = (\mathbf{A}^T\mathbf{B}_{u,v})/[(A^TA)^{1/2}(B^T_{u,v}B_{u,v})^{1/2}].$$

By definition, the numerator is unity when the two images represented by the vectors $\mathbf{A}$ and $\mathbf{B}_{uv}$ are in register. Computation of the variance of $R_S(u, v)$ when the images are out of register requires knowledge of the fourth-order joint movements of the two images. Again, this information is generally not available. The only recourse in such a situation is to resort to a simulation of the registration system, or actually implement the system and measure its performance.

As a preliminary step in assessing the performance of the statistical correlation measure, a computer simulation was performed for the registration of the image of Fig. 4 translated with respect to itself. In the simulation, a window of 16 X 16 pixels was employed in a 32 X 32 pixel search area. The image correlation matrix was assumed to be Markov with a correlation factor of $\rho = 0., 0.5, 0.8, 0.9, 0.95, 0.96, 0.97, 0.98, 0.99, 1.00$. In all ten cases the misregistration was successfully detected. An indication of the signal-to-noise ratio performance, and a verification of the utility of the statistical correlation measure, is given by Fig. 5. In the simulation one image was offset horizontally by three pixels and vertically by four pixels. The curves in Fig. 5 are the statistical correlation measure taken through the peak to illustrate the vertical shift. The correlation functions have been normalized to unity at their peak for purposes of comparison. It should be noted that, for $\rho = 0.$, the conventional correlation measure, it is relatively difficult to distinguish the peak of $R_S(u, v)$. For $\rho = 0.9$ or greater, $R_S(u, v)$ peaks sharply at the correct point.

VII. Extensions of Correlation Measure

A common criticism of the correlation measure form of image registration is the great amount of computation that must be performed if the window and search areas are large. With the correlation measure technique, no decision can be made until the correlation array $R_S(u, v)$ is computed for all u, v. Furthermore, the amount of computation of $R_S(u, v)$ is the same for all amounts of misregistration. These deficiencies of the standard correlation measure have led to the search for sequential algorithms which could inherently provide a misregistration estimate with fewer computations.

One method of sequential test that has been proposed is the sequential similarity detection algorithm [4]. The basic form of this algorithm is deceptively simple. The error

$$\xi_S = \sum_j \sum_k | f_1(j, k) - f_2(j-u, k-v) |$$

is accumulated for pixel values in a window area. If the error exceeds a predetermined threshold value before all $J \times K$ points in the window area are examined, then it is assumed that the test has failed for the particular window, and a new window is checked. If the error grows slowly, then the number of points examined when the threshold is finally exceeded is recorded and denoted as a rating of the test window. Eventually, when all test windows have been examined, the window with the largest rating is assumed to be the proper registration window. The highest rating possible is equal to the number of window points $J \times K$. It should be noted that only a relatively few number of computations are necessary for windows that correspond to a gross misregistration. There are several extensions to the sequential test algorithm that have been proposed to speed its convergence and improve its reliability [4]. The sequential test algorithm is a nonlinear processing algorithm, and does not lend itself to analysis. Simulation tests have been performed on the registration of cloud pictures with success. It appears that the system performance for other classes of imagery and sensor noise conditions must be evaluated by simulation or system tests. A hybrid system employing a sequential test algorithm to discard gross misregistration points followed by a statistical correlation measure for the remaining test points is a very favorable candidate for a system. Such a system could provide the performance advantages of the statistical correlation measure with the speed advantages of the sequential test.

VIII. Summary

An extension ov the basic correlation measure for translational image registration has been developed. This extension, which involves linear, spatial preprocessing of one of the images prior to the application of the basic correlation measure, has been shown to provide a considerable improvement in the detectability of image misregistration.

References

[1] P.F. Anuta, "Digital registration of multispectral video imagery," *Soc. Photo-Optical Instrum. Engs. J.*, vol. 7, pp. 168-175, September 1969.

[2] W.K. Pratt, "Generalized Wiener filtering computation techniques," *IEEE Trans. Computers*, vol. C-21, pp. 636-641, July 1972.

[3] A. Arcess, P.H. Mengert, and E.W. Trombini, "Image detection through bipolar correlation," *IEEE Trans. Information Theory*, vol. IT-16, pp. 534-541, September 1970.

[4] D.I. Barnea and H.F. Silverman, "A class of algorithms for fast digital image registration," *IEEE Trans. Computers*, vol. C-21, pp. 179-186, February 1972.

Image Registration Error Variance as a Measure of Overlay Quality

C. D. McGILLEM, FELLOW, IEEE, AND M. SVEDLOW

Abstract—When one image (the signal) is to be registered with a second image (the signal plus noise) of the same scene, one would like to know the accuracy possible for this registration. This paper derives an estimate of the variance of the registration error that can be expected via two approaches. The solution in each instance is found to be a function of the effective bandwidth of the signal and the noise, and the signal-to-noise ratio. Application of these results to LANDSAT-1 data indicates that for most cases registration variances will be significantly less than the diameter of one picture element.

I. INTRODUCTION

MANY INSTANCES arise in which one would like to register two different images of the same scene. When one attempts to accomplish this overlay of images, several problems are encountered. An important question that arises is, given images of a particular scene, to within what tolerance can the two images be aligned? This is the problem with which this paper deals.

Two models for the variance of the error in the registration of two different images of the same scene are developed. The method of solution employed is analogous to that used for the determination of the error in the measured delay time in a radar system. For purposes here the radar system model assumes that the returned signal is a delayed version of the original signal corrupted by additive noise. As adapted to the registration of two images, the noise is defined as the difference between the two images at the correct registration position, and is therefore additive. The time delay corresponds to a spatial translation or displacement.

Several analyses of the radar problem have been carried out based upon different premises [1], [2], [3]. These approaches may be categorized as those which use the probability density function of the noise directly and those which do not. The first case utilizes maximum *a posteriori* probability, maximum likelihood, or minimum mean square error estimates. All three estimators are based upon knowledge of the noise probability density function. The second case is based only upon the output of a filter which gives a maximum output at the correct time delay when the input is noise free.

The solution to the problem of the first case, in which the probability density function of the noise is directly involved, depends upon the cost function which is assigned to the error and the *a posteriori* distribution, $p_f[m(\tau)]$, of the signal as a function of a parameter, $m(\tau)$, given the received signal, f. A minimum mean square error estimate is the mean of $p_f[m(\tau)]$; an absolute value cost function gives the median of the probability function; the maximum *a posteriori* estimate yields the maximum of $p_f[m(\tau)]$. The maximum likelihood estimate may be viewed as the same as the maximum *a posteriori* estimate when there is no prior knowledge of the density function of the parameter, $p[m(\tau)]$, or $p[m(\tau)]$ is assumed uniform over the entire range of interest. All four of the above cost functions will yield the same solution when $p[m(\tau)]$ is uniform and the conditional density function $p_m[\tau(f)]$ is symmetric and unimodal [3]. A Gaussian distribution which has been assumed for $p_f[m(\tau)]$ in several analyses, is a member of this latter class. The reason for the use of the Gaussian distribution is the availability of a closed form analytical solution.

An analysis of this sort should prove useful in several respects. The results should give an indication of the best possible registration of two images given the models of the data and noise. Once the models of the parameters involved have been found or assumed, an optimum processor to implement the overlaying procedure may be developed. Comparison of existing registration systems with the results obtained herein may also be performed. However, one must keep in mind the assumptions the entire analysis will be based on, for different assumptions may yield different results.

It is assumed in the following investigation that the useful signal is present, reducing the problem to one of estimation only rather than detection as well as estimation. It is further

Manuscript received April 10, 1975; revised July 3, 1975. This work was sponsored by the National Aeronautics and Space Administration, in part under Grant NGL-15-005-112, and in part under Contract NAS9-14016.

The authors are with the Laboratory for Applications of Remote Sensing, Purdue University, West Lafayette, IN 47907.

Reprinted from *IEEE Trans. Geosci. Electron.*, vol. GE-14, pp. 44–49, Jan. 1976.

assumed that the signal shape is known and nonrandom, although the parameter that is to be measured is a random variable. Since the original signal is known, it does not have a probability density function. However, the second signal does contain noise and possibly other perturbations and is therefore a sample function of a random process. The problem will be approached with this in mind.

II. METHOD 1

This derivation of the variance of the registration error is an adaptation of the solution obtained by Zubakov and Wainstein [4]. In this problem one assumes that the additive noise is jointly Gaussian with zero mean. It is also assumed that the density function of the parameter (i.e., the misregistration or displacement of the images) is uniform in the range of interest.

With these assumptions one may construct the likelihood function and then find its peak to determine the optimum estimator.

$$\Lambda(\tau_x, \tau_y) = p_m(\tau_x, \tau_y) \frac{p_{m(\tau_x, \tau_y)}(f)}{p_0(f)} \tag{1}$$

where,

$\Lambda(\tau_x, \tau_y)$	likelihood function of the displacement parameters, τ_x and τ_y;
$p_m(\tau_x, \tau_y)$	density function of the parameters, τ_x and τ_y, given the known signal;
$p_{m(\tau_x, \tau_y)}(f)$	conditional density of $f(x,y)$ given $m(x,y,\tau_x,\tau_y)$ is present;
$p_0(f)$	conditional density of $f(x,y)$ given $m(x,y,\tau_x,\tau_y)$ is absent;
$m(x,y,\tau_x,\tau_y)$	known signal as a function of the spatial coordinates and the displacement parameters;
$f(x,y)$	$m(x,y) + n(x,y)$ = received signal;
$n(x,y)$	additive noise; assumed independent of the signal.

Since the data that are being analyzed are discrete, it is convenient to use integer subscripts rather than continuous spatial coordinates. A further notational savings is realized by combining the double subscripts into a single subscript. A two dimensional array $m_{ij}, i = 1, \cdots, p; j = 1, \cdots, q$, is converted to a one dimensional data set $m_h, h = 1, \cdots, pq$. This conversion loses nothing from the standpoint of the results to be derived.

In the discrete case a continuous function has been sampled and may be denoted,

$$\begin{aligned} m_h &= m(x_i, y_j) \\ n_h &= n(x_i, y_j) \\ f_h &= f(x_i, y_j) = m_h + n_h \\ h &= 1, \cdots, H \\ H &= pq = \text{total number of samples.} \end{aligned}$$

To arrive at an analytical result, the probability density function of the noise must be known. Because of the many independent contributions to the differences between images being registered, it is reasonable to approximate the density function as being Gaussian. The probability density function of the noise is therefore given by

$$p_n(n) = \frac{1}{(2\pi)^{H/2} |R|^{1/2}} \exp\left(-\frac{1}{2} n^T R^{-1} n\right) \tag{2}$$

where R is the covariance matrix of the noise, $R_{gh} = E[n_g n_h]$. The density functions in the likelihood equation then become,

$$\begin{aligned} p_{m(\tau_x, \tau_y)}(f) &= p_n(f - m(\tau_x, \tau_y)) \\ p(f) &= p_n(f) \\ f^T &= (f_1, \cdots, f_H) \\ m^T &= (m_1, \cdots, m_H). \end{aligned}$$

The likelihood function can be reduced to

$$\Lambda(\tau_x, \tau_y) = p_m(\tau_x, \tau_y) \cdot \exp\left(\sum_g^H \sum_h^H Q_{gh} f_g m_h(\tau_x, \tau_y) - \frac{1}{2} \sum_g^H \sum_h^H Q_{gh} m_g(\tau_x, \tau_y) m_h(\tau_x, \tau_y)\right)$$

$$Q_{gh} = gh^{\text{th}} \text{ element of } R^{-1}. \tag{3}$$

Since it is only the maximum of $\Lambda(\tau_x, \tau_y)$ which is desired, the problem can be reduced even further. Let $p_m(\tau_x, \tau_y)$ be a uniform distribution over a given area. This is a reasonable assumption since there is no *a priori* knowledge about the actual distribution. The question in point here is concerned only with a spatial delay, so that the summation term,

$$\mu = \sum_g^H \sum_h^H Q_{gh} m_g(\tau_x, \tau_y) m_h(\tau_x, \tau_y) \tag{4}$$

will also be a constant function of τ_x and τ_y. The only factor which is not a constant with respect to τ_x and τ_y is,

$$\phi = \sum_g^H \sum_h^H Q_{gh} f_g m_h(\tau_x, \tau_y). \tag{5}$$

Therefore the maximum of $\Lambda(\tau_x, \tau_y)$ is determined solely by the maximum of ϕ. The optimum processor is then the one which finds the maximum of ϕ. This type of processor may be viewed as a correlator which is weighted according to the inverse noise covariance function, Q_{gh}. For the case in which the noise is white with spectrum $N_0/2$, the covariance matrix becomes $(N_0/2)I$ (I = identity matrix), and the optimum processor is simply a correlator.

$$\phi = \frac{2}{N_0} \sum_h^H f_h m_h(\tau_x, \tau_y). \tag{6}$$

Given that the maximum point (this translation position is denoted by $(\hat{\tau}_x, \hat{\tau}_y)$) of the likelihood function has been found, a measure of the accuracy of the estimate is necessary so that the performance of the estimator may be evaluated. One such measure is the variance of the estimate about the maximum point of $\Lambda(\tau_x, \tau_y)$. For this analysis it is convenient to use $\ln[\Lambda(\tau_x, \tau_y)]$ which is a monotonic function of $\Lambda(\tau_x, \tau_y)$.

The logarithm of the likelihood function is expanded in a second order Taylor series as a function of the delay parameters about its peak in the x-axis and y-axis directions separately. It is assumed that $\ln [\Lambda(\tau_x, \tau_y)]$ can be approximated by a second order polynomial around its peak.

Only the results in the x-axis direction are given since the y-axis direction results are completely analogous.

$$\ln \Lambda(\tau_x, \hat{\tau}_y) \simeq \ln \Lambda(\hat{\tau}_x, \hat{\tau}_y) + \frac{\partial \ln \Lambda(\hat{\tau}_x, \hat{\tau}_y)}{\partial \tau_x}(\tau_x - \hat{\tau}_x) + \frac{1}{2}\frac{\partial^2 \ln \Lambda(\hat{\tau}_x, \hat{\tau}_y)}{\partial \tau_x^2}(\tau_x - \hat{\tau}_x)^2 \tag{7}$$

where

$$\frac{\partial \ln \Lambda(\hat{\tau}_x, \hat{\tau}_y)}{\partial \tau_x} = \left.\frac{\partial \ln \Lambda(\tau_x, \tau_y)}{\partial \tau_x}\right|_{\substack{\tau_x = \hat{\tau}_x \\ \tau_y = \hat{\tau}_y}}.$$

A necessary condition for the maximum point of $\ln \Lambda(\tau_x, \tau_y)$ is that,

$$\frac{\partial \ln \Lambda(\hat{\tau}_x, \hat{\tau}_y)}{\partial \tau_x} = 0 = \frac{\partial \ln \Lambda(\hat{\tau}_x, \hat{\tau}_y)}{\partial \tau_y}. \tag{8}$$

The Taylor series expansion may then be reduced to

$$\ln \Lambda(\tau_x, \hat{\tau}_y) \simeq \ln \Lambda(\tau_x, \hat{\tau}_y) + \frac{1}{2}\frac{\partial^2 \ln \Lambda(\hat{\tau}_x, \hat{\tau}_y)}{\partial \tau_x^2}(\tau_x - \hat{\tau}_x)^2. \tag{9}$$

Rearranging this equation one obtains,

$$\Lambda(\tau_x, \hat{\tau}_y) = \Lambda(\hat{\tau}_x, \hat{\tau}_y) \exp\left(-\frac{1}{2}\frac{(\tau_x - \hat{\tau}_x)^2}{\Delta_x^2}\right) \tag{10}$$

where

$$\Delta_x^2 = -\left[\frac{\partial^2 \ln \Lambda(\hat{\tau}_x, \hat{\tau}_y)}{\partial \tau_x^2}\right]^{-1} = \text{variance in the } x\text{-direction.} \tag{11}$$

Assuming $p_m(\tau_x, \tau_y)$ to be uniformly distributed,

$$\frac{1}{\Delta_x^2} = \sum_g^H \sum_h^H Q_{gh}[m_g(\hat{\tau}_x, \hat{\tau}_y) - f_g]\frac{\partial^2 m_h(\hat{\tau}_x, \hat{\tau}_y)}{\partial \tau_x^2} + \sum_g^H \sum_h^H Q_{gh}\frac{\partial m_g(\hat{\tau}_x, \hat{\tau}_y)}{\partial \tau_x}\frac{\partial m_h(\hat{\tau}_x, \hat{\tau}_y)}{\partial \tau_x}. \tag{12}$$

If one further assumes a large signal-to-noise ratio, then

$$\frac{1}{\Delta_x^2} = \sum_g^H \sum_h^H Q_{gh}\frac{\partial m_g(\hat{\tau}_x, \hat{\tau}_y)}{\partial \tau_x}\frac{\partial m_h(\hat{\tau}_x, \hat{\tau}_y)}{\partial \tau_x} \tag{13}$$

since $[m_g(\hat{\tau}_x, \hat{\tau}_y) - f_g]$ is dependent only upon the noise and is small compared to $m_g(\hat{\tau}_x, \hat{\tau}_y)$.

Greater insight into the solution may be obtained by looking at the result in the frequency domain as opposed to the spatial domain. This transformation yields an interesting answer. The variance becomes,

$$\Delta_x^2 = \frac{1}{\Delta W_x^2 \mu} \tag{14}$$

where

ΔW_x^2 effective bandwidth in the x-axis direction;
μ signal-to-noise ratio;

$$\mu = \sum_u^p \sum_v^q \frac{|M(u, v)|^2}{S_R(u, v)} \tag{15}$$

$$\Delta W_x^2 = \frac{4\pi^2 \sum_u^p \sum_v^q \frac{u^2 |M(u, v)|^2}{S_R(u, v)}}{\sum_u^p \sum_v^q \frac{|M(u, v)|^2}{S_R(u, v)}} \tag{16}$$

$M(u, v)$ Fourier transform of the known signal;
$S_R(u, v)$ noise spectrum.

In the spatial domain,

$$\mu = \sum_g^H \sum_h^H Q_{gh} m_g(\hat{\tau}_x, \hat{\tau}_y)\, m_h(\hat{\tau}_x, \hat{\tau}_y) \tag{17}$$

$$\Delta W_x^2 = \frac{\sum_g^H \sum_h^H Q_{gh} \frac{\partial m_g(\hat{\tau}_x, \hat{\tau}_y)}{\partial \tau_x}\frac{\partial m_h(\hat{\tau}_x, \hat{\tau}_y)}{\partial \tau_x}}{\sum_g^H \sum_h^H Q_{gh}\, m_g(\hat{\tau}_x, \hat{\tau}_y)\, m_h(\hat{\tau}_x, \hat{\tau}_y)}. \tag{18}$$

With the above assumptions the variance has been reduced to a function of the effective bandwidth and signal-to-noise ratio. This implies that if one can estimate the effective bandwidth and the signal-to-noise ratio in the x-axis and y-axis directions, then the variance of the registration error can be estimated.

Now consider the second derivation for the variance which is based upon different assumptions.

III. METHOD 2

A second derivation of the variance of the registration error is developed in this section. In this case, the only assumption about the signal and processor is that in the absence of noise, the output of the processor will be a maximum at the correct time delay [2]. No assumptions about the probability distribution of the noise are needed. As will be seen, the results of this derivation are similar to those obtained in the previous derivation, even though the two approaches are quite unalike.

The signal corresponding to the image to be overlayed is comprised of two components, the desired signal and additive noise. This signal is passed through a filter and the position where the maximum of the output signal occurs is taken to be the correct registration position. However, the filter is designed to yield a maximum at the correct delay only in the noise free case. The discrepancy between these two positions is the registration error.

First consider the parameters involved.

$f(x, y)$ signal;
$m(x, y)$ additive noise;
$f(x, y) + m(x, y)$ data set to be registered;

$h(x, y)$ — filter impulse response;
$g(x, y)$ — $f(x, y) * h(x, y)$ = output signal in the absence of noise;
$n(x, y)$ — $m(x, y) * h(x, y)$ = output due to the noise input;
$z(x, y)$ — $g(x, y) + n(x, y)$ = composite output signal used to estimate the correct registration position;
$(\tilde{x}, \tilde{y})$ — true registration position;
$(\bar{x}, \bar{y})$ — estimated registration position.

The derivation proceeds as follows. First expand $g(x, y)$ in a second order Taylor series about (x, y).

$$g(x,y) \simeq g(\tilde{x},\tilde{y}) + g_x(\tilde{x},\tilde{y})\,[x-\tilde{x}] + g_y(\tilde{x},\tilde{y})\,[y-\tilde{y}] + g_{xy}(\tilde{x},\tilde{y})\,[x-\tilde{x}]\,[y-\tilde{y}] + \tfrac{1}{2}g_{xx}(\tilde{x},\tilde{y})\,[x-\tilde{x}]^2 + \tfrac{1}{2}g_{yy}(\tilde{x},\tilde{y})\,[y-\tilde{y}]^2 + \cdots \tag{19}$$

where

$$g_x(\tilde{x},\tilde{y}) = \left.\frac{\partial g(x,y)}{\partial x}\right|_{x=\tilde{x},\, y=\tilde{y}}.$$

This subscript notation is used for the remainder of this section. Assume that $(x - \tilde{x})$ and $(y - \tilde{y})$ are small enough so that all higher order terms may be neglected.

Note that a necessary condition for a maximum is

$$\frac{\partial g(\tilde{x},\tilde{y})}{\partial x} = 0 = \frac{\partial g(\tilde{x},\tilde{y})}{\partial y}.$$

Substitute this result into the equation for $z(x, y)$.

$$z(x,y) = g(\tilde{x},\tilde{y}) + g_{xy}(\tilde{x},\tilde{y})\,[x-\tilde{x}]\,[y-\tilde{y}] + \tfrac{1}{2}g_{xx}(\tilde{x},\tilde{y})\,[x-\tilde{x}]^2 + \tfrac{1}{2}g_{yy}(\tilde{x},\tilde{y})\,[y-\tilde{y}]^2 + n(x,y). \tag{20}$$

Again use the necessary condition for an observed maximum, $\partial z(\bar{x}, \bar{y})/\partial x = 0 = \partial z(\bar{x}, \bar{y})/\partial y$,

$$z_x(\bar{x},\bar{y}) = 0 = g_{xy}(\tilde{x},\tilde{y})\,[\bar{y}-\tilde{y}] + g_{xx}(\tilde{x},\tilde{y})\,[\bar{x}-\tilde{x}] + n_x(\bar{x},\bar{y}) \tag{21}$$

$$z_y(\bar{x},\bar{y}) = 0 = g_{xy}(\tilde{x},\tilde{y})\,[\bar{x}-\tilde{x}] + g_{yy}(\tilde{x},\tilde{y})\,[\bar{y}-\tilde{y}] + n_y(\bar{x},\bar{y}). \tag{22}$$

Arrange these equations in terms of $(\bar{x} - \tilde{x})$ and $(\bar{y} - \tilde{y})$, the error in the registration.

$$(\bar{x}-\tilde{x}) = \frac{g_{xy}n_y - g_{yy}n_x}{g_{xx}g_{yy} - g_{xy}^2} \tag{23}$$

$$(\bar{y}-\tilde{y}) = \frac{g_{xy}n_x - g_{xx}n_y}{g_{xx}g_{yy} - g_{xy}^2} \tag{24}$$

where the arguments, $(\bar{x}, \bar{y})$ and $(\tilde{x}, \tilde{y})$ have been left out for notational convenience.

One can now find the variance of the error by taking the expectation of $(\bar{x} - \tilde{x})^2$ and $(\bar{y} - \tilde{y})^2$. It is assumed that $E[\bar{x} - \tilde{x}] = 0 = E[\bar{y} - \tilde{y}]$.

$$\text{Var}\,[\bar{x}-\tilde{x}] = E[(\bar{x}-\tilde{x})^2] = \overline{(\bar{x}-\tilde{x})^2} \tag{25}$$

$$\text{Var}\,[\bar{y}-\tilde{y}] = E[(\bar{y}-\tilde{y})^2] = \overline{(\bar{y}-\tilde{y})^2} \tag{26}$$

$$\overline{(\bar{x}-\tilde{x})^2} = \frac{g_{xy}^2\overline{n_y^2} - 2g_{xy}g_{yy}\overline{n_y n_x} + g_{yy}^2\overline{n_x^2}}{[g_{xx}g_{yy} - g_{xy}^2]^2} \tag{27}$$

$$\overline{(\bar{y}-\tilde{y})^2} = \frac{g_{xy}^2\overline{n_x^2} - 2g_{xy}g_{xx}\overline{n_y n_x} + g_{xx}^2\overline{n_y^2}}{[g_{xx}g_{yy} - g_{xy}^2]^2}. \tag{28}$$

One may use these equations to calculate the variance of the error, but in doing so, it is found that a filter function must be specified first. This is intrinsic in the parameters in these equations. This is seen more clearly if one writes these terms as a function of the filter (stationarity is assumed).

$$\overline{n_y^2(\bar{x},\bar{y})} = \iiiint h_y(\bar{x}-\alpha,\bar{y}-\beta)\,h_y(\bar{x}-\gamma,\bar{y}-\lambda) \cdot R_m(\alpha-\gamma,\beta-\lambda)\,d\alpha\,d\beta\,d\gamma\,d\lambda \tag{29}$$

$$\overline{n_y(\bar{x},\bar{y})\,m_x(\bar{x},\bar{y})} = \iiiint h_y(\bar{x}-\alpha,\bar{y}-\beta) \cdot h_x(\bar{x}-\gamma,\bar{y}-\lambda)\,R_m(\alpha-\gamma,\beta-\lambda) \cdot d\alpha\,d\beta\,d\gamma\,d\lambda \tag{30}$$

$$\overline{n_x^2(\bar{x},\bar{y})} = \iiiint h_x(\bar{x}-\alpha,\bar{y}-\beta)\,h_x(\bar{x}-\gamma,\bar{y}-\lambda) \cdot R_m(\alpha-\gamma,\beta-\lambda)\,d\alpha\,d\beta\,d\gamma\,d\lambda \tag{31}$$

$$g_{xx}(\tilde{x},\tilde{y}) = \iint h_{xx}(\tilde{x}-\alpha,\tilde{y}-\beta)\,f(\alpha,\beta)\,d\alpha\,d\beta \tag{32}$$

$$g_{yy}(\tilde{x},\tilde{y}) = \iint h_{yy}(\tilde{x}-\alpha,\tilde{y}-\beta)\,f(\alpha,\beta)\,d\alpha\,d\beta \tag{33}$$

$$g_{xy}(\tilde{x},\tilde{y}) = \iint h_{xy}(\tilde{x}-\alpha,\tilde{y}-\beta)\,f(\alpha,\beta)\,d\alpha\,d\beta \tag{34}$$

where

$$R_m(\alpha-\gamma,\beta-\lambda) = \overline{m(\alpha,\beta)\,m(\gamma,\lambda)}. \tag{35}$$

Equations (27) and (28) will allow one to find the variance of the error for any filter function; however, they seem to bear little resemblence to the results in the first section. To obtain a particular solution, a specific filter function must be chosen. The one that has been picked is intuitively pleasing in two ways: it is an optimum type filter in that it maximizes the signal-to-noise ratio; and it yields an answer in terms of the signal bandwidth and signal-to-noise ratio. This filter is the so called "matched filter."

Let

$$H(u,v) = \frac{F^*(u,v)\exp(-j2\pi(\tilde{x}u + \tilde{y}v))}{S_m(u,v)} \tag{36}$$

where,

$S_m(u, v)$ — Fourier transform of $R_m(x, y)$;
$F(v, u)$ — Fourier transform of $f(x, y)$;
$H(u, v)$ — Fourier transform of $h(x, y)$.

Substituting this filter function into equations (27) and (28), the results simplify to,

$$\overline{(\bar{x} - \tilde{x})^2} = \left[\frac{g_{xy}^2}{g_{yy}} - g_{xx}\right]^{-1} \tag{37}$$

$$\overline{(\bar{y} - \tilde{y})^2} = \left[\frac{g_{xy}^2}{g_{xx}} - g_{yy}\right]^{-1}. \tag{38}$$

This simplification is seen more easily if one first converts equations (29) through (34) to the frequency domain and then inserts the matched filter.

One obtains the final result by converting these last two equations to the frequency domain. They then become,

$$\overline{(\bar{x} - \tilde{x})^2} = \left[- \frac{g_{xy}^2}{B_y^2 \text{ SNR}} + B_x^2 \text{ SNR}\right]^{-1} \tag{39}$$

$$\overline{(\bar{y} - \tilde{y})^2} = \left[- \frac{g_{xy}^2}{B_x^2 \text{ SNR}} + B_y^2 \text{ SNR}\right]^{-1} \tag{40}$$

where

$$B_x = \left[\frac{4\pi^2 \iint u^2 \frac{|F(u, v)|^2}{S_m(u, v)} du\, dv}{\iint \frac{|F(u, v)|^2}{S_m(u, v)} du\, dv}\right]^{1/2} \tag{41}$$

B_x effective bandwidth of input signal in the x-axis direction;

$$B_y = \left[\frac{4\pi^2 \iint v^2 \frac{|F(u, v)|^2}{S_m(u, v)} du\, dv}{\iint \frac{|F(u, v)|^2}{S_m(u, v)} du\, dv}\right]^{1/2} \tag{42}$$

B_y effective bandwidth of input signal in the y-axis direction;

$$\text{SNR} = \iint \frac{|F(u, v)|^2}{S_m(u, v)} du\, dv \tag{43}$$

SNR output signal-to-noise ratio.

It is seen that the variance of the error is again expressable in terms of the effective signal bandwidth and signal-to-noise ratio. These results are similar to those obtained in the first section, but the relationships are not quite as simple.

A further simplification can be obtained by making some additional assumptions. The error variance expressions then will be the same as in the first method. These assumptions concern the term $g_{xy}(\tilde{x}, \tilde{y})$ in equations (39) and (40). If this term equals zero, then the desired result is obtained. Such a condition involves the quantity $[|F(u, v)|^2]/[S_m(u, v)]$ since $g_{xy}(\tilde{x}, \tilde{y})$ is a function of the quantity. Let $K(u, v) = [|F(u, v)|^2]/[S_m(u, v)]$ for notational convenience. Since $K(u, v)$ is an even function of u and v, in order for $g_{xy}(\tilde{x}, \tilde{y})$ to equal zero it is sufficient that,

$$K(u, v) = K(-u, v) \tag{44}$$

or necessary and sufficient that,

$$\int_0^\infty \int_0^\infty uv\, K(u, v)\, du\, dv = \int_0^\infty \int_0^\infty uv\, K(-u, v)\, du\, dv. \tag{45}$$

The expressions then become

$$\overline{(\bar{x} - \tilde{x})^2} = \frac{1}{B_x^2 \text{ SNR}} \tag{46}$$

$$\overline{(\bar{y} - \tilde{y})^2} = \frac{1}{B_y^2 \text{ SNR}}. \tag{47}$$

An example of when these last assumptions might apply is the following situation. Let $F(u, v)$ and $S_m(u, v)$ be band-limited to W_x and W_y in the respective axis directions. And let $[|F(u, v)|^2]/[S_m(u, v)]$ equal a constant. This would occur when the noise spectrum has a shape similar to the signal spectrum. In this case, it might be advantageous to model the two spectra as differing only by a constant factor for simplicity in estimating the variance to be expected. This may be written,

$$\frac{|F(u, v)|^2}{S_m(u, v)} = c, \text{ a constant.} \tag{48}$$

From Equation (43)

$$\text{SNR} = c \int_{-W_x}^{W_x} \int_{-W_y}^{W_y} du\, dv. \tag{49}$$

So,

$$c = \frac{\text{SNR}}{4W_x W_y}. \tag{50}$$

Then from equations (41), (42) and (43),

$$B_x^2 \text{ SNR} = 4\pi^2 c \left[\frac{2W_x^3}{3}\right](2W_y) \tag{51}$$

$$B_y^2 \text{ SNR} = 4\pi^2 c(2W_x)\left[\frac{2W_y^3}{3}\right]. \tag{52}$$

Substituting in the expressions for c, the variances are,

$$\overline{(\bar{x} - \tilde{x})^2} = \frac{3}{4\pi^2 W_x^2 \text{ SNR}} \tag{53}$$

$$\overline{(\bar{y} - \tilde{y})^2} = \frac{3}{4\pi^2 W_y^2 \text{ SNR}}. \tag{54}$$

The respective standard deviations then are,

$$\text{Standard deviation of } (\bar{x} - \tilde{x}) = \frac{1}{2\pi W_x} \sqrt{\frac{3}{\text{SNR}}} \tag{55}$$

$$\text{Standard deviation of } (\bar{y} - \tilde{y}) = \frac{1}{2\pi W_y} \sqrt{\frac{3}{\text{SNR}}}. \tag{56}$$

One may obtain a quantitative feel for the values of these expressions by using the sampling intervals for the LANDSAT-1 data in this example. The sampling interval is about 60 meters along the columns and about 80 meters along the lines. Substituting these values in equations (55) and (56), one finds that,

Standard deviation of error along the

$$\text{lines} = \frac{44.1}{\sqrt{\text{SNR}}} \text{ meters} \tag{57}$$

Standard deviation of error along the

$$\text{columns} = \frac{33.1}{\sqrt{\text{SNR}}} \text{ meters.} \tag{58}$$

These results indicate that with the chosen filter, the standard deviation of the registration error is quite small.

IV. Conclusion

An evaluation of the quality of the registration of two images is possible via an estimate of the variance of the error. This should prove useful in several respects. It may be a basis for the analysis of different registration systems by giving a way to estimate the expected accuracy of the system. It also provides a straightforward way of estimating this error.

The two approaches used are quite different even though the solutions are similar. The variance in each case was found to be a function of the effective bandwidth of the signal and noise, and the signal-to-noise ratio.

As a final consideration the basic assumptions needed for the two methods are listed. These assumptions are important and must be realized fully to be sure that they apply to the situation in which they will be utilized. For the first method these assumptions are: the noise is additive and independent of the signal; the joint probability density function of the noise is Gaussian; the *a priori* distribution of the delay parameters is uniform over the range of interest; the variance may be modeled in the x-axis and y-axis directions separately; the final result is dependent upon a large signal-to-noise ratio [cf. step from equation (12) to (13)]. The basic assumptions for the second method are: the noise is additive and independent of the signal; the noise spectrum must be known; the chosen filter is the "matched filter;" to obtain results completely analogous to the first method there is one further assumption that must be made about the ratio $[|F(u, v)|^2]/[S_m(u, v)]$ [cf. equations (44) and (45)].

References

[1] DiFranco, J. V., Rubin, W. L., *Radar Detection*, Prentice-Hall, Inc., New Jersey, 1968.

[2] McGillem, C. D., Thurman, L. A., *Radar Resolution Enhancement and Target Surface Reconstruction*, TR-EE 73-14, April 1973, Purdue University.

[3] Van Trees, H. L., *Detection, Estimation, and Modulation Theory, Part 1*, John Wiley & Sons, Inc., New York, 1968.

[4] Zubakov, V. D., Wainstein, L. A., *Extraction of Signals From Noise*, Prentice-Hall, Inc., New Jersey, 1962.

Part IV
Image Enhancement for Manual Interpretation

Organized by Paul E. Anuta, Associate Editors
and Azriel Rosenfeld

Recent Developments in Digital Image Processing at the Image Processing Laboratory at the Jet Propulsion Laboratory

DOUGLAS A. O'HANDLEY AND WILLIAM B. GREEN

Abstract—**Image processing of spacecraft images has been carried on at the Jet Propulsion Laboratory since 1964. The most recent advances in removal of geometric distortion and residual image effects along with various types of mapping projections are covered. The recent applications of image processing to the areas of biomedicine, forensic sciences, and astronomy are discussed. These treatments are of a tutorial nature and should serve as a guide to more complete discussions on the subjects.**

I. Introduction

IMAGE processing at the Jet Propulsion Laboratory started in 1964, when pictures of the moon transmitted by Ranger 7 were processed by a computer to correct various types of image distortion inherent in the on-board television camera [1]. Digital image enhancement and analysis techniques were applied to photographs from the five successful Surveyor missions to the moon [2] and to Mariner 4, the 1964 flyby mission to Mars [3]. These initial efforts were followed by the 1969 flyby missions of Mariners 6 and 7. The 201 complete television frames of Mars were processed by digital computers. Digital techniques were developed to remove the spacecraft encoder effects and to correct for camera distortions [4].

The intent of this paper is to present a summary of current development activities within the Image Processing Laboratory at Caltech's Jet Propulsion Laboratory. Because of the current broad activities and limited space, the areas are not presented in detail. It is hoped that the wide variety of subjects presented will be an indication of the varied applications of digital image processing. In the following sections specific activities are summarized and appropriate references to more detailed discussions indicated wherever possible.

The first three sections consider recent developments in the area of digital processing of images returned from JPL spacecraft. These include 1) a new method for removing camera-system-induced geometric distortion from images obtained utilizing vidicon tubes, 2) a recently developed method to perform first-order correction of vidicon images for residual image effects, and 3) the projection of flight imagery to standard mapping projections to support mapping activities.

The remaining three sections consider new applications of digital image processing to photographs taken on Earth, including those taken in the areas of biomedical application forensic sciences, and astronomy.

Manuscript received April 12, 1972; revised April 18, 1972. This paper presents the results of research carried out at the Jet Propulsion Laboratory, California Institute of Technology, Pasadena, Calif. This work was sponsored jointly by the National Aeronautics and Space Administration under Contract NAS 7-100 and the National Institutes of Health under Biotechnology Resources Grant RR-00443; and in part by a U. S. Public Health Service under Grant HL 14138-01, Specialized Center of Research—Atherosclerosis, and in part by the Caltech President's Fund under Grant to Dr. P. H. Richter.

The authors are with the Jet Propulsion Laboratory, California Institute of Technology, Pasadena, Calif. 91103.

II. Correction of Camera-Induced Geometric Image Distortion

The camera systems flown on recent JPL spacecraft (Mariners 6, 7, and 9) have been designed such that optical geometric distortion is negligible, and the main cause of geometric distortion is the vidicon tube. The large number of images returned from recent planetary missions that utilize vidicon tubes as the photosensitive component of the camera system made necessary the development of automated and reliable methods of removing camera-system-induced geometric distortion from space imagery. For support of the mapping and variable features requirements of the Mariner 9 orbital Mars mission, it was possible to develop such a method and to apply it routinely to approximately 7000 flight images.

In a vidicon camera the image is scanned from a photosensitive surface designed to accumulate charge on its back surface as a function of the light incident on the front surface; the scanning is performed utilizing an electron beam.

Geometric distortion occurs because of the way in which scanning of the photosensitive surface is performed. Magnetic coils are used to deflect the beam in a raster pattern and it is not practical to design and drive coils in such a way that the scan lines are straight, or that the beam is deflected at a uniform rate. Consequently, the raster deviates substantially from the ideal uniform rectangular array of straight lines. In addition to this fairly large effect, the magnetic environment of the vidicon tube may vary slightly due to changing camera orientation and the operation of other on-board science instruments. Another noticeable influence on electron beam deflection is the charge distribution itself that is being scanned. Local variations in the distribution will cause local variations in beam deflection. Thus the geometric distortion actually varies as a function of the scene being photographed.

Small metallic squares are deposited on the active surface of each vidicon tube and appear as black spots within each sampled image. These are referred to as reseau marks. The locations of these reseau marks within an image are used to determine the overall distortion of that image. A geometric transformation is generated in order to remove the camera-system-induced geometric distortion from that image. After the transformation is performed, the image represents the actual scene as viewed by the camera system.

The actual process of geometric calibration of a flight camera system is quite complex, as is the algorithm used to locate the reseau marks within each flight frame. This preflight calibration cannot be thoroughly treated here but the basic procedure is outlined. The calibration process begins

Reprinted from *Proc. IEEE*, vol. 60, pp. 821–828, July 1972.

by empirically determining the functional form of the intensity variation of the images of the reseau marks. This functional form is correlated with the observed reseau points in all subsequent calibration and flight frames in order to precisely locate the reseau position. Calibration concludes with a determination of where the reseau marks should appear in a geometrically correct image. This determination is based on the criterion that the geometrically corrected image of a grid test target conform as accurately as possible with the original target.

Upon receipt of pictures from space, each flight frame is corrected for geometric distortion by locating its reseau marks and applying this transformation.

In Fig. 1(a) a Mariner 9 calibration image of a uniform grid target is shown. Note the barrel distortion and the curvature of the black edge of the frame (the *mask*). The same image after geometric restoration is shown in Fig. 1(b).

III. Removal of Residual Image From Vidicon Systems

Several conflicting requirements have existed on recent planetary missions that cause the problem of residual image in pictures returned by the camera systems resulting from use of vidicon tubes as the photosensitive device. The desire to maximize the number of frames returned leads to the requirement that the vidicon tube store an image long enough to be scanned and then be capable of a rapid erasure before the next image is exposed. The Mariner 9 design was a significant improvement over past camera designs, but the stringent requirements for accurate photometry to support detailed surface analysis and especially for accurate image differencing required that vidicon residual image be first calibrated and then removed from flight frames wherever possible.

An attempt to calibrate the residual image properties of the vidicon tubes flown on Mariner 6 and 7 was only partially successful [4]. For Mariner 9 the attempts were much more successful, but much remains to be done before the phenomenon is understood well enough to be modeled analytically.

To appreciate the complexity of the problems, consider the following observations obtained from engineering measurements on vidicon systems: the amount of residual image from one or more previous images is a function of 1) the spatial position on the vidicon photosensitive material, 2) the intensity of the current and one or more previous images at each point on the vidicon, 3) the wavelength of incident light, and 4) the temperature.

Faced with these observations, a finite amount of financial resources, and a fixed launch period, the Image Processing Laboratory developed a first-order model for reducing residual image in the frames returned from Mariner 9. This model is described and, at best, represents only a start in completely characterizing this phenomenon.

Several compromises were made in developing this model. The major compromise was development of a model based only on the current image and the immediately preceding image recorded on the same vidicon tube. Thus no attempt was made to model long-term residual image effects which are known to exist (for example, a sharp black–white transition will persist for many frames).

For calibration purposes, a target was utilized consisting of alternating clear and opaque vertical bars. A sequence was devised utilizing a bar target exposure, followed by a uniformly illuminated exposure, followed by an overexposure (to remove any remaining bar target residual), followed by several frame erasures. This sequence was repeated for several combinations of intensity in both target and flat-field frames; thus the effect of light intensity in the previous and current image was modeled. The preflight calibration sequences were taken for two filters in the Mariner 9 wide-angle camera and for the narrow-angle camera, at one temperature. Extrapola-

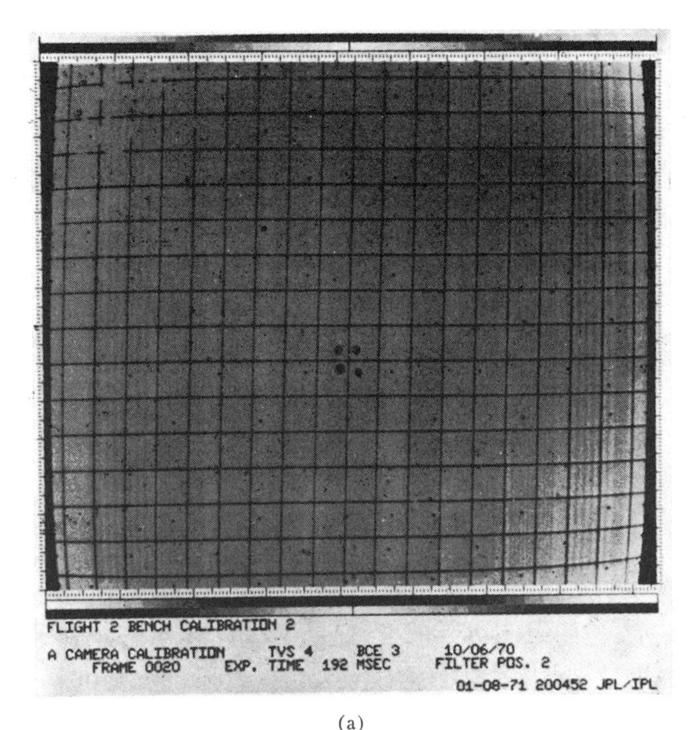

(a)

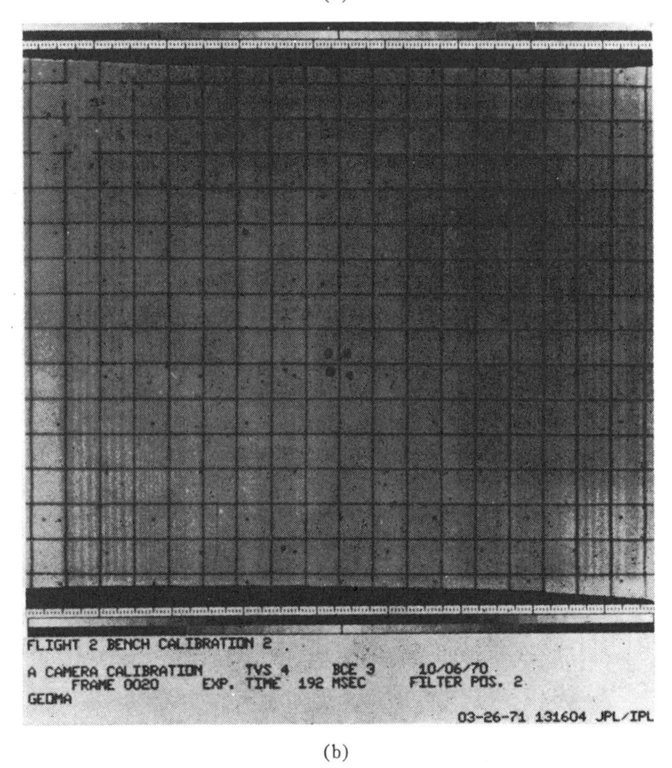

(b)

Fig. 1. (a) Mariner 9 grid target before removal of camera-system-induced geometric distortion. (b) Mariner 9 grid target after removal of camera-system-induced geometric restoration.

tion to other filter positions in the wide-angle camera and to the spacecraft operating temperature was performed using engineering measurements.

For each pair of bar-target and flat-field exposures, the amount of bar-target residual in the flat-field exposure as a function of position can be determined by computing the local modulation within the flat-field exposure; a table can be constructed at an array point throughout the vidicon that represents the amount of residual image from the previous frame in the current frame as a function of intensity of both the previous and current frame at each point. This set of tables is the basis for correcting flight frames.

Fig. 2(a) is a flat-field frame that was preceded by a bar-target exposure; the frame has been contrast-enhanced and the bar-target residual is clearly visible. Fig. 2(b) is the same frame after applying the residual image reduction algorithm and the same contrast enhancement as in Fig. 2(a). A significant reduction in residual image is apparent. For typical flight frames the residual image within a particular frame was reduced by approximately a factor of 5. The residual removal is thus not perfect, but this degree of correction was satisfactory for most scientific objectives of the Mariner 9 mission.

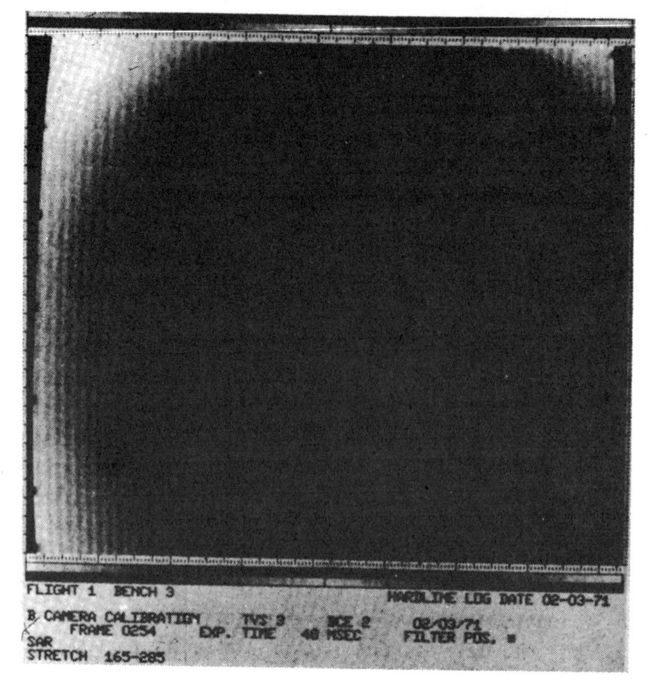

(a)

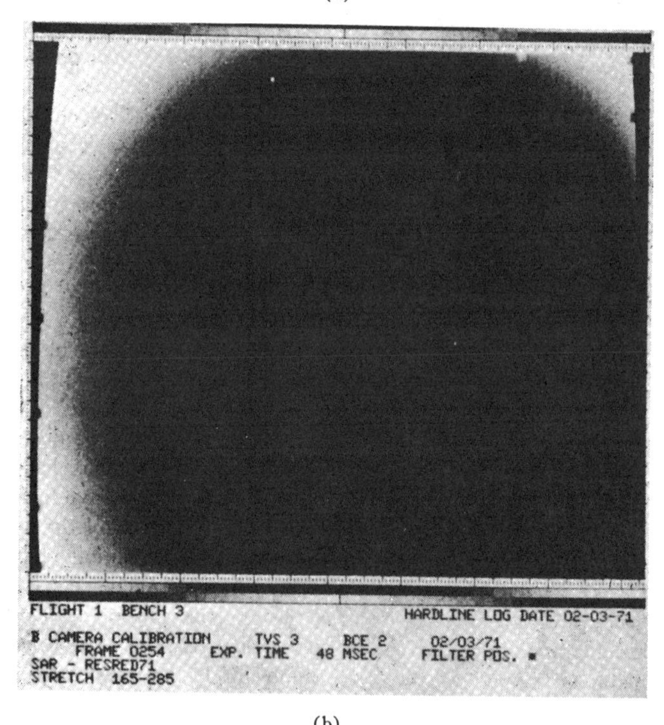

(b)

Fig. 2. (a) Flat field with bar target residual visible, before application of residual image reduction algorithm. (b) Flat field with bar target residual after application of residual image reduction algorithm.

IV. Mapping Projections of Planetary Images

With the increase in the number of frames returned from planetary missions, the goal of mapping our neighboring planets is becoming a realistic one. The Image Processing Laboratory has been developing algorithms that perform geometric transformations on flight images that transform the images to a variety of standard mapping projections. The computer has become an indispensable tool of the cartographer as he maps the surface of distant planets.

In addition to the cartographic requirements, there are additional reasons for transforming images to standard projections. Use of these projections is one means of correcting for camera viewing geometry to allow valid comparison of surface detail imaged under quite different viewing angles. Also, particularly for the Mariner 9 orbital mission, image differencing can be achieved only after the two frames to be differenced are brought into the same reference space.

The mapping projections currently used include orthographic, mercator, polar stereographic, and lambert conformal. The projections are performed using the same geometric transformation program that is used to remove camera-system-induced geometric distortion. The frame is projected to a standard projection using *tie points* obtained either from spacecraft telemetry data containing spacecraft position and camera pointing data or from cartographic control grids that establish particular planetary features as corresponding to particular longitude–latitude locations on the planet. Typically, several generations of projections are performed starting with projections based on initial spacecraft position data (which may be imprecise) and ending with projections based on accurate reduction and analysis of spacecraft position data or accurate cartographic control grids established many months after the images are returned.

For Mariner 6 and 7, final mapping projections were based on the first preliminary control net of Mars, using the data of Davies and Berg [5]. Using these data, the first computer mosaic of the south polar region of Mars was generated, using projection and mosaicking software developed in support of the postmission data analysis activity of the Mariner 6 and 7 flyby missions [6]. Fig. 3 shows a Mariner 7 frame after orthographic projection. This frame has been high-pass filtered before projection to enhance surface detail, and contrast-enhanced. The processing is intended to achieve maximum discriminability of surface features, while not necessarily preserving true surface brightness variations. The projected image thus represents an accurate view of the topography of the planet as seen from a vantage point vertically above the center of the projected frame.

V. Biomedical Applications

The application of image processing techniques to medical X-ray films has been the subject of study at the Jet Propulsion Laboratory for the last five years. Initially, the goal of the study was to develop methods for improving the image quality. Following the lead of the spacecraft image processing development, an attempt was made to restore high-frequency image [7], [8] components by linear filtering methods. While such methods were quite successful with spacecraft pictures, they were considerably less successful with X-ray films because of the inherently high noise levels, which tended to increase as a result of enhancement.

A partial solution to the noise problem was obtained through the use of processing techniques that are scene-dependent rather than camera-dependent as is the case with spacecraft picture processing [9], [10]. One application of this type under study involves the enhancement of fine blood vessels in cerebral angiograms.[1] In this case, the computer has been programmed to search for and enhance all short straight-line segments. Fine blood vessels can be represented as a connected sequence of such segments.

While the development of computer enhancement techniques is continuing, a particularly promising new area of development is being pursued that involves the use of computers to extract quantitative information from X-ray films. While human observers are poorly suited to perform repetitive precision measurements from visual data. it is an ideal job for a computer. At the current time, the Jet Propulsion Laboratory is collaborating with Dr. David H. Blankenhorn, Principal Investigator of the Specialized Center of Research—Atherosclerosis at the Los Angeles County/University of Southern California Medical Center, to develop a computer method for detecting and measuring the extent of atherosclerosis from arteriograms [11].

Atherosclerosis is a vascular disease that results in the deposit of fatty lesions (plaques) along the inside of the blood vessels. In advanced stages, these lesions partially or totally occlude vessels and are a common cause of heart failure.

Detection of large lesions in advanced cases can be made from X-ray/radio-opaque films of blood vessels that have been filled with radio-opaque dye. Deflection of this dye from the vessel wall by the plaques cause the vessel shadow on the film to look bumpy or pinched. By visual or manual techniques, small or early lesions cannot, in general, be seen reliably nor can changes in larger lesions that might indicate either advance or regression of the disease. The techniques being developed at the Jet Propulsion Laboratory are intended to detect these small changes.

An example of a processed film is shown in Fig. 4. A digital reconstruction of the 6-cm long artery is shown in Fig. 4(a) and a display of the vessel edges found by the computer is shown in Fig. 4(b). The edges were selected as the points of maximum intensity change along each line of samples crossing the vessel. A contrast-enhanced version of the vessel is shown in Fig. 4(c). Approximately one quarter of the distance from the left edge in Fig. 4(d) a dark area can be seen that is not visible in the unenhanced version in Fig. 4(a). This is probably caused by a lesion that does not cause much edge deflection because of its orientation relative to the film and X-ray source. Fig. 4(c) shows a plot of vessel width (multiplied by 2) and Fig. 4(e) shows a plot of relative mean film density between the detected edges.

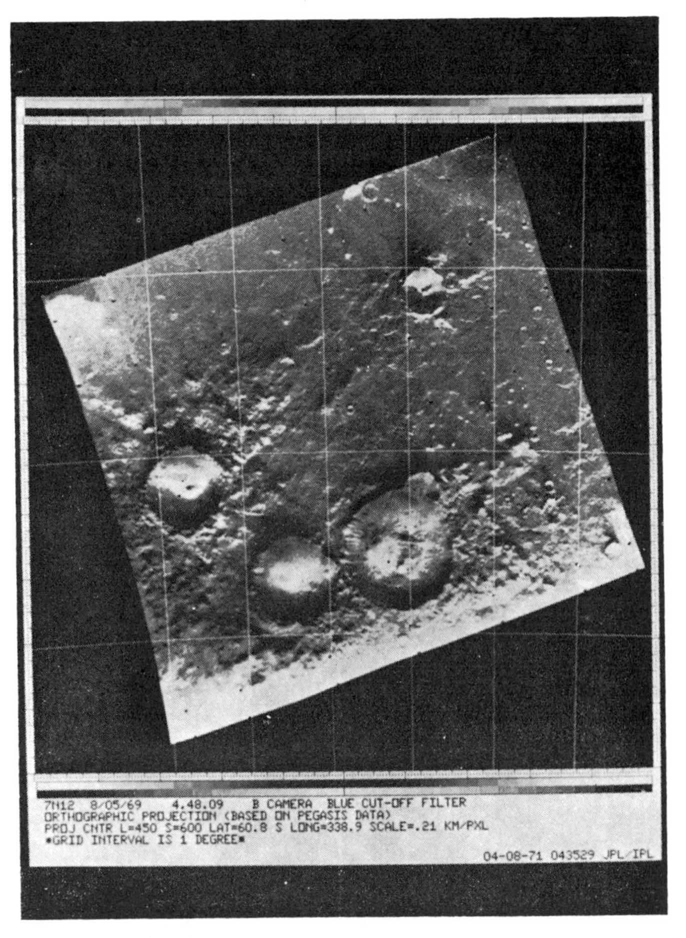

Fig. 3. Mariner 7 frame after orthographic projection and processing to enhance topographic detail.

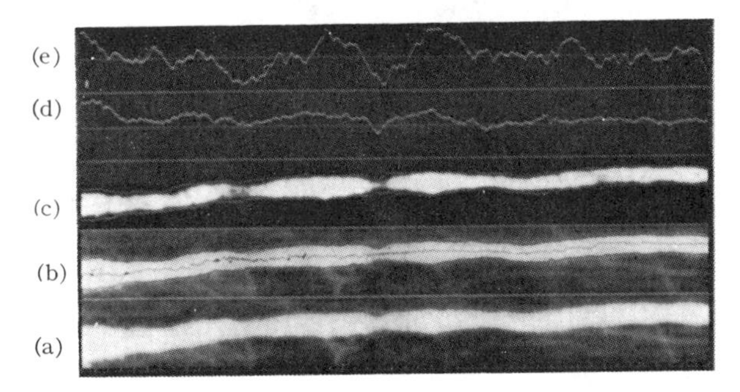

Fig. 4. Computer measurement of blood vessel plaque. (a) X-ray image of femoral artery filled with radio-opaque dye. Actual vessel length = 6 cm. (b) Computer-processed image showing detected vessel edges and center. (c) Contrast-enhanced image showing filling variations caused by plaque. (d) Vessel width variations as measured between detected edges. (e) Average grey-level variations as measured between detected edges.

From the edge data, a measure of vessel *roughness* can be derived and an estimate of relative volume variation can be obtained from the film density data. Early results using phantom studies and excised vessels that were available for visual examination have been highly encouraging.

[1] In collaboration with Dr. Calvin Rumbaugh, Department of Neuroradiology, Los Angeles County/University of Southern California Medical Center.

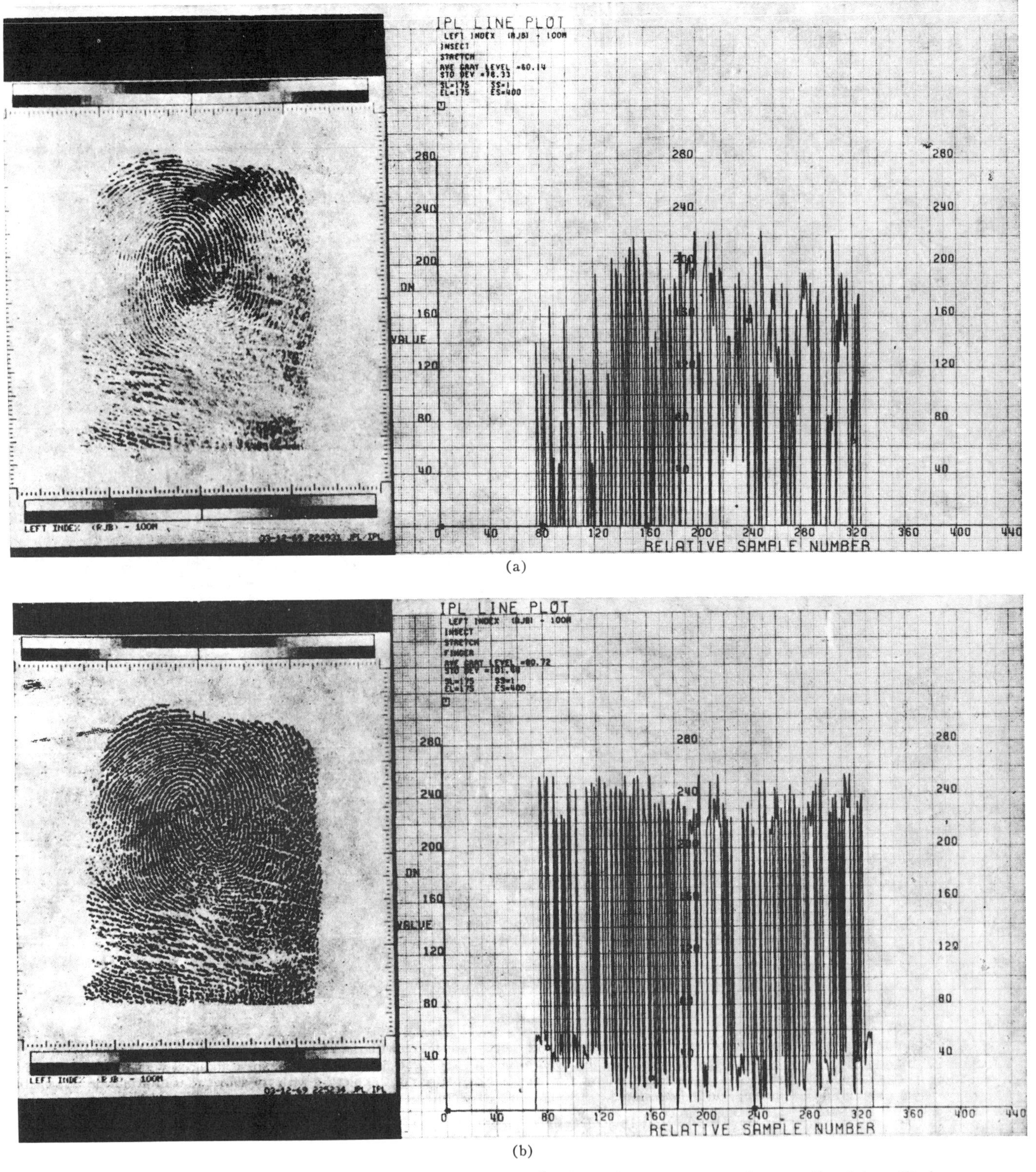

(a)

(b)

Fig. 5. Before and after photographs showing how the maxima/minima enhancement process increases the plot amplitudes.

VI. Forensic Science Application

In the forensic sciences one application of computer processing is fingerprint image enhancement. Digital image processing can be used to reduce the effects of low-quality recording techniques in acquiring latent fingerprints and standard inked fingerprints [12], [13]. In some cases, it has been found that the needed information is there but cannot be seen by the unaided human eye. In the area of automatic machine classification, the same problem exists because most automatic classifiers are not designed to accommodate fingerprints that are underinked or overinked.

The approach taken in this study was to develop an algorithm which would detect the sequential ridge structure, independent of orientation, in all fingerprints. Once this de-

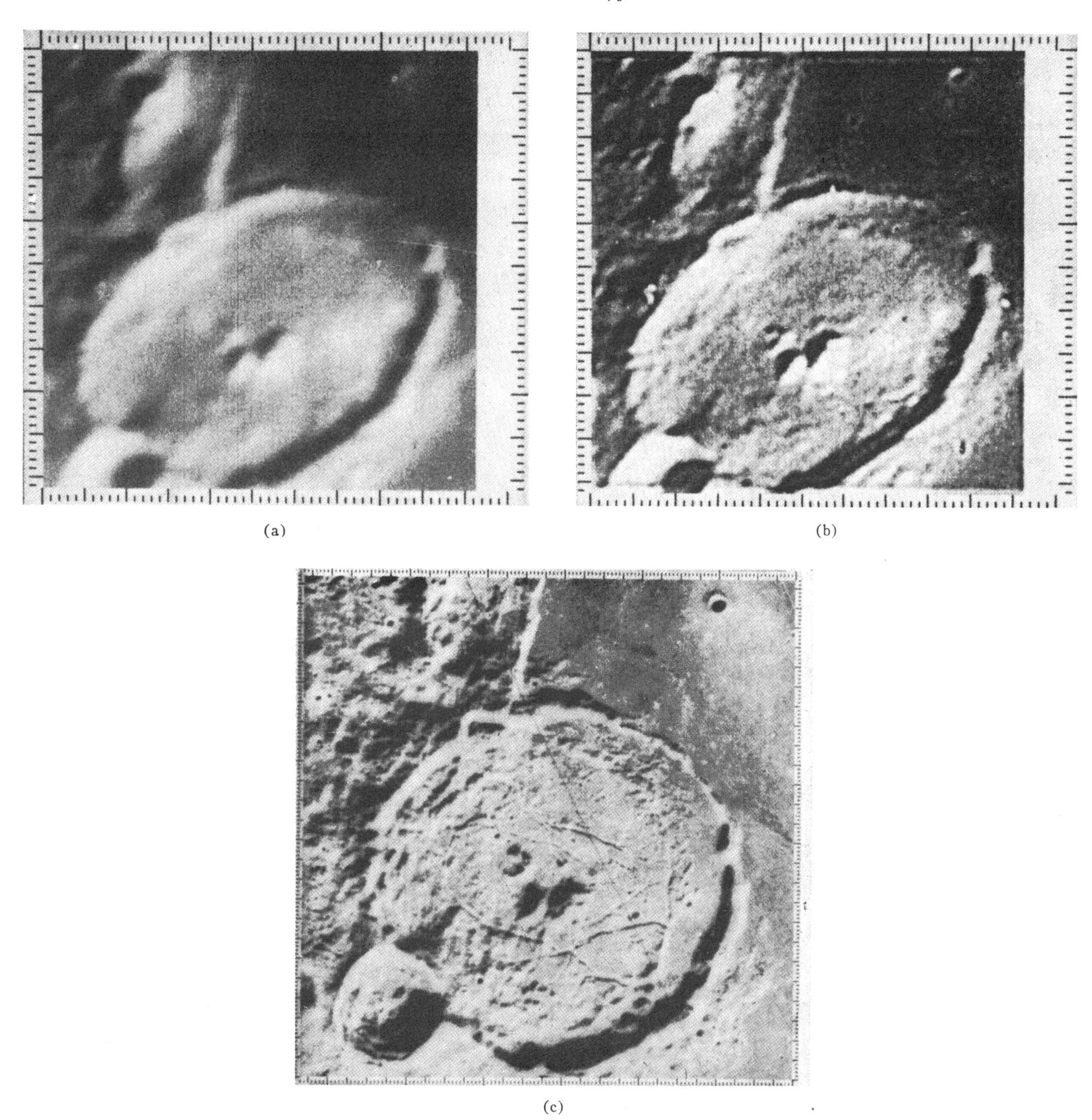

Fig. 6. (a) Original unenhanced picture of the lunar crater Gassendi. The picture was taken with the 24-in polar telescope at the Aerospace Corporation San Fernando Observatory, and is shown here in its digitized form with 256×256 picture elements. (b) Enhanced version of (a) showing reduction of resolution loss originally created by atmospheric turbulence. Comparison of this picture with (c) shows that information not visible in the original photograph has been accurately restored. (c) Lunar Orbiter picture of Gassendi. This picture is for comparison purposes only, and has not been used in the enhancement process.

tection is accomplished, then a rescaling of the gray scale numbers representing the fingerprints is produced to equally display all the information in the fingerprint.

In order to develop an algorithm which would enhance fingerprints, a closer look at the appearance of the resultant scanned image is required. Fig. 5(a) is a photograph of a fingerprint which was acquired by inking the finger with standard ink and recording it on a strip of clear acetate. The plot to the right of the photograph is a line plot from left to right across the middle of the fingerprint. It can be seen that the print exhibits areas of underinking and, consequently, poor contrast. It can also be seen that the areas of low contrast are identical to the areas in the plot where the individual amplitudes are low, whereas in the middle of the print, where the contrast is better, the amplitudes are higher. The objective of the computer program is to develop an algorithm which will recognize the successive peaks and valleys in the fingerprint and then stretch the amplitudes to improve the contrast. Restated, we wish to identify local maxima and minima, regardless of the orientation of the ridges or valleys.

Fig. 5(b) illustrates the application of the algorithm to the original fingerprint. The replot of the line trace across the fingerprint after enhancement has increased the amplitudes. It can be seen that there is a net overall increase in contrast, and the underinked areas are now more visible.

Application of this technique to overinked areas has also been carried out [14]. This process has shown the capability to recover latent fingerprint images. It would seem that it might have application as a preprocessing step taken prior to automatic or semiautomatic classification of fingerprints.

VII. Astronomical Application

Of all the sources of noise which ultimately limit the resolution in astronomical photography with large earth based telescopes, turbulence in the atmosphere causes the greatest degradation. For example, the Rayleigh limit for the 200-in Hale telescope on Mt. Palomar is 0.025″, whereas the very best photographs taken with that instrument do not exceed 0.25″, and a more common value is 1.0″ or greater [15]. Although part of this loss can be attributed to the figure of the mirror itself (and its mounting support), most is due to thermally induced refractive index variations in the atmosphere which produce random phase variations in the incoming electromagnetic waves.

Theoretical analysis of this problem [16] shows that the effect of such turbulence is to reduce image contrast at the higher spatial frequencies of the object so that the atmosphere resembles a low-pass filter. An important point, however, is that unlike an aperture which possesses a definite cutoff frequency linearly related to the aperture size, the turbulence produces a relatively constant attenuation beyond a certain spatial frequency, the amount of attenuation and the frequency at which it becomes constant depending upon the magnitude of the phase fluctuations and their power spectrum.

The foregoing suggests that enhancement of such images should be possible if the attenuation is not so severe that the higher frequencies are lost in the detector noise. The signal-to-noise ratio may be improved by reducing the noise and/or increasing the signal, but in the photographic case the two are competing. This is because the amount of attenuation produced by the atmosphere depends upon the integration time of the exposure, being less for shorter exposures, while the grain noise in film increases with increasing film sensitivity, which would permit shorter exposures.

This problem for various classes of astronomical objects including the sun, moon, planets, and their satellites is under study and has encountered some very promising results to date. Fig. 6(a) shows a photograph of the lunar crater Gassendi taken with the San Fernando Solar Observatory 24-in telescope operated by the Aerospace Corporation. Fig. 6(b) shows the enhanced version of this same picture. Also shown [Fig. 6(c)] for comparison, but not utilized in the enhancement process, is a Lunar Orbiter picture of the same crater. These examples show that an improvement in resolution from about 2.5″ to 0.8″ has been achieved and that details not visible in the original picture have been faithfully restored by the enhancement process.

This particular enhancement was secured by means of high-pass filtering in the Fourier domain of a single photograph recorded on a very fine grained emulsion, the exposure time being 0.25 s. The noise which is visible in the upper right corner of the enhanced picture is actually the result of insufficient intensity resolution in the scanning process in this case, and not to film grain. The exposure time of 0.25 s has resulted in an averaging of the random phase variations in the Fourier transform of the scene induced by the atmospheric turbulence so that a simple amplitude filter was adequate to restore the higher frequencies.

More complex filtering and averaging processes are also being studied and all of these are being applied to individual photographs and sequencies of short time exposures taken with large telescopes (Hale 200-in, Mt. Wilson 60-in, etc.) in an effort to achieve substantially better resolution with these instruments than has been possible up until now. In addition, high-resolution high-gain low-noise image intensifiers are being used to improve the overall signal-to-noise ratio and permit shorter time exposures.

Acknowledgment

The work described in this paper represents the dedicated efforts of many individuals at the Image Processing Laboratory at the Jet Propulsion Laboratory. The technique for locating reseaus in Mariner 9 flight frames was developed by J. Kreznar, the method of correcting flight frames for geometric distortion discussed in this paper was developed by J. Kreznar and J. Seidman, techniques for calibrating and removing residual image for the Mariner 9 mission were implemented by P. Jepsen and A. Schwartz and the original impetus for this effort was supplied by T. Rindfleisch, the Image Processing Laboratory software for cartographic projections was initially developed by A. Gillespie and J. Soha, and software development and data analysis support were supplied by R. Ruiz and S. Harami. The three nonspace efforts presented in this review are the result of the work of R. Selzer in the biomedical applications and R. Blackwell in the forensic sciences. Dr. P. H. Richter, Associate Professor of Physics, San Fernando Valley State College, contributed the section on the astronomical application. All of these individuals have given a great deal of time to their respective disciplines and the results reported here.

The Image Processing Laboratory computer facility support and software maintenance is provided by a group headed by J. Lindsley and hardware fabrication and maintenance is provided by a group headed by F. Billingsley. Finally, many of the techniques described in this paper owe their existence to the dedication and leadership of T. Rindfleisch, formerly of the Jet Propulsion Laboratory, who is now at Stanford University.

References

[1] R. Nathan, "Picture enhancement for the moon, Mars, and man," in *Pictorial Pattern Recognition*, G. C. Cheng, Ed. Washington, D. C.: Thompson, 1968, pp. 239–266.
[2] Mission Reports for Surveyors I, III, V, VI, VII, Jet Propulsion Laboratory Tech. Reps. TR 32-1023, TR-1177, TR-1246, TR-1262, and TR-1264.
[3] "Mariner Mars 1964 project report," Jet Propulsion Laboratory Tech. Rep. 32-884.
[4] T. C. Rindfleisch, J. A. Dunne, H. J. Frieden, W. D. Stromberg, and R. M. Ruiz, "Digital processing of the Mariner 6 and 7 pictures," *J. Geophys. Res.*, vol. 76, pp. 394–417, Jan. 1971.
[5] M. E. Davies and R. A. Berg, "A preliminary control net of Mars," *J. Geophys. Res.*, vol. 76, pp. 373–393, Jan. 1971.
[6] A. R. Gillespie and J. M. Soha, "An orthographic photomap of the

south pole of Mars," *Icarus*, in press.
[7] R. H. Selzer, "Digital computer processing of X-ray photographs," in *Proc. Rochester Conf. on Data Acquisition and Processing in Biology and Medicine* (Rochester, N. Y., July 27, 1966), vol. 5. New York: Pergamon, 1968, pp. 309–325.
[8] R. H. Selzer, "Use of computers to improve biomedical image quality," in *1968 Fall Joint Computer Conf., AFIPS Conf. Proc.*, vol. 33, Pt. I. Washington, D. C.: Thompson, 1968, pp. 817–834.
[9] R. H. Selzer, "Recent progress in computer processing of X-ray and radioisotope scanner images," *Biomed. Sci. Instrum.*, vol. 6, pp. 225–234, 1969.
[10] S. D. Rockoff and R. H. Selzer, "Radiographic trabecular quantitation of human lumbar vertebrae in situ," in *Proc. Conf. on Progress in Methods of Bone Mineral Measurements* (Bethesda, Md., 1968, NIH, NIAMS), pp. 331–351.
[11] S. H. Brooks, R. H. Selzer, D. W. Crawford, and D. H. Blankenhorn, "Computer image processing of peripheral vascular angiograms," in preparation.
[12] B. C. Bridges, *Practical Fingerprinting.* New York: Funk and Wagnalls, 1942, 374 pp.
[13] H. Cummins and M. Charles, *Fingerprints, Palms and Soles.* New York: Dover, 1961, pp. 319.
[14] R. J. Blackwell, "Fingerprint image enhancement by computer methods," presented at 1970 Carnahan Conf. Electron. Crime Countermeasures (Lexington, Ky., Apr. 17, 1970).
[15] I. S. Bowen, "The 200 inch Hale telescope," in *Telescopes*, vol. I, *Stars and Stellar Systems*, G. P. Kuiper and B. M. Middlehurst, Eds. Chicago, Ill.: Univ. Chicago, 1960, pp. 1–15.
[16] E. L. O'Niell, *Introduction to Statistical Optics.* New York: Addison-Wesley, 1963, pp. 86, 99.

Image Restoration: The Removal of Spatially Invariant Degradations

MAN MOHAN SONDHI

Abstract—This is a review of techniques for *digital* restoration of images. Optical and other analog processors are not discussed. Restoration is considered from the point of view of space-domain as well as of spatial-frequency-domain descriptions of images. Consideration is restricted to degradations arising from noise and spatially invariant blurring. However, many of the space-domain methods apply, with minor modifications, to spatially varying blur as well. Some examples of restoration are included to illustrate the methods discussed. Included also is a section on methods whose potential has not yet been exploited for image restoration.

I. Introduction

THE FIELD of image restoration in the modern sense of the term began in the early 1950's with the work of Maréchal and his co-workers [1]. Although the possibility of optical spatial filtering had been demonstrated by the experiments of Abbé and Porter some fifty years earlier, it was Maréchal who first recognized its potential for restoring blurred photographs. His success stimulated others to study image restoration from the point of view of optical compensation of the degradations. In the past few years the versatility of the digital computer has been brought to bear upon the problem, with promising results. With digital processing it is possible to overcome many inherent limitations of optical filtering and, indeed, to explore new approaches which have no conceivable optical counterparts.

In this paper we describe various digital techniques available for the restoration of degraded optical images. Except for references to various examples of optically restored images we exclude optical processing [2] from our discussion.

We consider imaging under incoherent illumination only and represent images by their intensity distributions. Let $p(x, y)$ represent the original undistorted picture image. We assume d to be the result of adding a noise intensity $n(x, y)$ to a blurred image $b(x, y)$ of p. We restrict our discussion to those situations where the blurring is equivalent to linear spatially invariant filtering. Thus

$$d(x, y) = b(x, y) + n(x, y) \tag{1}$$

where

$$b(x, y) = \int_{-\infty}^{\infty} dx' \int_{-\infty}^{\infty} dy' h(x - x', y - y') p(x', y'). \tag{2}$$

Here $h(x, y)$ (often called the point spread function) is the response of the blurring filter to a two-dimensional unit impulse $\delta(x)\delta(y)$.

In terms of the model of image degradation expressed by (1) and (2) we define the restoration task as follows: With d given, utilize the available *a priori* information about n, h, and p to make a good estimate $\hat{p}(x, y)$ of p. The various restoration schemes differ from each other in the assumed *a priori* information as well as in the criterion by which the goodness of the estimate is judged.

The assumption that d is available for processing is not strictly valid. Assuming instantaneous shutter action and negligible noise, the total exposure in the image plane is proportional to d. What is recorded, in general, is a nonlinear function of the exposure (e.g., the H–D curve [3] for photographic emulsions). Therefore, d may plausibly be assumed available only over a small range around the average exposure. It is possible to accurately measure the nonlinear function by using standard gray scales. Such a measurement can be used to recover d over a larger dynamic range. However, any attempt at extending this range must ultimately be frustrated by a drastic increase in the noise level.

Our assumption that noise is additive is also subject to criticism. Many of the noise sources (e.g., stray illumination, circuit noise, roundoff) may be individually modeled as additive. However, because they occur both before and after the nonlinear transduction previously mentioned their effect on d may be assumed additive only over a small dynamic range.

Manuscript received December 7, 1971; revised March 6, 1972.

The author is with Bell Telephone Laboratories, Inc., Murray Hill, N. J. 07974. Presently he is a Guest Scientist at the Department of Speech Communication, Royal Institute of Technology (KTH), Stockholm, Sweden, during the academic year 1971–1972.

Reprinted from *Proc. IEEE*, vol. 60, pp. 842–853, July 1972.

Nevertheless, because it makes the problem mathematically tractable, the assumption is common to almost all work on image restoration. In photographic recording granularity is often the controlling source of noise. Huang [4] has shown that it is much more accurate to model this noise as multiplicative. Some cases in which multiplicative noise can be handled are discussed in Stockham's paper in this issue.[1]

The assumption of linearity and spatial invariance of the blur is quite accurate for a great variety of configurations of practical interest. For blurring due to imperfections in the optical imaging system Dumontet [5] has shown that the assumption is accurate for objects of small enough angular subtense photographed in incoherent light. The exact calculation of the function h is usually very complicated. Frequently, however, the imaging may be assumed isoplanatic, i.e., a convolution may be assumed to relate the complex wave amplitudes in the object and image planes. (Dumontet's analysis shows that, in general, this is a much more restrictive assumption.) In such cases the computation of the impulse response for incoherent illumination is considerably simplified [3, pp. 120–125]. We give the general formula and some examples computed in this way in Appendix I.

Another type of degradation which is equivalent to spatially invariant filtering occurs due to motion of the image in its plane during exposure. The total exposure at any point of the film is obtained in this case by integrating the instantaneous exposure over the time interval during which the shutter is open. To isolate the effect of image motion let us assume that the shutter operates instantaneously and that the optical imaging is perfect. Then if $\hat{x}(t)$ and $\hat{y}(t)$ are, respectively, the x and y components of the displacement and T the duration of the exposure, we have

$$b(x, y) = \int_0^T p(x - \hat{x}(t), y - \hat{y}(t))\, dt. \tag{3}$$

To show that this relation can be put in the form of (2) it is convenient to take the Fourier transform of each member of (3). Let ω and σ represent the spatial frequency components in the x and y directions. Let $B(\omega, \sigma)$, $P(\omega, \sigma)$, $H(\omega, \sigma)$ represent the Fourier transforms of b, p, and h, respectively. Then if it is assumed that the order of integrations over x, y, and t may be reversed, we have from (3),

$$B(\omega,\sigma) = \iint e^{-j\omega(x+\sigma y)} b(x, y) dx dy$$

$$= P(\omega, \sigma) \int_0^T e^{j\omega\hat{x}(t)+j\sigma\hat{y}(t)} dt. \tag{4}$$

Comparison with the Fourier transform of (2) shows that

$$H(\omega, \sigma) = \int_0^T e^{j\omega\hat{x}(t)+j\sigma\hat{y}(t)} dt. \tag{5}$$

The function H is, of course, the transfer function of the blurring filter. If desired, the impulse response h can be obtained from H by inverse Fourier transformation. Expressions equivalent to (3) and (5) have been used by Lohmann [6], Shack [7], Levi [8], Som [9], and others to derive the filters for various types of motion. References [6]–[8] consider also the modification of (3) in order to include the effects of time-varying shutter action, scattering in the emulsion, and finite memory of the recording medium. (The last mentioned effect is of interest when a fluorescent phosphor is usd as a detector. For photographic recording the memory is for all practical purposes infinite.) We will not be concerned with these refinements. Except in extremely high speed photography, the shutter may be assumed instantaneous unless deliberately designed to be otherwise. With instantaneous shutter action the rest of the effects merely introduce subsidiary filters in cascade with the filter of (5), without destroying spatial invariance. In Appendix I we give some examples of impulse responses obtained from (5).

It is worth noting that in deriving (5) we have assumed that the image is invariant in time except for a displacement of origin. This is not necessarily true even when the camera moves perpendicular to the optic axis. For example the images of objects at different distances from the camera move by different amounts, thereby producing spatially varying blur. On the other hand time invariance of the image is not necessary for spatial invariance of the blur. Lohmann and Paris [10] show, for instance, that the periodic defocussing due to sinusoidal motion parallel to the optic axis gives rise to spatially invariant blur.

As a last example we mention blurring due to atmospheric turbulence. During a short exposure the atmosphere may be assumed stationary and the wave aberration due to its inhomogeneities adds to that due to the optical system. As before, therefore, the impulse response in this case is spatially invariant for objects of small angular subtense. Sometimes, due to the presence (intentional or unintentional) of a point source in the field of view, this impulse response can be estimated. In general, however, little is known about such an "instantaneous" impulse response except, possibly, some statistical properties of an ensemble of such impulse responses. With long exposures, on the other hand, the blurring filter is, as expected, much more stable. Indeed, Hufnagel and Stanley [11] have shown that $H(\omega, \sigma)$ may be approximated by the function $\exp[-c(\omega^2+\sigma^2)^{5/6}]$ where c is a constant depending upon the atmospheric turbulence.

With the foregoing examples we hope to have convinced the reader that the class of image degradations which may be approximated by (1) and (2) is large enough to be of considerable practical interest. We proceed now to a critique of the definition of the restoration task given earlier in this section.

Let us note first of all that the very term "restoration" categorically implies use of *a priori* information. In the absence of such information there is no possibility of even discovering that there is anything wrong with a given picture. It is *a priori* information that leads us to the conclusion that the graininess of a photograph is an artifact and not part of the scene; or that a streak in a photograph of the sky is due to a star whose image moved during exposure and not due to some hitherto unidentified filamentary object.

Frequently, of course, the *a priori* "information" is little more than a reasonsable assumption. Thus for instance, all that is usually known about the desired picture is that $P(\omega, \sigma)$ is negligible outside some region (in most cases around the origin) of the ω–σ plane. Likewise, *a priori* information about the noise is usually very meager and for the most part it is assumed that n has a known constant spectral density. The *a priori* knowledge about the impulse response on the other hand runs the gamut from near certainty (e.g., known track-

[1] See pp. 828–842.

ing or focussing errors) to near ignorance (e.g., short-term turbulence blur).

The manner in which this information is utilized depends upon the purpose of the restoration. We have tacitly assumed that the restored picture is to be viewed by a human observer. Ideally, therefore, we should require that the restoration should redisplay the information in the degraded image in such a way that it enables a human observer to identify the original objects with as much detail as possible. However, to do so we would need to know the processing limitations as well as the fidelity criterion of the human visual system. Lacking such knowledge we are forced to the more modest definition of restoration given above. In demanding that the restoration makes a good estimate of p we recognize one of the important processing limitations of the human visual system, namely, its inability to deconvolve images. In leaving unspecified the criterion by which the goodness of the estimate is to be judged we confess ignorance of the optimum tradeoff between resolution and noise.

Each of the restoration schemes which we will describe in the succeeding sections explicitly or implicitly assumes some objective intuitively reasonable criterion of goodness. Inverse filtering for instance, attempts perfect resolution without regard to noise; "optimum filtering" minimizes mean-squared error without regard to resolution; the method of Section IV attempts perfect resolution but in a manner less sensitive to noise than inverse filtering; the method of constrained deconvolution constructs a restoration consistent with prespecified tolerances and allows the viewer to change these tolerances to get any desired number of different restorations to choose from; the method of Backus and Gilbert (which we describe in Section VI and which to our knowledge has not yet been used for image restoration) allows the user to obtain a restoration with any prespecified tradeoff between resolution and noise, again allowing him to select the balance he judges to be best.

At present it is not possible to order these methods in superiority. Perhaps no such ordering exists and the best choice might inherently depend on the application. In any event, no detailed study appears to have been made to evaluate the relative merits and drawbacks of these methods from the point of view of a human observer. Considering our present uncertainty about the fidelity criterion of the human visual system, flexible methods like the last two previously mentioned might provide the most useful, though expensive, tools in the search for the optimum restoration.

II. Inverse Filtering

The basic idea of inverse filtering is extremely simple. As in Section I, let ω and σ represent the x and y spatial frequencies and let upper-case letters represent Fourier transforms of functions denoted by the corresponding lower-case letters. Then the Fourier transformation of (1) and (2) gives

$$D(\omega, \sigma) = B(\omega, \sigma) + N(\omega, \sigma) \tag{6}$$

$$B(\omega, \sigma) = P(\omega, \sigma)H(\omega, \sigma) \tag{7}$$

in the spatial frequency domain. Restoration by inverse filtering merely consists of dividing both sides of (6) by the transfer function H and taking the inverse transform. Thus

$$\begin{aligned} \hat{p}(x, y) &= F^{-1}(D/H) \\ &= p(x, y) + F^{-1}(N/H) \end{aligned} \tag{8}$$

provided H does not vanish at any point (ω, σ). Maréchal *et al.* [1], Tsujiuchi [12], Harris [13], McGlamery [14], and Mueller and Reynolds [15] have all published restorations obtained by this technique with minor modifications. The modifications are aimed at the two fairly obvious flaws in this method. First, H invariably decreases rapidly for large values of ω and σ. (All the examples in Appendix I, for instance, have this property.) Noise, on the other hand, has a fairly uniform spectral distribution. From (8) it is apparent, therefore, that the restoration enhances high-frequency noise. Second, very often H has zeros at spatial frequencies within the range of interest. Linear uniform motion, sinusoidal motion, defocussing are all examples of such blurring. Division by zero being undefined, (8) is meaningless in such cases and the method breaks down. (It may be argued that in a digital realization of (8), the sample points in the ω–σ plane can be always chosen to avoid the points or curves at which H is zero. Note, however, that if H has a zero, no sampling rate is strictly adequate for $1/H$. Thus exact reconstruction is impossible even in the noiseless case. In the presence of noise, any "reasonable" set of sampling points will include many at which the reciprocal of H is very large. Thus the restoration will include a large amount of noise at spatial frequencies in the neighborhood of a zero of H.)

The modifications suggested to overcome these drawbacks are all *ad hoc* and intuitive. As an example we might mention Harris's suggestion [13] of multiplying (6) by a function $H_0(\omega, \sigma)$ before dividing by H. Thus

$$\hat{p}(x, y) = F^{-1}[BH_0/H] + F^{-1}[NH_0/H]. \tag{9}$$

By choosing H_0 to be zero over the region of the ω–σ plane where B is dominated by N we can reduce high-frequency noise. By choosing H_0/H to be finite wherever H is zero we can eliminate the problem with infinities. The price we pay is that $\hat{p}$ is no longer a noisy version of p but a noisy version of p blurred through the filter H_0. The requirements imposed on H_0 are intuitively reasonable. However, there is a noncountable infinity of functions that satisfy these requirements and we must arbitrarily choose one.

The restorations given in [11]–[15] cited above prove, however, that in spite of all the limitations and arbitrariness inverse filtering is capable of yielding reasonably good restorations in many situations where noise is not the controlling degradation.

III. Minimization of Mean-Squared Error

One way to avoid the arbitrariness of the inverse filtering approach is to minimize the discrepancy between p and $\hat{p}$. The measure of this discrepancy should, ideally, correspond to that of the human visual system. Such a measure, even if it were known in detail, would be too complex to use. As an alternative it is possible to find the optimum restoration for some simpler objectively defined criterion and see how much improvement this affords. The minimum mean-squared-error (MSE) criterion is one of the few objective criteria for which the optimum restoration can be computed.

From the point of view of image restoration for a human observer the foregoing is, admittedly, a flimsy justification for minimizing MSE. We mention just two examples to demonstrate the gross inadequacy of the criterion. It is well known that the eye demands much more faithful reproduction of regions where the intensity changes rapidly than of regions with little change. It is also well known that the sensitivity of the eye to a given error in intensity depends strongly upon the intensity. The minimum MSE criterion, on the other hand, weights an error independently of the intensity (and intensity gradient) at which it occurs.

However, Harris [16] and Horner [17], [18] have shown that the method is capable of producing some dramatic restorations. In many of the examples given in their papers the degraded images appear to be shapeless blobs while the eye can discern well-defined objects in the minimum MSE restorations. The same objects are far less discernible in the inverse-filter restorations included as controls.

As a signal processing technique, minimization of MSE has been familiar to electrical engineers since the late 1940's. In the specific context of image processing the method was recently proposed by Helstrom [19]. We present here a brief outline of the mathematical derivation of the optimum restoration.

Let the picture and the noise be members of the random processes $\{p\}$ and $\{n\}$, respectively. Then the minimum MSE estimate of $p(x, y)$ is the function $\hat{p}(x, y)$ which, for every point (x, y) of the image plane, minimizes the quantity ϵ_1 given by

$$\epsilon_1 = E[(\hat{p} - p)^2]. \tag{10}$$

Here E denotes expectation over the $\{n\}$ and $\{p\}$ processes. Let e denote the error $\hat{p}-p$, and let ϵ_d denote the conditional expectation of e^2, given the degraded image d. Then ϵ_d is a random variable whose expectation over the $\{d\}$ ensemble gives ϵ_1. Thus ϵ_1 is minimized if ϵ_d is minimized. On the other hand it is easily verified that

$$\epsilon_d = E[(p - m)^2 \mid d] + E[(\hat{p} - m)^2 \mid d] \tag{11}$$

where m is the conditional expectation $E(p|d)$. Thus ϵ_d, and hence ϵ_1, is minimized if we chose

$$\hat{p} = m = E(p \mid d). \tag{12}$$

The simplicity of (12) is deceptive. The computation of the conditional expectation is extremely difficult for arbitrary random processes $\{n\}$ and $\{p\}$. The problem becomes much simpler if we assume these processes to be stationary and jointly Gaussian, and d to be available on the entire plane. In that case it can be shown that the optimum estimate of (12) can be obtained by spatially invariant filtering of d. This means that $\hat{p}$ must have the form $q*d$, where * denotes convolution and q is the (as yet undetermined) restoring filter.

Strictly speaking, of course, the Gaussian hypothesis cannot be valid because the $\{p\}$ and $\{n\}$ processes are ensembles of positive functions. However, even low-contrast scenes for which this one-sidedness is not significant, might not be well approximated as Gaussian. In such cases the "optimum" restoration derived in the following does not minimize MSE. However, of all estimates obtainable by spatially invariant filtering of d, it gives the least MSE. To keep the formulas simple we further assume $\{n\}$ and $\{p\}$ to be statistically independent and to have zero means. (The zero mean assumption merely means that the average values can be well estimated and subtracted before processing. This is usually possible).

The minimizing filter is easiest to describe in the spatial frequency domain. Note that ϵ_1 is also the integral over the frequency plane of the power spectral density of the error process $\{e\} = \{q*d-p\}$. Let Φ_p, Φ_n, Φ_d, and Φ_e (all understood to be functions of the spatial frequencies ω and σ) denote, respectively, the power spectral densities of the processes $\{p\}$, $\{n\}$, $\{d\}$, and $\{e\}$. Then in view of (1) and (2) and the assumption that the $\{n\}$ and $\{p\}$ processes are independent with zero means, we get

$$\epsilon_1 = \iint \Phi_e d\omega d\sigma \tag{13}$$

with

$$\Phi_e = \Phi_p + Q\bar{Q}\Phi_d - (QH + \bar{Q}\bar{H})\Phi_p \tag{14}$$

and

$$\Phi_d = H\bar{H}\Phi_p + \Phi_n. \tag{15}$$

Here the bar over a symbol denotes its complex conjugate. Note that since a power spectral density is nonnegative, the vanishing of Φ_d at some point (ω, σ) implies that $\Phi_e = \Phi_p$ at that point. Clearly at all such points Q may be chosen arbitrarily (say equal to zero). At all other points (14) can be rewritten as

$$\Phi_e = \left| Q\sqrt{\Phi_d} - \bar{H}\Phi_p/\sqrt{\Phi_d} \right|^2 + \Phi_p\Phi_n/\Phi_d. \tag{16}$$

Thus to minimize ϵ_1 we must choose

$$\begin{aligned} Q &= \Phi_p\bar{H}/\Phi_d \\ &= \bar{H}/(H\bar{H} + \Phi_n/\Phi_p). \end{aligned} \tag{17}$$

This completes our derivation of the optimum filter. Note that with $\Phi_n = 0$, Q becomes the inverse filter, and that when $H=0$, $Q=0$ (rather than $Q=\infty$ as is the case with the inverse filter). To determine Q, both H and the "noise-to-signal ratio" Φ_n/Φ_p must be known for all (ω, σ) where Φ_d and Φ_p are nonzero. For the most part Φ_n/Φ_p is assumed constant over the frequency range of interest, although for restoration of turbulence-blurred aerial reconnaisance photographs. Horner [18] suggests that a monotonically decreasing function (e.g., a negative exponential) of $\sqrt{\omega^2+\sigma^2}$ may be more appropriate.

For a digital realization of the minimum MSE filter we may approximate D and Q by their discrete Fourier transforms and compute the product QD from which the estimate $\hat{p}$ may be computed for any given values of x and y. Harris [16] has used this method of computation. There is an alternative approach which starts with samples of d on a grid in the x–y plane and yields estimates of $\hat{p}$ at points on a similar grid. Let $\boldsymbol{d}$ denote a column vector of discrete samples of d ordered in a suitable manner to produce a linear array. With similar interpretation of $\boldsymbol{p}$ and $\boldsymbol{n}$, the digital equivalent of (1) and (2) can be written as

$$\boldsymbol{d} = \boldsymbol{H}\boldsymbol{p} + \boldsymbol{n} \tag{18}$$

where the matrix $\boldsymbol{H}$ is the matrix of the weights obtained by approximating the integral in (2) according to some quadrature formula. For simplicity let us assume that $\boldsymbol{p}$ has the same number of components as $\boldsymbol{d}$ and $\boldsymbol{n}$ (although the theory can be modified to allow $\boldsymbol{p}$ to have a different dimensionality). We then seek a restoring matrix $\boldsymbol{Q}$ such that the estimate

$$\hat{\boldsymbol{p}} = \boldsymbol{Q}\boldsymbol{d} \tag{19}$$

minimizes the error measure

$$\epsilon_2 = E[(\hat{\boldsymbol{p}} - \boldsymbol{p})^T(\hat{\boldsymbol{p}} - \boldsymbol{p})]. \tag{20}$$

Here ^{T} denotes transposition and E is the expectation as before. (The error measure ϵ_2 can also be made more general without much difficulty.) Assuming $\boldsymbol{n}$ and $\boldsymbol{p}$ to be zero-mean uncorrelated random vectors we get from (18), (19), and (20)

$$\epsilon_2 = \mathrm{Tr}[(\boldsymbol{QH} - \boldsymbol{I})\boldsymbol{\Phi}_p(\boldsymbol{QH} - \boldsymbol{I})^T + \boldsymbol{Q\Phi}_n\boldsymbol{Q}^T]. \quad (21)$$

Here Tr denotes the trace of a matrix, $\boldsymbol{I}$ is the identity matrix, and $\boldsymbol{\Phi}_n$, $\boldsymbol{\Phi}_p$ are the correlation matrices $E(\boldsymbol{nn}^T)$ and $E(\boldsymbol{pp}^T)$, respectively. The minimum of ϵ_2 in (21) can be shown to occur for

$$\boldsymbol{Q} = \boldsymbol{\Phi}_p\boldsymbol{H}^T(\boldsymbol{H\Phi}_p\boldsymbol{H}^T + \boldsymbol{\Phi}_n)^{-1}. \quad (22)$$

Note the formal similarity between (17) and (22).

The theory of minimum MSE estimation can be modified and extended in many ways. We mention two variations that have been proposed in connection with image restoration.

One is due to Slepian [20] who considers the case when h is a member of a random process $\{h\}$. To describe his approach, let R_1 be the rectangle $|x| \leq a_1$, $|y| \leq b_1$, R_2 the rectangle $|x| \leq a_2$, $|y| \leq b_2$, and R_3 the rectangle $|x| \leq L_1$, $|y| \leq L_2$ with $L_1 \geq a_1 + a_2$ and $L_2 \geq b_1 + b_2$. Let p be zero outside R_1, h be zero outside R_2, and d be known everywhere on R_3. Let the process $\{n\}$ be stationary with zero mean and let the process $\{h\}$ be independent of $\{n\}$. Let the error measure be

$$\epsilon_3 = E\int (p - \hat{p})^2\, dxdy \quad (23)$$

where the expectation is over the $\{n\}$ and $\{h\}$ ensembles. Note that ϵ_3 is akin to the error measure ϵ_2 of (20) rather than to ϵ_1 of (10). With p deterministic and the estimate $\hat{p}$ restricted to be of the form $q*d$, Slepian shows that the optimum restoring filter is given by

$$Q = P\overline{P}E(\overline{H})/[P\overline{P}E(H\overline{H}) + \eta] \quad (24)$$

where

$$\eta(\omega, \sigma) = N(\omega, \sigma) * \left[\frac{\sin(\omega L_1)}{\omega}\cdot\frac{\sin(\sigma L_2)}{\sigma}\right]. \quad (25)$$

If p is a member of a random process $\{p\}$ independent of $\{n\}$ and $\{h\}$, then the expectation is over the $\{p\}$ process as well and $P\overline{P}$ in (24) is replaced by Φ_p. Again the formal resemblance of (24) to (17) is striking.

Slepian also considers the error measure

$$\epsilon_4 = \int_{-L_1}^{L_1} dx \int_{-L_2}^{L_2} dy\, (\hat{p} - p)^2 \quad (26)$$

which is much more reasonable in view of the fact that d is available only on R_3. In this case a rather complicated integral equation can be derived for Q; however this equation cannot be solved explicitly.

We close our discussion of least squares filtering with a brief description of a method of solution proposed by Rino [21]. Suppose H and N (and therefore D) to be limited in spatial frequencies to a square S, given by $|\omega| \leq W/2$, $|\sigma| \leq W/2$ and let d be available on a square S_2 given by $|x| \leq L/2$, $|y| \leq L/2$. (In principle S_1 and S_2 can be replaced by arbitrary bounded regions and the analysis carried through formally. In practice such extensions are formidable except for the trivial one of replacing the squares by rectangles.) In such a situation the restoring filter of (17) is not optimum. Indeed the optimum restoring filter is not necessarily spatially invariant, and must be obtained by solving an integral equation which poses computational difficulties. Rino utilizes the dual orthogonality property of prolate spheroidal wave functions $\psi_n(c, x)$ [22] to get around this difficulty. With $c = WL/4$ it is known that the $\psi_n(c, x)$ $(n = 0, 1, 2, \cdots)$ form a complete orthogonal basis for the class of functions which are square-integrable on $|x| \leq L/2$; they also form a complete orthogonal basis on the infinite line for the class of functions band-limited to $|\omega| < W/2$ (i.e., functions whose Fourier transforms vanish for $|\omega| > W/2$). It follows, then that the functions $\psi_m(c, x)\psi_n(c, y)$ $(m, n = 0, 1, 2, \cdots)$ have similar properties with respect to the squares S_2 and S_1. With these properties in mind the function $d(x, y)$ may be expanded as

$$d(x, y) = \sum_m \sum_n d_{mn}\psi_m(c, x)\psi_n(c, y) \quad (27)$$

where

$$d_{mn} = \int_{-L/2}^{L/2} dx\psi_m(c, x) \int_{-L/2}^{L/2} dy\psi_n(c, y) d(x, y). \quad (28)$$

However, since D vanishes outside S_1, the expansion of (27) gives d everywhere on the x–y plane with zero MSE. For d extrapolated in this manner the optimum restoring filter can be obtained from (17). We have

$$q(x, y) = \int_{-W/2}^{W/2} d\omega \int_{-W/2}^{W/2} d\sigma e^{j(\omega x + \sigma y)}\Phi_p\overline{H}/\Phi_d \quad (29)$$

and

$$\hat{p}(x, y) = \sum_m \sum_n d_{mn} q(x, y) * [\psi_m(c, x)\psi_n(c, y)]. \quad (30)$$

Substitution of d_{mn} from (28) shows, as expected, that $\hat{p}$ is obtained from d by a spatially varying filter. Rino shows, further, that if the series in (30) is truncated at $m, n = K$, then with $K > 2c/\pi$ the (spatially varying) restoring filter is well approximated for restorations limited to the square S_2. For situations where the WL product is not very large, (30) might be a computationally efficient way to synthesize the optimum filter.

IV. A Method for Removal of Motion-Blur

In Section II we observed that even in the absence of noise, exact reconstruction of p is impossible whenever H has zeros at values of ω and σ in the range of interest. Uniform linear motion is such a case with a transfer function $(\sin a\omega)/(a\omega)$ when the motion is in the x direction with the total displacement a. Slepian [23] has pointed out that in this case if p is zero (or known) outside the interval $0 \leq x \leq L$ then it can be uniquely reconstructed from a knowledge of $b(x)$ on $0 \leq x \leq L$. The method of reconstruction is extremely simple. Thus let $\hat{x}(t) = at/T$ and $\hat{y}(t) = 0$ in (3). Then

$$b(x) = \int_0^T p(x - at/T)dt, \qquad 0 \leq x \leq L \quad (31)$$

where we have suppressed the dependence of b and p on y. Substituting $\tau = x - at/T$ and ignoring a scale factor we get

$$b(x) = \int_{x-a}^{x} p(\tau)d\tau, \qquad 0 \leq x \leq L \quad (32)$$

and by differentiation

$$b'(x) = p(x) - p(x - a), \qquad 0 \le x \le L. \tag{33}$$

For convenience let us assume that $L = Ka$ with K an integer. (Otherwise we would have to repeatedly make special statements about the last fractional interval.) Then (33) may be written in the equivalent form

$$p(x + ka) = b'(x + ka) + p(x + (k - 1)a), \qquad 0 \le x < a, \quad k = 0, 1, \cdots, K - 1. \tag{34}$$

Let us denote by $\phi(x)$ the portion of the scene that moves into the field of view during exposure. Then

$$\phi(x) = p(x - a), \qquad 0 \le x < a \tag{35}$$

and (34) can be solved recursively in terms of $\phi(x)$. The solution is

$$p(x) = \sum_{k=0}^{m} b'(x - ka) + \phi(x - ma), \qquad 0 \le x \le L \tag{36}$$

where m is the integral part of x/a. Thus if $\phi(x)$ is known, the undistorted scene $p(x)$ can be recovered exactly. (Slepian shows that a similar argument can be carried through for rotation of the image about a point outside itself.)

If this method of inversion is used in the presence of noise, then b must be replaced by d in the above analysis. In that case a term $\sum_0^m n'(x - ka)$ must be subtracted from the right-hand side of (36). Thus in the interval $ka \le x \le (k+1)a$ the noise power in the restoration is k times the noise power in d', provided a is larger than the distance over which the input noise is correlated.

Cutrona and Hall [24] have pointed out that we can exactly reconstruct the estimate

$$\hat{p}(x) = \int_{x-\delta}^{x} p(\tau)d\tau \tag{37}$$

from $b(x)$ by an algorithm similar to Slepian's but one which avoids differentiation. It is easily verified that with $\phi(x) = 0$, say,

$$\hat{p}(x) = \sum_{k=0}^{m} [b(x - ka) - b(x - ka - \delta)] \tag{38}$$

with m defined as before to be the integral part of x/a. When used in the presence of noise this algorithm ought to give a restoration less noisy than Slepian's. If δ is chosen smaller than the resolution limit of the eye, the blurring given by (37) cannot be detected and the noise reduction is obtained at no cost.

In the preceding analysis we assumed the direction and extent of blurring to be known. These quantities need not be known *a priori*, however. They can be estimated from the degraded image. Thus if $\hat{x}(t) = at/T$, $\hat{y}(t) = ct/T$, then from (4)

$$|B(\omega, \sigma)| = |P(\omega, \sigma)| \left| \frac{\sin[(a\omega + c\sigma)/2]}{[(a\omega + c\sigma)/2]} \right|. \tag{39}$$

The parameters a and c may be obtained, as Slepian points out, from the regularly spaced lines $(\omega a + \sigma c)/2 = \pm k\pi$, $k = 1, 2, \cdots$, at which $|B|$ vanishes. In the presence of noise these lines may be obscured. A better method would be to take the (two-dimensional) cepstrum of b; i.e., the Fourier transform of log $|B|$. In the cepstrum a bright line will appear (followed by fainter equispaced lines), instead of the lines of zero intensity.

Slepian had proposed this method with star images in mind. In such an application $\phi(x)$ can be safely assumed to be zero. However if we try to apply the algorithm to a blurred aerial photograph, say, then $\phi(x)$ is not even approximately known. In collaboration with M. R. Schroeder we have found one way to estimate ϕ from the degraded image itself. The key is to note that ϕ is repeated K times across the restoration given by (36). Let us define

$$\tilde{p}(x) = \sum_0^m b'(x - ka) \tag{40}$$

and rewrite (36) as

$$\phi(x - ma) = p(x) - \tilde{p}(x). \tag{41}$$

If we evaluate the left and right sides of (41) for $ka \le x < (k+1)a$, and add the results for $k = 0, 1, \cdots, K-1$, we get

$$\phi(x) = \frac{1}{K}\sum_0^{K-1} p(x + ka) - \frac{1}{K}\sum_0^{K-1} \tilde{p}(x + ka), \qquad 0 \le x < a. \tag{42}$$

The first sum on the right-hand side of (42) is, of course, unknown. However, it is intuitively evident that for large values of K this sum approaches the mean value of p. We, therefore, assume this sum to be a constant p_m independent of x and take

$$\hat{\phi} = p_m - \frac{1}{K}\sum_0^{K-1} \tilde{p}(x + ka) \tag{43}$$

to be an estimate of ϕ. Substituting this into (41) we get the estimate

$$\hat{p}(x) = \tilde{p}(x) - \frac{1}{K}\sum_0^{K-1} \tilde{p}(x + ka) + p_m. \tag{44}$$

We close this section by noting that there are examples other than uniform linear motion for which the impulse response has no inverse on the entire plane, but has an inverse when restricted to a finite region. In principle, therefore, the basic idea of Slepian's method ought to carry over to these situations as well. In practice, however, the numerical solution of such integral equations is difficult and "exact" inversion is not possible. (See Section VI in this connection.) The success of the method for uniform motion is really due to the fact that after suitable restriction of the domain, the equation can be solved explicitly.

V. The Method of Constrained Deconvolution

We have emphasized throughout this paper that optimum restoration for a human observer is practically impossible due to the complexity of the fidelity criterion. In view of this, a restoration procedure recently proposed by MacAdam [25] is noteworthy. It allows the experimenter to search for (and hopefully converge on) the optimum restoration by using a digital computer in an interactive manner. The experimenter specifies certain input parameters and tolerances, on the basis of which the computer produces a restoration. If the result is not satisfactory, the experimenter may try a new set of specifications based on his previous observations and on his subjective evaluation of what he thinks is wrong with the restoration. In this manner he may obtain as many different restorations as he desires.

Only experimentation with this technique can tell whether it results in aimless wandering or a noticeable drift towards an optimum. However, such use of introspective evaluation in a feedback loop is effective in many contexts and certainly ought to be explored for image restoration.

The algorithm used to produce each restoration is interesting in its own right, independently of its use in a man–machine feedback loop. MacAdam calls it the method of constrained deconvolution for reasons that will become apparent from the outline that follows.

Let us assume that p, h, and b are all of finite extent. Let $p_{ij}(i, j=1, n_p)$, $h_{ij}(i, j=1, n_h)$, and $b_{ij}(i, j=1, n_b)$ be samples of these functions. We assume the sampling to be fine enough so that (2) is well approximated by the matrix convolution

$$b_{ij} = \sum_{l=1}^{n_p} \sum_{k=1}^{n_p} h_{i+1-k, j+1-l} p_{kl}, \qquad i, j = 1, \cdots, n_b. \quad (45)$$

(If a subscript on h lies outside the set $\{1, \cdots, n_h\}$, the corresponding sample is understood to be zero.) By suitably ordering the samples of b, h, and p and interspersing zeros it turns out[2] that (45) can be written as a convolution of *vectors*. Thus with

$$I = n_b(n_p - 1) + n_p, \quad J = n_b^2, \quad M = n_b(n_h - 1) + n_h \quad (46)$$

it is possible to form the vectors $\boldsymbol{p}$ (I-dimensional), $\boldsymbol{b}$ (J-dimensional), and $\boldsymbol{h}$ (M-dimensional) such that in terms of their components (45) becomes

$$b_j = \sum_{i=1}^{I} h_{1+j-i} p_i, \qquad j = 1, \cdots, J. \quad (47)$$

MacAdam introduces the effect of noise and *a priori* information in an interesting way by specifying lower and upper limits for each component of each of the vectors $\boldsymbol{b}$, $\boldsymbol{h}$, and $\boldsymbol{p}$. Suppose $\tilde{\boldsymbol{b}}$, $\tilde{\boldsymbol{h}}$, and $\tilde{\boldsymbol{p}}$ are vectors which satisfy these constraints as well as the equation

$$\tilde{b}_j = \sum_{i=1}^{I} \tilde{h}_{1+j-i} \tilde{p}_i, \qquad j = 1, \cdots, J. \quad (48)$$

Then $\tilde{\boldsymbol{p}}$ is a restortion of $\boldsymbol{p}$ consistent with the given constraints.

The inequality constraints along with (48) define $2(I+J)$ half-spaces (each bounded by a hyperplane) in the I-dimensional space of the components of $\boldsymbol{p}$. The region bounded by these half-spaces is a convex region—the solution solid. The coordinates of any point in this solid are the components of a valid restoration $\tilde{\boldsymbol{p}}$. MacAdam gives an iterative algorithm to reach one such point from the origin. He gives a second algorithm to improve the restoration $\tilde{\boldsymbol{p}}$ by moving it close to the center of the solution solid. This improved $\tilde{\boldsymbol{p}}$ is then the final estimate $\hat{\boldsymbol{p}}$.

One of the nice features of this method of restoration is that the constraints on $\boldsymbol{h}$, $\boldsymbol{b}$, and $\boldsymbol{p}$ enable us to ensure that these vectors never have negative components. None of the other methods discussed here can guarantee this in the presence of noise. The method also gives the experimenter a degree of control over the restoration which is unmatched by other methods.

[2] MacAdam attributes this discovery to R. W. Preisendorfer.

VI. Other Possibilities

In purely mathematical terms, the restoration of a degraded photograph is equivalent to the solution of a Fredholm integral equation of the first kind. This type of equation arises in many physically motivated problems. Many methods have been proposed, in various contexts, for numerically solving such equations. We present a few of these here in the hope that they may be profitably used for image restoration.

To keep the presentation simple let us consider only one-dimensional blurring, e.g., due to linear nonuniform motion. The extension to two dimensions is quite straightforward; only the equations are more cumbersome. Let us also assume p to be of finite extent. Thus suppressing the dependence on y, the integral equation

$$\int_0^L h(x - t) p(t) dt = d(x) - n(x) \quad (49)$$

is to be solved for p.

The main source of difficulty in solving (49) is that the inverse (if there is one) of the integral operator is not bounded. Clearly, for any integrable h, $\int_0^L h(x-t) \sin \omega t \, dt \to 0$ as $\omega \to \infty$. Hence a large high-frequency component in p might produce an insignificant change in d. Conversely, a small amount of noise n might give a very large error (usually a high-frequency oscillation) in the inverse.

With $n(x)$ unconstrained the problem is clearly meaningless. Constraints on n restrict the solution to a set of possible solutions. The problem thus reduces to one of choosing the "best" among these to be the estimate $\hat{p}$. Phillips [26] has proposed that $\hat{p}$ be chosen as the smoothest solution in the sense that for all possible solutions p,

$$\int_0^L [\hat{p}''(x)]^2 dx \le \int_0^L [p''(x)]^2 dx. \quad (50)$$

Twomey [27], [28] has given an improved and more general version of this idea as follows: Let

$$\boldsymbol{Hp} = \boldsymbol{d} - \boldsymbol{n} \quad (51)$$

be the digital equivalent of (49). Here $\boldsymbol{p}$, $\boldsymbol{d}$, and $\boldsymbol{n}$ are column vectors and $\boldsymbol{H}$ is a rectangular ($K \times J$) matrix of quadrature coefficients. For notational convenience let us assume $K \ge J$. Let $\boldsymbol{n}^T\boldsymbol{n}$ be constrained to have some given value. Then of the possible solutions of (51) choose the one that minimizes $c = \boldsymbol{p}^T \boldsymbol{C} \boldsymbol{p}$, where $\boldsymbol{C}$ is some square matrix that makes c a reasonable criterion of smoothness. With μ a Lagrangian multiplier the problem thus reduces to finding $\boldsymbol{p}$ such that the gradient

$$\nabla_p[(\boldsymbol{d} - \boldsymbol{Hp})^T(\boldsymbol{d} - \boldsymbol{Hp}) + \mu \boldsymbol{p}^T \boldsymbol{C} \boldsymbol{p}] = 0. \quad (52)$$

Straightforward manipulation of (52) gives

$$\boldsymbol{p} = (\boldsymbol{H}^T\boldsymbol{H} + \mu \boldsymbol{C})^{-1} \boldsymbol{H}^T \boldsymbol{d}. \quad (53)$$

The correct value $\hat{\mu}$ of μ is choesn by trial and error to give the desired value of $\boldsymbol{n}^T\boldsymbol{n}$. For this value of μ (53) gives the estimate $\hat{\boldsymbol{p}}$. Suppose (51) has a solution $\boldsymbol{p}_0$ in the absence of noise. Then $\boldsymbol{H}^T\boldsymbol{d} = \boldsymbol{H}^T\boldsymbol{H}\boldsymbol{p}_0$. By substituting this into (53),

Twomey shows that $\hat{p}$ is a "filtered" version of p_0 in that components of p_0 along the eigenvectors of $H^T H$ corresponding to small eigenvalues are attenuated.

A similar type of smoothing can also be obtained by approximating the matrix $H^T H$ by one of lower rank [29].[3] Note that the symmetric $J \times J$ matrix $H^T H$ has the dyadic expansion

$$H^T H = \sum_1^J \lambda_i u_i u_i^T \quad (54)$$

where (λ_i, u_i), $i = 1, \cdots, J$, are the (nonnegative) eigenvalues and orthonormal eigenvectors of $H^T H$. Let $\lambda_1 \geq \lambda_2 \cdots \geq \lambda_J$. Then truncating the series in (54) at $M < J$ we can get a smoothed estimate $\hat{p}$ by expanding $H^T d$ in terms of u_i, $i = 1, \cdots, M$. Thus

$$\hat{p} = \sum_1^M (\alpha_i / \lambda_i) u_i \quad (55)$$

where $\alpha_i = u_i^T H^T d$.

A result analogous to (55) can be obtained also via the singular value decomposition and the pseudo inverse. The matrix H has a decomposition.

$$H = U \Lambda V^T \quad (56)$$

where U and V are unitary matrices whose columns are the orthonormal eigenvectors of HH^T and of $H^T H$, respectively, and Λ is a $K \times J$ matrix whose elements are zero except on the main diagonal with entries $\sqrt{\lambda_i}$, $i = 1, \cdots, J$. Let $\tilde{\Lambda}$ be obtained from Λ by setting $\lambda_i = 0$ for $i > M$ and let $\tilde{\Lambda}^I$ be the pseudoinverse (obtained by replacing each *positive* entry in $\tilde{\Lambda}^T$ by its reciprocal). Then the estimate [30]

$$\hat{p} = U^T \tilde{\Lambda}^I V d \quad (57)$$

has many interesting properties, among them the elimination of spurious oscillation for proper choice of M.

In all of these methods it is clear that resolution is sacrificed in order to reduce the effects of noise. However it is not easy to get a quantitative measure of the extent of residual blurring in the restoration. Also the tradeoff between resolution and noise can be controlled only indirectly through specifications (of μ, C, M in the above examples) whose influence is not easy to foresee.

Backus and Gilbert [31] have recently proposed a method (for geophysical modeling of the earth!) which alleviates this problem. They give a reasonable quantitative measure of spatial spread due to the residual blur. For any given value of spread they derive the restoration which minimizes the noise variance, thus allowing a direct and simple control over the tradeoff between noise and resolution. Conceptually this is the most reasonable approach to restoration with a flexible objective criterion [32].[4]

To formulate their approach let us define the vectors n, d, and $h(t)$ with components $n(x_i)$, $d(x_i)$, and $h(x_i - t)$, respectively, where x_i, $i = 1, 2, \cdots, K$ are the points at which $d(x)$ is known. Then (49) gives the vector equation

$$d = \int_0^L h(t) p(t) dt + n. \quad (58)$$

We assume that the estimate of the picture at a point x is to be given by a linear combination of the $d(x_i)$. Thus with $a(x)$ an arbitrary vector, the allowed estimates are of the type

$$p_a(x) = a^T(x) d$$

$$= \int_0^L A(x, t) p(t) + n^T a(x) \quad (59)$$

where

$$A(x, t) = a^T(x) h(t). \quad (60)$$

In the sequel the dependence of quantities on x will be suppressed. The kernel $A(t)$ provides a local *average* of $p(x)$ so that we impose the constraint

$$\int_0^L A(t) dt = 1 \quad (61)$$

and define the spread of A from x as s_a given by

$$s_a = \int_0^L (x - t)^2 A^2(t) dt. \quad (62)$$

Defining

$$u = \int_0^L h(t) dt \quad (63)$$

and

$$S = \int_0^L (x - t)^2 h(t) h^T(t) dt \quad (64)$$

(61) becomes

$$u^T a = 1 \quad (65)$$

and (62) becomes

$$s_a = a^T S a. \quad (66)$$

Under the assumption that the $h_i(t)$ are linearly independent on $(0, L)$, it is easily shown that S is a positive definite matrix. Let N be the covariance matrix $E(nn^T)$ also assumed positive definite. Then the variance of the noise in the estimate of (59) is

$$\sigma_a^2 = a^T N a. \quad (67)$$

We may now state the problem as follows: For a given spread s, minimize σ_a^2 given by (67) over all a that satisfy (65) and the inequality

$$s_a \leq s. \quad (68)$$

Let $\epsilon(s)$ denote the minimum value of σ_a^2 under these constraints. It is then evident that $\epsilon(s)$ in nonincreasing.

In the space of the components of a, the inequality (68) defines the interior of an ellipsoid centered at the origin. There is clearly a minimum value of s below which the ellipsoid does not reach the hyperplane given by (65). This value can

[3] The continuous analog of this is the approximation by degenerate kernels. The effects of such approximation on the different types of noise encountered in image restoration are discussed in [29].

[4] The method described in [32] also allows tradeoff between noise and resolution. The data are assumed continuous and the restoring filter is restricted to be spatially invariant. The transfer function of the filter is the solution of a differential equation similar to the Schrödinger equation of quantum mechanics.

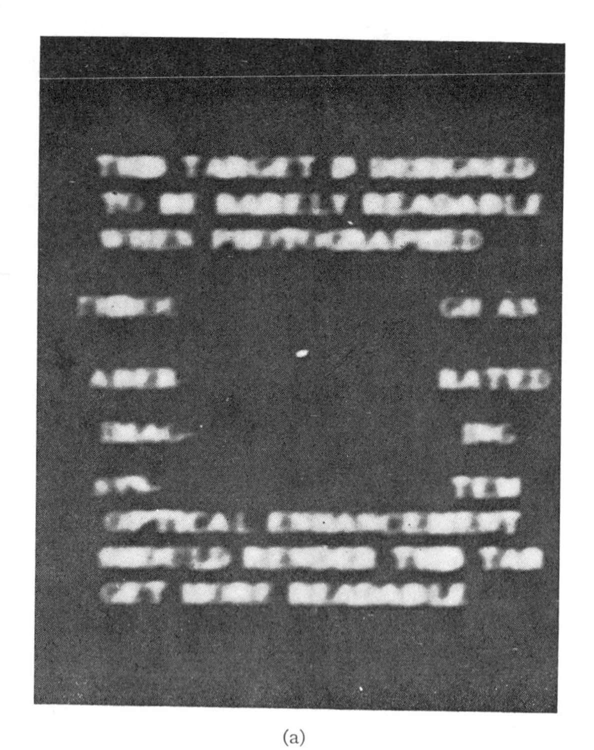

(a)

(b)

(c)

Fig. 1. Optical restoration of long-term turbulence blur by inverse filtering (Mueller and Reynolds). (a) The degraded image. (b) The inverse filter transfer function. (c) The restored image.

be shown to be

$$s_{\min} = 1/(\boldsymbol{u}^T S^{-1} \boldsymbol{u}) \tag{69}$$

and occurs for $\boldsymbol{a} = \boldsymbol{a}_s$ given by

$$\boldsymbol{a}_s = (S^{-1}\boldsymbol{u}) \cdot s_{\min}. \tag{70}$$

In view of the monotonicity of $\epsilon(s)$, the value of $\sigma_a{}^2$ for $\boldsymbol{a} = \boldsymbol{a}_s$ may be called $\epsilon_{\max}$.

Likewise, there is a minimum value of ϵ given by

$$\epsilon_{\min} = 1/(\boldsymbol{u}^T N^{-1} \boldsymbol{u}) \tag{71}$$

which occurs for $\boldsymbol{a} = \boldsymbol{a}_\epsilon$ given by

$$\boldsymbol{a}_\epsilon = (N^{-1}\boldsymbol{u})\epsilon_{\min}. \tag{72}$$

The value of s_a for $\boldsymbol{a} = \boldsymbol{a}_\epsilon$ may be called $s_{\max}$ since there is no point in choosing $s > s_{\max}$. Backus and Gilbert show by an interesting rigorous argument that for $s_{\min} < s < s_{\max}$ $\epsilon(s)$ is strictly decreasing, and there is a unique vector $\hat{\boldsymbol{a}}_s$ which solves the minimization problem. This vector satisfies (65) and the equations

$$(S + \alpha N)\boldsymbol{a} = \lambda \boldsymbol{u} \tag{73}$$

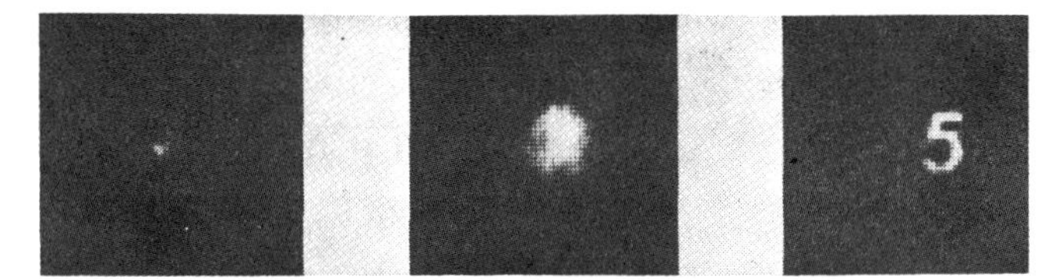

Fig. 2. Digital restoration of short-term turbulence blur by inverse filtering (McGlamery). From left to right: the impulse response, the degraded image, and the restored image.

Fig. 3. Optical restoration of long-term turbulence blur by minimizing MSE (Horner). For the restoration the transfer function of the blurring filter was approximated by a Gaussian curve and the noise power was assumed to be 0.001 of the signal power at all frequencies.

and

$$a^T S a = s \tag{74}$$

with $\alpha > 0$, and some real λ.

An algorithm for computing $\hat{a}_s$, as well as an extension of the theory to minimization of relative noise (rather than absolute noise) are given in [31].

VII. Examples of Restoration

In this section we present examples of restorations achieved by various investigators. The examples illustrate the methods outlined in Sections II–IV. We had hoped to include an example of constrained deconvolution also. However, in the opinion of its inventor, no photographs are available at present, which adequately illustrate the method.

Figs. 1 and 2 illustrate inverse filtering. The restoration in Fig. 1 was produced optically by Dr. P. F. Mueller and Dr. G. O. Reynolds, from a photograph blurred by long-term turbulence. The restoration in Fig. 2, due to Dr. B. L. McGlamery, is an example of digital inverse filtering applied to short-term turbulence blur. In each of these examples the turbulence was generated by a heater in the optical path, the impulse response was photographed along with the scene

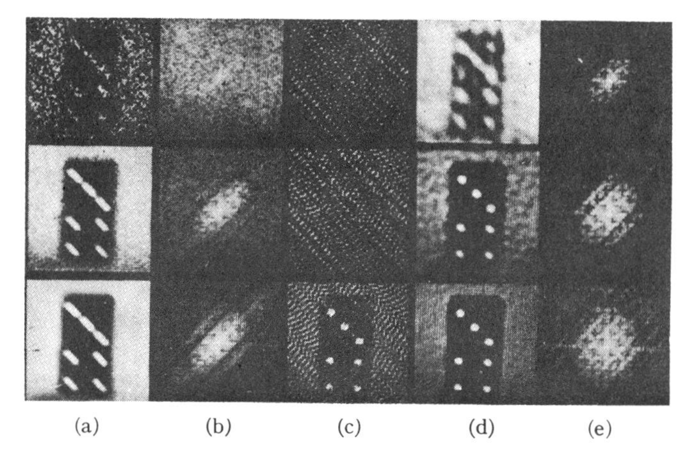

(a) (b) (c) (d) (e)

Fig. 4. Digital restoration of degradation due to uniform motion and noise by minimizing MSE (Harris). (a) The blurred image with maximum signal/noise ratios of 1, 10, and 100, respectively, from top to bottom. (b) The corresponding spectra. (c) Inverse-filter restorations. (d) Minimum MSE restorations. (e) Spectra of the restorations in column (d).

and the noise in the degraded image was mainly due to the emulsion. (See [14] and [15], respectively, for the details concerning the experimental setup.)

The next two figures are examples of minimum MSE restorations. Fig. 3, due to Dr. J. L. Horner, is an optical restoration of a turbulence-degraded image. As discussed in [18] the measured transfer function of the long-term turbulence was approximated by a Gaussian function of spatial frequency and restoring filters were synthesized for various assumed values of signal/noise ratio. This ratio has a value 1000:1 for the restoration in Fig. 3.

Fig. 4, due to Dr. J. L. Harris, Sr., illustrates minimum MSE restoration of blurring due to uniform linear motion in the presence of noise. The degraded images were themselves simulated on a digital computer, as were the restorations. (See [16] for details.)

The restoration in Fig. 5 represents processing indicated by (44) of Section IV. The blurring and restoration were produced digitally; however no noise was added. The figure, therefore, does not indicate the limitations of the method. It does indicate, however, that the distortion of the scene due to the error in estimating ϕ is not intolerable.

Appendix

In this Appendix we collect some examples of spatially invariant blurring. We give all transfer functions or impulse responses to within scale factors which can be absorbed into the units used for the measurement of light intensity.

A. Blurring due to Optical Imperfections

In these cases the filter is easiest to describe in terms of $H(\omega, \sigma)$. Let us define a complex function $A(x, y)$ as follows:

$$A(x, y) = |A(x, y)| e^{-jB(x,y)} \tag{A-1}$$

where

$$\begin{aligned} |A(x, y)| &= 1, \quad \text{for } (x, y) \text{ within the exit pupil} \\ &= 0, \quad \text{outside} \end{aligned} \tag{A-2}$$

and $B(x, y)$ is the wave aberration. When imaging a point source the optical system should ideally produce, at the exit pupil, a spherical wave which converges to a point in the image

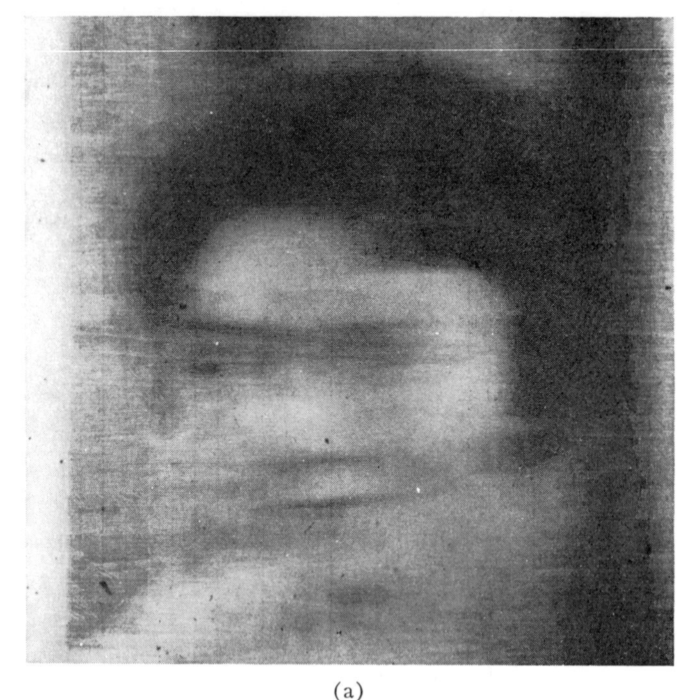

(a)

(b)

Fig. 5. Digital restoration of uniform motion blur by (44) of Section IV. Blurring was simulated on the digitized picture and no noise was added. The blurring distance was about $\frac{1}{8}$ the width of the photograph.

plane. Then $B(x, y)$ is the departure of the phase from that of the ideal wave. In terms of A we have [3, pp. 120–125]

$$H(\omega, \sigma) = \iint A(\omega' - \alpha\omega, \sigma'' - \alpha\sigma) \cdot \bar{A}(\omega' + \alpha\omega, \sigma' + \alpha\sigma)d\omega' d\sigma'. \quad \text{(A-3)}$$

Here $\alpha = \lambda l/2$, where λ is the wavelength of the light and l the distance of the image from the exit pupil. From (A-3) the following examples can be computed:

1) Circular aperture of radius r; $B \equiv 0$. In this case H is circularly symmetric. With $\rho = \sqrt{\omega^2 + \sigma^2}$ we get

$$H(\omega, \sigma) = \cos^{-1}(\alpha\rho/r) - (\alpha\rho/r)\sqrt{1 - (\alpha\rho/r)^2}. \quad \text{(A-4)}$$

2) Square aperture of side s, $B \equiv 0$. In this case $H(\omega, \sigma) = G(\omega)G(\sigma)$ where

$$G(\omega) = (1 - |\alpha\omega/s|). \quad \text{(A-5)}$$

3) Square aperture of side s; $B = c(x^2 + y^2)$. This aberration represents focusing error whose severity increases with c. The transfer function is again of the form $G(\omega)G(\sigma)$ with

$$G(\omega) = \text{sinc}\left[\frac{cs^2}{\pi} \cdot \frac{\alpha\omega}{s}\left(1 - \left|\frac{\alpha\omega}{s}\right|\right)\right] \quad \text{(A-6)}$$

where $\text{sinc}(x) = \sin x/x$.

4) Computation of H even for a simple aberration like defocusing is complicated for other shapes of the aperture. However, for extreme defocusing the image of a point source spreads into a patch of the same shape as the aperture. Approximately, therefore,

$$h(x, y) = |A(\beta x, \beta y)| \quad \text{(A-7)}$$

where β is proportional to c, and determines the size of the patch.

B. Motion Blur

5) In cases where the image moves in two dimensions, rigidly, without change of orientation, (5) of Section I applies. For example [6] with $\hat{x}(t) = r\cos(2\pi t/T)$, $\hat{y}(t) = r\sin(2\pi t/T)$ we get circular motion. In this case

$$H(\omega, \sigma) = J_0(2\pi r\sqrt{\omega^2 + \sigma^2}) \quad \text{(A-8)}$$

where J_0 is the zeroeth-order Bessel function.

6) For linear motion (not necessarily uniform) the description is easier in terms of the impulse response. Thus with $\hat{y}(t) = 0$ in (3) of Section I

$$b(x, y) = \int_0^T p(x - \hat{x}(t), y)dt. \quad \text{(A-9)}$$

Let $\hat{x}(t) = s$, $\hat{x}(T) = a$, and $(d/dt)\hat{x}(t) = v(s)$. Then

$$b(x, y) = \int_0^a p(x - s, y)ds/|v(s)|. \quad \text{(A-10)}$$

Here a is the total distance moved during exposure, and $v(s)$ is the velocity as a function of distance. The limits in (A-10) can be made $\pm\infty$ by defining $v(s)$ to be zero outside the range $0 \le s \le a$. Then a comparison with (2) shows that

$$h(x, y) = \delta(y)|1/v(x)|, \quad 0 \le x \le a$$
$$= 0, \quad \text{otherwise.} \quad \text{(A-11)}$$

The impulse response for uniform acceleration, oscillation, and various other linear motions can be obtained directly from (A-11).

7) An example where the image retains a constant orientation but suffers periodic defocussing due to a motion $z(t) = z_0 \sin \nu t$. Let the lens be diffraction limited (i.e., $B \equiv 0$) of radius r and let $\omega_0 = 2r/\lambda l$ where λ is the wavelength of light and l the image distance. Then for an integral number of cycles of mo-

tion [10]

$$H(\omega, \sigma) = \int_0^{1-\mu} J_0(\beta x)\sqrt{1-(x+\mu)^2}\, dx \qquad \text{(A-12)}$$

where

$$\mu = \sqrt{\omega^2 + \sigma^2}/\omega_0$$

and

$$\beta = \lambda \omega_0{}^2 \mu z_0.$$

8) An example where the orientation changes but the image moves rigidly occurs when the camera rotates about the optic axis. (Note, this is quite different from example 5.) In this case the blurring becomes "spatially invariant" in *polar coordinates*. Thus if $b(r, \theta)$ and $p(r, \theta)$ are the blurred image and original picture in polar coordinates centered on the axis of rotation, then [23]

$$b(r, \theta) = \int_0^T p(r, \theta + \phi(t))dt \qquad \text{(A-13)}$$

where $\phi(t)$ is the angular displacement at time t. This equation has exactly the same form as (A-9), and so the impulse response can be derived in a similar manner.

Acknowledgment

The author wishes to thank the authors mentioned in Section VII for kindly supplying him with the various illustrative restorations; the editor of *Applied Optics* for permission to use Fig. 3; and the editor of the *Journal of the Optical Society* for permission to use Figs. 1 and 2. He also thanks S. Felicetti for editorial help in preparing this manuscript.

References

[1] A. Maréchal, P. Croce, and K. Dietzel, "Amélioration du contraste des détails des images photographiques par filtrage des frequences spatiales," *Opt. Acta*, vol. 5, pp. 256–262, 1958. Also references in this paper to earlier work of the group.

[2] For a recent survey of optical processing techniques, see A. Vander Lugt, "A review of optical data-processing techniques," *Opt. Acta*, vol. 15, pp. 1–33, Feb. 1968.

[3] J. W. Goodman, *Introduction to Fourier Optics*. San Francisco, Calif.: McGraw-Hill, 1968, pp. 154–158.

[4] T. S. Huang, "Some notes on film grain noise," in Woods Hole Summer Study Rep. on *Restoration of Atmospherically Degraded Images Vol. 2*, Alexandria, Va.: Defense Documentation Center, July 1966, pp. 105–109.

[5] P. Dumontet, "Sur la correspondance objet-image en optique," *Opt. Acta*, vol. 2, pp. 53–63, July 1955.

[6] A. Lohmann, "Aktive Kontrastübertragungstheorie," *Opt. Acta*, vol. 6, pp. 319–338, Oct. 1959.

[7] R. V. Shack, "The influence of image motion and shutter operation on the photographic transfer function," *Appl. Opt.*, vol. 3, pp. 1171–1181, Oct. 1964.

[8] L. Levi, "Motion blurring with decaying detector response," *Appl. Opt.*, vol. 10, pp. 38–41, Jan. 1971.

[9] S. C. Som, "Analysis of the effect of linear smear," *J. Opt. Soc. Amer.*, vol. 61, pp. 859–864, July 1971.

[10] A. W. Lohmann and D. P. Paris, "Influence of longitudinal vibrations on image quality," *Appl. Opt.*, vol. 4, pp. 393–397, Apr. 1965.

[11] R. E. Hufnagel and N. R. Stanley, "Modulation transfer function associated with image transmission through turbulent media," *J. Opt. Soc. Amer.*, vol. 54, pp. 52–61, Jan. 1964.

[12] J. Tsujiuchi, "Correction of optical images by compensation of aberrations and by spatial frequency filtering," in *Progress in Optics*, vol. 2. New York: Wiley, 1963, pp. 131–180.

[13] J. L. Harris, Sr., "Image evaluation and restoration," *J. Opt. Soc. Amer.*, vol. 56, pp. 569–574, May 1966.

[14] B. L. McGlamery, "Restoration of turbulence-degraded images," *J. Opt. Soc. Amer.*, vol. 57, pp. 293–297, Mar. 1967.

[15] P. F. Mueller and G. O. Reynolds, "Image restoration by removal of random media degradations," *J. Opt. Soc. Amer.*, vol. 57, pp. 1338–1344, Nov. 1967.

[16] J. L. Harris, Sr., "Potential and limitations of techniques for processing linear motion-degraded imagery," in *Evaluation of Motion-Degraded Images*. Washington, D. C.: U. S. Gov. Printing Office, 1968, pp. 131–138.

[17] J. L. Horner, "Optical spatial filtering with the least-mean-square-error filter," *J. Opt. Soc. Amer.*, vol. 59, pp. 553–558, May 1969.

[18] ——, "Optical restoration of images blurred by atmospheric turbulence using optimum filter theory," *Appl. Opt.*, vol. 9, pp. 167–171, Jan. 1970.

[19] C. W. Helstrom, "Image restoration by the method of least squares," *J. Opt. Soc. Amer.*, vol. 57, pp. 297–303, Mar. 1967.

[20] D. Slepian, "Linear least-squares filtering of distorted images," *J. Opt. Soc. Amer.*, vol. 57, pp. 918–922, July 1967.

[21] C. L. Rino, "Bandlimited image restoration by linear mean-square estimation," *J. Opt. Soc. Amer.*, vol. 59, pp. 547–553, May 1969.

[22] D. Slepian and H. O. Pollak, "Prolate spheroidal wave functions, Fourier analysis and uncertainty—I," *Bell Sys. Tech. J.*, vol. 40, pp. 43–64, Jan. 1961.

[23] D. Slepian, "Restoration of photographs blurred by image motion," *Bell Syst. Tech. J.*, vol. 46, pp. 2353–2362, Dec. 1967.

[24] L. J. Cutrona and W. D. Hall, "Some considerations in post-facto blur removal," in *Evaluation of Motion-Degraded Images*. Washington, D. C.: U. S. Gov. Printing Office, 1968, pp. 139–148.

[25] D. P. MacAdam, "Digital image restoration by constrained deconvolution," *J. Opt. Soc. Amer.*, vol. 60, pp. 1617–1627, Dec. 1970.

[26] D. L. Phillips, "A technique for the numerical solution of certain integral equations of the first kind," *J. Assoc. Comp. Mach.*, vol. 9, pp. 84–97, Jan. 1962.

[27] S. Twomey, "On the numerical solution of Fredholm integral equations of the first kind by the inversion of the linear system produced by quadrature," *J. Assoc. Comp. Mach.*, vol. 10, pp. 97–101, Jan. 1963.

[28] ——, "The application of numerical filtering to the solution of integral equations encountered in indirect sensing measurements," *J. Franklin Inst.*, vol. 279, pp. 95–109, Feb. 1965.

[29] C. K. Rushforth and R. W. Harris, "Restoration, resolution, and noise," *J. Opt. Soc. Amer.*, vol. 58, pp. 539–545, Apr. 1968.

[30] G. H. Golub, "Least squares, singular values, and matrix approximations," *Aplikace Matematiky*, vol. 13, pp. 44–51, 1968.

[31] G. Backus and F. Gilbert, "Uniqueness in the inversion of inaccurate gross earth data," *Phil. Trans. Roy. Soc. London*, vol. 266, ser. A, pp. 123–192, Mar. 1970.

[32] A reviewer has drawn our attention to the following interesting paper: H. A. Smith, "Improvement of the resolution of a linear scanning device," *SIAM J. Appl. Math.*, vol. 14, pp. 23–40, Jan. 1966.

Digital Image Processing

B. R. HUNT, MEMBER, IEEE

Invited Paper

Abstract—A review of the field of digital image processing is presented, with concentration upon image formation and recording processes, digital sampling and digital image display, and with in-depth coverage of image coding and image restoration. New results in image restoration are also presented, covering restoration by use of an eye-model constraint and nonlinear restoration by maximization of the posterior density function.

I. Introduction

THE FRONT PAGES of papers around the world have shown the images of the surface of the moon and the face of Mars, transmitted to earth by NASA's unmanned lunar and planetary exploration vehicles. Yet the remarkable wealth of sharp detail visible in the photos was not the product of the vehicles alone; most of the images released to the public have been digitally processed at the Jet Propulsion Laboratory (JPL) of the California Institute of Technology [1], [2].

The work of JPL in processing space images is probably the most visible part of digital image processing to the general public. It is, however, only a single segment of the growing volume of research and applications in digital image processing. Vision being the most powerful of the five human senses, it is natural that the digital computer be applied, in an increasing number of ways, to images [3]-[13].

What factors account for this growth in the popularity of digital image processing? One important factor has been *necessity;* there is often no substitute for an image in order to convey the full range of information that is relevant to a specific task. But necessity does not solely explain the growth of *digital* image processing. Images can also be processed by *optical* means, and the development of the laser has, indeed, resulted in much activity in optical image processing [14]-[17]. Digital image processing has grown for the same reasons that all other applications of digital signal processing have grown, namely, as follows.

1) Hardware advances. The continued development of solid-state integrated-circuit technology.

2) Software advances. The discovery of fast convolution algorithms [18], [19].

3) Flexibility. The possession of much greater flexibility by digital systems to carry out nonlinear processing algorithms, iterative processes, and processes requiring tests and decision making, when compared to optical processors.

The third of the said points, flexibility, is the one that may be the most important in future applications of image processing. The great quantity of information in an image, and inherent nonlinearities combine to make digital image processing potentially one of the most complex branches of the field of digital signal processing.

Manuscript received September 3, 1974; revised October 24, 1974. This work was supported by the U.S. Atomic Energy Commission under Contract W-7405-ENG-36.

The author is with Los Alamos Scientific Laboratory, University of California, Los Alamos, N.Mex. 87544.

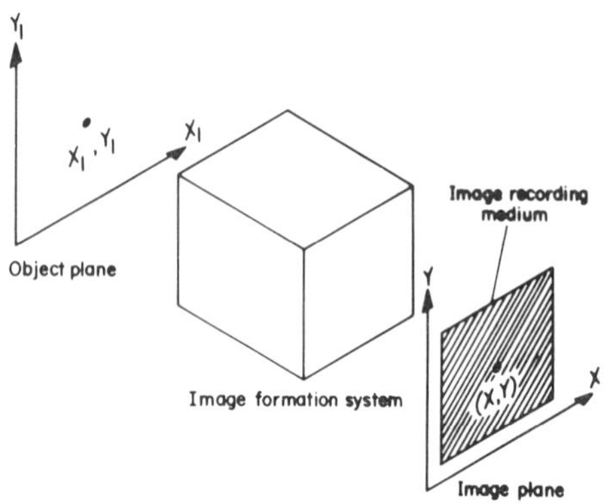

Fig. 1. Schematic of image formation system.

In this article we shall try to review some of the more important concepts in digital image processing and then concentrate upon two problem areas (coding and restoration) which have been among the most active. We thus neglect the important theoretical work in areas such as pattern recognition and artificial intelligence, and the practical applications being done in medicine, geosciences, physics, etc. Space limitations make it impossible to do justice to these many other topics.

II. Basic Concepts in Digital Image Processing

Every visual scene is an image, specifically, the image formed by the human eye upon the retina. The eye is, of course, not the only image formation system. Images are formed by optical methods and by penetrating nuclear radiation. We can abstract the principal elements of an image formation system into a diagram such as Fig. 1. The "black box" in Fig. 1 generates the image by acting upon a radiant energy component of the object. Thus image formation is inherently involved with ascertaining the state of a remote region. In optical systems, the radiant light intensity reflected or emitted by the object is transformed by a set of lenses and apertures. In a radiographic image formation system, the nuclear radiation passing through an object is passed through radiation-opaque apertures and/or pinholes [20]. The physical form of the devices in the black box of Fig. 1 is less important than the equations which describe the transformation process.

A. Image Formation Processes and Equations

The basic nature of realistic image formation systems is that they possess *global* behavior in the processes by which they form images. Consider a point in the image plane, point (x, y) in Fig. 1. The image at point (x, y) is a function of contributions in a (possibly infinite) neighborhood of (x_1, y_1). Representing the image radiant energy distribution as $g(x, y)$ and the

Reprinted from *Proc. IEEE*, vol. 63, pp. 693-708, Apr. 1975.

object radiant energy distribution as $f(x, y)$, the most general description of g is

$$g(x, y) = \int_{-\infty}^{\infty} \int_{-\infty}^{\infty} h(x, y, x_1, y_1, f(x_1, y_1))\, dx_1\, dy_1. \quad (1)$$

The function h involving the object distribution as an argument allows for nonlinear processes in the formation of the image, since we have assumed nothing about linearity. The function h is referred to as the *point-spread function.*

A simplifying assumption is that the point-spread function merely *weights* the object distribution as a scalar multiplier. In this case the point-spread function takes the form:

$$h(x, y, x_1, y_1, f(x_1, y_1)) = h(x, y, x_1, y_1)\, f(x_1, y_1)$$

and the resulting image formation equation is linear

$$g(x, y) = \int_{-\infty}^{\infty} \int_{-\infty}^{\infty} h(x, y, x_1, y_1)\, f(x_1, y_1)\, dx_1\, dy_1. \quad (2)$$

The next step in simplifying assumptions is to make the point-spread function *invariant* as a function of position. Thus

$$h(x, y, x_1, y_1) = h(x - x_1, y - y_1) \quad (3)$$

which leads to the two-dimensional convolution

$$g(x, y) = \int_{-\infty}^{\infty} \int_{-\infty}^{\infty} h(x - x_1, y - y_1)\, f(x_1, y_1)\, dx_1\, dy_1. \quad (4)$$

Finally, the assumption that the point-spread function is *separable*,[1]

$$h(x, y, x_1, y_1) = h_1(x, x_1)\, h_2(y, y_1) \quad (5)$$

$$h(x - x_1, y - y_1) = h_1(x - x_1)\, h_2(y - y_2) \quad (6)$$

leads to two one-dimensional operations on the object distribution

$$g(x, y) = \int_{-\infty}^{\infty} h(x, x_1) \cdot \left[\int_{-\infty}^{\infty} h(y, y_1)\, f(x_1, y_1)\, dy_1\right] dx_1 \quad (7)$$

$$g(x, y) = \int_{-\infty}^{\infty} h(x - x_1) \cdot \left[\int_{-\infty}^{\infty} h(y - y_1)\, f(x_1, y_1)\, dy_1\right] dx_1 \quad (8)$$

and the order of integration in (7) and (8) is arbitrary. Equation (2) is image formation under a *space-variant* point-spread function; (4) represents *space-invariant* image formation. A simplified nomenclature for representing these various cases has been developed by Andrews [22].

A well-developed body of work on optical image formation systems is available (see [17], [23], for example). Under the assumptions of ideal (aberration-free) lenses it can be shown

[1] In texts on integral equations, the kernel h is said to be *degenerate* [21].

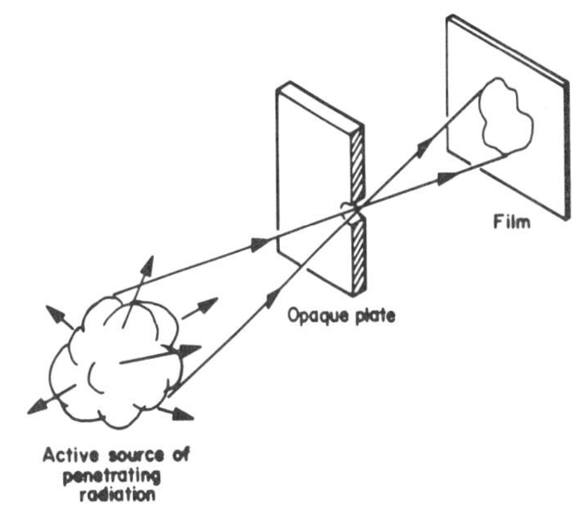

Fig. 2. Active penetrating radiation image formation.

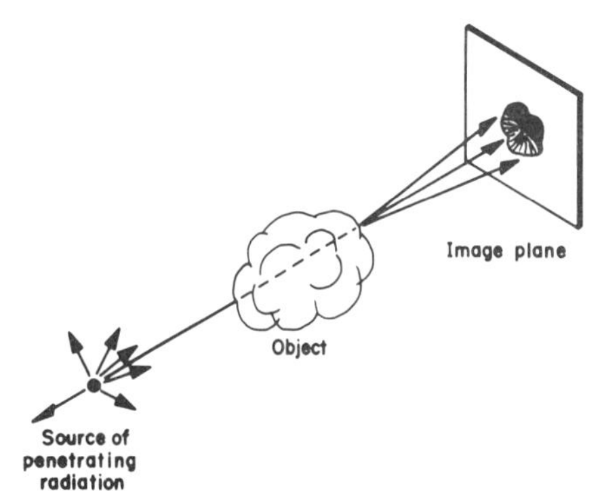

Fig. 3. Passive penetrating radiation image formation.

that:

1) lenses have the ability to perform Fourier transforms;
2) an optical image formation system can be described with a space-invariant model (4);
3) the point-spread function h, in (4), is the Fourier transform of the exit pupil for coherent light, and is the Fourier transform of the autocorrelation function of the exit pupil in incoherent light.

The body of work available on images formed by penetrating (nuclear) radiation is not as extensive or as well-known as for optical systems. Penetrating radiation systems for image formation are basically of two types: active and passive. In an active system the object to be imaged is itself a source of some nuclear radiation, such as neutrons, gamma rays, or X-rays. To form an image of object it is necessary to utilize a pinhole, as in Fig. 2. The principle is the same as the basic pinhole camera in optics. The image which is formed can be described by an equation such as (4). Calculation of the point-spread function h associated with the pinhole is made difficult, however, by the penetrating nature of the radiation [24], [25]. The use of Fresnel zone plates with active penetrating radiation image systems is also possible, and has led to preliminary studies of three-dimensional (holographic) radiography [20], [26].

In a passive penetrating radiation image formation system, the object to be imaged is illuminated by a source of radiation (neutrons, X-rays, gamma rays) which penetrates the object. The relative density of material in the object attenuates the radiation beam and a shadow-image, a projection through the the object, is observed at the image plane; see Fig. 3. Images such as those represented by Fig. 3 are referred to generically as *radiographs.*

The basic equations to describe radiographic image formation are the most complex of all. This complexity is a function of the number of different sources of image degradation phenomena in radiographic imagery and the nature of the phenomena themselves. We list the more important ones.

1) Source size. For perfect resolution the source must be infinitesimally small. In reality it is of finite size. The resulting degradation is modeled by a convolution [9].

2) Geometric effects. Given two sharp parallel edges, one edge lying on a straight-line radius to the source and the other not, the shadow-image nature of radiography will give a sharp image of the edge on the radius, whereas the other will be blurred out. This is a response that must be modeled as a space-variant point-spread function.

3) Radiation scatter. When a quantum of radiation interacts with the atoms in an object, two things occur: the radiation is attenuated in intensity as it collides with the atoms, and it is scattered from a straight-line path by the collisions. The net effect is to produce an image of a point of matter which is blurred by the deviation from a straight line by the scattering (see Fig. 3). Since the amount of scattering is a function of the object which is being radiographed, the formal equation of image formation is nonlinear, as in (1). Approximations can be made by assuming an average density for the object; the required scattering integrals to estimate the point-spread-function can be evaluated numerically without Monte Carlo calculations to simulate the radiation scatter [9].

The pragmatic approach to radiographic image formation systems has been dominant in past efforts in this field; namely, a linear space-invariant model, of the form of (4), is assumed and the point-spread-function estimated by approximations.

B. *Image Recording Processes*

The system capable of forming an image is of no utility without the means to sense and record the image. The recording process is a necessary step in digital image processing. Current computing technology is not capable of performing digital image processing operations of interest on pictures even of the moderate quality of commercial TV, at the bandwidth of commercial TV. Thus digital image processing does not take place in real time, as many one-dimensional digital signal processing operations do.

Image recording is divided into two major technologies: photochemical and photoelectric. Photochemical technology is exemplified by photographic film, photoelectric by the television[2] camera. The utilization of both film and television camera is extremely widespread; both film and television have been used to record images for input to digital image processing.

Photographic film relies upon the properties of halide salts of silver to record images. Silver halides are changed by exposure to light (the exact nature of the change is still not definitely known) so that the action of mild reducing agents (called developers) result in the deposition of free silver [27]. It was found experimentally by Hurter and Driffield [27] that the mass of silver deposited is linearly proportional to the logarithm of the total exposure E, where *exposure* E is defined as the time integral of light intensity over the interval of exposure. There is, in addition, a region of saturation, where

[2]We are using "television" in the generic sense of electronic image recording, rather than implying a specific sense such as systems for commercial broadcast TV.

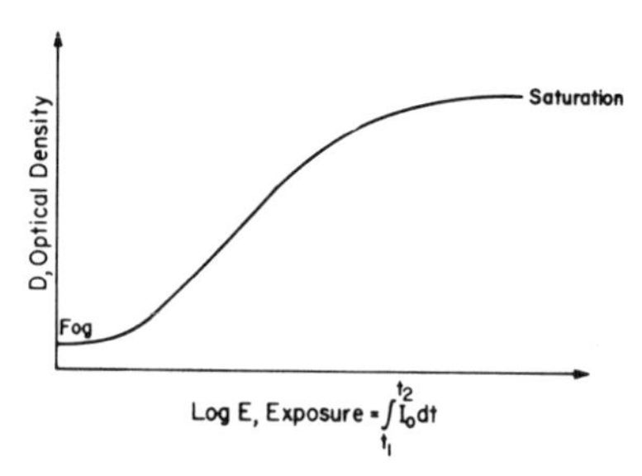

Fig. 4. Typical film response curve.

all silver is deposited, and a region of "fog," where some silver is deposited even in the absence of light. Hurter and Driffield found also that this relation could be displayed in a curve of D-log E, where D is *optical density* and is defined as

$$D = \log_{10}\left(\frac{I_1}{I_2}\right) \tag{9}$$

where I_1 is the intensity of a reference source of light and I_2 is the intensity of light transmitted through (or reflected from) a photographic film when illuminated by the reference source; $I_1 \geqslant I_2$ always. A typical D-log E curve appears as in Fig. 4, and is called the *characteristic curve.*

Assume an exposure recorded in the linear region of this curve; the film is developed and then examined by shining light through it. The intensity of light passing through the film is governed by a form of Bouger's law [27], [28]

$$I_2 = I_1 \exp(-k_1 m_{ag}) \tag{10}$$

where I_1 and I_2 are defined as before, k_1 is an absorption coefficient, and m_{ag} is the mass density of silver per unit area deposited by development. Hurter and Driffield showed, for suitable definition of k_1 and m_{ag}, that

$$D = k_1 m_{ag}$$

and thus substituting into (10)

$$I_2 = I_1 \exp(-D). \tag{11a}$$

But Hurter and Driffield showed, as visible in Fig. 4, that in the linear region

$$D = \gamma \log E - D_0 \tag{11b}$$

where γ is the slope of the linear region and D_0 is the offset due to the linear region not passing through the origin. We assume that the intensity was constant during the exposure interval, so that $E = I_0 t$, where I_0 is the original intensity creating image. Without loss of generality we let $t = 1$. Thus we substitute (11b) into (11a) and have:

$$I_2 = I_1 \exp(-\gamma \log I_0 + D_0) = k_2 (I_0)^{-\gamma}. \tag{12}$$

Depending upon whether γ is positive or negative (a negative film has positive γ, a positive film has negative γ) the representation of intensity in the recorded image is a highly nonlinear mapping of the original intensity or a less nonlinear representation of original intensity. For $\gamma = -1$ the representation is exact.

Equation (12) shows that film is a means (although nonlinear) of recording light intensities. Any recording system, however, is subject to noise. In photographic film the noise is a property of randomness in silver grain formation. First, the size and shape of silver grains are a random quantity. Second, two seemingly identical grains will not respond to exposure and

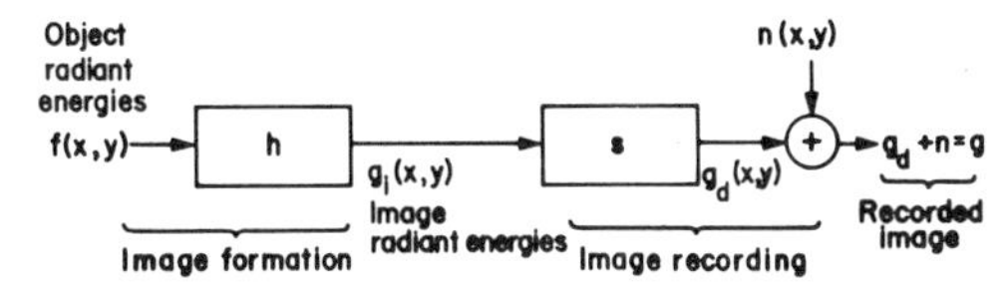

Fig. 5. Block diagram representation of image formation and recording.

development in an identical way. This randomness in film grains results in a randomness in the local silver density in a microscopic region of the developed image. *Film-grain noise*, as this randomness has been called, has been studied extensively. The random fluctuations in silver density due to film grain can be approximated by an additive Gaussian random variable distributed about the mean density [29], [30]. Further, it is usually describable as a very high-frequency or white-noise random process.

The importance of film-grain-noise is that it is additive in film density, *not* intensity. Thus, given an intensity image I_0, the image plus noise is recorded in density as (neglecting the offset D_0):

$$D = \gamma \log (I_0) + n \tag{13}$$

where n is a zero-mean Gaussian random variable with a high-pass or white-noise power spectrum. The noisy version of (12) becomes

$$I_2 = k_2 (I_0)^{-\gamma} \exp(-n) = k_2 (I_0)^{-\gamma} n_1 \tag{14}$$

where n_1 is a *multiplicative noise process* (and would obey log-normal statistics under the aforementioned Gaussian assumption). As before, in regions of saturation or fog, this simple relation is not valid.

The multiplicative nature of film-grain noise is a consistent property of images recorded by photographic film.[3] In photoelectric systems the recorded image is scanned by electron beam, and the camera output current is derived from the beam current. Like any current, the camera output is subject to noise sources (shot-noise, amplifier noise, etc.) which can be treated as additive random fluctuations. In most photoelectric cameras there is a power-law relation between the input and output [31], which is usually described in terms of a plot of the logarithm of camera output current versus the logarithm of camera input light intensity. The power-law relation appears as a straight line in such a plot, the slope of the line being referred to as the *gamma* of the image sensor. An equation such as (12) can be derived to determine the intensity representation of the photoelectric image in terms of the output signal (usually current) of the photoelectric image sensor [31]-[33]. Fortunately, a number of photoelectric systems can be designed or made to operate with a slope of unity in a positive image fashion (equivalent to $\gamma = -1$), and noise is additive to intensity raised to the power gamma.

The importance of the aforementioned properties of image formation and recording processes can be summarized in a model such as Fig. 5. The system h has "memory"; it is a linear or nonlinear operation in the form of (1), (2), (4), (7), or (8). The system s has no memory but represents nonlinearities in image recording. Finally, the term n added to s is a random process, to account for the unavoidable presence of noise that occurs as a consequence of recording the image. Typically, n is assumed to be a Gaussian white-noise process for film-grain noise.

For many purposes it is sufficient to assume that s is the logarithm function. This assumption of the logarithm frames one of the most important distinctions in digital image processing; the distinction between *intensity images* and *density images*. An intensity image we define to be an image represented by values which are linearly proportional to the intensity of the original radiant energy component involved in the image formation. A density image we define to be an image in which the values are proportional to the logarithm of the intensity of the original radiant energy component involved in the image formation. Distinctions between density and intensity processing are important, as we shall see in Section IV.

C. Image Sampling and Quantization

Given a recorded image, the image must be sampled and the resulting samples quantized into a finite number of bits. Quantization effects and sample spacing can be analyzed by extensions from one-dimensional digital signal processing. Quantization errors[4] are equivalent to the addition of a random process to the original data [34]. Likewise, sample spacings must satisfy the Nyquist theorem, which can be proved in two dimensions for images [17].

An image is usually not band limited; therefore, aliasing can be expected in sampling an image. In one-dimensional digital signal processing, aliasing is prevented by the use of a low-pass filter which is rolled-off to zero at the Nyquist frequency of the sampler. In digital image processing this analog prefiltering of the data is not so easily accomplished. Making a slightly out-of-focus copy of the original image is sometimes possible; however, the copy introduces additional wide-band film-grain noise.

Systems to sample and quantize images are modifications of known equipment for microdensitometry. Such systems project a spot of light upon film, and the amount of light I_2 transmitted through (or reflected from) the film is compared (by photomultiplier and associated electronics) to the amount of light I_1 observed in the absence of film. The transmittance T is defined as

$$T = \frac{I_2}{I_1} \tag{15}$$

where $I_2 \leqslant I_1$ always. Analog circuitry computes T or the optical density D (9). The spot of light projected on the film is moved in a raster scanning sequence, e.g., left-to-right, top-to-bottom, and the output sampled at given coordinate spacings. The mathematical model that describes this is

$$g_1(x,y) = \int_{-\infty}^{\infty} \int_{-\infty}^{\infty} h_a(x - x_1, y - y_1) g(x_1, y_1)\, dx_1 dy_1 \tag{16}$$

where g_1 is the output that is sampled at $x = j\Delta x$, $y = k\Delta x$. The matrix of samples $g_1(j\Delta x, k\Delta y)$ for $j = 0, 1, 2, \cdots, N-1$ and $k = 0, 1, 2, \cdots, M-1$, is referred to as the sampled or *digital image*. g is the image recorded on photographic film, and h_a is a function that describes the intensity profile of the spot of light which is projected onto the film. The function

[3] One exception is radiography at high energies, where film density is a linear function of exposure, and grain noise is, therefore, additive in intensity.

[4] Quantization noise could be considered as another noise term to include in Fig. 5. We discount such effects under the assumption that in a good sampling system such effects will be of second order compared to the inherent film-grain noise.

h_a is the effective *aperture* through which the film is being observed.

As (16) indicates, the image which is sampled is not the recorded image g but an image g_1 which is modified by the aperture. In the spatial frequency domain:

$$\mathcal{G}_1(f_x, f_y) = \mathcal{H}_a(f_x, f_y)\mathcal{G}(f_x, f_y) \tag{17}$$

where script letters indicate Fourier transforms. If the aperture projected on film is infinitesimally small, i.e., a Dirac impulse, then $\mathcal{G}_1 = \mathcal{G}$. However, conventional microdensitometry systems cannot produce Dirac impulse apertures, hence $\mathcal{G}_1 \neq \mathcal{G}$. From the viewpoint of aliasing, a Dirac aperture would be undesirable. Equation (17) shows that a suitable aperture can accomplish an analog prefiltering to prevent high-frequency noise aliasing. By simple variations of the aperture dimensions and sample spacing it is possible to achieve almost any desired attenuation of high-frequency noise in order to control aliasing [35].

Aliasing of images is most troublesome when the visual structures are periodic. Moiré patterns result from displaying pictures aliased in insufficient sampling of a periodic structure; the effects are usually annoying, but seldom disastrous, and are easily detected so a more fine sampling may be chosen. The overall effects of even moderate aliasing can often not be detected in either a processed or unprocessed picture after redisplay; one estimate has been made that if a large fraction, say 95 percent, of the spatial frequency energy of the image lies below the Nyquist frequency, then aliasing probably will not cause noticeable distortion [36].

The distinction having been previously made between density and intensity images, which should be quantized: a density image or an intensity image? The answer can be a function of the dynamic range of the image, the number of bits to be retained after quantization, and the signal-to-noise ratio of the image [37]. One is well-served to take the advice of Stockham [38] against fixed "rules of thumb" for a number of bits and choice of intensity versus density representation.

A complete analysis of microdensitometry and the image sampling process can be quite detailed. A recent paper contains probably the most comprehensive survey of the topic that is readily available [39].

D. Reconstruction and Display of Digital Images

The recovery of an analog signal from digital data is achieved, in one-dimensional digital signal processing, by the use of the ideal low-pass filter (sinc function impulse response) in the Shannon–Whittaker sampling/reconstruction theorem [17]. Image reconstruction systems cannot be made that represent a simple extension to two dimensions of the Shannon–Whittaker theorem. The sinc impulse response of the ideal low-pass filter has negative values, which implies negative light in the context of image reconstruction (an impossibility).

Image reconstruction systems are basically the same as sampling systems. A spot of light is projected onto unexposed film (or light response is generated from a CRT phosphor by electron beam), and the intensity of the *display spot* is modulated as the digital data values require. The spot is scanned across the film in a raster fashion. It is clear that, following (16), we can describe image reconstruction by

$$g_2(x,y) = \int_{-\infty}^{\infty}\int_{-\infty}^{\infty} h_d(x - x_1, y - y_1) g_1(x,y)\, dx_1 dy_1. \tag{18}$$

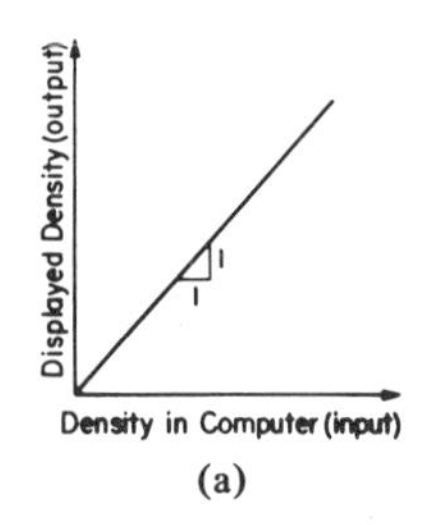

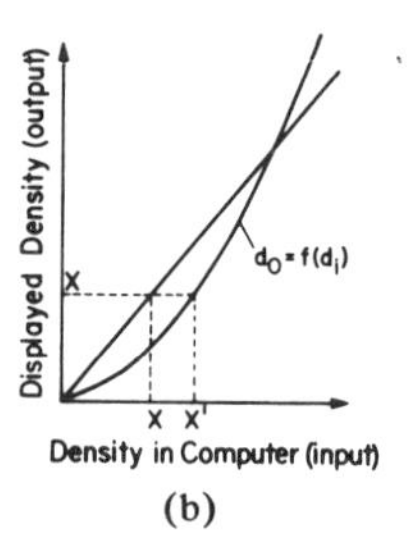

Fig. 6. (a) Ideal display characteristic. (b) Nonideal display characteristic.

g_2 is the reconstructed image; g_1 is a function consisting of Dirac impulses, spaced Δx, Δy apart; h_d is a function that describes the distribution of light intensity in the display spot that reconstructs the image. We thus have in the frequency domain

$$\mathcal{G}_2(f_x, f_y) = \mathcal{H}_d(f_x, f_y)\, \mathcal{G}_p(f_x, f_y) \tag{19}$$

where $\mathcal{G}_p$ is the transform $\mathcal{G}_1$ in (17) after having been sampled and aliased

$$\mathcal{G}_p(f_x, f_y) = \sum_{n=-\infty}^{\infty} \sum_{m=-\infty}^{\infty} \mathcal{G}_1\left(f_x + \frac{m}{\Delta x}, f_y + \frac{n}{\Delta y}\right). \tag{20}$$

In (19) the Fourier transform of the display spot plays the role of the filter in the Shannon–Whittaker theorem which filters out the baseband from the higher ordered aliases.

The properties of display spots in conventional use govern the reconstruction process. In virtually all systems of which the author is aware, the display spot is of a simple form: a Gaussian spot (in CRT and flying-spot scanner systems); or a circular or square spot of uniform intensity within the spot, and of zero intensity outside the spot. The use of simple display spots does not fully attenuate the higher ordered aliases, which will be present to some extent in the reconstructed image [35]. The fact that such displays are usually not considered undesirable demonstrates the validity of the judgment that a limited amount of aliasing in a displayed image is usually acceptable [36].

Equations (17) and (19) show that the process of sampling and displaying an image distorts the spectrum of the image. Given a knowledge of the scanning and display apertures, these effects can be corrected as a problem in image restoration (Section IV), [35].

We make a final remark concerning fidelity in image displays. If a number exists in the computer and represents a particular value of transmittance or density, then a perfect fidelity display is one which creates an image with the same value of transmittance or density as in the computer. In terms of input–output transfer characteristics, the display has the behavior shown in Fig. 6(a). Unfortunately, the author has yet to encounter a display that behaves in this ideal fashion. An actual display characteristic might be such as in Fig. 6(b), in which the true characteristic deviates from the 45° line of the ideal. The display can be *linearized*, however, in the following steps.

1) Collect data on the input/output characteristics of the display by generating, in the computer, fixed steps of transmittance or density; the steps are sent to the display and the true response of the display to a given step is measured.

2) On Fig. 6(b), the characteristic is $d_0 = f(d_i)$. The linear characteristic is given by $d_i = f^{-1}(d_0)$. This transformation

can be constructed empirically by least-squares polynomial fits of the data determined in step 1) or by a table look-up scheme.

3) Before displaying any image it is first transformed by the f^{-1} function. Techniques such as those cited for display linearization have been applied successfully at several installations, including the author's laboratory.[5] They represent a calibration step in digital-to-analog data conversion which is not usually found in one-dimensional digital signal processing with circuits operated under linear conditions.

III. Coding and Bandwidth Compression of Digital Images

Digital data communications have grown very rapidly in the past ten years [40], [41]. Consequently, the transmission and storage of images by digital techniques will increase proportionately. This raises the general problem referred to as *picture coding and bandwidth compression.*

As discussed by Wintz [42], it is possible to conceive of a code which assigns a unique code word to every possible N by N digital image; but the possible number of pictures is so enormous (even "astronomical" is not a sufficient adjective) that the task is impossible even for images of trivial size. Such a scheme fails because of its emphasis on representing all fine structure in all possible pictures. Results achievable with current (and foreseeable future) technology require deemphasizing fine structure and analyzing structure of a more global nature, i.e., statistical structure.

A. Coding Considerations

Suppose that a digital image $g(j, k)$ is created by sampling an image as a matrix of N by N samples and then arbitrarily quantizing each sample picture element (pixel) to R bits. If all the N^2 pixels so sampled were statistically independent, then N^2R bits would nominally be required to represent the resulting digital image. Of course, pictures of interest are not associated with statistical independence between adjacent pixels; there is a great amount of statistical predictability between adjacent pixels. A way to describe this is to use the *image covariance matrix:*

$$R_g = E[(g - E(g))(g - E(g))^T] \tag{21}$$

where g is an N^2 by 1-column vector formed by lexicographically ordering the rows of the image sample matrix, i.e., row 1 of $g(j, k)$ occupies elements 1 through N of g, row 2 is in elements $N + 1$ through $2N$, etc. The expectation operators in (21) are, of cource, ensemble operations.

Cole has shown that all pictures look basically the same in a covariance or power spectrum sense [43]; the use of a simple Markov process model to develop image covariance properties has also been successful [44]. One concludes that covariance statistics are not adequate to describe image structure and that the ensemble average in (21) is probably meaningful over ensembles of pictures which look basically the same, but differ in minor random details.

The matrix R_g describes the degree of statistical dependence between pixels in the digital image. It is possible to represent the matrix R_g in terms of the eigenvalues and eigenvectors:

$$R_g = \Phi\Lambda\Phi^T = \sum_{i=1}^{N} \lambda_i \phi_i \phi_i^T \tag{22}$$

[5] **Dr. Thomas G. Stockham, Jr., University of Utah, has been the most exacting practitioner of linearization techniques, and Dr. Stockham's aid in introducing the author to the subject is gratefully acknowledged.**

where Φ is the N^2 by N^2 collection of orthogonal column eigenvectors and Λ is the diagonal matrix of (real) eigenvalues.

As is well known, (22) describes a rotation of the data vector g into a space with uncorrelated coordinates. Further, it has been shown that in the rotated (eigenvalue) space a great amount of the total statistical variability in the image data can be accounted for with a small number of coordinates in the transformed space and with minimum-mean-square-error (MMSE) [45]-[47]. Such behavior is a general property of the representation of (22), the discrete *Karhunen-Loeve expansion* [48].

Besides the examination of statistical correlations between pixels, the occupancy of quantization levels must also be considered. In either the original or the transformed space, the assignment of an equal number of R bits to each pixel will lead to inefficiency, since not all of the 2^R quantization levels will be present with equal probability. Therefore, a number of schemes have been devised to make best use of bits assigned to quantization of the data, including uniform and nonuniform quantization and quantization based upon objective and subjective criteria. The paper by Wintz contains an excellent summary of quantization schemes [42].

A final consideration in digital image coding is the characteristic behavior of the ultimate judge of imagery. The human visual system has known limitations and some (partially) known idiosyncrasies. The use of particular properties of the human visual system in image coding is referred to as *psychovisual coding.* The human visual system is more sensitive to image spatial frequencies which lie in the "mid-range" rather than in the low or high frequencies; consequently, coding schemes can be developed which use error criteria that are weighted corresponding to the natural weighting of the human visual system. Other considerations of the behavior of the human visual system in image coding have been in the use of contrast sensitivity, spatial resolution, and temporal response [49]. The human visual system also possesses behavior equivalent to logarithmic transformation; the incorporation of the logarithmic eye response into an image coding model results in image coding schemes which are less sensitive to quantization errors and more immune to channel noise, as demonstrated by Stockham [38], [50]. Another property of the human visual system is neural inhibition in the eye retina, which leads to high-pass filter-like behavior [51]. Such behavior results in making the high-frequency edges in images of more importance, and is utilized in a coding scheme known as synthetic highs [52].

The exploitation of visual system properties in image coding is limited by knowledge of more detailed properties of the visual system. Research in analytical models of the visual system indicates many remaining difficulties in a more complete description to encompass known behavior [53].

B. Transform Image Coding

Principles used in the coding of one-dimensional signals have been successfully applied to image coding, e.g., differential PCM and delta modulation, [44], [54], [55]. However, one coding technique which seems to have been more important in the image coding area, and which is of interest in the general context of digital signal processing, is transform image coding.

The discovery of fast transform techniques and their applications have been of great importance to digital signal processing. In image coding, fast transforms have been used to approximate the coding efficiency of Karhunen-Loeve expansions while avoiding the computational requirements of Karhunen-

Loeve. Given the eigenvalues and eigenvectors from (22), we order the eigenvalues and their associated eigenvectors

$$\lambda_1 \geqslant \lambda_2 \geqslant \lambda_3 \geqslant \cdots \geqslant \lambda_{N^2}. \tag{23}$$

Then the coded image is derived by the transformation

$$g_1 = \Phi g. \tag{24}$$

The vector g_1 is in a space where the principal axes of data variation are orthogonal and uncorrelated, and the ordering of (23) means that the components of g_1 represent variations in decreasing order. By selecting and saving components of g_1, one can generate a vector of length $M < N^2$; saving the largest M components of g_1 is an approach used. The M components saved are transmitted, along with "bookkeeping" information to indicate the components actually saved, and a reduction in bandwidth is achieved. At the receiver, the vector of length M is made into a vector g_2 of length N^2 by inserting zeroes for the components not transmitted, and the image is reconstructed by the transformation:

$$g_3 = \Phi^T g_2 \tag{25}$$

the mean-square-error in coding is thus

$$\|g_3 - g\|^2 = \epsilon \tag{26}$$

and is minimal over all possible linear transformations.

Equations (24) and (25) indicate that the Karhunen–Loeve expansion transformations require on the order of N^4 computations; for typical N (say $N = 500$) this is excessive even with the best of modern computing technology. (Even more difficult can be the computation of the eigenvalues and and eigenvectors of the original N^2 by N^2 covariance matrix.) Experimental studies have shown that the covariance between two pixels rapidly approaches zero as the distance between the pixels increases. Consequently, it is possible to code the image by breaking it up into subblocks of size P by P; covariance and coding computations are performed just as already mentioned in P by P regions, and the total computing time is then proportional to Q^2P^4, where $Q = N/P$. If $P \ll N$ an appreciable reduction in computing time can be achieved. $P = 8$ to 16 is typical. Even for subblock coding, however, the amount of computation is substantial.

In transform coding the eigenvector transformation of (24) is replaced by one of the known fast transform algorithms. The fast Fourier, Hadamard, Haar, and slant transforms have all been tried [46]. The purpose of the transform step is to create a domain in which the data is uncorrelated and signal energy is compacted into a small number of components. The above fast transform algorithms are known for their energy compaction properties; in addition, the Fourier can be shown to generate uncorrelated components as the value of N goes to infinity [56].

The gain in computation by the use of fast transforms is great. Since the aforementioned fast transforms are separable into one-dimensional operations on rows and columns, then to transform an entire image is of the order of $N^2 \log_2 N^2$ computing operations. Finally, coding the picture in P by P subblocks using fast transforms produces even further savings in computation.

Given a pragmatic motivation, it is satisfying to note that transform coding works. Comparisons are difficult to make, but Wintz [42] presents data showing the performance of various picture coding schemes. Transform coding methods can produce coded pictures with requirements as small as 1 bit per pixel. Since it has been stated that studies of images show entropies on the order of 1 bit per pixel [57], transform coding approaches such rates and is definitely superior to various PCM schemes [42]. Comparisons between fast transforms for image coding and Karhunen–Loeve coding show that little is lost in fast transform coding, either in objective or subjective criteria [45], [46].

The most recent and interesting result in image coding is the work of Tescher [58], who has demonstrated the importance of phase. It has been known for some time that the phase of the Fourier transform of an image is more important in representing the image than is the amplitude [59]. By including phase in an adaptive image coding procedure Tescher was able to demonstrate coding requiring $\frac{1}{3}$ to $\frac{1}{2}$ bit per pixel, with results that equal subjective and objective error criteria on coding without phase. The results of Tescher demonstrate the potential for bandwidth reduction on the order of 16 to 24 to 1.

IV. Digital Image Restoration

Digital image restoration is a field that has seen as much activity as digital image coding; however, the overall state of image restoration is not as satisfying aesthetically and intellectually as that of digital image coding, because the general image restoration problem is made more difficult by the nonlinearities inherent in various image recording processes.

A. Basic Formulations of the Restoration Problem

The image restoration problem can be described in terms of Fig. 5: take the recorded image and process it in some fashion to obtain an estimate of the object radiant energy distribution f. The estimate $\hat{f}$ of the object radiant energy distribution is referred to as the *restored image*. In the case of an ideal image recording medium, the functional block s in Fig. 5 would be the identity mapping and the noise term n would be everywhere zero. In such an ideal case the image restoration problem reduces to solving an integral equation such as (1), (2), or (4) for the function f. Even this ideal case is difficult, however, for the underlying equation is *ill-conditioned*. It can be shown by the Rieman–Lebesque lemma that in (2) an arbitrarily small perturbation in g can correspond to arbitrarily large perturbations in the function f [60].

The digital image restoration problem is posed in terms of a discrete approximation to either (2) or (4) (we are avoiding discussion of the nonlinear problem because of its even greater difficulty). A discrete approximation of either (2) or (4) requires the use of *quadrature* rules from numerical analysis; as discussed in previous sections, satisfying the Nyquist criterion is a prime objective in sampling, and the resulting equal spacing leads to equal-interval or Newton–Cotes quadrature rules [61]. The simplest quadrature rule of this class for approximating (2) or (4) is the rectangular rule, which gives under the ideal assumptions in Fig. 5,

$$g_i(j\Delta x, k\Delta y) \cong \sum_{n=a_2}^{b_2} \sum_{m=a_1}^{b_1} h(j\Delta x, k\Delta y, m\Delta x_1, n\Delta y_1) \cdot f(m\Delta x_1, n\Delta y_1) \tag{27}$$

and also

$$g_i(j\Delta x, k\Delta y) \cong \sum_{n=a_2}^{b_2} \sum_{m=a_1}^{b_1} h((j-m)\Delta x, (k-n)\Delta y) \cdot f(j\Delta x, k\Delta y). \tag{28}$$

Note the approximation in (27) and (28). The left- and right-hand sides are equal if we include an error term in the integral quadratures. If images of infinite extent are dealt with, then: $a_1 = a_2 = -\infty$ and $b_1 = b_2 = \infty$. Such images are not practical for digital restoration, of course, and we assume the images to be of finite extent. Then, without loss of generality, we assume: $a_1 = a_2 = 0$ and $b_1 = M$, $b_2 = N$.

Equations (27) and (28) make explicit the linear nature of image restoration. If we use a lexicographic order to create vectors (recall (21)) from the matrices g and f, then the sums in (27) and (28) can be represented as a matrix product. The elements of the matrix correspond to the coefficients h in (27) and (28), and appear in the appropriate positions in the matrix to multiply the lexicographically ordered elements of f. Thus (27) and (28) can be written in the form

$$g_i = H_V f \tag{29}$$

$$g_i = H_T f \tag{30}$$

where the approximation has been dropped for simpler notation but should be remembered as implicit. Equation (29) is *linear space-variant* digital restoration and (30) is *linear space-invariant* digital restoration.

The linear systems in equations (29) and (30) possess the same behavior of being ill-conditioned as do the underlying integral equations. In fact, it can be shown that the more finely one samples the data to improve the approximation, the more ill-conditioned is the system of linear equations [63]. Further, the size of the system is extremely large.

Considering the computational problem, equations (29) and (30) differ by the nature of the matrices H_V and H_T. Matrix H_V, space-variant restoration, has the maximum number of degrees of freedom: $(MN)^2$ total degrees of freedom. However, matrix H_T is of special form. It is a vector or block Toeplitz matrix, i.e., H_T is made up of a total of M^2 partitions, each partition being of size N by N; the partitions are arranged in Toeplitz matrix fashion and each partition is itself Toeplitz [64]. Consequently, the matrix H_T has only MN degrees of freedom.

The Toeplitz structure of matrix H_T can be readily exploited. First, the block Toeplitz structure of (28) and (30) can be described as a subpartition of a circular convolution structure by using a vector or block circulant matrix, and the associated block circulant matrix can be inverted and the linear system solved by use of the two-dimensional discrete Fourier transform [62]. The discrete Fourier transform computations can be implemented by the fast Fourier transform (FFT) to place the computational requirements in the realm of existing computers. Second, the circulant inversion is only an approximation to the true restoration which requires a Toeplitz inversion; the approximation can be improved by iteration if necessary [64]. In general, however, circulant approximations to Toeplitz forms improve as the number of points increase [65] (a rare case in which Mother Nature is going in the same direction as the analyst desires to go).

The matrix H_V cannot be treated in a simplified fashion, and the general space-variant image restoration problem remains computationally intractable. However, there are special cases of space-variant restoration which have recently been solved, such as space-variant motion degradation [66] and optical coma [67]. Finally, the use of singular-value decompositions in space-variant restoration is being investigated [68], and may be of utility in the future.

An alternative point of view to the linear equations already discussed is in filtering. Equations (4) and (28) describe a linear spatial filter, and the image restoration problem can be described, in a formal fashion at least, as the choice of the filter *inverse* to the point-spread-function, so as to yield the original data when the degraded image is processed by the filter. The filtering viewpoint does contribute insight into a case where the simple inverse filter approach will work. If the original image blur is so slight that the corresponding spatial filter remains appreciably greater than zero amplitude response throughout the image bandwidth, and if the signal-to-noise ratio is very high over the image bandwidth, the inverse filter can be expected to achieve good results [69].

Equations (28) and (30) represent a formulation of the restoration problem under the assumption of an ideal image recording media. A realistic model must include a more complex form for the function s plus the addition of noise (the basic model of Fig. 5). The recorded image is

$$g_d(x, y) = s(g_i(x, y)) + n(x, y).$$

Using the discrete representations and lexicographic-ordered vectors as before:

$$g_d = s(H_V f) + n \tag{31}$$

and

$$g_d = s(H_T f) + n. \tag{32}$$

The notation $s(x)$ means each component of the vector x is transformed by s.

For computational purposes, we again concentrate upon (32). Assuming exposure in the linear region of film response, the function s is the base 10 logarithm, and the distinction between density images and intensity images becomes important. Assuming a photographic image is measured and digitized in optical density units, then the density image restoration problem is

$$g_d = \gamma \log_{10} (H_T f) + n \tag{33}$$

where γ is the slope of the characteristic curve in the linear region;[6] in (33) the original object radiant energies f are to be estimated. If the recorded image is measured and digitized in film transmittance units (proportional to intensity of the original object radiant energies), then s is the gamma power and the intensity restoration problem becomes

$$g_i = N_1 (H_T f)^{\gamma} \tag{34}$$

where N_1 is a diagonal matrix whose diagonal entries are

$$\{N_1\}_{jj} = (10)^{n_j}$$

and it is necessary to account for the multiplicative nature of the noise in the intensity domain.

The aforementioned complexities have mostly been avoided in digital image restoration. For restoration of intensity images, the multiplicative nature of film-grain noise is usually neglected and an additive model assumed:

$$g_i = H_T f + n_1 \tag{35}$$

is the assumed intensity restoration, with n_1 an additive process (the relation between n_1 and n in (33) is usually not accounted for, and a convenient assumption is made for n_1, say Gaussian statistics). For density image restorations, the *low contrast* assumption can be made; if the intensity span of

[6] We neglect the offset D_0 from (11b) without loss of generality. Also, we can let $\gamma = 1$ in (33).

the original image is small, then the logarithm of a small variation is approximately linear. The corresponding density image restoration is

$$g_d \cong \gamma H_T f + n. \tag{36}$$

Equations (35) and (36) both result in additive signal and noise, and known theory is applicable. The author's personal preference is for density domain restoration of images, even when the low-contrast approximation may not be valid, because of the superior visual quality of the results; similar observations have been noted by others [70], [71].

Sources of image degradation for which digital restoration is often sought are: motion blur, out-of-focus optics, and blur due to atmospheric turbulence [23], [72].

B. Image Restoration Techniques

The FFT makes feasible the numerical solution of the large linear systems presented in the space-invariant digital restoration problem. The ill-conditioned behavior of the systems is not affected by the employment of the FFT for solutions. The ill-conditioned behavior is controlled from assumptions made to develop a particular restoration technique. Given the ill-conditioned behavior and the existence of noise, there is no unique solution. Instead, there is an infinite family of solutions, and one must choose a criterion that determines the solution which is optimal in some sense.

Equations (35) and (36) are the same in structure (though different in meaning) so we use the structure as a paradigm

$$g = H_T f + n. \tag{37}$$

A criterion of solution is the MMSE criterion:

$$\min E(f - \hat{f})$$

where $\hat{f}$ is the restored image estimate. Since H_T in (37) is block Toeplitz, the block circulant approximation may be employed. Assuming the covariance functions of both image and noise are stationary and decay to zero in a finite interval, then the resulting covariance matrices of the lexicographic-ordered vectors f and n are also block Toeplitz and block circulant approximations may also be used. The resulting restoration algorithm is a linear digital spatial filter which can be described in the frequency domain as

$$\mathcal{H}_W(m, n) = \frac{\overline{\mathcal{H}(m, n)}}{|\mathcal{H}(m, n)|^2 + \dfrac{\Phi_n(m, n)}{\Phi_f(m, n)}} \tag{38}$$

and the restored-image Wiener estimate is described in the frequency domain as

$$\hat{\mathcal{F}}(m, n) = \mathcal{H}_W(m, n)\, \mathcal{G}(m, n) \tag{39}$$

where in (38) and (39) the script letters are two-dimensional discrete Fourier transforms of corresponding lower case quantities, the Φ's are power spectra of noise and image, and the overbar denotes complex conjugate. The point-spread function must be known so that its Fourier transform may be computed for the filter. The application of Wiener filter methods to digital image restoration has been made in a number of different problems. [15], [72]-[76].

Wiener filter restorations possess the shortcoming of requiring extensive *a priori* information, namely, the point-spread function and detailed knowledge of image and noise autocovariance functions. Constrained least squares estimation is a technique that eliminates the requirement of covariance information [62]. The constrained least squares estimate is obtained by solving the minimization problem

$$\begin{aligned} &\text{minimize:} \quad f^T C^T C f \\ &\text{subject to:} \quad [H_T f - g]^T [H_T f - g] = e \end{aligned} \tag{40}$$

where C is a constraint matrix, e is proportional to the noise variance, and T superscript denotes matrix transpose. Given the previous assumptions on H_T and assuming that C is expressible in Toeplitz form, this problem may be solved by the discrete Fourier transform also, and the frequency domain filter is [62]:

$$\mathcal{H}_c(m, n) = \frac{\overline{\mathcal{H}(m, n)}}{|\mathcal{H}(m, n)|^2 + \lambda |\mathcal{C}(m, n)|^2} \tag{41}$$

where λ is a parameter determined by iteration. The similarity to (38) is evident; it is possible to generate a family of filters of which (38) is a special case [77].

Of special interest are the homomorphic techniques of Stockham, Cole, and Cannon, in which the point-spread function is assumed to be unknown and then estimated from the degraded image by taking averages of image segments in the log-spectral domain. Cole's work concentrated upon point-spread functions with no phase component [43] and was extended to estimation of phase in specific point-spread functions by Cannon [71]. The basis of both techniques is the concept of homomorphic systems [70]. The homomorphic filter can be described in the frequency domain as

$$\mathcal{H}_H(m, n) = \sqrt{\frac{1}{|\mathcal{H}(m, n)|^2 + \dfrac{\Phi_n(m, n)}{\Phi_f(m, n)}}} \tag{42}$$

where the symbols are as already discussed. For a point-spread function with zero phase, the homomorphic filter is the geometric mean between the Wiener filter (38), and the inverse filter:

$$\mathcal{H}_I(m, n) = \frac{1}{\mathcal{H}(m, n)}. \tag{43}$$

The aforementioned techniques have been applied to realistic digital pictures, i.e., pictures with sufficient resolution in picture elements to convey appreciable visual structure.[7] Most image restoration work on practical images of which the author is aware utilize one or more of the aforementioned techniques. A number of other digital image restoration methods have been proposed, but not applied to realistic images, chiefly because the computing requirements were beyond the resources of the researcher.

Related to constrained estimation is a restoration technique based upon multiple linear equality and inequality constraints on least squares problems [78]. Linear inequality constraints require solution of a quadratic programming problem, and imply extensive computing when digital images are involved. A different approach to inequality constraints has also encountered computing limitations, even though the method embodies the constraints in a heuristic search technique [79].

[7] It is the author's personal prejudice that resolution of 256 by 256 pixels is a minimum for a realistic digital picture, with 512 by 512 and 1024 by 1024 being, respectively, preferable. This prejudice may be nurtured by the marvelous computing facilities available to the author. Research groups with lesser computing power no doubt see the virtue (of necessity) in digital pictures with less resolution. But some published work with 10 by 10 pictures seems to be stretching the point.

The work of Frieden [80] and Hershel [81] is of interest as a different approach. A random grain model is assumed. Restoration is accomplished by maximizing the likelihood function of grain allocation and simultaneously satisfying image formation equations in the form of (35) or (36). The maximization process generates a nonlinear equation which assures positive restoration. The properties of existence and uniqueness of solutions to the nonlinear equation have not been investigated. Approximate solutions are known but reduce to Wiener filter processing; exact solutions require very large amounts of computation [81]. More recent work of Frieden has been to solve the restoration problem by Monte Carlo allocation of grains subject to image formation equations [82].

Recent Bayesian formulation of image restoration as a problem in state-space analysis is another hopeful direction [83]. Earlier attempts at formulating a Bayesian restoration were constrained by computing requirements [84], but did indicate potential merit.

C. Comparison of Image Restoration Techniques

Given the diversity in image restoration techniques, the natural question is: how do the techniques compare, and which is best? Parameters which affect the outcome of a restoration are: signal-to-noise ratio; shape and extent of the point-spread function; correlation properties of both image and noise; the criterion of optimality chosen for restoration; the extent and the nature of the available *a priori* information. A study has been made to compare the four simplest restoration techniques (inverse filter, Wiener filter, constrained least squares estimate, homomorphic filter) [77]. Some of the results of that study can be summarized as follows.

a) Signal-to-noise ratio. If the signal-to-noise ratio is great enough there is no preference, the techniques converging to the inverse filter.

b) Image and noise correlation. Images are basically low pass in behavior, so high-pass noise presents the simplest restoration problem, and low pass and white noise present equal (but different quality) difficulties.

c) *A priori* information. The homomorphic filter can construct the restoration information from the degraded image itself, hence requires the least *a priori* information. The constrained least squares estimate is next, requiring the point-spread function; noise variance is also required but can be estimated from the degraded image. The Wiener filter requires the most *a priori* information: the point-spread function, plus correlation functions of signal and noise.

d) Visual quality of restored image. For the limited number of cases explored, homomorphic filters produced the restorations with best visual quality, with constrained least squares estimates close in ranking, and Wiener filter a distant third. This preference ranking was most pronounced for lower signal-to-noise ratios (in neighborhood of 10 dB); similar results have been noted in comparing Wiener and homomorphic filters [43].

It may be initially surprising that the Wiener filter performs so poorly in comparison to the other image restoration techniques. The optimality of the Wiener filter is derived on the basis of linear theory, however, and the human visual system perceives images with a number of known nonlinear effects [38], [69], and probably a number of others that are unknown. The logarithmic structure of the homomorphic filter makes its preferability seem reasonable in view of logarithmic eye response [69]; the desirable performance of constrained least squares is not explainable by current theory.

We have not mentioned techniques which result in bandwidth extrapolation of bandwidth-limited images, or "super-resolution". The techniques of Frieden possess such properties [80]. In theory, at least, general properties of Fourier transforms and bandlimited functions can be exploited for super-resolution [85], [86]. In practice, however, the existence of noise and problems of computational stability limit the practical utility of current techniques for super-resolution.

There are some general guidelines to be used in carrying out image enhancement, but no formulation of a specific mathematical problem, as in restoration. Linear filtering, with a high-pass filter, is useful for revealing additional fine structure and detail in an image [9], [10], [87], [88], for example; low-pass filtering is often used to suppress noise; notch filtering is useful to remove the effects of a periodic noise. Histogram equalization is a scheme that has particular utility in enhancement, and can be related in an information-theory sense to the generation of an image which possesses maximum entropy [59], [89], [90]. Other image enhancement techniques that have proven useful include: homomorphic processing for dynamic range compression with linear filtering enhancement processing, the "crispening" technique of Stockham [70]; nonlinear processing of Fourier transform magnitude with Fourier transform phase left unaltered, the "alpha processing" of Andrews [59]; image intensity or density mappings into a color display, usually referred to as "pseudocolor" [59]; rectification of known geometric distortions [88]; generation of image contours; discarding high-order bits before display [22]; etc. Since enhancement is an *ad hoc* process, the list will, no doubt, keep growing as the experience and ingenuity of image enhancers increases.

V. Utilization of a Simple Visual Model in Image Restoration

In digital image restoration, many applied problems which are likely to be presented to the analyst can be characterized as lacking detailed *a priori* information. In the lack of any *a priori* information the homomorphic methods of Stockham, Cole, and Cannon [43], [71] can be used to generate satisfactory restorations. In the presence of complete *a priori* information the optimal filter (Wiener) can be used (although its utility is questionable, in view of the comparisons discussed in the previous section). In between these two extremes is a region in which constrained least squares estimates function, requiring only the shape of the point-spread function to be implemented. In the following we will show restoration based upon combining constrained least squares estimates with *a priori* information which is always available: the response of the human visual system.

A. Human Visual System Response

Determining the response of the human visual system is made difficult by its inaccessible and distributed nature. One cannot isolate a working component of this system and measure its properties. Instead, one must infer the visual system behavior on the basis of indirect evidence. It is on the basis of such indirect data that we know of the visual phenomena associated with Mach bands, Cornsweet's illusion, etc. [38].

Experiments show that the human visual system perceives the logarithm of incident light intensity entering the eye [38]; this result may seem surprising but is consistent with what the

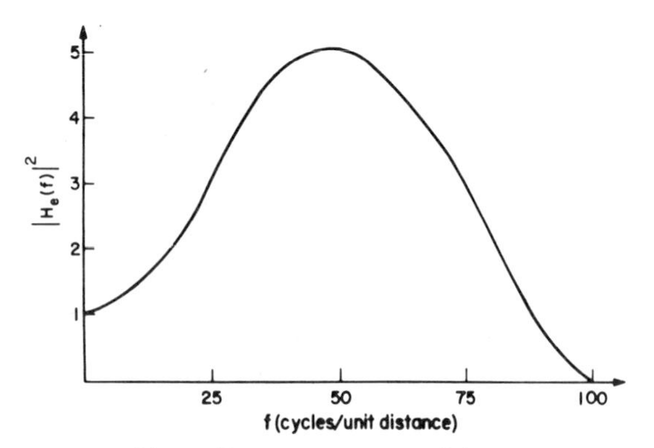

Fig. 7. Frequency response of the eye.

eye can perceive in dynamic range of light intensity. Visual perception takes place from illumination levels of moonless night to high noon on a snowfield, an intensity range of eight orders of magnitude [91].

The human visual system also acts as a spatial frequency filter. No person can see with infinite resolution, so there must be a limit to the maximum spatial frequency which the eye can perceive, at a given viewing distance. In such terms, the eye must behave as a low-pass filter, with no resolution beyond a specific high-frequency limit. More surprising, however, is the spatial frequency response of the eye in the low- and mid-range spatial frequencies. The visual system weights the mid-range frequencies more heavily than either the low or high frequencies. A plot of the visual system response as a function of frequency appears as in Fig. 7 as a result [92]. The response is assumed to be radially (circularly) symmetric, and Fig. 7 is thus an intersection on a plane of the response. The rising low-frequency characteristic corresponds to differentiator-like behavior, with mid-range emphasis. The response in Fig. 7 can be described by an equation[8] [92]

$$H_e(f)^2 = (1 + 0.005\, f^2) \exp\left[-(f/50.0)^2\right], \quad \text{for } 0 \leqslant f \leqslant 100 \text{ cycles/unit distance.} \tag{44}$$

In Fig. 7 the frequency axis is normalized so that 100 cycles per unit distance corresponds to the frequency cutoff of the visual system. In practice, the frequency cutoff is a function of viewing distance, so that at a given viewing distance the frequency axis is rescaled so that the 100 cycles per unit distance cutoff maps into the observed cutoff frequency at that distance.

B. Constrained Least Squares and Visual System Effects

The frequency response characteristic of Fig. 7 has determinable effects in the perception of images by the visual system. Assume that the incident intensities entering the eye originated from an image which was correlated, i.e., the covariance function of the image was not a Dirac function. The image perceived by the brain would not have the same correlation. The linear filter in eye response would induce new correlation on the image, and the power spectrum of the image perceived by the brain would have been multiplied by Fig. 7 (multiplied by proportionality constants for the logarithmic transformation of the spectral power of the original uncorrelated image).

[8] This is (69) of [92] except for the coefficient of the f^2 term, which was reported as 0.05; there is a decimal point mislocated, the response of Fig. 9 being obtained by 0.005 as the coefficient of f^2.

We recall (40), which defines the constrained-least-squares estimate. Examine the side condition first. Since the least squares solution is not exact, we wish to constrain the residual to a meaningful quantity. If $e \cong N^2\sigma_n^2$, then the residual is constrained to represent the noise, where σ_n^2 is the noise variance. This is reasonable since substitution of the exact value of the original object f into the side condition would result in a residual value equal to $N^2\sigma_n^2$. If we next assume a choice of C in (40) such that

$$C^T C = R_f^{-1} \tag{45}$$

where R_f is the covariance matrix of the original object to be restored, then the transformation Cf yields an uncorrelated or "whitened" representation of f [48], and the problem becomes one of minimizing the length of the whitened object vector f subject to the given side constraint. The absence of a weight matrix in the constraint of (40) is equivalent to an assumption that the noise is uncorrelated ("white" noise) [77]. The choice of matrix C such that (45) is satisfied is required for pre-whitening in the statistical theory of least squares estimation. The estimate derived by solution of (40) is necessarily biased. However, in a Bayesian approach, optimal properties for the estimator, i.e., smallest mean-square error for a given bias, result from the use of the whitening transformation based upon the inverse of the covariance matrix R_f [93], [94], and one can again relate such approaches to the optimal theory of Wiener [77].

The use of the covariance matrix of f is undesirable, since the covariance matrix is usually unobtainable as *a priori* information. Instead we make the following assumption: that the covariance of f is a Dirac function, i.e., the original object intensities f are samples from a "white" random process. This can be characterized as the *maximum ignorance assumption*. In such an event, the correlation of the image perceived by the brain is the correlation induced by the visual system. For the maximum ignorance assumption, therefore, the covariance matrix R_f is that of the visual system with the corresponding power-spectrum description given by (44). Following (41), therefore, the restoration filter in the discrete frequency domain is

$$\mathcal{H}_v(m, n) = \frac{\overline{\mathcal{H}(m, n)}}{|\mathcal{H}(m, n)|^2 + \lambda/(|\mathcal{H}_e(m, n)|^2)} \tag{46}$$

where $\mathcal{H}(m, n)$, as before, is the Fourier transform of the point-spread function, and $\mathcal{H}_e(m, n)$ is a two-dimensional formulation of the radially symmetric visual system response in (44).

C. Experimental Results

Fig. 8 is a digital display of an image sampled at 200-μm spacing in a square raster for an experimental study of the effects of the eye-model assumptions discussed earlier in a constrained least squares restoration. The image of Fig. 8 was degraded by using existing subroutines at Los Alamos Scientific Laboratory for digital spatial filtering [87], [95]. A Gaussian-shaped point-spread function was used to simulate image blur by long-term observation through turbulent atmosphere. The point-spread function was circularly symmetric with a full width at half maximum of 1400 μm (7 samples). After the image was blurred, noise was added in the density domain to simulate film-grain noise. Since it is known from other work [43] that restorations with high noise present the greatest difficulty, even for small or moderate degrees of image blur,

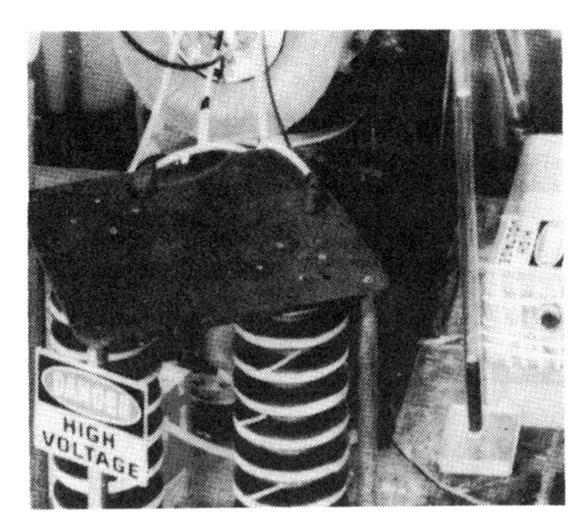

Fig. 8. Original digitized image.

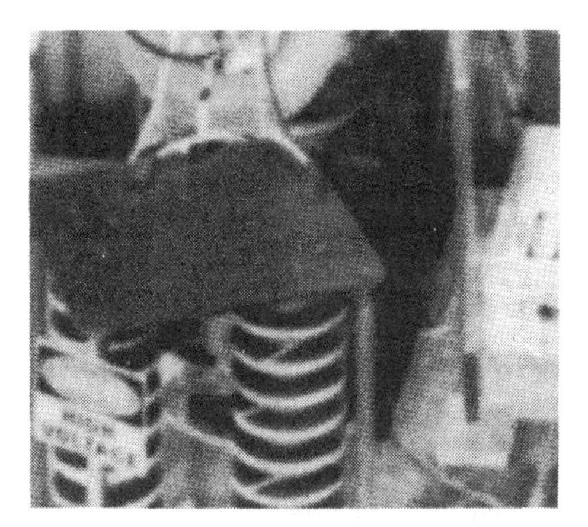

Fig. 9. Blurred image with noise added.

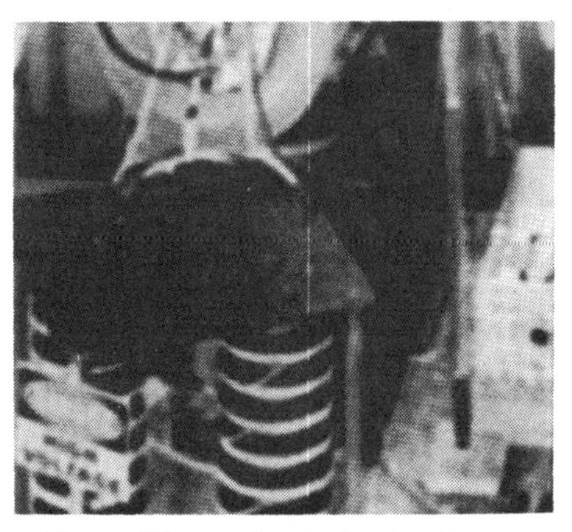

Fig. 10. Wiener estimate of restored image.

the noise was chosen so that the variance in the density domain of the image in Fig. 8 was only ten times the variance of the added noise. The amplitude statistics of the noise were Gaussian, and noise was correlated by a mild high-pass filtering with a discrete frequency domain characteristic of

$$\exp\left[-\left(\frac{\sqrt{m^2+n^2}-2.5}{2.0}\right)^2\right] \tag{47}$$

which is circularly symmetric, and the quantities are in units of cycles per millimeter. The image resulting from the blur and noise is in Fig. 9. The image blur and noise contribute roughly equally to the overall degradation of the image quality observed.

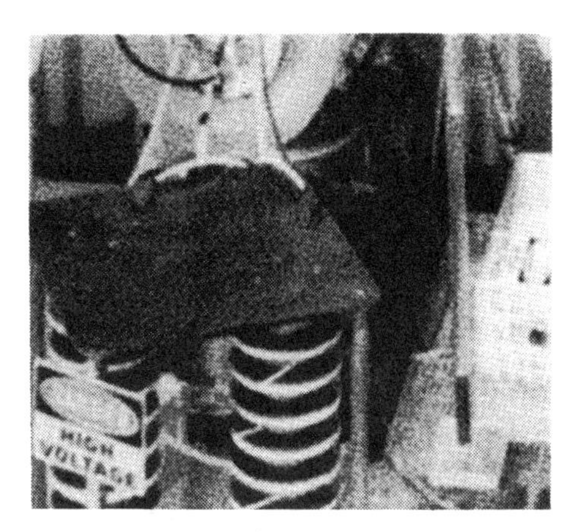

Fig. 11. Eye-constraint estimate of restored image.

Restoration based upon the inverse filter was attempted. The ill-conditioned nature of the problem and the high noise resulted in no discernable image; a random noise field was the visual appearance of the restoration. Next a Wiener filter restoration was computed. The power spectrum of the original image in Fig. 8 was computed and a piecewise exponential-polynomial model fitted to the measured power spectrum. The noise power spectrum was computed from (47) and the Wiener restoration carried out. The small-signal assumption (linear approximation to the logarithm) was made and the processing was carried out in the density domain. The result is presented in Fig. 10. The noise has been greatly suppressed, but no gain in resolution is discernible. In fact, the overall impression is that the suppression of noise has resulted in an overall *loss* of fine structure and resolution of detail from the degraded image of Fig. 8.

The constrained least squares restoration was next carried out, using the maximum ignorance assumptions about both signal and noise, i.e., the original image and noise were both assumed to be uncorrelated (white) random processes. Thus the only covariance statistics under this maximum ignorance assumption would be that induced by the visual system and the resulting filter for constrained least squares estimation would be that induced by (46), with λ varied to satisfy the side condition in (40) for $e \cong N^2 \sigma_n^2$. Restoration could not be carried out until frequency cutoff in Fig. 7 is specified. For the experiment the following rationale was adopted: set the frequency cutoff of the visual system model to be equal to the Nyquist frequency of the sampled image in Fig. 8, which is 2.5 cycles per millimeter for the 200-μm raster (1 mm = 100 μm). Since resolution is a function of the distance from eye to object, the choice of visual system frequency cutoff to be located at the Nyquist frequency implies that the visual system model will be valid only at a certain distance and the resulting restored images should be viewed at that distance. Visually this distance may be determined by holding the original sampled image at the distance where the sampled raster just becomes indistinguishable.[9] With the function $\mathcal{H}_e(m, n)$ in (46) scaled to a frequency cutoff 2.5 cycles per millimeter, the image restoration was carried out, with processing again in the density domain for the small-signal assumption. See Fig. 11.

[9] For the author, this distance is 2 to 3 ft from the original image; obviously this viewing distance is a function of each individual's eyes. Since the size reduction in the printed version of this paper is unknown, and the effect of half-tone reproduction of the image unknowable *a priori*, it is impossible to say what the reader's viewing distance should be of the images in the published version of this paper.

When Fig. 11 is viewed at the proper distance, so that the visual system cutoff is properly located, the superiority of Fig. 11 to Fig. 10 is very evident. Further, the superiority of Fig. 11 to Fig. 9 is also evident; there is a definite impression of overall increased sharpness to Fig. 11 as compared to Fig. 9, and a number of fine details are more visible, as comparison to Fig. 8 will show.

VI. Maximum a Posteriori Estimation for Digital Image Restoration

The eye model work in the previous section illustrates the development of digital image restoration techniques to be used in the presence of maximum ignorance assumptions. The opposite side of the coin is to develop restoration techniques that can function in the presence of the maximum amount of *a priori* information. The reader's first response is that this is the nature of restoration techniques based on the Wiener filter. But such is not true; the Wiener filter is developed from a linear model such as (37), which is obtained by simplifying approximations that do not account for the inherent nonlinear response of film or photoelectric sensors. In this section we present new restoration results that are free of linearity assumptions. Detailed exposition of these results will be presented in a forthcoming paper [96]; in this final section we expose the elements so as to alert the reader of the nature of current and anticipated research in this area [96].

A. Basic Model Assumptions

We adopt the general model of (32) as the basis of development

$$g_d = s(H_T f) + n. \tag{32}$$

Our strategy will be to develop a probability density for the vector f conditioned upon the recorded image g_d and then maximize the conditional density to generate the estimate.

The vector n is a sample from the additive noise field; in film-grain-noise or photoelectronic system noise it is common to treat the noise as a Gaussian random process. The resulting sample vector n can be described by the multivariate normal probability density

$$p(n) = ((2\pi)^{N/2} |R_n|^{1/2})^{-1} \exp(-(n - \bar{n})^T R_n^{-1} (n - \bar{n})) \tag{48}$$

where $\bar{n}$ and R_n are mean and covariance of the sampled noise, respectively. Zero-mean noise is assumed, $\bar{n} = 0$. The assumption of noise stationarity is also made, so that the matrix R_n is a Toeplitz form.

The vector f, representing the object intensities to be restored, is also assumed to be a sample from an underlying two-dimensional random process. We choose as the model for the intensities a Gaussian process fluctuating about a positive mean. This is partially unrealistic; the Gaussian process generates negative values, no matter how large the positive mean. However, if the variance about the mean is small compared to the mean, a very large fraction of the process values are positive and the model is a reasonable approximation. Histograms of images show that intensities can sometimes be modeled by Gaussian statistics, and sometimes not; it is an assumption we adopt for mathematical tractability. The sampled intensities f we describe by the multivariate normal density

$$p(f) = ((2\pi)^{N/2} |R_f|^{1/2})^{-1} \exp(-(f - \bar{f})^T R_f^{-1} (f - \bar{f})) \tag{49}$$

where $\bar{f}$ and R_f are mean and covariance matrices, respectively. In both (48) and (49) we are assuming N-point samples; for two-dimensional descriptions, the number of samples N would be the product of sampled image dimensions.

B. Derivation of Estimator Equations

Given the recorded image, sampled as g_d, the density of f conditioned on g_d is determined from Bayes' law:

$$p(f|g_d) = \frac{p(g_d|f)p(f)}{p(g_d)}. \tag{50}$$

The *maximum a posteriori* (MAP) estimate is derived by differentiating with respect to f and equating the result to zero. We can first take the logarithm of both sides and maximize, and the result is

$$\frac{\partial \ln [p(f|g_d)]}{\partial f} = \frac{\partial \ln [p(g_d|f)]}{\partial f} + \frac{\partial \ln [p(f)]}{\partial f} = 0. \tag{51}$$

The term involving $p(f)$ is available from (49). The conditional of g_d on f is describable in terms of the noise density from (48) as

$$p(g_d|f) = ((2\pi)^{N/2} |R_n|^{1/2})^{-1} \exp(-(g_d - s(H_T f))^T \cdot R_n^{-1}(g_d - s(H_T f))). \tag{52}$$

Substituting from (49) and (52) into (51), one can show that

$$\hat{f} = \bar{f} + R_f H_T^T S_b R_n^{-1} [g_d - s(H_T \hat{f})] \tag{53}$$

is the equation which the MAP estimate satisfies; the matrix S_b is a diagonal matrix of derivatives given by

$$S_b = \begin{bmatrix} \left.\frac{\partial s(u)}{\partial u}\right|_{u=b_1} & & & \emptyset \\ & \left.\frac{\partial s(u)}{\partial u}\right|_{u=b_2} & & \\ & & \ddots & \\ \emptyset & & & \left.\frac{\partial s(u)}{\partial u}\right|_{u=b_N} \end{bmatrix} \tag{54}$$

and

$$b = H_T \hat{f}.$$

Equation (53) has several interesting features which we comment upon.

1) The *a priori* mean image is explicitly part of the estimate. If little *a priori* knowledge is available the mean $\bar{f}$ can be assumed to be a constant or zero. If more substantial knowledge of the restored image is available, however, this knowledge could be used as the *a priori* mean.

2) If the function s is a linear transformation, then the matrix S_b becomes an identity matrix, and (53) can be manipulated into a form which describes the space-domain form of the discrete Wiener filter. The use of circulant approximations for R_f, H_T, and R_n then allows the MAP estimate to be computed in the discrete frequency domain as in (38). The equivalence of MAP and MMSE estimation for linear systems and symmetric densities is already known [97].

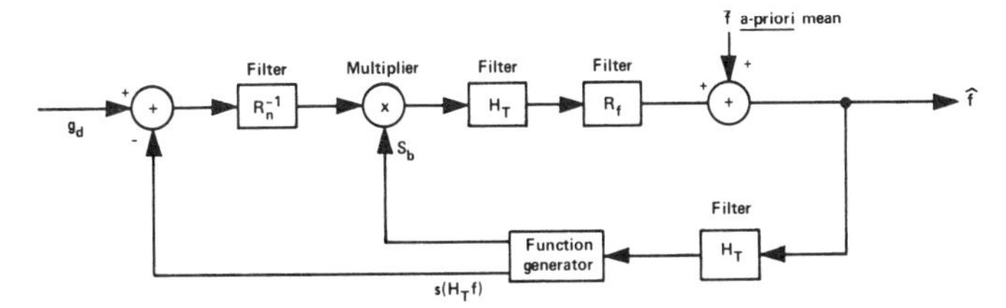

Fig. 12. Feedback structure of the MAP estimate.

3) The estimate of (53) can be described in a feedback structure. See Fig. 12. The structure is, in fact, equivalent to the nonlinear feedback system derived in Van Trees for one-dimensional continuous time-variable systems [97, p. 435]. An equation in the form of (53) can be derived directly from corresponding equations in Van Trees [97, p. 435] by making discrete variable approximations to the equivalent integral equations. The approach of Van Trees uses the Karhunen–Loeve expansion for one-dimensional signals. The approach already given with discrete data is much simpler, and the use of lexicographically ordered image vectors and block matrices makes the two-dimensional estimator direct, without having to resort to a modification of the derivation of Van Trees.

C. Solution of MAP Equations

Equation (53) is a nonlinear matrix equation, with the unknown vector appearing on both sides of the equation. The solution of the equation for the MAP estimate is not known in general, just as the corresponding nonlinear integral equation from Van Trees has not been solved in general. Current efforts are focusing on methods of solving (53) that are consistent with the computational requirements of digitized imagery. Two methods are currently under investigation, which we now describe.

1) Fast Transform Iteration: Equation (53) can be written in the functional form

$$\hat{f} = \Phi(\hat{f}) \tag{55}$$

where Φ is a nonlinear vector function of $\hat{f}$. An iterative scheme for solving (55) is based upon successive substitutions, given an initial guess $\hat{f}_0$:

$$\hat{f}_1 = \Phi(\hat{f}_0)$$
$$\hat{f}_2 = \Phi(\hat{f}_1)$$
$$\vdots$$
$$\hat{f}_n = \Phi(\hat{f}_{n-1}).$$

Under known conditions on the partial derivatives [98], the process converges, i.e.,

$$\lim_{n\to\infty} \hat{f}_n \to \hat{f}.$$

Only the nonlinear function s and the products with S_b contribute to making up the nonlinearities in (53). The remaining operations are linear matrix products; by the use of circulant approximations, the linear matrix products can be carried out by the use of discrete frequency transforms. Thus the FFT can be used to carry out the iterative solution process. Given a guess $\hat{f}_0$, the quantity $H_T\hat{f}_0$ is computed by FFT, transformed to $s(H_T\hat{f}_0)$ and then subtracted from the recorded digital image g_d. Fast transforms are used to compute the product with R_n^{-1}. The product with S_b is a scalar quotient operation, followed by fast transform computation of the product $R_f H_T^T$. Then $\bar{f}$ is added to generate $\hat{f}_1$. Iterations proceed in this fashion. The conditions under which the process converges are under study in the hope of demonstrating conclusively that the fast transform iteration process will converge [98], since if convergence is demonstrated the fast transform process is readily within the computational requirements of current computer technology.

2) Gradient Methods: Maximization of the posterior density function is equivalent to minimization of the quadratic forms associated with the Gaussian assumptions. Equation (53) is, in fact, the gradient of the quadratic forms evaluated at the maximum. Under the assumptions of Toeplitz forms for the point-spread function and covariance matrices, the gradient can be calculated by FFT techniques. Using standard iterative schemes for gradient optimization, e.g., steepest ascent, a fast iterative algorithm can be employed to solve the MAP equations. Actual computer solutions of the MAP equations have been achieved using this method, and will be reported [96].

From the viewpoint of special purpose hardware, the MAP estimate is particularly intriguing; the feedback structure of Fig. 12 should be realizable in hardware. Structures such as Fig. 12 cannot be implemented directly in MAP estimates of time functions because of the nonrealizable nature of the filters, i.e., anticipatory response and noncausal filters are required [97]. However, causality is not a restriction in image processing, and the indicated filters can be implemented optically since causality does not arise as an issue. Likewise, feedback is also possible in optical systems. Finally, there now exist digitally programmable light modulators which can be used to create the nonlinear operations associated with the function s and its derivatives. Thus the possibility exists that a hybrid digital/optical processor can be built to serve as a special-purpose hardware device for image restoration by MAP estimation.

D. Discussion

As stated at the beginning of Section IV, the field of digital image restoration is not as satisfying intellectually as that of digital image coding. The entry of more sophisticated processing techniques, beginning with MAP estimation, is the first step in bringing intellectual unity to this field, the author believes. The MAP estimation technique offers, for the first time, an equation for the restored image which keeps the nonlinear characteristics of image recording, the function s, explicit in the formulation. Further, the relations between MAP estimates and the MMSE estimates used in previous restoration work are known in terms of the mode and mean of the conditional densities. For nonlinear systems the MAP and MMSE estimates are different, but asymptotic equivalence is often possible (and is being examined for the forementioned equations). Finally, the generalized Cramer–Rao bound (the information matrix of Van Trees [97]) should make it possible to quantify the performance of the different techniques.

Much remains to be done, of course. The basic equation (53) must be implemented in efficient and optimized computer programs. Relationships to other estimation techniques (constrained least squares, maximum likelihood, etc.) must be explored. The derivation of (53) was based upon the assumption of a Gaussian description of the image, and a more appropriate always-positive description, such as the log-normal density, should be examined. Finally, encompassing eye-model knowledge, such as in Section V, into the MAP method should be explored. When all the tasks are complete we should have a better basis on which to describe digital image restoration in terms of complete formulations and without the simplifying approximations noticeable heretofore. Until these things are done, digital image restoration should remain one of the more interesting and challenging branches of the larger field of digital signal processing.

Acknowledgment

The author's thanks go to the management of the Los Alamos Scientific Laboratory (LASL) for continued support in image processing research, software, and hardware, particularly, Dr. F. E. Dorr, Dr. D. H. Janney, Dr. A. F. McGirt, and Dr. R. K. Ziegler of LASL. Thanks also go to all the staff of the Image Analysis Group at LASL, who keep things going so that I may have the time to draft a paper such as this. And to be singled out for special thanks, Dr. H. C. Andrews of USC and Dr. T. G. Stockham, Jr., of the University of Utah, two individuals who have contributed so much to digital image processing, to the author's intellectual development and appreciation of the area, and to the author personally.

References

[1] R. Nathan, "Picture enhancement for the moon, Mars, and man," in *Pictorial Pattern Recognition*, G. C. Cheng, Ed. Washington, D.C.: Thompson, 1968.

[2] T. C. Rindfleisch, J. A. Dunne, H. J. Frieden, W. D. Stromberg, and R. M. Ruiz, "Digital processing of the Mariner 6 and 7 pictures," *J. Geophys. Res.*, vol. 76, pp. 394-417, Jan. 1971.

[3] E. A. Smith and D. R. Phillips, "Automated cloud tracking using precisely aligned digital ATS pictures," *IEEE Trans. Comput.*, vol. C-21, pp. 715-729, July 1972.

[4] W. E. Shenk, "Meteorological uses of automated techniques with satellite measurements," in *Proc. Computer Image Processing and Recognition Conf.*, Univ. Missouri, Aug. 24-25, 1972.

[5] C. B. Shelman and D. Hodges, "Fingerprint research at Argonne National Laboratory," in *Proc. Carnahan Conf. on Elec. Crime Countermeasures*, Apr. 25-27, 1973.

[6] R. N. Sutton and E. L. Hall, "Texture measures for automatic classification of pulmonary disease," *IEEE Trans. Comput.*, vol. C-21, pp. 667-676, July 1972.

[7] R. P. Kruger, W. B. Thompson, and A. F. Turner, "Computer diagnosis of pneumoconiosis," *IEEE Trans. Syst., Man, Cybern.*, vol. SMC-4, pp. 40-49, Jan. 1974.

[8] D. A. Ausherman, S. J. Dwyer, and G. S. Lodwick, "Extraction of connected edges from knee radiographs," *IEEE Trans. Comput.*, vol. C-21, pp. 753-757, July 1972.

[9] B. R. Hunt, D. H. Janney, and R. K. Zeigler, "An introduction to restoration and enhancement of radiographic images," Los Alamos Scientific Lab., Los Alamos, N.Mex., Rep. LA-4305, 1970.

[10] B. R. Hunt, D. H. Janney, and R. K. Zeigler, "Radiographic image enhancement by digital computers," *Mater. Eval. J. ASNT*, vol. 31, pp. 1-5, Jan. 1973.

[11] G. W. Wecksung and K. Campbell, "Digital image processing at E G & G," *Comput.*, vol. 7, pp. 63-71, May 1974.

[12] D. C. Chu, "Spectrum shaping for computer generated holograms," Ph.D. dissertation, Dep. Elec. Eng., Stanford Univ., Stanford, Calif., 1974.

[13] J. W. Brault and O. R. White, "The analysis and restoration of astronomical data via the fast Fourier transform," *Astron. & Astrophys.*, vol. 13, pp. 169-189, 1971.

[14] J. L. Horner, "Optical spatial filtering with the least-mean-square-error filter," *J. Opt. Soc. Amer.*, vol. 59, pp. 297-303, May 1969.

[15] J. L. Horner, "Optical restoration of images blurred by atmospheric turbulence using optimal filter theory," *Appl. Opt.*, vol. 9, pp. 167-171, Jan. 1970.

[16] A. R. Shulman, *Optical Data Processing*. New York: Wiley, 1970.

[17] J. W. Goodman, *Introduction to Fourier Optics*. New York: McGraw-Hill, 1968.

[18] J. W. Cooley and J. W. Tukey, "An algorithm for the machine calculation of complex Fourier series," *Math. Comput.*, vol. 19, pp. 297-301, 1965.

[19] T. G. Stockham, Jr., "High speed convolution and correlation," in *1966 Spring Joint Computer Conf., AFIPS Conf. Proc.*, vol. 28. Washington, D.C.: Spartan, 1966, pp. 229-233.

[20] W. L. Rogers, K. S. Han, L. W. Jones, and W. H. Beierwaltes, "Application of a Fresnel zone plate to gamma-ray imaging," *J. Nucl. Med.*, vol. 13, p. 612, 1972.

[21] T. L. Saaty, *Modern Nonlinear Equations*. New York: McGraw-Hill, 1967.

[22] H. C. Andrews, "Digital image restoration: A survey," *Comput.*, vol. 7, pp. 36-45, May 1974.

[23] E. L. O'Neil, *Introduction to Statistical Optics*. Reading, Mass.: Addison-Wesley, 1963.

[24] R. L. Mather, "Gamma-ray collimator penetration and scattering effects," *J. Appl. Phys.*, vol. 28, pp. 1200-1207, Oct. 1957.

[25] F. M. Tomnovec and R. L. Mather, "Experimental gamma-ray collimator sensitivity patterns," *J. Appl. Phys.*, vol. 28, pp. 1208-1211, Oct. 1957.

[26] K. S. Han, G. J. Berzins, and S. T. Donaldson, "Pulsed X-ray imaging by incoherent holographic techniques," in *Proc. Electro-Optical Systems Design Conf.*, Sept. 18-20, 1973.

[27] C. E. K. Mees, *The Theory of the Photographic Process*. New York: Macmillan, 1954.

[28] *American Institute of Physics Handbook*. New York: McGraw-Hill, 1972.

[29] E. W. H. Selwyn, "A theory of graininess," *Phot. J.*, vol. 75, p. 571, 1935.

[30] E. W. H. Selwyn, "Experiments on the nature of graininess," *Phot. J.*, vol. 79, p. 513, 1939.

[31] L. M. Biberman and S. Nudelman, *Photoelectronic Imaging Devices*, vols. 1 and 2. New York: Plenum Press, 1971.

[32] G. M. Glasford, *Fundamentals of Television Engineering*. New York: McGraw-Hill, 1955.

[33] R. W. Drury, Ed., *Electronic Imaging Systems Symp.* Washington, D.C.: SPSE Pub., 1970.

[34] B. Gold and C. M. Rader, *Digital Processing of Signals*. New York: McGraw-Hill, 1969.

[35] B. R. Hunt and J. R. Breedlove, "Scan and display considerations in processing images by digital computer," IEEE Computer Society Repository, Paper R-74-174, May 1974.

[36] R. Legault, "Aliasing problems in two-dimensional sampled imagery," in *Perception of Displayed Information*, L. Biberman, Ed. New York: Plenum Press, ch. 7, 1973.

[37] D. J. Connor, R. C. Brainard, and J. O. Limb, "Intraframe coding for picture transmission," *Proc. IEEE*, vol. 60, pp. 779-791, July 1972.

[38] T. G. Stockham, Jr., "Image processing in the context of a visual model," *Proc. IEEE.*, vol. 60, pp. 828-842, July 1972.

[39] R. E. Swing, "The optics of microdensitometry," *Opt. Eng.*, vol. 12, pp. 185-198, Nov. 1973.

[40] J. W. Bayless, S. J. Campanella, and A. J. Goldberg, "Digital voice communications," *IEEE Spectrum*, vol. 10, pp. 28-34, Oct. 1973.

[41] C. G. Bell, "More power by networking," *IEEE Spectrum*, vol. 11, pp. 40-45, Feb. 1974.

[42] P. A. Wintz, "Transform picture coding," *Proc. IEEE*, vol. 60, pp. 809-820, July 1972.

[43] E. R. Cole, *The Removal of Unknown Image Blurs by Homomorphic Filtering*, Univ. Utah, Comput. Sci. Dep. Rep. UTEC-CSc-029, June 1973.

[44] A. Habibi, "Comparison of nth-order DPCM encoder with linear transformations and block quantization techniques," *IEEE Trans. Commun. Technol.*, vol. COM-19, pp. 948-956, Dec. 1971.

[45] W. K. Pratt and H. C. Andrews, "Transform image coding," Univ. Southern California, Los Angeles, USCEE Rep. 387, Mar. 1970.

[46] W. H. Chen, "Slant transform image coding," Univ. Southern California, Los Angeles, USCEE Rep. 441, May 1973.

[47] J. Pearl, H. C. Andrews, and W. K. Pratt, "Performance measures for transform data coding," *IEEE Trans. Commun.*, vol. COM-20, pp. 411-415, June 1972.

[48] K. Fukanaga, *An Introduction to Statistical Pattern Recognition*. New York: Academic Press, 1973.

[49] W. F. Schreiber, "Picture coding," *Proc. IEEE*, vol. 55, pp. 320-330, Mar. 1967.

[50] T. G. Stockham, Jr., "Intra-frame encoding for monochrome images by means of a psychophysical model based on nonlinear filtering of multiplied signals," in *Proc. 1969 Symp. on Picture Bandwidth Reduction*. New York: Gordon and Breach, 1972.

[51] H. K. Hartline, "Visual receptors and retinal interaction," *Science*, vol. 164, pp. 270-278, Apr. 1969.

[52] D. N. Graham, "Image transmission by two-dimensional contour coding," *Proc. IEEE*, vol. 55, pp. 336-346, Mar. 1967.

[53] P. Colas-Baudelaire, "Digital picture processing and psychophysics: A study of brightness perception," Univ. Utah, Comput. Sci. Dep. Rep. UTEC-CSc-74-025, Mar. 1973.

[54] J. B. O'Neal, Jr., "Predictive quantizing systems (differential pulse code modulation) for the transmission of television signals," *Bell Syst. Tech. J.*, vol. 45, May 1966.

[55] A. Habibi, "Data modulation and DPCM coding of color signals," in *Proc. Int. Telemetry Conf.*, vol. 8, pp. 333-343, Oct. 1972.

[56] W. B. Davenport and W. L. Root, *An Introduction to the Theory of Random Signals and Noise*. New York: McGraw-Hill, 1958.

[57] A. Habibi and G. Robinson, "A survey of digital picture coding," *Comput.*, vol. 7, pp. 22-34, May, 1974.

[58] A. G. Tescher, "The role of phase in adaptive image coding," Univ. Southern California, Los Angeles, USCIPI Rep. 510, Dec. 1973.

[59] H. C. Andrews, A. G. Tescher, and R. P. Kruger, "Image processing by digital computer," *IEEE Spectrum*, vol. 9, pp. 20-32, July 1972.

[60] D. L. Phillips, "A technique for the numerical solution of certain integral equations of the first kind," *J. Ass. Comput. Mach.*, vol. 9, pp. 84-97, 1962.

[61] A. Ralston, *A First Course in Numerical Analysis*. New York: McGraw-Hill, 1965.

[62] B. R. Hunt, "The application of constrained-least-squares-estimation to image restoration by digital computer," *IEEE Trans. Comput.*, vol. C-22, pp. 805-812, Sept. 1973.

[63] B. R. Hunt, "A theorem on the difficulty of numerical deconvolution," *IEEE Trans. Audio*, vol. AU-20, pp. 94-95, Mar. 1972.

[64] M. P. Ekstrom, "An iterative-improvement approach to the numerical solution of vector Toeplitz systems," *IEEE Trans. Comput.*, vol. C-23, pp. 320-325, Mar. 1974.

[65] R. M. Gray, "Toeplitz and circulant matrices: A review," Stanford Univ., Stanford, Calif., Rep. SU-SEL-71-032, June 1971.

[66] A. S. Sawchuk, "Space-variant image motion degradation and restoration," *Proc. IEEE*, vol. 60, pp. 854-861, July 1972.

[67] G. M. Robbins and T. S. Huang, "Inverse filtering for linear shift-variant imaging systems," *Proc. IEEE*, vol. 60, pp. 862-872, July 1972.

[68] H. C. Andrews and C. L. Patterson, "Outer product expansions and their uses in digital image processing," Aerospace Corp., El Segundo, Calif., Rep. ATR-74(8139)-2, Jan. 1974.

[69] M. M. Sondhi, "Image restoration: the removal of spatially invariant degradations," *Proc. IEEE*, vol. 60, pp. 842-853, July 1972.

[70] A. V. Oppenheim, R. W. Schafer, and T. G. Stockham, Jr., "Nonlinear filtering of multiplied and convolved signals," *Proc. IEEE*, vol. 56, pp. 1264-1291, Aug. 1968.

[71] T. M. Cannon, "Digital image deblurring by nonlinear homomorphic filtering," Ph.D. dissertation, Univ. Utah, Comput. Sci. Dep., Aug. 1974.

[72] B. L. McGlamery, "Restoration of turbulence degraded images," *J. Opt. Soc. Amer.*, vol. 57, pp. 293-297, Mar. 1967.

[73] C. W. Helstrom, "Image restoration by the method of least-squares," *J. Opt. Soc. Amer.*, vol. 57, pp. 297-303, Mar. 1967.

[74] C. K. Rushforth and R. W. Harris, "Restoration, resolution, and noise," *J. Opt. Soc. Amer.*, vol. 58, pp. 539-545, Apr. 1968.

[75] J. L. Harris, "Potential and limitations of techniques for processing linear motion-degraded imagery," in *Evaluation of Motion Degraded Images*. Washington, D.C.: U.S. GPO, 1968, pp. 131-138.

[76] W. K. Pratt, "Generalized Wiener filtering computational techniques," *IEEE Trans. Comput.*, vol C-21, pp. 636-641, July 1972.

[77] B. R. Hunt and H. C. Andrews, "Comparison of different filter structures for image restoration," in *Proc. 6th Annual Hawaii Int. Conf. on Systems Sciences*, Jan. 1973.

[78] N. D. Mascarenhas, "Digital image restoration under a regression model—the unconstrained, linear equality and inequality constrained approaches," Univ. Southern California, Los Angeles, USCIPI Rep. 520, Jan. 1974.

[79] D. P. McAdam, "Digital image restoration by constrained deconvolution," *J. Opt. Soc. Amer.*, vol. 59, pp. 748-752.

[80] B. R. Frieden, "Restoring with maximum likelihood," Univ. Arizona, Optical Sciences Center, Tucson, Ariz., Tech. Rep. 67, Feb. 1971.

[81] R. S. Hershel, "Unified approach to restoring degraded images in the presence of noise," Univ. Arizona, Optical Sciences Center, Tucson, Ariz., Tech. Rep. 72, Dec. 1971.

[82] B. R. Frieden, "2-D restoration by decision-rule allocation of pseudograins," in *Proc. Spring Meet.* Washington, D.C.: Optical Soc. Amer., Apr. 1974.

[83] A. K. Jain and E. Angel, "Image restoration, modeling, and reduction of dimensionality," *IEEE Trans. Comput.*, vol. C-23, pp. 470-476, May 1974.

[84] W. H. Richardson, "Bayesian-based iterative method of image restoration," *J. Opt. Soc. Amer.*, vol. 62, pp. 55-59, Jan. 1972.

[85] J. L. Harris, "Diffraction and resolving power," *J. Opt. Soc. Amer.*, vol. 56, pp. 569-574, May 1966.

[86] B. R. Frieden, "Bandlimited reconstruction of optical objects and spectra," *J. Opt. Soc. Amer.*, vol. 57, pp. 1013-1019, Aug. 1967.

[87] B. R. Hunt, "Data structures and computational organization in digital image enhancement," *Proc. IEEE*, vol. 60, pp. 884-887, July 1972.

[88] D. A. O'Handley and W. B. Green, "Recent developments in digital image processing at the Image Processing Laboratory at the Jet Propulsion Laboratory," *Proc. IEEE*, vol. 60, pp. 821-827, July 1972.

[89] E. L. Hall, "Almost uniform distributions for computer image enhancement," *IEEE Trans. Comput.*, vol. C-23, pp. 207-208, Feb. 1974.

[90] E. L. Hall *et al.*, "A survey of preprocessing and feature extraction techniques for radiographic images," *IEEE Trans. Comput.*, vol. C-20, pp. 1032-1044, Sept. 1971.

[91] C. S. Williams and O. A. Becklund, *Optics*. New York: Wiley-Interscience, 1972.

[92] D. J. Sakrison and V. R. Alagazi, "Comparison of line-by-line and two-dimensional encoding of random images," *IEEE Trans. Inform. Theory*, vol. IT-17, pp. 386-398, July 1971.

[93] A. E. Hoerl and R. W. Kennard, "Ridge regression: Biased estimation for non-orthogonal problems," *Technometrics*, vol. 12, pp. 55-68, Feb. 1970.

[94] H. Raiffa and R. Schlaiffer, *Applied Statistical Decision Theory*. Boston, Mass.: Harvard Univ. Press, 1961.

[95] B. R. Hunt and D. H. Janney, "Digital image processing at Los Alamos Scientific Laboratory," *Comput.*, vol. 7, pp. 57-62, May 1974.

[96] B. R. Hunt, "Bayesian methods in digital image restoration," to be published.

[97] H. L. VanTrees, *Detection, Estimation, and Modulation Theory*, vol. I. New York: Wiley, 1968.

[98] T. L. Saaty and J. Bram, *Nonlinear Mathematics*. New York: McGraw-Hill, 1964.

Image Processing in the Context of a Visual Model

THOMAS G. STOCKHAM, JR., MEMBER, IEEE

Abstract—**A specific relationship between some of the current knowledge and thought concerning human vision and the problem of controlling subjective distortion in processed images are reviewed.**

I. Introduction

IMAGE QUALITY is becoming an increasing concern throughout the field of image processing. The growing awareness is due in part to the availability of sophisticated digital methods which tend to highlight the need for precision. Also there is a developing realization that the lack of standards for reading images into and writing images out of digital form can bias the apparent effectiveness of a process and can make uncertain the comparison of results obtained at different installations. Greater awareness and the desire to respond to it are partially frustrated, because subjective distortion measures which work well are difficult to find. Part of the difficulty stems from the fact that physical and subjective distortions are necessarily different.

Manuscript received January 31, 1972; revised April 20, 1972. This research was supported in part by the University of Utah Computer Science Division monitored by Rome Air Development Center, Griffiss Air Force Base, N. Y. 13440, under Contract F30602-70-C-0300, ARPA order number 829.

The author is with the Computer Science Division, College of Engineering, University of Utah, Salt Lake City, Utah 84112.

The ideas presented here spring from our reevaluation of the relationship between the structure of images and 1) the problem of quantitative representation, 2) the effect of desired processing and/or unwanted distortion, and 3) the interaction of images with the human observer. They provide a framework in which we think about and perform our image processing tasks. By adding to our understanding of what is to be measured when dealing with images and by strengthening the bridge between the objective (physical) and the subjective (visual) aspects of many image processing issues, these ideas have clarified the meaning of image quality and thus have enhanced our ability to obtain it. We offer them with the hope that they may aid others as well.

In the course of the discussion it is noted that image processors which obey superposition multiplicatively instead of additively, bear an interesting resemblance both operationally and structurally to early portions of the human visual system. Based on this resemblance a visual model is hypothesized, and the results of an experiment which lends some support to and provides a calibration for the model are described. This tentative visual model is offered only for its special ability to predict approximate visual processing characteristics. (See footnote 11.)

In recent years there has been a large amount of quantitative work done by engineers and scientists from many fields

Reprinted from *Proc. IEEE*, vol. 60, pp. 828–842, July 1972.

in support of a model for human vision. While many of these works are not referenced explicitly here, we have attempted to reference papers and texts which do a good job of collecting these references in a small number of places while providing a unifying interpretation [1]–[5].

II. Some Philosophy about Image Processing

The notion of processing an image involves the transformation of that image from one form into another. Generally speaking, two distinct kinds of processing are possible. One kind involves a form of transformation for which the results appear as a new image which is different from the original in some desirable way. The other involves a result which is not an image but may take the form of a decision, an abstraction, or a parameterization. The following discussion limits itself primarily to the first kind of processing.

The selection of a processing method for any particular situation is made easier when the available processes have some kind of mathematical structure upon which a characterization of performance can be based. For example, the bulwark for most of the design technology in the field of signal processing is the theory of linear systems. The fact that the ability to characterize and utilize these systems is as advanced as it is, stems directly from the fact that the defining properties of these systems guarantee that they can be analyzed. These analyses, based on the principle of superposition, lead directly to the concepts of scanning, sampling, filtering, waveshaping, modulation, stochastic measurement, etc.

Equally important, however, is the idea that the mathematical structure of the information being processed be compatible with the structure of the processes to which it is exposed. For example, it would be impossible to separate one radio transmission from another if it were not for the fact that the linear filters used are compatible with the additive structure of the composite received signal.

In the case of images the selection of processing methods has often been based upon tradition rather than upon a consideration of the ideas given above. In fields such as television and digital image processing where electrical technology is a dominating influence, the tradition has centered around the use of linear systems.

This situation is a very natural one since the heritage of electrical image processing stems from those branches of classical physics which employ linear mathematics as their foundation. Specifically, it is interesting to follow the development from electromagnetic field theory to electric measurements, circuit theory, electronics, signal theory, communications theory, and eventually to digital signal processing. The situation is similar when considering the role of optics in image processing, the laws of image formation and degradation being primarily those determined from linear diffraction theory.

The question that arises is whether this tradition of applying linear processing to images is in harmony with the ideas given above. The major point at issue cannot be whether the processors possess enough structure, because linear systems certainly do. The issue is then whether that structure is compatible with the structure of the images themselves. To clarify this issue the question of image structure must be elaborated upon.

III. The Structure of Images

As an energy, signal light must be positive and nonzero. This situation is expressed in (1)

$$\infty > I_{x,y} > 0 \tag{1}$$

where I represents energy, or intensity as it is commonly called, and x and y represent the spatial domain of the image. Furthermore, since images are commonly formed of light reflected from objects, the structure of images divides physically into two basic parts. One part is the amount of light available for illuminating the objects; the other is the ability of those objects to reflect light.

These basic parts are themselves spatial patterns, and like the image itself must be positive and nonzero as indicated in (2) and (3)[1]

$$\infty > i_{x,y} > 0 \tag{2}$$

$$1 > r_{x,y} > r_{\min} \approx 0.005. \tag{3}$$

These image parts, called the illumination component and the reflectance component, respectively, combine according to the law of reflection to form the image $I_{x,y}$. Since that law is a product law, (2) and (3) combine as in (4)

$$\infty > I_{x,y} = i_{x,y} \cdot r_{x,y} > 0 \tag{4}$$

which is in agreement with (1).

It follows from (4) that two basic kinds of information are conveyed by an image. The first is carried by $i_{x,y}$, and has to do primarily with the lighting of the scene. The second is carried by $r_{x,y}$, and concerns itself entirely with the nature of the objects in the scene. Although they are delivered in combination, these components are quite separate in terms of the nature of the message conveyed by each.

So far it has been assumed that the process of forming an image is carried out perfectly. Since ideal image forming methods do not exist and can only be approached, a practical image will only approximate that given in (4). Because most image forming methods involve linear mechanisms such as those which characterize optics, a practical image can be regarded as an additive superposition of ideal images. This fact is expressed in (5)

$$\infty > \tilde{I}_{x,y} = \int_{-\infty}^{\infty} I_{X,Y} \, h_{x,X;y,Y} \, dX dY > 0 \tag{5}$$

where $\tilde{I}_{x,y}$ represents a practical image and $h_{x,X;y,Y}$ represents the so-called point spread function of the linear image forming mechanism. In other words $h_{x,X;y,Y}$ is the practical image that an ideal image consisting of a unit intensity point of light located at $x=X$ and $y=Y$ would produce. Obviously h must be nonnegative.

If the point spread function is the same shape for all points of light in the ideal image, then the superposition integral (5) becomes a convolution integral (6)

$$\infty > \tilde{I}_{x,y} = \int_{-\infty}^{\infty} I_{X,Y} \, h_{x-X;y-Y} \, dX dY > 0 \tag{6}$$

[1] It is almost impossible to find a material that reflects less than about 1 percent of the incident light.

which is conventionally expressed using a compact notation as in (7)

$$\infty > \tilde{I}_{x,y} = I_{x,y} * h_{x,y} > 0. \tag{7}$$

Combining (4) and (7) we obtain (8)

$$\infty > \tilde{I}_{x,y} = (i_{x,y} \cdot r_{x,y}) * h_{x,y} > 0 \tag{8}$$

which under the assumption of a position invariant point spread function summarizes the essential structure of practical images as they are considered in most current efforts.

The expression (8) places in evidence the three essential components of a practical image. If $h_{x,y}$ is sufficiently small in its spatial extent, the practical image can be taken as an adequate approximation to the ideal. If $h_{x,y}$ fails in this respect, the practical image can be processed by any one of a variety of methods in an attempt to remedy the situation.[2]

Since the objective of the present discussion focuses primarily on the structure of an ideal image, it will be assumed in the following that the effect of $h_{x,y}$ can be neglected.[3] Primary concern here is thus redirected to (4).

We now return to the issue posed at the end of Section II as to whether or not the mathematical structure of linear processors is compatible with the structure of the images themselves. Since (4) indicates that the image components are multiplied to form the composite, and further since linear systems are compatible with signals possessing additive structure, it follows that there exists basic incompatibility. However, this incompatibility depends in a basic way upon some implicit assumptions which have been imposed upon the structure as described in (4).

An essential ingredient to the structure of images as expressed in (4) is the assumption that an image is an energy signal. This assumption really amounts to a choice of a representation for an image. The nature of that choice can be extremely important. To clarify this concept the question of representation must be elaborated upon.

IV. The Representation of Images

A key question in the transmission, storage, or processing of any information is that of representation. The reason that the choice of representation is important is that the problems of transmission, storage, and processing can be substantially effected by it.

If an ideal physical image is considered as a carrier of information, it follows that nature has already chosen a representation. It takes the form of light energy. Furthermore, if one takes nature literally when sensing an optical image, one will continue that representation by creating a signal proportional to the intensity of that light energy. Indeed this representation seems like a very natural one, and in fact as already indicated, it is commonly used in television and digital image processing.

Strangely enough representation by light intensity analogy is a relatively new practice in image technology. The process of photography, now over a century old, does not use it. It has only been with the advent of electrical imaging methods that it has received attention.

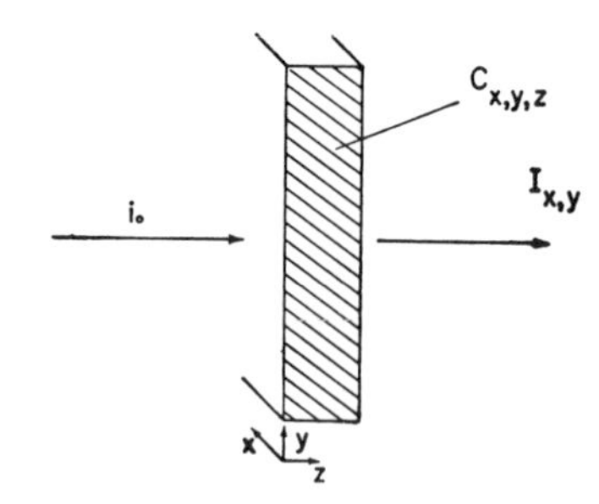

Fig. 1. An intensity image $I_{x,y}$ as reproduced by the transmission of light through a volume concentration of amorphous silver $C_{x,y,z}$.

In order to clarify this point, imagine a black and white photographic transparency which portrays some optical image. In order to see the reproduction one must illuminate the transparency uniformly with some intensity i_0 and somehow view the transmitted pattern of light intensity $I_{x,y}$. The quantities of light which are transmitted are determined by the volume concentrations of amorphous silver suspended in a gelatinous emulsion. Thus it is these concentrations which represent the image in its stored form. Let these concentrations be expressed as $C_{x,y,z}$.

Physically the situation is as depicted in Fig. 1. In order to derive the relationship between the reproduced image $I_{x,y}$ and $C_{x,y,z}$ we must consider the transmission of light through materials. The physics of the situation is given in (9)

$$\frac{di}{dz} = -kC_{x,y,z}i \tag{9}$$

where i is the intensity of the light at any point in the transmitting material and k is a constant representing the attenuating ability of a unit concentration of amorphous silver. Integration of (9) according to standard methods yields (10)

$$\int_{i_0}^{I_{x,y}} \frac{di}{i} = -k \int_0^{z_t} C_{x,y,z} dz \tag{10}$$

where z_t represents the thickness of the emulsion. Since the integral in the right-hand side of (10) represents the total quantity of silver per unit area of the transparency independent of how that silver is distributed in the z dimension, (10) can be rewritten as in (11)

$$\ln (I_{x,y}/i_0) = -kd_{x,y}. \tag{11}$$

A solution of (11) for $I_{x,y}$ yields (12)

$$I_{x,y} = i_0 e^{-kd_{x,y}}. \tag{12}$$

From (11) it can be seen that in the case of a photographic transparency, the physical representation of the image is actually $d_{x,y}$ which is proportional to the logarithm of the reproduced intensity image. In turn (12) reveals that the physical representation $d_{x,y}$ is exponentiated during its conversion to light intensity. Further, it follows that if $I_{x,y}$ is a faithful reproduction of the original intensity image from which the transparency was made, then the quantities of silver used to form the representation $d_{x,y}$ must have been

[2] For an excellent and recent summary, bibliography, and set of references representative of the many interesting efforts in this area, see Section II of a recent article by Huang *et al.* [1].

[3] There is still much to be learned both practically and theoretically about restoring practical images to the point where this is possible. Such restoration methods are very important; and since they attempt in part to compensate for distortions caused by linear mechanisms, linear processing is used extensively and often with great success.

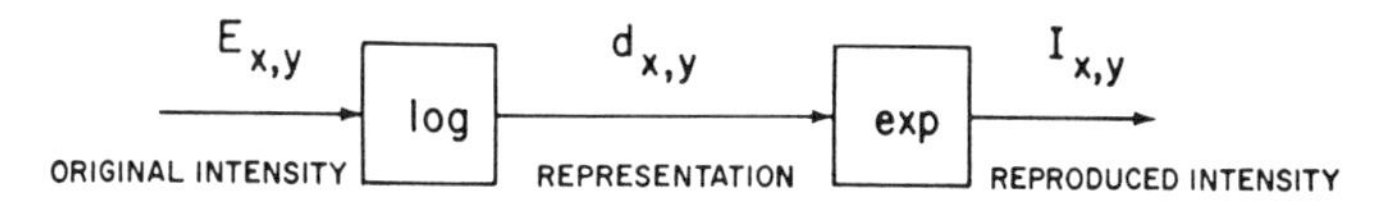

Fig. 2. In photography an image is represented by the total quantity $d_{x,y}$ of amorphous silver per unit image area. For faithful reproduction $d_{x,y}$ must be proportional to the logarithm of the image intensities.

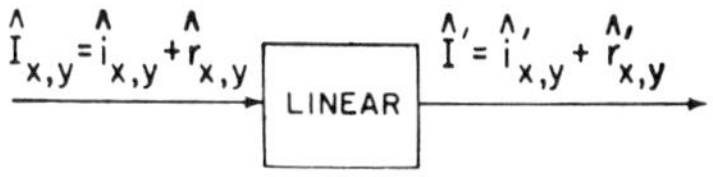

$\hat{I}_{x,y}=\hat{i}_{x,y}+\hat{r}_{x,y}$ → LINEAR → $\hat{I}'=\hat{i}'_{x,y}+\hat{r}'_{x,y}$

Fig. 3. A density image as processed by a linear system. Note that the basic structure of the image is preserved. The output is a processed illumination plus a processed reflectance regardless of what the process may be.

deposited in the emulsion by a process which was logarithmically sensitive to light energy.

This situation is summarized in Fig. 2 where the logarithmic and exponential transformations which mechanize the formation of a photographic image are placed in evidence. The variables i_0 and k which appear in (11) and (12) have been omitted for convenience since they are only scaling constants.[4]

The relationship of (12) is well known in photography but is usually presented in a somewhat altered form as in (13).

$$\log_{10}(i_0/I_{x,y}) = D_{x,y}. \tag{13}$$

Here the quantity $D_{x,y}$, called density, is proportional to $d_{x,y}$ but related directly to the common logarithm in a manner similar to that used in the definition of the decibel. Because $d_{x,y}$ and $D_{x,y}$ are both related to the popular notion of density it is reasonable to call any logarithmic representation of an image a density representation. As indicated above, all such representations are the same except for the choice of the two constant parameters.

Taking this into account (11) and (12) may be generalized to (14) and (15)

$$\hat{I}_{x,y} = \log(I_{x,y}) \tag{14}$$

$$I_{x,y} = \exp(\hat{I}_{x,y}) \tag{15}$$

where the hatted variables represent density and the unhatted variables represent intensity. All density representations are the same except for a scale factor and an additive constant.

V. Relationships Between Processing, Structure, and Representation

A study of the use of a density representation for images leads to a chain of interesting observations. These observations begin with the introduction of density representations into the previous discussion concerning the structure of ideal images. This introduction changes (1)–(4)[5]

$$\infty > \hat{I}_{x,y} = \log(I_{x,y}) > -\infty \tag{16}$$

$$\infty > \hat{i}_{x,y} = \log(i_{x,y}) > -\infty \tag{17}$$

$$0 > \hat{r}_{x,y} = \log(r_{x,y}) > \hat{r}_{\min} \tag{18}$$

and

$$\infty > \hat{I}_{x,y} = \hat{i}_{x,y} + \hat{r}_{x,y} > -\infty \tag{19}$$

where $\hat{i}_{x,y}$ and $\hat{r}_{x,y}$ represent illumination[6] and reflection densities, respectively.

It is obvious from these equations that a change from an energy representation to a density representation has introduced some interesting changes in the apparent structure of images. There is no longer a restriction upon the range of the representation. To see this fact compare (1) with (16). The manner in which the basic components of the scene are combined has been changed from multiplication to addition (compare (4) and (19)). Finally, the scene components themselves have been changed from an energy representation to a density representation.

In the case of the reflection component the transformation to a density representation is a very satisfactory one. This is so, because to a great extent the physical properties of an object which determine its ability to reflect light are the densities of the light blocking materials from which it is formed. The situation is similar to that of the photographic transparency as described in (9)–(12). Thus by using (19) the physical properties of an object are represented more directly than in (4).

The single most important effect of using a density representation is that it makes the structure of images compatible with the mathematical structure of linear processing systems. This fact is true, because linear systems obey additive superposition and from (19) we see that the basis for the structure of a density representation of an image is additive superposition.

To build upon this observation consider Fig. 3 in which a density image is being processed by a linear system. The input of the system is given as in (19). It follows from the property of superposition in linear systems that the output must be given in (20)

$$\infty > \hat{I}_{x,y}' = \hat{i}_{x,y}' + \hat{r}_{x,y}' > -\infty \tag{20}$$

where the primes indicate processed quantities. But (21) is in the same form as (19). What (20) says is that the basic structure of a density image is preserved by any linear processor. More specifically the illumination component of the processed image *is* the processed illumination component and the reflection component of the processed image *is* the processed reflection component.

For comparison consider the effect of a linear system upon an intensity image. The input is given in (4). It is clear that the notion of structure preservation cannot be maintained in this case. What is even more embarrassing is the fact that there is little guarantee that the output will be positive and nonzero which it must if it is to be regarded as an image at all.

Because an image carries information, and because information can be measured using concepts of probability, it is interesting to consider the probability density functions

[4] Actually i_0 is just a constant of proportionality on the image intensity and can be neglected if one considers normalized images only. Also k can be absorbed into the logarithmic and exponential transformations by adjusting the base being used.

[5] The minimum reflection density using the common logarithm would almost never exceed 2.0. See footnote 1.

[6] The concept of an illumination density may seem strange at the outset but proves to be an important mathematical concept even though it may be difficult to assign it any physical significance.

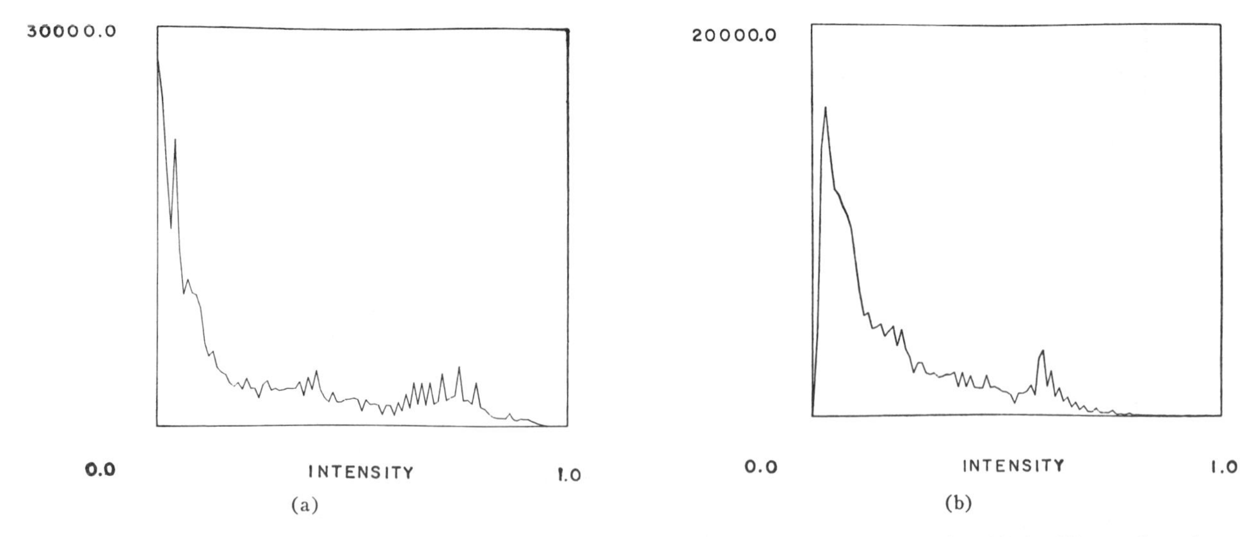

Fig. 4. Intensity histograms of 100 bins each obtained from high quality images carefully digitized to 340 by 340 samples using 12 bit/sample. (a) Three wide dynamic range scenes. (b) Two scenes of less dynamic range (approx. 30:1).

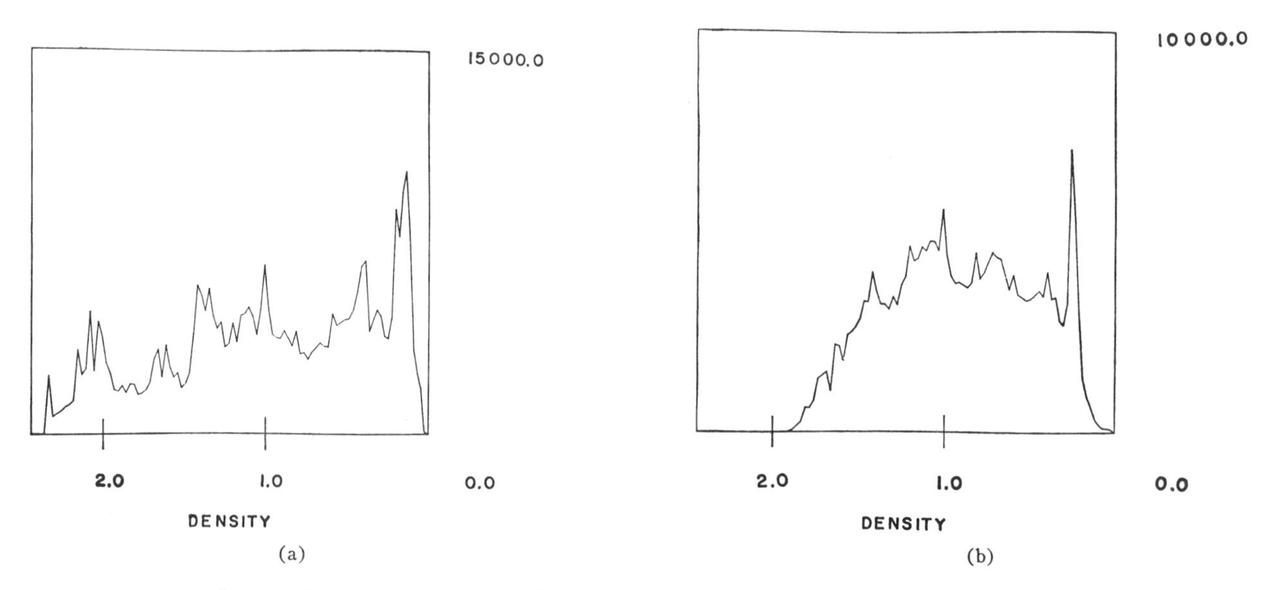

Fig. 5. Density histograms of 100 bins each obtained from the same images as in Fig. 4.

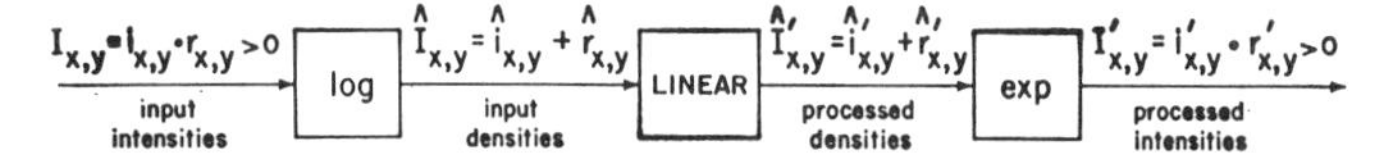

Fig. 6. An intensity image as processed by a multiplicative system. Again the basic structure of the image is preserved and the output is a processed illumination times a processed reflectance.

which are associated with both forms of representation. To this end Fig. 4 shows histograms for images which were represented by intensities and Fig. 5 shows histograms for the same images as represented by densities. These images were obtained using very careful methods from very high quality digital images.

It is instructive to compare the highly skewed distributions of Fig. 4 with the more nearly symmetric ones of Fig. 5. The fact that a density representation of an image tends to fill the representation space more uniformly than an intensity representation implies some important advantages for the former. For example, consider the problem of digitizing either representation by means of a quantizer using a binary code. The nearly symmetric distributions of Fig. 5 imply a more efficient use of the information carrying capacity of the binary code, a rectangular distribution being ideal in this respect. In addition, the symmetric distributions are more nearly aligned with the conventional assumptions associated with signals in many theoretical studies.

VI. Multiplicative Superposition in Image Processors

For some purposes it is important to be able to think of an image as represented by intensities. It is absolutely essential to do so when sensing an image to begin with or when reproducing an image for observation. In these cases it is possible to retain the match between the structure of images and the structure of processors by combining the concepts embodied in Figs. 2 and 3. This situation is depicted in Fig. 6. The input is given as in (4). It follows from (20) and (15) that

$$\infty > I_{x,y}' = \exp(\hat{I}_{x,y}') = \exp(\hat{i}_{x,y}' + \hat{r}_{x,y}') > 0 \quad (21)$$

which by the properties of the exponential function becomes

Fig. 7. Two grayscales.[8] (a) Linear intensity steps. (b) Linear density steps.

$$\infty > I_{x,y}' = \exp(\hat{\imath}_{x,y}') \cdot \exp(\hat{r}_{x,y}') > 0. \quad (22)$$

But in analogy with (21) we have

$$i_{x,y}' = \exp(\hat{\imath}_{x,y}') \quad (23a)$$

and

$$r_{x,y}' = \exp(\hat{r}_{x,y}'). \quad (23b)$$

So substituting (23) into (22) we get

$$\infty > I_{x,y}' = i_{x,y}' \cdot r_{x,y}' > 0 \quad (24)$$

which is in the same form as (4).

Again the basic structure of the image is preserved. However, this time the multiplicative superposition which characterizes the structure of an intensity image is compatible with the mathematical structure of the processor of Fig. 6. It follows that Fig. 6 depicts a class of systems which obey multiplicative superposition [2]. Besides demonstrating the preservation of structure for intensity images (24) also reveals the fact that a multiplicatively processed image is itself positive and nonzero and thus realizable. This later observation transcends the fact that the system used to process the input densities in Fig. 6 is linear, because the processed intensities are formed by exponentiating the processed densities regardless of how those densities were produced. The result of exponentiating a real density is always positive and nonzero. This property of density processing is called the realizable output guarantee.

VII. Multiplicative Superposition in Vision

Although a great deal of sophisticated and elaborate knowledge has been gained in the last several decades about the problem of communicating electrically between various sorts of automatic mechanisms, dissappointingly little has been done to match the ultimate source and receiver, namely the human being, to this body of knowledge and these systems. The basic obstacles have been a lack of understanding of the human mechanisms in terms describable by the available theory and the difficulty in studying the human mechanisms which are involved.

The philosophy that any communications system, whether man-made or natural, has structure and that that structure should be matched to the communications task at hand, seems to provide a stepping stone for understanding the operation of some of these systems. In this regard we would like to take the concept of a multiplicative image processor and explore its possible relationship to the known properties of early portions of the human visual system.

In many respects the multiplicative image processors previously described and their canonic form as represented in Fig. 6 bear an interesting resemblance to many operational characteristics of the human retina.[7] The presence of an approximately logarithmic sensitivity in vision has been known for some time [3]. Even more readily evident, and mechanized through the process of neural interaction, is the means for linear filtering [3], [4].

A. Logarithmic Sensitivity

The fact that light sensitive neurons fire at rates which are proportional to the logarithm of the light energy incident upon them has been measured for simple animal eyes [3, pp. 246–253]. Similar experiments with human beings are inconvenient to say the least, but there are some interesting experiments that serve as a partial substitute. The most convincing of these is the so called "just noticeable difference" experiment [5]. In this experiment an observer is asked to adjust a controllable light patch until it is just noticably brighter or darker than a reference light patch. The experimenter then steps his way through the gamut of light intensities from very bright to very dark. The step numbers are then plotted as a function of the intensity of the reference light. The resulting curve is very close to logarithmic over several orders of magnitude of intensity.

For a direct but less objective demonstration of this relationship consider the gray-scale steps[8] presented in Fig. 7. In Fig. 7(a) the scale consists of equally spaced intensity steps. In Fig. 7(b) the scale consists of exponentially spaced intensity steps which is the same as equally spaced density steps. The scale in Fig. 7(b) appears as a more nearly equally spaced scale than that of Fig. 7(a) so that the eye appears to respond more nearly to densities than to intensities.

B. Linear Filtering through Neural Interaction

The mechanism for linear spatial processing in vision is observed in the Hartline equations [4, pt. I, ch. 3], [3, ch. 11, pp. 284–310]. The effect of this processing can be observed by means of a number of simple optical illusions.

The simplest of these illusions is known as the illusion of simultaneous contrast[9] and can easily be observed in Fig. 8. In this image we observe two small squares surrounded by larger rectangles, one light, one dark. In fact the two small

[7] A recent, lucid, and elaborate discussion of these characteristics is presented by Cornsweet [3]. See especially chs. XI and XII.

[8] This and several other test images shown here should be presented using a calibrated display or calibrated photography. An uncertain but considerable distortion will have taken place during the printing of this paper. The reader must take this into account and estimate the possible degradation for himself.

[9] For a more complete discussion see [3, pp. 270–284].

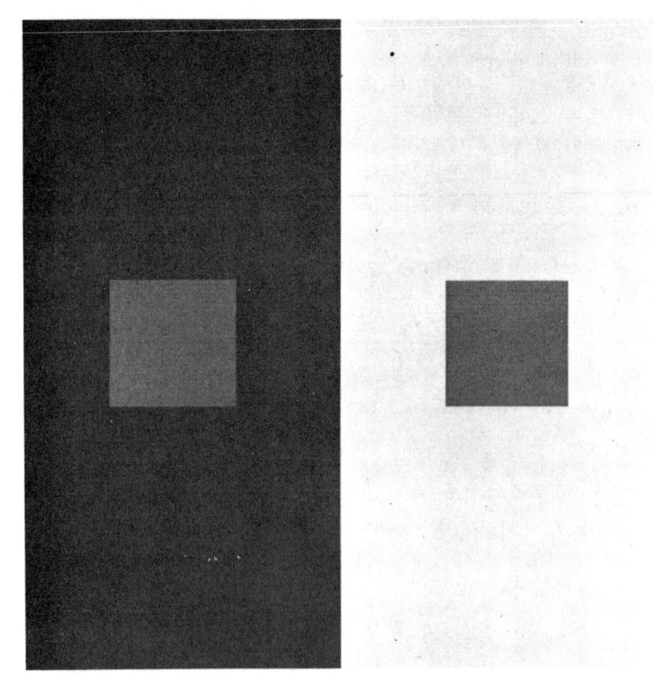

Fig. 8. The illusion of simultaneous contrast. The two small squares are of exactly the same intensity.

squares are exactly the same shade of gray. They appear different, however, due to their surroundings. This illusion can be explained at least qualitatively by assuming that the image has been subject to linear spatial filtering in which low spatial frequencies have been attenuated relative to high spatial frequencies. Filters of this type cause the averages of different areas in one image to seek a common level. Since in Fig. 8 the area of the left has a darker average, it will be raised, making the left square brighter. Likewise, since the area on the right has a lighter average, it will be lowered, making the right square less bright.

Another illusion can be observed by returning attention to Fig. 7(b). Each rectangle in this gray scale is one uniform shade of gray. However, each rectangle appears to be darker near its lighter partner and lighter near its darker partner. Again the phenomenon can be explained at least qualitatively by the assumption of linear spatial filtering.[9]

The final illusion to be discussed here is presented in Fig. 9. It is known as the illusion of Mach bands [3, pp. 270–284], [4]. In this image[8] there are two large areas, one light and one dark but each of a uniform shade. These two areas are coupled by a linearly increasing density wedge (exponentially increasing intensity wedge) as indicated in Fig. 9(b). The observer will notice that immediately at the left and at the right of this wedge are a dark and light band as implied by Fig. 9(c). These bands, known as Mach bands, can also be explained at least qualitatively by linear processing.[10]

C. *Saturation Effects*

So far this discussion has implied that the linear spatial processing of densities can explain a number of visual phe-

[10] Quantitative studies of this illusion are common. Unfortunately, almost all of them employ a matching field or light which in turn perturbs the measurement considerably. Mach himself warned of this problem [4, pp. 50–54, 262, 305, 322] and suggested that there is no solution. The psychophysical experiment to be described later is offered as a possible counter example to this suggestion.

(a)

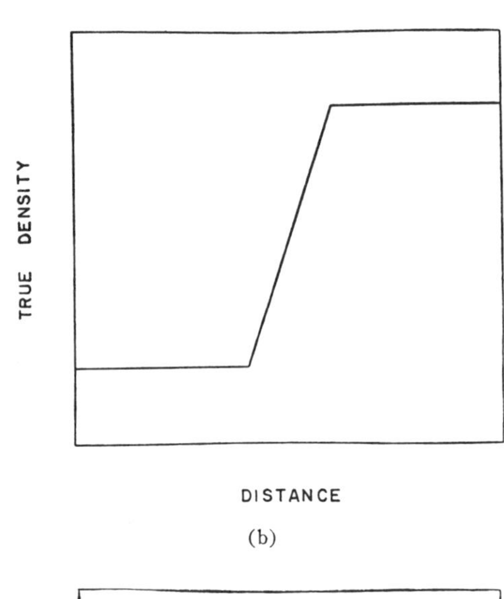

(b)

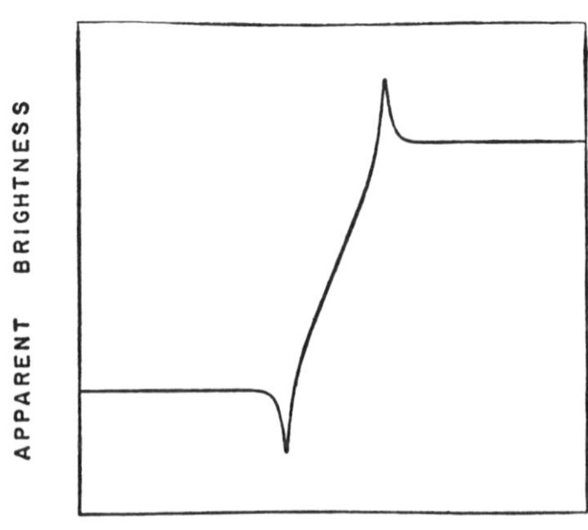

(c)

Fig. 9. The illusion of Mach bands. (a) Observe the dark and light bands which run vertically at the left and right of the ramp, respectively. (b) The true density representation of the image. (c) The approximate apparent brightness of the image.

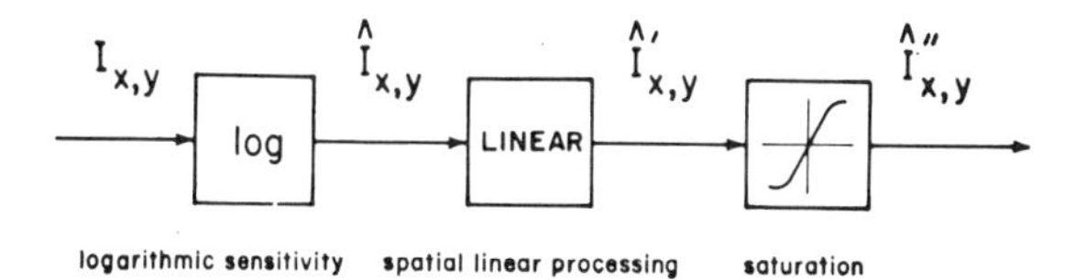

Fig. 10. A possible approximate model for the processing characteristics of early portions of the human visual system.

nomena. It is clear that these visual phenomena are only observable if there is a proper amount of light available for their presentation. It is common knowledge that below certain illumination levels one cannot see well if at all. The same is true if illumination levels become too great.

The physical limitations of any visual mechanism guarantee that saturation or threshold effects will occur if intensity levels are raised or lowered far enough. In this respect any consideration of the relationship between the processing of densities and properties of vision must eventually include the effects of saturation.

D. A Process Model for Early Portions of the Human Visual System

The preceding discussions suggest a model for the processing characteristics of early portions of the human visual system.[11] This model is shown in Fig. 10. The output $\hat{I}_{x,y}''$ is a saturated version of a linearly processed density representation. The linear processing is presumably of the form in which low spatial frequencies are attenuated relative to high spatial frequencies.

The most useful implications of this model do not come from its relationship to the optical illusions which we have already discussed as much as from the operational characteristics it embodies. The operational characteristics in question center around the ability of the human visual system to maintain its sensitivity to patterns of relatively low contrast in the context of a total image in which intensities are spread across a very large dynamic range,[12] and its ability to preserve an awareness of the true shades of an object in spite of huge differences in illumination. Moreover, these abilities are embodied without sacrificing the basic structure of images with respect to the separate physical components of illumination and reflectance!

If the illumination component of an image did not vary in space, (4) would become

$$I_{x,y} = i \cdot r_{x,y}. \tag{25}$$

In this case[13] the dynamic range of an image would be limited to about 100:1, because it would be determined by the reflection component[1] alone. Problems with saturation effects would be relieved if not avoided altogether. In addition the true shade of an object would be reproduced directly by $I_{x,y}$.

[11] This model is representative of approximate processing characteristics at early stages only. It is not intended as a biophysical or anatomical model for any specific visual mechanism or as an exact or complete processing representation. In image processing some such model must be assumed even if it is by default. The classical default assumption is that of fidelity reproduction namely that like an ideal camera the eye "sees" what it sees.

[12] The dynamic range of an image is the ratio of the greatest to the least intensity value therein contained. Ratios in excess of 1000:1 are often encountered by the eye or camera.

[13] This configuration, often sought at great expense in photographic and television studios, is called flat lighting.

Unfortunately, the illumination component of an image varies a great deal, often more than the reflectance component. For example a black piece of paper in bright sunlight will reflect more light than a white piece of paper in shadow. In the proper environment both situations could occur in the same image at the same time, but an observer would always call the white paper "white" and the black paper "black" in spite of the fact that the black paper would be represented by a higher intensity than the white paper. This visual phenomenon is called brightness constancy. Moreover, if there were low contrast markings on either sheet of paper they could be read in spite of their insignificance with respect to the total intensity scale.

With these facts in mind it is interesting to note that the system of Fig. 10 tends to produce an output in which the variations in illumination are indeed reduced. This is so, because the illumination component dominates the Fourier spectrum of a density image at low spatial frequencies while the reflectance component dominates at high spatial frequencies. As a result, the spatial linear filtering previously described reduces the illumination variations, because it attenuates low frequencies relative to high frequencies. At the same time the basic structure of images is preserved because the model operates linearly on a density representation.

The detailed consequences of this situation are described in more detail in [2, sec. V]. There the use of multiplicative processors for the purpose of simultaneous dynamic range reduction and detail contrast enhancement is discussed and demonstrated. An example of an image possessing some serious dynamic range problems is shown in Fig. 11 before and after such processing. Notice how the illumination is extremely variable from the outside to the inside of the building. In the unprocessed image, details within the room though present in the original are obscured by the limited dynamic range capabilities of the printing process you are now viewing. In the processed image these details are present in spite of this limitation.

E. Model and Process Compatibility

When the image of Fig. 11(b) is observed, the total processing system including the approximate visual model is that shown in Fig. 12 which combines Figs. 6 and 10. In Fig. 12(a) the two linear systems which characterize the processor and the visual system are labeled H and V, respectively. Fig. 12(b) shows the simplified exact equivalent system in which as much merging of subprocesses as is possible has been performed. The new composite linear system labeled $H \cdot V$ is merely the cascade of the two previous ones.

Fig. 12(b) demonstrates the compatibility of the visual model and the multiplicative image processor. It does so by placing in evidence the fact that within the validity of the model the experience of viewing a processed image is indistinguishable from that of viewing an unprocessed image except that it is possible to alter the linear processing performed through the manipulation of the linear system labeled H.

F. Model Testing and Calibration

The approximate visual model of Fig. 10 has been motivated in the above by studying certain illusions, noting certain aspects of neural structure and neural measurement, and by concentrating attention upon certain desirable and available performance characteristics. This motivation can be sup-

(a)

(b)

Fig. 11. A large dynamic range scene. (a) Before processing. (b) After processing with a multiplicative processor adjusted to attenuate low and to amplify high frequency components of density. (Note: These and all other images in this paper are digital.)

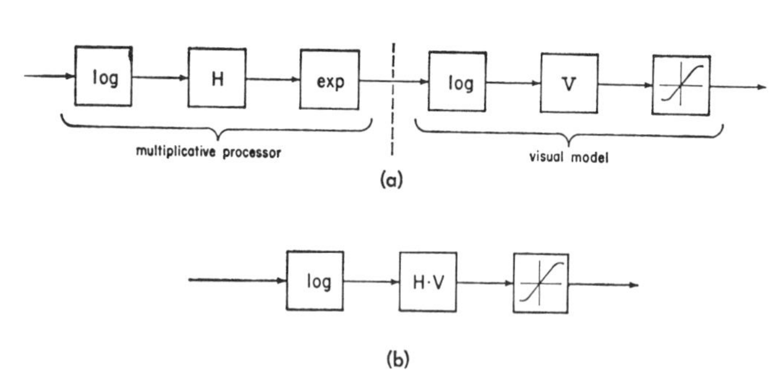

Fig. 12. Total processing system including visual model when viewing Fig. 11(b). (a) Unsimplified system. Processed intensities appear at the vertical dotted line. (b) Simplified system with processors merged.

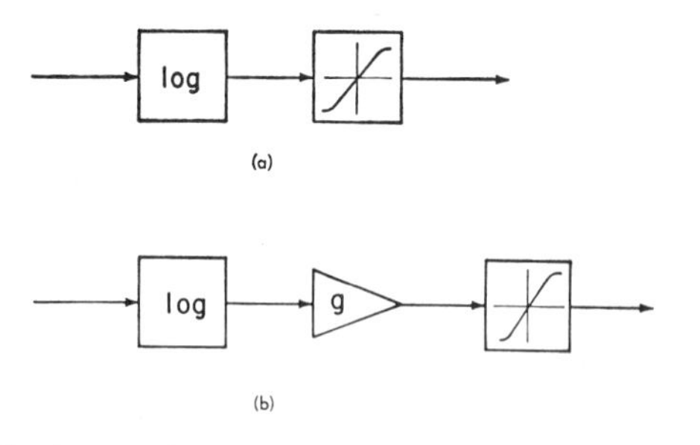

Fig. 13. Total processing system when viewing an image which has been subject to a multiplicative processor the linear component of which has been adjusted to be the inverse of the linear component of the visual model. (a) H is exactly the inverse of V. (b) H is the inverse of V except for a constant of proportionality g.

ported by a testing experiment which is suggested by the situation depicted in Fig. 12. If the system H were adjusted to become the inverse of the system V, the system of Fig. 12(b) could be further simplified as shown in Fig. 13. In this situation it should not be possible to observe the optical illusions described above and portrayed in Figs. 8 and 9.

An experiment designed to find an H which would simultaneously cancel the optical illusions described above can be carried out with significant success [6]. By comparing the pattern of Fig. 14 with Figs. 8 and 9 one can see that this pattern strongly induces the illusions in question.[8] If one processes this pattern by means of a multiplicative processor with the system H adjusted according to (26)

$$H = V^{-1} \tag{26}$$

one obtains a pattern which appears to have little remaining illusion phenomena.

Such a processed pattern[14] is shown in Fig. 15. The illusions have been significantly suppressed, and the apparent brightness of Fig. 15 follows the profile of true density of Fig. 14 remarkably well. The degree to which the illusions have been suppressed provides additional support for the model of Fig. 10. In addition an estimate of the system V results as a byproduct since (26) can be solved for V in terms of the actual H used in the experiment.

It should be noted that the above results support the logarithmic component of the model and its position in the system because the cancellation of the illusions depends upon the neutralization of the exponential component of the multiplicative processor. Without this neutralization Fig. 12(a) could not be reduced to Fig. 12(b).

Although one might find a system H that would cancel the illusions for a single fixed pattern, it has been shown that the experiment succeeds about equally well for all patterns such

[14] Here the comments of footnote 8 must be considered most seriously since the illusion cancelling experiment is a sensitive one and gray-scale distortions can upset it easily. The calibrated print sent to the publisher appears as described in the text. A limited number of such calibrated prints are available to readers with sufficient interest and requirements. As published here the pattern should be viewed approximately at arms length.

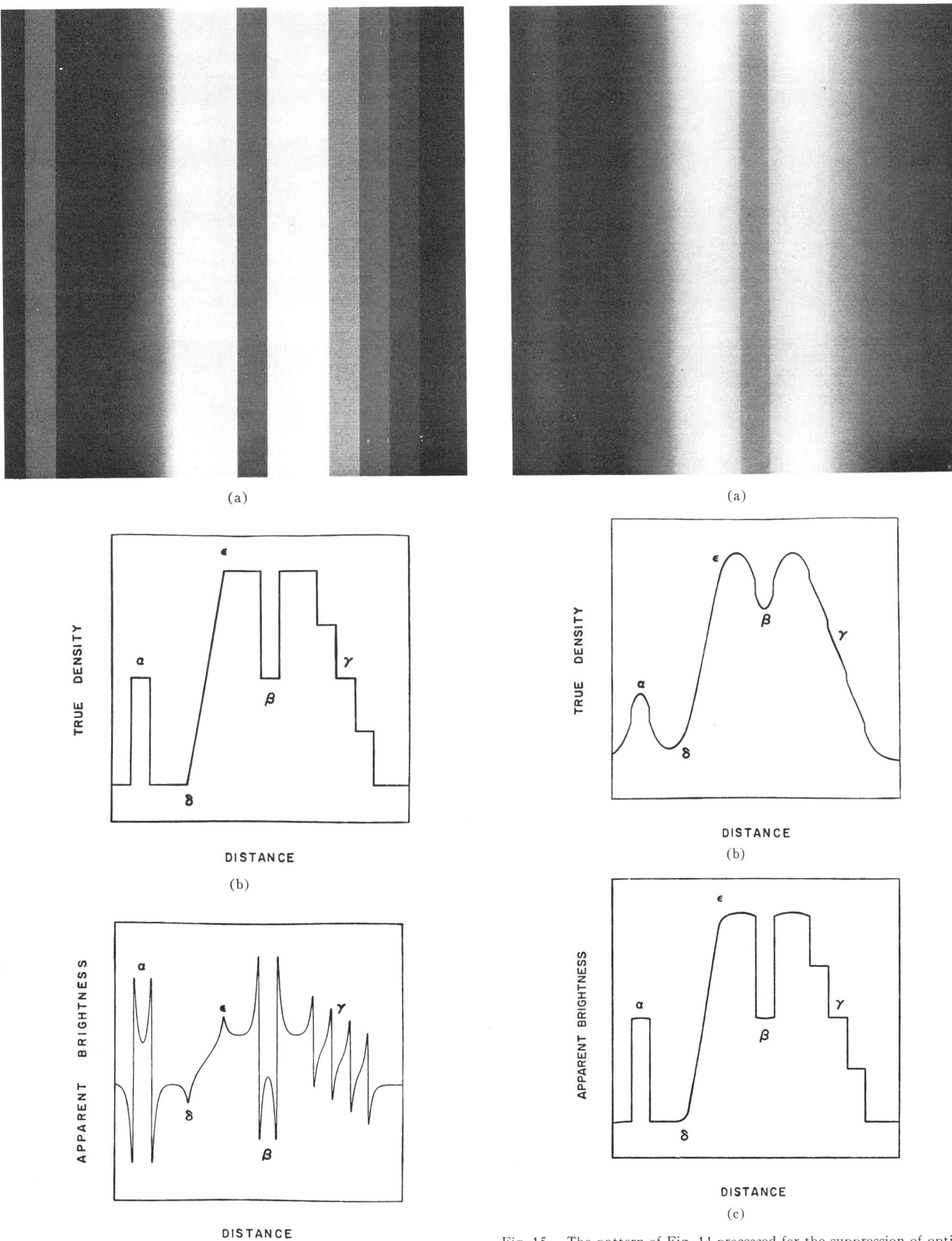

Fig. 14. Pattern for use in testing and calibrating the visual model. (a) Observe the illusions of simultaneous contrast α, β, γ, and Mach bands δ, ϵ. (b) The true density representation of the image. (c) The approximate apparent brightness of the image.

Fig. 15. The pattern of Fig. 14 processed for the suppression of optical illusions. Compare with Fig. 14. (a) Appraise the amounts of remaining simultaneous contrast α, β, γ, and Mach bands δ, ϵ. (b) The true density representation of the processed image. (c) The approximate apparent brightness of the processed image as observed from a calibrated print. Curve taken as a subjective consensus from five knowledgeable observers.

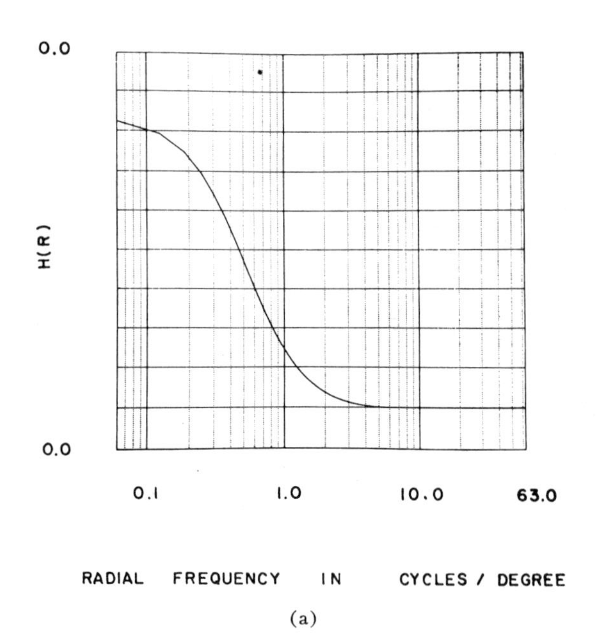

(a)

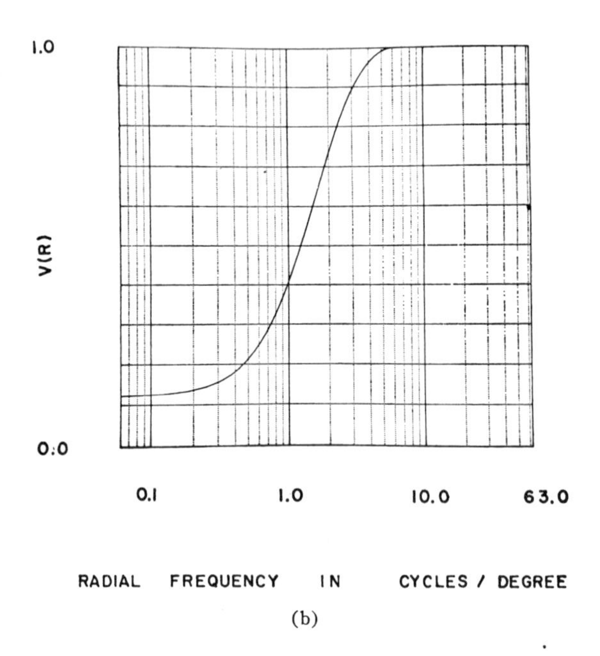

(b)

Fig. 16. Frequency response of one-dimensional systems used in test of eye model. (a) Response of system H for cancelling illusions. (b) Relative response of system V as estimated from H.

as Fig. 14 not just the one shown here. Alternately, it has been shown that the cancellation of Fig. 15 holds across a wide range of the constant of proportionality g in which the processed patterns have enough dynamic range to be clearly visible and not so much dynamic range so as to produce saturation effects.[15]

The actual linear system H used in the experiment described above was found by a cut-and-try procedure wherein an initial estimate was refined through successive rounds of processing, visual evaluation, and system redesign.

[15] Since the cancellation of these illusions requires only that the apparent brightnesses of Fig. 15 take on a profile of a certain *relative* shape, the true value of g in (26) and in Fig. 13(b) cannot be determined. Thus V can only be estimated to within an unknown constant of proportionality.

Since the test patterns varied only in one dimension, the development of a one-dimensional linear system for H was all that was required.[16] The one-dimensional frequency response of that system is shown along with its inverse in Fig. 16. It follows from two-dimensional Fourier analysis that under the assumption that the two-dimensional frequency response of the eye model has circular symmetry, the curve of Fig. 16(b) represents a radial cross section of that two-dimensional frequency response. Specifically

$$V(R) = V(X). \tag{27}$$

In addition the two-dimensional point spread function of the system V can be determined either from the Bessel transform of $V(R)$ or from the two-dimensional Fourier transform of the surface of revolution generated by $V(R)$.

It is interesting to compare the frequency response characteristics obtained here with those determined elsewhere. An excellent summary discussion and associated references are available [3, ch. 12, pp. 330–342]. In this respect there is a marked similarity between the approach taken here and the work of Davidson [3, ch. 12, pp. 330–342] in which problems with both logarithmic sensitivity and spatial interference between test patterns and matching fields are avoided.[17]

One might wonder what the world would look like if the eye did not create the illusions that we have been discussing. In this regard consider Fig. 17 which bears the same relation to Fig. 11(a) as Fig. 15(a) bears to Fig. 14(a).

VIII. Image Quality and the Visual Model

Image quality is a complicated concept and has been studied in a variety of ways and contexts. In most situations a final measure of quality can be defined only in the subjective sense. It can be measured only approximately and with difficulty by means of slow and expensive tests involving human observers. As the understanding of the human visual mechanism grows, objective measures become more feasible. So it is that with the aid of the visual model of Fig. 10 it is possible to define such a measure of image quality. By virtue of the discussions presented in Section VII one expects this measure to be related to some basic subjective considerations. An objective measure is defined by measuring the difference between a distorted image and its reference original, only after each has been transformed by the model. An example of such a definition based on a mean-square error measure is given in (28)

$$E^2 = \iint [V_{x,y} \circledast (\log I_{x,y} - \log R_{x,y})]^2 \, dx dy \tag{28}$$

[16] For the purpose of this experimental effort the linear system portion of the eye model was assumed to be position invariant. Since peripheral and central (foveal) vision possess quite different resolution properties, this assumption falls short of reality and leaves room for further refinements. For this reason and because the cancellation of illusions as shown in Fig. 15 might be improved we have not given an analytic expression for our present best estimate for $V(R)$ as part of (27). Tentatively we are using

$$V(R) = 742/(661 + R^2) - 2.463/(2.459 + R^2)$$

where R is the radial spatial frequency in cycles per degree. See Fig. 16(b). See also [7].

[17] One can still find fault with these methods, because the test patterns used do not fill the visual field and so there is still interaction between them and the surround which is uncontrolled. See also footnote 16.

Fig. 17. The scene of Fig. 11(a) processed for the suppression of optical illusions. Compare with Fig. 11(a).

where E is the objective measure, $V_{x,y}$ is the two-dimensional point spread function of the visual model, $I_{x,y}$ is the image being measured, and $R_{x,y}$ is the reference original. For examples of the use of such an objective measure see Sakrison and Algazi [7] and Davisson [8]. Since the model emphasizes certain aspects of an image and deemphasizes certain others in a manner approximately the same as early portions of the human visual system, distortions which are important to the observer will be considered heavily while those which are not will be treated with far less weight. This will be so even though the important distortions may be physically small and the unimportant ones physically large, which is frequently the case.

With the above ideas in mind it becomes clear that when an image is to be distorted as a result of the practical limitations which characterize all transmission, storage, and processing mechanisms it makes sense to allow such distortions to take place after the image has been transformed by the model. The image can then be transformed back again just before it is to be viewed. For example if an image bandwidth compression scheme is to be implemented it probably makes much better sense to invoke that scheme upon the model-transformed image than upon the physical intensity image. The motivations for this argument are not entirely subjective. Since the model transformation emphasizes the reflectance components and deemphasizes the illumination components of a scene, it renders that scene more resistant to disturbing influences on certain physical grounds as well, because it can be argued that the reflectance component is the more important one.

For some applications it may be inconvenient to transform an image by means of the complete visual model before exposing it to disturbing influences, because the processing power required to mechanize the linear portion of the model might be somewhat high in terms of the present technology. However, for a variety of reasons it is at least desirable to employ a density representation to provide part of the resistant effect. One reason is that no disturbance can violate the property of density processing which guarantees a realizable output. Another is that since the eye is logarithmically sensitive, it considers errors on a percentage basis. Because disturbances and distortions tend to distribute themselves uniformly throughout the range of a signal, they represent extremely large percentage distortions in the dark areas of an intensity image. To make matters worse, as can be seen from the intensity histograms of Fig. 4, dark areas are by far the most likely in intensity images.

These effects can be observed most readily when images are quantized in preparation for digital processing. The classically familiar quantization contours are most visible in the dark areas of intensity represented images but distribute nearly uniformly in density represented images. As a result, the use of a given number of bits to represent an image produces more readily observable quantization distortion in the form of contouring when an intensity rather than a density representation is employed. Indeed, for images of large dynamic range the disparity can be very great.[18]

As an illustration of the issues presented in this section consider Figs. 18 and 19. Fig. 18 shows the digital original of Fig. 11(a) in combination with white noise with a rectangular probability density function. In each of the three different combinations shown the peak signal to peak noise ratio was exactly the same namely 8:1. The noise disturbs an intensity representation in Fig. 18(a), a density representation in Fig. 18(b), and a model-processed image in Fig. 18(c). For additional discussion and examples see [6].

Fig. 19 shows another image quantized to 4 bit (i.e., 16 equally spaced levels exactly spanning the signal range). The quantization disturbs an intensity representation in Fig. 19(a), and a density representation in Fig. 19(b).

IX. Summary and Conclusions

The discussions presented in this paper concentrated upon the structure of images and the compatibility of that structure with the processes used to store, transmit, and modify them. The harmony of density representation and multiplicative processing with the physics of image formation was emphasized and special attention was drawn to the fact that early portions of the human visual system seem to enjoy that harmony. A visual model based upon these observations was introduced and a test yielding a calibration for the model was presented. Finally, an objective criterion for image quality based upon that model was offered and some examples of the use of the model for protecting images against disturbances were given.

During the past five years these concepts have been developed and employed in a continuing program of digital image processing research. Their constant use in guiding the

[18] The number of bits needed to represent an image cannot properly be determined without specifying at least the quality and character of the original, the kind of processing contemplated, the quality of the final display, the representation to be used, and the dynamic range involved. Similarly, the number of bits to be saved by using a density instead of an intensity representation given a fixed subjective distortion depends at least on the dynamic range in question. In the light of the quality obtainable with present technology the "rules of thumb" which have been popularly used in the past should be regarded with caution.

(a)

(b)

(c)

Fig. 18. Noisy disturbance in the context of three different representations. Peak signal to peak noise is 8:1 in all cases. (a) Disturbed intensities. (b) Disturbed densities. (c) Disturbed model-processed image. Compare with Fig. 11(a).

basic philosophy of the work has resulted in an ability to obtain high and consistent image quality and to enhance and simplify image processing techniques as they were proposed. Their ability to provide engineering insight and understanding complementary to existing ideas has been an invaluable aid in planning and in problem solving.

Continuing research is attempting to include within the model the aspects of color and time and to enlarge upon the model in the context of visual processes which take place at points farther along the visual pathway. It is hoped that enlargements and refinements of the model will continue to suggest useful image processing techniques and that digital

(a)

(b)

(c)

Fig. 19. Quantization distortion in the context of two different representations. In both cases 16 equally spaced levels exactly spanning the signal range were used. (a) Quantized intensities. (b) Quantized densities. (c) Original.

signal processing methods will continue to permit the investigation of those techniques which might be too complex to be explored without them.

ACKNOWLEDGMENT

I wish to thank the people who have helped me in the course of the image processing research which has led to the ideas presented here. I am grateful to A. V. Oppenheim for his theory of homomorphic filtering, which for me is the *sine qua non* of these views. Many thanks are also due to C. M. Ellison, D. M. Palyka, D. H. Johnson, P. Baudelaire, G. Randall, R. Cole, C. S. Lin, R. B. Warnock, R. W. Christensen, M. Milochik, Kathy Gerber, and to the many too numerous to name who have given encouragement, interest, and ideas. Special appreciation goes to my wife Martha who has given me unceasing support.

REFERENCES

[1] T. S. Huang, W. F. Schrieber, and O. J. Tretiak, "Image processing," *Proc. IEEE*, vol. 59, pp. 1586–1609, Nov. 1971.

[2] A. V. Oppenheim, R. W. Schafer, and T. G. Stockham, Jr., "Nonlinear filtering of multiplied and convolved signals," *Proc. IEEE*, vol. 56, pp. 1264–1291, Aug. 1968.

[3] T. N. Cornsweet, *Visual Perception*. New York: Academic Press, 1970.

[4] F. Ratliff, *Mach Bands: Quantitative Studies on Neural Networks in the Retina*. San Francisco, Calif.: Holden-Day, 1965.

[5] L. M. Hurvich and D. Jameson, *The Perception of Brightness and Darkness*. Boston, Mass.: Allyn and Bacon, 1966, pp. 7–9.

[6] T. G. Stockham, Jr., "Intra-frame encoding for monochrome images by means of a psychophysical model based on nonlinear filtering of multiplied signals," in *Proc. 1969 Symp. Picture Bandwidth Compression*, T. S. Huang and O. J. Tretiak, Eds. New York: Gordon and Breach, 1972.

[7] D. J. Sakrison and V. R. Algazi, "Comparison of line-by-line and two-dimensional encoding of random images," *IEEE Trans. Inform. Theory*, Vol. IT-17, pp. 386–398, July 1971.

[8] L. Davisson, "Rate-distortion theory and applications," this issue, pp. 800–808.

Part V
Information Extraction by Machine Processing

Organized by Paul E. Anuta and Azriel Rosenfeld, Associate Editors
and Ruzena Bajcsy

Techniques for Change Detection

ROBERT L. LILLESTRAND

Abstract—The problem of change detection presents itself for imaging systems that view the same scene repeatedly. Current research programs based on the processing of side-looking radar imagery show that spatial alignment of the various parts of the image must be highly accurate if noise in the difference picture is to be reduced to acceptably low levels. Typically, the spatial alignment accuracy must be better than one-fourth of the diameter of the smallest resolvable feature in the imagery, and this often requires several hundred degrees of freedom in the performance of the map warp for images that are of the order of 10^7 picture cells (pixels) in size. Gray scale rectification of conjugate sampling points is less difficult, requiring typically only 10 to 20 percent as many degrees of freedom. Point by point adjustment for differences in mean transparency and contrast is employed. Recently developed equipment provides a continuous pipeline processing capability. With this equipment, each picture element of the second image is transformed with four degrees of freedom (two spatial and two gray scale). The digital correlator is capable of processing 4×10^5 six-bit picture elements per second when used in conjunction with a CDC 1700 computer.

Index Terms—Digital correlator, geometric image distortion, image superposition, side-looking radar, subregion correlation, transparency rectification.

Manuscript received November 16, 1971; revised March 6, 1972. This work was supported by the U. S. Air Force and by Control Data Corporation, Minneapolis, Minn. A preliminary version of this paper was presented at the IEEE, UMC, Two-Dimensional Digital Signal Processing Conference, Columbia, Mo., October 6–8, 1971.

The author is with the Research Division, Control Data Corporation, Minneapolis, Minn. 55440.

System Definition

AT THE present time most efforts at the detection of changes in images of the same scene involve manual comparisons. Usually these are done with some form of image superposition equipment that provides alternate viewing of one image and then the other in quick succession. Rather than expose the human viewer to all of the information contained in both images, significant increases in the quantity and quality of his work can be achieved by presenting only the changes. Typically, in the case of side-looking radar, increases in the speed of change detection ranging from 10^1 to 10^3 result from the use of automatic techniques, the higher level being typical of urban scenes that contain a high concentration of features that have large radar cross sections.

In one of the early papers on change detection, given by Rosenfeld [1], the basic problems are defined. This paper includes a consideration of various possible coefficients of correlation as measures of the quality of image registration as well as the description of steps necessary for the implementation of an automatic change detection system. Shepard [2] discussed the need for an automatic system that detects changes between two sets of aerial photographs and points out

Reprinted from *IEEE Trans. Comput.*, vol. C-21, pp. 654–659, July 1972.

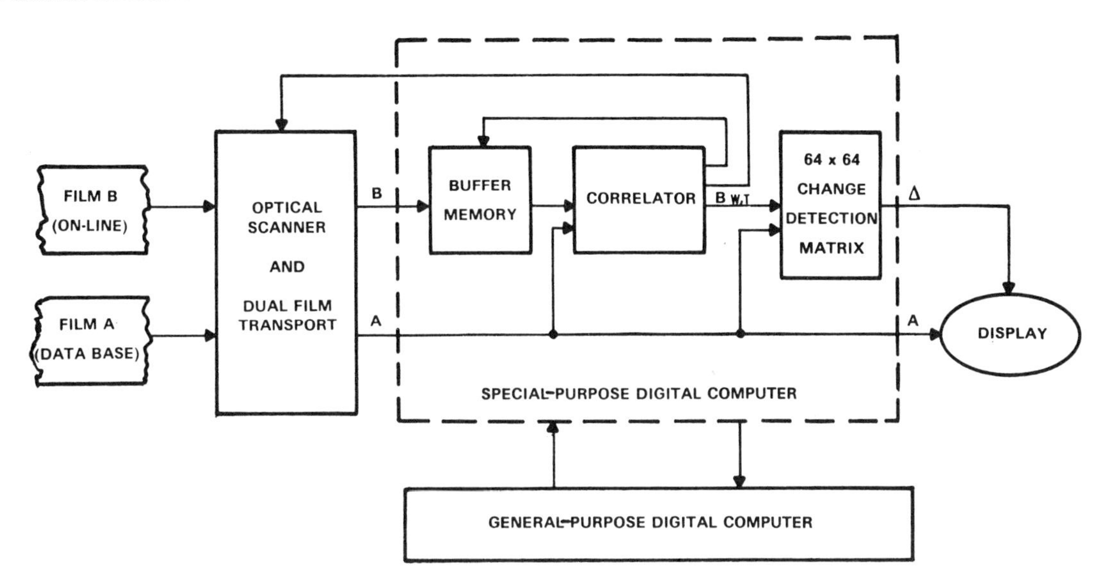

Fig. 1. System block diagram.

that manual procedures are too slow. Several methods for automating the process of change detection are given by Paolantonio [3]. Kawamura [4] describes a system used to automatically detect meaningful changes in photographic data for purposes of city planning. In this paper he considers the problems of both change detection and pattern recognition. Typical geometric image distortions derived from sensor-related and external errors are described in the paper by Bernstein and Silverman [5]. They also describe a general class of sequential algorithms for image registration.

The block diagram of a high-speed change detection system constructed at Control Data Corporation is shown in Fig. 1. Since the SLR imagery is continuous, rather than appearing a frame at a time, as is the case for most photography, a dual film continuous transport optical scanner is used for the A/D conversion. A small general-purpose computer (CDC 1700) is paired with a special-purpose hardwired computer for the data processing. This latter equipment is a multiple strip (up to 16 channels) digital correlator that has an auxiliary digital memory that serves as an input buffer and picture cell interpolator. By means of the buffer memory, the requirement for the precise servo control of one of the film scanner beams relative to the other is eliminated. The small general-purpose computer shown in Fig. 1 contains about 30 alterable parameters and this makes it possible to optimize the processing for various types of imagery.

The digital processor is capable of correlating 4×10^5 six-bit pixels per second. Using appropriate conversions to reduce the processing load of the special-purpose digital computer to equivalent six-bit adds, a value of about 40×10^6 six-bit adds per second is obtained. The correlated output of the B channel ($B_{W,T}$) is compared with the A channel gray scale value in the change detection matrix to provide whatever transformation function is required prior to display. This equipment performs such tasks as shadow suppression and thresholding.

For an image that is 1024 pixels wide, one image frame moves past the viewer every 2.5 s when correlating 4×10^5 pixels. By adding additional computing equipment, the image is modularly expandable in steps of 2^{10} pixels, with an upper limit of about 2^{14} pixels imposed by the resolution of the scanner. For this image width, the expanded system can correlate up to 6.4 million six-bit pixels per second.

The automatic digital change detection may be divided into the following steps: 1) subregion correlation; 2) spatial rectification; 3) transparency rectification; 4) change presentations. This is the order of presentation for the succeeding sections of the present paper.

The image research work conducted in our laboratory involves a CDC 6600 computer that is used to perform these functions when new imagery is received, during which time the algorithms are developed and the processing parameters are selected. Typically, for types of imagery not previously processed, this may involve 100 to 200 h of image processing. At this stage, no particular effort is made to achieve a high processing speed, but rather, emphasis is placed on the achievement of the desired processing result through the use of experimental methods. Having completed this work, a special-purpose system of the type shown in Fig. 1 is constructed (in this case, for SLR imagery). This latter equipment is a pipeline processor with a high degree of internal parallelity and is capable of processing imagery at speeds that are between one and two orders of magnitude faster than the 6600 processing used in the initial development of the algorithms.

Method of Correlation

The method of correlation consists of dividing the image into a series of small subregions and searching for maximum values of the correlation coefficient by displacing one subregion x and y relative to its conjugate. The fraction of the total image area that these subregions cover, as well as the number of subregions that are used, depend upon the complexity of the spatial warp of the imagery and the distribution of features, but typical values range from 5 to 50 percent. The number of pixels contained in one subregion area can be varied from $(2^5)^2$ to $(2^7)^2$.

In the case of the continuous processor shown in Fig. 1, a series of more or less parallel correlation tracks are employed. As shown in Fig. 2, these tracks creep along the film strip in the direction of film transport. The interpolated location of the maximum value of the correlation coefficient is computed each time the subregion advances by a one-pixel step along the track and a synthetic image of conjugate points of film B is computed. This is illustrated in Fig. 3. The location of the maximum of the correlation coefficient can be predicted with a high degree of accuracy since the subregions creep along the track incrementally. Resultantly, not more than nine trial subregion positions need to be used to compute the location of the maximum of the correlation coefficient. Since the correct conjugate point on film B will not, in general, correspond with any specific point sampled by the scanner, an interpolated value based on the measured transparency of the four pixels surrounding it is used.

Should one or more subregions encounter a featureless area, as would be the case for tracks 4, 5, and 6 of Fig. 3, a cross-track linkage has been built into the servo system. This "elastic harness" permits tracks that are highly correlated with the data base imagery to carry along adjacent strips for which the values of the correlation coefficient are low. This feature is desirable at the startup of the image tracking since lock-on for any one track propagates laterally as the film is transported, eventually resulting in lock-on over the full width of the imagery.

In general, the assessment of the quality of subregion correlation for the operation of the cross-track linkage requires more than just a high value of the correlation coefficient. For example, if the features contained in a given subregion consist largely of parallel edges, a highly elliptical shape will be obtained for the contours of the correlation coefficient shown in Fig. 3. In this case, even though the correlation coefficient is large, there will be almost no subregion positional constraint in the direction of the semimajor axis of the ellipse. Because of this problem, tests must be made for both the magnitude of the correlation coefficient at the central maximum and for the manner in which it varies in the neighborhood of this maximum. In the case of side-looking radar imagery, the ratio of semimajor to semiminor axes seldom exceeds 3:1 since there are usually a significant number of point symmetric features.

A second order servo is used to slave the film B track to its film A conjugate. The values assigned to the feedback loop parameters are derived from simulations on

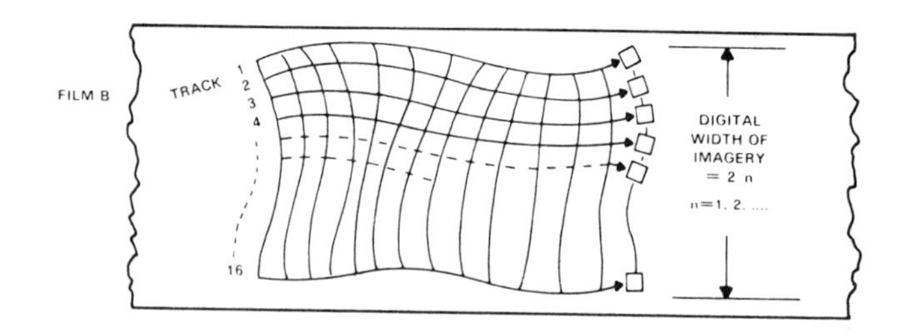

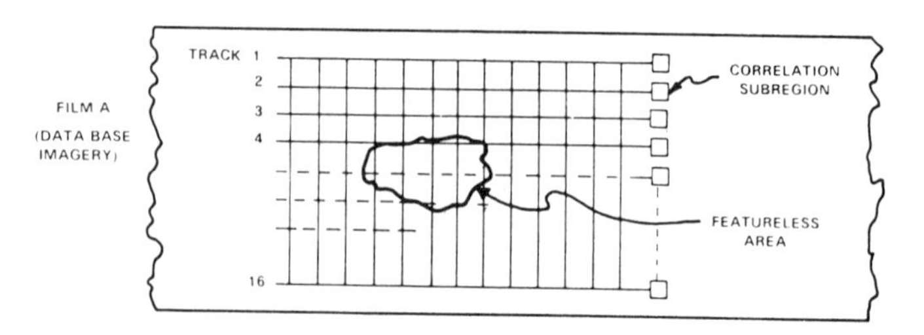

Fig. 2. Film strip correlation showing individual tracks.

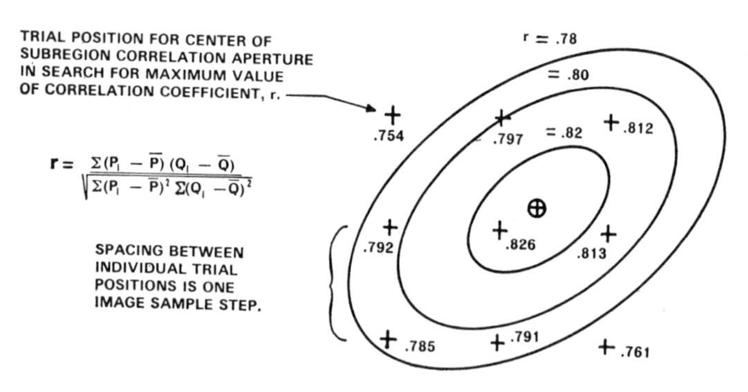

Fig. 3. Interpolation between image sample cells to determine location of maximum of correlation coefficient.

the 6600 computer and depend on the pattern of statistical and systematic deviations of the conjugate subregions for the type of imagery in question.

Spatial Rectification

Experience gained during the past three years in the digital processing of SLR imagery has repeatedly shown the importance of precise image registration if low noise levels in the difference image are to be achieved. There are many possible reasons for correlating imagery, and stable lock-on at the principal maximum of the correlation coefficient can often be achieved acceptably without employing the elaborate techniques that we have found necessary for change detection.

Fig. 4 illustrates three techniques that have been used for the implementation of the map warp so as to achieve the required precision in image registration. For imagery consisting of, say 10^6 pixels, a 16×16 array of correlation subregions might typically be used in studying the map warp problem. We then use a least square curve-fitting procedure for bivariate polynomials of increasing order and then examine the size of the residuals. This is shown as σ_{FIT} in Fig. 5 where an 8×16 subregion array was used. The remaining 8×16 sub-

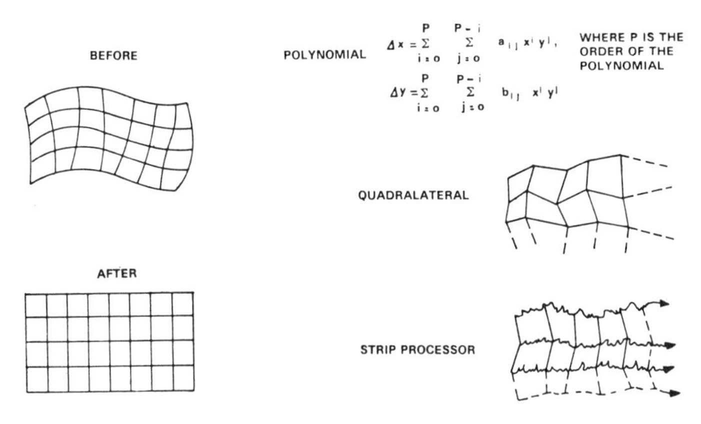

Fig. 4. Spatial rectification of imagery.

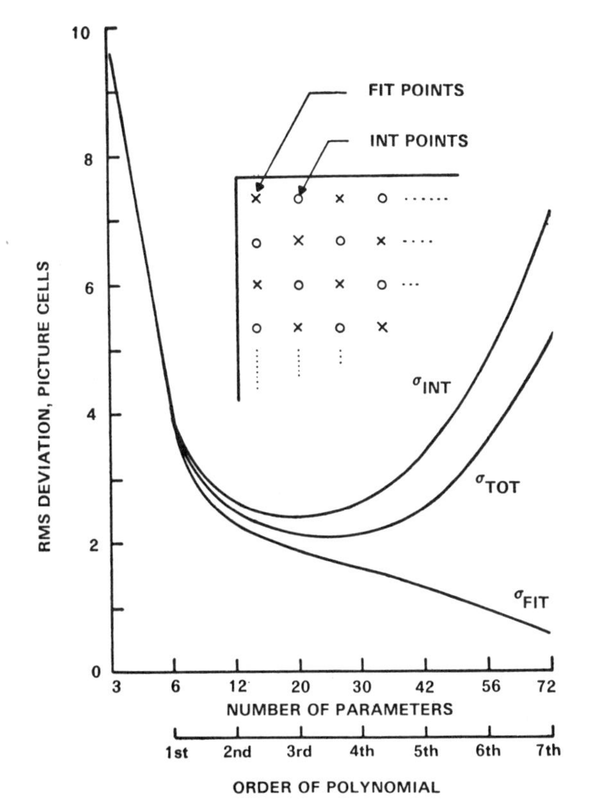

Fig. 5. Residuals after spatial rectification as a function of order of approximating polynomial.

regions form a checkerboard pattern with the others and are used to check the quality of the fit in terms of the interpolation problem. The rms residual of these intermediate points is designated σ_{INT}.

The three parameter solution consists of Δx, Δy, and $\Delta\theta$ and this forms the reference case for the study of spatial rectification since this is indicative of the superposition accuracy that can be achieved by merely sliding the two images relative to one another. By adding one more parameter, a difference in the scale of the two images can be corrected, whereas if three more parameters are added to the reference case, one can correct for anamorphosis, in which the differing scales are at right angles to one another, but at some unknown orientation relative to the x-y axes.

As one would expect when using increasingly higher order polynomial approximations, the values of σ_{FIT} approach zero. This advantage is lost, however, because of the poor interpolatory properties of the higher order polynomials, as is shown by the increase in the value of σ_{INT}. The value of σ_{TOT}, which is the rms error for the total collection of points, thus shows a broad minimum with the best values obtained for this particular type of imagery for the third- and fourth-order solutions.

Beyond fifth order, the check with intermediate points becomes so bad that a different map warp scheme must be employed. Experience with the quadrilaterals shown in Fig. 4, in which each corner corresponds to the center of a correlation subregion, shows improved performance for the 16×16 array, both in terms of the value of the correlation coefficient for the overall image and in terms of the noise in the difference image. A logical extension of the quadrilateral technique involves the use of bicubic spline fitting to remove the discontinuity in the first derivative of the map warp at the corners of the quadrilaterals. While this is consistent with physical intuition about the nature of the warp, this technique has not proved necessary in the spatial rectification of SLR imagery since a nearly equivalent effect can be produced by simply using more quadrilaterals.

The strip processor shown at the lower right portion of Fig. 4 resembles the quadrilateral solution, except that the interpolation between the strips is recomputed each time the tracker subregion is advanced by one pixel.

Experimental data such as that shown in Fig. 5 indicates that the map warp is often highly nonlinear. This, coupled with the nonuniform distribution of features in the various subregions, suggests that iterative processing may be worthy of investigation. When the feature content is low, the size of the correlation subregions must be large. If this condition occurs in combination with a large map warp, the images being processed are likely to benfit from iterative map warp processing.

Transparency Rectification

If the six-bit gray scale values for conjugate pixels are plotted against one another, as shown on the diagrams on the left of Fig. 6, systematic differences in mean transparency and contrast will usually be found. To the extent that the gray scale values of conjugate points are similar, the values will lie along a 45° regression line. Deviations from this line are a measure of nonrepeatability of the two spatially rectified images, and $\sigma_{\perp}$ is the rms deviation of the individual pixels in a direction perpendicular to the rectified regression line. The pixels along (1) represent a feature added to film B that is embedded in a uniformly low background level of film A; pixels along (2) represent this same case except that the background is strong.

Since the corrections for mean transparency and contrast vary from one part of the image to another, the

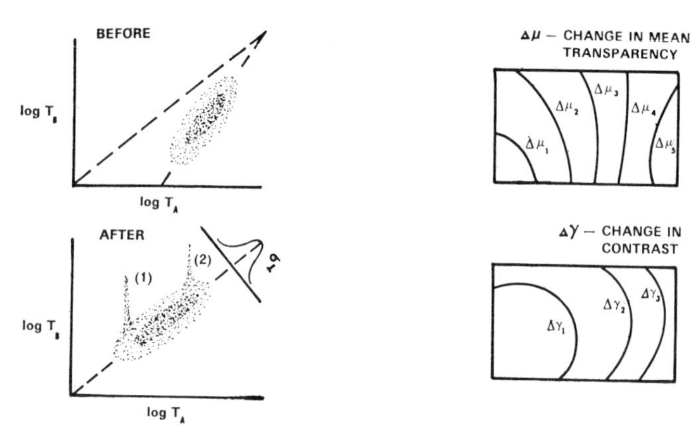

Fig. 6. Transparency rectification of imagery.

digital processor must be capable of varying these corrections over the area of the image, as shown by the contours at the right of Fig. 6. Thus the combined spatial and transparency rectifications require that each pixel of film B be modified with four degrees of freedom: Δx, Δy, $\Delta\mu$, $\Delta\gamma$. Although some of the imagery processed has shown evidence of nonlinearities in the scatter diagram of Fig. 6, the values normally obtained for $\sigma_\perp$ are sufficiently large to make the use of a more complicated nonlinear regression line of questionable advantage.

From the point of view of change detection, the individual pixels may be classified as shown in Fig. 7. Computer programs have been developed that automatically plot data in this form for various subregions or for an entire picture. This type of feature classification diagram is normally made only after completing the spatial and transparency rectifications.

Points falling on different regions of a scatter diagram of this type represent differing feature classifications. Typically, the diagonal threshold bounds are set at values that are three to four times $\sigma_\perp$. Features that are added to film B fall in the upper left-hand corner, while features that have been removed from film B fall in the lower right-hand corner. Shadows are regions of no radar return and correspond to transparent areas on the film. By studying diagrams of the type shown in Fig. 7, shadow suppression thresholds can be established to avoid the creation of changes in the difference picture that are the result of shadows of varying length.

A feature classification diagram for a subregion consisting of about 10^4 pixels is shown in Fig. 8. Spatial and transparency rectification were performed prior to making this plot. The distribution of gray scale values is bimodal, as can be seen from the camelback shape. That is, there is a preponderance of the film area that is either very light or very dark. This particular subregion contains no significant feature changes or shadows.

Large values of the correlation coefficient are obtained either because the value of $\sigma_\perp$ is small or because the scene contrast is large, as is the case for the data shown

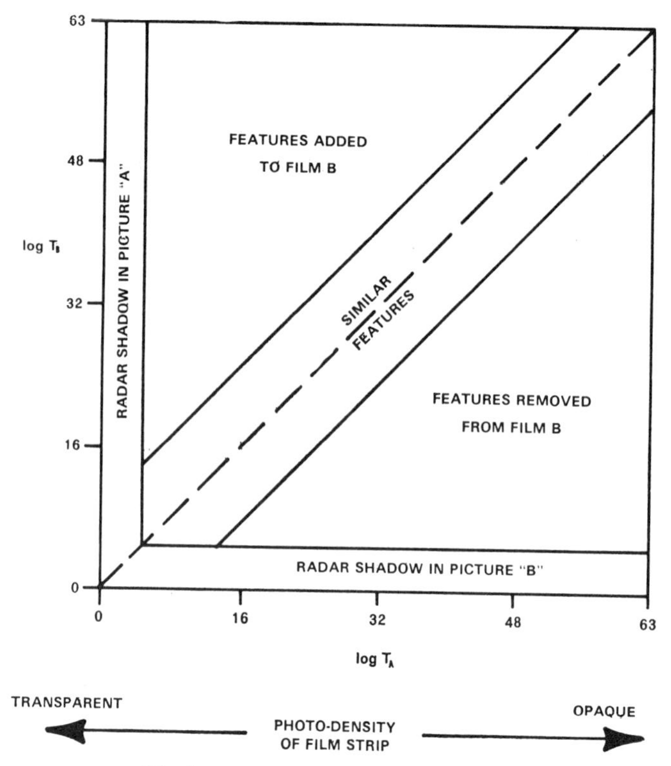

Fig. 7. Feature classification diagram.

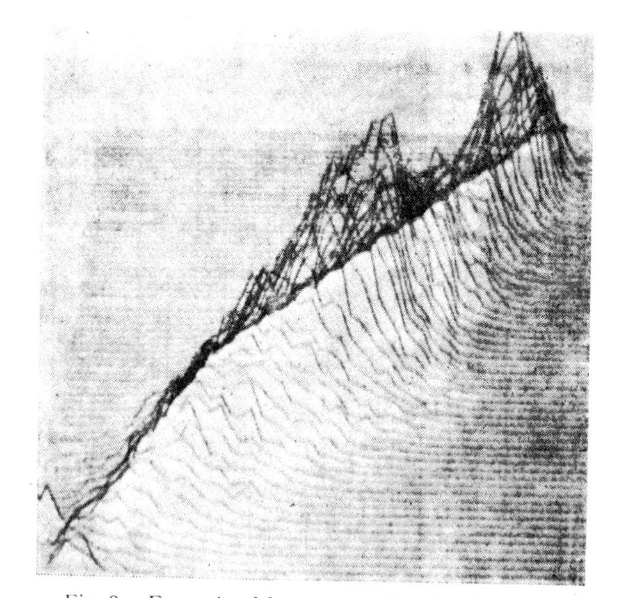

Fig. 8. Example of feature classification diagram.

in Fig. 8. When evaluating the result obtained for the superposition of an individual subregion as compared with others on the same image, a large value of the correlation coefficient should not be used as the sole criterion of image repeatability and it is the value of $\sigma_\perp$ that is used to automatically set the thresholds for change detection.

Illustrative Imagery

There are many ways of displaying the results of automatic digital change detection, particularly if color

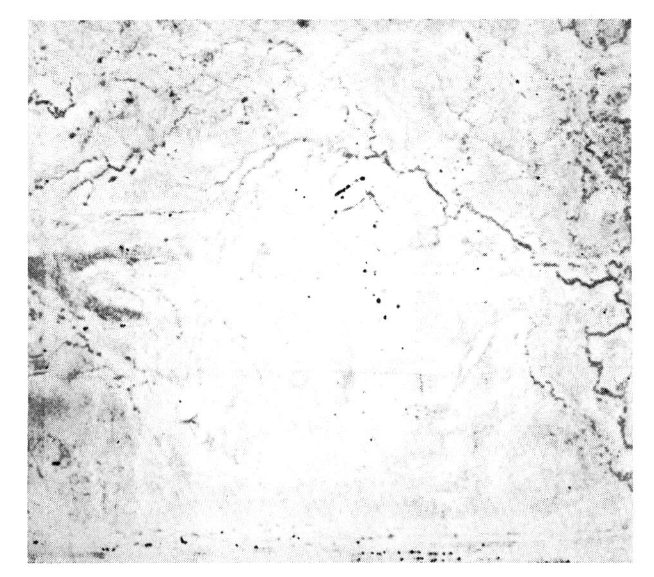

Fig. 9. Data base imagery. Features added are enhanced.

Fig. 10. Data base imagery. Features removed are enhanced.

displays are available. The two black and white renditions given in Figs. 9 and 10 are for an image that consists of about 10^6 pixels, each of which is encoded to six bits of gray scale information. A digitally generated presentation of the data base imagery (film A) is shown in the background. In the case of Fig. 9, the features that have been added to film B are shown in enhanced form. For Fig. 10, the features that have been removed are shown in enhanced form. (Some of the contrast in the original imagery has been lost in the offset printing, and consequently the changes as shown are somewhat more difficult to detect than is normally the case.) When making the original digital images, the background scene was reduced to one-fourth of its original value, thus providing the viewer with some contextual information to aid in the evaluation of the changes that have taken place.

Acknowledgment

This paper has presented a very abbreviated description of image processing techniques being developed at Control Data Corporation, Minneapolis, Minn., and contributions were made by many people on the staff of the Research Division.

References

[1] A. Rosenfeld, "Automatic detection of changes in reconnaissance data," in *Proc. 5th Conv. Mil. Electron.*, 1961, pp. 492–499.

[2] J. R. Shepard, "A concept of change detection," *Photogrammetr. Eng.*, vol. 30, pp. 648–651, July 1964.

[3] A. Paolantonio, "Difference measurements in automatic photo-interpretation of surveillance maps," *Inform. Display*, vol. 6, pp. 41–44, Mar./Apr. 1969.

[4] D. G. Kawamura, "Automatic recognition of changes in urban development from aerial photographs," *IEEE Trans. Syst., Man, Cybern.*, vol. SMC-1, pp. 230–239, July 1971.

[5] R. Bernstein and H. Silverman, "Digital techniques for earth resource image data processing," presented at the 8th Annu. AIAA Meeting, AIAA Paper 71-978, Oct. 25–28, 1971.

Digital Image-Processing Activities in Remote Sensing for Earth Resources

GEORGE NAGY, SENIOR MEMBER, IEEE

Abstract—**The United States space program is in the throes of a major shift in emphasis from exploration of the moon and nearby planets to the application of remote sensing technology toward increased scientific understanding and economic exploitation of the earth itself. Over one hundred potential applications have already been identified. Since data from the unmanned Earth Resources Technology Satellites and the manned Earth Resources Observation Satellites are not yet available, the experimentation required to realize the ambitious goals of these projects is carried out through approximation of the expected characteristics of the data by means of images derived from weather satellite vidicon and spin-scan cameras, Gemini and Apollo photographs, and the comprehensive sensor complement of the NASA earth resources observation aircraft.**

The extensive and varied work currently underway is reviewed in terms of the special purpose scan and display equipment and efficient data manipulation routines required for high-resolution images; the essential role of interactive processing; the application of supervised classification methods to crop and timber forecasts, geological exploration, and hydrological surveys; the need for nonsupervised classification techniques for video compaction and for more efficient utilization of ground-control samples; and the outstanding problem of mapping accurately the collected data on a standard coordinate system.

An attempt is made to identify among the welter of "promising" results areas of tangible achievement as well as likely bottlenecks, and to assess the contribution to be expected of digital image-processing methods in both operational and experimental utilization of the forthcoming torrent of data.

I. Introduction

THE OBJECT of this survey is to give an account of experimental developments in digital image processing prompted by the major environmental remote sensing endeavors currently underway, such as the already operational weather satellite program of the National Oceanographic and Atmospheric Agency (NOAA), the projected Earth Resources Technology Satellite (ERTS) and Skylab experiments, the NASA Earth Resources Aircraft Program (ERAP), and the Department of the Interior's Earth Resources Observation System (EROS).

Sources of Information: The most comprehensive and readily accessible source of material in this area is the seven volumes published so far of the *Proceedings of the International Symposium on Remote Sensing of Environment*, held annually under the auspices of the Center for Remote Sensing Information and Analysis of the University of Michigan.

Other useful sources of information are the *NASA-MSC Annual Earth Resources Program Reviews*, the *Proceedings of the Princeton University Conference on Aerospace Methods for Revealing and Evaluating Earth's Resources*, the publications of the American Society of Photogrammetry and of the Society of Photo-Optical Instrumentation Engineers, the *Journal of Applied Meteorology*, the *Proceedings of the IEEE* (pertinent special issues in April 1969 and in July 1972), the *IEEE Transactions on Computers* and the *IEEE Transactions on Man, Machines, and Cybernetics*, the Journals of *Remote Sensing of Environment* and of *Pattern Recognition*, and the proceedings of several symposia and workshops on picture processing and on pattern and target recognition. Previous introductory and survey articles include Shay [185], Colwell and Lent [37], Leese *et al.* [120], Park [167], Dornbach [48], and George [67].

As is the case with most emerging fields of research, the assiduous reader is likely to encounter considerable redundancy, with many experiments republished without change in the electrical engineering and computer literature, in the publications dealing with aerial photography and photogrammetry, in the various "subject matter" journals (agronomy, meteorology, geophysics), in the pattern recognition press, and in the increasing number of collections devoted exclusively to remote sensing.

A depository of relevant published material, government agency reports, and accounts of contractual investigations is maintained by NASA at the Earth Resources Research Data Facility at the Manned Spaceflight Center in Houston, Texas (Zeitler and Bratton [223]). The Facility also maintains a file of most of the photographs obtained by the NASA satellites and earth observation aircraft, and by other cooperating agencies, institutions, and organizations. Provisions are made for convenient browsing through both the printed material and the vast amounts of photography. The Center publishes *Mission Summary Reports* and detailed *Screening and Indexing Reports* of each data-collection operation and acts in principle as a clearinghouse for the exchange of such material. All of its holdings are cataloged by subject, location, and author, but in its periodically published computer compiled *Index* [155]; documents cannot, unfortunately, be located by either author or title. An annotated list of references to the literature is, however, also available [154].

For background information, the book *Remote Sensing*, embodying the report of the Committee on Remote Sensing for Agricultural Purposes appointed by the National Academy of Sciences, is recommended as much for its comprehensive coverage (the chapters on "Imaging with Photographic Sensors," "Imaging with Nonphotographic Sensors," "Applications," and "Research Needs," are particularly interesting) as for the quality of its photographic illustrations [161]. The reports of the other Committees are also available [185].

The International Geographic Union is compiling a survey of current work, including a list of participating scientists, in geographic data sensing and processing. The long-range plans of the United States, as presented to the Committee on Science and Astronautics of the U. S. House of Representatives, are set forth in [197], [38], and [60].

Contents of the Paper: Although much of the current ac-

Manuscript received January 31, 1972; revised June 30, 1972.

The author was with IBM Thomas J. Watson Research Center, Yorktown Heights, N.Y. 10598. He is now with the Department of Computer Science, University of Nebraska, Lincoln, Neb. 68508.

Reprinted from *Proc. IEEE*, vol. 60, pp. 1177–1200, Oct. 1972.

tivity is sponsored by NASA, most of the early work in remote sensing was initiated by military intelligence requirements; in particular, the development of imaging sensors was greatly accelerated by the deployment of high-altitude photoreconnaissance aircraft and surveillance satellites. Very little information is, however, available in the open literature about the actual utilization of the collected imagery. The few published experiments for instance, in the *Proceedings of the Symposium on Automatic Photo Interpretation* (Cheng *et al.* [31]), deal almost exclusively with idealized target recognition or terrain classification situations far removed from presumed operational requirements. In view of the scarcity of up-to-date information, this aspect of remote sensing will be discussed here only in passing despite its evident bearing and influence even on strictly scientific and economic applications.

We shall also largely avoid peripheral application of digital computers to the collection or preparation of pictorial material intended only for conventional visual utilization, as in the calculation of projective coefficients in photogrammetry or the simulation of accelerated transmission methods independent of the two-dimensional nature of the imagery. Nor shall we be concerned with statistical computations arising from manually derived measurements, as in models of forest growth and riparian formations based on aerial photographs, or in keys and taxonomies using essentially one-dimensional densitometric cross sections or manual planimetry.

Omitted too is a description of the important and interesting Sideways Looking Airborne Radar all-weather sensors. Such equipment will not be included in the forthcoming satellite experiments. Its potential role in remote sensing is discussed by Simonett [187], Hovis [94], and Zelenka [224].

The diffuse and unstructured nature of terrestrial scenes does not lend itself readily to elegant mathematical modeling techniques and tidy approximations; an empirical approach is well-nigh unavoidable. The first ERTS vehicle is not, however, expected to be launched until the second half of 1972, and the Skylab project is scheduled for 1973, hence, preparatory experimentation must be based on other material. Although none of the currently available sources of imagery approximates closely the expected characteristics of ERTS and Skylab, some reflect analogous problems, and several are of interest on their own merits as large scale data-collection systems. These sensor systems, including both spaceborne and airborne platforms, are described in Section II.

A large portion of the overall experimental effort has been devoted to developing means for entering the imagery into a computer, for storing and retrieving it, and for visual monitoring—both of the hardware available for scanning and displaying high-resolution imagery, and of the software packages necessary for efficient manipulation of large amounts of two-dimensional (and often multiband) imagery in widely disparate formats. These matters are discussed in Section III.

Section IV is devoted to image registration, the difficult problem of superimposing two different pictures of the same area in such a way that matching elements are brought into one-to-one correspondence. This problem arises in preparing color composites from images obtained simultaneously through separate detectors mounted on the same platform, in constructing mosaics from consecutive overlapping pictures from a single sensor, in obtaining a chronological record of the variations taking place in the course of a day or a year, and in comparing aspects of the scenery observed through diverse sensor systems. The most general objective here consists of mapping the images onto a set of standard map coordinates.

Section V is concerned with the application of automatic classification techniques to the imagery. The major problem is the boundless variability of the observed appearance of every class of interest, due to variations inherent in the features under observation as well as in atmospheric properties and in illumination. The difficulty of defining representative training classes under these circumstances has led to renewed experimentation with adaptive systems and unsupervised learning algorithms. From another point of view, the classification of observations into previously undefined classes is an efficient form of data compression, an objective of importance in its own right in view of the quantity of data to be collected.

By way of conclusion, we attempt to gauge the progress accomplished thus far in terms of what still remains to be done if automatic digital image processing is to play a significant part in the worthwhile utilization of the remote sensing products about to become widely available.

The remainder of this Introduction lists some of the proposed applications for ERTS and Skylab, outlines the functional specifications for the image collection systems designed for these platforms, and describes the central data processing facility intended to accelerate widespread utilization of the ERTS image products.

A. Objects of the United States Remote Sensing Program

It is too early to tell whether expectations in dozens of specific application areas are unduly optimistic [185], [38], [60]. Certainly, few applications have emerged to date where satellite surveillance has been conclusively demonstrated to have an economic edge over alternative methods; it is only through the combined benefits accruing from many projects that this undertaking may be eventually justified.

Typical examples of proposed applications are crop inventory and forecasting, including blight detection, in agriculture [61], [169]; pasture management in animal husbandry [97], [32]; watershed management and snow coverage measurement in hydrology [135], [22]; ice floe detection and tracking in oceanography [93], [196]; demarcation of lineaments and other geographic and geomorphological features in geology and in cartography [219], [59]; and demographic modeling [209].

Much of the digital image processing development work to date has been directed at removing the multifarious distortions expected in the imagery and in mapping the results on a standard reference frame with respect to the earth. This process is a prerequisite not only to most automatic classification tasks but also to much of the conventional visual photointerpretation studies of the sort already successfully undertaken with the Apollo and Gemini photographs [37].

The pattern recognition aspects of the environmental satellite applications are largely confined to terrain classification based on either spectral characteristics or on textural distinctions. Object or target recognition as such is of minor importance since few unknown objects of interest are discernible even at the originally postulated 300 ft per line-pair resolution of the ERTS-A imaging sensors.

B. Plans for ERTS and Skylab

The ERTS satellites will be launched in a 496-nmi 90-min near-polar (99°) sun-synchronous orbit. The total payload is about 400 lb.

The two separate imaging sensor systems on ERTS-A (the first of the two Earth Resources Technology Satellites) consist 1) of three high-resolution boresighted return-beam vidicons sensitive to blue–green, yellow–red, and deep-red solar infrared regions of the spectrum, and 2) of an oscillating-mirror transverse-sweep electromechanical multispectral scanner with four channels assigned to blue–green, orange, red, and reflective infrared (IR) bands. ERTS-B will carry a fifth MSS channel in the thermal infrared.

The target of the vidicon tubes is exposed for a period of 12 ms/frame; the readout takes 5 s. This design represents a compromise between the requirements of minimal motion smear, sufficient illumination for acceptable signal-to-noise ratio, and low bandwidth for transmission or recording. In the oscillating-mirror scanner high signal-to-noise ratio is preserved through the use of multiple detectors for each band.

The field of view of both types of sensors will sweep out a 100-nmi swath of the surface of the earth, repeating the full coverage every 18 days—with 10-percent overlap between adjacent frames. Every 100-nmi square will thus correspond to seven overlapping frames consisting of approximately 3500 by 3500 picture elements for each vidicon and 3000 by 3000 elements for each channel of the mirror scanner, digitized at 64 levels of intensity.

The resolution on the ground will be at best 160 m/line-pair for low-contrast targets in the vidicon system and 200 m/line-pair in the mirror scanner [126], [159], [160], [12]. A comparison of the various resolution figures quoted for the Gemini/Apollo photography and for the ERTS/Skylab sensors, and more pessimistic estimates of the resolving power of the ERTS sensors, can be found in [34].

The pictures will be either transmitted directly to receiving stations at Fairbanks, Alaska, Mojave, Calif., and Rosman, N. C., if within range, or temporarily stored on video tape. The vidicon data will then be transmitted in frequency-modulated form in an analog mode while the scanner information is first digitized and then transmitted by pulse-code modultation (PCM) [67]. Canadian plans to capture and utilize the data are described in [198].

The center location of each picture will be determined within one half mile from the ephemeris and attitude information provided in the master tracking tapes which will also be made available to the public.

The sources of geometric and photometric distortion and the calibration systems provided for both sensors are described in some detail in Section IV, where digital implementation of corrective measures is considered. We note here only that estimates for digital processing on an IBM 360/67 computer of a single set of seven ERTS images ranges from 2 min for geometric distortion correction only to 136 min for complete precision processing including photometric correction [217].

The Skylab program will utilize a combined version of the Apollo command-and-service module and a Saturn third stage with a total vehicle weight of 130 000 lb in a low (250-nmi) orbit permitting observation of the earth between latitudes 50° N and 50° S.

The major imaging systems of the Skylab EREP (Earth Resources Experimental Package) consist of a 13-band multispectral scanner covering the ranges 0.4–2.3 μ and 10–12 μ, and of six 70-mm cartographic cameras having suitable film–filter combinations for four bands between 0.4 and 0.9 μ. The instantaneous field of view of the multispectral scanner will be 80 m^2 with a 78-km swath. The low-contrast resolution of the camera system will be 30 m/line-pair with a 163-km^2 surface coverage. A number of nonimaging sensors, such as a lower resolution infrared spectrometer, microwave radiometer/scatterometer altimeter, will also be on board, as well as an optical telescope [215], [168].

The multispectral data will be recorded on board in PCM on 20 000 BPI 28-track tape and returned with the undeveloped film at the end of each manned period of Skylab.

C. Throughput Requirements

Only the relatively well-defined processing load of the centralized NASA facility for the ERTS imagery will be considered here, since it is clear that the quantity of data required for each application ranges from the occasional frame for urban planning [165], to the vast quantities needed for global food supply forecasts [71]. The expected requirements of the user community are discussed in some detail in [72] and [146]. The coverage extended for the North American continent is of the order of

$$\frac{3000 \text{ nmi} \times 3000 \text{ nmi (area)}}{100 \text{ nmi} \times 100 \text{ nmi (frame size)} \times 18 \text{ days (period)}} = 50 \text{ sets}$$

of seven pictures per day. Each set of pictures contains approximately 10^8 bits of data, thus each day's output is the equivalent of 125 reels of 1600 bit/in magnetic tape. This estimate neglects the effects of cloud cover, which is discussed in [190] and [68].

At the NASA Data Processing Facility all of the imagery will be geometrically corrected to within at most 0.5 nmi in linearity and at most 1 nmi in location, and distributed in the form of 70-mm annotated black-and-white transparencies prepared by means of a computer-controlled electronic-beam recorder. In addition, about 5 percent of the images will undergo precision processing designed to reduce registration and location errors with both sensors to within 200 ft (to allow the preparation of color composites), and to reduce photometric degradation to under 1 percent of the overall range. All of the precision-processed data, 5 percent of the raw MSS data, and 1 percent of the raw RBV data will be made available on standard digital tape [220], [217], [138]. The current plan is to use the ephemeris and tracking data for the bulk processing and analog cross correlation against film chips of easily observable landmarks for the precision processing [138].

This is, of course, only the beginning; the subcontinent represents but 15 percent of the total area of the globe. While nations other than the U. S. and Canada may eventually obtain the data by direct transmission from the satellite [62], much of the original demand will be funneled through the NASA facility. The initial capacity of the photographic laboratories is to be 300 000 black-and-white and 10 000 color prints or transparencies per week; it is clear that the major emphasis is *not* on the digital products.

II. Characteristics of the Data Currently Available for Experimentation

At the initial stages of an image-processing experiment, the actual content of the pictorial data under investigation is sometimes less important than its format, resolution, distortion, and grey-tone characteristics, and its relation to other pictorial coverage of the same area. Fortunately, a large variety of data, much of it already digitized, is available to the tenacious investigator, and the supply is being replenished perhaps faster than it can be turned to profitable use.

The sources covered in this section include the vast collection of the National Environmental Satellite Center, the photography from the Gemini and Apollo missions, and both

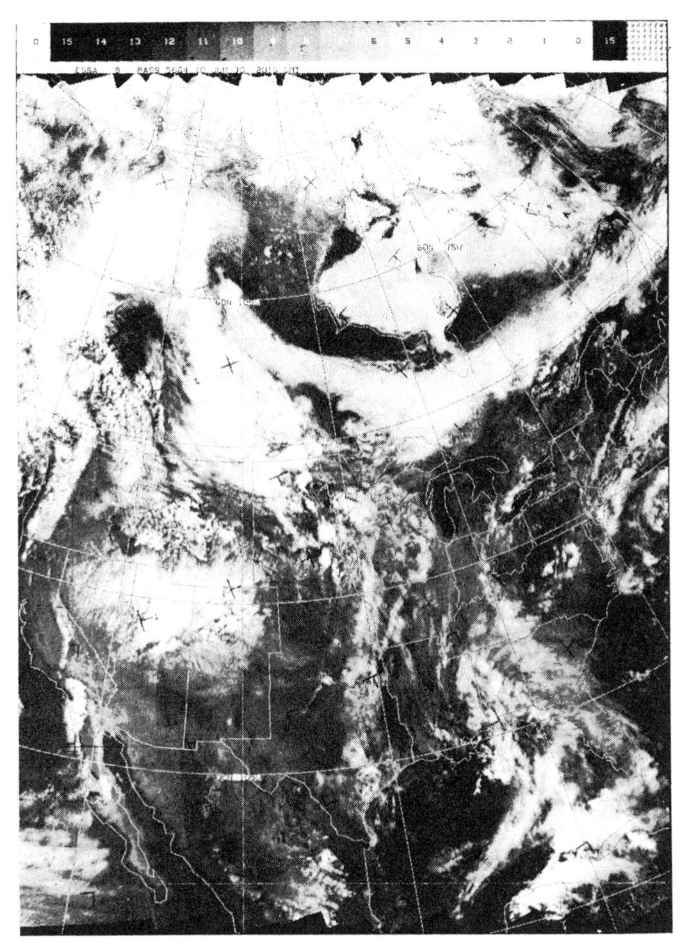

Fig. 1. ESSA-9 mosaic of North America. Traces of the reconstruction from the separate video frames are evident from the fiducial marks. The deviation of the overlay from the true coast lines shows a registration error carried into the mapping program. A programming error, since corrected, may be seen in the checkerboard in NW corner. Note gray wedge and annotation.

photographic and multispectral coverage of over two hundred specially selected test sites obtained by the NASA earth observation fleet.

A. Weather Satellites

Data have been obtained so far from 25 individual satellites beginning in 1959 with Vanguard and Explorer and continuing in the early 1960's with the ten satellites of the TIROS series and later with the Environmental Survey Satellites (ESSA) of the Tiros Operational Satellite System. The current operational series (ITOS) has been delayed because of the premature failure of ITOS-A. Data have also been collected by the Applications Technology Satellites in high geosynchronous orbits and by the experimental Nimbus series.

At present the major meteorological function of these systems is to provide worldwide cloud and wind-vector information for both manual and automated forecasting services, but extensions to other atmospheric characteristics are also underway [222], [41]. The newer satellites provide, for instance, accurate sea-level temperatures in cloud-free regions [123], [174], cloud-height distributions (through the combination of infrared sensor information with ground-based National Meteorological Center pressure and temperature observations [47]), and somewhat less accurate altitude-temperature and humidity profiles (based on the differential spectral absorption characteristics of the atmosphere). Other applications are mapping snow and ice boundaries and observation of sea state [135]. In addition, over 4000 storm advisories have been issued as a result of satellite observed disturbances [120].

The individual frames of Advanced Vidicon Camera System video, obtained from the latest operational polar-orbiting satellites, contain approximately 800 by 800 points at a resolution varying from 1.5 mi at the subsatellite point to 3.0 mi at the edges. The video is quantized to 64 levels of grey, with nine fiducial marks (intended to allow removal of geometric camera distortion), appearing in black on white. The overlap between successive frames is about 50 percent in the direction of the orbit and about 30 percent laterally at the equator; each frame covers about 1700 by 1700 nmi [23]. The two-channel Scanning Radiometer operates at about half the resolution of the AVCS [29].

Digital mosaics are available on a daily basis in either Universal Transverse Mercator or Polar Stereographic projections (Fig. 1). Each "chip" contains 1920 by 2238 points digitized at 16 levels, covering an area of about 3000 by 3000 mi^2. Multiday composites including average, minimum, and maximum brightness charts for snow, ice, and precipitation studies, are also issued periodically. The positional accuracy of any individual point is usually good to within 10 mi. On the high-quality facsimile output provided by the National Environmental Satellite Service geodesic gridlines and coastlines are superimposed on the video to facilitate orientation, but the only extraneous signals in the actual digital data are the fiducial marks from the vidicon camera [26], [24].

The geosynchronous ATS's are equipped with telescopic spin-scan cloud cameras which take advantage of the spin of the satellite itself to provide one direction of motion. The signal from these sensors can be monitored with relatively simple equipment; currently over 600 receiving stations throughout the world take advantage of the wide-angle coverage provided of the Atlantic and Pacific Oceans (Fig. 2). The average altitude is of the order of 20 000 mi, but the high angular resolution of the spin-scan camera allows ground definition comparable to that of the ITOS vidicons. Each frame consists of approximately 2000 by 2000 points. The maximum repetition rate is one frame every 24 min [28].

The Nimbus satellites are used mainly for experimentation with instrumentation to be eventually included in operational systems. Nimbus III, for instance, launched in 1969, carries a triad of vidicon cameras, a high-resolution infrared radiometer, an infrared spectrometer, an ultraviolet monitor, an image dissector camera system, and an interrogation, recording, and location system for data collection from terrestrial experimental platforms.

At present, the imagery from the various satellites is archived at the original resolution only in graphic form, but the last few days' coverage is usually available from the National Environmental Satellite Service Center at Suitland, Md., on digital magnetic tape. Medium-scale archival data tapes going back to January 1962 are maintained by the National Weather Record Center in Asheville, N. C. [23].

An excellent summary of the history, status, and prospects of meteorological satellites data processing, including an extensive bibliography, is contained in [120] and updated in [41].

B. Gemini and Apollo Photography

Most of the 2000 photographs collected on the six Gemini missions between 1964 and 1968 were obtained with hand-held cameras. The astronauts appear to have favored high-oblique

(a)

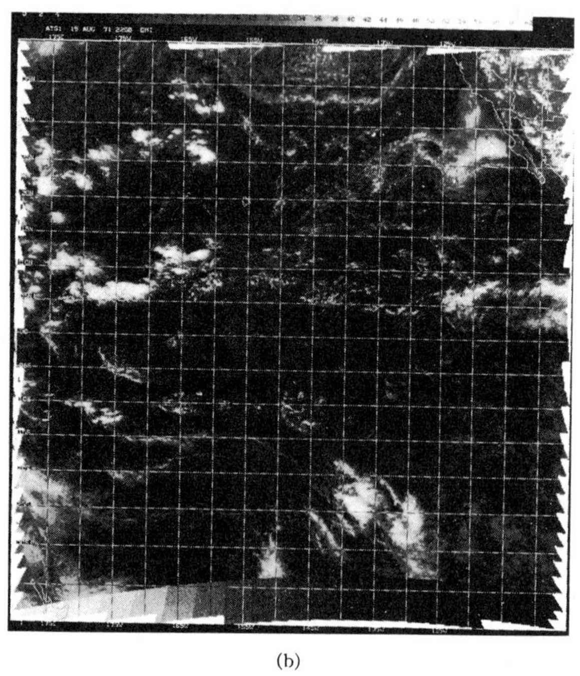

(b)

Fig. 2. ATS image. Raw video and Mercator projection of the Pacific Ocean from the first ATS. These illustrations were obtained through the courtesy of the National Environmental Satellite Center.

Fig. 3. Digitized Apollo photograph. This photograph of central Arizona was digitized at a resolution corresponding to 4200 lines on a drum scanner and was then recorded on film using the same device. The fiducial marks were introduced in the digital data for testing certain reseau detection algorithms [10], [11].

shots, with considerable variation in the scale and orientation of photographs of the same area [151].

For registration experiments, three Gemini photographs of Cape Kennedy obtained within a three-year interval are quite suitable. Two of these photographs are almost at the same scale and show little foreshortening, while the third is an oblique view extending clear across Florida [10].

In addition to hand-held cameras, some of the Apollo missions were equipped with a bank of four boresighted 70-mm Hasselblad cameras. The SO-65 project on the Apollo-9, in particular, was designed to assess the capabilities and limitations of multiband photography in a variety of applications. The satellite photography was carefully coordinated with aerial photography from almost a dozen aircraft flying at altitudes ranging from 3000 to 60 0000 ft, with airborne multispectral scanner coverage, and with the simultaneous collection of terrain information ("ground truth") from several test sites [152].

For agricultural purposes, the most popular test site appears to be the Imperial Valley of California. Extensive coverage is available for this area, with overlapping frames both in the same orbit of the Apollo-9 and in successive orbits several days apart. Each photo-quadruplet consists of three black-and-white negatives in the green, red, and infrared, and of a false-color composite including all three of these bands. The scale of the vertical photography is about 1:1 500 000 [153].

The photographs are obtainable in the form of third-generation prints, negatives, or 35-mm slides from the Technology Application Center of the University of New Mexico. Several dozen photographs, including some of the Imperial Valley and Cape Kennedy pictures mentioned above, have been digitized by Fairchild Camera and Instrument Corporation, Optronics International, Inc., and IBM (among others), at a resolution corresponding to 4000 lines/frame (Fig. 3). The quality of the pictures gives little justification for higher resolution [5].

C. High-Altitude Photography

The MSC Earth Resources Aircraft Program operates half-a-dozen specially equipped airplanes gathering data over some 250 NASA designated test sites [48]. The particular missions flown are decided largely on the advice of 200 or so principal investigators of diverse affiliations appointed for specific research tasks involving remote sensing. About 500 000 frames are collected annually, and are available from NASA "by special request."

Detailed descriptive material is available for each mission, including flight log summaries, charts showing flight lines, lists of camera characteristics (some missions fly as many as a dozen different cameras (see Fig. 4), film and filter combinations, roll and frame numbers, and plots of the earth loca-

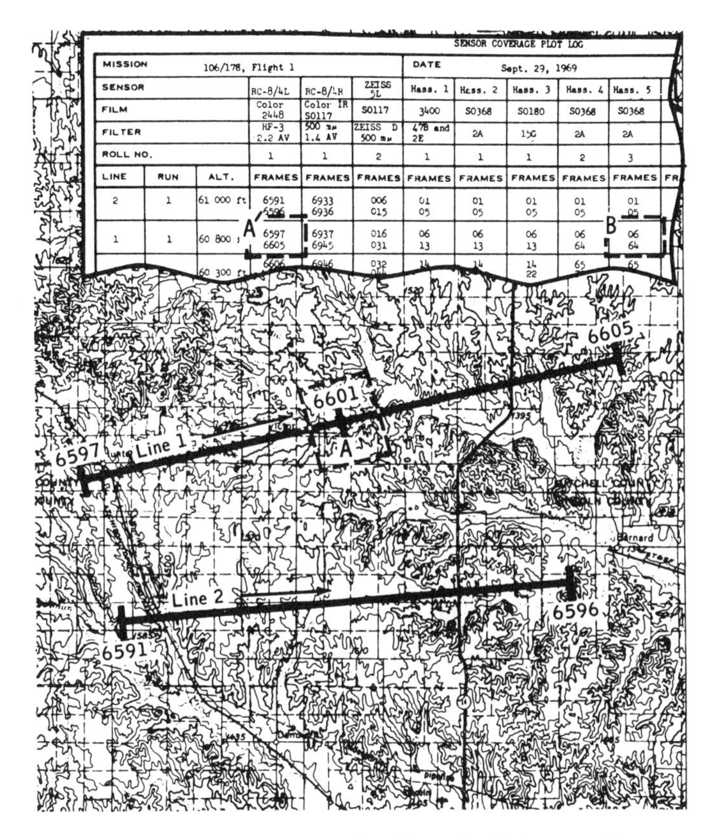

Fig. 4. NASA–ERAP documentation. Example of photographic coverage plot and corresponding plot for a high-altitude flight of the Earth Resources Observation Program (excerpted from NASA–MSC Screening and Indexing Report, Mission 123, Houston, Tex., July 1970).

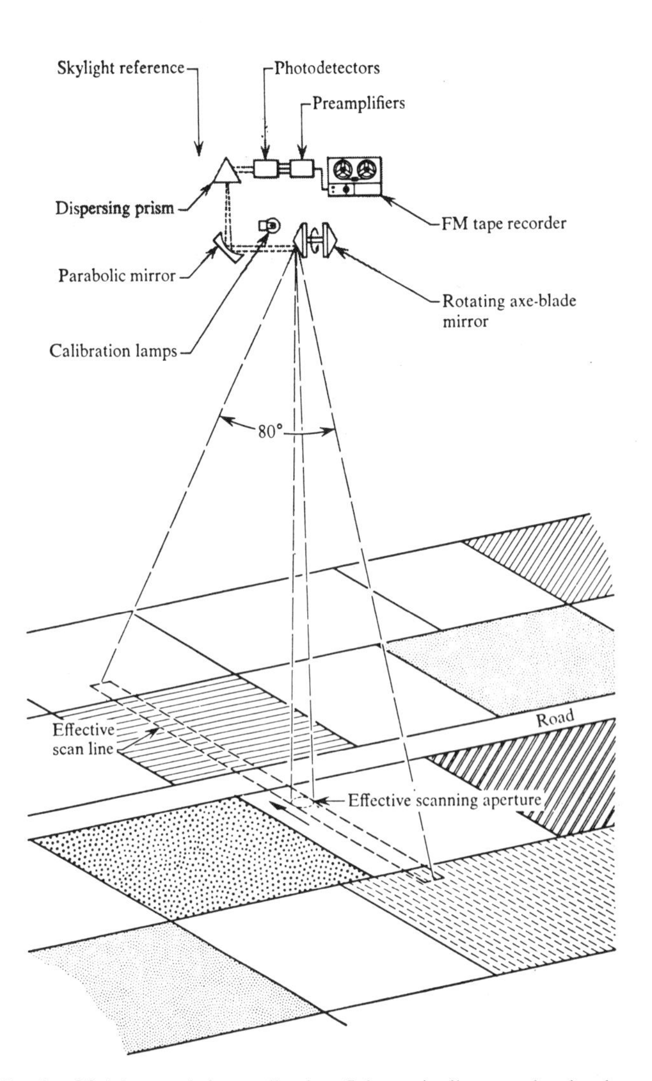

Fig. 5. Multispectral data collection. Schematic diagram showing how the light reflected or radiated from the ground is decomposed into its spectral components, converted into an electric signal, and recorded on board in analog form for subsequent digitization.

tion of selected frames. For simulation of future satellite data, the most suitable imagery is probably that obtained by the 60 000-ft ceiling RB-57 twin jet reconnaissance aircraft and by the recently acquired "ERTS-simulator" U-2's equipped with four multiband cameras (red, green, pan-IR, color-IR) with 40-mm lenses imaging the earth at only twice the resolution of the expected ERTS coverage.

Several of the aircraft are equipped with line scanners of various types. Such instruments will be described in the next section.

D. *Airborne Multispectral Scanners*

In instruments of this type, one scan direction is provided by the forward motion of the aircraft and the other by the rotation or oscillation of a prism or mirror (Fig. 5). Emitted or reflected radiation from the ground is imaged onto an array of detectors sensitive to various bands in the spectrum; the recorded output is an array of multidimensional vectors where each vector represents a specific position on the ground and each component corresponds to a spectral channel [92].

Scanners in operation today have anywhere from one to twenty-four spectral channels of varying bandwidths; cover all or part of the spectrum from the ultraviolet to the thermal infrared; offer a spatial resolution of 2 mrad down to about 0.1 rad; and differ widely with respect to calibration sources, attitude control and tracking accuracy, and method of recording the information. In general, most of the instruments can be flown in various configurtions in conjunction with other sensors including photographic cameras, nonimaging probes, and radar [126].

Most of the published work is based on data obtained by means of the scanners mounted on an unpressurized (10 000-ft ceiling) C-47 aircraft operated for NASA by the University of Michigan. Since 1966, over 150 missions have been completed, with such varied purposes as the study of soil distribution, arctic ice, bark attack on ponderosa pine, water depth, sinkhole-prone conditions, water-fowl habitats, urban features, and most recently and extensively, corn blight. Only a small fraction of the collected information has been automatically analyzed; the remainder is printed out in analog form for visual examination [148].

Digitized data from one of the Michigan flights are typically in the form of an array of 12-dimensional vectors (there are additional channels available but due to separate mounts they are not all in spatial register), with 220 samples perpendicular to the flight direction and up to several thousand samples along the flight line. Ground resolution at 5000-ft flight altitude is of the order of 60 ft.

Several calibration sources are viewed and recorded during the period that neither surface of the rotating axe-blade mirror is looking at the ground. These sources include lamps filtered to match the solar spectrum as closely as possible, black-body thermal references, and background illumination collected through a diffuser [112], [124].

After the flight, the analog signal recorded on FM tape is

digitized, corrected for roll angle, unskewed, and normalized with respect to the calibration signals. Noise bursts, out-of-sync conditions, and other anomalies are detected by elaborate preprocessing programs which also provide appropriate coordinate labels for subsequent identification with respect to photographs or other sources of independent terrain information.

Michigan is currently testing the new M7 scanner which is designed for recording wavelengths from 0.34 to 12 μ [39]. In addition, NASA is testing a 24-channel scanner on a C-130 Hercules (30 000-ft) aircraft equipped with an elaborate automatic data annotation system for facilitating earth location of the imagery [111], [221], [218], [69]. Other scanners with a smaller number of channels and less flexible arrangements are flown on smaller aircraft (such as Bendix's Beechcraft, and Colorado State's Aerocommander) for special missions.

The design of a 625-line color television system intended for airborne service is described in [142].

III. Image-Processing Systems

Although the availability of a large general-purpose digital computer is a valuable asset for experimentation with processing techniques, special hardware is required for converting the raw data into computer-readable form, for monitoring the results of the processing operation and for entering ground truth or other ancillary data pertaining to observed features of the image.

Many experimenters feel that parallel processors must also be available for performing the calculations at the speed necessary to evaluate the results on a significant enough variety of imagery. With conventional processors a large amount of programming work of a rather stultifying nature is required to decompose two-dimensional arrays containing up to 10^7 bytes into fragments of a size suitable for manipulation within the constraints imposed by core size, sequential tape access, blocked disk formats, and, possibly, multiprogrammed operating systems, and to reconstitute these arrays after transformations which may involve changes in the relations between, as well as within, the segments. Some alternatives to sequential digital processing are discussed by Preston elsewhere in this issue.

The large initial investment necessary to begin coming to grips with the more interesting problems offered by *real* data (as opposed to mathematical abstractions or shreds of hand-picked and selected imagery) accounts for the domination of this area of research by a few relatively large institutions, as reflected in the bibliography accompanying this paper.

It is not, as mentioned before, indispensable for each institution to provide means for digitizing the raw data. In principle, this is a one-time operation. In practice, however, it is a very ticklish procedure, and shortcomings may be discovered only after considerable experience with the digitized imagery. For this reason, many experimenters believe it necessary to develop their own scanners and other analog-to-digital conversion equipment, a challenge usually outlasting their purse and patience.

The equipment required for precision conversion of multispectral FM recordings and directly captured weather satellite pictures is so specialized that it will not be described here. Some of the calibration and synchronization problems encountered are described in [148] and [28]. Optical scanners suitable for diverse applications are, however, commercially available and will be briefly discussed.

There is no question that adequate grey-tone output must be conveniently available for experimentation, but opinion seems divided as to whether a television-type screen display with some interactive capability or a high-quality hardcopy output is preferable. A common compromise is a low-resolution CRT display with higher resolution (because flicker-free operation is not required) Polaroid recording capability.

Other less easily defined aspects of processing *large* pictures discussed in this section are the operating systems and utility programs necessary for any sort of coherent experimentation, special processors for anticipated quantity production, and the role of man–machine interaction in both experimental and operational systems.

Some examples of digital-computer-oriented remote sensing facilities, and of the equipment they contain are as follows:[1]

NOAA's National Environmental Satellite Center at Suitland, Md.: three CDC 6600's, CDC 160A, two ERM 6130's and 6050's, CDC 924, three 5000 by 5000 element Muirhead recorders, Link 35-mm archival microfilm unit [28], [29];

NASA–MSC's complex for the analysis of multispectral recordings and photography: 160 by 111 digital color display, closed-circuit television display, Xerox hardcopy output, Grafacon tablet, keypack, analog tape drives, IBM 360/44 processor [56], [58];

Caltech's JPL operation: film-scanner, CRT display, FM tape conversion, facsimile hardcopy, IBM 360/75 [18], [162];

University of Michigan multispectral facility: SPARC analog computer, drum scanner, analog film recorder, CRT display, FM tape conversion, CDC 3600 [131];

Purdue University's Laboratory for Applications of Remote Sensing: 577 by 768 element flicker-free 16-level digital TV display, light pen, continuous image motion, selective Polaroid or negative hardcopy without obstruction of display, FM tape conversion, IBM 360/67 [116], [117], [203];

University of Kansas KANDIDATS (Kansas Digital Image Data System) and IDECS (Image Discrimination, Enhancement, and Combination System): three flying-spot scanners for transparencies (25 mm to 3 by 4 in) and a vidicon camera controlled by a PDP 15/20, electronic congruencing unit (rotation, translation, and change of scale), 20 by 20 element linear processor and level selector, 24-channel digital disk storage, monochrome and color displays with built-in crosshatch generator, film output, GE-635 computer [77];

Computing Science Center of the University of Maryland: flying-spot film scanner, drum scanner/recorder, CRT display, vidicon, Univac 1108 [171];

University of Southern California Image Processing Laboratory: IER 1000 by 1000 element flying-spot color scanner and display, Muirhead rotating-drum color scanner and recorder, digital color television and display, Adage vector display with joystick and light pen, IBM 360/44, IBM 370/155, and HP 2100 computers connected to ARPA net [173];

Perkin-Elmer's Sampled Image Laboratories: drum scanner/recorder, flying-spot scanner, high-resolution traveling-stage microscope image-plane scanner, CRT and storage tube displays, precision plotting table, linked IBM 360/67, XDS 930, H-516, and Varian 620-i computers [212];

[1] Some of this information, obtained through personal communication, is more recent than the references would indicate.

IBM Research Division's facility at Yorktown Heights: film scanner, CRT output, image dissector, digital color TV display, graphic tablets, 360/91, 360/67, and 1800 computer net [85].

Another group of facilities is dedicated primarily to automatic photointerpretation, with only fragmentary information available about the work. Examples of this group are as follows:

McDonnell Douglas Astronautics: a compact vidicon scanner with 70-mm film scanner and a minicomputer, and a larger interactive system with a 1024-line (nominal) image dissector, scan converter, rear-projection viewer, 16-level digital TV display, alphanumeric display, joystick, XDS 930 computer [104], [105];

SOCRATES (Scope's Own Conditioned-Reflex Automatic Trainable Electronic System), a 20 by 20 photodiode and threshold logic array, the successor to Conflex I [164], [211];

SARF, General Motors' phoenix-like interactive Signature Analysis Research Facility [192];

MULTIVAC, Hughes Research Laboratories' 10 by 10 element binary array processor [8];

Litton Industries' Automatic Target Recognition Device, a hybrid system with a programmable CRT scanner, Recomp II process control computer, and interactive operation [205];

Cornell Aeronautical Laboratories' adaptive image processing operation using 35/70-mm CRT scanner, storage scope output, PDP-9, IBM 360/65 and 370/165 [134], [143];

ASTRID, Ohio State's Automatic Recognition and Terrain Identification Device, a hybrid computer system oriented toward processing line segments [163].

A good review of image enhancement facilities for remote sensing throughout the country is available in [37]. Among the systems discussed are the following: the NASA-USDA Forestry Remote Sensing Laboratory Optical Color Combiner at Berkeley, Calif.; the University of Kansas IDECS system; the two-band 1000-line Philco-Ford Image-Tone Enhancement System; and the Long Island University Multispectral Camera-Viewer. Abroad, we know of sustained activity only at the Institute for Information Processing and Transmission of Karlsruhe University [106], [89], [107], though some earlier European work is described in [21].

A. Optical Scanners

Since the data most closely resembling the expected ERTS and Skylab imagery in terms of resolution are available in the form of photography, optical scanners are necessary to translate the grey-tone (or color) information into computer-readable code.

CRT flying-spot scanners and television cameras (image dissectors or vidicons) are the most inexpensive and fastest devices available, but beyond a degree of resolution corresponding to about 500 by 500 picture elements quantized at 16 levels of intensity, the nonlinearities introduced by such scanners tend to exceed the degradation and distortion present in the photography itself. Owing to the nonuniform sensitivity, scanning the pictures section by section introduces even graver problems in juxtaposing adjacent sections.

Mechanical-drum and flatbed microdensitometers (50 000 dollars and up) are easily capable of the accuracy required for almost any type of photography, with even the less expensive digitizers (15 000 dollars) producing 2000 by 2000 arrays with up to eight bits of grey-scale quantization. Owing to the narrow spectral range of the source of illumination and matching detector configuration, most of these machines cannot be readily converted to color work. For this purpose, one must turn to scanners specially designed for the simultaneous production of color-separation plates in the printing industry, suitably modified by the addition of an analog-to-digital interface. The low speed of operation of such scanners (of the order of $\frac{1}{4}$ s/scan-line) generally requires off-line operation or an elaborate interrupt structure [141], [210], [171], [212].

Due to the lower contrast of opaque prints, positive or negative transparencies constitute the preferred medium for scanning. Because of multiple surface reflections. transparencies cannot be scanned with a reflection scanner by simply providing a uniform reflective background. Drum scanners designed for film have either a glass drum or some self-supporting arrangement with edge guides.

B. Grey-Tone Output Devices

Many flying-spot scanners and drum microdensitometers can be modified to operate in a write mode. This is a particularly convenient arrangement since the format conversion problems are altogether eliminated, and compatible resolution and sensitivity characteristics are guaranteed. The only drawback is that closely controlled wet processing with attendant time delay is usually required for consistent grey-tone reproduction. Film recorders may also be used with Ozalid foil overlays to produce high-quality color transparencies [130].

Facsimile recorders are less expensive and can provide 16 levels of grey on a 4000 by 4000 array (Fig. 2). Programmable flying-spot recorders with special character masks, such as the widely used Datagraphics 4020, provide about 500 by 500 distinct elements, but elaborate programming is necessary, with frequent recalibration, to secure even eight reasonably uniformly distributed intensity levels on either paper or film. Spatial resolution may, however, be traded off for grey-scale resolution by resorting to halftone techniques [79], [183].

Although all of these devices are generally used in a fixed-raster mode of operation, control of the beam deflection in an electron-beam recorder being developed by CBS Laboratories is the intended mechanism for the correction of "bulk-processed" video tapes at the NASA-ERTS Data Processing Facility [138]. This is an essentially analog system under control of an XDS Sigma 3 digital computer. A laser-beam recorder with a 10-μ spot over a 20 by 20-cm area has been developed at ITEK Corporation [125].

Line-printer overstrike programs are still useful for quick turn around, particularly with elongated formats such as that of the Michigan MSS. The visual qualities of this form of output are greatly improved by judicious use of watercolors and transparent overlays; modifications intended for on-line terminal use are, however, agonizingly slow. Isometric, perspective, and isodensity Calcomp plots offer another alternative for the impecunious investigator.

Television-type grey-scale displays are generally refreshed either from a high-speed core buffer or from a digital video disk. A single-line buffer is sufficient to fill the video disk, but a full-frame buffer (typically 520 by 600 bytes or picture elements) renders it much easier to change only parts of a picture without regenerating the entire frame.

Color displays need three times as much buffer storage as black-and-white pictures, but offer no particular difficulty if registered color separations are available in digital form. The calibration procedures necessary to produce color composites from *multiband* photography (filtered black-and-white exposures) are discussed in [19].

It is possible to circumvent the need for an image buffer by resorting to the now available grey-scale storage scope. Such a device requires a much simpler interface than a refreshed system, but the saving is to some extent illusory since most of the cost resides in the programming system necessary to select, retrieve, edit, and otherwise manipulate the displayed pictures.

Without a sophisticated programming system, the display can be used only to show the coarsest changes in the picture or as a conversation piece with lay visitors. For meaningful experimentation, it is desirable to be able to show two or more versions of the same picture simultaneously, to form overlays, to label specified features, to display intensity histograms and other computed functions, to vary the density modulation to bring into prominence regions of different intensity or to compensate for the amplitude nonlinearities of both the data collection system and the display itself, and to to perform many other more specific functions quickly and effortlessly.

Just how good a display must be to prove useful is a moot point and depends largely on the ingenuity of the user in casting the relevant information in a form compatible with the available display capabilities. So far, there is insufficient evidence to evaluate, in realistic terms, the contribution of display systems to the development of specific image-processing algorithms.

C. Operating Subsystems and Utility Programs

As already mentioned in several different connections, the major part of the programming effort required to cope effectively with large image arrays (four orders of magnitude larger than binary character arrays and about two orders of magnitude larger than most biomedical pictures) must be devoted to conceptually trivial matters such as: the decomposition and reconstitution of pictures; edge effects; efficient packing and unpacking routines for the various modes of point representation; variations in the basic byte and word sizes between different machines; aspect-ratio and other format changes among scanner, analog-to-digital converter, internal processor, and output devices; tape and disk compatibility; storage-protect and filing devices to ensure the preservation of valuable "originals" without undue accumulation of intermediate results; diagnostic routines permitting inspection of the actual values of relatively small segments at given image coordinate locations; left–right, up–down, black–white confusions; and myriad other frustrating details.

To avoid having to reprogram all these ancillary routines for each new function to be performed on the pictures, it is desirable to set up a procedure-oriented language or system within the framework of which new programs can be readily incorporated. Such a system may provide the necessary interface with the special-purpose hardware, allow access to a library of subroutines, supervise extensive runs in the batch mode on large computers, and offer special image-oriented debugging and diagnostic facilities in a time-shared mode of operation [125], [85].

Since most functions to be performed on a picture are local operations in the sense that the values of only a small subset of all the picture elements need be known in order to compute the value of an element in the output picture, the provision of a generalized storage policy is essential for the efficient performance of arbitrary window operations. For instance a window of size $n \times m$ may take n disk accesses if the image is stored line by line. A worst case example is a horizontal edge-finding operation on an image stored by vertical scans. With large images in such cases it is usually worth rewriting the array in an appropriate format before proceeding with the calculation; flexible means for accomplishing the reshuffling must be a part of any image-processing system worthy of the name unless an entire image can be accommodated in the fast random-access memory. A valuable discussion of the computational aspects of two-dimensional linear operations on very large arrays (up to 4000 by 4000 elements) is presented in [99].

Among the best known image-processing systems are the various versions of PAX developed in conjunction with the Illiac III, and later rewritten for the CDC 3600, IBM 7094 and System/360, and also for the Univac 1108. In PAX, images are treated as stacks of two-dimensional binary arrays. Arithmetic operations on integer-valued image elements are replaced by logical operations performed in parallel on the corresponding components (such as the 2^3 level) of several picture elements. Planes are defined in multiples of the word size of the machine, but aside from a few such restrictions, PAX II is conveniently imbedded in Fortran IV, with the debugging facilities of the latter available in defining new subroutines. Both a conversational mode and batch demand processing have been implemented at the Computing Center of the University of Maryland. The major subroutines are designed for the definition of planes, masks, and windows, logical functions of one or more planes or windows, neighborhood operations, area and edge determination, preparation of mosaics, tracing of connectivity, creation of specific geometric figures such as circles and disks, distance measurements, grey-scale overprint, grey-scale histograms and normalization, superimposition of grids, noise generation, moment-of-inertia operations, translation, rotation and reflections, and include as well a number of basic "macro" operations intended to facilitate expansion of the program library [149], [170], [30], [101], [102].

Other examples of comprehensive programming systems described in the literature are Purdue's LARSYSAA [116], [202], [203], General Motors' SARF [192], and the University of Kansas KANDIDATS [76]. Such systems are generally difficult to evaluate because the publication of interesting research results developed with the system (as opposed to contrived examples) tends to lag indefinitely behind the system description. Furthermore, very seldom is there any indication of the breakdown between the amount of effort expended in the development of the system and the time required to conduct given experiments.

D. Special-Purpose Digital and Hybrid Systems

At a time of continually waning interest in special-purpose processors for pattern recognition, there are two main arguments for their use in remote sensing. The first is the inability of even the largest digital computers to cope with element-by-element classification at a speed approaching the rate of collection of the data; an airborne MSS typically spews out 10^8 samples/h while the 360/44 can ingest only about 10^5 bytes/h on a ten-class problem [133]. The second argument is based on the need for on-board processing owing to the excessive bandwidth requirements for transmitting data from a spaceborne platform.

An example of a hybrid classification system is the SPARC machine at the Infrared and Optics Laboratory of the University of Michigan [130]. SPARC has 48 analog multipliers operating in parallel, and performs quadratic maximum-likelihood decisions on 12-component vectors at the rate of one every 10 μs.

Because of the difficulty of calibrating the machine, requiring manual setting of the potentiometers corresponding to the entries in up to four previously computed 12×12 class-covariance matrices, exact duplication of results is almost impossible and the machine is therefore of marginal utility in strictly experimental investigations. A successor featuring direct digital control and an interactive display capability is on the drawing boards [132], [133].

The proposed NASA–ERTS data processing facility makes use of optical correlation techniques against chips containing easily identifiable landmarks for registering the data, and special digital hardware for point-by-point correction of vignetting in the vidicons and other systematic errors [71], [138].

So far, no on-board satellite image processor has been installed, but a feasibility study based on several hundred photographs of clouds and diverse lunar terrain features concludes that an acceptable classification rate can be attained [45].

Another study, using photographs of six "typical" terrain features, proposes a simple adaptive processor based on coarsely quantized average intensity levels, spatial derivatives, and bandpass spatial filter output [100]. Electrooptical preprocessing techniques using image intensifier tubes are described in [83].

The highly circumscribed test material used in these experiments leaves some doubt as to their relevance to the output of currently available spaceborne sensors.

In spite of the commercial availability of relatively inexpensive FFT hardware and long shift-register correlators, there appears to have been no attempt so far to apply these devices to digital image processing for remote sensing.

E. Interactive Processing

Interactive processing in remote sensing does not necessarily imply the kind of lively dialogue between man and machine envisioned by early proponents of conversational systems and already realized to some extent in computer-aided design and information retrieval, and in certain areas of pattern recognition (see paper by Kanal elsewhere in this issue).

The prime objective of on-line access to the imagery is to provide an alternative to laborious and error-prone off-line identification, by row and column counts on printouts or interpolation from measurements on hardcopy output, of features which are easily identifiable by eye yet difficult to describe algorithmically without ambiguity. Examples of such features are landmarks—such as mountains, promontories, and confluences of rivers—for use in accurate registration of photographs, and the demarcation of field boundaries for crop-identification studies based on multispectral recordings.

Even a system without immediate visual feedback, such as a graphic tablet on which a facsimile rendition of the digitized image can be overlaid, is considerably superior to keying in the measured coordinate values. One step better is the displacement of a cursor on the display under control of a tablet, joystick, or mouse. The ideal is direct light-pen interaction, but this is not easy to implement on a high-resolution digital display. For really accurate work, a zoom option on the display is necessary for accurate location of the features of interest, but the amount of computation required is prohibitive in terms of response time [84].

A thoroughly tested system for locating nonimaging sensor data in relation to a closed-circuit television display of simultaneously obtained photography is described in [55] and [58]. The accuracy of the computer-generated overlays is shown to be better than 0.5° by reference to salient landmarks [54].

There has also been discussion of on-line design of spatial filters, decision boundaries, and compression algorithms. Here again, however, the waiting time between results is lengthy, and the operator must transmit so little information to the machine—typically just a few parameter values—that batch processing with high-quality hardcopy output is preferable in many instances. With multiprogrammed systems, the difference between on-line and off-line operation tends to blur in any case, with the distinction sometimes reduced to whether one enters the necessary commands at a terminal in the office or at a nearby remote job entry station.

Another possible desideratum is a display browsing mode, allowing inspection of large quantities of images. Here also, however, reams of hardcopy output, with on-line operation, if any, confined to pictures of interest, may be preferable.

It would thus appear that the most appropriate applications of interactive concepts, in the context of remote sensing, are 1) the debugging of program logic, where small image arrays may be used to keep the response time within acceptable limits, and 2) the entry of large quantities of positional information, where practically no computation is required and no viable off-line alternatives exist.

IV. Image Restoration and Registration

The need for exact (element-by-element) superimposition of two images of the same scene upon one another arises in the preparation of color composites, chronological observations, and sensor-to-sensor comparisons. The spatial, temporal, and spectral aspects of image congruence are discussed in [3]. Here we shall attempt to categorize the types of differences which may be encountered between two pictures of the same scene on the basis of the processing requirements necessary to produce a useful combined version. Only digital techniques are presented; the advantages and disadvantages of optical techniques are discussed elsewhere in this issue by Preston, and in [165], [166].

Geometric distortions in electronically scanned imagery are due to changes in the attitude and altitude of the sensor, to nonlinearities and noise in the scan-deflection system, and to aberrations of the optical system.

Photometric degradation (occasionally also referred to as "distortion," with questionable propriety) arises due to modulation transfer-function defects including motion blur, nonlinear amplitude response, shading and vignetting, and channel noise.

The atmospheric effects of scattering and diffraction, and variations in the illumination, also degrade the picture, but these effects are in a sense part of the scene and cannot be entirely eliminated without ancillary observations.

Once the pictures to be matched have been corrected for these sources of error, resulting in the digital equivalent of perfect orthophotos, the *relative* location of the pictures must still be determined before objective point-by-point comparisons can be performed. In reality, this is a chicken-or-egg problem, since the pictures cannot be fully corrected without locating a reference image, but the location cannot be determined accurately without the corrections.

Tracking and ephemeris data usually provide a first approximation to the position of the sensor at the time of exposure, but for exact registration more accurate localization is required. In operator-aided systems, such as the operational NESC mapping program, the landmarks are located by eye, while in fully automatic systems some correlation process is usually employed. A compromise is preliminary location of the landmarks by the operator, with the final "tuning" carried out by computer [191] in a manner analogous to the widely used track measurement programs for bubble-chamber photographs, described by Strand elsewhere in this issue.

A major difficulty in multispectral correlation or matched filtering of perfectly corrected images is the existence of *tone reversals*, or negative correlations, between spectral bands. This phenomenon, however, constitutes the very essence of most of the spectral discrimination techniques described in Section V.

Many of the current image restoration and enhancement techniques are intended to facilitate the task of visual interpretation. These matters are discussed in detail in the July 1972 special issue of the *Proceedings of the IEEE*, and in [96]. Difficulties arise because the transformations required to reveal or emphasize one set of features may, in fact, degrade features desirable for another purpose, yet such techniques have consistently produced visually startling results in Mars images [176]. An excellent discussion of the necessary compromises is offered by Billingsley [20].

A. Mathematical Formulation

The registration problem is cumbersome to state mathematically in its entire generality, but the following formulation may help in understanding the work currently in progress.

The scene under observation is considered to be a two-dimensional intensity distribution $f(x, y)$. The recorded image is another (digitized) two-dimensional distribution $g(u, v)$. The image is related to the "true" scene $f(x, y)$ through an unknown transformation T:

$$g(u, v) = T(f(x, y)).$$

Thus in order to recover the original information from the recorded observations, we must first determine the nature of the transformation T, and then execute the inverse operation T^{-1} on the image.

When independent information is available about T, such as calibration data on distortion and degradation, or a model of atmospheric effects, or attitude data concerning the angle of view, then the two operations may be separated.

Often, however, only indirect information about T is available, usually in the form of another image or a map of the scene in question. In this case, our goal must be to transform one of the pictures in such a manner that the result looks as much as possible like the other picture. The measure of similarity is seldom stated explicitly, since even if the two pictures are obtained simultaneously, the details perceptible to the two sensors may be markedly different. Thus for instance, in registering photographs of the same scene obtained simultaneously through different color filters we would want shorelines and rivers, but not the intensity levels, to correspond. On the other hand, if the photographs are obtained years apart with the purpose of observing the erosion of the shoreline or the shift in drainage patterns, then we must expect changes in the location of such features. Seasonal variations also give rise to problems of this type.

In some studies it is assumed that except for the effect of some well-defined transformation of interest, the image of a given scene is produced either by the addition of independently distributed Gaussian white noise, or by multiplication by exponentially distributed noise. While these assumptions lead to the expected two-dimensional generalization of the familiar formulas of detection, estimation and identification theory, they bear little relation to the observed deviations in many situations of practical interest.

The case of known (or derivable) T is sometimes known as image restoration, as opposed to the classical registration problem where T must be obtained by repeated comparison of the processed image with some standard or prototype. This dichotomy fails, however, when the parameters of T are obtained by visual location of outstanding landmarks followed by automatic computation of the corrected image.

B. Single-Point Photometric Corrections

To make any headway on either problem, at least the form of the unknown transformation must be known. We can then parametrize the transformations and write $g(x, y) = T_{\mathbf{c}}(f(x, y))$ to indicate that the true value (grey level) of a point with coordinates (x, y) depends only on the observed value at (x, y). The components of $\mathbf{c}$ specify the regions where a given correction factor is applicable.

Examples of such degradation are the vignetting due to the reduced amount of light reaching the periphery of the image plane in the sensor, and the shading due to sun-angle in the TIROS and ESSA vidicon data. Since the combined degradation is quite nonlinear with respect to both intensity and position, the appropriate correction factors are prestored for selected intensity levels on a 54 by 54 reference lattice, and the individual values in the 850 by 850 element picture are interpolated by cubic fit. Camera warmup time through each orbit, as well as the sawtooth effect owing to the non-uniformly reciprocating focal-plane shutter, are taken into account, but contamination by the residual images on the photocathode is neglected. The final output is claimed to be photometrically accurate (or at least consistent) within 5 percent, which is sufficient for the production of acceptable montages for visual inspection [26], [24].

If the preflight calibration does not yield a sufficiently accurate description of the response of the post-launch system, as was the case with the early Mariner pictures, then the correction may be based on the average grey-scale distribution of many pictures on the assumption that the true distribution is essentially uniform [156].

Spectral calibration of digitized aerial color photography can be performed on the basis of the measured reflectance characteristics of large ground calibration panels [52].

Single-point photometric corrections have also been extensively applied to airborne multispectral imagery. A comprehensive discussion of the various factors contributing to variations in the output of the multispectral scanner, including the crucial non-Lambertian reflectance characteristics of vegetative ground cover, is contained in [112]. This study also offers an experimental evaluation of various normalization methods based on *relative* spectral intensities, and a formula for eliminating channel errors resulting in "unlikely" observations. A followup study [113] describes interactive techniques based on visual examination of certain amplitude averages. Good examples of the importance of amplitude preprocessing in extending the range of multispectral recognition are shown in [88], [188].

A more theoretical approach to automatic derivation of the complicated relation between sun-angle and look-angle is presented in [43] and an analysis of scattering phenomena in different layers of the atmosphere in [172].

C. Multipoint Photometric Correction

This class of operations may be symbolized as

$$g(x, y) = T_c(f(n(x, y)))$$

where $n(x, y)$ denotes a *neighborhood* of the point (x, y). In the simplest case, the corrected value at (x, y) depends on the observations at two adjacent points:

$$g(x, y) = T_c(f(x-1, y), f(x, y)).$$

Such linear filtering operations (the properties of the filter are characterized by $\mathbf{c}$) are common in correcting for motion blur (the convolution of the true video with a rectangular pulse corresponding to the length of the exposure), for loss of resolution due to modulation transfer function (MTF) rolloff, for periodic system noise, and for scan-line noise [18], [156], [179], [157], [50], [10], [217], [184], [177].

The desired filtering operation may be performed directly in the space domain as a local operation [179], [10], [11], [156], with typical operators ranging from 3 by 3 neighborhoods for motion blur to 21 by 41 elements for more complex sources of low-frequency noise, by convolution with the fast Fourier transform [80], [50], [2], [186], or through optical techniques [96]. In processing speed the local operators show an advantage as long as only very-high-frequency effects are considered and they are also less prone to grid effects in the final results [184], [7]. Optical processing has not yet been used on an operational basis on nonphotographic imagery, principally because of the difficulty of interfacing the digital and optical operations [95].

D. Geometric Distortion

Geometric distortion affects only the position rather than the magnitude of the grey-scale values. Thus

$$f(u, v) = f(T_c(x, y))$$

where T_c is a transformation of the coordinates.

If the transformation is linear, the parameter vector contains only the six components necessary to specify the transformation, i.e., $\mathbf{c} = (A, B, C, D, E, F)$ where

$$u = Ax + By + C$$
$$v = Dx + Ey + F$$
$$f(u, v) = f(Ax + By + C, Dx + Ey + F).$$

Important subcases are pure translation ($A = E = 1$, $B = D = 0$), pure rotation ($C = F = 0$, $A^2 + B^2 = D^2 + E^2 = 1$), and change of scale ($A/B = D/E$, $C = F = 0$). From an operational point of view, the transformation is specified by the original and final location of three noncollinear points. In executing a linear transformation on the computer, it is sufficient to perform the computations for a small segment of the image in high-speed storage, and transform the remainder, segment by segment, by successive table lookup operations. Aside from the saving in high-speed storage requirements, this technique results in approximately a tenfold decrease in computation over direct implementation of the transformation.

Along the same lines, a projective transformation is specified by eight parameters, which may be derived from the location of four pairs of picture elements. The execution of this transformation is, however, more complicated, since the relative displacement of the picture elements is not uniform throughout the frame.

It is important to note that owing to the quantization of the coordinate axes, the actual computation of the corrected image is usually performed in reverse in the sense that the program proceeds by determining the *antecedent* of each element or set of elements in the new image. Were the transformation performed directly, one would be faced with the possibility of the occurrence of gaps in the case of dilations and the superposition of several elements in the case of contractions. Since the computed coordinates of the antecedents do not in general fall on actual grid points, it is customary to adopt a nearest neighbor rule for selecting the appropriate element, though local averages are sometimes used instead.

Translation, rotation, and perspective transformations occur in practice owing to changes in the position and attitude of the sensor platform, while scale changes are frequently required by format considerations in input–output. The most bothersome distortions cannot, however, be described in such a simple form. Properties of the transducer itself, such as pincushion distortion, barreling, optical aberrations, and noise in the deflection electronics are best characterized by their effect on a calibration grid scanned either prior to launching or during the course of operation. In addition, fiducial marks are usually etched on the faceplate of the camera in order to provide *in situ* registration marks.

The expected sources of distortion in ERTS-A RBV imagery are discussed in quantitative terms in [136] with particular reference to correction by means of analog techniques such as optical projection and rectification, line-scan modulation, orthophoto correlation, and analytically (digital computer) controlled transformation of incremental areas. Because of the nonuniformity of the distortions over the entire format, and the possibility of tone reversals from object to object in the spectral bands, only the last system is accorded much chance of success.

The precise measurement of the location of the 81 (9×9 array) fiducial marks (also called reseau marks) on the faceplate as well as on the output image of a number of return-beam vidicons destined for the ERTS-A satellite is described in detail in [137]. It is shown that to provide a frame of reference for eventual correction of the imagery to within one-half resolution element, the vidicon parameters must be established to the accuracy shown in Table I.

The detection of the fiducial marks in the ERTS vidicon images, with experiments on simulated pictures derived from digitized Apollo photographs, is described in [11], [16], [129]. The basic technique is "shadow casting" of the intensity distributions on the x and y axes of the picture. This is shown to correspond, for the selected fiducial-mark geometry, to matched-filter detection.

The actual correction, by interpolation between the grid coordinates of the distortion on extraterrestrial images, is reported in [156], on TIROS vidicon data in [26], and on ATS spin-scan cloud-camera pictures in [28]. The correction of geometrical distortions may be efficiently combined with the production of rectified orthophotos (equivalent to a 90° angle of view) [156] and with the generation of standard cartographic products such as Mercator and Polar Stereographic projections [27], [23], [29].

TABLE I

CALIBRATION REQUIREMENTS FOR PARAMETERS OF THE RETURN-BEAM VIDICON FOR ERTS-A (from [137])

Parameter	System Accuracy	Calibration Accuracy
Reseau coordinates	$\pm$ 10 μm	$\pm$ 3 μm
Lens:		
Focal length	+270 μm −980 μm	$\pm$20 μm
Principal point	$\pm$ 65 μm	$\pm$ 5 μm
Radial distortion	$\pm$ 30 μm	$\pm$ 5 μm
Electronic distortion	$\pm$750 μm[a] (about 125 TVL of 6 μm each)	$\pm$ 5 μm
Orientation between cameras	$\pm 7.0\times10^{-4}$ rad (about 2.4 ft)	$\pm 1.5\times10^{-5}$ rad (about 3 in)

[a] Recent data indicate that the 750-μm value may be too high. However, the complete RBV–transmission–EBR system has not been tested yet.

A fast algorithm suitable for digital computers equipped with MOVE BYTE-STRING instructions has been reported in [217], [10], [11]. This algorithm is intended for the correction of small distortions, such as those due to the camera characteristics, and is based on the fact that relatively large groups of adjacent picture elements retain their spatial relationship in the corrected picture. The program computes the maximum number of adjacent elements that may be moved together without exceeding a preset error (typically one coordinate increment). Experimental results [17], [129] show that the boundaries between such groups are not visually detectable.

When the imagery is intended mainly for visual use rather than for further computer processing, the technique of "gridding" offers an expedient alternative to mapping [27]. With this method the picture elements are left in their original location and the positional information is inserted by the superposition of latitude and longitude coordinates over the image. Since only the locations of selected image points need be computed for this purpose, the method is quite rapid and is extensively used with the specially modified output devices of the National Environmental Satellite Center.

E. Automatic Determination of Anchor Points

Finding corresponding points in two pictures of the same scene can be accomplished by correlating a "window function" from one picture with selected portions of the other picture. If the two images differ only by a shift or translation, then only a two-dimensional search is involved. If, however, a rotation and a change of scale are also necessary, then the time required for exhaustive search exceeds all practicable bounds.

Once one window function has been adequately matched, the location of the maximum value of the correlation function usually gives a good idea of where the search for the next window function should be centered, even if the transformation is not quite linear.

In terms of our earlier formulation, we are trying to evaluate the parameter vector $\boldsymbol{c}$ under the assumption that both $f(x, y)$ and $g(x, y)$ are available and that parameters representative of the entire image can be derived by establishing the correspondence of selected subimages. This usually involves maximization of a similarity function subject to the constraints imposed by the postulated transformation.

The similarity function to be maximized may take on several forms. If, for instance, we take two $m\times m$ images Y_1 and Y_2 considered as vectors in an m^2-dimensional space, then an $n\times n$ window function V_1 and the trial segment V_2 may be considered as projections of Y_1 and Y_2 onto the n^2-dimensional subspaces spanned by the coordinates corresponding to the elements in the two subimages. Reasonable choices for the similarity function are as follows

1) $$V_1\cdot V_2$$

2) $$\frac{V_1\cdot V_2}{Y_1\cdot Y_2}$$

3) $$\frac{V_1\cdot V_2}{|V_1|\cdot|V_2|}.$$

The first of these functions suffers from the defect that false maxima may be obtained by positioning the window function over some high-intensity region of the target picture. Normalizing the entire image 2) does not circumvent this difficulty completely. Alternative 3) is thus usually chosen, representing the angle between the window function and the trial segment considered as vectors, in spite of the fact that this procedure requires renormalization of the trial segment for each displacement.

A good account of the problems encountered in an experimental investigation of this problem, viewed as "one of determining the location of matching context points in multiple images and alteration of the geometric relationship of the images such that the registration of context points is enhanced," is given in [3], [4]. Both 14-channel multispectral images obtained one month apart and digitized Apollo-9 photographs were tested. Window functions ranging in size from 4 by 4 to 24 by 24 picture elements and located at the vertices of a grid were used to obtain least squares estimates of a "generalized spatial distance" incorporating the translational, rotational, and scale parameters of the required transformation. The information derived at each vertex was used to center the search space at the next vertex at the most likely values of the parameters. The actual correlation function was the correlation coefficient[2]

$$\phi(k, l) = \frac{\sum_{i=1}^{n}\sum_{j=1}^{n} f(i+k+s, j+l+s)g(i,j)}{\sum_{i=1}^{n}\sum_{j=1}^{n} f^2(i+k, j+l)\sum_{i=1}^{n}\sum_{j=1}^{n} g^2(i,j)}$$

where the image f is of size $m\times m$; the window is of size $n\times n$; $s=(m-n)/2$; and k and l range from $-s$ to s.

The average values were previously subtracted from f and g to yield zero-mean functions; $\phi(k, l)$ is then bounded by -1 and $+1$. In order to circumvent the problem of tone reversals, experiments were also conducted on gradient enhancement techniques designed to extract significant edge information from the images, but this is of less value than might be expected, owing to the noisy nature of the data. The computation was carried out by means of FFT routines, which are shown to yield an average improvement of an order of magnitude in speed over the direct method. The displacements determined by the program are of the order of 2–3 elements in the multispectral data and 10 elements in the photographs; safeguards are included to reject maxima obtained under certain suspect circumstances (for instance on patches of uniform intensity).

[2] The formula is given in this form in [4]; the more customary formulation has a square root in the denominator.

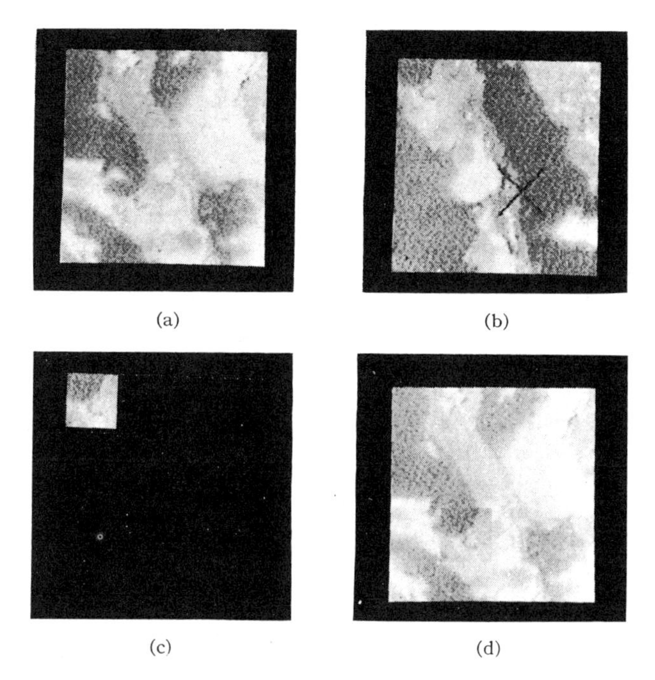

Fig. 6. ITOS-1 registration experiment. A 32 by 32 window function (c) extracted from the top left corner of picture #1 (b) is tried in every possible position of the search area (a) by means of the sequential similarity detection algorithm. The position of the best match determined by the algorithm is shown in (d), where the window function is inserted into the search area. Note the discontinuities due to the change in cloud coverage between the times of exposure of the two pictures.

Similar work has been reported on a pair of black-and-white Gemini photographs of Cape Kennedy, obtained thirty months apart, and on a SO 65 color-separation triplet [10]. On the black-and-white pictures, windows of size 16 by 16 and 32 by 32 elements gave fairly consistent results, but on the SO-65 IR-red, and IR-green pairs, 128 by 128 elements were necessary even with preprocessing by means of an edge-enhancement technique using the coefficient of dispersion or mean-to-variance ratio over 3×3 subregions. These experiments were carried out on a terminal-based system in a time-shared environment.

Impressive savings in processing time can be demonstrated on the basis of the observation that when two windows do not match, the cumulative distance function rises on the average much more rapidly with each pair of picture elements examined than when the windows do match [13]. The class of *sequential similarity detection algorithms*, as the method is called, was analyzed on the assumption that the deviation between two pictures in register is exponentially and independently distributed, and optimal stopping rules were derived under various conditions. The number of operations required was tabulated according to the size of the search space and the size of the window. It is shown, for example, that for a search area of dimensions 2048 by 2048 and a window size of 256 by 256 elements (much larger than in the experiments actually carried out), the direct method would require about 10^{12} operations, the FFT, 10^{10} operations, and the best of the sequential algorithms, 2×10^{7} operations. The test vehicle in this instance consisted of pairs of ITOS AVCS frames obtained two days apart over the Baja California and Gulf Coast regions (Fig. 6).

Another shortcut for multiple template matching, based on prior selection of "rare" configurations, is compared to the conventional systematic search in [147].

F. Cloud Motion

An interesting twist on the registration problem occurs when the camera remains stationary and the scene shifts. This is precisely the case in attempting to determine wind-velocity vectors from changes in the location of cloud masses in the 30-min intervals between successive readouts from the geostationary ATS spin-scan cloud cameras.

An excellent account of the importance of this problem in the context of the Global Atmospheric Research Program, which requires wind-velocity measurements accurate to within 3 knots, is presented in [191]. An interactive computer facility (WINDCO) for tracing cloud motion by means of 32 by 32 element FFT correlation on a Univac 1108 computer is also described. A highlight of this paper is the thorough treatment of the computations necessary to determine the satellite orbit and attitude from least squares approximations of the measured location of landmarks appearing in the pictures. It is shown that the parameters necessary for mapping operations can be determined to an accuracy superior to the resolution of the individual picture elements.

Correlation techniques using the FFT were also applied to 64 by 64 element windows extracted from preprocessed (i.e., projected on standard map coordinates) versions of the large ATS pictures by Leese and his colleagues. They report an approximately twentyfold improvement over direct calculation of the lagged product. The metric selected was the correlation coefficient; the observed peak value was usually about 0.7. The results are claimed to be comparable to or better than visually obtained values when only a single layer of clouds is present or when the wind velocity is constant with altitude. The major source of inaccuracy, in addition to the gradual deformation of the cloud masses, is the occurrence of relative mapping errors between the two pictures [121], [122].

Essentially similar results were obtained with a binary matching technique operating on edge configurations extracted from ATS cloud pictures quantized to only two levels. This method is a factor of two or more faster than the FFT method, but is also unable to cope with multilayer cloud patterns drifting in different directions. The human observer, however, has little difficulty in separating such motion components on the basis of accelerated "loop movies" showing twenty or more successive exposures [121].

Yet another sustained attempt at generating wind-velocity maps from ATS pictures is taking place at the Stanford Research Institute. In this work the bright points (clouds) are aggregated by means of a clustering algorithm, and the ground location is determined through successively finer-grained cross correlation against preselected 20 by 20 element templates. At the last stage a directed-search technique is used, as if the cross-correlation function were monotonic, but the results are checked by means of several independently chosen starting points. The major remaining difficulties are said to be connected with changes in sun angle [73], [53].

G. Parallax Measurements

The determination of parallax from stereo photographs differs from the problems just discussed in that the displacement function may vary quite rapidly from point to point; therefore, the use of window functions based on uniform displacements is inappropriate.

Good results in estimating parallax on a single digitized low-altitude stereo pair were obtained with the assumptions that the difference in the grey levels of corresponding trial pairs of points in the two pictures is a zero-mean Gaussian process and that the departure of the displacement function from its average value over a small neighborhood is also of the same form. Since in this formulation the grey-level difference distribution is conditional upon the parallax, for which the Gaussian continuity condition gives an *a priori* distribution, Bayes' rule may be invoked to estimate the *a posteriori* dependence of the parallax on the observed video distribution.

The computation of the altitude contour lines from the parallax information is a straightforward problem in analytic geometry. Some results of applying the complete procedure to the estimation of forest-tree heights are shown in [9].

It should be noted that there exist automatic and semiautomatic photogrammetric image-correlating devices (*stereoplotters*) which recognize corresponding subsections on a stereopair of photographs and simultaneously transform and print one of the images in an orthographic projection or generate altitude contour lines. Several types of equipment are available, with electronic (scanned) or direct optical image transfer, digital or mechanical implementation of the projective parameters, manual or automatic correlation, and on-line or off-line printout [207], [98]. These sophisticated and expensive precision instruments are not, however, considered to be sufficiently adaptable to cope with the types of distortion expected from the various satellite sensors [136].

V. Classification Techniques

Experimental work in automatic classification through remote sensing may be neatly dichotomized according to whether the primary features consist of spectral or spatial characteristics. Examples of the former are largely based on the output of the Michigan multispectral scanner, while examples of the latter include aerial photographs, high-resolution photographs of the moon, and satellite cloud pictures.

The reason why the expected development of methods based on both spatial *and* spectral characteristics has not yet taken place to any significant degree is that the spatial resolution of the airborne multispectral scanner is too low to allow discrimination of most objects of interest. With multiband aerial photography, on the other hand, automatic registration techniques have not yet come far enough along to permit the preparation of digital color composites in sufficient numbers for significant experimentation.

The typical classification experiment in either domain is open to criticism on several counts. The data, collected in a single region under favorable conditions by airborne or spaceborne sensors, are *examined in their entirety by the experimenter*, who decides which areas are most representative of the region as a whole. The samples from these areas are assembled to form the training set, which is characterized by ground-truth information delineating certain categories of interest. A statistical categorizer or decision box is constructed on the basis of the statistical parameters extracted from the training set.

The classification performance is evaluated on another portion of the data (the test set), also carefully *selected to enhance the probability of correct classification*, where the location and extent of the different types of cover is known to the experimenter. If the error rate is unacceptably high, then offending portions of the test set may be included in the training set for another iteration through the whole cycle!

The details of the various experiments differ with respect to the source of data, the method of labeling the training and test sets, the number and nature of the classes to be identified, the degree of statistical sophistication of the categorizer, and the method of evaluating the results, but the general scheme of classifying samples of the test set according to their similarity to the training set remains the same.

Since it is impossible to review in detail the many dozen pattern-recognition experiments reported in the literature, we shall confine our attention to a few tasks which have been the object of sustained efforts and will examine other contributions in terms of their deviation from the specimen tasks. Rather than offer a description of each classification problem, method of collecting the data, and decision algorithm, we shall concentrate on the experimental procedure, the manner of evaluating the results, and the validity of the derived inferences. For background, the reader cannot do better than turn to the excellent discussion of these matters in relation to conventional photointerpretation in [36], which includes many fine examples drawn from Gemini pictures and coordinated low-altitude oblique aerial photography.

A. Crop Recognition

In the United States, the recognition of crop species by means of multispectral *signatures*, or distinctive spectral characteristics, has been extensively studied since 1966 with a view to providing timely information for fertilization practices, blight control, harvesting schedules, and yield forecasts [61], [90], [124]. There has been much discussion of the importance of systematic coverage throughout the growing season [167], [195], but most of the experiments reported use only data collected in a single day.

Almost all of the earlier work in automatic classification is based on data collected by the Michigan scanner at altitudes ranging from 3000 to 10 000 ft in intensively farmed areas of the Midwest, though a number of missions were also flown in the Imperial Valley agricultural test site of southern California in conjunction with the Apollo-9 experiments. As a rule only the twelve bands spanning the visible and near-IR range of the spectrum are used, because the other channels of the Michigan scanner do not produce data in register with the first twelve channels and cannot be readily transformed into the same spatial coordinate system [25].

Ground truth is generally obtained either through ground surveys or by means of color, IR, and black-and-white photography obtained during the overflights. The performance of skilled photointerpreters using the various types of imagery has been most recently evaluated in [14], with still valid ideas on the most appropriate roles for manual and automatic techniques discussed in [35] and [118].

Most of the classification experiments were performed by investigators associated with the Infrared and Optics Laboratory of the University of Michigan and with the Laboratory for Applications of Remote Sensing of Purdue University. The primary instruments for experimentation were a CDC 3600 at Michigan and IBM 360/44 and 360/67 computers at Purdue. A number of results were also obtained with the Michigan SPARC analog processor.

Extensive programming systems were developed at both installations to allow data editing, data normalization, inspection of statistical attributes of spectral distributions, selection of spectral bands, derivation of categorizer parameters, execution of classification procedures, and evaluation of results [115], [202], [112], [148], [87]. Both organizations take pride in the accessibility and ease of assimilation of their software packages as attested, for instance, in [206] and [208].

The ground swath covered in each flight by the Michigan scanner is normally twice as wide as the flight altitude. i.e.,

$\frac{1}{2}$ to 2 mi, with the flight lines varying in length from 2 to 20 mi. With a few notable exceptions [88], however, not all of the data collected are run through the training and classification algorithms, and even when all of the data are actually classified, the determination of the accuracy of classification over the entire region is usually hampered by the difficulty of entering into the computer the complete ground-truth description.

The specific fields chosen for training and for classification are usually selected for uniform appearance and for nearness to the centerline of the swath since recognition tends to be worse at oblique look-angles [113]. Sometimes the central portions of certain fields are chosen for the training set, with the remainder of the field used for checking the "generalization" capability of the categorizer [63].

The training areas are frequently augmented by including areas where "preliminary" classification runs show poor recognition rates [202], [131]. In view of the variability of widely spaced samples, this procedure usually requires the specification of several subclasses for each category of interest [131], [113]. Automatic mode determination by means of clustering algorithms has also been attempted with fair success [201], [63], [6], [65], [66].

The ratio of the size of the training set to the size of the test set has been steadily decreasing as it is discovered that in multispectral recognition the quantity of samples collected in any given field is much less important than the judicious inclusion of small "representative" regions along, and particularly across, the flight line. Though excellent results have been obtained under favorable conditions over large areas up to 90 mi away from the training fields, even these experiments show that the experimenter must ever be on the alert to change the configuration of training and test fields should an inopportune fluff of cloud intrude on the original experimental design [88].

Much effort has been devoted to the selection of suitable subsets of the spectral channels in order to increase throughput. Among the feature selection methods tested are principal component analysis, divergence, and minimax pairwise linear discriminants. All reports agree that four to six channels perform as well as the full set of ten or twelve (and sometimes better, owing to the use of suboptimal classifiers), but, as expected, the best subset varies from application to application. Linear combinations of channels have also been used, with similar results, but the necessary computation can be justified only for problems involving a large number of categories [116], [117], [140], [64]–[66].

Theoretical attempts to relate the laboratory-observed spectral characteristics of plants to remote observations through the atmosphere are discussed in [51] and [91].

Michigan and Purdue seem to agree that quadratic decision boundaries derived from the maximum likelihood ratio based on Gaussian assumptions constitute the preferred method of classification. Almost all of the experiments discussed in this section made use of this technique, though linear decisions, sequential tree logic, potential functions, and nearest neighbor classification have been shown to yield very similar results, at least on small samples [193], [63], [158], [65], [66]. The quadratic decision rule is to choose the class (or subclass) i which exhibits the largest value

$$g(x) = (x - M_i)^T \tilde{K}_i (x - M_i) + C_i$$

where x is a multispectral measurement vector; M_i is the class mean vector estimated from the training samples; $\tilde{K}_i$ is the class covariance matrix; and C_i is a constant related to the *a priori* probability of class i.

This decision function can be computed rather rapidly (0.3 ms per six-dimensional sample, eight classes, on an IBM 360/44) by making use of table lookup and sequential search procedures [57].

One means of improving the recognition results achieved by such methods is "per-field" classification, where the individual sample vectors are replaced by the average values of the spectral components in each field. Not only does this ease the computational requirements, but it also reduces the contribution to the error rate of weedy patches, irrigation ditches, and other irregularities [6], [14], [65]. Efforts are underway to develop automatic methods for deriving the field boundaries in order to give real meaning to these per-field figures, but while the majority of the boundary segments may be readily obtained by "local operators," linking them up in the desired topological configuration is a difficult matter [4], [213], [214].

Recognition results range from 90 to 98 percent on a single crop (wheat) versus "background" [131], [88], through 91 percent on eight classes in the same flight line, dropping to 65 percent with only four classes across flight lines [63], to 30 percent on really difficult-to-distinguish crops with training and test samples selected from different fields [70].

In comparing the classification performance of a given system to how well one might be expected to do by chance alone, the *a priori* probabilities which are obtained from the ground truth associated with the training set should be taken into account. On the above four-class problem, for instance, almost half the samples were drawn from soybeans, hence one would reach approximately 50-percent correct recognition without any reference at all to the multispectral information.

Crop-recognition experiments have also been conducted on digitized samples of color and color-IR photography. These experiments are based on sample sizes ranging from minuscule to small: 59 samples used for a comparison of several different recognition algorithms [193]; a few hundred samples from carefully chosen densitometric *transects* (single scans) with six classes (including "low-reflectance," "medium-reflectance," and "high-reflectance" water!), and no differentiation between training and test data [175]; 6000 pixels of training data from 60 000-ft RB-57 multiband and multiemulsion photography, divided into four classes (corn, soybeans, pasture, and trees) with 95-percent accuracy [87]; seven fields from a dissector-digitized photograph (81-percent accuracy on dry cotton, 33-percent accuracy on wet cotton) [70]; and 50 000 samples (about 1 percent of one frame) of automatically digitized Apollo SO-65 photography [127], [5], [6]. Recognition rates on the latter material range from 30 to 70 percent, with about 3 percent of the carefully selected specimen areas used for training, and the rest for test. In comparing the results on satellite data to the performance obtained by airborne sensors on similar tasks, it must be observed that the photographs were scanned at a resolution corresponding to about 200 ft on the ground, and no photometric calibration was attempted.

The largest series of experiments to date are the Summer 1971 corn-blight surveys flown by the Michigan (and other) aircraft and divided up for processing between Michigan and Purdue. Strong United States Department of Agriculture (USDA) logistic support for these experiments was presumably based on the promising results obtained in August and September of 1970 [128], [14]. At the time of writing no published information was available on the 1971 experiments, but the results will be eventually released by the Corn Blight Information Center in Washington, D. C.

B. Terrain Classification Based on Spectral Characteristics

The entry of ground-truth information for crop classification, difficult as it is, is a relatively straightforward matter compared to the problems encountered in other types of terrain classification, because the geometry of cultivated fields tends to be quite simple, and it can be safely assumed that each field corresponds to one crop type or to a standard mixture.

In the absence of similar simplifying features in other problem areas there has been no attempt so far to enter ground-truth information into the computer with a view to objective quantitative evaluation of the results. Instead, classification maps are generated which are visually compared to maps or aerial photographs of the area. The results of such comparisons are invariably "promising."

Among applications which have been tackled so far with the help of the Michigan scanner are: the location of areas of potential sink-hole activity in Florida [130], [33]; a beautifully illustrated and ecologically captivating study of hydrobiological features within the Everglades [110]; and soils mapping in Indiana (where difficulties were encountered in attempting to extend the classification to samples located 4 km from the training set) [114], [206]. In addition, there have been innumerable other projects where prints of the unprocessed MSS output, or simple thresholded versions thereof, have so far proved sufficient for the purpose at hand, but where automatic processing may be eventually required as research findings are translated into operational requirements.

Terrain classification has also been attempted by means of digitized aerial photographs. Here also the difficulty of preparing detailed ground-truth maps for the entire area under investigation presents an insurmountable obstacle to quantitative evaluation of the results. In one study, for example, the investigators had to resort to marking isolated patches containing "marked vegetative zonation or relatively homogeneous stands of a dominant species" with 18-in plastic strips which were discernible in the photographs. This particular experiment is also distinguished by conscientious spectral calibration by means of large colored panels displayed on the ground at flight time and subsequently scanned with the remainder of the imagery [52].

For some other applications of interest, and some novel insights, the reader may consult [40], [109].

C. Shape Detection

The major methodological difference between research aimed at automating the production of land usage maps and research concerned with the classification and recognition of objects on the basis of their geometrical properties (including texture) is an inevitable consequence of the relative scarcity of objects of interest. In multispectral crop classification, for instance, every decision results in a positive contribution, but if we are looking for houses, roads, tornadoes, or eagle's nests, then many more pictures will have to be examined to obtain statistically significant results [78].

To gain some idea of the preliminary (and often unreported) manipulation necessary to obtain acceptable results in this area, we shall examine in some detail a rather extensive feasibility study intended to explore alternative designs for a satellite-based recognition system. Various aspects of this study have been reported in outside publications [44], [45], [103], [46], but most of the information presented here is obtained from an internal contract report [49].

The pictures originally selected by the contracting agency, as representative of the material the satellite-borne classifier would encounter, consisted of 311 black-and-white prints containing 198 lunar features, such as craters, rimas, and rilles, and 323 Nimbus cloud samples of various types.

One hundred and 47 of the better prints were selected for digitization on the basis of *visual examination*. In the cloud pictures, fiducial marks and certain "long black lines" were eliminated by means of a water-color retouching kit. The prints were then rotated *by the experimenters* to align the shadow directions to reduce variation due to changes in the solar illumination. The resolution of the slow-scan television camera used to convert the pictures to an intermediate analog tape was set *by the operator* on each pattern in such a way that the maximum size variation was kept below 1.5:1. The resolution was then further reduced by averaging to 50 by 50 picture elements for the lunar patterns, and 75 by 75 elements for the clouds. The dynamic range was also individually adjusted *by the operator* to let each pattern "fill" the 3-bit digital grey scale.

A file of 1000 training samples and 200 test samples was produced for both moonscapes and clouds by a digital editing process consisting of replicating individual patterns by means of translations of up to 15 percent of the effective frame size and rotations of up to 15°. To avoid letting "difficult" patterns predominate, not all patterns were replicated the same number of times. Due care was taken, however, to avoid including replicas of the same pattern in both the training and test set.

Each pattern was characterized by the output of *property filters* consisting either of intuitively designed measurements or of features derived by statistical means. Eight different classification schemes, gleaned from a literature search resulting in 167 titles, were tried on a subset of the data, but none of the eight methods (forced adaptive learning, "error correction," Madaline, another piecewise-linear method, mean-square error criterion, "iterative design," Bayes weights, and direct estimation of the distributions) proved notably superior to the others. Of greatest benefit to accurate classification, it turned out, was the "reduced aperture" technique, whereby insignificant portions of the picture are eliminated prior to the feature-extraction stage! The best combination of features, decision methods, and reduced aperture consistently achieved better than 90-percent correct classification on a half-dozen different ways of grouping the patterns into classes.

Experiments were also conducted on pattern segmentation. The data for these tests consisted of side-by-side montages of cloud patterns which had been correctly classified in isolated form. In this way it was possible to demonstrate that the *input* to the decision units on successive segments generally gives a reliable indication of the location of the boundary between the different cloud types (at least as long as the progression of segments is limited to the direction perpendicular to the boundary)!

Hardware configurations were evaluated for parallel/analog, sequential/hybrid, and sequential/digital methods of implementation. While all three configurations were capable of meeting the desired real-time operation criterion under the particular choice of assumptions, the last design showed a definite advantage in terms of *weight*.

Studies in a similar vein, but with more emphasis on parallel methods of implementation, have been conducted for several years by Hawkins and his colleagues [81]–[83]. The objects of interest here, extracted from aerial photographs, are orchards, oiltank farms, woods, railroad yards, roads, and lakes. The photographs were scanned with a flying-spot scanner under program control (PDP-7), with examples of various local feature-extraction operations, such as matched filters, gradient detection, contrast enhancement, and thresholding demonstrated by means of electrooptical techniques.

Another point of view altogether, but surprisingly similar conclusions regarding the feature-extraction stage, are represented by the M.I.T. pilot study of a semiautonomous Mars Rover. The input to this low-resolution system is provided by a stereo television system under control of a PDP-9, but the output, following a hierarchical multilayer model of the vertebrate visual subsystem, remains in the realm of conjecture [200].

A good description of the application of Golay rotation-invariant hexagonal-neighborhood logic to both local and global feature extraction, including measurements of object diameter, area, perimeter, curvature, and particle counts, is given in [108]. Experimental results are presented on two 500 by 500 arrays extracted from aerial photographs.

Diverse analog and digital methods for implementing texture measurements are considered in [204]. Haralick in a very thorough series of experiments, has classified 54 scenes (each $\frac{1}{8}$ by $\frac{1}{8}$ inch in area) from 1:20000 scale photographs into 9 classes on the basis of the statistical dependence of the grey levels in adjoining picture elements. He compared the results obtained (average 70 percent correct) to those of five trained photointerpreters using the entire 9 by 9 in photograph (81 percent correct) [76].

An early experiment to detect broken and continuous cloud cover in Tiros imagery on the basis of texture is reported in [178]. Local operators containing 5 by 5 picture elements were used for purely textural characteristics, with 15 by 15 element operators for distinction based on size and elongation. A follow-up study introduces additional criteria such as amount of edge per unit area and number of grey-level extrema per unit area [181], [182].

A good review of cloud-cover work, emphasizing the difference between classifying a given segment and producing a cloud map, can be found in [199]. This paper also presents new heuristic algorithms for delineating cloudy regions.

D. Nonsupervised Classification

Nonsupervised learning or *cluster seeking* are names applied to methods of data analysis where *only* the observed values are used explicitly to group samples according to some intrinsic measure of similarity. It is true that closer examination inevitably reveals that some additional information, such as the expected number of groups and some suitable metric, was used by the experimenter to achieve the desired results, but the nature of the ground-truth information associated with the samples enters the algorithm only in a circuitous manner.

In remote-sensing experiments this approach has been used 1) to alleviate the problem of multimodal probability distributions in supervised classification methods, 2) to circumvent the need for *a priori* selection of training samples, 3) to extract the boundaries between homogeneous regions in multispectral arrays, and 4) to condense the amount of information stored or transmitted.

Experiments showing the application of nonparametric clustering techniques to discover the "natural constituents" of the distributions characterizing the various terrain classes are described in [63] and [65] on multispectral scanner observations and in [5], [6], [87] on digitized multiband photographs. We learn, for example, that *mode estimation* yields a good approximation to the mean values of the spectral components in individual fields, or that seven different subclasses of "bare soil" were found in a certain batch of data, but there is not sufficient information presented to determine how much clustering reduces the error rate in comparison to using a single quadratic boundary per class, or alternatively, whether clustering permits the substitution of simpler boundary equations. In each of the papers mentioned the clustering procedure is conducted on an even smaller number of samples than the main supervised classification experiments, because the algorithms used so far tend to be very complicated and time-consuming, frequently involving repeated merging and partitioning of the tentative cluster assignments.

When the clustering technique is used on all of the available observations without any regard to the class descriptions, then it may be considered equivalent to a stratified or two-stage sampling design insuring the efficient collection of the ground-truth information [119], [1]. In other words, the classifier is allowed to designate the appropriate areas for the collection of class-identifying information on the basis of the data itself, rather than on *a priori* considerations. The necessary ground truth may then be collected after the analysis rather than before, reducing the risk that the areas sampled may not be typical of significant portions of the data. One of the few complete experiments testing this idea is described in [193], but the test data unfortunately contained only a few dozen digitized samples. A somewhat larger experiment on nine classes extracted from aerial color photographs by means of a trichromatic microdensitometer is reported in [189]; here the clustering was performed by inspection on two-dimensional projections of the three-dimensional measurement space pending the completion of suitable programmed algorithms. Finally, 2500 samples of high-altitude photography were classified by clustering in [87], but since the study was oriented toward other objectives, these results were not evaluated in detail.

The extraction of boundary information for registration or other purposes is even more difficult on noisy multispectral data than on black-and-white images. A comparison of gradient techniques with clustering is reported in [214], with both methods applied to congruencing data obtained at different times in [4]. A related problem is the conversion of gray-scale imagery to binary arrays by thresholding; here a clustering technique oriented towards the joint occurrence of certain gray-level values in adjacent picture elements has been shown to yield better results than thresholds based on the intensity histogram alone [180].

Both iterative and single-pass clustering algorithms have been thoroughly investigated by Haralick and his colleagues for the purpose of discovering data structures in remote-sensing observations that may reduce storage and transmission requirements for such data. The single-pass algorithms, tried on 80×80 microdensitometric samples of three-band aerial photography, are based on a chaining technique operating in either the image space or in the measurement space [74]. The iterative technique depends on preliminary mapping of the data in a high-dimensional binary space and adjusting the coefficients of prototype vectors to minimize the least square deviation; it proved relatively unsuccessful in reaching an acceptable error rate in compacting 27 000 twelve-dimensional multispectral observations of Yellowstone National Park [75]. An exhaustive experimental investigation of various aspects of the performance of the iterative clustering algorithm on five distributions (three arbitrary, and two tenuously connected with remote observations) is described at length in [46], and a theoretical analysis of the same family of algorithms is presented in [42].

VI. Conclusions

It would appear that few of the results demonstrated to date warrant our sanguine expectations regarding *automatic* processing of the output of the first generation of earth resources survey satellites. Even making allowances for the ever-accelerating march of science, some of the current preparations are uncomfortably reminiscent of the early attempts at automatic translation and "universal" pattern recognition.

The first grounds for scepticism concern the question of instrumentation. Is it really possible to measure consistently the reflectance of 4000 by 4000 distinct picture elements with satellite-borne vidicon cameras and multispectral sensors when meticulously tuned electronic scanning equipment (as opposed to mechanical microdensitometers) in image processing laboratories throughout the country yield only about one quarter as many usable points? The Michigan scanner has an effective transverse resolution of 220 lines (though the optical system is nearly five times as good), the ITOS camera system has barely 800 lines; improvements in this area have not, historically, taken order-of-magnitude jumps. Even if specifications are met, how much can we learn from 200 by 200-ft picture elements?

In regard to the crucial question of automatic registration, very little can be said until the accuracy of the satellite attitude and ephemeris information and the magnitude of the geometric and photometric degradation are firmly established. It should be noted, however, that so far no completely automatic techniques have been developed even for the registration of weather satellite imagery, where far greater errors could be tolerated. Although it is frequently implied that the major obstacle to the implementation of digital registration methods based on correlation is the inordinate amount of computation required, none of the work performed to date indicates that the required accuracy could be attained at *any* cost.

Many of our troubles can, of course, be attributed to the lack of representative raw data. Whatever will be the significant characteristics of the ERTS and Skylab coverage, it is safe to predict that it will bear little resemblance to the digitized photographs and low-altitude multispectral observations which are currently available for experimentation. The weather satellite pictures, which are in many ways most representative, are not nearly detailed enough for most of the suggested earth resources applications. Although there have been attempts at simulating the scale and resolution of the expected data [15], these simulations did not include some of the essential features of the electronic data-collection systems, and were, in any case, largely ignored by the image-processing community.

A related source of frustration is due to our failure to take advantage of the economies of scale resulting from large well-coordinated efforts. The tendency seems to be for each research group to attempt to build its own "complete" image-processing system independently of what may be available elsewhere. Since such an undertaking exceeds the resources of most laboratories, a small part of the overall problem is selected for special attention, with the remainder treated in cursory fashion without regard for previous work on the subject. If, for instance, the selected topic is feature extraction, then the feature extraction algorithms are exercised with minimal preprocessing and normalization of the data, and without any attempt at relating the performance achieved to the manner in which the results are to be utilized. Yet chances are that, as in any large programming effort, the "interface" problems will in the end dwarf the contributions necessary to develop the individual modules.

Leaving aside for the moment the essential but so far tremulous infrastructure in image registration and restoration, we come to the more glamorous aspects of automatic data classification. Without further insistence on the many sins in experimental design (i.e., lack of separate training and test sets, and failure to select the test data *independently* of the training data), and on the widespread and arrant disregard for the statistical rules of inference governing small-sample behavior, we shall take at face value the 60-, 75-, and 90-percent recognition rates achieved on favorable test sites on a half-dozen carefully selected terrain categories or crop species. Studies of the error rates tolerable in various applications would leave little doubt that such performance levels, though perhaps not totally useless, can be reached far more economically by means of other than satellitic surveillance, but seldom in the pattern-recognition literature is there any reference to the minimum acceptable classification rates.

Perhaps an absolute standard is too much to ask. Yet, with only a handful of major test sites under study for automatic classification, are there any published comparisons of relative recognition rates obtained by *different* investigators under similar circumstances? The same photographs and printouts appear in publication after publication, but a sufficient number of conditions are changed in each study to render all comparison meaningless.

Significant progress in this area cannot be expected until sophisticated interactive systems are developed for entering ground-truth information directly in the reference frame of the digitized images. This, in turn, requires high-resolution gray-scale display devices, with attendant software facilities for pinpointing landmarks and outlining boundaries, implementing projective and other transformations, preparing mosaics, comparing the images with digitized maps, serial photographs and other prestored information, and in general, easing the burden of programming the sundry necessary details.

These observations are not meant to imply that digital methods will not play an important role in processing the torrent of data that will soon be released by the survey satellites. It is likely, however, that the computer will for a time continue to be relegated to such relatively unexciting tasks as format conversion, "cosmetic" transformations, merging the image stream and the ephemeris and attitude information, correction of systematic distortions, keeping track of the distribution list and other bookkeeping chores, while interpretation of the images is left to the ultimate user. The prototype operation is likely to be modeled on that of the successful and efficient National Environmental Satellite Center, where over one hundred frames are smoothly processed and assembled daily, with dozens of computer products distributed on a routine basis to the four corners of the globe.

The situation is more encouraging with regard to data collected by relatively low-flying aircraft. Here the scattering effects of the atmosphere are not quite so debilitating, the sensor configuration can be optimized for specific objectives, the flights can, in principle, be timed for the best combination of illumination and terrain conditions, and bore-sighted cameras and navigational subsystems can be used to ensure adequate ground location of the electronic images. The yield of usable pictures should then be sufficiently high to allow fairly sophisticated automatic analysis of the data in selected applications. An example of a plausible if modest candidate for digital processing is the bispectral airborne forest-fire detection system described in [86].

In spite of the many difficulties, the long-range objective must of course remain the eventual combination of spaceborne and airborne sensor systems with conventional ground observations in a grand design for a worldwide hierarchical data-collection system enabling more rational utilization of our planet's plentiful but by no means unlimited natural resources.

We conclude with a plea for more cohesive exploitation of the talent and funds already committed, increased exchange of data sets in a readily usable form, more collaboration and standardization in image-processing software, formulation of realistic goals, and persistence.

References

[1] R. C. Aldrich, "Space photos for land use and forestry," *Photogrammetric. Eng.* (*Special Issue on Remote Sensing*), vol. 37, pp. 389–401, 1971.

[2] H. C. Andrews, *Computer Techniques and Image Processing.* New York: Academic Press, 1970.

[3] P. E. Anuta, "Digital registration of multispectral video imagery," *SPIE J.*, vol. 7, pp. 168–175, 1969.

[4] ——, "Spatial recognition of multispectral and multitemporal digital imagery using fast Fourier transform techniques," *IEEE Trans. Geosci. Electron.*, vol. GE-8, pp. 353–368, Oct. 1970.

[5] P. E. Anuta, S. J. Kristof, D. W. Levandowski, R. B. MacDonald, and T. L. Phillips, "Crop, soil and geological mapping from digitized multispectral satellite photography," in *Proc. 7th Int. Symp. Remote Sens. Environ.* (Ann Arbor, Mich.), pp. 1983–2016, 1971.

[6] P. E. Anuta and R. B. MacDonald, "Crop surveys from multiband satellite photography using digital techniques," *J. Remote Sens. Environ.*, vol. 2, pp. 53–67, 1971.

[7] R. J. Arguello, H. R. Sellner, and J. A. Stuller, "Transfer function compensation of sampled imagery," in *Proc. Symp. Two-Dimensional Image Processing* (Columbia, Mo.), pp. 11-2-1–11-2-12, 1971.

[8] R. H. Asendorf, "The remote reconnaissance of extraterrestrial environments," in *Pictorial Pattern Recognition*, Cheng *et al.*, Eds. Washington, D. C.: Thompson, 1968, pp. 223–238.

[9] R. Bakis and P. G. Langley, "A method for determining the height of forest trees automatically from digitized 70 mm color aerial photographs," in *Proc. 7th Int. Symp. Remote Sens. Environ.* (Ann Arbor, Mich.), pp. 705–714, 1971.

[10] R. Bakis, M. A. Wesley, and P. M. Will, "Digital image processing for the Earth Resources Technology Satellite data," in *Proc. IFIP Congress 71* (Ljubljana, Yugoslavia), 1971.

[11] ——, "Digital correction of geometric and radiometric errors in ERTS data," in *Proc. 7th Int. Symp. Remote Sens. Environ.* (Ann Arbor, Mich.), pp. 1427–1436, 1971.

[12] G. Barna, "Frame type imaging sensors," in *Proc. Princeton Univ. Conf. Aerospace Methods for Revealing Earth's Resources* (Princeton, N. J.), p. 1.1, June 1970.

[13] D. Barnea and H. F. Silverman, "A class of algorithms for fast digital image registration," *IEEE Trans. Comput.*, vol. C-21, pp. 179–186, Feb. 1972; also IBM Res. Rep. RC-3358, 1971.

[14] M. E. Bauer, R. M. Swain, R. I. Mroczynski, P. E. Anuta, and R. B. MacDonald, "Detection of southern corn leaf blight by remote sensing techniques," in *Proc. 7th Int. Symp. Remote Sens. Environ.* (Ann Arbor, Mich.), pp. 693–704, 1971.

[15] D. J. Belcher, E.E. Hardy, and E.S. Phillips, "Land use classification with simulated satellite photography," *USDA Econ. Res. Serv.*, Agriculture Information Bull. 352, 1971.

[16] R. Bernstein, H. Branning, and D. G. Ferneyhough, "Geometric and radiometric correction of high resolution images by digital image processing techniques," presented at the IEEE Annu. Int. Geoscience Electronics Symp., Washington, D. C., 1971.

[17] R. Bernstein and H. F. Silverman, "Digital techniques for Earth resource image data processing," presented at the AIAA 8th Annu. Meeting, Sept. 1971; also IBM-FSD Tech. Rep. FSC 71-6017, Gaithersburg, Md.

[18] F. C. Billingsley, "Digital image processing at JPL," in *SPIE Computerized Imaging Techniques Seminar* (Washington, D. C.), pp. II 1–10, 1967.

[19] F. C. Billingsley, A. F. H. Goetz, and J. N. Lindsley, "Color differentiation by computer image processing," *Photogr. Sci. Eng.*, vol. 14, pp. 28–35, Jan.–Feb. 1970.

[20] F. C. Billingsley, "Photo reconnaissance applications of computer processing of images," in *Proc. Symp. Two-Dimensional Image Processing* (Columbia, Mo.), pp. 7-2-1–7-2-12, 1971.

[21] W. T. Blackband, Ed., "Advanced techniques for aerospace surveillance," Advisory Group for Aerospace Research and Development (AGARD), U. S. Dep. Commerce, NADO ASTIA Doc. AD 73887.

[22] P. Bock, "Space acquired data: Hydrological requirements," in *Proc. Princeton Univ. Conf. Aerospace Methods for Revealing Earth's Resources* (Princeton, N. J.), p. 14.1, June 1970.

[23] A. L. Booth and V. R. Taylor, "Meso-scale archive and products of digitized video data from ESSA satellite," *Bull. Amer. Meteorol. Soc.*, vol. 50, pp. 431–438, 1969; also U. S. Dep. Commerce Tech. Memo. NESCTM-9.

[24] R. E. Bradford and J. F. Gross, "Conditioning of digitized TIROS and ESSA satellite vidicon data," in *SPIE Computerized Imaging Techniques Seminar* (Washington, D. C.), pp. VII 1–8, 1967.

[25] J. Braithwaite, "Airborne multispectral sensing and applications," *SPIE J.*, vol. 8, pp. 139–144, 1970.

[26] C. L. Bristor, W. M. Callicott, and R. E. Bradford, "Operational processing of satellite cloud pictures by computer," *Mon. Weather Rev.*, vol. 94, pp. 515–527, 1966.

[27] C. L. Bristor, "Computer processing of satellite cloud pictures," U. S. Dep. Commerce, Washington, D. C., Tech. Memo. NESCTM-3, 1968.

[28] ——, "The earth location of geostationary satellite imagery," *Pattern Recogn.*, vol. 2, pp. 269–277, 1970.

[29] C. L. Bristor and J. A. Leese, "Operational processing of ITOS scanning radiometer data," presented at IEEE Int. Conf. Telemetry, 1971, NESC Preprint.

[30] E. B. Butt and J. W. Snively, "The Pax II picture processing system," Univ. of Md. Comput. Sci. Cen., Rep. 68-67, 1969.

[31] G. C. Cheng, R. S. Ledley, D. K. Pollock, and A. Rosenfeld, Eds., *Pictorial Pattern Recognition.* Washington, D. C.: Thompson, 1968.

[32] L. J. Cihacek and J. V. Drew, "Infrared photos can map soils," *Farm, Ranch and Home Quart.*, Univ. of Nebr., pp. 4–8, Spring 1970.

[33] A. E. Coker, R. E. Marshall, and N. S. Thomson, "Application of multispectral data processing techniques to the discrimination of sinkhole activity in Florida," in *Proc. 6th Int. Symp. Remote Sens. Environ.* (Ann Arbor, Mich.), pp. 65–78, 1969.

[34] A. P. Colvocoresses, "Image resolution for ERTS, SKYLAB, and GEMINI/APOLLO," *Photogrammetr. Eng.*, vol. 37, 1972.

[35] R. N. Colwell, "The extraction of data from aerial photographs by human and mechanical means," *Photogrammetria*, vol. 20, pp. 211–228, 1965.

[36] ——, "Determining the usefulness of space photography for natural resource inventory," in *Proc. 5th Int. Symp. Remote Sens. Environ.* (Ann Arbor, Mich.), pp. 249–289, 1968.

[37] R. N. Colwell and J. D. Lent, "The inventory of Earth resources on enhanced multiband space photography," in *Proc. 6th Int. Symp. Remote Sens. Environ.* (Ann Arbor, Mich.), pp. 133–143, 1969.

[38] R. N. Colwell, "The future for remote sensing of agricultural, forest, and range resources," presented before the Comm. Science and Astronautics, U. S. House of Representatives, Jan. 26, 1972.

[39] J. J. Cook and J. E. Colwell, "Michigan M7 scanner," Cent. for Remote Sensing Informat. and Analysis, Willow Run Labs., Univ. of Mich., Ann Arbor, Mich., Newsletter, 1971.

[40] M. Cook, "Remote sensing in the Isles of Langerhans," in *Proc. 7th Int. Symp. Remote Sens. Environ.* (Ann Arbor, Mich.), pp. 905–918, 1971.

[41] D. S. Cooley, "Applications of remote sensing and operational weather forecasting," in *Proc. 7th Int. Symp. Remote Sens. Environ.* (Ann Arbor, Mich.), pp. 941–950, 1971.

[42] P. W. Cooper, "Theoretical investigation of unsupervised learning techniques," NASA Contract NAS 12-697, Final Rep., 1970.

[43] R. B. Crane and M. M. Spencer, "Preprocessing techniques to reduce atmospheric and sensor variability in multispectral scanner data," in *Proc. 7th Int. Symp. Remote Sens. Environ.* (Ann Arbor, Mich.), pp. 1345–1356, 1971.

[44] E. M. Darling and R. D. Joseph, "An experimental investigation of video pattern recognition," in *Pictorial Pattern Recognition*, Cheng *et al.*, Eds. Washington, D. C.: Thompson, 1968, pp. 457–469.

[45] ——, "Pattern recognition from satellite altitudes," *IEEE Trans. Syst. Sci. Cybern.*, vol. SSC-4, pp. 38–47, Mar. 1968.

[46] E. M. Darling and J. G. Raudseps, "Non-parametric unsupervised learning with application to image classification," *Pattern Recog.*, vol. 2, pp. 313–336, 1970.

[47] A. G. DeCotiis and E. Conlan, "Cloud information in three spatial dimensions using thermal imagery and vertical temperature profile data," in *Proc. 7th Int. Symp. Remote Sens. Environ.* (Ann Arbor, Mich.), pp. 595–606, 1971.

[48] J. E. Dornbach, "Manned systems for sensing Earth's resources," in *Proc. Princeton Univ. Conf. Aerospace Methods for Revealing Earth's Resources* (Princeton, N. J.), p. 6.1, June 1970.

[49] Douglas Aircraft Co., "Research on the utilization of pattern recognition techniques to identify and classify objects in video data," NASA Contract Rep. CR-999, 1968.

[50] J. A. Dunne, "Mariner 1969 television image processing," *Pattern Recogn.*, vol. 2, pp. 261–268, 1970.
[51] D. L. Earing and I. W. Ginsberg, "A spectral discrimination technique for agricultural applications," in *Proc. 6th Int. Symp. Remote Sens. Environ.* (Ann Arbor, Mich.), pp. 21–32, 1969.
[52] W. G. Egan and M. E. Hair, "Automated delineation of wetlands in photographic remote sensing," in *Proc. 7th Int. Symp. Remote Sens. Environ.* (Ann Arbor, Mich.), pp. 2231–2252, 1971.
[53] R. M. Endlich, D. E. Wolf, D. J. Hall, and A. E. Brain, "Use of pattern recognition techniques for determining cloud motion from sequences of satellite photographs," *J. Appl. Meteorol.*, vol. 10, pp. 105–117, 1971.
[54] W. G. Eppler, "Accuracy of determining sensor boresight position during aircraft flight test," in *Proc. 6th Int. Symp. Remote Sens. Environ.* (Ann Arbor, Mich.), pp. 205–226, 1969.
[55] W. G. Eppler and R. D. Merrill, "Relating remote sensor signals to ground-truth information," *Proc. IEEE*, vol. 57, pp. 665–675, Apr. 1969.
[56] W. G. Eppler, D. L. Loe, E. L. Wilson, S. L. Whitley, and R. J. Sachen, "Interactive display/graphics systems for remote sensor data analysis," in *3rd Annu. Earth Resources Program Review* (NASA-MSC, Houston, Tex.), p. 44-1, Dec. 1970.
[57] W. G. Eppler, C. A. Helmke, and R. H. Evans, "Table look-up approach to pattern recognition," in *Proc. 7th Int. Symp. Remote Sens. Environ.* (Ann Arbor, Mich.), pp. 1415–1426, 1971.
[58] W. G. Eppler, D. L. Loe, E. L. Wilson, S. L. Whitley, and R. J. Sachen, "Interactive system for remote sensor data analysis," in *Proc. 7th Int. Symp. Remote Sens. Environ.* (Ann Arbor, Mich.), pp. 1293–1306, 1971.
[59] W. A. Fischer, "User needs in geology and cartology," in *Proc. Princeton Univ. Conf. Aerospace Methods for Revealing Earth's Resources* (Princeton, N. J.), p. 11.1, June 1970.
[60] J. C. Fletcher, "NASA's long-range earth resources survey program," presented before the Comm. Science and Astronautics, U. S. House of Representatives, Jan. 26, 1972.
[61] H. T. Frey, "Agricultural application of remote sensing—The potential from space platforms," USDA Agricultural Information Bull. N-328, Washington, D. C.
[62] A. W. Frutkin, "Status and prospects of international Earth Resources Satellite programs," in *Proc. Princeton Univ. Conf. Aerospace Methods for Revealing Earth's Resources* (Princeton, N. J.), p. 17.1, June 1970.
[63] K. S. Fu, D. A. Landgrebe, and T. L. Phillips, "Information processing of remotely sensed agricultural data," *Proc. IEEE*, vol. 57, pp. 639–653, Apr. 1969.
[64] K. S. Fu, P. J. Min, and T. J. Li, "Feature selection in pattern recognition," *IEEE Trans. Syst. Sci. Cybern.*, vol. SSC-6, pp. 33–39, Jan. 1970.
[65] K. S. Fu, "On the application of pattern recognition techniques to remote sensing problems," Purdue Univ. School of Elect. Eng. Rep. TR-EE 71-73, 1971.
[66] ——, "Pattern recognition in remote sensing," in *Proc. 9th Allerton Conf. Circuits and Systems Theory* (Univ. of Illinois), Oct. 1971.
[67] T. A. George, "Unmanned spacecraft for surveying Earth's resources," in *Proc. Princeton Univ. Conf. Aerospace Methods for Revealing Earth's Resources* (Princeton, N. J.), p. 7.1, June 1970.
[68] F. A. Godshall, "The analysis of cloud amount for satellite data," in *Trans. N. Y. Acad. Sci.*, Ser. 2, vol. 33, pp. 436–453, 1971.
[69] I. L. Goldberg, "Design considerations for a multispectral scanner for ERTS," in *Proc. Purdue Centennial Year Symp. Information Theory* (Lafayette, Ind.), pp. 672–680, 1969.
[70] V. W. Goldsworthy, E. E. Nelson, and S. S. Viglione, "Automatic classification of agricultural crops from multispectral and multi-camera imagery," presented at Symp. Automatic Photointerpretation and Recognition, Washington, D. C., 1971; also McDonnell Douglas Astronautics Co., Rep. WD.
[71] H. M. Gurk, C. R. Smith, and P. Wood, "Data handling for Earth Resources Satellite data," in *Proc. 6th Int. Symp. Remote Sens. Environ.* (Ann Arbor, Mich.), pp. 247–259, 1969.
[72] H. M. Gurk, "User data processing requirements," in *Proc. Princeton Univ. Conf. Aerospace Methods for Revealing Earth's Resources* (Princeton, N. J.), p. 9.1, June 1970.
[73] D. J. Hall, R. M. Endlich, D. E. Wolf, and A. E. Brain, "Objective methods for registering landmarks and determining cloud motions from satellite data," in *Proc. Symp. Two-Dimensional Image Processing* (Columbia, Mo.), pp. 10-1-1–10-1-7, 1971.
[74] R. M. Haralick and G. L. Kelly, "Pattern recognition with measurement space and spatial clustering for multiple images," *Proc. IEEE*, vol. 57, pp. 654–665, Apr. 1969.
[75] R. M. Haralick and I. Dinstein, "An iterative clustering procedure," presented at EIA Symp. Automatic Photointerpretation and Recognition, Baltimore, Md., Dec. 1970; also in *Proc. 9th IEEE Symp. Adaptive Processes* (Austin, Tex.), Dec. 1970.
[76] R. M. Haralick and D. E. Anderson, "Texture-tone study with application to digitized imagery," presented at Symp. Automatic Photointerpretation and Recognition, Washington, D. C.; also Univ. of Kansas, CRES Tech. Rep. 182 2, 1971.
[77] R. M. Haralick, "Data processing at the University of Kansas," presented at the *4th Ann. Earth Resources Program Review* (NASA-MSC, Houston, Tex.) 1972.
[78] T. J. Harley, L. N. Kanal, and N. C. Randall, "System considerations for automatic imagery screening," in *Pictorial Pattern Recognition*, Cheng *et al.*, Eds. Washington, D. C.: Thompson, 1968, pp. 15–32.
[79] L. D. Harmon and K. C. Knowlton, "Picture processing by computer," *Science*, vol. 164, pp. 19–29, 1969.
[80] J. C. Harris, "Computer processing of atmospherically degraded images," in *SPIE Computerized Imaging Techniques Seminar* (Washington, D. C.), pp. xv 1–5, 1967.
[81] J. K. Hawkins, G. T. Elerding, K. W. Bixby, and P. A. Haworth, "Automatic shape detection for programmed terrain classification," in *Proc. Soc. Photo-Optical Instrument. Eng., Filmed Data and Computers Seminar*, Boston, (Mass.), 1966.
[82] J. K. Hawkins and G. T. Elerding, "Image feature extraction for automatic terrain classification," in *SPIE Computerized Imaging Techniques Seminar* (Washington, D. C.), pp. vi 1–8, 1967.
[83] J. K. Hawkins, "Parallel electro-optical picture processing," in *Pictorial Pattern Recognition*, Cheng *et al.*, Eds. Washington, D. C.: Thompson, 1968, pp. 373–386.
[84] N. M. Herbst and P. M. Will, "Design of an experimental laboratory for signal and image processing," in *Advanced Computer Graphics*, Parslow and Green, Eds. New York: Plenum, 1971.
[85] ——, "An experimental laboratory for pattern recognition and signal processing," *Commun. Ass. Comput. Mach.*, vol. 15, pp. 231–244, 1972.
[86] S. M. Hirsch, F. A. Madden, and R. F. Kruckeberg, "The bispectral forest fire detection system," in *Proc. 7th Int. Symp. Remote Sens. Environ.* (Ann Arbor, Mich.), pp. 2253–2272, 1971.
[87] R. M. Hoffer, P. E. Anuta, and T. L. Phillips, "Application of ADP techniques to multiband and multiemulsion digitized photography," in *Proc. Amer. Soc. Photogrammetry 37th Ann. Meeting* (Washington, D. C.), 1972.
[88] R. M. Hoffer and F. E. Goodrick, "Variables and automatic classification over extended remote sensing test sites," in *Proc. 7th Int. Symp. Remote Sens. Environ.* (Ann Arbor, Mich.), pp. 1967–1982, 1971.
[89] F. Holderman and H. Kazmierczak, "Generation of line drawings from grey-scale pictures," Tech. Rep. Res. Group for Information Processing and Transmission of the Univ. of Karlsruhe, under German Ministry of Defense Contract T 0230/92340/91354.
[90] R. A. Holmes, "An agricultural remote sensing information system," in *EASCON '68 Rec.*, pp. 142–149, 1968.
[91] R. A. Holmes and R. B. MacDonald, "The physical basis of system design for remote sensing in agriculture," *Proc. IEEE*, vol. 57, pp. 629–639, Apr. 1969.
[92] M. R. Holter and W. L. Wolfe, "Optical-mechanical scanning techniques," *Proc. IRE*, vol. 47, pp. 1546–1550, Sept. 1959.
[93] R. Horvath and D. S. Lowe, "Multispectral survey in the Alaskan Arctic," in *Proc. 5th Int. Symp. Remote Sens. Environ.* (Ann Arbor, Mich.), pp. 483–496, 1968.
[94] W. Hovis, "Passive microwave sensors," in *Proc. Princeton Univ. Conf. Aerospace Methods for Revealing Earth's Resources* (Princeton, N. J.), p. 3.1, June 1970.
[95] T. S. Huang and H. L. Kasnitz, "The combined use of digital computer and coherent optics in image processing," in *Computerized Imaging Techniques Seminar* (Washington, D. C.), pp. xvii 1–6, 1967.
[96] T. S. Huang, W. F. Schreiber, and O. J. Tretiak, "Image processing," *Proc. IEEE*, vol. 59, pp. 1586–1609, Nov. 1971.
[97] H. F. Huddlestone and E. H. Roberts, "Use of remote sensing for livestock inventories," in *Proc. 5th Int. Symp. Remota Sens. Environ.* (Ann Arbor, Mich.), pp. 307–323, 1968.
[98] T. A. Hughes, A. R. Shope, and F. S. Baxter, "The USGS automatic orthophoto system," in *Proc. Amer. Soc. Photogrammetry 37th Ann. Meeting* (Washington, D. C.), pp. 792–817, 1971.
[99] B. R. Hunt, "Computational considerations in digital image enhancement," in *Proc. Symp. Two-Dimensional Image Processing* (Columbia, Mo.), pp. 1-4-1–1-4-7, 1971.
[100] J. M. Idelsohn, "A learning system for terrain recognition," *Pattern Recogn.*, vol. 2, pp. 293–302, 1970.
[101] E. G. Johnston, "The PAX II picture processing system," in *Picture Processing and Psychopictorics*, B. S. Lipkin and A. Rosenfeld, Ed. New York: Academic Press, 1970, pp. 427–512.
[102] E. G. Johnston, M. N. Fritz, and A. Rosenfeld, "A study of new image processing techniques," Univ. of Maryland Comput. Sci. Cent. Rep. AFCRL-71-0434, 1971.
[103] R. D. Joseph, R. G. Burge, and S. S. Viglione, "Design of a satellite-borne pattern classifier," in *Proc. Purdue Centennial Year Symp. on Information Theory* (Lafayette, Ind.), 1969, pp. 681–700.
[104] ——, "A compact programmable pattern recognition system," presented at the Symp. Application of Reconnaissance Technology to Monitoring and Planning Environmental Change, Rome, N. Y., 1970; also McDonnell Douglas Astronautics Co. Rep. WD 1409.
[105] R. D. Joseph and S. S. Viglione, "Interactive imagery analysis," in *Proc. Symp. on Two-Dimensional Image Processing* (Columbia, Mo.), pp. 7-1-1–7-1-6, 1971; also McDonnell Douglas Astro-

nautics Co. Paper WD 1754, Aug. 1971.
[106] H. Kazmierczak and F. Holderman, "The Karlsruhe system for automatic photointerpretation," in *Pictorial Pattern Recognition*, Cheng *et al.*, Eds. Washington, D. C.: Thompson, 1968, pp. 45–62.
[107] H. Kazmierczak, "Informationsreduktion und Invarianz-Leistung bei der Bildverarbeitung mit einen Computer," in *Pattern Recognition in Biological and Technical Systems*. New York: Springer, 1971.
[108] S. J. Kishner, "Feature extraction using Golay logic," in *Proc. Perkin-Elmer Symp. on Sampled Images* (Norwalk, Conn.), pp. 7-1-7-8, 1971.
[109] M. C. Kolipinski and A. L. Higer, "On the detection of pesticidal fallout by remote sensing techniques," in *Proc. 6th Int. Symp. Remote Sens. Environ.* (Ann Arbor, Mich.), p. 989, 1969.
[110] M. C. Kolipinski, A. L. Higer, N. S. Thomson, and F. J. Thomson, "Automatic processing of multiband data for inventorying hydrobiological features," in *Proc. 6th Int. Symp. Remote Sens. Environ.* (Ann Arbor, Mich.), pp. 79–96, 1969.
[111] C. L. Korb, "Physical considerations in the channel selection and design specification of an airborne multispectral scanner," in *Proc. Purdue Centennial Year Symp. Information Theory* (Lafayette, Ind.), pp. 646–657, 1969.
[112] F. J. Kriegler, W. Malika, R. F. Nalepka, and W. Richardson, "Preprocessing transformations and their effects on multispectral recognition," in *Proc. 6th Int. Symp. Remote Sens. Environ.* (Ann Arbor, Mich.), pp. 97–132, 1969.
[113] F. J. Kriegler, "Implicit determination of multispectral scanner variation over extended areas," in *Proc. 7th Int. Symp. Remote Sens. Environ.* (Ann Arbor, Mich.), pp. 759–778, 1971.
[114] S. J. Kristof, A. L. Zachary, and J. E. Cipra, "Mapping soil types from multispectral scanner data," in *Proc. 7th Int. Symp. Remote Sens. Environ.* (Ann Arbor, Mich.), pp. 2095–2108, 1971.
[115] D. A. Landgrebe and T. L. Phillips, "A multichannel image data handling system for agricultural remote sensing," in *SPIE Computerized Imaging Techniques Seminar* (Washington, D. C.), pp. xii 1–10, 1967.
[116] D. A. Landgrebe, P. J. Min, P. H. Swain, and K. S. Fu, "The application of pattern recognition techniques to a remote sensing problem," Purdue Univ., Lafayette, Ind., 1968, LARS Information Note 080568.
[117] D. A. Landgrebe, "The development of machine technology processing for Earth resource survey," in *3rd Ann. Earth Resources Program Review* (NASA-MSC, Houston, Tex.), p. 40-1, Dec. 1970.
[118] P. G. Langley, "Automating aerial photointerpretation in forestry —How it works and what it will do for you," in *Proc. Ann. Meet. Soc. Amer. Forestry* (Detroit, Mich.), 1965.
[119] ——, "New multistage sampling techniques using space and aircraft imagery for forest inventory," in *Proc. 6th Int. Symp. Remote Sens. Environ.* (Ann Arbor, Mich.), pp. 1179–1192, 1969.
[120] J. A. Leese, A. L. Booth, and F. A. Godshall, "Archiving and climatological application of meteorological satellite data," U. S. Dep. Commerce, Environ. Sci. Serv. Admin., Rockville, Md., Rep. to the World Meteorol. Org. Comm. for Climatol.; also ESSA Tech. Rep. NESC 53, July 1970.
[121] J. A. Leese, C. S. Novak, and V. R. Taylor, "The determination of cloud pattern motion from geosynchronous satellite image data," *Pattern Recogn.*, vol. 2, pp. 279–292, 1970.
[122] J. A. Leese, C. S. Novak, and B. B. Clark, "An automated technique for obtaining cloud motion for geosynchronous satellite data using cross-correlation," *J. Appl. Meteorol.*, vol. 10, pp. 110–132, 1971.
[123] J. A. Leese, W. Pickel, B. Goddard, and R. Bower, "An experimental model for automated detection, measurement and quality control of sea-surface temperature from ITOS-IR data," in *Proc. 7th Int. Symp. Remote Sens. Environ.* (Ann Arbor, Mich.), pp. 625–646, 1971.
[124] R. R. Legault, "Summary of Michigan Multispectral Investigations program," in *3rd Ann. Earth Resources Program Review* (NASA-MSC, Houston, Tex.) p. 37-1, Dec. 1970.
[125] S. H. Lerman, "Digital image processing at ITEK," in *Proc. Symp. on Two-Dimensional Image Processing* (Columbia, Mo.), pp. 5-5-1–5-5-4, 1971.
[126] D. S. Lowe, "Line scan devices and why use them," in *Proc. 5th Int. Symp. Remote Sens. Environ.* (Ann Arbor, Mich.), pp. 77–100, 1968.
[127] R. B. MacDonald, "The application of remote sensing to corn blight detection and crop yield forecasting," in *3rd Ann. Earth Resources Program Review* (NASA-MSC, Houston, Tex.), p. 28-1, Dec. 1970.
[128] ——, "The application of automatic recognition techniques in the Apollo IX SO-65 experiment," in *3rd Ann. Earth Resources Program Review* (NASA-MSC, Houston, Tex.), p. 39-1, Dec. 1970.
[129] H. Markarian, R. Bernstein, D. G. Ferneyhough, C. E. Gregg, and F. S. Sharp, "Implementation of digital techniques for correcting high resolution images," presented at the ASP-ACSM Fall Convention, Sept. 1971; also IBM-FSD, Gaithersburg, Md., Tech. Rep. FSC 71-6012.
[130] R. E. Marshall, "Application of multispectral recognition techniques for water resources investigations in Florida," in *Proc' Purdue Centennial Year Symp. Information Theory* (Lafayette Ind.), pp. 719–731, 1969.
[131] R. E. Marshall, N. S. Thomson, and F. J. Kriegler, "Use of multispectral recognition techniques for conducting wide area wheat surveys," in *Proc. 6th Int. Symp. Remote Sens. Environ.* (Ann Arbor, Mich.), pp. 3–20, 1969.
[132] R. E. Marshall and F. J. Kriegler, "Multispectral recognition system," in *Proc. 9th IEEE Symp. Adaptive Processes, Decision, and Control* (Austin, Tex.), p. xi-2, 1970.
[133] ——, "An operational multispectral survey system," in *Proc. 7th Int. Symp. Remote Sens. Environ.* (Ann Arbor, Mich.), pp. 2169–2192, 1971.
[134] H. M. Maynard, L. F. Pemberton, and C. W. Swonger, "Application of image processing research to the retrieval of filmed data," in *Proc. Soc. Photo-Optical Instrument. Eng., Filmed Data and Computers Seminar* (Boston, Mass.), pp. xvii 1–10, 1966.
[135] E. P. McClain, "Potential use of Earth satellites for solving problems in oceanography and hydrology," presented at the Nat. Meeting of the Amer. Astronaut. Soc., Las Cruces, N. Mex., 1970.
——, "Application of environmental satellite data to oceanography and hydrology," U. S. Dep. Commerce, Washington, D. C., Tech. Memo. NES CTM 19.
[136] R. B. McEwen, "An evaluation of analog techniques for image registration," U. S. Geol. Survey Prof. Paper 700-D, pp. D305–D311, 1970.
[137] ——, "Geometric calibration of the RBV system for ERTS," in *Proc. 7th Int. Symp. Remote Sens. Environ.* (Ann Arbor, Mich.), pp. 791–808, 1971.
[138] E. P. McMahon, "Earth Resources Technology Satellite system data processing," in *Proc. EASCON 70*, pp. 24–27, 1970.
[139] T. M. Mendel and K. S. Fu., Eds., *Adaptive, Learning and Pattern Recognition Systems*. New York: Academic Press, 1971.
[140] P. J. Min, D. A. Landgrebe, and K. S. Fu, "On feature selection in multiclass pattern recognition," in *Proc. 2nd Ann. Princeton Conf. Information Sciences and Systems*, pp. 453–457, 1968.
[141] R. T. Moore, M. C. Stark, and L. Cahn, "Digitizing pictorial information with a precision optical scanner," *Photogrammetr. Eng.*, vol. 30, pp. 923–931, 1964.
[142] M. Mozer and P. Serge, "High resolution multispectral TV camera systems," in *Proc. 7th Int. Symp. Remote Sens. Environ.* (Ann Arbor, Mich.), pp. 1475–1482, 1971.
[143] J. L. Muerle and D. C. Allen, "Experimental evaluation of techniques of automatic segmentation of objects in a complex scene," in *Pictorial Pattern Recognition*, Cheng *et al.*, Eds. Washington, D. C.: Thompson, 1968, pp. 3–14.
[144] J. L. Muerle, "Some thoughts on texture discrimination by computer," in *Picture Processing and Psychopictorics*. New York: Academic Press, 1970, pp. 371–379.
[145] A. H. Muir, "Some shifting relationships between the user community and the Earth Resources Program," in *Proc. 7th Space Cong.* (Cocoa Beach, Fla.), pp. 2-11–2-14, 1970.
[146] L. E. Mumbower, "Ground data processing considerations for Earth resources information," in *Proc. Princeton Univ. Conf. Aerospace Methods for Revealing Earth's Resources* (Princeton, N. J.), p. 8.1, June 1970.
[147] R. Nagel and A. Rosenfeld, "Ordered search techniques in template matching," *Proc. IEEE* (Lett.), vol. 60, p. 242, Feb. 1972; also Univ. of Maryland Comput. Sci. Cent. Rep. TR-166.
[148] R. F. Nalepka, "Investigation of multispectral discrimination techniques," Infrared and Optics Lab., Inst. Sci. and Technol. of the Univ. of Michigan, Ann Arbor, Mich., Rep. 2264-12-F, 1970.
[149] R. Narasimhan, "Labeling schemata and syntactic description of pictures," *Inform. Contr.*, vol. 7, pp. 151–179, 1964.
[150] NASA-MSFC, "Skylab in-flight experiments," Skylab Program Summary Description, June 1971.
[151] NASA, "Earth photographs from Gemini VI through XII," Washington, D. C., NASA SP-171, 1969.
[152] NASA, "Apollo-9 photographic plotting and indexing report," NASA-MSC, Mapping Sci. Lab. for Earth Resour. Div., Sci. and Appl. Directorate, Houston, Tex., 1969.
[153] NASA, NASA-MSC Earth Resour. Aircraft Program, Houston, Tex., Mission 89 Summary Rep., 1969.
[154] NASA, "Remote sensing of Earth resources: A literature survey with indexes," NASA Publ. SP-7036, 1970.
[155] NASA, "Earth resources research facility index, vol. I—Documentary data, vol. II—Sensor data," Houston, Tex., NASA-MSC-02576, 1971.
[156] R. Nathan, "Picture enhancement for the Moon, Mars and Man," in *Pictorial Pattern Recognition*, Cheng *et al.*, Eds. Washington, D. C.: Thompson, 1968, pp. 239–266.
[157] ——, "Spatial frequency filtering," in *Picture Processing and Psychopictorics*, B. S. Lipkin and A. Rosenfeld, Eds. New York: Academic Press, 1970, pp. 151–163.
[158] R. Newman and B. Reisine, "Practical applications of sequential pattern recognition techniques," in *Proc. 7th Space Cong.* (Cocoa Beach, Fla.), pp. 2.1–2.9, 1970.

[159] V. T. Norwood, "Optimization of a multispectral scanner for ERTS," in *Proc. 6th Int. Symp. Remote Sens. Environ.* (Ann Arbor, Mich.), pp. 227–236, 1969.

[160] ——, "Scanning type imaging sensors," in *Proc. Princeton Univ. Conf. Aerospace Methods for Revealing Earth's Resources* (Princeton, N. J.), p. 2.1, June 1970.

[161] National Research Council (Comm. on Remote Sensing for Agricultural Purposes), *Remote Sensing with Special Reference to Agriculture and Forestry*. Washington, D. C.: Natl. Acad. Sci., 1970.

[162] D. O'Handley and W. B. Green, "Recent developments in digital image processing at the Image Processing Laboratory at the Jet Propulsion Laboratory," *Proc. IEEE*, vol. 60, pp. 821–828, July 1972.

[163] K. W. Olson, "An operational pattern recognition system," in *Pictorial Pattern Recognition*, Cheng *et al.*, Eds. Washington, D. C.: Thompson, 1968, pp. 63–73.

[164] R. F. Owens, R. W. VanDuinen, and L. D. Sinnamon, "The Socrates adaptive image processor," in *SPIE Computerized Imaging Techniques Seminar* (Washington, D. C.), pp. v 1–5, 1967.

[165] J. J. O. Palgen, "Applicability of pattern recognition techniques to the analysis of urban quality from satellites," *Pattern Recogn.*, vol. 2, pp. 255–260, 1970.

[166] ——, "Application of optical processing for improving ERTS data," Allied Research Associates Tech. Rep. 16, NASA Contract NAS-10343, 1970.

[167] A. B. Park, "Remote sensing of time dependent phenomena," in *Proc. 6th Int. Symp. Remote Sens. Environ.* (Ann Arbor, Mich.), pp. 1227–1236, 1969.

[168] ——, "The role of man in an observatory/laboratory spacecraft," in *Proc. EASCON 70*, pp. 21–23, 1970.

[169] ——, "User needs in agriculture and forestry," *Proc. Princeton Univ. Conf. Aerospace Methods for Revealing Earth's Resources* (Princeton, N. J.), p. 13.1, June 1970.

[170] J. L. Pfaltz, J. W. Snively, and A. Rosenfeld, "Local and global picture processing by computer," in *Pictorial Pattern Recognition*, Cheng *et al.*, Eds. Washington, D. C.: Thompson, 1968, pp. 353–372.

[171] A. Pilipchuk, "Mechanical scanner for off-line picture digitization," Comput. Sci. Cent., Univ. of Maryland Rep. 69-92, 1969.

[172] J. F. Potter, "Scattering and absorption in the Earth's atmosphere," in *Proc. 6th Int. Symp. Remote Sens. Environ.* (Ann Arbor, Mich.), pp. 415–430, 1969.

[173] W. K. Pratt, "Semi-annual technical report of the USC Image Processing Laboratory," Univ. of South. California Rep. 411, Mar. 1972.

[174] P. K. Rao and A. E. Strong, "Sea surface temperature mapping off the eastern United States using NOAA's ITOS satellite," in *Proc. 7th Int. Symp. Remote Sens. Environ.* (Ann Arbor, Mich.), pp. 683–692, 1971.

[175] A. J. Richardson, R. J. Torline, and W. A. Allen, "Computer identification of ground patterns from aerial photographs," in *Proc. 7th Int. Symp. Remote Sens. Environ.* (Ann Arbor, Mich.), pp. 1357–1376, 1971.

[176] T. C. Rindfleisch, "Getting more out of Ranger pictures by computer," *Astronaut. Aeronaut.*, vol. 7, 1969.

[177] T. C. Rindfleisch, J. A. Dunne, H. F. Frieden, and W. D. Stromberg, "Digital processing of the Mariner 6 and 7 pictures," *J. Geophys. Res.*, vol. 76, pp. 394–417, 1971.

[178] A. Rosenfeld, C. Fried, and J. Ortan, "Automatic cloud interpretation," *Photogrammet. Eng.*, vol. 31, p. 991, 1965.

[179] A. Rosenfeld, *Picture Processing by Computer*. New York: Academic Press, 1969.

[180] A. Rosenfeld, H. Huang, and V. B. Schneider, "An application of cluster detection to text and picture processing," *IEEE Trans. Inform. Theory*, vol. IT-15, pp. 672–681, Nov. 1969; also Univ. of Maryland Comput. Sci. Cent. Rep. TR-68-68.

[181] A. Rosenfeld and E. B. Troy, "Visual texture analysis," Comput. Sci. Cent., Univ. of Maryland, Tech. Rep. TR70-116.

[182] A. Rosenfeld and M. Thurston, "Edge and curve enhancement for visual scene analysis," *IEEE Trans. Comput.*, vol. C-20, pp. 562–569, May 1971.

[183] L. Rossol, "Generation of halftones by computer-controlled microfilm recorder," *IEEE Trans. Comput.* (Short Note), vol. C-20, pp. 662–664, June 1971.

[184] H. R. Sellner, "Transfer function compensation of sampled imagery," in *Proc. Perkin-Elmer Symp. on Sampled Images* (Norwalk, Conn.), pp. 4-1–4-14, 1971.

[185] J. R. Shay, "Systems for Earth resources information," in *Proc. Purdue Centennial Year Symp. Information Theory* (Lafayette, Ind.), pp. 635–643, 1969.

[186] H. F. Silverman, "On the use of transforms for satellite image processing," in *Proc. 7th Int. Symp. Remote Sens. Environ.* (Ann Arbor, Mich.), 1971, pp. 1377–1386.

[187] D. S. Simonett, "Potential of radar remote sensors as tools in reconnaissance geomorphic, vegetation and soil mapping," in *9th Int. Cong. Soil Sci. Trans.*, vol. IV, Paper 20, pp. 271–280, 1968; also Univ. of Kansas, CRES Rep. 117-2.

[188] H. W. Smedes, M. M. Spencer, and F. J. Thomson, "Processing of multispectral data and simulation of ERTS data channels to make computer terrain maps of a Yellowstone National Park test site," in *3rd Annu. Earth Resources Program Review* (NASA-MSC Houston, Tex.), p. 10-1, Dec. 1970.

[189] H. W. Smedes, H. J. Linnerud, S. G. Hawks, and L. B. Woolamer, "Digital computer mapping of terrain by clustering techniques using color film on a three-band sensor," in *Proc. 7th Int. Symp. Remote Sens. Environ.* (Ann Arbor, Mich.), pp. 2057–2072, 1971.

[190] C. R. Smith and T. F. Shafman, "Cloud cover limitations on satellite photography of the United States," in *Trans. 49th Annu. Meeting of the Amer. Geophys. Union* (Washington, D. C.), p. 188, 1968.

[191] E. A. Smith and D. R. Phillips, "Automated cloud tracking using precisely aligned digital ATS pictures," in *Proc. Symp. Two-Dimensional Image Processing* (Columbia, Mo.), pp. 10-2-1–10-2-26, 1971.

[192] G. L. Stanely, W. C. Nienow, and G. G. Lendaris, "SARF—An interactive signature analysis research facility," in *Proc. Purdue Centennial Year Symp. Information Theory* (Lafayette, Ind.), pp. 436–445, 1969.

[193] D. Steiner, "A methodology for the automated photo-identification of rural land use types," in *Automatic Interpretation and Classification of Images*, A. E. Grasselli, Ed. New York: Academic Press, 1969, p. 235.

[194] D. Steiner, K. Baumberger, and H. Maurer, "Computer processing and classification of multi-variate information from remote sensing imagery," in *Proc. 6th Int. Symp. Remote Sens. Environ.* (Ann Arbor, Mich.), pp. 895–908, 1969.

[195] D. Steiner, "Time dimension for crop surveys from space," *Photogrammetr. Eng.*, vol. 36, pp. 187–194, 1970.

[196] R. E. Stevenson, "Oceanographic data requirements for the development of an operational satellite system," in *Proc. Princeton Univ. Conf. Aerospace Methods for Revealing Earth's Resources* (Princeton, N. J.), p. 12.1, June 1970.

[197] H. G. Stever, Keynote Address to the 13th Meeting of the Panel on Sci. and Technol., Comm. Sci. and Astronaut., U. S. House of Representatives, Jan. 25, 1972.

[198] W. M. Strome, "Canadian data-handling facility for ERTS," in *Proc. Canadian Computer Conf.*, pp. 423101–423118, 1972.

[199] J. P. Strong, "Automatic cloud cover mapping," Univ. of Maryland Comput. Sci. Cent. Rep. TR-163.

[200] L. L. Sutro and W. S. McCulloch, "Steps toward the automatic recognition of unknown object," in *Proc. Inst. Elec. Eng. NPL Conf. Pattern Recognition* (Teddington, U. K.), pp. 117–133, 1968.

[201] P. H. Swain and K. S. Fu, "On the application of a nonparametric technique to crop classification problems," in *Proc. Nat. Electronics Conf.*, 1968.

[202] P. H. Swain and D. A. Germann, "On the application of man-machine computing systems to problems in remote sensing," in *Proc. 11th Midwest Symp. Circuit Theory* (Notre Dame, Ind.), May, 1968; also *Software Age*, vol. 2, pp. 13–20, 1968.

[203] P. H. Swain, "Data processing I: Advancements in machine analysis of multispectral data," Purdue Univ. LARS Informat. Note 012472, to be included in the *4th Annu. Earth Resources Program Review* (NASA-MSC, Houston, Tex.), 1972.

[204] G. D. Swanlund, "Design requirements for texture measurements," in *Proc. Symp. Two-Dimensional Image Processing* (Columbia, Mo.), pp. 8-3-12–8-3-19, 1971.

[205] W. Swoboda and J. W. Gerdes, "A system for demonstrating the effects of changing background on automatic target recognition," in *Pictorial Pattern Recognition*, Cheng *et al.*, Eds. Washington, D. C.: Thompson, 1968, pp. 33–42.

[206] M. G. Tanguay, R. M. Hoffer, and R. D. Miles, "Multispectral imagery and automatic classificaton of spectral response for detailed engineering soils mapping," in *Proc. 6th Int. Symp. Remote Sens. Environ.* (Ann Arbor, Mich.), pp. 33–64, 1969.

[207] M. M. Thompson, Ed., *Manual of Photogrammetry*. Falls Church, Va.: Amer. Soc. of Photogrammetry, 1966.

[208] F. J. Thomson, "User oriented data processing at the University of Michigan," in *3rd Annu. Earth Resources Program Review* (NASA-MSC, Houston, Tex.), p. 38-1, Dec. 1970.

[209] W. R. Tobler, "Problems and prospects in geographical photointerpretation," in *Pictorial Pattern Recognition*, Cheng *et al.*, Eds. Washington, D. C.: Thompson, 1968, pp. 267–274.

[210] D. R. Tompson, "An IBM special cartographic scanner," in *Proc. ASP-ACSM Conv.* (Washington, D. C.), 1967.

[211] M. R. Uffelman, "Target detection, prenormalization, and learning machines," in *Pictorial Pattern Recognition*, Cheng *et al.*, Eds. Washington, D. C.: Thompson, 1968, pp. 503–521.

[212] R. M. Vesper and D. N. Rosenzweig, "Sampled image laboratories and facilities," in *Proc. Perkin-Elmer Symp. Sampled Images* (Norwalk, Conn.), pp. 10-1–10-8, 1971.

[213] A. G. Wacker, "A cluster approach to finding spatial boundaries in multispectral imagery," LARS Informat. Note 122969, 1969.

[214] A. G. Wacker and D. A. Landgrebe, "Boundaries in multispectral images by clustering," in *Proc. 1970 9th IEEE Symp. Adaptive Processes* (Austin, Tex.), pp. 322–330, 1970.

[215] R. N. Watts, "Skylab scheduled for 1972," *Sky Telesc.*, pp. 146–148, 1970.
[216] G. R. Welti and S. H. Durrain, "Communication system configuration for the Earth Resources Satellite," in *Communication Satellites for the 70's Systems*, N. E. Feldman and C. M. Kelly, Eds. (vol. 26 of *Progress in Astronautics and Aeronautonics*). Cambridge, Mass.: M.I.T. Press, 1971, pp. 289–314.
[217] P. M. Will, R. Bakis, and M. A. Wesley, "On an all-digital approach to Earth Resources Satellite image processing," Yorktown Heights, N. Y.,IBM Tech. Rep. RC-3027, 1970.
[218] C. L. Wilson, M. M. Beilock, and E. M. Zaitzeff, "Design considerations for aerospace multispectral scanning systems," in *Proc. Purdue Centennial Year Symp. on Information Theory* (Lafayette, Ind.), pp. 658–671, 1969.
[219] F. J. Wobber, "Space age prospecting," *World Mining*, p. 25, June 1968.
[220] P. Wood, "User requirements for Earth Resources Satellite data," in *Proc. EASCON 70*, pp. 14–20, 1970.
[221] E. M. Zaitzeff, C. L. Wilson, and D. H. Ebert, "MSDS: An experimental 24-channel multispectral scanner system," *Bendix Tech. J.*, pp. 20–32, 1970.
[222] B. Zavos, "Application of remote sensing to the World Weather Watch and the Global Atmospheric Research Program," in *Proc. 7th Int. Symp. Remote Sens. Environ.* (Ann Arbor, Mich.), pp. 207–220, 1971.
[223] E. O. Zeitler and R. O. Bratton, "Uses of Earth Resources data in the NASA/MSC Earth Resources research data facility," in *Proc. 7th Int. Symp. Remote Sens. Environ.* (Ann Arbor, Mich.), pp. 925–936, 1971.
[224] J. S. Zelenka, "Imaging radar techniques in remote sensing applications," in *Proc. 9th Annu. Allerton Conf. Circuit and System Theory*, pp. 975–986, 1971.

Digital Image Processing for Information Extraction

FRED C. BILLINGSLEY

Jet Propulsion Laboratory,
Pasadena, California, U.S.A.

(Received 31 May 1972)

In recent years the modern digital computer has been used to process images, to emphasize details, to sharpen pictures, to modify the tonal range, to aid picture interpretation, to remove anomalies, and to extract quantitative information. A price to be paid for this extreme flexibility in handling linear and non-linear operations is that a number of anomalies caused by the camera, such as geometric distortion, MTF roll-off, vignetting, and non-uniform intensity response must be taken into account or removed to avoid their interference with the information extraction process. Once this is done, computer techniques may be used to emphasize details, perform analyses, classify materials by multi-variate analysis (usually multi-spectral), detect temporal differences, etc. Digital processing may also be used to modify various aspects of pictures to enhance the ability of the human photo interpreter in extracting information. A number of these processes are illustrated in this paper.

Introduction

The traditional method for extracting information from pictures has been by eye. The recording of the information extracted must of necessity be by manual processes and is thus time consuming and subject to the many vagaries of human operations. Incoherent optical processing has been used to enhance the presentation to the analyst by means of such darkroom techniques as photographic contrast manipulation, dodging, photogrammetric rectification, multi-film sandwiching operations and the like. With the advent of the laser, coherent optical processing has been utilized to give access to the Fourier transform plane for such operations as spatial frequency filtering and correlation.

The modern digital computer has made practical processing techniques for handling non-linear operations in both the geometrical and intensity domains, various types of non-uniform noise clean-up, and the numerical analysis of pictures. The types of processing operations which are possible

Reprinted with permission from *Int. J. Machine Perception of Patterns and Pictures*, Inst. Physics, pp. 337–362, 1972.

are limited only by the ingenuity of the analyst: affine or non-affine geometrical transformation, application of camera calibration data, intensity shading and contrast manipulations, one- or two-dimensional spatial filtering, coherent or non-coherent noise removal, numerical data extraction, multi-picture comparisons, etc.

Our primary concern will be with the digital processing, using as illustrations pictures produced by the Image Processing Laboratory of the Jet Propulsion Laboratory. The techniques for image processing have been developed primarily for processing pictures as returned from the NASA space vehicles (for example, Ranger, Surveyor and Mariner) to the Jet Propulsion Laboratory. However, these same techniques and to a very large extent even the same specific programs have been utilized to process pictures from many other sources, such as electron and light microscopes, industrial and medical radiographs, Apollo pictures of the Earth and Moon, and various aerial photographs. This paper will review the digital image processing developed at JPL, and will discuss briefly the multi-spectral processing developed at LARS, Purdue University.

It is convenient to consider the image processing techniques and processes to be grouped into three general areas, depending upon the use to which the processed image is to be put:

Enhancement—The modification of subjective features of a picture to alter the impact of that picture on the viewer. In general, the enhancement processing will not produce a picture with numerically accurate data.

Quantitative Restoration—The application of camera calibration or other numerical data to the picture content so that the resulting picture has point-for-point numerical meaning. This group would also include the set of processes involved in the actual camera calibrations. The primary effort here is the application of camera calibration data to remove camera artifacts, although the removal of other contaminants such as atmospheric effects belongs in this category.

Information Extraction—The conversion of the picture data into the information or decision required by the analyst, perhaps with a reduction in the quantity of information required. In this category would be the various pattern recognition techniques used to process multi-spectral images for analysis and classification.

Processing Examples

One of the early applications for digital processing was the analysis of the roughness of the lunar surface. In this analysis altitude contours of the lunar surface were derived from the (monoscopic) pictures taken by the Ranger spacecraft. This process utilized the known photometric function (reflected brightness vs. viewing angle) to derive the surface slope, from which the surface altitude may be calculated by integrating along the zero phase line (Rindfleisch, 1966). This set of altitude data calculated for the entire picture may then be computer contoured. One of the Ranger pictures, its photogrammetrically rectified equivalent, and the altitude contours are shown in Plate 1.

Since this analysis used the reflected brightness to determine the surface slope, it is important that the brightness be accurate. Since television camera systems are not sufficiently uniform in their intensity response over the face of the picture corrections must be applied. For this purpose pre-flight calibration is obtained by photographing a series of uniformly illuminated flat fields at different brightnesses and from these determining a brightness response function for every pixel (picture element) in the picture. A demonstration of the variation in camera response across the test field is shown in Plate 2, in which the images of the flat fields at four different brightnesses have been contoured (Rindfleisch, *et al.* 1971). The variation in the brightness response at different light levels is apparent.

The true ground coverage plot of the contours shown in Plate 1 was produced by geometrically rectifying the incoming picture. This and other geometric manipulations to be described below is done digitally by defining to the computer the new desired locations of known points. The locations of all the intermediate points are generated in the output picture by interpolation between the defined points.

Another similar trapezoidal stretching is shown in Plate 3, in which a low angle photograph of one of the JPL buildings has been stretched to produce an approximation to a frontal view.

Another major use of geometrical stretching is for the registration of one picture to another, as would be required, for instance, in matching the different colour separation framelets of a given area, for registering multispectral scanner images to photographs (for example, the ERTS television camera and multi-spectral scanner images) and the like. An example of the stretching required to register color framelets from the multi-Hasselblad experiment S-158 is shown in Plate 4. In this experiment, four Hasselblad cameras with 80 mm lenses were ganged, and used to simultaneously photograph the lunar surface through different colour filters. Photography was obtained during the Apollo 12 mission from the orbiting command module. The black and white grids in this figure represent the rubber sheet stretching required to cause registration of two of the color components, and represent the cumulative distortions from the camera lenses, film flatness effects, film stretching, and scanner variations.

Another situation encountered in the S-158 experiment was the variation in brightness fall-off away from image center (vignetting) at different f: stops. The cameras were calibrated by taking an array of photometer measurements in the focal plane. These data have been interpolated to match the picture sizing, and a typical set is shown in Plate 5. The contour lines represent 5% decrease in brightness relative to the center of the frame. This effect, while tolerable for human eyeball analysis, will severely hamper certain types of numerical analysis when intercomparisons of pictures taken at different f:stops is required. These effects, once measured, may of course be removed during the digital processing.

The effect of this vignetting can be clearly seen in Plate 6. The upper left framelet is an area of Frau Mauro photographed by S-158. The remaining framelets are density slices taken over the total brightness range. The circular aspect of the slices is caused by the approximately circularly symmetric vignetting.

A major use for digital filtering at JPL has been for the enhancement of the high spatial frequency content (fine details) of the pictures. An example of this process is given in Plate 7, which shows the appreciable sharpening of a picture due to this type of filtering. Our approach to this process is discussed by Nathan (1971) and is illustrated in Fig. 1. Consider the image of a star. Due to the geometry involved this image should be essentially a delta function having zero dimension. In practice, however, this image will have a finite diameter due to light which should have impinged at the delta function point arriving instead in the nearby area. The job of the high frequency enhancement filter is to gather this nearby light and to reinsert it at the delta function point. The filter function must, therefore, have a ring of negative values around the central pixel to remove the unwanted light and a high value at the center pixel to reinsert the light there. In Fig. 1 is shown the relationship between this point spread function correction and the relative spatial frequency enhancement (which are two equivalent ways of considering the same process), together with the numerical values of a practical digital filter.

The difference in processing required for different uses of a picture is shown in Plates 8 and 9. Plate 8 is one of the Mariner '69 pictures which has been correctly photometrically rectified (i.e., photometric calibration has been applied to restore numerical accuracy). The large difference in brightness between the south polar cap of Mars and the surrounding area

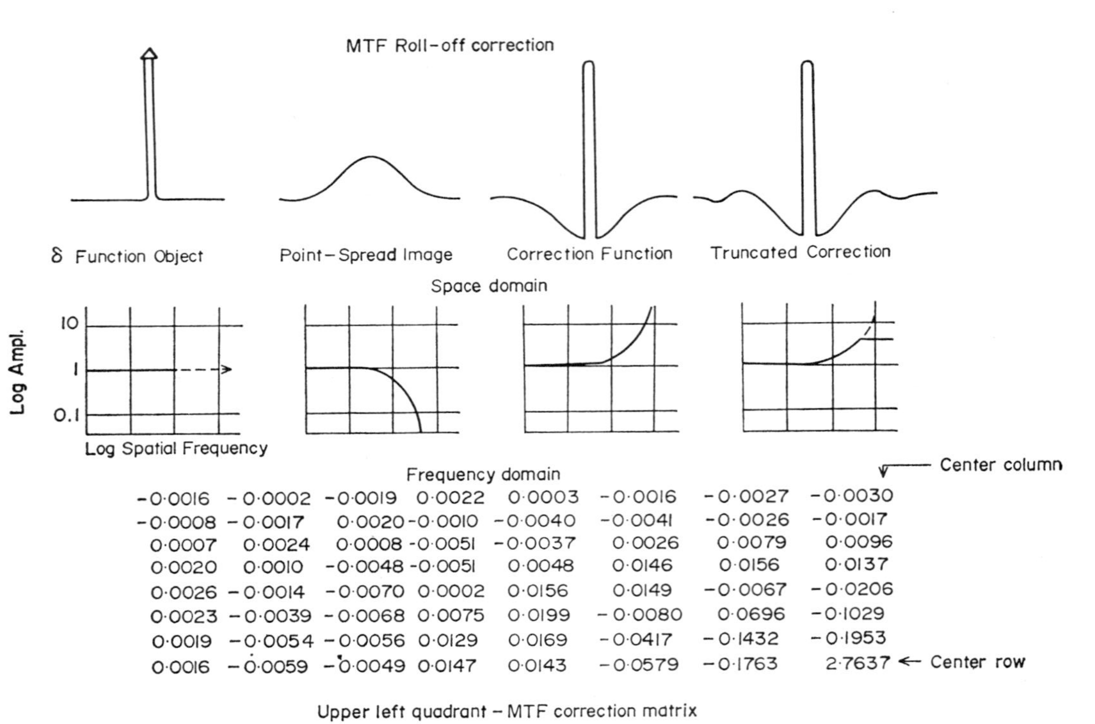

-0·0016	-0·0002	-0·0019	0·0022	0·0003	-0·0016	-0·0027	-0·0030
-0·0008	-0·0017	0·0020	-0·0010	-0·0040	-0·0041	-0·0026	-0·0017
0·0007	0·0024	0·0008	-0·0051	-0·0037	0·0026	0·0079	0·0096
0·0020	0·0010	-0·0048	-0·0051	0·0048	0·0146	0·0156	0·0137
0·0026	-0·0014	-0·0070	0·0002	0·0156	0·0149	-0·0067	-0·0206
0·0023	-0·0039	-0·0068	0·0075	0·0199	-0·0080	0·0696	-0·1029
0·0019	-0·0054	-0·0056	0·0129	0·0169	-0·0417	-0·1432	-0·1953
0·0016	-0·0059	-0·0049	0·0147	0·0143	-0·0579	-0·1763	2·7637

FIG. 1. Top row: the point spread function and its compensation as shown in real picture space. Middle row: equivalent correction as expressed in the spatial frequency domain. Bottom row: a practical digital filter (upper left quadrant) for high frequency compensation.

fairly thoroughly obliterates detail in both areas, although the numbers in the digital form of the picture are correct and useful for photometric analyses. To render more of the detail visible the lower spatial frequency photometry was deliberately destroyed by using a filter which greatly reduces the low spatial frequency content (Dunn *et al.*, 1971). This reduces the large black-to-white excursions and superimposes all of the fine detail on a more or less medium gray background. In this form the contrast of the resulting picture may be increased to increase the visibility of this fine detail as shown in Plate 9. This "maximum discriminability" picture has proven quite useful in the analysis of surface features where photometry was of secondary consideration. A further use of geometric rubber sheet stretching is for the projection of various pictures onto a common scale to allow mosaicking. A number of frames of the southern hemisphere of Mars have been so projected and mosaicked by computer (Gillespie & Soha, 1972). The resulting orthophoto centered on the Martian south pole is shown in Plate 10.

Another example of the processing for enhancement of visual effects is shown in Plate 11. This is a radiograph of an arm bone containing a tumor. The desire is to be able to see the blood vessel structure both in the bone area and in the flesh area surrounding it. In the original, this detail is masked by the large black-to-white variation between the bone and the flesh. Complete removal of all the low spatial frequency content and enhancement of the high frequencies removes this shading, allowing contrast stretching to be applied to enhance the fine (high frequency) detail.

The previous enhancements utilized completely generalized filter processes with parameters chosen to perform the types of filtering desired. For some types of enhancements more specialized techniques must be used. Plate 12 illustrates the improvement in finger print images possible. In this enhancement a set of processes invoking filtering, maximum gradient analysis and sequential ridge structure detection were combined (Blackwell, 1970).

Some Diagnostic Processing

The recent Mars flyby missions contained in each picture a coherent noise pattern consisting of a diagonal structure of small vertical lines. This pattern was relatively constant in a given picture but was quite different from picture to picture, so that a single set of filtering parameters could not be derived. Because of this, the noise pattern had to be individually located for each picture. This was done by taking the two-dimensional Fourier transform of the picture (Rindfleisch *et al.*, 1971). The coherent noise structure appears in the Fourier plane as a geometrically arrayed series of star-like spots representing energy concentrations at specific spatial frequency locations.

A program was developed to locate these spots and to remove them from the Fourier plane. Retransformation of the picture-sans-noise restores the original image, but without the coherent noise. Plate 13 shows an original frame, its Fourier transform, the noise pattern isolated from the series of spots and the retransformed cleaned up picture.

Multi-image processing may be used to average together a group of images for the reduction of, for instance, film grain noise. Plate 14 shows one of a series of original images of a spiral nebula and the average of 2, 4, and 8 of the images. The reduction in system noise is apparent. This process requires, of course, precise registration, since misregistration will also result in removal of the star images or other fine detail.

Information Extraction

Time lapse photography for the detection of temporal changes may be conveniently analysed by a computer. Plate 15(a) is one frame of a time sequence showing the micro-circulation of the blood. The circulation pattern itself is revealed by subtracting time-adjacent photographs. In this subtraction all stationary material is cancelled leaving only the moving components visible, as shown in Plate 15(b).

A similar process is shown in Plate 16, in which the two left framelets are time sequence images of a freeway. Computer registration of the freeway and subtraction results in cancelation of all stationary vehicles and the freeway structure itself, leaving as plus-minus pairs each moving vehicle. Translation of one freeway image with respect to the other to correspond to some particular speed such as the speed limit and subtraction will result in all cars going at that speed being cancelled, and the cars going below or above that particular speed showing as image pairs.

Multi-spectral Analysis

Multi-spectral analysis for the analysis of images which are being obtained from the ERTS spacecraft has been developed at several centers in the United States. The examples given here are from the Laboratory for the Application of Remote Sensing (LARS) at Purdue University (Landgrebe, 1971). In this analysis advantage is taken of the fact that different materials will have different reflectivities in different spectral ranges, such that by suitable selection of the ranges utilized various different materials may be located in photographs. Plate 17 illustrates the multispectral response of four typical materials photographed in three spectral bands. The process may be briefly described as follows: consider a plot of the spectral reflectance

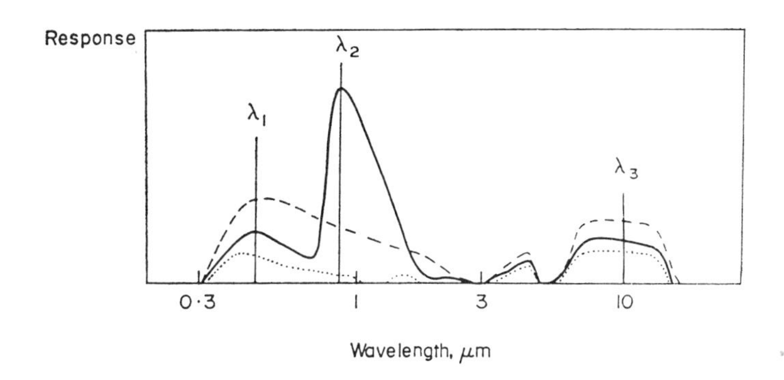

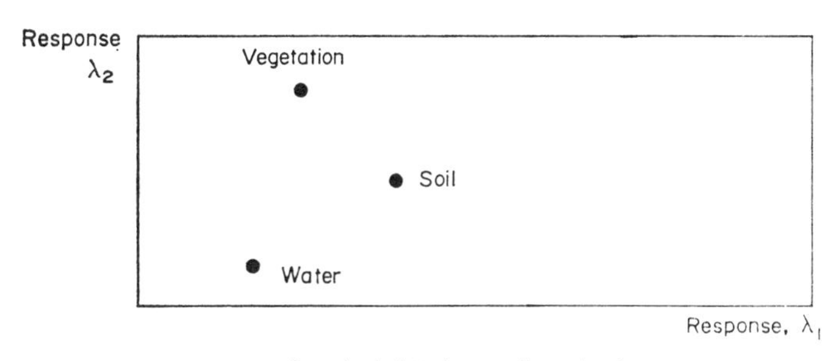

Spectral data in two-dimensional feature space

FIG. 2. Typical spectral reflectance curves of several materials. The relative differences in the reflectances at different wavelengths form a basis for discriminating between the materials. —— Vegetation; – – – – Soil; · · · · Water.

of different materials as a function of wavelength (Fig. 2). Typically each material will have a different spectral response such that a set of samples of the spectrum at different wavelengths will uniquely describe each material. Consider for simplicity the case of two samples at λ1 and λ2. If the responses at these two wavelengths are used as Cartesian co-ordinates of a decision plane it will be found that in general all of the samples of a given material will tend to cluster near the same area of the plane and that for separable materials each material sample group will be located at a different area. Advantage is taken of this tendency by manually defining a number of training areas in the picture in which the material type is known. The measurements from each of these materials are used to determine the locale in the decision space to be assigned to that material. All of the different materials visible in the photograph are so treated; this results in a decision space which is completely defined. Each unknown sample in the picture is then compared in

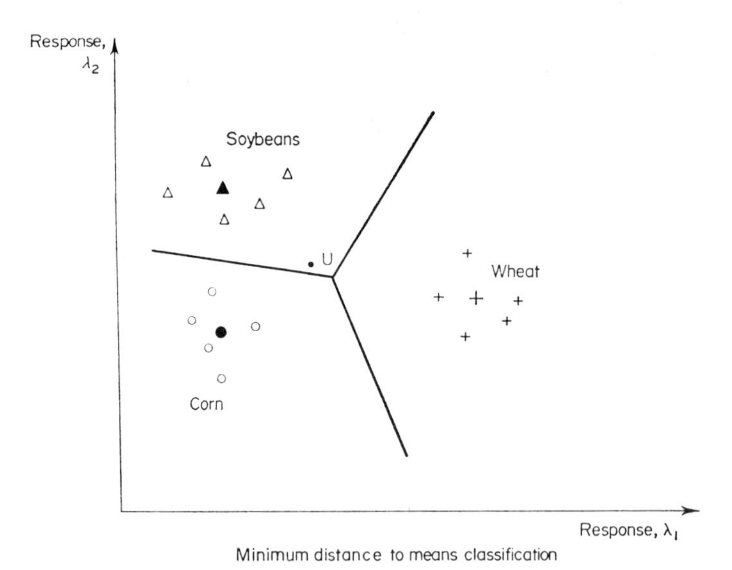

FIG. 3. Decision space, showing classification of unknown sample point U as soybeans since it falls in the soybeans space.

this decision space to the set of known training samples and is classified as being that material which it most nearly matches (Fig. 3).

A different type of color analysis has been used at JPL for the analysis of lunar photographs. Lunar color and its variations across the surface has interested planetary astronomers for many years. This interest has heightened with the growing weight of evidence obtained from accurate Earth based photoelectric photometry which points towards a positive correlation between color and compositional differences (Goetz, Billingsley, Yost & McCord, 1971). For this color analysis pictures of the same area were taken through a Mt Wilson telescope on black-and-white film through two color filters (Plate 18(a)). After registering these color components in the computer the data was intensity corrected to log exposure, after which the pictures were subtracted one from the other (Billingsley, Goetz & Lindsley, 1970). The conversion to log exposure and the subsequent subtraction produces a picture whose numerical value at a given point is a measure of the red/blue ratio at that point. The resultant picture thus represents the red/blue ratio rather than the brightness for the area covered, for the digital processing has eliminated the brightness component. Plates 18(b), (c), (d) shows the red, grey, and blue areas, respectively, each brought to saturation to indicate their respective locations. Color versions of these pictures are given in Billingsley *et al.* (1970). For the S-158 experiment, three-color analysis was performed in much the same manner by producing red/green and blue/green photographs. These two ratios may be used together to define a unique color which is representative of the ratio combination, so that for each picture point an output color may be assigned representing the red, green and blue triplet from the input pictures. A detailed explanation of this process and color pictures are given in Billingsley (1972).

Biomedical Analysis

Since the introduction of a method allowing microscopic examination of individual human chromosomes the karyotype has emerged as a tool of increasing diagnostic value. Under microscopic examination the chromosomes from a somatic cell in the metaphase stage of cell division appear in a scattered disarray. The karyotype is a systematic grouping of metaphase chromosomes from a single cell. This grouping is conceived to assist the geneticist in the identification of individual chromosomes. The diagnostic value of the karyotype is predicated upon a consistent pattern in normal persons and the correlation of certain chromosomal aberrations with specific clinical observations. At present manual karyotyping is so tedious and expensive that its general application is usually limited to those situations involving a suspected abnormality. Reduction in cost of the karyotyping process through computer processing would allow a much more wide spread clinical use of this tool (Castleman, 1970).

Plate 19 is a digitized version of a typical good chromosome spread in which the chromosomes are isolated and do not touch or overlap. The process consists in scanning the spread on a microscope slide under computer control, following which the individual chromosomes are isolated. The isolation is accomplished by a brightness thresholding technique which first analyzes the histogram of brightness of the complete picture to establish a threshold for separation of dark objects, and which then collects together each group of connected points above the threshold. These groups of connected points, which are the chromosomes, may be displayed as shown in Plate 20. The chromosome in each cell is then rotated until it is upright, as determined by the minimization of the area of an enclosing rectangle, following which the total length, area, and position of the centromere (the central constriction) are measured. Classification is by location of the measurements in a two-dimensional decision plane shown in Fig. 4. The resultant classified chromosomes are then arranged for display and labeled with the result as shown in Plate 21. The pictorial output which may be produced preserves all the structural detail of the chromosomes, allowing

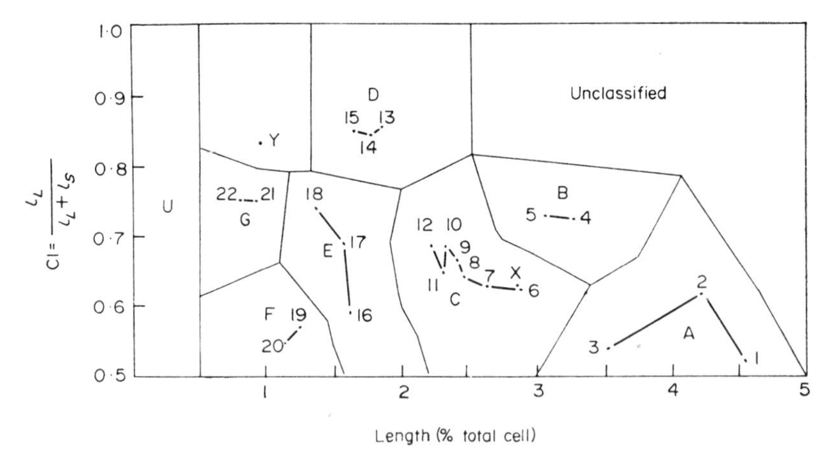

FIG. 4. Typical decision plane used to classify the chromosomes into groups.

the clinician to verify the classification visually and thereby correct errors. Also, the pictorial data can be enhanced by processes such as those described in this paper to show clearly the presence of secondary constrictions which provide clues for the identification of individual chromosomes.

Conclusion

The image processing techniques developed for enhancement and calibration of the JPL imaging experiments have not only been proven useful for the original application but have also been successfully applied to images from a variety of sources. As computer processing becomes more widely available, as the processing costs decrease, and as digital enhancement techniques are improved to speed up the processing and provide more useful results, the use of computer processing for enhancement of images will become increasingly important and widespread. It will prove particularly advantageous in those cases requiring analysis and classifications of the images where no simple photographic techniques are possible, such as karyotyping and the classification of multi-spectral images.

This research presents the results of one phase of research carried out at the Jet Propulsion Laboratory, California Institute of Technology, under Contract No. NAS 7-100, sponsored by the National Aeronautics and Space Administration.

All of the above work with the exception of the multi-spectral analysis was performed by personnel of the Image Processing Laboratory of the Jet Propulsion Laboratory. In particular I would like to acknowledge the work of Dr Robert

Nathan who fathered the original image processing development, Tom Rindfleisch, who developed the Ranger photoclinometery and guided the Mariner '69 processing, and Robert Selzer and Dr Kenneth Castleman for the biomedical processing. The multi-spectral work discussed here has been developed at LARS under the guidance of Dr David A. Landgrebe.

This paper was originally presented at the conference on Machine Perception of Patterns and Pictures held at the National Physical Laboratory in April 1972 and published by the Institute of Physics as Conference Series No. 13: *Machine Perception of Patterns and Pictures*.

References

Blackwell, R. J. (1970). Fingerprint image enhancement by computer methods, 1970 Carnahan Conference on Electronic Crime Countermeasures, 17 April, 1970, Lexington, Ky.

Billingsley, F. C. (1972). Computer-generated color image display of lunar spectral reflectance ratios. *Photogr. Sci. Engng*, **16**, 51.

Billingsley, F. C., Goetz, A. F. H. & Lindsley, J. N. (1970). Color differentiation by computer image processing. *Photogr. Sci. Engng*, **14**, 28.

Castleman, K. R. (1970). In *Pictorial Output for Computerized Karyotyping in Perspectives in Cytogenetics*, pp. 316–323. Eds S. W. Wright, B. F. Crandall & L. Boyer. Springfield, Ill.: C. C. Thomas.

Dunn, J. A., Stromberg, W. D., Ruiz, R. M., Collins, S. A. & Thorpe, T. E. (1971). Maximum discriminability versions of the near encounter Mariner pictures. *J. geophys. Res.*, **76**, 438.

Gillespie, A. R. & Soha, J. M. (1972). An orthographic photomap of the south pole of Mars. *Icarus* **16**, 522.

Goetz, A. F. H., Billingsley, F. C., Yost, E. & McCord, T. B. (1971). Apollo 12 multispectral photography experiment. Proc. of the Second Lunar Science Conference, Suppl. 2. *Geochim. cosmochim. Acta*, **3**, 2301. MIT Press.

Landgrebe, D. A. (1971). Systems approach to the use of remote sensing, LARS Information Note 041571. Laboratory for the Application of Remote Sensing, Purdue University, Lafayette, Ill.

Nathan, R. (1971). Image processing for electron microscopy: I. Enhancement procedures. *Advances in Optical and Electron Microscopy* 4, pp. 85–125. New York: Academic Press.

Rindfleisch, T. C. (1966). *Photogramm. Engng*, **32**, 262.

Rindfleisch, T. C., Dunne, J. A., Frieden, H. J., Stromberg, W. D. & Ruiz, R. M. (1971). Digital processing of the Mariner 6 and 7 pictures. *J. geophys. Res.*, **76**, 394.

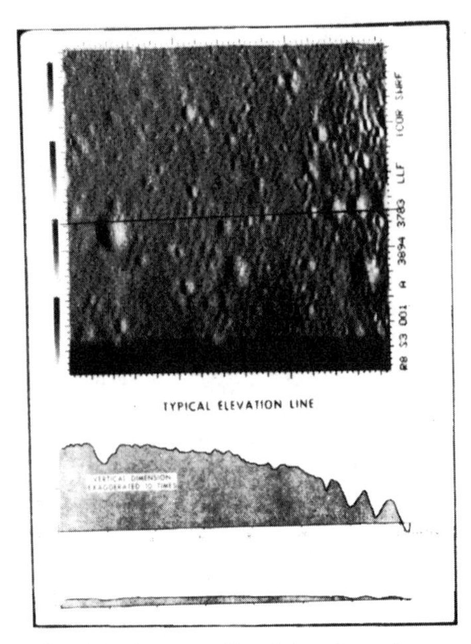

PLATE 1. (a) A photograph of the lunar surface taken by a Ranger spacecraft, with a zero-phase line and derived altitude profile.

PLATE 1. (b) Photogrammetrically rectified image.

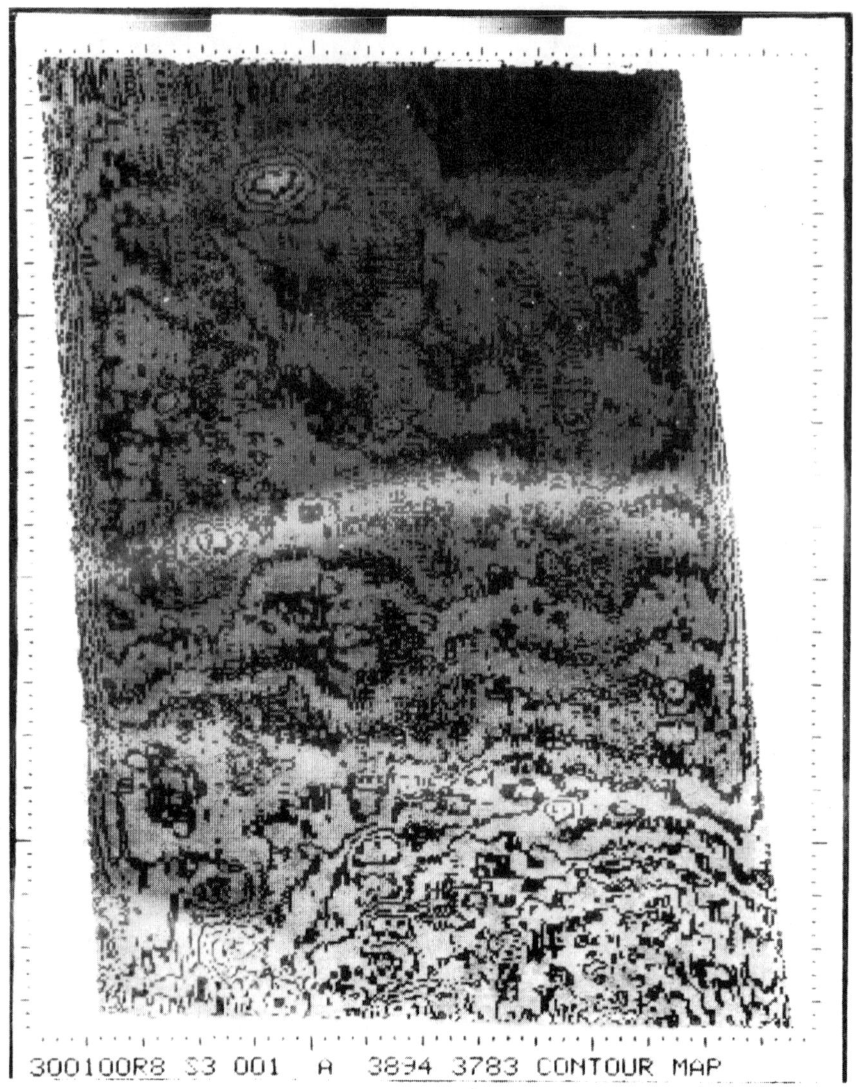

PLATE 1. (c) Altitude contours derived from a series of altitude profiles.

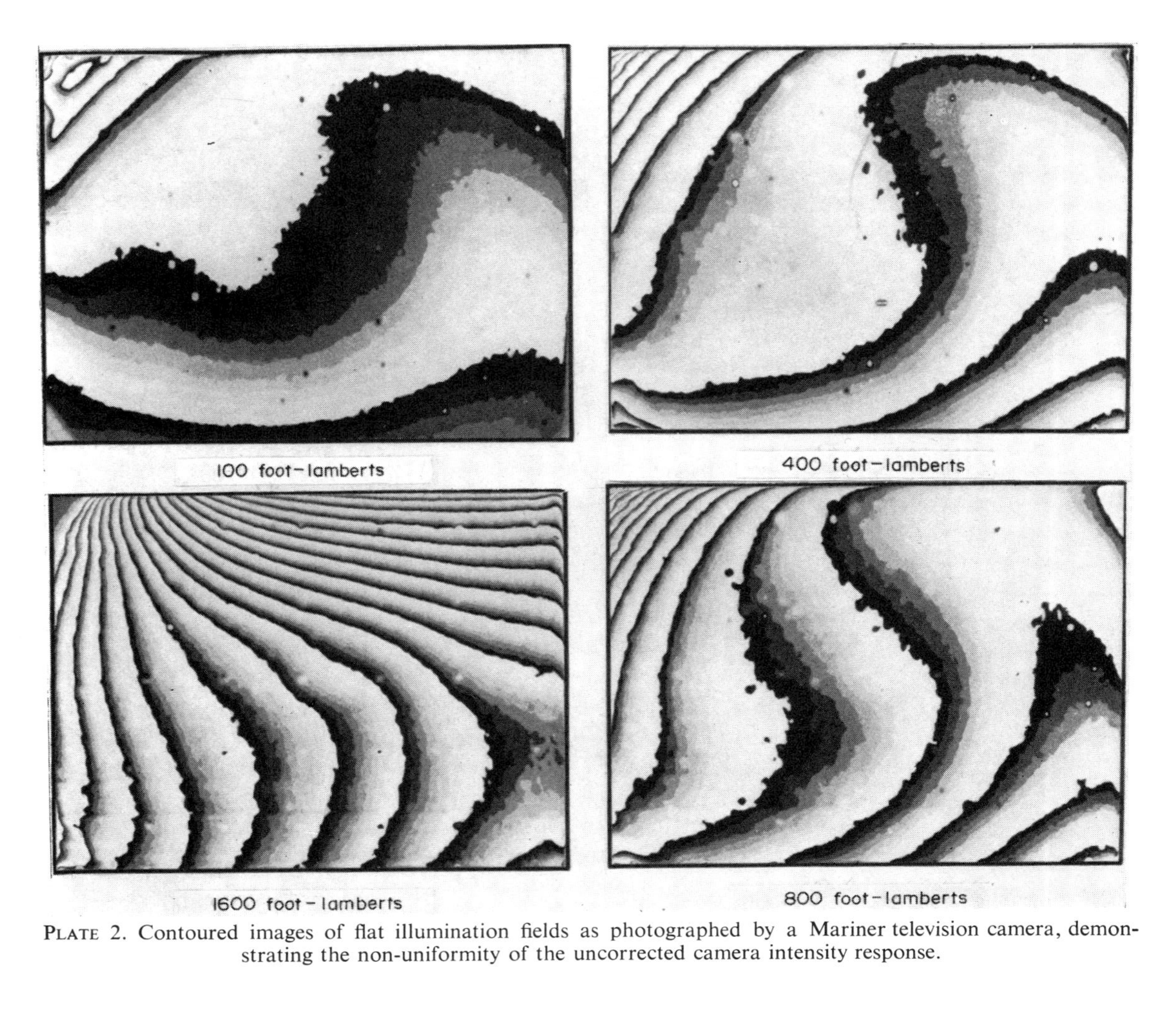

PLATE 2. Contoured images of flat illumination fields as photographed by a Mariner television camera, demonstrating the non-uniformity of the uncorrected camera intensity response.

PLATE 3. (a) A JPL building photographed from a low angle.

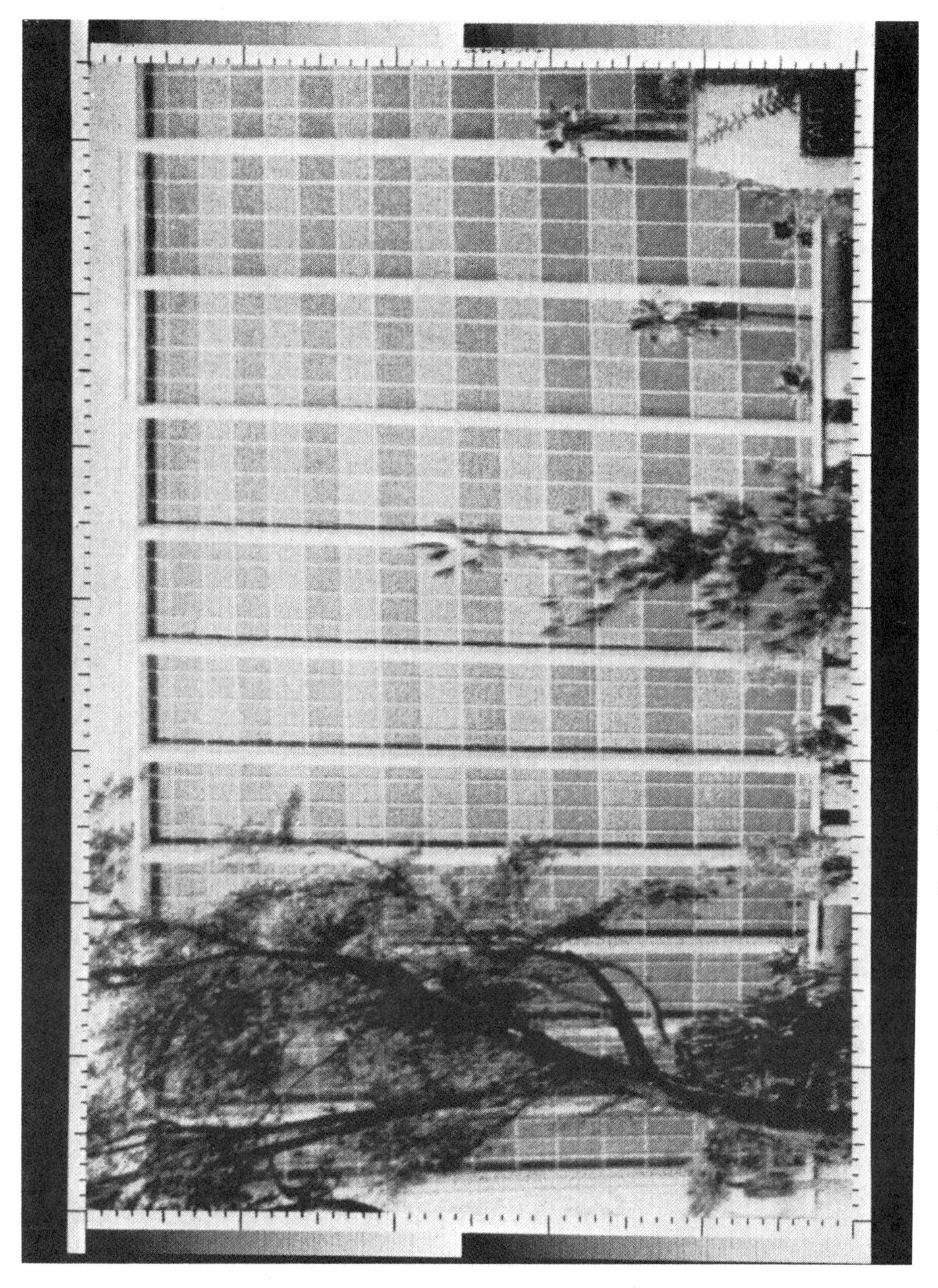

PLATE 3. (b) Rectified version, using computer stretching.

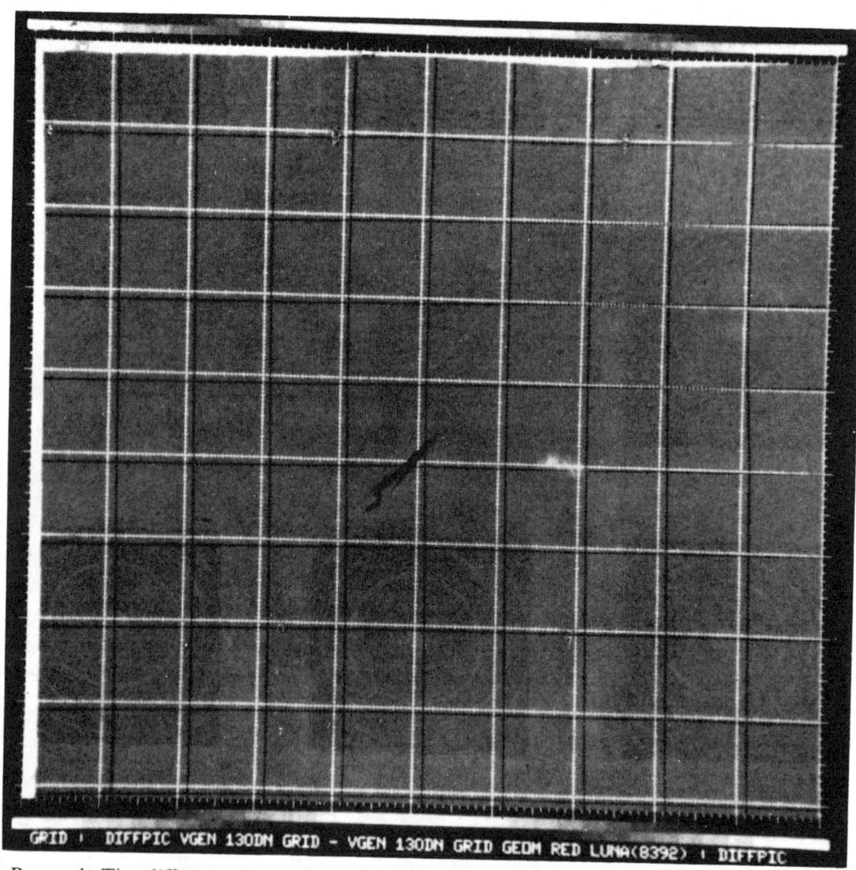

PLATE 4. The difference in position of the black and white grids represents the rubber sheet stretching required to register two color separation framelets to each other. The pictures are from the S-158 Lunar Multispectral Photography Experiment flown on Apollo 12.

Contouring maps of camera vignetting S-158 lunar multispectral photography experiment.

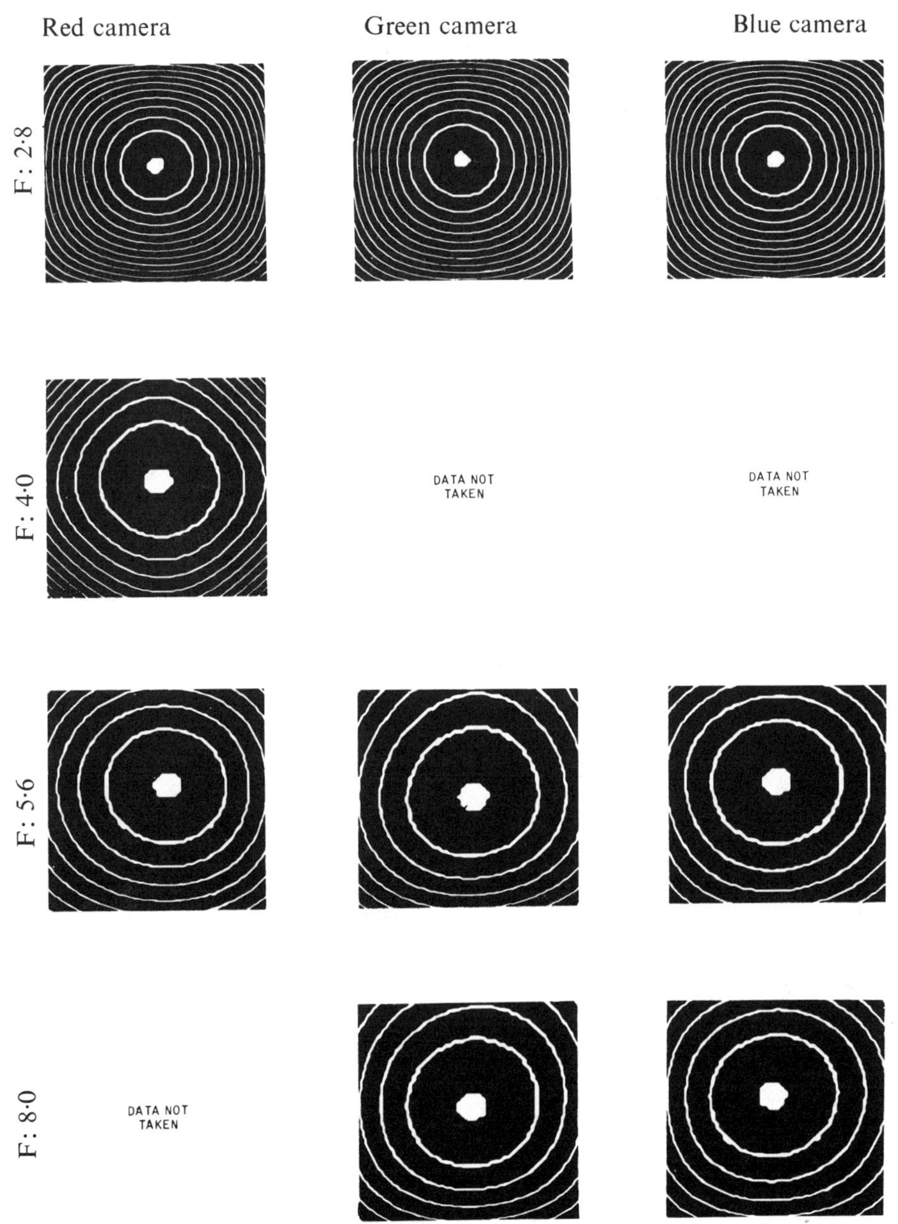

Black area covers total image frame. All vignetting frames are upright, looking at rear of camera, with camera upright. Contours are 5% transmittance differences, relative to 100% transmittance at center.

Plate 5. A set of brightness contours as obtained by calibration of the S-158 cameras at various f:stops.

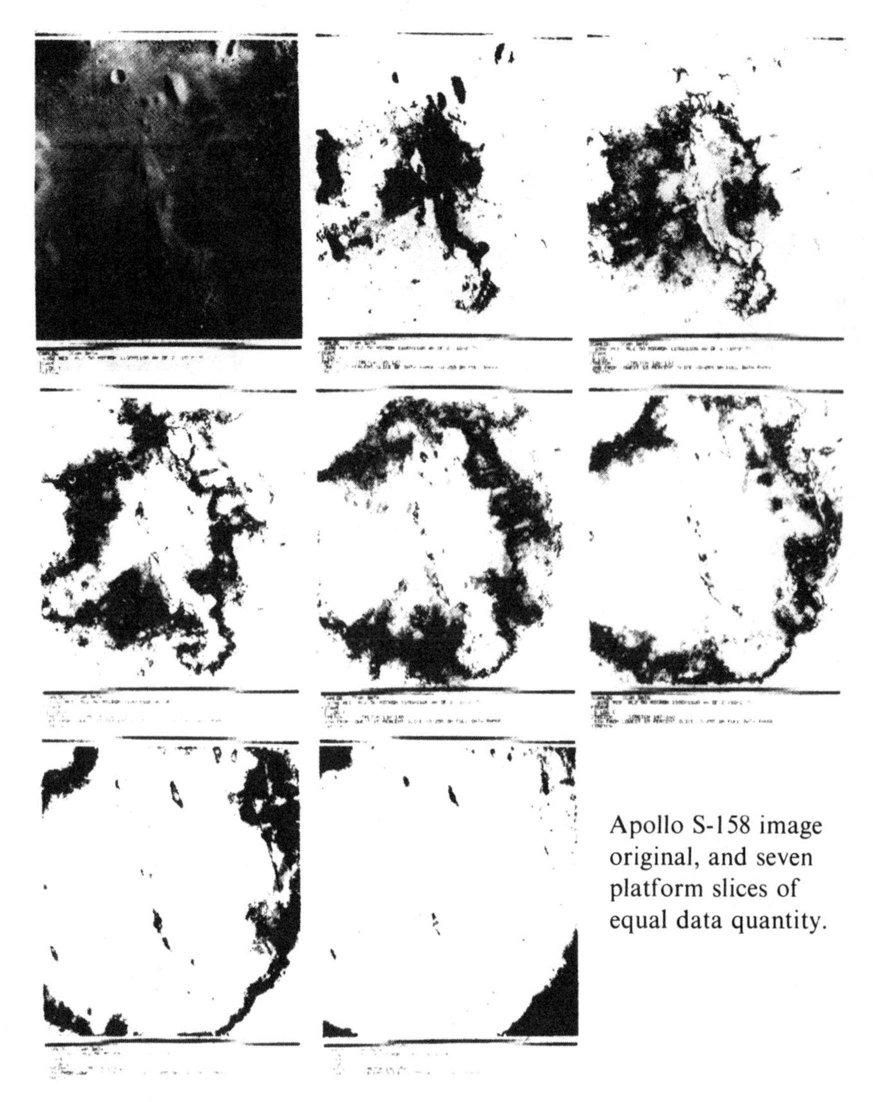

PLATE 6. A photograph of Frau Mauro by S-158, and seven slices of density taken to cover the density range. Note the circularly symmetric appearance of the slices, caused by lens vignetting.

PLATE 7. (a) A photograph of a footpad of the Surveyor spacecraft resting on the Moon, before and after digital sharpening. Before enhancement.

PLATE 7. (b) As for Plate 7 (a) but after enhancement.

PLATE 8. A photometrically correct picture of the edge of the Martian south polar cap.

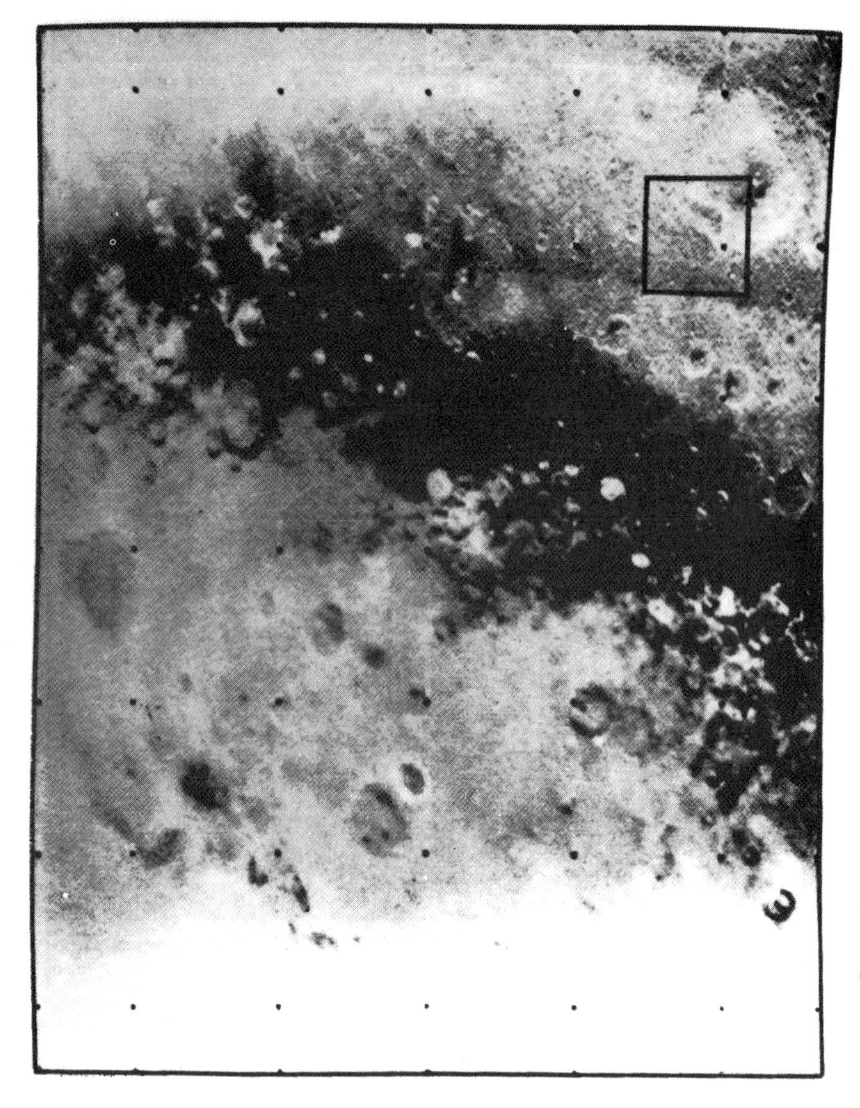

PLATE 9. A maximum discriminability version of Plate 8, in which the large brightness variation has been removed to allow the detail to be visible.

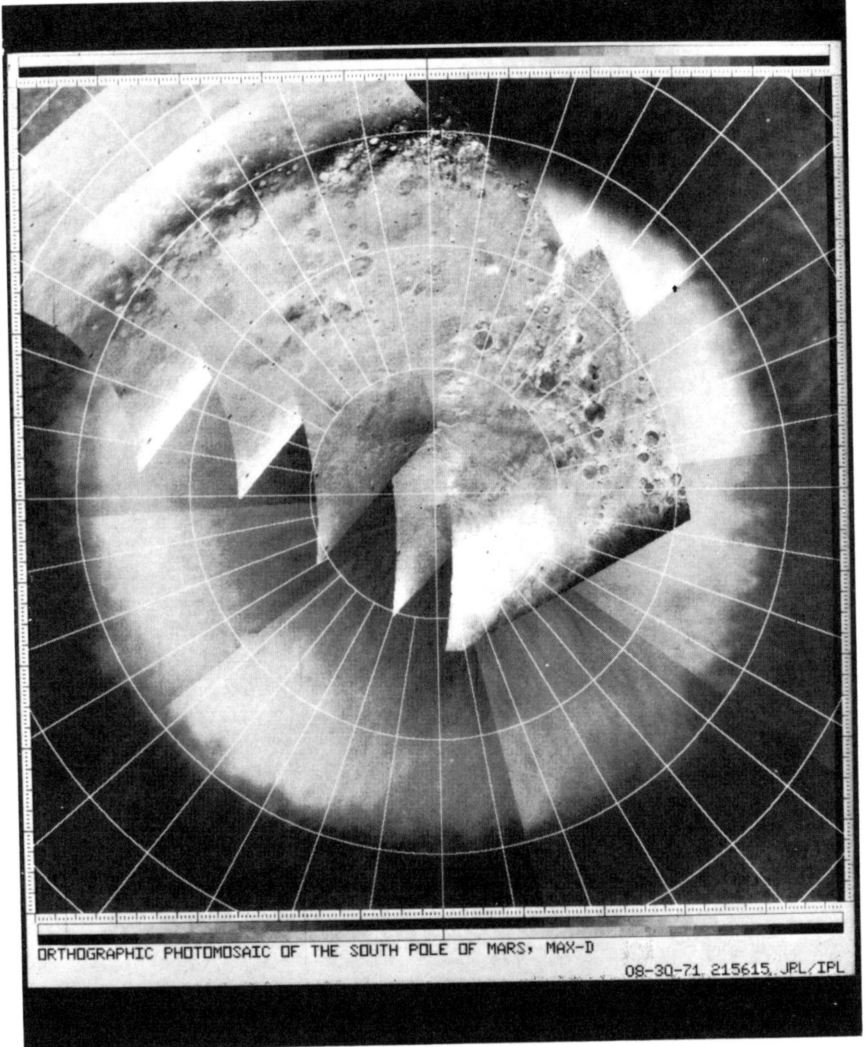

PLATE 10. Orthophoto mosaic of the Martian south polar cap, assembled by computer.

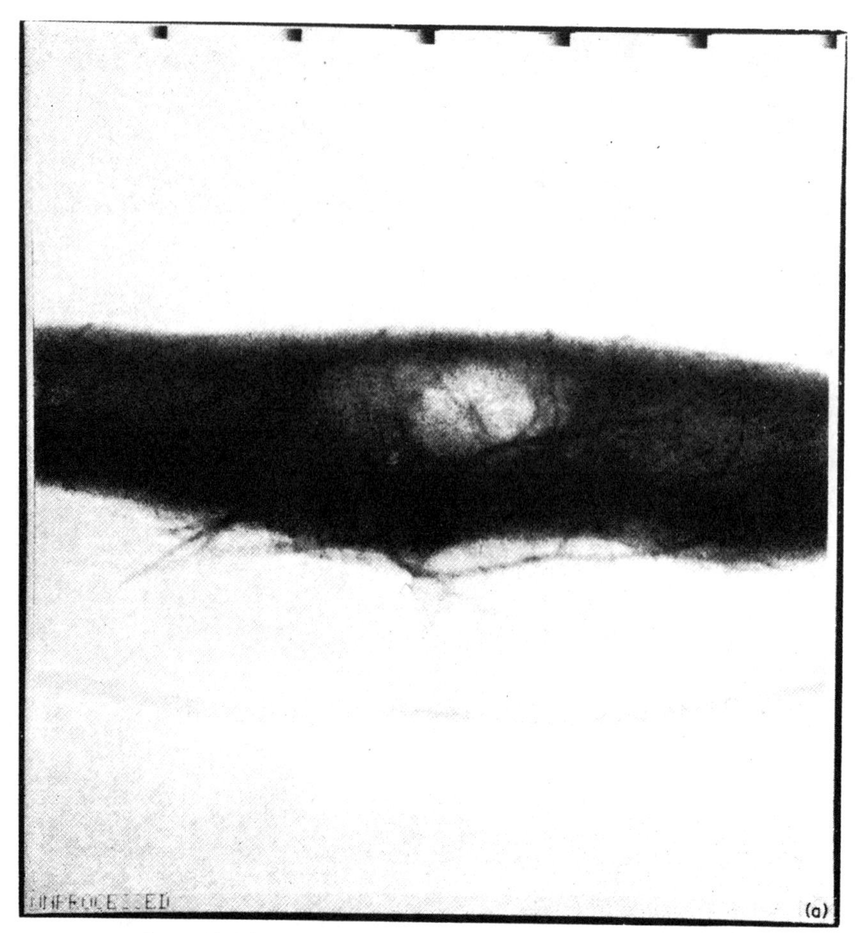

PLATE 11. (a) Radiograph of an arm bone containing a tumor.

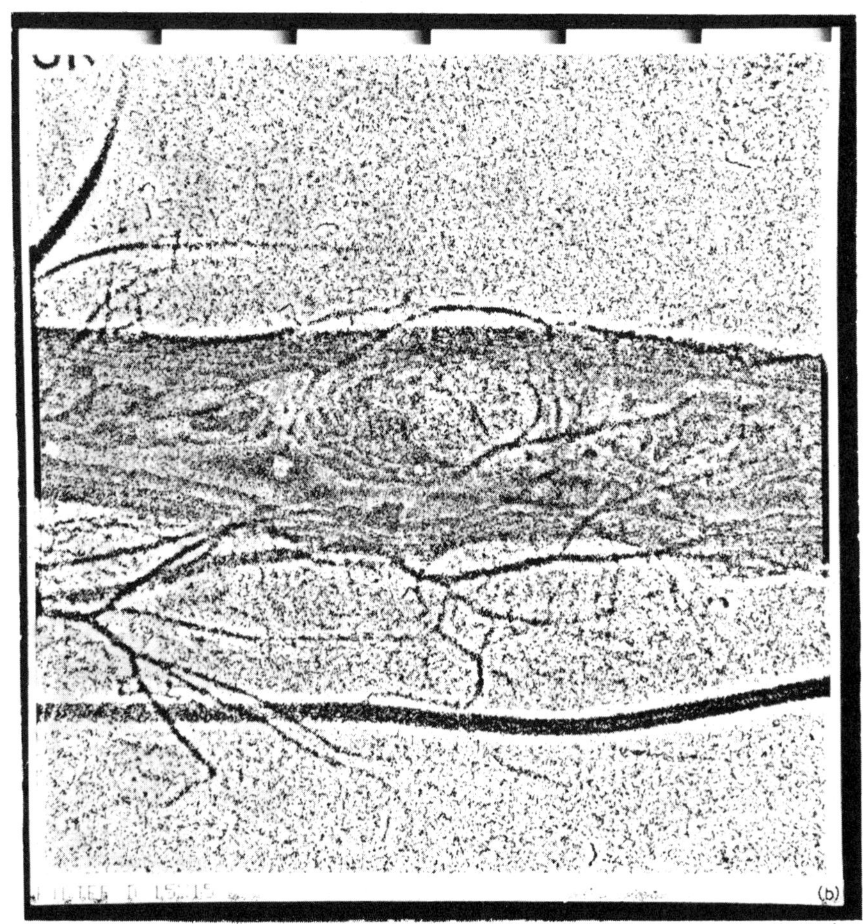

PLATE 11. (b) By computer processing to remove gross shading, blood vessel detail may be better seen.

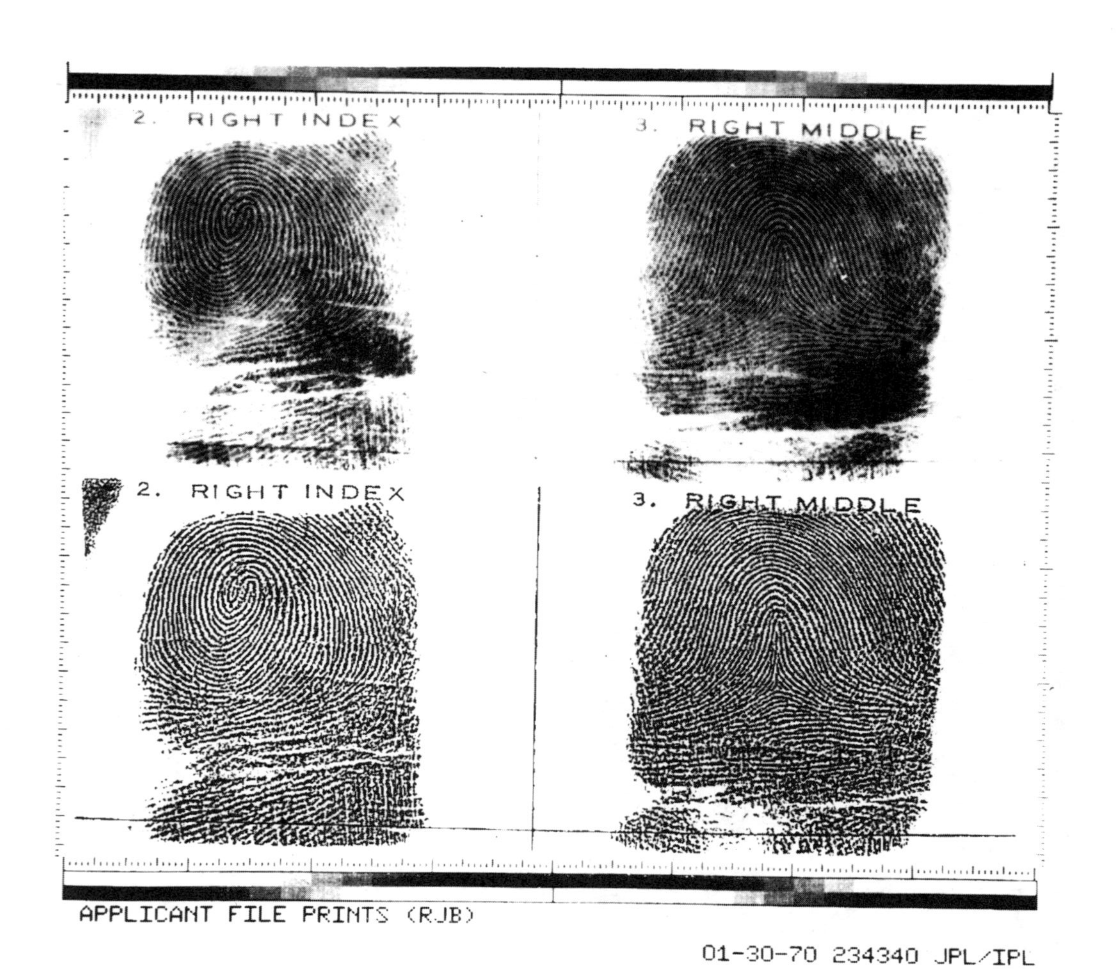

PLATE 12. Restoration of fingerprint detail by computer processing.

Two-dimensional Fourier spectrum of a picture

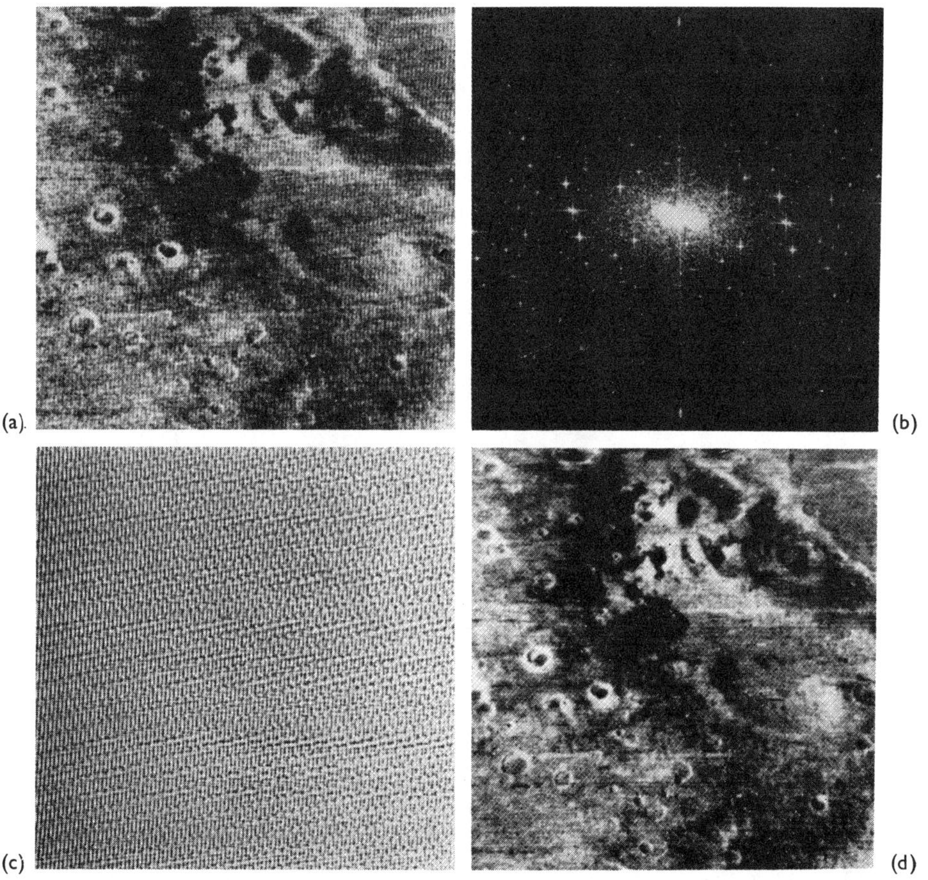

PLATE 13. (a) A picture from the Mariner 7 mission, showing a coherent noise interference. (b) The Fourier transform of this picture, showing the noise spikes as a geometric array of points. (c) The noise alone as reconstructed from the noise spikes from the Fourier plane. (d) The cleaned picture with the noise removed.

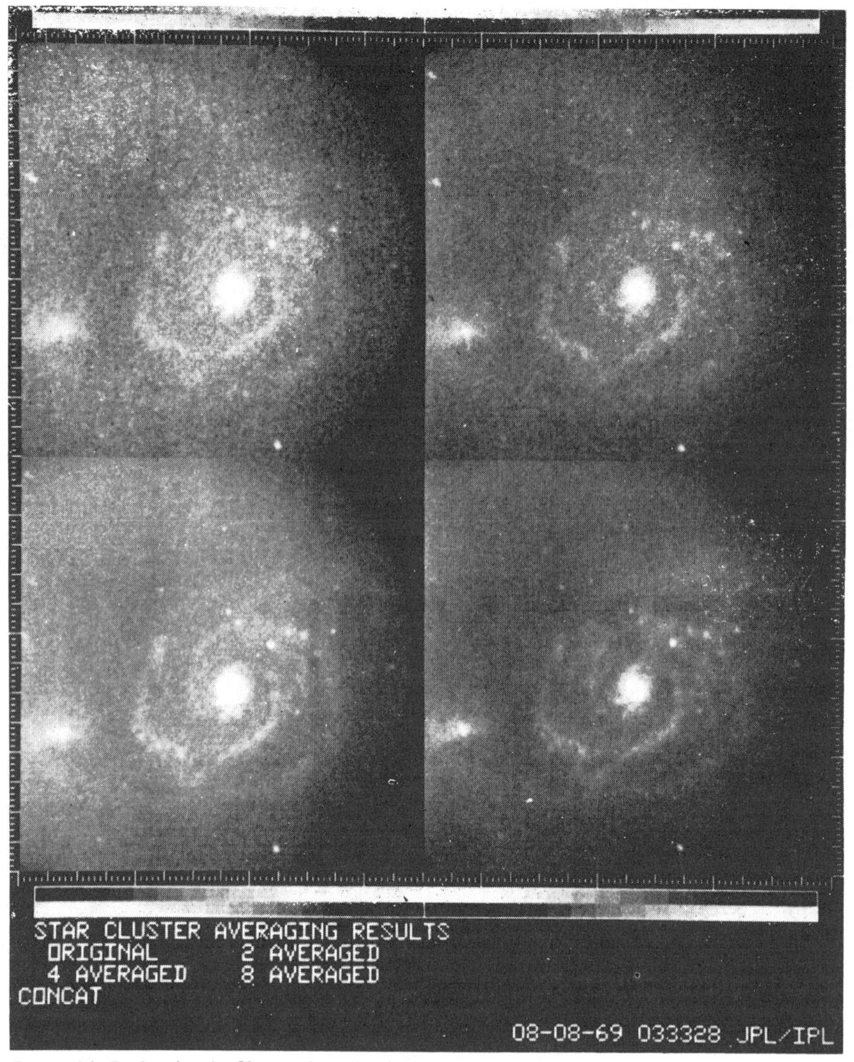

PLATE 14. Reduction in film grain noise by averaging of two, four and eight images of a spiral nebula.

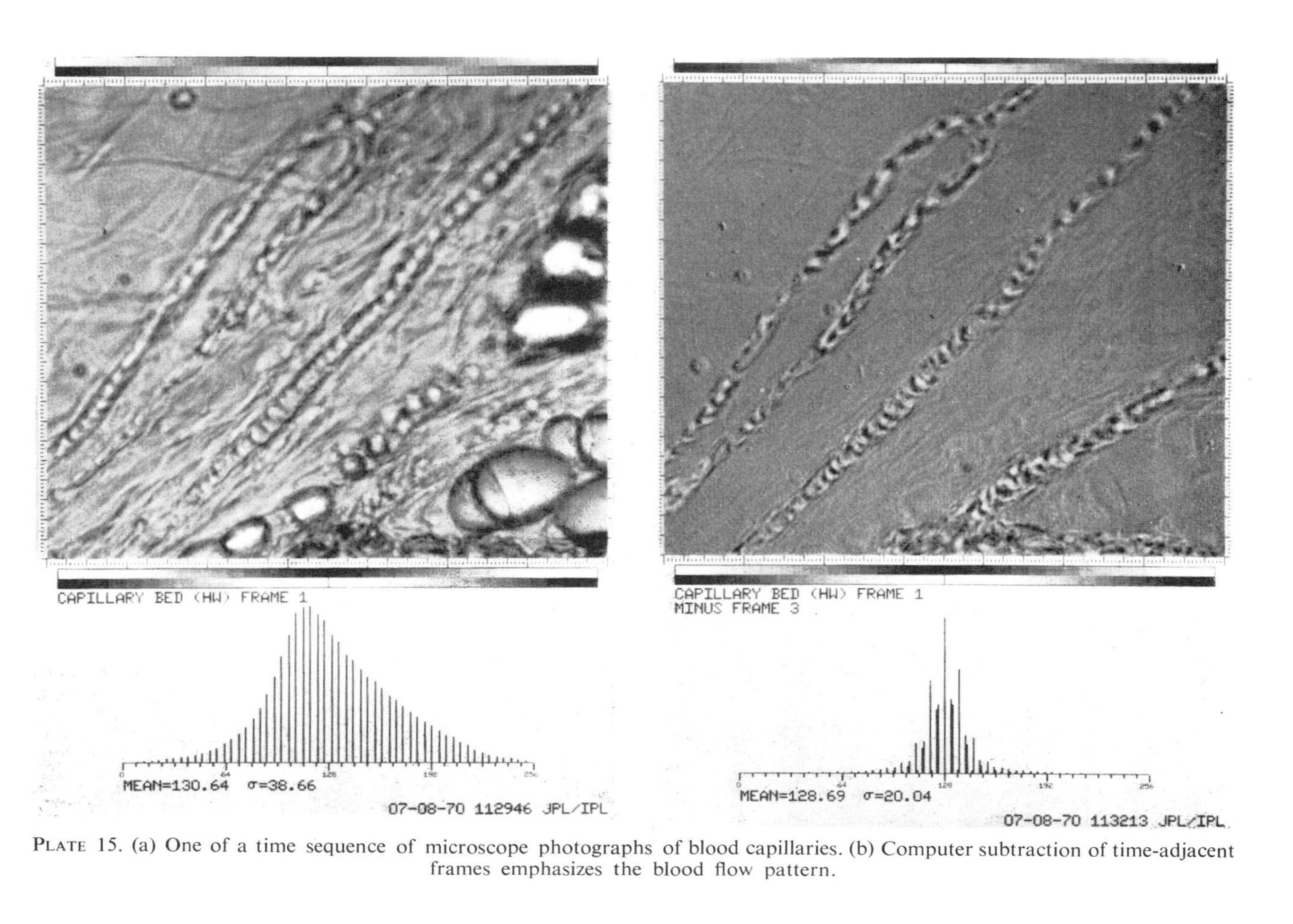

PLATE 15. (a) One of a time sequence of microscope photographs of blood capillaries. (b) Computer subtraction of time-adjacent frames emphasizes the blood flow pattern.

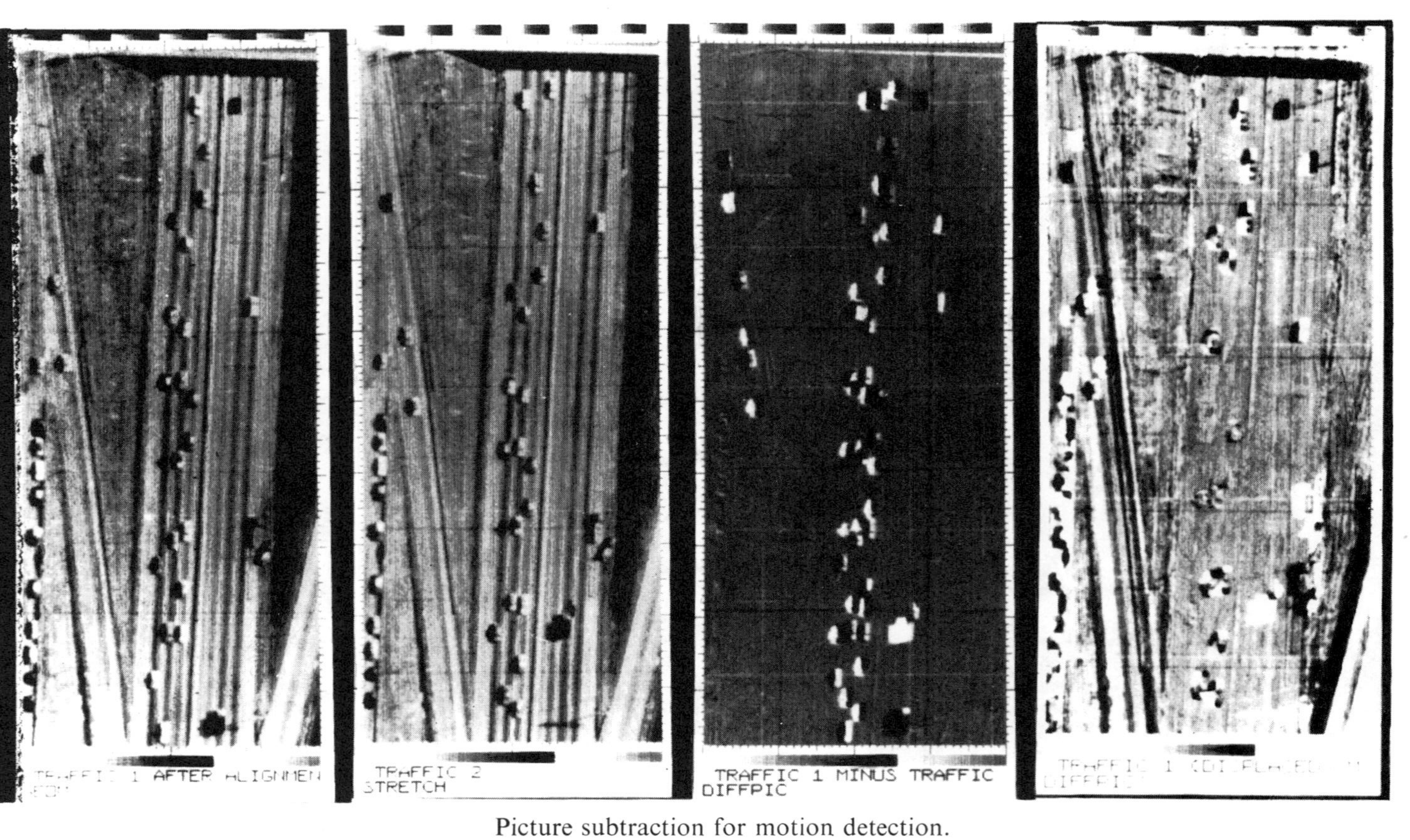

Picture subtraction for motion detection.

PLATE 16. Time sequence photography of a freeway. (a) Aerial photograph of freeway traffic. (b) Same as (a), after a time lapse. (c) Resultant of subtraction of (b) from (a). (d) Subtraction of displaced image. All vehicles corresponding to displacement velocity cancel.

PLATE 17. Multispectral response photographs of corn, alfalfa, stubble and bare soil as seen in three spectral bands.

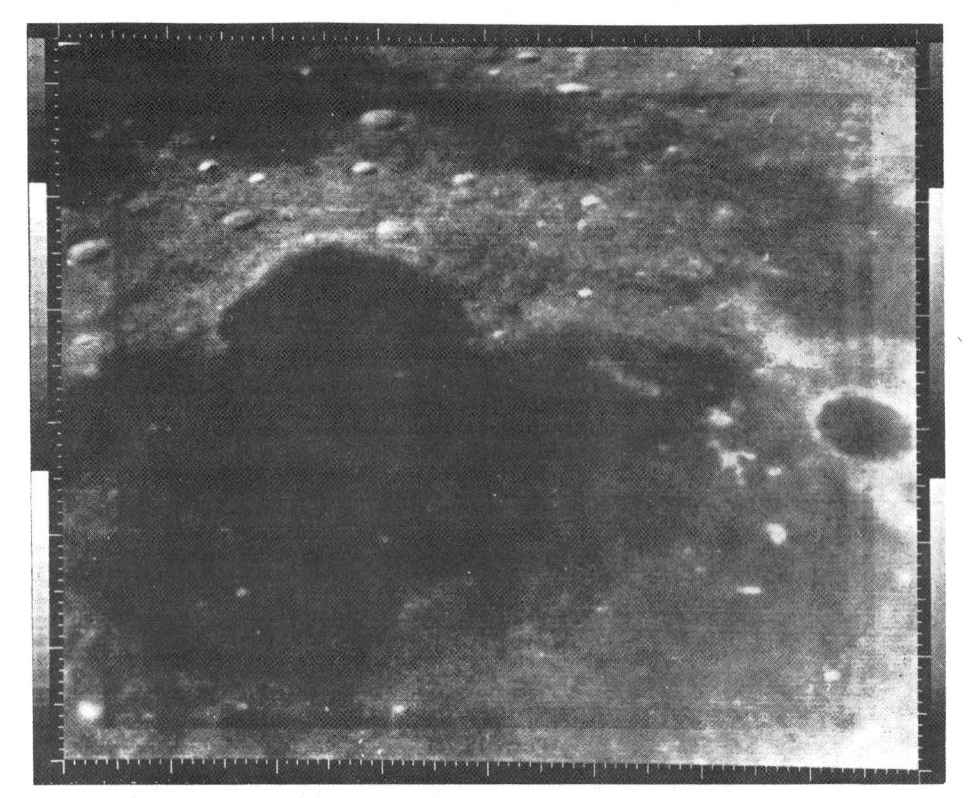

PLATE 18. (a) Mare Imbrium area of the Moon used for color analysis.

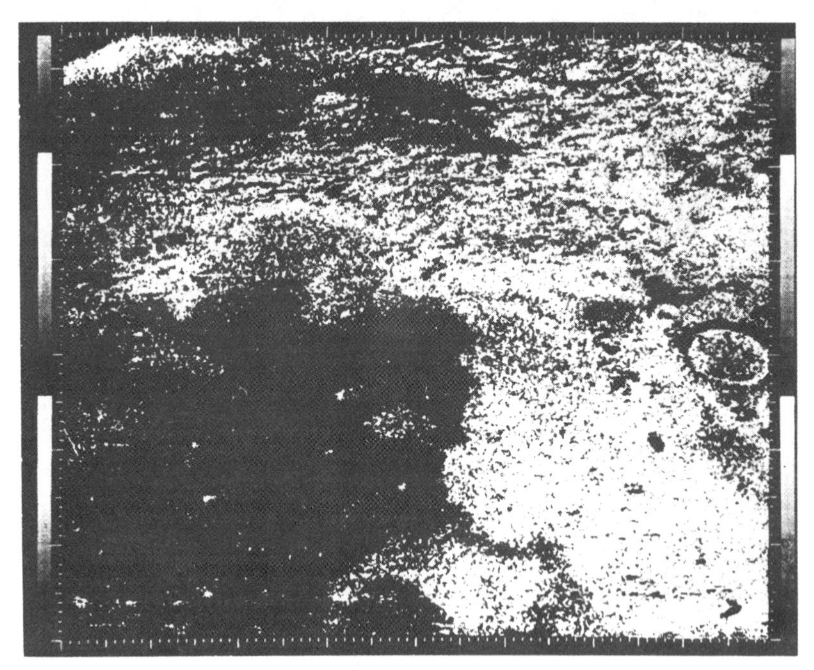

PLATES 18. (b), (c), (d) Red, grey and blue areas (referenced to a point in Plato which is defined as grey) are white in these separations. Note the color boundary which is almost invisible in Plate 18 (a).

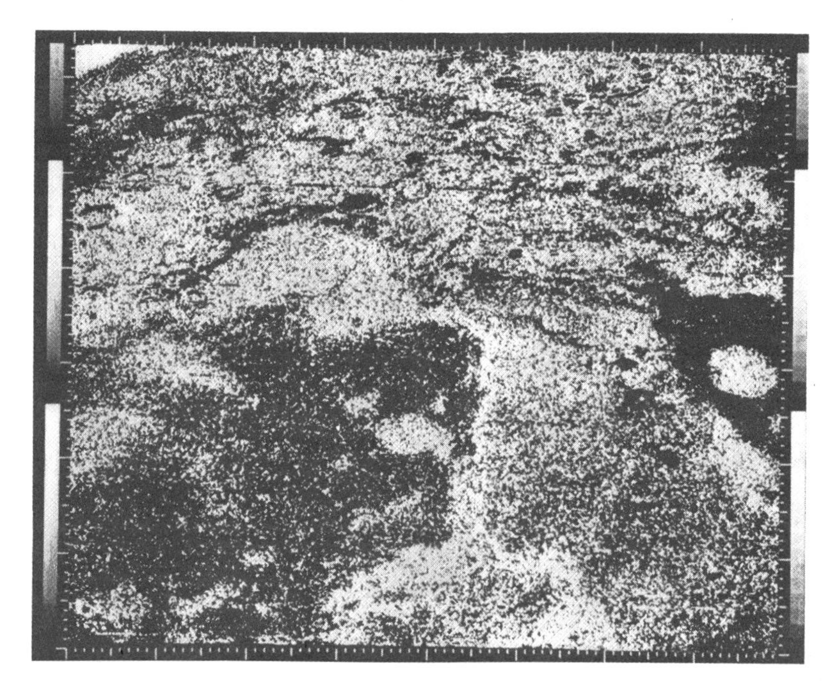

Plate 18. (c)

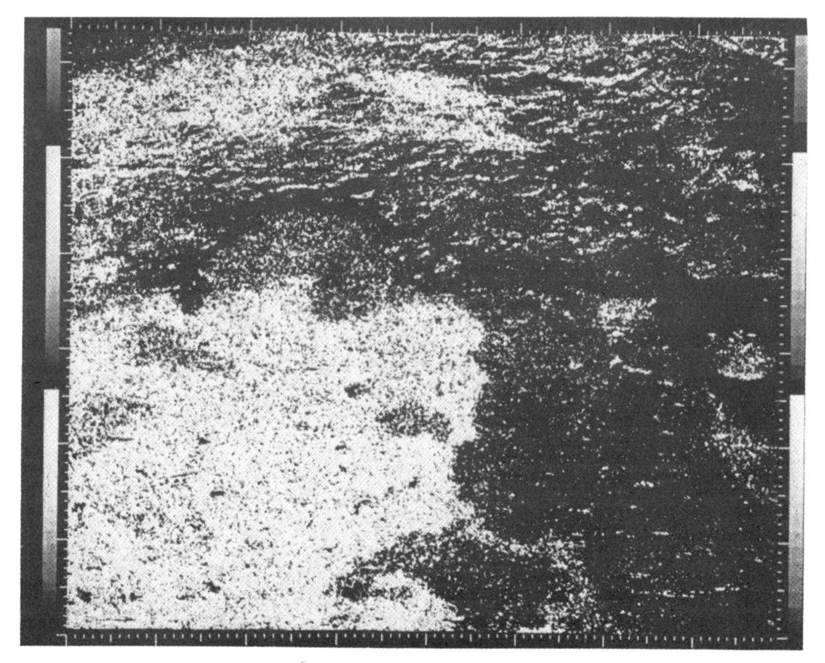

Plate 18. (d)

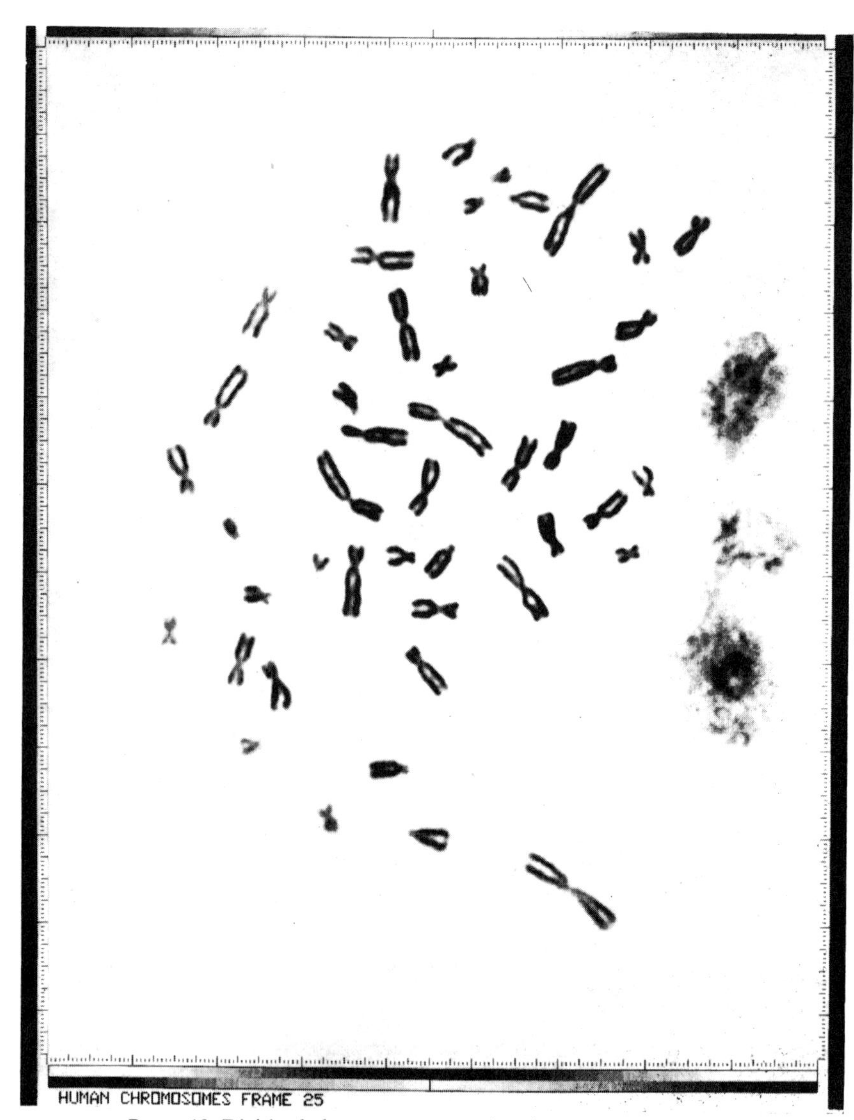

PLATE 19. Digitized chromosome spread to be computer analyzed.

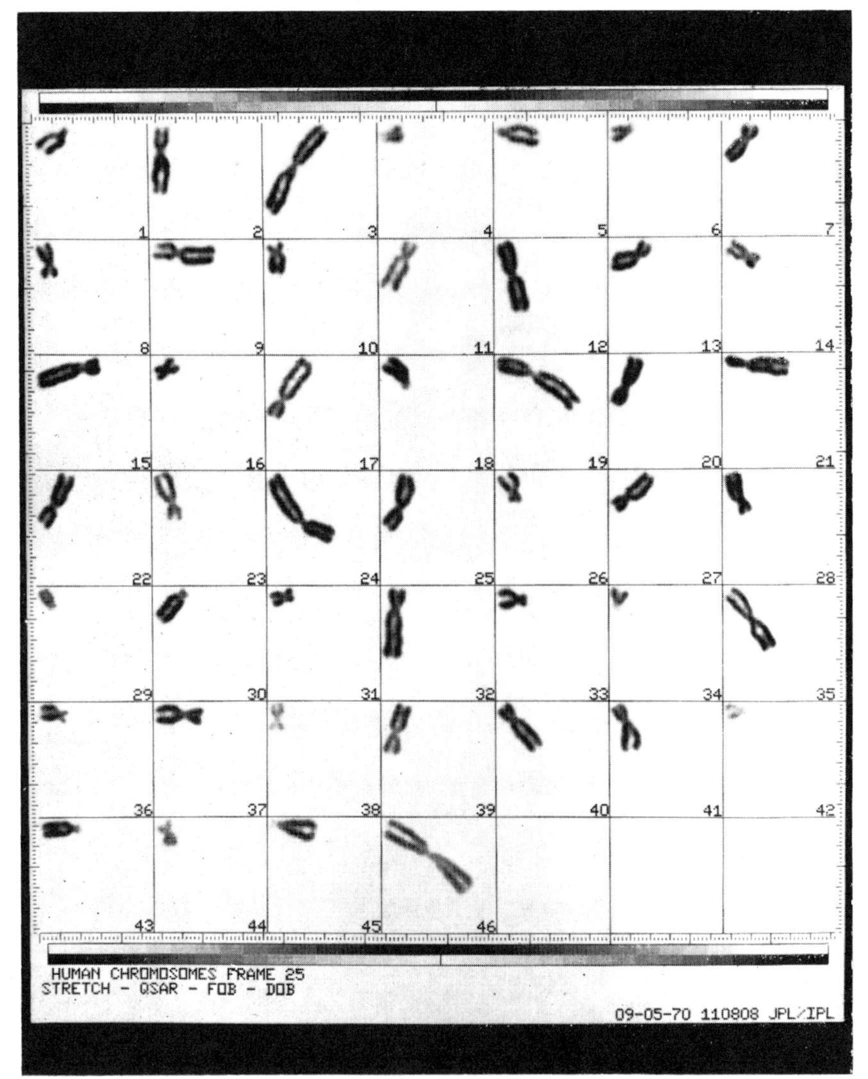

PLATE 20. Isolated chromosomes from Plate 19.

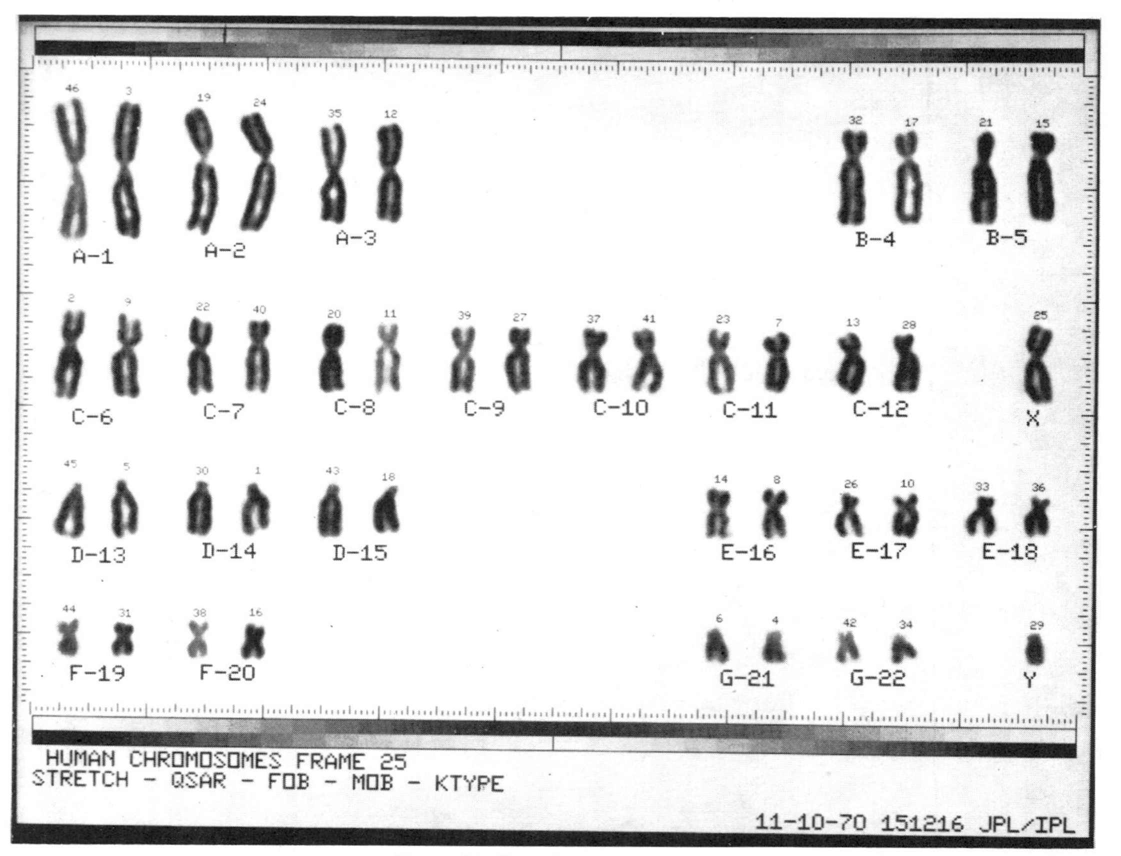

PLATE 21. Completed karyogram.

Pattern Recognition in Remote Sensing of the Earth's Resources

K.-S. FU, FELLOW, IEEE

(Invited Paper)

Abstract—**This paper reviews some recent topics in pattern recognition applied to remote sensing problems. In decision-theoretic pattern recognition, four topics are presented: per-field classifications, cluster analysis and sequential partitioning procedure, feature selection, and estimation of misclassification. The syntactic approach to pattern recognition is then introduced and its application to remote sensing problems illustrated. Problems for further research are discussed.**

I. Introduction

ONE important problem in remote sensing of the earth's resources is the characterization and classification of (spectral) measurements taken from various distances above the earth. For example, based on certain spectral measurements, we may want to classify various crops planted in a particular region. Or, we would like to identify one or several specific soil types in a region from the spectral measurements. This problem of analyzing data with respect to a specific goal could be considered as one falling into the general problem of pattern recognition [1]-[12]. The many different pattern recognition methods may be grouped into two general approaches; namely, the decision-theoretic (or statistical) approach and the syntactic (or structural or linguistic) approach. In the decision-theoretic approach, a set of characteristic measurements, called features, are extracted from the patterns; the recognition of each pattern (assignment to a pattern class) is usually made by partitioning the feature space. Most of the pattern recognition methods applied to remote sensing problems are decision-theoretic methods [13]-[16]. Linear and piecewise linear discriminant functions, and the Bayes classification rule have been used for data classification. Linear and some nonlinear feature-space transformations, and (statistical) distance measures have been applied to problems of feature extraction and selection. This paper discusses several selected topics in decision-theoretic pattern recognition methods and then introduces the syntactic approach to pattern recognition in remote sensing problems. For basic pattern recognition methods with remote sensing applications, refer to [14].

In the decision-theoretic approach, four topics will be discussed; per-field classifications, cluster analysis, feature selection, and estimation of misclassification. The syntactic approach will then be introduced, and its application to remote sensing problems illustrated. These topics are selected with the intention of introducing additional pattern recognition tools to the remote sensing community.

Manuscript received July 23, 1975. This work was supported by the National Science Foundation under Grant ENG 74-17586.

The author is with the School of Electrical Engineering, Purdue University, West Lafayette, IN 47907.

II. Per-Field Classification

In some remote sensing problems, classification can be carried out on the per-field or per-region basis. For example, in accordance with general crop planting practice, if an agricultural field has a certain majority of resolution elements classified as "ω_i," then one can feel reasonably sure that the entire field as a whole is in class "ω_i." Based on this observation, we shall formulate a pattern classification procedure as follows [17]. Let $F_1, F_2, \cdots, F_m$ be the probability distributions of the feature vectors corresponding to the pattern classes $\omega_1, \omega_2, \cdots, \omega_m$, respectively. Then, given a set of points (patterns) $\{X_1, X_2, \cdots, X_n\} = \{X\}$ all from the same field, the problem is to decide to which F_i they belong. A distance measure $d(F_i, F_j)$ is proposed as a suitable metric of the separation between any two distributions F_i and F_j, and this metric will be used as a rule for decision making. Let $G(X)$ be the distribution governing $\{X\}$. We shall compare the magnitudes of $\{d(G, F_i)\}$; the distribution F_i which minimizes the distance $d(G, F_i)$ shall be presumed to be the distribution which contains $G(X)$. Therefore, the vector set $\{X\}$ is classified into one of the known classes.

Let $F_i(X)$ and $F_j(X)$ be the distributions of the pattern classes ω_i and ω_j, respectively. Then the distance between distributions F_i and F_j is defined as follows:

$$d^2(F_i, F_j) = \int_{\Omega_x} \left(\sqrt{p(X/\omega_i)} - \sqrt{p(X/\omega_j)}\right)^2 dX. \tag{2.1}$$ [1]

If we define

$$\rho(F_i, F_j) = \int_{\Omega_x} \sqrt{p(X/\omega_i)} \cdot \sqrt{p(X/\omega_j)}\, dX \tag{2.2}$$

we have

$$d^2(F_i, F_j) = 2 - 2\rho(F_i, F_j). \tag{2.3}$$

The quantity $\rho(F_i, F_j)$ expresses the correlation between distributions F_i and F_j, and we can use $\rho(F_i, F_j)$ in decision making. It is important to note that to minimize the $d(F_i, F_j)$ is the same as to maximize the $\rho(F_i, F_j)$. Hence, the decision (classification) can be made by choosing the class which maximizes the correlation function ρ.

The distance $d(F_i, F_j)$ defined in (2.1) is applicable to any

[1] It should be noted that $d(F_i, F_j)$ is the Matusita distance between F_i and F_j.

Reprinted from *IEEE Trans. Geosci. Electron.*, vol. GE-14, pp. 10-18, Jan. 1976.

TABLE I
CLASSIFICATION RESULTS OF AIRCRAFT DATA TAKEN FROM AREA C-4 (JULY 1966)

Class	No. of Fields	No. of Field Classified into			
		Soybean	Corn	Pasture	Stubble
Soybean	19	17	2	0	0
Corn	31	1	30	0	0
Pasture	18	0	0	18	0
Stubble	13	0	0	3	10

Total No. of Fields Classified = 81

No. of Fields Correctly Classified = 75

Percent Correct Classification of Per-Field Classifier = 92.6

Percent Correct Classification of MLDR Classifier = 79.7

pair of distributions. We shall now turn to a specific example of multivariate Gaussian case. Let F_i and F_j be N-dimensional Gaussian distributions with mean vectors M_i, M_j, and covariance matrices K_i, K_j, respectively.

$$\rho(F_i, F_j) = \frac{|K_i^{-1}K_j^{-1}|^{1/4}}{|\frac{1}{2}(K_i^{-1}+K_j^{-1})|^{1/2}} \exp\left[\tfrac{1}{4}(K_i^{-1}M_i + K_j^{-1}M_j)^T \cdot (K_i^{-1}+K_j^{-1})^{-1}(K_i^{-1}M_i + K_j^{-1}M_j) - \tfrac{1}{4}(M_i^T K_i^{-1} M_i + M_j^T K_j^{-1} M_j)\right]. \tag{2.4}$$

When $K_i = K_j = K$,

$$\rho(F_i, F_j) = \exp\left[-\tfrac{1}{8}(M_i - M_j)^T K^{-1}(M_i - M_j)\right]$$

and when $M_i = M_j = M$,

$$\rho(F_i, F_j) = \frac{|K_i^{-1}K_j^{-1}|^{1/4}}{|\frac{1}{2}(K_i^{-1}+K_j^{-1})|^{1/2}}. \tag{2.5}$$

In order to evaluate the proposed per-field classification scheme 81 agriculture fields in the LARS Flight Line C4 area were used as test fields. Based on the histogram study, five subclasses were chosen to represent the soybean, the corn and the pasture. Four subclasses were used to represent stubble. This was due to the multimodal distribution of each of the four classes. The classification results are summarized in Table 1. When the same data were classified by the Maximum-Likelihood Classification Rule (MLDR) with the same set of training statistics, the classification accuracy was 79.7%.

In addition to the distance measure in (2.1), other distance measures and the maximum-likelihood classification rule have also been applied to the per-field classification [18], [19]. The problem now is the determination of field boundaries. Merging or bottom-up method, splitting or top-down method, and a mixed split-and-merge method have been proposed for the determination of field or region boundaries in remote sensing images [19]-[22].

III. CLUSTER ANALYSIS AND SEQUENTIAL PARTITION PROCEDURE

One of the goals of pattern recognition study is to find meaningful descriptions to adequately characterize a set of data. The descriptions can then be used for classification purposes. Parametric classification techniques can be used when the parametric form of each class distribution is known. On the other hand, if the parametric form is unknown, nonparametric techniques must be used. Two nonparametric techniques are discussed in this section.

A. Cluster Analysis

A cluster seeking or simply a clustering procedure is a nonparametric method to determine a type of structure describing a set of empirical data. Data which are similar in some sense are lumped together and isolated or partitioned from the remaining data set. A set of patterns contained in a region of feature space where the density of patterns is large compared to the density of patterns in the surrounding regions is usually called a cluster. Many cluster seeking procedures have been proposed [23], [24]. In general, they may be described in terms of the following steps.

Step 1: Partition the data set using an appropriate similarity criterion. Patterns which are considered similar, usually in terms of comparing a prespecified similarity measure with one or more thresholds, are grouped into the same cluster.

Step 2: Test the partition to determine whether it is significant; that is, whether or not the subsets of patterns are sufficiently distinct (for example, in terms of distances or correlations between clusters). If not, merge any subsets which are not sufficiently distinct.

Step 3: Repeatedly partition the subsets created, testing at each step as in Step 2, until no further permanent subdivisions result, or until some other stopping criterion is satisfied. The stopping criterion, for example, may be that the clusters established have satisfied the *a priori* information about the total number of classes or that a certain limitation on computing time and/or data storage is reached.

A specific clustering procedure results from the general procedure when a similarity measure, distinctness test, and stopping rule are supplied. Let X_i and X_j be any two given samples (pattern vector) with components $x_{i1}, \cdots, x_{iN}$ and $x_{j1}, \cdots, x_{jN}$. A list of several commonly used similarity measures is given in the following:

(i) Dot product

$$X_i \cdot X_j = |X_i||X_j| \cos(X_i, X_j). \tag{3.1}$$

(ii) Similarity ratio

$$S(X_i, X_j) = \frac{X_i \cdot X_j}{X_i \cdot X_i + X_j \cdot X_j - X_i \cdot X_j}. \tag{3.2}$$

(iii) Weighted Euclidean distance

$$d(X_i, X_j) = \sum_{k=1}^{N} w_k (x_{ik} - x_{jk})^2. \tag{3.3}$$

(iv) Unweighted Euclidean distance

$$d(X_i, X_j) = \sum_{k=1}^{N} (x_{ik} - x_{jk})^2. \tag{3.4}$$

(v) Boolean "and" (or weighted Boolean "and")

$$\sum_{k=1}^{N} x_{ik} \cap x_{jk}. \tag{3.5}$$

(vi) Normalized correlation

$$\frac{X_i \cdot X_j}{\sqrt{(X_i \cdot X_i)(X_j \cdot X_j)}} \tag{3.6}$$

where

$$X_i \cdot X_j = \sum_{k=1}^{N} x_{ik} x_{jk}.$$

Because of the lack of a unified theoretical treatment for clustering procedures, it is usually rather difficult to evaluate the performance of a clustering procedure except empirically. Many parameters such as similarity measure, merging and splitting criteria, and thresholds are dependent upon the designer's choice. In many cases, man-machine interaction capability is required to facilitate the on-line adjustments of some of the design parameters. Nevertheless, from a practical application point of view, clustering analysis is considered as a practical and effective approach in pattern recognition. Applications of clustering procedures to remote sensing problems can be found in [25]-[27].

B. A Sequential Partitioning Procedure

A nonparametric sequential procedure which seeks to partition the feature space into successively finer regions for classification purpose has recently been applied to remote sensing [28]. The resultant partition of the feature space from the training samples can be easily implemented by a layered structure of threshold devices. The procedure is summarized as follows: For a two-class one-dimensional case, let $T_1 = \{X_1, \cdots, X_{n_1}\}$ be a set of n_1 independent training samples from class ω_1 where X_i is a one-dimensional feature measurement of the ith training sample. Similarly, let $T_2 = \{Y_1, \cdots, Y_{n_2}\}$ be a set of n_2 independent training samples from class ω_2. Let K be a prespecified parameter[2] determined from the combined sample size $n = n_1 + n_2$. Denote c_1 and c_2 to be the relative cost associated with misclassifying patterns from class ω_1 and class ω_2, respectively, and c_0 to be cost of indecision (pattern class ω_0).

Step 1: Order the combined sample set $T_1 \cup T_2$ according to increasing numerical value to form an ordered set T. Partition the set T into successive groups or regions of K samples (refer to Fig. 1(a)-(c)).

Step 2: For each group count the total number of X's and number of Y's and assign a pattern class according to the following: If

[2] The particular choice of K is not of concern in the concepts and the mathematical analysis of the procedure. Since the purpose of partitioning the training set into groups of K samples is to reduce the number of partitions which need to be considered, K should not increase as fast as n does. That is, mathematically, $\lim_{n\to\infty} K/n = 0$.

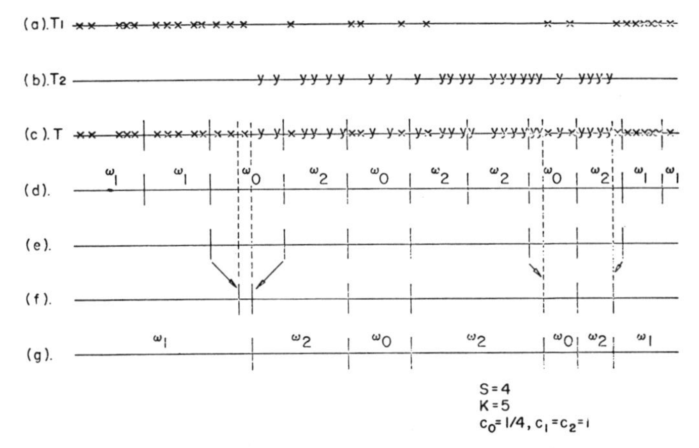

Fig. 1. Illustration of basic partitioning procedure.

$$\min_i \left\{ \sum_{\substack{j=1 \\ \neq i}}^{m=2} c_j \text{ (number of samples from } \omega_j) \right\} < c_0 K$$

$$\text{assign class } \omega_i (i = 1 \text{ or } 2) \tag{3.7}$$

otherwise, assign class ω_0. Then combine consecutive regions which have been assigned to the same class (refer to Fig. 1(d)-(e)).

Step 3: Adjust the partitioning boundaries by perturbing them a maximum of $K/2$ samples in either direction and relocating them at positions where the most improvement in classification accuracy is obtained (Fig. 1(f)).

Step 4: If less than $K/2$ samples remain in any one region, dissolve that region and place its samples among the neighboring two regions so as to yield least increase in misclassifications (Fig. 1(g)).

Step 5: Repeat Step 2. For this final partition compute the empirical classification statistic or the total cost S as

$$S = \sum_{i=1}^{m=2} c_i \text{ (number of samples misclassified from } \omega_i)$$
$$+ c_0 \text{ (number of unclassified samples).} \tag{3.8}$$

The implementation of the above classification algorithm can be carried out easily by using threshold devices. For the example in Fig. 1, six thresholds will be required for the implementation. The structure of the classifier for the example in Fig. 1 is shown in Fig. 2. For an unknown input pattern X (one-dimensional in this case), if

$$\begin{array}{lll} X < t_1 & \text{then} & X \sim \omega_1 \\ t_1 < X < t_2 & \text{then} & X \sim \omega_2 \\ t_2 < X < t_3 & \text{then} & X \sim \omega_0 \\ t_3 < X < t_4 & \text{then} & X \sim \omega_2 \\ t_4 < X < t_5 & \text{then} & X \sim \omega_0 \\ t_5 < X < t_6 & \text{then} & X \sim \omega_2 \\ t_6 < X & \text{then} & X \sim \omega_1. \end{array}$$

Notice that this algorithm is easily extended for the multiclass case. The only modification necessary is the formation of additional class labels used in Step 2.

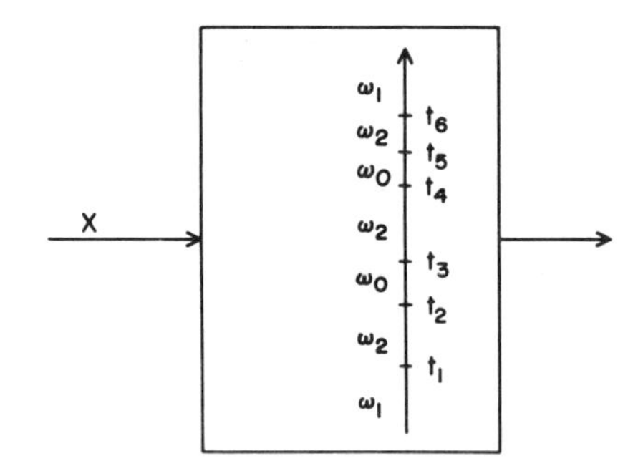

Fig. 2. Structure of classifier for the sample in Figure 1.

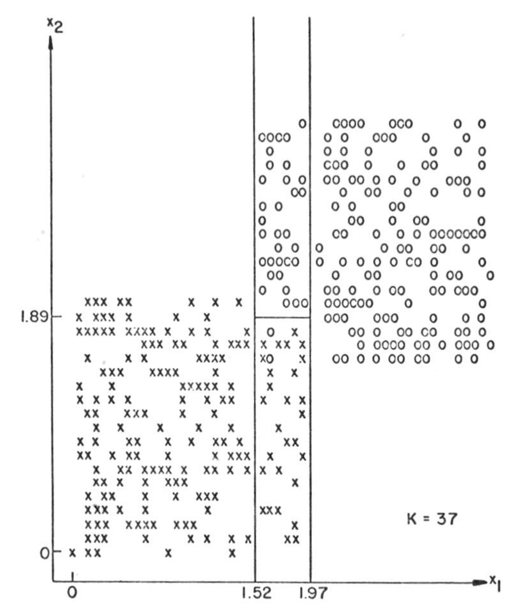

Fig. 3. Partition of a two-dimensional feature space.

Additional information would be required to classify the samples falling into the ω_0 regions; that is, as in the sequential classification procedures, the second feature measurement will be needed. In such a situation, the feature space becomes two-dimensional, and a second layer of threshold devices will be used to implement the partition on the second dimension or feature. A typical two-class two-dimensional example is given in Fig. 3. The basic idea in the extension to multidimensional feature space is to apply the algorithm to each feature or dimension separately. The empirical classification statistic S obtained from Step 5 of the algorithm is computed for each of the features or dimensions. The dimension associated with the lowest value of S is selected as the dominant dimension; in other words, the corresponding feature measurement will be taken first. The feature space is then partition by parallel hyperplanes which have the dominant dimension as the common normal vector and intersect it at the boundaries determined by the algorithm for this dimension (Fig. 3). For each new region ω_0 formed by the above partitions, the procedure is repeated until no ω_0 regions are produced.

The following example illustrates the application of the sequential partitioning procedure to crop classification problems. Only features 1, 10 and 12, corresponding to the spectral wavelength bands of 0.40–0.44 μ, 0.66–0.72 μ and 0.80–1.00 μ were used. In order to demonstrate the nonparametric characteristic of the procedure, a relatively smaller number of samples were drawn from each class (corresponding to each crop) compared with the usually large number of samples required to accumulate the parametric information (that is, the Gaussian assumption). For each of the examples to be discussed, a comparison to the maximum-likelihood classification rule (MLDR) using Gaussian assumption was also made for each set of data.

TABLE II
CLASSIFICATION OF 400 TWO-CLASS (SOYBEAN AND CORN) TRAINING SAMPLE (FEATURES 1, 10 AND 12)

Method	Correct (percent)	Incorrect (percent)	Undecided (percent)
Sequential Partitioning Procedure	326 (81.5)	59 (14.5)	15 (4)
MLDR with Gaussian assumption	328 (82)	72 (18)	0 (0)

Two hundred (three-dimensional) samples were drawn from each of the soybean and corn crops. The classification results of the training set by the nonparametric partitioning procedure and by the maximum likelihood classification rule using Gaussian assumption are given in Table 2. The structure of the

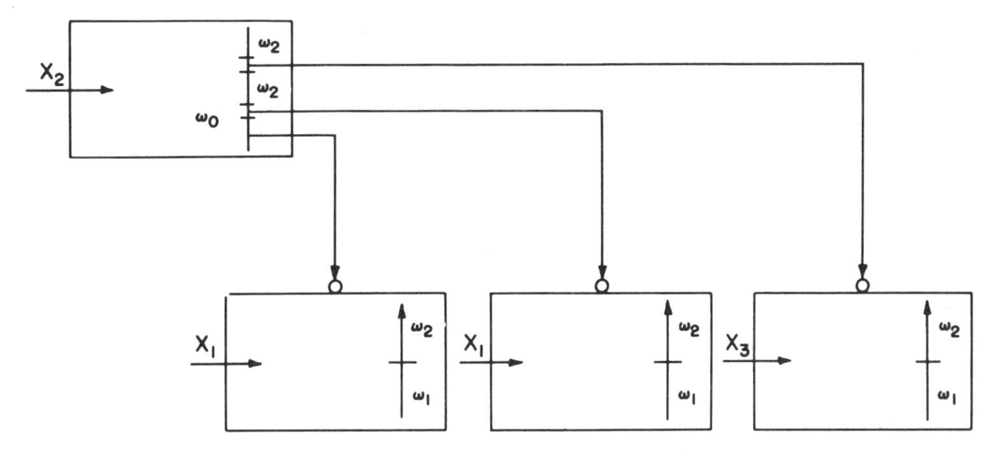

Fig. 4. Classifier structure for example 3.

TABLE III
CLASSIFICATION OF 912 TWO-CLASS (SOYBEAN AND CORN) TEST SAMPLES

Method	Correct (percent)	Incorrect (percent)	Undecided (percent)
Sequential Partitioning Procedure	705 (77.5)	185 (20)	22 (215)
MLDR with Gaussian assumption	698 (76.5)	214 (23.5)	0 (0)

classifier determined by the nonparametric partitioning procedure is given in Fig. 4. Table 3 gives a summary of the classification results for a test set of 912 samples, including the 400 training samples [28].

IV. FEATURE SELECTION

A. Distance Measure for Feature Selection

The main goal of feature selection is to select a subset of l features from a given set of N features ($l < N$) without significantly degrading the performance of the recognition system, that is, the probability of misrecognition, or more generally, the risk of decision. Unfortunately, a direct calculation of the probability of misrecognition is often impossible or impractical partially due to the lack of general analytic expressions which are simple enough to be treated. One approach is to find some other indirect criteria to serve as a guide for feature selection.

The most common approach is to define a distance or separability measure between the probability distributions corresponding to the classes under investigation. The distance measures suggested in the existing literature can be put in one of the following two categories: (i) those derived by semiheuristic reasoning, (ii) those based on information theory and statistics. The heuristic measures usually have no explicit relation to the probability of error; however, their relative simplicity to evaluate is very attractive when the statistical model of the problem is not known. In this section, we will focus our interest on measures derived from information theory and statistics. In recent years, there have been an increasing interest in distance measures and their corresponding error bounds in the literatures on statistical decision theory, communication and pattern recognition. A detailed historical review of the development of this type of measures (with references going back to 1912) and the application of the divergence and Bhattacharyya distance to signal selection can be found in Kailath [29]. The relation between Bayes error P_e and the equivocation for the case of m-classes is explored in Hellman and Raviv [30]. Lainiotis and Park [31]-[32] used pair-wise Bhattacharyya distances to derive upper and lower bounds on P_e for the M-class case. Inter-relations between several separability measures are discussed by Lainiotis and Park [33], Kobayashi and Thomas [34] and more extensively by Toussaint [35]. (Toussaint also presents a more general version of the distance by Patrick and Fischer from [36].) More recently, Ito [37] and Devijver [38] suggested distance measures which are similar to the one proposed by Vajda [39].

Assuming that the most important characteristic of the distance measure is its upper bound on the error P_e, we can arrange the measures listed above in an increasing order of importance. Consider the m-class problem, denote the upper bound on P_e when using Bhattacharyya's distance U_B, Matusita's distance U_M, equivocations U_E, Vajda's entropy U_V, Devijver's Bayesian distance U_D, Ito's measure (for $n = 0$) U_1, Kolmogorov's variational distance U_K and M_0 - distance by Toussaint U_T, we get for $m = 2$

$$P_e = U_K \leqslant U_V = U_D = U_1 = U_T \leqslant U_E \leqslant U_B = U_M \tag{4.1}$$

and for $m > 2$, U_K is no more equal to P_e and not necessarily smaller than U_V, although according to many examples we run on the computer, it appears that $U_K < U_V$ even if $m > 2$ for $P_e \leqslant 50\%$. The divergence and Kullback-Leibler numbers, which are simply related to each other, are excluded from the ordering in (4.1) due to the lack of a known upper bound except for normal distributions where its bound is larger than U_B. The distance measure by Patrick and Fischer $d^2(\omega_1, \omega_2)$ and its generalized form by Toussaint induce upper bounds which are larger than U_V, but their relation to U_E and U_B was not yet explored. It should be noted that the inequality relation in (4.1) should be interpreted as "pointwise" relations rather than "in expectation" as they are stated in (4.1). This observation is somewhat damaging the glamour of the bounds on the probability of misrecognition, and even more disappointing is the fact that the best bound on the error (except U_K which is nothing but the error itself) derived from distance measures is equal to U_{1NN} the asymptotic error of the simple first nearest neighbor classifier. This implies that for moderately large data sets, in a nonparametric problem, one should not attempt to estimate any of the discussed distance measures since the nearest neighbor or the edited nearest neighbor [40] classifier will be much easier to apply and bear better or equal result.

Lissack and Fu [41] have recently proposed a family of distance measures

$$J_\alpha(\omega_i, \omega_j) = E[\,|p(\omega_i\,|X) - p(\omega_j\,|X)\,|^\alpha], \quad \alpha \geqslant 0 \qquad (4.2)$$

and derive the corresponding lower and upper bounds on the probability of misrecognition. It is shown that if $\alpha = 1$ the bounds exactly coincide with the error and an increase or decrease of α will loosen the bounds. Then it is observed that U_2, the bound derived for $\alpha = 2$, is equal to U_V and hence U_α for $1 \leqslant \alpha \leqslant 2$ is better or equal to any bound in (4.1). For $0 \leqslant \alpha \leqslant 1$ the upper bounds are not pointwise (as in the discussion above), and, therefore, could not be compared in general with previous results as in (4.1). However, for all the problems they have investigated, they have obtained that, for example, $\alpha = 0.5$ bears a tighter upper bound than $\alpha = 2$, $U_{0.5} \leqslant U_2$. This result should certainly limit the use of common distance measures and promote the application of J_α with $1 \leqslant \alpha \leqslant 2$.

Some applications of distance measures to feature selections in remote sensing can be found in [42]-[44].

B. Adaptive Feature Selection

In some practical applications, the number of features N and the number of pattern classes m could be both very large (e.g., fifty spectral measurements for fifty types of ground cover). In such cases, it would be advantageous to use a multi-level recognition system based on the decision tree scheme. A typical decision tree scheme is shown in Figure 5. At the first level, m classes are classified into i groups using only N_1 features. Here, $i << m$ and $N_1 << N$, and the N_1 features selected are the best features to classify these i groups. In an extreme case, $i = 2$ so a two-class classifier can be used, or $N_1 = 1$ so a one-dimensional (thresholding) classifier can be used. The same procedure will then be repeated at the second level for each of the i groups. For example, the group of m_1 classes at the second level can be classified into j_1 groups using only N_{21} features, the group of m_2 classes can again be classified into j_2 groups using N_{22} features, etc. Continue in this fashion for the third level, fourth level, etc., until each of the original m classes can be separately identified. Now, following each tree path in the decision tree, we should be able to recognize each of the m classes.

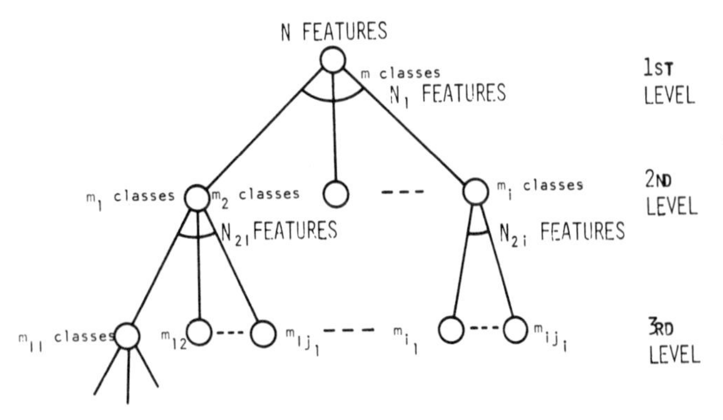

Fig. 5. A decision tree scheme for classification.

The idea of adaptively selecting a smaller number of features at different levels of classification appears to be very attractive in applications. An optimal design of such a tree scheme may be computationally quite complex. However, several heuristic design techniques have recently been suggested [45], [46]. The decision tree scheme for classification results in a layered classifier structure. It should be interesting to compare the layered classifier using the decision tree scheme with that using the sequential partitioning procedure discussed in Section III-B.

V. Estimation of Probability of Misclassification

The performance of a classifier is usually expressed in terms of its probability of misclassification. Consider that we have a Bayes classifier. If the class *a priori* probabilities, $P(\omega_i)$, and the class conditional probability density functions, $p(X\,|\,\omega_i)$ are known, we can calculate exactly the probability of error. On the other hand, if $p(X\,|\,\omega_i)$ are not completely known, we can only estimate the probability of misclassification from a (finite) set of pattern samples. There are two approaches to the estimation of misclassification: the resubstitution approach and the cross-validation approach [47], [48]. In the resubstitution method, the pattern samples used for training the classifier are also used as test samples. It is quite easy to realize that the resubstitution method often gives a (optimistically) biased estimate of the actual performance of the classifier since it provides no measure of the generalization capability of the "learning" machine.

In cross-validation methods, the available pattern samples are partitioned in various ways such that an estimate is finally obtained which reflects the actual performance on samples not used for training. One method, called "hold-out" method, suggests that some (usually half of the) samples are held out to test the classifier. Another method is the "leave-one-out" method. In this method, given n samples, a classifier is trained on $(n - 1)$ samples, tested on the remaining sample, and then the results of all such partition of size $(n - 1)$ for the training set and one for the test set are averaged. Except in some special cases, this method requires n times the computation of the hold-out method. Unfortunately, an attempt to analytically determine the optimal partitioning in order to minimize the variance of the estimate has not been very successful. Although it is desirable to use an unbiased estimator of the probability of misclassification, it is also important to use an estimator with small variance. It has been shown that the leave-one-out method has much greater variance than the resubstitution method for the case of discrete class distributions, and experimental results indicate similar behavior in the continuous case.

A compromise between the hold-out method and the leave-one-out method is the "rotation" or π method. The π method consists of partitioning the n samples into a test set T_s of size k, $1 \leqslant k \leqslant n/2$, n/k an integer, and a training set T_r for the remaining samples. The classifier is then trained on T_r and tested on T_s to get an estimate of the probability of misclassification. The procedure is repeated with additional disjoint test sets and corresponding training set, and the average over the various disjoint test sets results is used for the final estimate. With $k = 1$, this method is the leave-one-out method, and with $k = n/2$ this gives a version of the hold-out method.

The average probability of misclassification using the resubstitution method gives a lower bound on the true probability of misclassification while the other methods yield upper bounds. This leads to the suggestion of using the estimate

$$\hat{P}_e = \beta E[\hat{P}_e(\pi)] + (1 - \beta)\hat{P}_e(R) \tag{5.1}$$

where $\hat{P}_e(\pi)$ = the estimate of misclassification using the π method

$\hat{P}_e(R)$ = the estimate of misclassification using the resubstitution method and $0 \leqslant \beta \leqslant 1$ is a constant depending upon the sample size n, the feature size N, and the test set size k.

Recently, Lissack and Fu [49] have proposed a method, called the F method, for the estimation of misclassification through their proposed distance measure (4.2). The estimation of $J_1(\omega_i, \omega_j)$ is reduced to the estimation of a scalar function $|p(\omega_i|X) - p(\omega_j|X)|$ at the given sample points. For the class distributions being the family of exponential distributions (which includes all normal distributions), the separation at a sample point is shown to be $\tan h |\frac{1}{2} d(X)|$ where $d(X)$ is the value of the logarithm of the likelihood ratio. This result relates the distance of a sample from the decision boundary to its contribution to the probability of misclassification. Experimental results on Gaussian data, using the F method have been reported. The F method has been found to have less bias and smaller variance than the leave-one-out method and the resubstitution method.

VI. Syntactic Pattern Recognition Applied to Remote Sensing Problems

The approach of using hierarchical structures and grammar rules to describe the structures of patterns has recently received increasing attention [11]. This approach is often called the structural or syntactic approach to distinguish it from the decision-theoretic or statistical approach. Practical applications include the description of chromosome images, the recognition of characters, spoken digits, electrocardiograms, and two-dimensional mathematical expressions, the identification of bubble chamber and spark chamber events, and the recognition of fingerprint patterns. In the syntactic approach, each pattern is described in terms of its parts, i.e., subpatterns. Each subpattern can again be described in terms of its parts. The simplest subpatterns are called the pattern primitives, and they constitute the basic symbols (the set of terminals) of the pattern language. The description of each primitive can be either deterministic or statistical and the recognition of primitives is often based on the decision-theoretic approach. Each class of patterns is now described by a set of sentences consisting of the primitives, and it can be generated by a pattern grammar. With the above description, it might be said that in the syntactic approach we often use the decision-theoretic approach for primitive recognition; however, the emphasis will be on the use of syntactic rules to describe the structure of patterns (the compositions rules of the primitives and subpatterns). Recognition of patterns is often performed in terms of a parsing algorithm with respect to the syntactic rules. A pattern is classified as from the class of which the pattern grammar will generate the pattern under consideration. When a pattern is classified, its structure in terms of a parse tree is also given. In addition to the use of one-dimensional string grammars, high dimensional grammars (e.g., tree grammars, graph grammars, web grammars, etc.) have been recently proposed for pattern description [50]-[52].

Multi-spectral signals measured by LANDSAT over Marion County (Indianapolis), Indiana were analyzed using clustering analysis. Fourteen clusters were found and the data from the metropolitan area were accordingly classified using a Bayes classifier. The result of the Bayes classifier provides the basic pattern primitives. The structure of the data classified (Fig. 6) can be expressed hierarchically as shown in Figure 7. Linguistic

Fig. 6. Photograph of Marion County imagery from digital display.

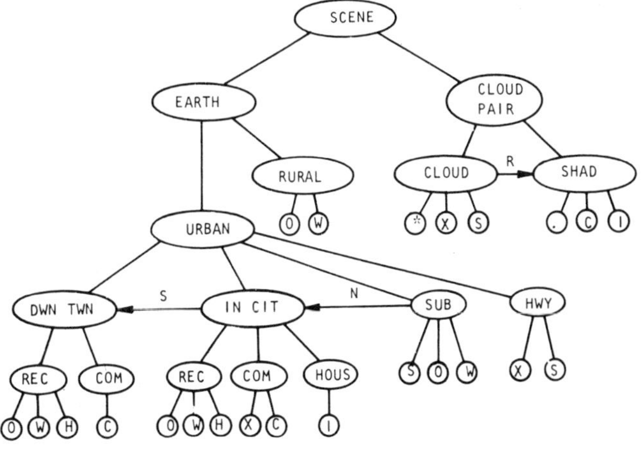

Fig. 7. A hierarchical graph model of the scene.

rules in terms of web grammars were constructed to describe cloud-shadow pairs and highways, respectively [53]. These linguistic rules essentially describe the spatial relations among the pattern primitives in the formation of cloud-shadow pairs and highways. The recognition of clouds and highways was performed in terms of a parsing algorithm. Preliminary results have shown that the recognition accuracy can be significantly improved using the information of spatial relations expressed syntactically. A similar viewpoint has also been implied in the processing of satellite pictures for the recognition of roads [54].

VII. Remarks

When the exact functional relationships between the ground cover types and the remotely sensed (spectral) measurements are not known, they can only be observed through experiments. This makes pattern recognition methods very useful for the data analysis in remote sensing. Depending upon the availability of ground truth, supervised or nonsupervised techniques are used to learning the functional relationships and, consequently, the classifier design. Depending upon the parametric knowledge about the distributions of spectral measurements for each class (each type of ground cover of interest) is known or not, parametric or nonparametric statistical pattern recognition methods can be applied. It should be kept in mind that a Gaussian assumption of the class distributions will result in a quadric decision boundary between two classes, and an additional assumption of equal covariance matrices will reduce the decision boundary to a hyperplane. Sufficient training samples are always needed to estimate the statistical parameters even in a parametric case. Otherwise, due to the poor estimates, the classifier's performance may become poorer even with the increase of features, or dimensionality.

In order to simplify the classifier structure and to make the computations involved more efficient, the problem of feature selection is important. This paper gives a brief overview of using distance measures for feature selection. The applicability of this approach to remote sensing problems certainly requires more extensive tests in the future. An improvement of classification performance could be obtained by sequentially (adaptively) selecting the features on the basis of a decision-tree scheme. In order to have a reliable performance measure of the classifier designed, an estimate of misclassification is necessary. The estimate of misclassification using all the training samples also as test samples is obviously optimistic and, therefore, biased. Several other estimation techniques are introduced in this paper. It is quite possible that a fair comparison of the performance of different classifiers may require the information of more than one estimate of misclassification.

In terms of measurement resolution and computation efficiency, per-field or per-region classification appears to be an attractive approach in remote sensing of earth's resources. However, an efficient and effective segmentation or boundary detection technique of multrispectral pictures is needed before each field and region can be accurately determined. The spatial relations among various regions in a given scene can be described using the syntactic approach. Contextual information can also be treated in a limited fashion using a statistical approach [55]. The syntactic approach, however, may provide a more direct connection to the problem of storage and retrieval of the information obtained from the data classified and interpreted. This data base problem, though not yet receiving much attention in remote sensing, will become more important when a large amount of analyzed data is accumulated.

References

[1] K. S. Fu, *Sequential Methods in Pattern Recognition and Machine Learning*, Academic Press, New York, 1968.

[2] N. J. Nilsson, *Learning Machines*, McGraw-Hill, New York, 1965.

[3] J. M. Mendel and K. S. Fu, eds., *Adaptive, Learning and Pattern Recognition Systems: Theory and Applications*, Academic Press, New York, 1970.

[4] K. Fukunaga, *Introduction to Statistical Pattern Recognition*, Academic Press, New York, 1972.

[5] E. A. Patrick, *Fundamentals of Pattern Recognition*, Prentice-Hall, Englewood Cliffs, New Jersey, 1972.

[6] W. Meisel, *Computer Oriented Approaches to Pattern Recognition*, Academic Press, New York, 1972.

[7] H. C. Andrews, *Introduction to Mathematical Techniques in Pattern Recognition*, Wiley, New York, 1972.

[8] C. H. Chen, *Statistical Pattern Recognition*, Hayden, Washington, DC, 1973.

[9] R. O. Duda and P. E. Hart, *Pattern Classification and Scene Analysis*, Wiley, New York, 1973.

[10] T. Y. Young and T. W. Calvert, *Classification, Estimation, and Pattern Recognition*, American Elsevier, New York, 1974.

[11] K. S. Fu, *Syntactic Methods in Pattern Recognition*, Academic Press, New York, 1974.

[12] J. T. Tou and R. C. Gonzalez, *Pattern Recognition Principles*, Addison-Wesley, Reading, MA, 1974.

[13] K. S. Fu, D. A. Landgrebe and T. L. Phillips, "Information Processing of Remotely Sensed Agricultural Data," *Proc. IEEE*, Vol. 57, No. 4, April 1969, pp. 639–653.

[14] K. S. Fu, "On the Application of Pattern Recognition Techniques to Remote Sensing Problems," Technical Report, TR-EE 71-13, School of Electrical Engineering, Purdue University, West Lafayette, Indiana, U.S.A.

[15] Proceedings of the Symposium of Machine Processing of Remotely Sensed Data, 1973 and 1975, Purdue University, West Lafayette, Indiana.

[16] Proceedings of the Symposium on Remote Sensing of Environment, 1964–1975, University of Michigan, Ann Arbor, Michigan.

[17] T. Huang, "Per-Field Classification of Remotely Sensed Agricultural Data," *Proc. 1970 Allerton Conference on Circuit and System Theory*.

[18] A. G. Wacker and D. A. Landgrebe, "The Minimum Distance Approach for Classification," LARS Information Note 100771, Purdue University, W. Lafayette, Indiana, 1971.

[19] R. L. Kettig and D. A. Landgrebe, "Classification of Multispectral Image Data by Extraction and Classification of Homogenous Objects," Proc. 197.

[20] T. V. Robertson, P. H. Swain, and K. S. Fu, "Multispectral Image Partitioning," Tech. Rept. TR-EE 73-26, School of Elec. Eng., Purdue University, West Lafayette, Indiana, August 1973.

[21] E. M. Rodd, "Closed Boundary Field Selection in Multispectral Digital Images," IBM Publication No. 320.2420, January, 1972.

[22] T. Pavlidis and S. L. Horowitz, "Picture Segmentation by a Directed Split-and-Merge Procedure," *Proc. Second International Joint Conference on Pattern Recognition*, August 1974, Copenhagen, Denmark.

[23] N. Jardine and R. Sibson, *Mathematical Taxonomy*, Wiley, New York, 1971.

[24] M. R. Anderberg, *Cluster Analysis for Application*, Academic Press, New York, 1971.

[25] P. H. Swain and K. S. Fu, "On the Application of a Nonparametric Technique to Crop Classification Problem," *Proc. 1968 National Electronics Conference*.

[26] G. C. Gustafson, "Numerical Classification Procedures in Fluvial Geomorphology," *Proc. 1973 Symp. on Machine Processing of Remotely Sensed Data*, LARS, Purdue University, W. Lafayette, Indiana 47907.

[27] E. P. Pan, W. A. Holley and H. D. Parker, Jr., "The JSC Clustering Program ISOCLS and Its Applications," *Proc. 1973 Symp. on Machine Processing of Remotely Sensed Data*, LARS, Purdue University, W. Lafayette, Indiana 47907.

[28] E. G. Henrichon and K. S. Fu, "A Nonparametric Partitioning Procedure for Pattern Classification," *IEEE Trans. on Computers*, Vol. C-18, No. 7, July 1969.

[29] T. Kailath, "The Divergence and Bhattacharyya Distance Measures in Signal Selection," *IEEE Trans. on Communication Technology*, Vol. COM-15, pp. 52–60, February 1967.

[30] M. E. Hellman and J. Raviv, "Probability of Error, Equivocation and the Chernoff Bound," *IEEE Trans. on Inform. Theory*, Vol. IT-16, pp. 368–372, July 1970.

[31] D. G. Lainiotis, "A Class of Upper Bounds on Probability of Error for Multihypothesis Pattern Recognition," *IEEE Trans. on Information Theory*, Vol. IT-15, pp. 730–731, November 1969.

[32] D. G. Lainiotis and S. K. Park, "Probability of Error Bounds," *IEEE Trans. on Systems, Man and Cybernetics*, Vol. SMC-1, No. 2, pp. 175–178, April 1971.

[33] D. G. Lainiotis and S. K. Park, "Feature Extraction Criteria: Comparison and Evaluation," 5th Hawaii Conf. on System Science, University of Hawaii, January 1972.

[34] H. Kobayashi and J. B. Thomas, "Distance Measures and Related Criteria," in *Proc. 5th Annual Allterton Conf. on Circuit and System Theory*, October 1967, pp. 491–500.

[35] G. T. Toussaint, "Feature Evaluation Criteria and Contextual Decoding Algorithms in Statistical Pattern Recognition," Ph.D. Thesis, University of British Columbia, August 1972.

[36] E. A. Patrick, and F. P. Fischer II, "Nonparametric Feature Selection," *IEEE Trans. on Information Theory*, Vol. IT-15, pp. 577–584, September 1969.

[37] T. Ito, "Approximate Error Bounds in Pattern Recognition," Machine Intelligence Workshop, March 1972.

[38] P. A. Devijver, "On a New Class of Bounds on Bayes Risk in Multihypothesis Pattern Recognition," Report R199 M.B.L.E. Brussels, December 1972.

[39] I. Vajda, "Bounds of the Minimal Error Probability on Checking a Finite or Countable Number of Hypothesis," *Problems of Information Transmission*, Vol. 26, No. 1, 1967.

[40] D. L. Wilson, "Asymptotic Properties of Nearest Neighbor Rules Using Edited Data," *IEEE Trans. Systems, Man and Cybernetics*, Vol. SMC-2, pp. 408–420, July 1972.

[41] T. Lissack and K. S. Fu, "A Separability Measure for Feature Selection and Error Estimation in Pattern Recognition," TR-EE 72-15, Tech. Rept. School of Electrical Engineering, Purdue University, W. Lafayette, Indiana, May, 1972.

[42] K. S. Fu, P. J. Min and T. J. Li, "Feature Selection in Pattern Recognition," *IEEE Trans. on Systems Science and Cybernetics*, Vol. SSC-6, No. 1, January 1970.

[43] P. H. Swain and R. C. King, "Two Effective Feature Selection Criteria for Multispectral Remote Sensing," *Proc. International Joint Conference On Pattern Recognition*, November 1973, Washington, DC.

[44] H. P. Decell Jr. and J. A. Wuirein, "An Iterative Approach to the Feature Selection Problem," *Proc. 1973 Symp. on Machine Processing of Remotely Sensed Data*, LARS, Purdue University, W. Lafayette, Indiana 47907.

[45] C. L. Wu, "The Decision Tree Approach to Classification," Ph.D. thesis, School of Elec. Eng., Purdue University, W. Lafayette, Indiana, 1975.

[46] H. Hauska and P. H. Swain, "The Decision Tree Classifier: Design and Potential," *Proc. 1975 Symp. on Machine Processing of Remotely Sensed Data*, LARS, Purdue University, W. Lafayette, Indiana 47907.

[47] G. T. Toussaint, "Bibliography on Estimation of Misclassification," *IEEE Trans. on Information Theory*, Vol. IT-20, July 1974.

[48] G. T. Toussaint, "Recent Progress in Statistical Methods Applied to Pattern Recognition," *Proc. Second International Joint Conference on Pattern Recognition*, August 1974, Copenhagen, Denmark.

[49] T. Lissack and K. S. Fu, "Error Estimation in Pattern Recognition via L^{α}-Distance Between Posterior Density Functions," *IEEE Trans. on Information* Theory (to appear).

[50] K. S. Fu and B. K. Bhargava, "Tree Systems for Syntactic Pattern Recognition," *IEEE Trans. on Computers*, Vol. C-22, December 1973.

[51] T. Pavlidis, "Linear and Context-Free Graph Grammars," *J. ACM*, January 1972.

[52] J. L. Pfaltz and A. Rosenfeld, "Web Grammars," *Proceedings International Joint Conference on Pattern Recognition*, Washington, DC, 1969.

[53] J. M. Brayer and K. S. Fu, "Web Grammars and Their Application to Pattern Recognition," Tech. Rept. TR-EE-75-1, School of E.E., Purdue University, W. Lafayette, Indiana, January 1975.

[54] R. Bajcsy and M. Tavakoli, "Computer Recognition of Roads from Satellite Pictures," *Proc. Second International Joint Conference on Pattern Recognition*, August 1974, Copenhagen, Denmark.

[55] J. R. Welch and K. G. Salter, "A Context Algorithm for Pattern Recognition and Image Interpretation," *IEEE Trans. on Systems, Man and Cybernetics*, Vol. SMC-1, January 1971.

Picture Recognition*

A. ROSENFELD and J. S. WESZKA

With 17 Figures

This chapter reviews methods of measuring properties of pictures, and extracting objects from pictures, for purposes of picture recognition and description. Subjects covered include

1) Properties of regions in pictures—in particular, textural properties.
2) Detection of objects in pictures—template matching, edge detection.
3) Properties of detected objects—projections, cross-sections, moments.
4) Object extraction—thresholding, region growing, tracking.
5) Properties of extracted objects—connectedness and counting, area and perimeter, compactness, convexity, elongatedness.
6) Object and picture representation—boundaries, skeletons, relational structures.

5.1 Introduction

This chapter describes methods of measuring properties of pictures, and extracting objects from pictures, for purposes of pattern recognition. Properties of pictures, or of objects that have been extracted from pictures, can serve as features for statistical picture classification; while the extracted objects can serve as primitives for linguistic pattern recognition of the pictures.

Section 5.2 discusses properties that are appropriate to measure for uniform regions in a picture. These are primarily properties that characterize the texture of the region. The dependence of these properties on the picture's grayscale, and methods of normalizing the grayscale, are also treated.

In Section 5.3, we consider how to detect objects in a picture, by matching parts of the picture against templates. Objects of known shape can be detected using templates that have the given shape; objects of unknown shape, that contrast with their backgrounds, can be detected using edge-or curve-like templates. The techniques in this section detect positions where objects are likely to be present, but do not explicitly "extract" the objects from the picture; the latter problem will be treated in Section 5.5.

Section 5.4 discusses properties that are appropriately measured for regions in a picture that contain an object on a background. These properties include moments, coefficients in orthonormal expansions, as well as properties derived from projections or cross-sections of the regions. The dependence of these prop-

* The support of the Information Systems Branch, Office of Naval Research, under Contract N00014-67A-0239-0012, is gratefully acknowledged, as is the help of SHELLY ROWE in preparing this paper.

Reprinted with permission from *Digital Pattern Recognition*, K. S. Fu, W. D. Keidel, and H. Wolter, Eds. New York: Springer, 1976, pp. 135–166.

erties on the positions, orientations, and size of the objects, and methods of normalizing with respect to these geometrical parameters, are also considered.

Object extraction from a picture is treated in Section 5.5. This process produces explicit "overlays", usually in the form of two-valued pictures having value 1 at object points and 0 at background points. The methods used include thresholding (preceded by suitable processing of the picture, if necessary), tracking, and region growing.

In Section 5.6, we discuss geometrical properties of objects, including connectedness (and methods of counting connected components), area and perimeter, extent, compactness, convexity, and elongatedness.

Section 5.7 is devoted to methods of representing objects, in particular using their borders or their "skeletons". The representation of a picture by a relational structure, involving objects, their properties, and relationships among them, is also briefly discussed.

Techniques for the analysis of three-dimensional scenes are not covered in this paper; see [5.1] for an introduction to this topic. We also do not cover picture processing for purposes other than recognition—image coding, enhancement, restoration, etc.

The subject of picture recognition has a large literature; only selected topics are treated, on an expository level, in this chapter. Additional details can be found in textbooks [5.1–3], while a large collection of references to the English-language literature can be found in a continuing series of survey papers [5.4–7] by the first author of this chapter. References on standard methods of picture analysis will not be given here; see [5.1–7]. We will, however, give a few references to recent developments of special interest, to which the reader can refer for detailed discussions of methods that could not be covered in the present brief paper.

5.2 Properties of Regions

In many cases, pictures can be regarded as made up of more or less uniformly textured regions, or as containing objects on a background, where the objects differ in texture from the background. Thus textural properties of regions in a picture are often of importance for picture description. This section discusses some of the commonly used textural properties.

Visual texture is a difficult concept to define, but it is commonly attributed to the repetitive occurrence of local patterns in the given region. Thus one can describe a texture by describing these local patterns and the rules for their arrangement. A good collection of papers on the analysis, synthesis, and perception of texture can be found in [5.8]. Selected approaches will be briefly reviewed in the next three subsections.

5.2.1 Analysis of the Power Spectrum

One way of analyzing local pattern arrangement in a region is to examine the power spectrum of the region, i.e., the squared magnitude of the region's two-dimensional Fourier transform. If the gray level at position (x, y) in the region is $f(x, y)$, then this Fourier transform is defined by

$$F(u, v)=\mathscr{F}(f(x, y))=\int_{-\infty}^{\infty}\int e^{2\pi i(ux+vy)}f(x, y)dxdy$$

so that the power spectrum is $|F(u, v)|^2=F(u, v)F^*(u, v)$, where $F^*(u, v)$ is the complex conjugate of $F(u, v)$. If the arrangement of local patterns over the region is periodic, say with period (u_0, v_0), then the power spectrum will have a high value at $(s/u_0, s/v_0)$, where s is the diameter of the region. Thus if the region is "busy", i.e., the patterns are fine-grained and closely spaced, the high values in the power spectrum will be spread out far from the origin, while for a coarsely textured region, the high values in the spectrum will be concentrated close to the origin. If the patterns, or their arrangement, are directionally biased, the spread of high values in the spectrum will be biased in the perpendicular direction; for example, horizontal streaks in the region will give rise to a vertical streak in the spectrum. These phenomena are illustrated in Fig. 5.1a–b.

A useful set of textural properties can be defined that take advantage of these properties of the power spectrum. Specifically, let (r, θ) be polar coordinates in the (u, v) plane, and suppose that we integrate $|F(u, v)|^2$ with respect to r and with respect to θ, i.e., we compute

$$F_1(r)=\int_0^{2\pi}|F(u, v)|^2 d\theta \quad \text{and} \quad F_2(\theta)=\int_0^{\infty}|F(u, v)|^2 dr\,.$$

If we wish, we can approximate these integrals by summing the $|F(u, v)|^2$ values over a set of thin rings centered at the origin, and over a set of narrow angular sectors emanating from the origin, respectively. The results, for the pictures of Fig. 5.1a, are shown in graph form in Fig. 5.1c–d. (These results were obtained using the discrete Fourier transform of f, which can be regarded as an approximation to the ordinary Fourier transform.) The degree of spread of the high values in $|F(u, v)|^2$ appears in the $F_1(r)$ graph as the rate of falloff from the peak at the origin; while directional biases in the high values appear as peaks in the $F_2(\theta)$ graph.

The values of $F_1(r)$ and $F_2(\theta)$, for specific choices of r and θ, can be used as textural properties. Alternately, we can compute measures of the spread or peakedness of these functions, e.g., the moment of inertia of $F_1(r)$ about the origin [i.e., $\int_0^{\infty} r^2 F_1(r)dr$], or the variance of $F_2(\theta)$.

5.2.2 Analysis of Local Property Statistics

An alternative approach to analyzing texture is to examine the frequency distribution of values of various local properties over the given region. A useful type of local property, in this connection, is a directional difference of averages taken over adjacent, non-overlapping neighborhoods in the picture. Let $A^{(r)}(x, y)$ denote the average of the gray levels ($f(x, y)$'s) in a neighborhood of radius r centered at (x, y); then a difference of non-overlapping A's in direction θ is defined by

$$D^{(r,\theta)}(x, y)=A^{(r)}(x+r\cos\theta, y+r\sin\theta)-A^{(r)}(x-r\cos\theta, y-r\sin\theta)$$

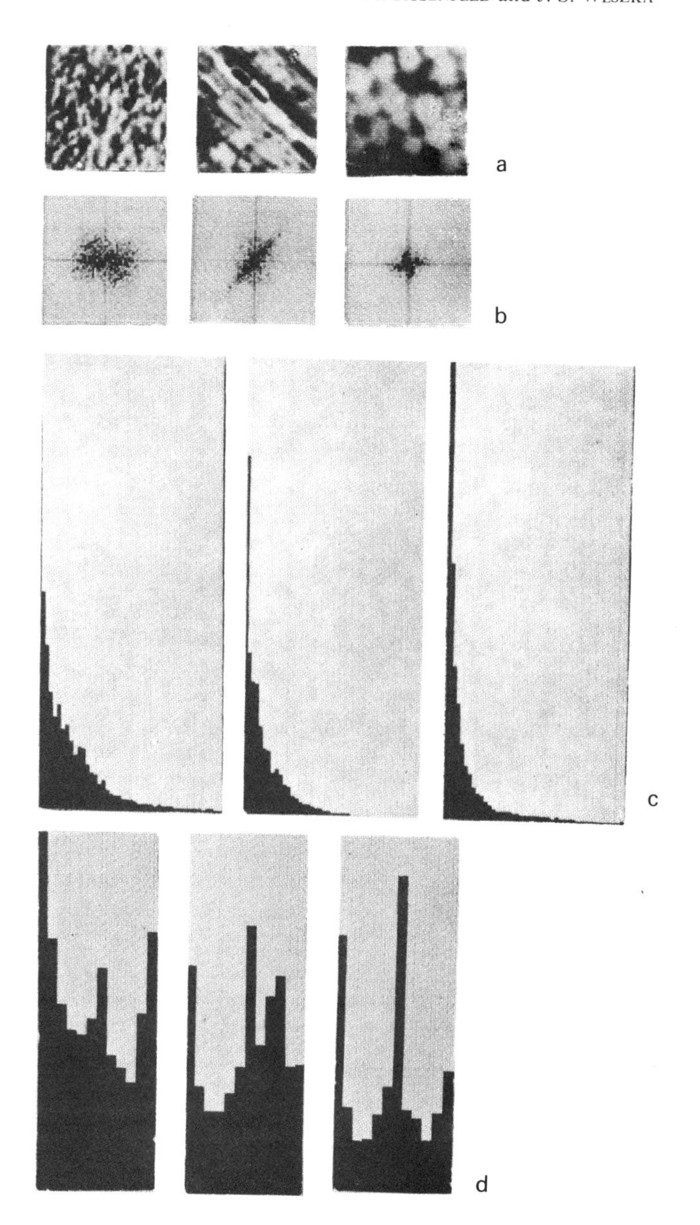

Fig. 5.1a–d. Texture analysis based on the power spectrum.

a) Input pictures, $f(x, y)$.

b) Power spectra, $\left|\iint_{-\infty}^{\infty} e^{2\pi i(ux+vy)} f(x, y)dxdy\right|^2 \equiv |F(u, v)|^2$, for the pictures in a).

c) Ring sums, $\int_0^{2\pi} |F(u, v)|^2 d\theta$, for a discrete set of rings (one unit wide).

d) Sector sums, $\int_0^{\infty} |F(u, v)|^2 dr$, for a discrete set of sectors (15° wide, starting at $-7\frac{1}{2}°$)

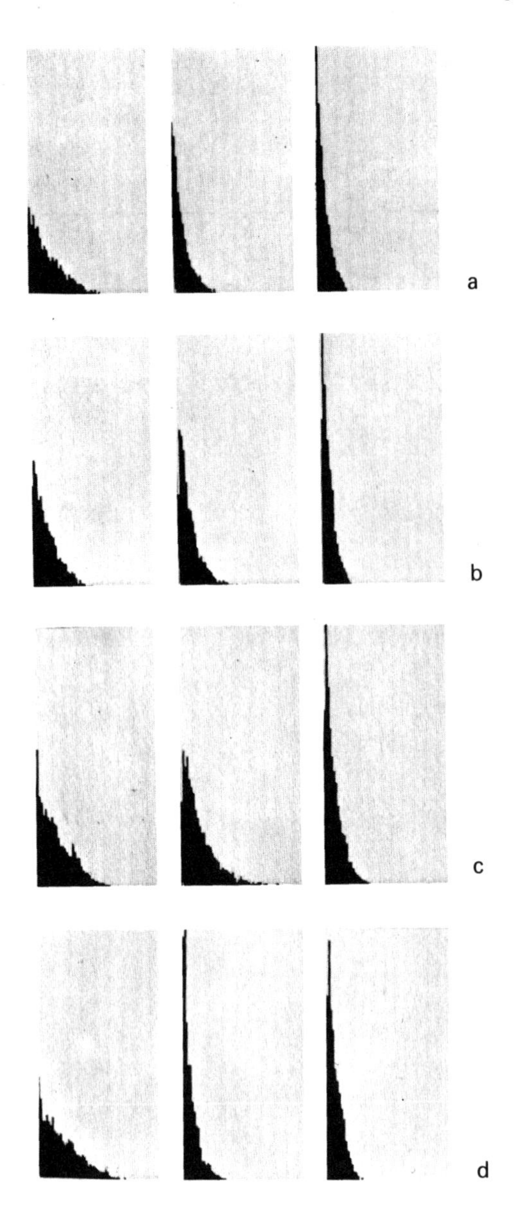

5.2a–d. Texture analysis based on local property statistics. Histograms of $D^{(r,\theta)}$, for the pictures in Fig. 5.1a, using four values of (r, θ):

Part	(r, θ)	Mean values, for the three pictures		
a)	$(1, 0)$	9.5	5.2	3.4
b)	$(1, \pi/2)$	7.2	6.1	3.2
c)	$(\sqrt{2}, \pi/4)$	12.3	4.8	4.2
d)	$(\sqrt{2}, 3\pi/4)$	10.4	9.4	4.7

If the texture in the region is "busy", $D^{(r,\theta)}$ should tend to have high values for small values of r; while if the texture is coarse, small r's should yield low values of D. If the texture has directional biases, the values of D will be higher for some θ's than for others; for example, if the texture is horizontally streaked, and we take r to be half the width of the streaks, then $D^{(r,\pi/2)}$ should be greater, on the average, than $D^{(r,0)}$.

Histograms of $D^{(r,\theta)}$, for $(r,\theta)=(1,0)$, $(1,\pi/2)$, $(\sqrt{2},\pi/4)$, and $(\sqrt{2},3\pi/4)$, are shown in Fig. 5.2a–d for the pictures in Fig. 5.1a. Statistics of these histograms,

Fig. 5.3a

	0–7	8–15	16–23	24–31	32–39	40–47	48–55	56–63
0– 7	290	214	122	49	11	5	1	0
8–15	214	145	223	173	66	33	12	1
16–23	122	223	106	180	130	97	39	5
24–31	49	173	180	94	180	131	90	23
32–39	11	66	130	180	118	191	132	64
40–47	5	33	97	131	191	129	221	88
48–55	1	12	39	90	132	221	150	215
56–63	0	1	5	23	64	88	215	304

	0–7	8–15	16–23	24–31	32–39	40–47	48–55	56–63
0– 7	398	165	28	5	7	0	0	0
8–15	165	257	206	91	18	14	1	0
16–23	28	206	215	235	83	16	10	0
24–31	5	91	235	181	240	64	16	1
32–39	7	18	83	240	194	219	55	4
40–47	0	14	16	64	219	220	246	21
48–55	0	1	10	16	55	246	245	183
56–63	0	0	0	1	4	21	183	394

	0–7	8–15	16–23	24–31	32–39	40–47	48–55	56–63
0– 7	367	224	15	6	0	0	0	0
8–15	224	257	221	36	4	0	0	0
16–23	15	221	268	230	22	0	0	0
24–31	6	36	230	241	229	18	1	0
32–39	0	4	22	229	273	212	2	0
40–47	0	0	0	18	212	285	204	8
48–55	0	0	0	1	2	204	336	136
56–63	0	0	0	0	0	8	136	437

5.3a–d. Texture analysis based on joint gray level statistics. Matrices $M^{(\Delta x,\Delta y)}$, for the pictures in Fig. 5.1a, using four values of $(\Delta x, \Delta y)$:

Part	$(\Delta x, \Delta y)$	Moment of inertia about main diagonal, for the three pictures (scaled)		
a)	(1, 0)	9.7	3.5	1.4
b)	(0, 1)	5.7	4.9	1.2
c)	(1, 1)	11.7	10.7	2.6
d)	(1, − 1)	15.6	3.5	2.0

such as their means or standard deviations, can be used as textural properties; the means are tabulated in Fig. 5.2. Other local properties can also be used in place of $D^{(r,\theta)}$; examples may be found in [5.8].

5.2.3 Analysis of Joint Gray Level Statistics

Still another method of texture analysis which has recently received attention [5.9] is based on examining the *joint* frequency distribution of pairs of gray levels, at various separations $(\Delta x, \Delta y)$ over the region. If we divide the grayscale into n intervals, we can represent such a distribution by an n-by-n matrix $M^{(\Delta x, \Delta y)}$ whose (h, k) element m_{hk} is the number of times that a point having gray level in the kth interval occurs in position $(\pm\Delta x, \pm\Delta y)$ relative to a point having gray level in the hth interval. For example, in the case of the picture

112
344
411

Fig. 5.3b

	0–7	8–15	16–23	24–31	32–39	40–47	48–55	56–63
0–7	342	217	43	14	2	1	0	0
8–15	217	193	240	117	41	9	2	0
16–23	43	240	154	238	114	51	13	0
24–31	14	117	238	122	220	122	47	6
32–39	2	41	114	220	126	230	125	32
40–47	1	9	51	122	230	148	249	66
48–55	0	2	13	47	125	249	185	210
56–63	0	0	0	6	32	66	210	353

	0–7	8–15	16–23	24–31	32–39	40–47	48–55	56–63
0–7	347	188	45	17	14	7	1	1
8–15	188	231	225	82	38	15	3	0
16–23	45	225	185	247	92	20	11	2
24–31	17	82	247	170	214	76	36	3
32–39	14	38	92	214	175	224	67	14
40–47	7	15	20	76	224	191	262	34
48–55	1	3	11	36	67	262	213	200
56–63	1	0	2	3	14	34	200	382

	0–7	8–15	16–23	24–31	32–39	40–47	48–55	56–63
0–7	398	190	9	0	0	0	0	0
8–15	190	295	212	29	1	0	0	0
16–23	9	212	287	211	18	0	0	0
24–31	0	29	211	267	223	21	0	0
32–39	0	1	18	223	278	218	7	0
40–47	0	0	0	21	218	288	194	4
48–55	0	0	0	0	7	194	316	141
56–63	0	0	0	0	0	4	141	425

Fig. 5.3c

	0–7	8–15	16–23	24–31	32–39	40–47	48–55	56–63
0–7	249	221	104	71	13	9	4	0
8–15	221	139	205	146	85	46	16	3
16–23	104	205	103	167	143	106	50	10
24–31	71	146	167	89	171	138	95	32
32–39	13	85	143	171	92	187	149	71
40–47	9	46	106	138	187	111	213	103
48–55	4	16	50	95	149	213	137	201
56–63	0	3	10	32	71	103	201	290

	0–7	8–15	16–23	24–31	32–39	40–47	48–55	56–63
0–7	272	198	77	46	38	23	14	4
8–15	198	170	189	136	73	34	17	12
16–23	77	189	143	223	113	59	34	15
24–31	46	136	223	117	183	104	60	19
32–39	38	73	113	183	129	197	102	39
40–47	23	34	59	104	197	136	230	97
48–55	14	17	34	60	102	230	156	215
56–63	4	12	15	19	39	97	215	295

	0–7	8–15	16–23	24–31	32–39	40–47	48–55	56–63
0–7	332	234	40	12	1	0	0	0
8–15	234	218	231	81	14	1	0	0
16–23	40	231	222	235	69	5	0	0
24–31	12	81	235	182	239	63	2	0
32–39	1	14	69	239	216	223	35	1
40–47	0	1	5	63	223	236	205	32
48–55	0	0	0	2	35	205	278	167
56–63	0	0	0	0	1	32	167	395

if we divide the gray levels into two interwals, (1,2) and (3,4), we have

$$M^{(1,0)} = \begin{pmatrix} 6 & 1 \\ 1 & 4 \end{pmatrix}; \quad M^{(0,1)} = \begin{pmatrix} 0 & 5 \\ 5 & 2 \end{pmatrix};$$

and so on for other values of $(\Delta x, \Delta y)$. The matrices for $(\Delta x, \Delta y) = (1, 0)$, $(0, 1)$, $(1, 1)$, and $(1, -1)$, using eight gray level intervals, are shown in Fig. 5.3a–d for the pictures in Fig. 5.1a.

If the given region is coarsely textured, then for small values of $\sqrt{(\Delta x)^2 + (\Delta y)^2}$ we can expect to see the entries in the $M^{(\Delta x, \Delta y)}$ matrix concentrated near the main diagonal, since pairs of points separated by $(\Delta x, \Delta y)$ will tend to have similar gray levels. For a busy texture, on the other hand, these entries will be more spread out. For a texture with directional biases, the amount of spread will depend on the direction $\tan^{-1}(\Delta y / \Delta x)$. These remarks suggest that the moment of inertia

Fig. 5.3d

	0–7	8–15	16–23	24–31	32–39	40–47	48–55	56–63
0– 7	224	189	134	88	32	19	9	1
8–15	189	108	202	171	103	69	37	13
16–23	134	202	88	135	135	114	70	25
24–31	88	171	135	81	143	142	107	51
32–39	32	103	135	143	87	158	158	99
40–47	19	69	114	142	158	96	205	125
48–55	9	37	70	107	158	205	114	188
56–63	1	13	25	51	99	125	188	249

	0–7	8–15	16–23	24–31	32–39	40–47	48–55	56–63
0– 7	376	133	26	15	12	5	1	1
8–15	133	286	201	59	24	5	4	0
16–23	26	201	234	218	57	21	5	1
24–31	15	59	218	200	236	56	17	4
32–39	12	24	57	236	212	190	54	6
40–47	5	5	21	56	190	255	211	18
48–55	1	4	5	17	54	211	289	113
56–63	1	0	1	4	6	18	113	424

	0–7	8–15	16–23	24–31	32–39	40–47	48–55	56–63
0– 7	345	222	33	4	1	0	0	0
8–15	222	230	261	52	2	0	0	0
16–23	33	261	220	240	43	7	0	0
24–31	4	52	240	208	240	44	1	0
32–39	1	2	43	240	233	238	24	0
40–47	0	0	7	44	238	242	211	17
48–55	0	0	0	1	24	211	289	152
56–63	0	0	0	0	0	17	152	410

of $M^{(\Delta x, \Delta y)}$ about its main diagonal, i.e., $\sum_{h,k}(h-k)^2 m_{hk}$, may be a useful textural property. It is shown, for the matrices mentioned above, in Fig. 5.3. A variety of other textural properties derived from $M^{(\Delta x, \Delta y)}$ matrices are discussed in [5.9].

5.2.4 Grayscale Normalization

The values of the textural properties defined above are sensitive to overall changes in the given picture's grayscale. For example, if the gray levels of the picture are multiplied by a constant c, then the power spectrum values, and the features derived from them, are multiplied by c^2. Similarly, the $D^{(r,\theta)}$ values are multiplied by c, while the $M^{(\Delta x, \Delta y)}$ matrices should also tend to be spread away from (or compressed toward) the main diagonal to a degree that depends on c. Other types of changes in the picture's grayscale will have more complicated effects on textural property values.

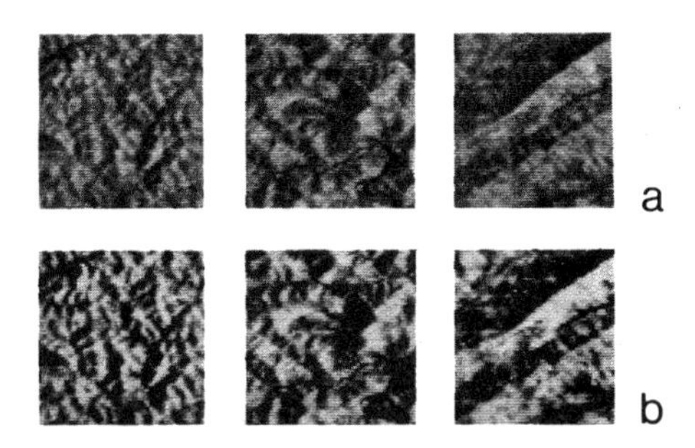

Fig. 5.4a and b. Grayscale normalization: Forcing each gray level to occur equally often. a) Input pictures. b) Results of normalization

If we want our textural properties to be invariant under grayscale changes, we can "normalize" the picture's grayscale before measuring the properties. One common way of doing this is to force the picture to have some standard frequency distribution of gray levels, e.g., a uniform distribution, in which all gray levels occur equally often. If there are n gray levels, we can do this by taking the n^{-1}th of the picture points having lowest gray level, and giving them all level 1; then taking the next lowest n^{-1}th and giving them level 2; and so on, with exact ties being resolved arbitrarily. (A more detailed description of this method can be found, e.g., in [5.9].) The results of forcing a set of pictures to have uniform gray level distributions are shown in Fig. 5.4. It is seen that the textures of these transformed pictures are essentially unchanged, though their grayscales are now "harsher".

5.3 Detection of Objects

Textural properties are appropriate descriptors for a uniform region in a picture, but not for a region that consists of an object on a background, or that overlaps two or more differently textured parts of the picture. In this section we discuss methods of detecting the presence of objects, or of edges between differently textured regions.

5.3.1 Template Matching

An object of known shape (and size) can be detected by matching the picture against a template having the given shape. This must be done for every possible position and orientation of the object. (Methods of speeding up the matching process will be discussed below.) The following are some standard measures of the degree of match between a picture $f(x, y)$ and a template T whose gray level at (x, y) is $t(x, y)$:

$$\max_{x,y \text{ in } T} |f(x, y)-t(x, y)|, \tag{5.1}$$

$$\int\int_T |f(x, y)-t(x, y)| dxdy, \tag{5.2}$$

$$\int\int_T [f(x, y)-t(x, y)]^2 dxdy. \tag{5.3}$$

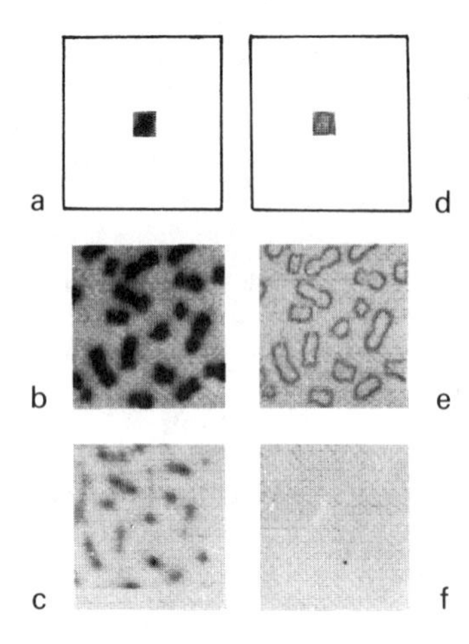

Fig. 5.5a–f. Template matching using the correlation coefficient. a) Template. b) Picture. c) Values of the correlation coefficient (nonlinearly scaled) for all possible shifts of a) relative to b). d) Outline template. e) Differentiated picture (see Subsect. 5.3.2). f) Same as c), using d) and e) in place of a) and b)

These measures are all zero for a perfect match, and have high values for poor matches. They differ, however, in the types of errors to which they are sensitive. For example, if $f(x, y)=t(x, y)$ except at a few points, where $|f(x, y)-t(x, y)|$ is large, measure (5.1) will show a large amount of mismatch, whereas measures (5.2) and (5.3) will show negligible mismatch. On the other hand, if $|f(x, y)-t(x, y)|$ is everywhere non-zero but small, e.g., $f(x, y)=t(x, y)+\varepsilon$, then measure (5.1) yields a small mismatch, ε, while measures (5.2) and (5.3) yield large mismatches, $\varepsilon|T|$ and $\varepsilon^2|T|$, respectively, where $|T|$ is the area of the template T.

A useful match measure that can be derived from (5.3) is the correlation coefficient

$$\frac{\int\int_T f(x, y)t(x, y)dxdy}{\sqrt{\int\int_T f^2(x, y)dxdy \int\int_T t^2(x, y)dxdy}}. \tag{5.4}$$

It can be shown that this always has value between 0 and 1, with value 1 being achieved if and only if $f(x, y)\equiv ct(x, y)$ for some positive constant c. In fact, measure (5.4) is evidently unaffected if the picture's grayscale is multiplied by a constant, unlike measures (5.1)–(5.3), which should be used only in conjunction with some form of grayscale normalization (see Section 5.2.4).

A template, a picture, and the values of the correlation coefficient (5.4) for all positions of the template relative to the picture, are shown in Fig. 5.5a–c. It will be noted that there are many near-misses—i.e., relatively high values appear even at positions where the match of template to picture is not very close. The matches are also not very sharply localized; match values drop off relatively slowly around a position of perfect match. These problems can be solved by using a template that resembles the outline of the desired object, rather than the solid object. Such a template is shown in Fig. 5.5d, and its match values with a derivative of the picture (Fig. 5.5e) are shown in Fig. 5.5f.

The process of matching a template against a picture in all possible positions is computationally costly. To reduce its cost, one can attempt to eliminate positions in which a match is unlikely, e.g., by measuring some simple property of the picture f (a textural property, say) in every position, and ignoring positions where this value differs too greatly from the value of the given property for the template t. Further reduction can be achieved by using a mismatch measure that can be expected to grow rapidly, as f and t are compared point by point, unless f actually matches t; thus a position where there is no match can be rejected, by the mismatch measure exceeding a threshold, after only a partial comparison has been made. More details on these methods of reducing the cost of template matching can be found in [5.10–11].

It is often unrealistic to assume that the shapes of the objects being looked for are known exactly; it would be more appropriate to use "templates" that could tolerate limited amounts of geometrical distortion. Two approaches to this problem have recently been investigated. In one of these [5.12], parameters that control the shape of the template are varied in an attempt to improve the degree of match. In the other [5.13], the template is regarded as made up of subtemplates joined by "springs", and one attempts to find positions of these subtemplates which maximize their degrees of match with the picture, while at the same time minimizing the tensions in the springs.

5.3.2 Edge Detection

Even if we do not know the shapes of the objects, we often do know something about how they differ from their background, e.g., that there should be a more or less abrupt change in gray level, in color, or in texture, at the boundary between objects and background. Under these circumstances, we should be able to detect the presence and location of an object by detecting such abrupt changes, or "edges". In the following paragraphs we discuss some basic methods of edge detection.

Abrupt changes in gray level can be detected by applying some type of derivative operation to the given picture; the result should have high values where edges are present, and low values elsewhere. If we do not know the directions of the desired edges, this operation should be isotropic (i.e., direction-independent). A simple isotropic derivative operation is the *gradient*; for the picture $f(x, y)$, this is the vector-valued function whose magnitude and direction are

$$\sqrt{(\partial f/\partial x)^2+(\partial f/\partial y)^2} \quad \text{and} \quad \tan^{-1}[(\partial f/\partial y)/(\partial f/\partial x)]\,.$$

(For digital pictures, these derivatives must be replaced by differences, and we can approximate the magnitude by using the sum or maximum of the absolute values, rather than the square root of the sum of the squares, if desired.) Another useful isotropic operation is the *Laplacian*

$$\frac{\partial^2 f}{\partial x^2}+\frac{\partial^2 f}{\partial y^2},$$

but this usually does not respond as strongly to edges as does the gradient. The magnitudes of the gradient and Laplacian, for the same set of pictures as in Fig. 5.1a, are shown in Fig. 5.6.

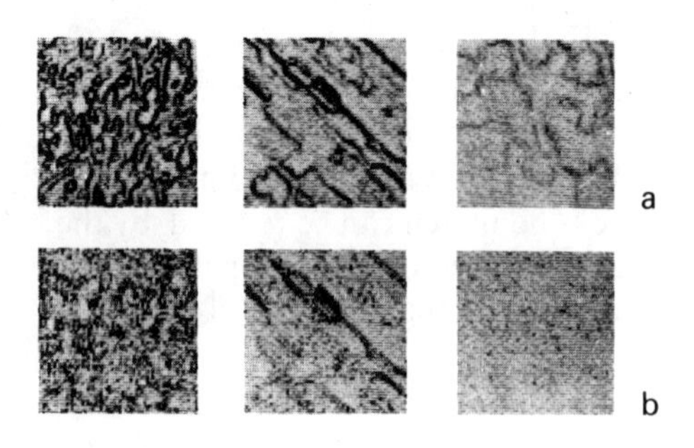

Fig. 5.6a and b. Magnitudes of derivative operations (using digital approximations) for the pictures in Fig. 5.1a. a) Gradient (multiplied by 2). b) Laplacian (multiplied by 4)

a

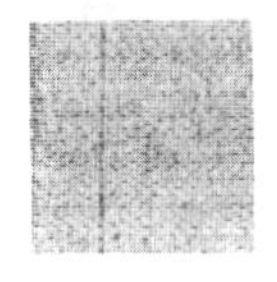

b

Fig. 5.7a and b. Vertical line detection. a) Input picture: Vertical dark line in noise. b) Output of vertical line detection operator for a)

When the picture is noisy, simple derivative operations may not be very useful in detecting edges. Under such circumstances, it may be more advisable to treat edge detection as a classification problem: given the observed gray levels in some neighborhood on the picture, is it more likely that they arise from an edge or from a uniform region? A detailed treatment of this approach may be found in [5.14]. Another possibility is to determine a step edge that best fits the observed gray levels, and to decide in favor of an edge being present if the height of this step exceeds some threshold; on this approach see [5.15–16].

Edges of known orientations and shapes can best be detected using "edge templates". For example, if we are looking for straight, vertical edges, we can sum the gray levels in two parallel, adjacent vertical strips, and subtract the results; the magnitude of this difference will be high if a straight vertical edge is present in that position. The same approach is commonly used to detect lines in a picture; we can sum the gray levels in three adjacent strips along the direction of the desired line, and subtract the sum along the center strip from the average of the sums along the flanking strips, obtaining a high magnitude of difference when a line that contrasts with its background runs down the center strip. Smooth curves can also be detected in this way (if we use strips in many orientations at each point), since locally they resemble lines, provided that they are not too sharply curved. An example of vertical line detection by this method is shown in Fig. 5.7. A disadvantage of this template approach is that it will give higher values for a high-contrast edge, or isolated point, than it will for a low-contrast line. On nonlinear line-detection "templates" that are not subject to this disadvantage see [5.17].

Objects that differ texturally from their backgrounds have edges that are characterized by changes in local property statistics, rather than by simple steps

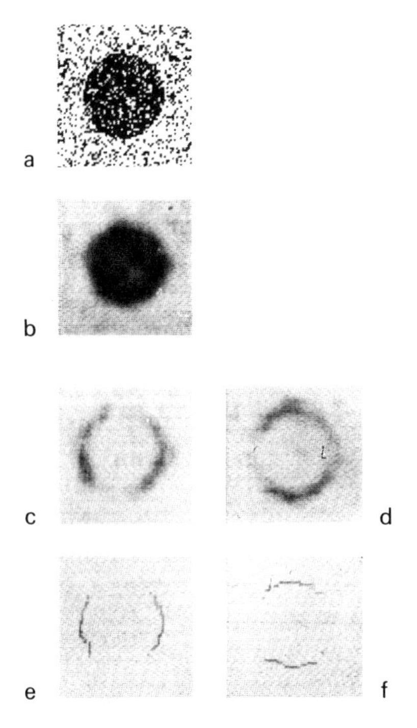

Fig. 5.8a–f. Detection of texture edges. a) Input picture: Noisy dark region on a noisy light background. b) Result of averaging a), using a 8-by-8 square averaging neighborhood at each point. c–d) Horizontal and vertical differences of adjacent, nonoverlapping averages that touch at each point of a). e–f) Horizontal and vertical maxima (respectively) of c–d)

in gray level. A straightforward approach to detecting such edges is to compute the appropriate local property at each point of the picture, and then look for pairs of adjacent neighborhoods in which the local property values differ significantly. For example, Fig. 5.8a shows a picture containing an object that has different gray level statistics from its background; this object can be detected by averaging the gray level over a neighborhood of every point (i.e., blurring the picture; Fig. 5.8b), and then taking differences of these averages in various directions (Fig. 5.8c–d), analogous to directional derivatives. As these figures show, edge detection based on differences of averages yields thick edges, since the edge is detected at many nearby positions; if desired, thin edges can be obtained by a nonmaximum suppression process (Fig. 5.8e–f). Further details on this approach may be found in [5.17–18].

5.4 Properties of Detected Objects

Suppose that a region in a picture contains an object on a background; to find such regions, we can look for places where there are good matches of templates with the picture. In this situation, we can obtain useful information about the object even without explicitly extracting it from the background, provided that we have at least a general idea about how the object and background differ. This section discusses properties that it may be useful to measure under such circumstances.

Input pictures	Moments for the pictures				
	m_{20}	m_{11}	m_{02}	m_{30}	m_{03}
0	6.96	0	16.25	0	0
1	1.41	−0.96	19.61	−0.10	10.18
2	8.69	1.88	21.50	2.34	9.00
3	8.03	0	20.82	−7.31	0
4	6.07	0.06	10.80	−9.27	7.38
5	9.19	−1.88	20.51	0	−27.56
6	8.69	1.16	17.36	2.34	14.80
7	6.80	4.00	18.76	4.36	−53.01
8	9.19	0	17.66	0	0
9	7.97	−0.69	15.61	−3.96	−18.18

Fig. 5.9. Moments for a set of numerals

5.4.1 Moments

Suppose, for concreteness, that the object has generally higher gray level than the background. Then we can get a good idea of the position of the object in the region, and of its extent in various directions, by computing *moments* of the picture's gray levels $f(x, y)$ over the region. The (i, j) *moment* of $f(x, y)$ over the region R is defined as

$$m_{ij} = \int\int_R x^i y^j f(x, y) dx dy .$$

The coordinates $(\bar{x}, \bar{y})$ of the *centroid* of R (its "center of gravity", if we think of gray level as corresponding to mass) are $\bar{x} = m_{10}/m_{00}$, $\bar{y} = m_{01}/m_{00}$. If we choose a coordinate system with $(\bar{x}, \bar{y})$ as the origin, the moments $\bar{m}_{ij}$ in this coordinate system (which are called "central moments") indicate how the gray levels in R are distributed relative to the centroid. For example, $\bar{m}_{20}$ and $\bar{m}_{02}$ are the moments of inertia of R around vertical and horizontal lines through the centroid, respectively; if $\bar{m}_{20}$ is greater than $\bar{m}_{02}$, the object is likely to be horizontally elongated, since the high gray levels are more spread out horizontally than they are vertically. The asymmetry of the object about these vertical and horizontal lines can be measured by the magnitudes of the moments $\bar{m}_{30}$ and $\bar{m}_{03}$, which are zero if there is perfect symmetry. The first few moments for a set of numeric characters are shown in Fig. 5.9.

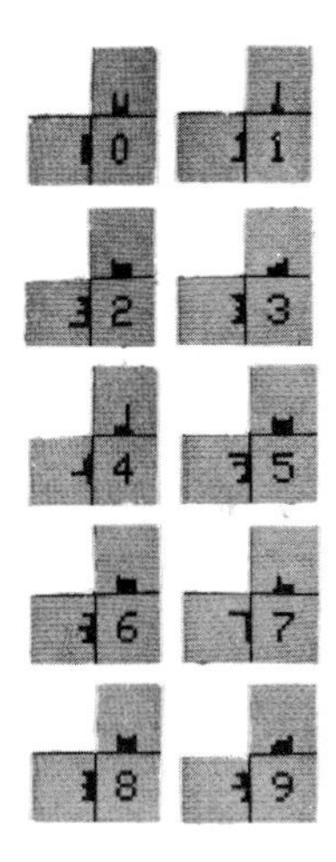

Fig. 5.10. Horizontal and vertical projections of the pictures in Fig. 5.9a

Other types of information about the arrangement of gray levels in a region can be obtained by computing the Fourier coefficients of $f(x, y)$, i.e.,

$$\int\int_R f(x, y) \sin(ux + vy) dx dy \quad \text{or} \quad \int\int_R f(x, y) \cos(ux + vy) dx dy .$$

If a particular coefficient is high, the gray level should tend to be periodically distributed (with a given period, phase, and orientation). Coefficients in other orthogonal expansions of f—e.g., Walsh coefficients—can also be useful. The orthogonality of the basis functions insures that properties measured in this way will be uncorrelated.

5.4.2 Projections and Cross-Sections

We can also get an idea of how the gray levels in R are distributed by examining the *projections* of $f(x, y)$ in various directions. For example, the projections of f in the x and y directions are

$$\int_R f(x, y) dx \quad \text{and} \quad \int_R f(x, y) dy .$$

These projections, for the same set of characters used in Fig. 5.9, are shown in graph form in Fig. 5.10. (Projections in a sufficient number of directions contain enough information, in principle, to reconstruct the picture; but we shall not pursue this subject here.) It is seen that, for objects having higher gray levels than their backgrounds, peaks in the projections can indicate the locations of major parts of the objects. Numerical properties of projections, such as their (one-dimensional) moments, Fourier coefficients, etc., can be useful as object descriptors.

More detailed information about the arrangement of gray levels in the region R can be obtained by examining *cross-sections* of $f(x, y)$ in various directions; e.g., the cross-sections in the x direction are just the functions $f(x, y_0)$ for par-

Character	Section	Number of runs of black points	Average run length
0	T	1	2
	M	2	1
	B	1	2
1	T	1	1
	M	1	1
	B	1	3
2	T	1	3
	M	1	3
	B	1	5
3	T	1	3
	M	1	2
	B	1	3
4	T	1	1
	M	2	1
	B	1	1
5	T	1	5
	M	1	4
	B	1	3
6	T	1	3
	M	1	4
	B	1	3
7	T	1	5
	M	1	1
	B	1	1
8	T	1	3
	M	1	3
	B	1	3
9	T	1	3
	M	1	4
	B	1	2

Fig. 5.11. Characteristics of three horizontal cross-sections (T = top, M = middle, B = bottom) of the pictures in Fig. 5.9a: number of runs of black points, and average run length

ticular values of y_0. Here again, for an object having high gray level, peaks in the cross-sections correspond to object parts. Some features of three horizontal cross-sections of each character in Fig. 5.9a are shown, in table form, in Fig. 5.11. Comparison of successive cross-sections can give useful information about object shape, in terms of how the peaks shift, expand, shrink, merge, and split as we track them from one cross-section to the next.

5.4.3 Geometrical Normalization

The values of moments, projections, etc. all depend on the position and orientation of the object in the given region. We can obtain properties that are invariant under translation or rotation of the object by using suitable combinations of

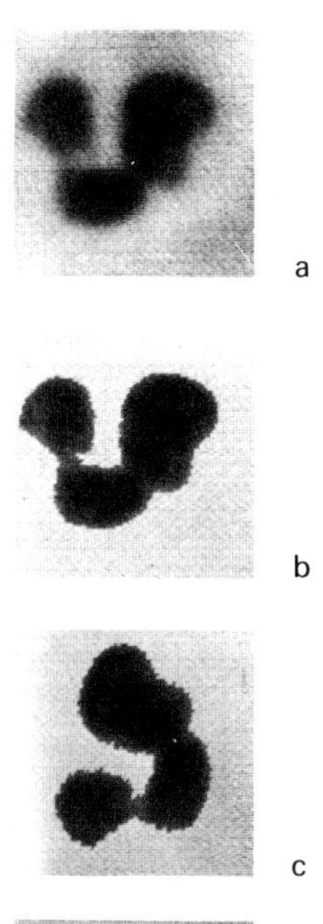

Fig. 5.12a–d. Geometrical normalization. a) Input picture (gray level range 0 to 63). b) Result of setting the gray levels (<32) of background points in a) to zero. c) Result of normalizing b) by rotating the object to make its principal axis vertical. d) Result of normalizing b) by rotating the object to make its least-area circumscribed rectangle upright

moments. For example, the central moments $\bar{m}_{ij}$ are invariant under translation; and the moment of inertia of the region R around its centroid, $\bar{m}_{20}+\bar{m}_{02}$, is invariant under rotation.

We can also use moments to "normalize" the position and orientation of the object, so that its projections, cross-sections, etc. can also be used to derive invariant properties of it. In fact, by using a coordinate system with origin at the centroid, we immediately obtain translation invariance. To get rotation invariance, we can find the line $y=x\tan\theta$, through this origin, about which the moment of inertia of R is least; i.e., we find the θ (which will usually be unique) such that

$$\int\int (x\sin\theta - y\cos\theta)^2 f(x,y)dxdy = \bar{m}_{20}\sin^2\theta - 2\bar{m}_{11}\sin\theta\cos\theta + \bar{m}_{02}\cos^2\theta$$

is as small as possible. (This line is called the *principal axis* of R.) We can then choose a rotated coordinate system in which this line is, say, vertical; this will tend to align the object so that it is elongated in the y direction. An example of a picture that has been normalized in this way is shown in Fig. 5.12a–c.

If the object has been explicitly extracted from the picture, another method of geometrical normalization can be used. We can circumscribe rectangles (say) around the object, and find an orientation for which the area of the circumscribed rectangle is least (again, this will usually be unique). We can then choose a rotated coordinate system in which the long sides of this minimum-area rectangle are, say, vertical. This too will tend to align the object so that it is vertically elongated; but for bent objects, it does not always yield the same orientation as the principal axis method, as we see in Fig. 5.12d.

Still another approach to geometrical normalization is to use a transform of the picture that is invariant under geometrical operations on the picture. For example, the autocorrelation and the power spectrum of f,

$$f \otimes f = \iint_{-\infty}^{\infty} f(x, y) f(x+h, y+k) dx dy \quad \text{and} \quad \left| \iint_{-\infty}^{\infty} f(x, y) e^{-2\pi i(ux+vy)} dx dy \right|^2$$

both remain invariant if the original picture f is translated. Rotation and scale invariance can be obtained in analogous ways, using polar-coordinate transforms.

5.5 Object Extraction

In this section we consider methods of explicitly extracting objects from a picture. The output of the extraction process must be a specific decision as to which points of the picture belong to the objects, and which to the background. (More generally, one can *segment* a picture into meaningful parts (not necessarily objects and background); here the result must explicitly indicate which points belong to each part.) The results of this decision can be expressed in the form of an "overlay", in register with the original picture; this overlay can be thought of as a two-valued picture that has value 1 at points belonging to the objects, and value 0 at points of the background.

5.5.1 Thresholding

If the objects occupy a distinctive gray level range—e.g., they have higher gray levels than the background—then they can be extracted by *thresholding;* in other words, for some threshold value θ, we can create an overlay $o(x, y)$ defined by

$$o(x, y) = 1 \quad \text{if} \quad f(x, y) \geqq \theta; \quad = 0 \text{ otherwise}$$

where $f(x, y)$ is the gray level of the given picture at (x, y). If we want to preserve information about the gray levels of the objects, while still distinguishing them from the background, we can perform "semi-thresholding" rather than thresholding—i.e., we can create a new picture $s(x, y)$ such that

$$s(x, y) = f(x, y) \quad \text{if} \quad f(x, y) \geqq \theta; \quad = 0 \text{ otherwise}.$$

The results of thresholding and semithresholding the picture in Fig. 5.12a, using various threshold levels, are shown in Fig. 5.13.

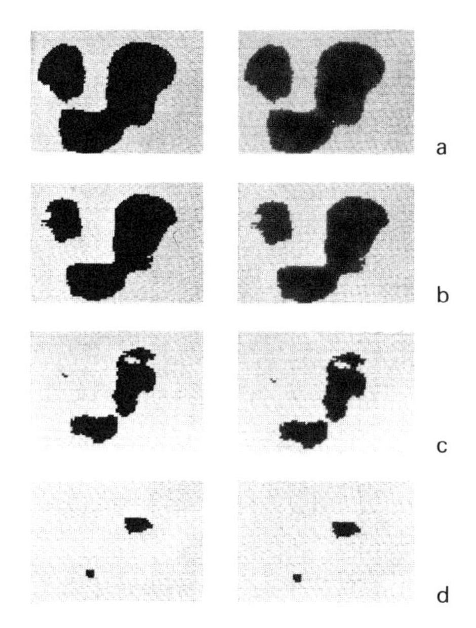

Fig. 5.13a–d. Thresholding, Results of thresholding and semithresholding the picture in Fig. 5.12a at a) 35, b) 40, c) 45, d) 50. (Fig. 5.12b showed the same picture semithresholded at 32)

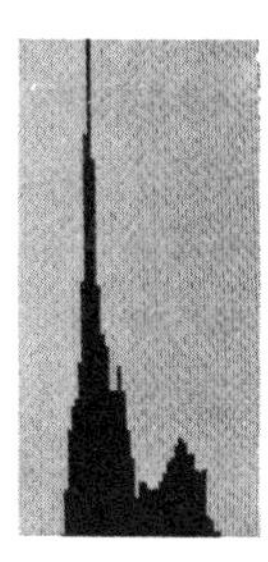

Fig. 5.14. Histogram of the gray levels in the picture in Fig. 5.11a

As Fig. 5.13 shows, proper choice of threshold is very important if one wants to extract objects correctly. A commonly used method of threshold selection is based on examining the histogram of the picture's gray levels. If the object and background gray level ranges are different, there should be peaks on this histogram corresponding to these ranges, separated by a valley corresponding to the intermediate gray levels that occur only rarely in the picture. The histogram for the picture in Fig. 5.12a is shown in Fig. 5.14. Note that the thresholds 32 and 35 used in Figs. 5.12b and 5.13a are at or near the valley bottom on this histogram.

The proper choice of threshold may vary from place to place in a picture; it may be preferable to threshold a picture piecewise. In fact, if we examine a small window of a picture, and discover that it has a strongly bimodal histogram, we can conclude that the window contains an object, or that an edge cuts across it [5.19]. When suitable thresholds have been chosen, say at the valley bottoms, for the windows that have bimodal histograms, we can assign thresholds to the non-bimodal windows by interpolation. It should be pointed out that if a picture has

a

b

c

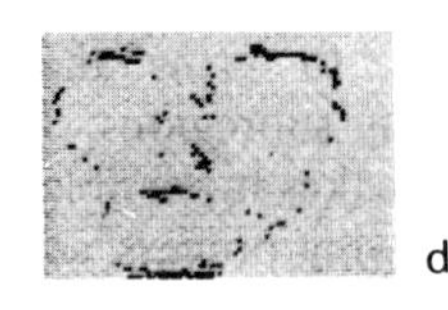
d

Fig. 5.15a–d. Outline extraction by thresholding. a) Result of displaying gray levels 35 to 38 in the picture in Fig. 5.12a as black, and all other levels as white. b) Same, using only levels 35 to 37. c) Result of displaying points of this picture at which the gradient has values $\geqq 3$ as black, and all other points as white. d) Same, using only values $\geqq 4$

a multimodal histogram, it is not in general safe to segment it by using thresholds at the bottoms of all of the valleys between the peaks, since there will be no way of distinguishing points that belong to a region of intermediate gray level from points that belong to transition zones between high and low gray level regions.

Outlines of objects, rather than the objects themselves, can be extracted in two ways. One approach is to "threshold" so as to extract points having a narrow range of gray levels, such as might be expected to occur just on the border between objects and background. Examples of this method are shown in Fig. 5.15a–b. The other approach is to apply an edge-sensitive operation to the picture, e.g., a gradient operation, and threshold the results; this should yield points on or near edges of objects, where the gradient has high values. Results obtained using this method are shown in Fig. 5.15c–d.

If an object differs texturally from its background, rather than occupying a distinctive range of gray levels, it cannot be extracted by simple gray level thresholding. For example, in the picture shown in Fig. 5.8a, both the object and the background have the same set of gray levels; they differ only in the probabilities of these gray levels. In cases like this, however, one can often convert the picture into a new form in which the objects and background do have different gray level ranges. Thus, in the case of Fig. 5.8a, if we simply blur the picture (as shown in Fig. 5.8b), we obtain a new picture in which the object is now generally darker than the background. The histogram of gray levels of this picture is shown in Fig. 5.16a; it now has peaks corresponding to the object and background gray level ranges. If we threshold at the bottom of the valley between these peaks, we obtain the reasonable result shown in Fig. 5.16b.

a

b

Fig. 5.16a and b. Extraction of a textured object by averaging and thresholding. a) Histogram of the gray levels in the picture in Fig. 5.8b. b) Result of thresholding this picture at 32 (near the bottom of the valley on the histogram)

5.5.2 Region Growing

A more flexible approach to object extraction or segmentation is to "grow" the objects, or the regions, by successively adding points, or merging subregions, if appropriate acceptance conditions are satisfied. This approach makes it possible to adjust the acceptance criteria in the course of the region growing process, so that they can depend on the textures and shapes of the growing regions, if desired. Object extraction by thresholding, on the other hand, applies a single fixed acceptance criterion to all points of the picture simultaneously.

A very simple example of object growing uses a high threshold to find "cores" of objects, and they are then allowed to grow by repeatedly adding adjacent points to them provided that these points exceed a certain lower threshold. This technique can be illustrated by referring to Figs. 5.12 and 5.13. By taking the high threshold as 45 or 50, we extract points belonging to two of the three objects. The growing process, using a lower threshold of 35 or 40, then extracts the two objects, but not the third, even though it contains many points whose gray levels exceed the lower threshold.

As already mentioned, the criteria used to accept new points into an object can vary as the object grows (and not stay the same throughout the growth process, as in the example just given). The criteria can thus be made to depend on the object's shape and texture. As an example, suppose that we want to extract objects that have fairly constant gray level and fairly simple shape. We can start with "object cores" that are connected components of constant gray level; and we can merge two adjacent components if this will result in a more compact shape (see Subsection 5.6.2 on measures of shape compactness), provided that the gray levels of the components do not differ too greatly. This process can be repeated, merging unions of components provided the shape becomes more compact and the average gray levels do not greatly differ. A region growing procedure of this kind is described in [5.20].

Another way of segmenting a picture is to start with a standard partition of it, say into grid squares. We split a square (say into quadrants) if its gray level is too highly variable; and we merge neighboring squares if their gray levels are similar. This process can be used to obtain a partition into pieces that are unions of squares, where the gray level (or some other property, if desired) is uniform, within some tolerance, on each piece. On procedures of this kind see [5.21].

5.5.3 Tracking

A special type of "region growing" is *tracking*, where one begins at a point lying on an edge or curve, and successively accepts neighboring edge or curve points until the entire object border, or the entire curve, has been traversed. Here again, the acceptance criteria can depend both on the contrast of the edge or curve and on its shape. For example, one can look for high-contrast neighboring points, subject to the constraint that the curvature of the resulting edge or curve be small. If following the path of highest contrast leads to high curvature, one can backtrack. Some recent examples of edge and curve tracking can be found in [5.22, 23].

Edge and curve detection operations (Subsection 5.3.2) often yield incomplete or broken results, since there may not be high contrast with the background at all points of the edge or curve. Edge and curve tracking, on the other hand, can be designed to bridge gaps, by looking for high-contrast points in a zone that extends the portion of the edge or curve already found.

A special case of tracking is *raster tracking*, where the edges or curves are tracked from row to row of the picture (so that the tracking is based on a row-by-row scan resembling a TV raster). Here the positions where the current row hits an edge or curve are noted, and hits are looked for in nearby positions on the next row (or within the next few rows, if we want to be able to bridge gaps). It is relatively straightforward to track any number of edges or curves simultaneously in this way, provided that they do not become very oblique (at the worst: tangential) to the raster lines.

5.6 Properties of Extracted Objects

Once we have explicitly extracted objects from a picture, we can analyze their geometrical properties. These include connectedness, size (area, perimeter, extent), and shape (compactness, convexity, elongatedness, etc.). These and other geometrical properties are the subject of this section. In what follows, S denotes an object, or more generally, an arbitrary finite set of digital picture points.

5.6.1 Connectedness

We say that two points (x_0, y_0) and (x_n, y_n) of S are *connected* in S if there exists a sequence of points of S, $(x_0, y_0), (x_1, y_1), \ldots, (x_n, y_n)$, such that (x_k, y_k) is a neighbor of (x_{k-1}, y_{k-1}), $1 \leqq k \leqq n$. Such a sequence of points is called a *path* in S. A maximal set of mutually connected points of S is called a *connected component* of S. If S has only one connected component, it is called *connected*.

There are two versions of the above definitions (and of many of the other definitions in this section), depending on whether or not diagonally adjacent points are considered to be neighbors. (Horizontally or vertically adjacent points, i.e., with $|x_k - x_{k-1}| + |y_k - y_{k-1}| = 1$, are always regarded as neighbors, so that a point always has four neighbors; but we may or may not allow the additional four points, with $(x_k - x_{k-1}, y_k - y_{k-1}) = (\pm 1, \pm 1)$, to be called neighbors also.) For example, if S consists of two diagonally adjacent points, $\begin{matrix} p & \\ & p \end{matrix}$, it is connected in the 8-neighbor sense, but not in the 4-neighbor sense. If we used a hexagonal rather than a square grid for our digital pictures, this ambiguity would not arise, and a point would always have six neighbors.

Let $\bar{S}$ be the complement of S; this includes all the points that lie outside the picture. If $\bar{S}$ has more than one connected component, then all but one of its components are *holes* in S; the remaining component, which contains the points outside the picture, is called the *background* of S. It turns out that if we want certain basic properties of connectedness to be valid, and if we want certain algorithms to work, we must use opposite types of connectedness for S and for $\bar{S}$—i.e., if we use 4-neighbor connectedness for S, then we must use 8-neighbor connectedness for $\bar{S}$, and vice versa. Thus if S consists of the four points $\begin{matrix} & p & \\ p & & p \\ & p & \end{matrix}$ it is connected in the 8-neighbor sense, and has a hole (since we use 4-neighbor connectedness for $\bar{S}$); but it is not connected in the 4-neighbor sense, and has no holes, if we use 8-neighbor connectedness for $\bar{S}$. On the other hand, $\begin{matrix} p & p & p \\ p & & p \\ p & p & p \end{matrix}$ is connected, and has a hole, in both senses. (In the case of a hexagonal grid, we can use 6-neighbor connectedness for both S and $\bar{S}$.) On the theory of digital connectedness see [5.24–26].

The *border* of S consists of those points of S that are neighbors of points of $\bar{S}$; the remainder of S is called its *interior*. For example, in the nine point object $\begin{matrix} b & b & b \\ b & i & b \\ b & b & b \end{matrix}$ we have labelled the border points b and the interior point i. A border point of S is called a *simple* point if deleting it from S would not disconnect the part of S inside its neighborhood; in other words, the point (x, y) of S is simple provided that, for any two neighbors of (x, y) that lie in S, there is a path in S, consisting of neighbors of (x, y), that joints the two given neighbors. For example, if the neighborhood of p is $\begin{matrix} n & \\ & p \\ & n \end{matrix}$, then p is simple in the 4-neighbor sense (since the diagonal neighbor can be ignored), but not in the 8-neighbor sense; while if the neighborhood is $\begin{matrix} & n \\ n & p \end{matrix}$, then p is simple in the 8-neighbor sense but not in the 4-neighbor sense. We will need to use simple points in defining "thinning" operations in Subsection 5.6.3.

One often wants to *count* the connected components of a given S, since this is a useful property of S. If none of the components has holes, this can be done in a single row-by-row scan of the picture, as follows: On the top row, count 1 for every run of points of S. On succeeding rows, for each run of points of S that is adjacent to k such runs on the preceding row (where $k \geq 0$), add $1-k$ to the count. When the scan is complete, it can be shown that the count will equal the total number of components of S. (If the components do have holes, the count will instead equal the number of components minus the number of holes.)

To count the components when they may have holes, one must in effect assign a unique "label" to the points of each component. On the first row, each run of points of S gets its own label; on subsequent rows,

a) if a run is adjacent to no runs on the previous row, it gets a new label;
b) if it is adjacent to just one run on the previous row, it gets that run's label;
c) if it is adjacent to several runs on the previous row, it gets one of their labels, and the other labels, if any, are noted to be equivalent to that label.

When the scan is complete, the number of inequivalent labels used is equal to the number of components. The areas of the components can be determined at the same time, by keeping count of how many times each label was used, and when labels are found to be equivalent, combining their counts.

5.6.2 Size, Compactness, and Convexity

As just indicated, the area of a set S of digital picture points is just the number of points in S. The *perimeter* of S can be defined as the number of border points of S (i.e., as the area of S's border). Alternatively, it can be defined as the number of adjacencies between points of S and points of $\bar{S}$, i.e., the number of pairs of neighboring points such that the first point is in S and the second in $\bar{S}$. This latter definition is essentially the number of moves that one must make in order to follow the border of S completely around; it is always greater than the area of the border. For example, if S is $\begin{matrix} ppp \\ ppp \\ ppp \end{matrix}$, its perimeter in the first sense is 8, in the second sense 12; if S is ppp, its perimeters are 3 and 8, respectively; while $\begin{matrix} p \\ p \\ p \end{matrix}$ has perimeters 3 and 12, respectively.

The *compactness* of S is often measured by A/P^2, where A is its area and P its perimeter. In the real plane, this is greatest $(=4\pi)$ for circles, and is smaller for all other figures. In the digital case, however, it is greatest for certain squares or octagons, depending on how we measure perimeter [5.27].

The *extent* of S in a given direction is the length of its projection perpendicular to that direction. For example, the *height*, or vertical extent, of S, is the number of rows of the picture that S occupies; while its *width*, or horizontal extent, is the number of columns that it occupies. The *diameter* of S is its greatest extent in any direction—or, equivalently, the greatest distance between any two points of S.

S is called *convex* if its cross-section along any line consists of at most a single line segment—or, equivalently, if any line segment whose endpoints are in S must lie entirely in S. (The definition of convexity in the digital case is somewhat more complicated, and will not be given here.) It is easily seen that if S is convex, it must be connected, and can have no holes. For example, $\begin{matrix} & p \\ p & p \end{matrix}$ is convex, but $\begin{matrix} p & & \\ p & & \\ p & p & p \end{matrix}$ is not. The smallest convex set containing S is called the *convex hull* of S.

The theory of digital convexity was first discussed in [5.28]. Algorithms have been developed for constructing the convex hull of S, for breaking an arbitrary S up into convex parts, and for finding concavities in S or "shadows" of S (e.g., points of $\bar{S}$ that are behind points of S when we take cross-sections in various directions); the details will be omitted here.

5.6.3 Arcs, Curves, and Elongatedness

In the digital case, S is called a *closed curve* if it is connected, and every point
 p
of S has exactly two neighbors in S. For example, p p is a closed curve in the
 p
ppp
8-neighbor sense, while p p is a closed curve in the 4-neighbor sense (but not
ppp
vice versa). S is called an *arc* if it is connected, and all but two of its points have exactly two neighbors in it, while the two exceptional points (the *endpoints*) each directions to the neighbors, is always a multiple of 45° (or 90°, if points are only allowed to have four neighbors). For many purposes, it is more useful to define slope in terms of the directions to points several steps away along the arc or curve; the farther away we go, the more values become possible, and our choice of how far away to go depends on how fine a level of detail we want in our description of the curve. Similarly, the *curvature* at a point can be defined as the difference between the slopes on the two sides of the point.

Angles on a curve can be defined as local maxima of the curvature [5.29]. Points of *inflection* can be defined as zero-crossings of the curvature; these points separate the curve into successive convex and concave pieces. For an arc to be have exactly one neighbor. The *slope* of an arc or curve at a point, if defined by the a *straight line* segment, its slope must be approximately constant; it turns out that this implies that the directions between successive pairs of neighbors along the arc take on at most two values, differing by 45°, where at least one of these values occurs only in runs of length 1, while the other value occurs in runs whose
 pp
lengths are as equal as possible [5.30]. Thus ppp is a straight line segment
 ppp
 p
 p
(in the 8-neighbor sense), but ppp and pp are not.
 p p

S is *elongated* if its greatest extent is much larger than its least extent. However, this definition does not cover all the situations in which we would want to call S elongated, since a coiled snake is elongated even though its overall shape is circular. We can develop a better definition by introducing the concepts of *shrinking* and *expanding* the set S. Let $S^{(1)}$ be the result of adding to S all points of $\bar{S}$ that are neighbors of points of S, and let $S^{(k)}$ be the result of repeating this
 p
process k times ($k=1, 2, \ldots$). Thus if S consists of a single point p, $S^{(1)}$ is ppp,
 p

```
  p
 ppp
ppppp
 ppp
  p
```

and $S^{(2)}$ is the pattern above, in the 4-neighbor sense, while in the 8-neighbor sense they are

```
ppp        ppppp
ppp  and   ppppp
ppp        ppppp
           ppppp
           ppppp
```

respectively. Similarly, let $S^{(-1)}$ be the result of deleting from S all points that are neighbors of points of $\bar{S}$, and let $S^{(-k)}$ be the result of repeating this process k times ($k = 1, 2, \ldots$). Thus if S is

```
  p
 ppp
ppppp
 ppp
  p
```

$S^{(-1)}$ in the 4-neighbor sense is

```
 p
ppp
 p
```

while in the 8-neighbor sense it is just p.

If we shrink and then re-expand, e.g., $(S^{(-k)})^{(k)}$, it is not hard to see that we may not get our original S back; but it can be shown that we must get a subset of it, i.e., we can never get points that were not in S. Let S_k^* be the set of points that we do not get back, and let C be a connected component of S_k^*. Since every point of C must have disappeared under k-step shrinking, the "width" of C is at most $2k$ (i.e., every point of C is at distance at most $2k$ from the complement $\bar{C}$). Suppose that C has area, say, $12k^2$; then its "length", defined as area/width, is at least $6k$. Since this length is three times the width, we can legitimately call C "elongated". Thus if we shrink and re-expand S, the large connected components of points that we lose (large in comparison with the amount of shrinking that we used) are *elongated parts* of S. For example, if S is

```
pppp
 pppppppppp
 pppppppppp
 pppp
 p
```

then $S^{(-1)}$ is

```
pp
pp
```

(in the 8-neighbor sense)

and $(S^{(-1)})^{(1)}$ is

```
pppp
pppp
pppp
pppp
```

We have thus lost three components, two of them consisting of only one point each, while the third consists of 12 points, and is elongated. Further discussion of how this method can be made to work in the presence of noise, and of how a similar scheme involving expansion and re-shrinking can be used to detect clusters of points in a set S, may be found in [5.31–32].

If S is everywhere elongated, it can be *thinned* by removing border points from it, provided that the points removed are simple (see Subsection 5.6.1) and have more than one neighbor in S; these conditions guarantee that the thinning process will not disconnect S, and will not shrink arcs that are already thin. Even if all such border points on one side of S (e.g., the points whose right-hand neighbors are in $\bar{S}$) are removed at once, it can be shown that S will not disconnect; but if points are removed on all sides at once, S may even vanish completely, e.g., if S is $\begin{matrix} ppppp \\ ppppp \end{matrix}$. More complex thinning algorithms can be used to remove border points on two sides at a time; see, e.g., [5.33]. It should be noted that thinning will not necessarily produce a "thin" object; for example, if S is

$$\begin{matrix} p\ \ p\ \ p \\ ppp \\ ppppp \\ ppp \\ p\ \ p\ \ p \end{matrix}$$

then thinning has no effect on it.

5.7 Representation of Objects and Pictures

In this section we discuss methods of representing objects that have been extracted from a picture; in particular, using borders or "skeletons" as representations. We also discuss the representation of pictures by relational structures involving objects or regions, their properties, and relationships among them.

5.7.1 Borders

An object is completely determined if we know its *borders*, provided that we also know which side of each border is inside the object and which is outside. Note that an object may have more than one border, if it has holes. Note also that two borders may have points in common, if a hole is near the outside edge of the object; and that a border may pass through the same point twice, if the object is "thin" at that point. Nevertheless, for any given border, simple algorithms exist that will "follow" or track it completely around, starting from any point of it (and given a neighbor of that point which lies outside the object); see, e.g., [5.24].

The sequence of moves that we make while following a border take us repeatedly from a point to one of its neighbors; if 8-neighbor adjacency is used in defining "border", this neighbor must be one of the four horizontal and vertical neighbors, while if 4-neighbor adjacency is used, successive points on the border can also be diagonal neighbors. For concreteness, let us make the latter assumption. Thus the moves can be specified by 3-bit numbers indicating which one of the eight neighbors of the current border point is the next border point; e.g., we can use the convention

$$\begin{matrix} 321 \\ 4p0 \\ 567 \end{matrix}$$

to designate p's eight neighbors by the numbers $0, \ldots 7$. The sequence of moves around the border is thus completely determined by specifying a string of 3-bit numbers. Conversely, any such string defines an (8-neighbor) path of points in the picture, if we are given the starting point. This method of representing object borders—or more generally, arbitrary paths—by strings of 3-bit numbers is called *chain coding*. For example, the path

```
  ppp
  p p
pppp
  p
  p
```

starting from its left end, has the chain code 00012446666. A detailed review of chain coding can be found in [5.34].

If we are given the chain codes of all the borders of an object, as well as a starting point for each chain, and a neighbor of that point which lies outside the object, we can reconstruct the object completely. To do this, we move as directed by each chain code and put down 1's in the (initially blank) picture, while putting down 0's at neighboring points that lie outside the object (these are always determinable, since the initial such point is given). When all the object's borders have been "painted" in this way, we can fill in the interior of the object by letting 1's expand into blanks (but not into 0's).

The border chain codes give a translation-invariant description of an object, since the position of the set of starting points can be chosen arbitrarily. However, this description is not rotation-invariant; in fact, if we rotate a curve, its chain code changes in nontrivial ways (unless the rotation is by a multiple of 90°, in which case we need only add a multiple of 2, modulo 8, to each code). For example, the line segment

```
     p
    p
   p
  p
 p
p
```

has chain code 11111, but if we rotate it by $-45°$, it becomes (approximately) *ppppppp*, which has chain code 0000000; this increase in length is needed to compensate for the fact that horizontal and vertical steps are only $1/\sqrt{2}=0.7$ as long as diagonal steps.

We can obtain a rotation-invariant description of a closed curve by using Fourier methods. (Of course, this description is no longer digital.) For example, let $\theta(s)$ be the slope of the curve at position s (measured along the curve, from an arbitrary starting point). Thus $\theta(s)$ is a periodic function, whose period is the perimeter of the curve. If we expand $\theta(s)$ in a Fourier series, the (squared) magnitudes of the series coefficients are rotation-invariant properties of the curve. For a more detailed discussion of this approach, see [5.35]; other types of Fourier descriptors can also be defined. Approximations to the shape of the curve can be obtained by truncating the series.

5.7.2 Skeletons

A very different way of describing an object is in terms of its "skeleton", which can be defined as follows: At each point p of the object S, there is a largest circle C_p centered at p that is entirely contained in the object. (If p is on the border of S, C_p has radius 0.) The circle C_p may be contained in another circle C_q; we can

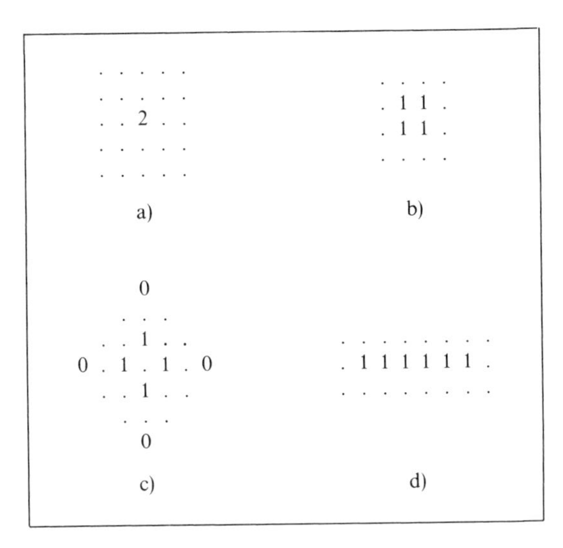

Fig. 5.17. Skeletons of some simple objects. Each skeleton point p is labeled with its associated radius r_p; the remaining points of the original objects are marked by dots

discard C_p in that case. It is easily seen that S is the union of the remaining circles. The set of centers p and radii r_p of these maximal circles C_p is called the *skeleton* of S; it is clear that when the skeleton is specified, S is completely determined.

In the digital case, it is convenient to use squares (S_p) instead of circles; the reasoning in the preceding paragraph remains valid when we do this. It is easily seen that if S_p is contained in S_q, there is some neighbor p' of p such that S_p is contained in $S_{p'}$; thus the points of the skeleton are just those points for which r_p is a local maximum. Unfortunately, this implies that the skeleton is quite disconnected; two neighboring points p and p' cannot both be skeleton points unless $r_p = r_{p'}$. Thus the skeleton is a somewhat less economical way of representing an object than is the border chain code. The skeleton points of some simple objects are shown in Fig. 5.17.

The most natural way of finding the skeleton of a given object is by shrinking and re-expanding it. In the notation used in Subsection 5.6.3, the skeleton points of radius k are just the points that are in $S^{(-k)}$ but not in $(S^{(-k-1)})^{(1)}$ [5.36]. On a method of finding the skeleton using two raster scans of the picture, in opposite directions, see [5.37]. Given the skeleton, the object can be reconstructed by simply expanding each skeleton point a number of times equal to its associated radius. The 8-neighbor expansion (and shrinking) should be used here, if the squares S_p are supposed to have horizontal and vertical sides; if we prefer to use diagonally oriented squares S_p (i.e., diamonds), 4-neighbor expansion and shrinking should be used. Alternating 4 and 8 yields octagons instead of squares [5.31].

5.7.3 Relational Structures

When objects have been extracted from a picture, or the picture has been segmented into regions, it becomes possible to contruct *descriptions* of the picture. Such descriptions can include

a) Representations of the objects (or regions);
b) Properties of the objects—color, texture, shape, etc.;
c) Relations among objects.

Under this last heading, geometrical relations are of particular interest. Some of these are easy to define, e.g., "is adjacent to" or "is inside of" (S is inside T if any path from S to the edge of the picture must pass through T). Others are much fuzzier, e.g., "is near" or "is to the left of". (This last example does not seem to have any simple definition, though it can be roughly defined by a combination of conditions such as the following: S is to the left of T if the centroid of S is farther left than the left-most point of T, and no point of S is as far to the right as the right-most point of T [5.38].)

A relational structure can be represented by a labeled directed graph in which each node represents an object (or region), and is labeled with a list of property values for that object; and arcs join pairs of related regions, and are labeled with lists of relation values. Note that properties and relations can have numerical values (e.g., area or distance); literal or nominal values (e.g., color = red, blue, etc.); or truth values (e.g., convexity, adjacency). Alternatively, the nodes of the graph can represent either objects, properties, or instances of relations; the arc between an object node and a property node is labeled with the value of that property for that object, while the arcs between a pair of object nodes and a relation instance node are labeled with the value of that relation for that pair of objects. (A node is needed for each instance of a relation, if we want to be able to determine which pairs of objects are related.) Relational structures can also be represented in other ways; for some examples, as well as a discussion on how to detect matches between such structures, see [5.39]. More specialized classes of relational structures can be defined for representing line drawings; such structures are widely used in computer graphics [5.40].

"Generalized" relational structures, in which the information is only partially specified, can be used as *models* for classes of pictures [5.38]. Such models play a very important role in analyzing pictures of the given class, since they determine what type of segmentation is wanted, what types of properties should be measured, and so on. Moreover, when the picture analysis process is guided by a model, it becomes possible to adjust the operations used, depending on which part of the picture is being analyzed; for example, when we have identified something in one part of the picture, the model can tell us where to look for other things. For practical scene analysis purposes, it is generally more effective to embody such knowledge about the given class of pictures in the picture analysis programs themselves, rather than in a formal model represented by a data structure.

References

5.1 R.O. Duda, P.E. Hart: *Pattern Classification and Scene Analysis* (Wiley, New York 1973)
5.2 A. Rosenfeld: *Picture Processing by Computer* (Academic Press, New York 1969)
5.3 A. Rosenfeld, A.C. Kak: *Digital Picture Processing* (Academic Press, New York 1976)
5.4 A. Rosenfeld: Computing Surveys **1**, 147 (1969)
5.5 A. Rosenfeld: Computing Surveys **5**, 81 (1973)
5.6 A. Rosenfeld: Computer Graphics and Image Processing **1**, 394 (1972)
5.7 A. Rosenfeld: Computer Graphics and Image Processing **3**, 178 (1974)

5.8 B. S. LIPKIN, A. ROSENFELD: *Picture Processing and Psychopictorics* (Academic Press, New York 1970) pp. 289–370
5.9 R. M. HARALICK, K. SHANMUGAM, I. DINSTEIN: IEEE Trans. Systems, Man, and Cybernetics SMC-**3**, 610 (1973)
5.10 D. I. BARNEA, H. F. SILVERMAN: IEEE Trans. Computers C-**21**, 179 (1972)
5.11 R. N. NAGEL, A. ROSENFELD: Proc. IEEE **60**, 242 (1972)
5.12 B. WIDROW: Pattern Recognition **5**, 175 (1973)
5.13 M. A. FISCHLER, R. A. ELSCHLAGER: IEEE Trans. Computers C-**22**, 67 (1973)
5.14 A. K. GRIFFITH: J. Assoc. Computing Machinery **20**, 62 (1973)
5.15 M. HUECKEL: J. Assoc. Computing Machinery **18**, 113 (1971)
5.16 M. HUECKEL: J. Assoc. Computing Machinery **20**, 634 (1973)
5.17 A. ROSENFELD, M. THURSTON: IEEE Trans. Computers C-**20**, 562 (1971)
5.18 A. ROSENFELD, M. THURSTON, Y-H. LEE: IEEE Trans. Computers C-**21**, 677 (1972)
5.19 C. K. CHOW, T. KANEKO: In *Frontiers of Pattern Recognition*, ed. by S. WATANABE (Academic Press, New York 1972) pp. 61–82
5.20 C. R. BRICE, C. L. FENNEMA: Artificial Intelligence **1**, 205 (1970)
5.21 S. L. HOROWITZ, T. PAVLIDIS: Proc. 2nd Intern. Joint Conf. Pattern Recognition **1974**, 424
5.22 U. MONTANARI: Comm. Assoc. Computing Machinery **14**, 335 (1971)
5.23 A. MARTELLI: Computer Graphics and Image Processing **1**, 169 (1972)
5.24 A. ROSENFELD: J. Assoc. Computing Machinery **17**, 146 (1970)
5.25 A. ROSENFELD: J. Assoc. Computing Machinery **20**, 81 (1973)
5.26 A. ROSENFELD: Information and Control **26**, 24 (1974)
5.27 A. ROSENFELD: IEEE Trans. Systems, Man, and Cybernetics SMC-**4**, 221 (1974)
5.28 J. SKLANSKY: Pattern Recognition **2**, 3 (1970)
5.29 A. ROSENFELD, E. JOHNSTON: IEEE Trans. Computers C-**22**, 875 (1973)
5.30 A. ROSENFELD: IEEE Trans. Computers C-**23**, 1264 (1974)
5.31 A. ROSENFELD, J. L. PFALTZ: Pattern Recognition **1**, 33 (1968)
5.32 A. ROSENFELD, C. M. PARK, J. P. STRONG: EASCON '69 Record **1969**, 264
5.33 R. STEFANELLI, A. ROSENFELD: J. Assoc. Computing Machinery **18**, 255 (1971)
5.34 H. FREEMAN: Computing Surveys **6**, 57 (1974)
5.35 C. T. ZAHN, R. Z. ROSKIES: IEEE Trans. Computers C-**21**, 269 (1972)
5.36 J. C. MOTT-SMITH: In *Picture Processing and Psychopictorics*, ed, by B. S. LIPKIN and A. ROSENFELD (Academic Press, New York 1970) pp. 267–278
5.37 A. ROSENFELD, J. L. PFALTZ: J. Assoc. Computing Machinery **13**, 471 (1966)
5.38 P. H. WINSTON: "Learning structural descriptions from examples", AI TR-231 (MIT, Cambridge, Mass. 1970)
5.39 H. G. BARROW, R. J. POPPLESTONE: In *Machine Intelligence* 6, ed. by B. MELTZER and D. MICHIE (University Press, Edinburgh 1971) pp. 377–396
5.40 R. WILLIAMS: Computing Surveys **3**, 1 (1971)

TREE SYSTEM APPROACH FOR LANDSAT DATA INTERPRETATION†

R. Y. Li and K. S. Fu
School of Electrical Engineering
Purdue University
West Lafayette, Indiana 47907

ABSTRACT

This paper describes a tree system approach which interpretates highways and rivers from LANDSAT pictures. The basic definitions of tree grammars and tree automaton and a grammatical inference procedure are first introduced. The interpretation process is conceived as a process of continuous verification of the hypothesized descriptions of objects in the picture. The LANDSAT imagery map of Lafayette, Indiana is used as a training data set and tree grammar is inferred from the interpretation process. The versatility of this set of syntactic rules is tested on a different data set and the initial results are reported.

I. INTRODUCTION

As the ability of satellites to gather data for the purpose of survey and monitoring of earth resources grows, the need to fully automate the recognition process of a large number of pictures obtained by satellite photography is also becoming more evident. In the past, the use of pattern recognition techniques has been very successful in the classification and interpretation of the data taken from agriculture fields, vegetation, water, soil, etc. However, these methods usually employ only spectral and/or temporal properties of the objects and neglect the spatial relationships among classes in the picture. Difficulties could then arise when one is dealing with smaller objects such as bridges, highway, river, etc. because the surrounding environment changes greatly the expected reflectance of those objects due to the resolution size. For instance, the gray level of a segment of the highway is digitized from a combined reflectance of concrete surfaces, grasses, trees, etc. Sometimes it is impossible to distinguish this class from, say, suburban scenes where similar features dominate. In cases like this, one has to extract a certain geometric feature from the data in order to interpret them more accurately. In other words, properties such as shape, size, and texture must be used to delineate one from the other among classes of similar spectral properties.

†This work was supported by the National Science Foundation Grant ENG 74-17586 and the ARPA Contract F30620-75-C-0150.

Often, the spatial relationships such as "surrounded by," "near by," and directional references can also be explored to locate classes of large areas where no definite shapes exist, such as those found in land use classification. For instance, in the study by Todd & Baumgardner[1] on land use classification of the Marion County (Indianapolis), Indiana, an overall accuracy of about 87 percent is reported using only the spectral information. Difficulties were encountered in the spectral separation of grassy (open country, agriculture) area and multi-family (older) housing. One solution to this problem consists of spatially dividing the data into urban and rural land use prior to classification. Over 95 percent accuracy of recognition may be achieved by this manual preprocessing step in their analysis. The use of syntactic methods to describe the spatial relationship has recently been suggested[2]. Brayer and Fu went further by constructing a hierarchical or tree graph model to contain the spatial distributions of all classes in the entire scene[3]. For instance, the earth scene consists of urban and rural area, and the urban area consists of the downtown area surrounded by the inner city area with near-by suburban area and a system of highways. These classes are then classified by utilizing their spatial relationships which are expressed in terms of syntactic rules; namely, those of a web grammar. The study undertaken here is similar to this approach, but a tree system is used as the main tool to interpret LANDSAT data where traditional approaches have not achieved satisfactory results.

II. BASIC DEFINITIONS OF TREE SYSTEM

The use of formal linguistics in describing physical patterns have received increasing attention recently[3]. The string representation has been used very often due to the availability of existing results in formal languages. But it is inadequate and sometimes inconvenient for descriptions of high-dimensional patterns or multiconnected graphs, so there is a need of developing higher dimensional pattern description languages. Recently, Fu and Bhargava have proposed the use of tree grammars for pattern description[5]. Tree grammars are generalizations of string grammars.

Reprinted from *Proc. IEEE Symp. on Machine Processing of Remotely Sensed Data*, June 29–July 1, 1976, pp. 2A-10–2A-16.

A tree grammar becomes a string grammar when the ranks of all variables are one or zero. We shall see that the use of tree grammars is justified because of their ability to describe easily the recursive nature of the physical patterns under consideration. Furthermore, a tree automaton can be easily constructed from a given tree grammar to recognize the trees generated.

A regular tree grammar over $<V_T, \gamma>$ is a four-tupled $G_t = (V, \gamma, P, S)$ where $<V, \gamma'>$ is a finite ranked alphabet with $V_T \subseteq V$ and $\gamma'/V_T = \gamma$, $V_N = V - V_T$, the set of non-terminals. P is the production rules of the form $A \rightarrow B$, if there is a production $\phi \overset{a}{\rightarrow} \psi$ in P such that ϕ is a subtree of A at a, and B is obtained by replacing the occurance of ϕ at a by ψ. S is a finite subset of T_V, called axioms, where T_V is the set of trees over alphabet V.

The tree language generated by a tree grammar G_t is defined as

$L(G_t) =$
$\{\alpha \in T_{V_T} \mid \text{there exists } Y \in S \text{ such that } Y \Rightarrow \alpha\}$

where T_{V_T} is the set of trees containing only terminal symbols.

A tree grammar $G_t = (V, \gamma, P, S)$ is expansive if each production in P is of the form

$$Y_o \rightarrow \begin{array}{c} x \\ / \quad \backslash \\ Y_1 \ldots\ldots Y_n \end{array}$$

where Y_o, $Y_1, \ldots\ldots Y_n$ are non-terminals, x is a terminal.

We also know that for every regular tree grammar G_t, one can effectively construct a tree automaton M_t such that $T(M_t) = L(G_t)$ where $T(M_t)$ is the set of trees accepted by M_t. We are interested in knowing the relation between tree automata and tree grammars, since the patterns will be described by a tree grammar and a tree automaton can be used to recognize these patterns.

Let $<V_T, \gamma>$ be a ranked alphabet and $V_T = \{x_1, \ldots\ldots x_n\}$. A tree automaton over V_T is a system $M_t = (Q, f_1, \ldots\ldots f_k, F)$ where

1. Q is a finite set of states
2. for each i, $1 \leqslant i \leqslant k$, f_i is a relation for $Q^{\gamma(x_i)} \rightarrow Q$
3. $F \subset Q$ is a set of final states

If each f_i is a function, f_i: $Q^{\gamma(x_i)} \rightarrow Q$, then M_t is determinable. Otherwise, it is non-deterministic.

The response relation p of a tree automaton M_t is defined as follows:

1. If $x \in V_{T_O}$, then $p(x) \rightarrow q$ iff $q \in Q$
2. If $x \in V_{T_n}$, $n > o$, then $p\left(\begin{array}{c} x \\ / \quad \backslash \\ t_1 \ldots t_n \end{array}\right) \rightarrow q$

   iff there exists $q_1, \ldots\ldots q_n \in Q$ such that $f_x(q_1, \ldots\ldots q_n) \rightarrow q$ and $p(t_i) \rightarrow q_i$, for $1 \leqslant i \leqslant n$, $t_i \in T_{V_T}$.

The language accepted by M_t is defined as

$T(M_t) = \{t \in T_{V_T} \mid \text{there exists } q \in F \text{ such that } p(t) \rightarrow q\}$. M_{t_1} and M_{t_2} are equivalent iff $T(M_{t_1}) = T(M_{t_1})$.

We summarize the construction procedures of tree automaton for a regular tree grammar as follows:

1. To obtain an expansive tree grammar (V', γ, P', S) for the given regular tree grammar (V, γ, P, S) over alphabet V_T.
2. The equivalent nondeterministic tree automaton is $M_t = (V' - V_T, f_1, \ldots. f_n, \{S\})$, where $f_x\ (q_1, \ldots. q_n) =$ q_o if $q_o \rightarrow x\ q_1, \ldots. q_n$ is a rule in P'. If f_i, $1 \leqslant i \leqslant k$, is a function, them M_t is deterministic; otherwise, M_t is non-deterministic.

As an example, if we denote a→, b↘, c↓, then the following multi-connected graph

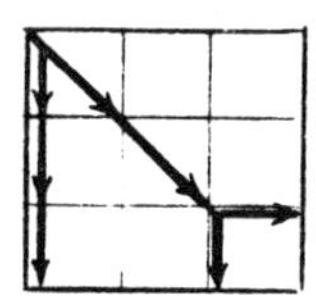

can be written in tree form as

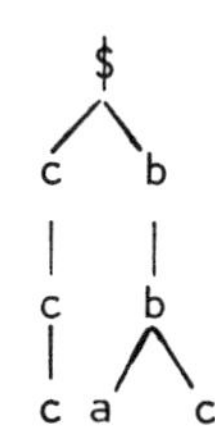

Sometimes when the patterns are not quite linear due to noise or distortion, we can apply a transformational grammar to linearize them.

III. INFERENCE OF TREE GRAMMAR

When the physical shape of the class under consideration is completely known and fixed, it is possible to write down the syntactic rules directly to describe its structure. If this is not the case, we have to construct a set of grammatical rules by examining a set of sample patterns known to come from that class in order to describe that particular class. This set of inferred rules should be able to describe and predict other sample patterns which are of the similar nature as the original training samples and presumably in the same class. Bhargava and Fu have suggested an inference procedure for tree grammars[6]. The basic idea consists of the following three steps:

1. Try to discover the syntactic structure of each given tree sample by looking for repetitions and dependent relationships, called repetitive substructures (RSS).

2. Decide what sublanguages make up the language and generate nonterminals for each sublanguage.

3. Combine equivalent nonterminals which have almost the same sublanguages and determine the appropriate relationships among sublanguages. The flow chart implementing the inference procedure is shown in Figure 1.

To start the inference process, we first find the types of terminals or primitives that will fit the subparts of the picture patterns for a given window size. After this initial extraction process, we have to decide the most probable combinations of primitives which occur as neighbors of each other in the set of observed training samples. These combinations are then applied to the training data set to test their recognition effectiveness. When the result appears to be satisfactory after some additions and deletions of the combinations, we can choose this set of patterns to represent the training samples. The appropriate grammar can then be inferred from these samples by following those three basic steps of grammatical inference. This process of learning can be repeated for higher levels if we are dealing with patterns of larger size. To prove the acceptance of the inferred grammar by other non-training sample patterns, a set of test data should be used. The success of this final step should prove that the spatial relationships among data samples of a particular class can be utilized in a broad sense.

IV. EXPERIMENTAL RESULTS

In the actual implementation of the above procedures, we first choose the LANDSAT imagery map of Lafayette, Indiana as the training data set. The original 17 clusters are further combined into seven ground cover types. They are general agriculture area, pasture with wheat dominant, forests, commercial area, residential area, highways and rivers. Among them, rivers and highways could serve as excellent examples for syntactic pattern recognition because of their simple shapes and the relative failures of statistical approaches. Our purpose would be to separate the lake or pond from the river and highway from any spectral similar features. Although we might expect that highways are usually built as a straight connection between two locations, in reality this is not true. The highways are built as straight lines only locally, but not globaly, in order to avoid the fatigue of the drivers. However, highways occasionally curve locally for directional changes when some natural obstacles such as rapid elevations occur. As a result, certain geometric requirements of a highway must be satisfied:

1. The width of a highway has an upperbound.

2. The local curvature of a highway has an upperbound to follow the requirement of maximal speed of automobiles.

The river, on the other hand, has a less rigid upperbound than highway in terms of local curvature. In other words, the river could make a sharper turn. In general, we can expect the river to exhibit the same linear pattern as the highway does. A small creek branching out from a river can be interpreted as the entrance or the exit road from the superhighway. For simplicity, we shall not write separate grammars for rivers and highways in our present study.

The lowest level or the primitives selected for both river and highway are based on a 2 X 2 pixel window of the following patterns:

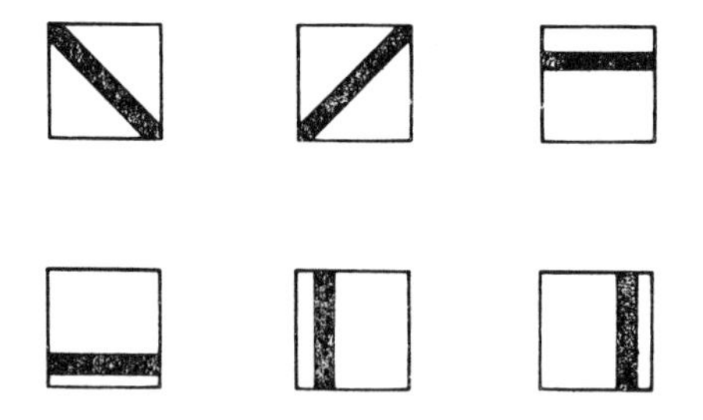

In short, this first-level extraction will eliminate all isolated points. Its main purpose, however, is to generate the terminals for further learning. The next step is to find the most probable combinations of primitives which occur as neighbors of each other in the river and highway data samples. For the sake of convenience, we choose a set of 4-tuple patterns which are more representative of suburban highways than, say, streets in commercial areas or any other features which reflect like a mixture of concrete and grass, like those appearing quite extensively in the new residential area in south Lafayette. Those 4-tuple patterns are shown in Figure 2.

After a series of trials and errors, we deduce a set of 26 combinations which give us a good result in terms of showing the Wabash River and Interstate Highway 65 in the Lafayette area. The ground truth in this case is provided by an infrared photography of the Lafayette area. The pointwise classified data of the Lafayette area is shown in Figure 3. The result from the syntactic method with selected pattern combinations is shown in Figure 4 for both highways and river. Since the 4-tuples can be applied in both directions, we really learn the highway and river structures from the 13 combinations shown in Figure 5.

It is possible for us to go one level further but these 13 patterns are probably sufficient for us to infer a tree grammar based on their structural information. Five of them are just straight lines, meaning no directional changes. The other eight have directional changes of no more than 45 degrees. It is true that these patterns only represent a segment of the highway and the river structures but their repetitive natures are certainly valid in the general context. Thus, we have completed the step (1) of the inference procedures.

The next step is to discover what subtrees make up the tree language and generate nonterminals for each subtree. We can divide those 13 patterns into three categories; they are shown as the three rows in Figure 5. If we denote a → (horizontal line segment), b ↘ (diagonal line segment), and c ↓ (vertical line segment), then the tree representation of the following superhighway pattern

will be

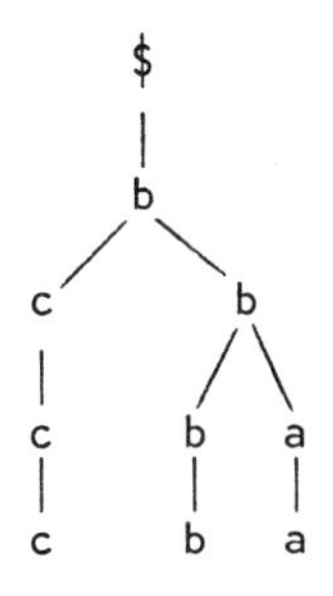

The subtrees of depth one within this tree can be expressed in terms of the following representation

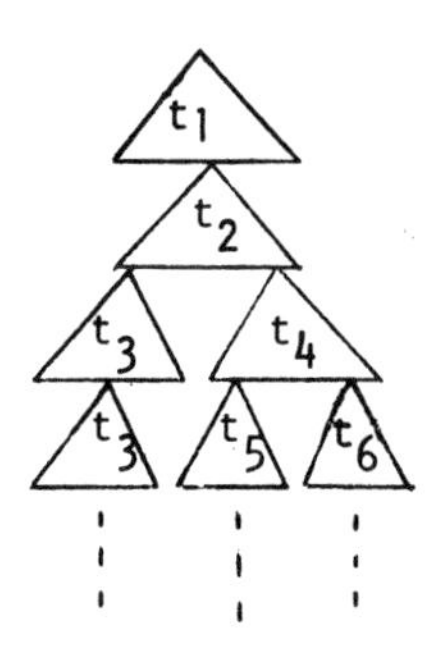

where the repetitive substructures (RSS) for the sublanguage are

```
t1 = $      t2 =  b      t3 = c
     |           / \          |
     b ,        c   b ,       c

t4 =  b     t5 = b       t6 = a
     / \         |            |
    b   a ,      b ,          a
```

Continuing in this fashion and following the flow chart in Figure 1, we obtain the following tree grammar for highway (or river) patterns:

$G_t = (V, \gamma, P, S)$

$V = \{s, a, b, c, \$, A_1, A_2, A_3, A_4, A_5, A_6\}$

$V_T = \{a, b, c\}$

$\gamma(a) = \{2, 1, 0\}$, $\gamma(b) = \{2, 1, 0\}$, $\gamma(c) = \{2, 1, 0\}$

$\gamma(\$) = \{2, 1\}$

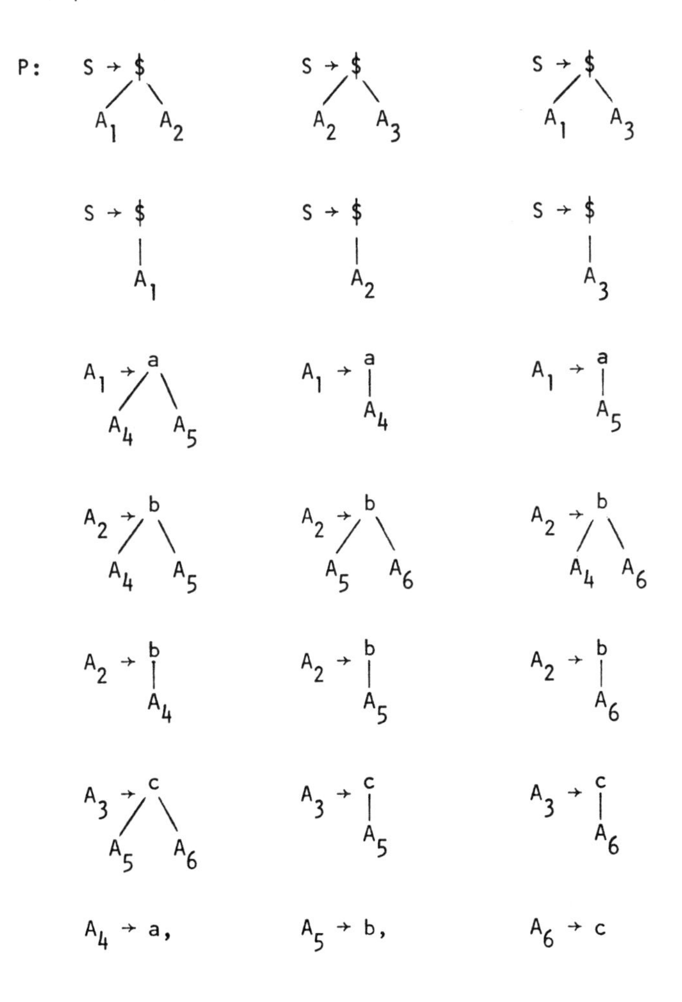

Corresponding to this tree grammar, we can then construct a tree automaton

$$M_t = (Q, f_\$, f_a, f_b, f_c, F) \quad \text{over } V_T$$

$$Q = \{A_1, A_2, A_3, A_4, A_5, A_6, q_F\}, \; F = \{q_F\},$$

$$V_T = \{\$, a, b, c\}$$

$$f_\$ (A_1, A_2) = q_F$$

$$f_\$ (A_2, A_3) = q_F$$

$$f_\$ (A_1, A_3) = q_F$$

$$f_\$ (A_1) = q_F$$

$$f_\$ (A_2) = q_F$$

$$f_\$ (A_3) = q_F$$

$$f_a (A_4, A_5) = A_1$$

$$f_a (A_4) = A_1$$

$$f_a (A_1) = A_1$$

$$f_b (A_4, A_5) = A_2$$

$$f_b (A_5, A_6) = A_2$$

$$f_b (A_4, A_6) = A_2$$

$$f_b (A_4) = A_2$$

$$f_b (A_5) = A_2$$

$$f_b (A_6) = A_2$$

$$f_c (A_5, A_6) = A_3$$

$$f_c (A_5) = A_3$$

$$f_c (A_6) = A_3$$

$$f_a = A_4$$

$$f_b = A_5$$

$$f_c = A_6$$

After an input tree extracted from a picture window is applied, if the tree automaton is in q_F then the picture contains a highway (or river) pattern. If the tree automaton reaches any other state we conclude that this particular picture does not have what we are looking for.

The tree automaton is tested on a new data set, that of Grand Rapids, Michigan. The total number of pixels being studied are about 57940, half of them mainly in the suburb, the other half mainly in the inner city. However, due to its poor resolution highway data has to be preprocessed using a local region expansion algorithm. In other words, a proper preprocessing algorithm can connect up those missing points in the data set which are due to inadequate reflections of highway surfaces whose ground covers are only a fraction of the pixel size ($\sim 79 \times 56\ m^2$). The method of preprocessing as employed is illustrated in Figure 6. This process essentially has the effect of lengthening and thicking the data samples.

The results on Grand Rapids, Michigan show that with appropriate preprocessing the highways in suburban areas can be detected as a road-like feature. In urban areas, there are too many streets and concrete parking lots confused as highways. On the other hand, the river, which is usually easier to find due to good resolution, is not so obvious in the lower portion of the urban-area data set due to the confusion with the shadow class. However, these rivers have been successively traced out in our syntactic approach. Figures 7, 8, 9, and 10 give the pointwise classification and the syntactic interpretation of highways and rivers respectively in the Grand Rapids area.

V. CONCLUDING REMARKS

There are some observations that we have obtained from these experiments on LANDSAT data:

1. Syntactic approach, and specifically the tree system approach here, can be very useful in picture recognition by analyzing the geometric patterns of the classes under investigation.

2. The spatial patterns of rivers and highways can be described by tree grammars.

3. The analysis of tree languages by tree automata is a simple and efficient procedure compared with other high-dimensional languages.

4. Preprocessing can be very helpful in handling the resolution problem when the continuity of the feature is very important.

More extensive tests on real data are certainly needed to justify the complete effectiveness and efficiency of the proposed tree system approach for LANDSAT data interpretation.

REFERENCES

1. Todd, W. J. and Baumgardner, M. F., "Land Use Classification of Marion County, Indiana, by Spectral Analysis and Digitized Satellite Data," LARS Information Note 101673, LARS, Purdue University, 1973.

2. Fu, K. S., "Pattern Recognition in Remote Sensing of the Earth's Resources," *IEEE Transactions on Geoscience Electronics*, Vol. GE-14, January, 1976.

3. Brayer, J. M. and Fu, K. S., "Web Grammar and Its Application to Pattern Recognition," Tech. Rept. TR-EE 75-1, Purdue University, West Lafayette, Indiana, December, 1975.

4. Fu, K. S., *Syntactic Methods in Pattern Recognition*, Academic Press, 1974.

5. Fu, K. S. and Bhargava, B. K., "Tree Systems for Syntactic Pattern Recognition," *IEEE Transactions on Computers*, Vol. C-22, December, 1973.

6. Bhargava, B. K. and Fu, K. S., "Transformations and Inference of Tree Grammars for Syntactic Pattern Recognition," Proc. 1974, International Conference on Systems, Man, and Cybernetics, October 2-4, Dallas, Texas.

7. Tavakoli, Mohamad and Bajcsi, Ruzena, "Computer Recognition of Roads from Satellite Pictures," University of Pennsylvania, Philadelphia, Pennsylvania, 1975.

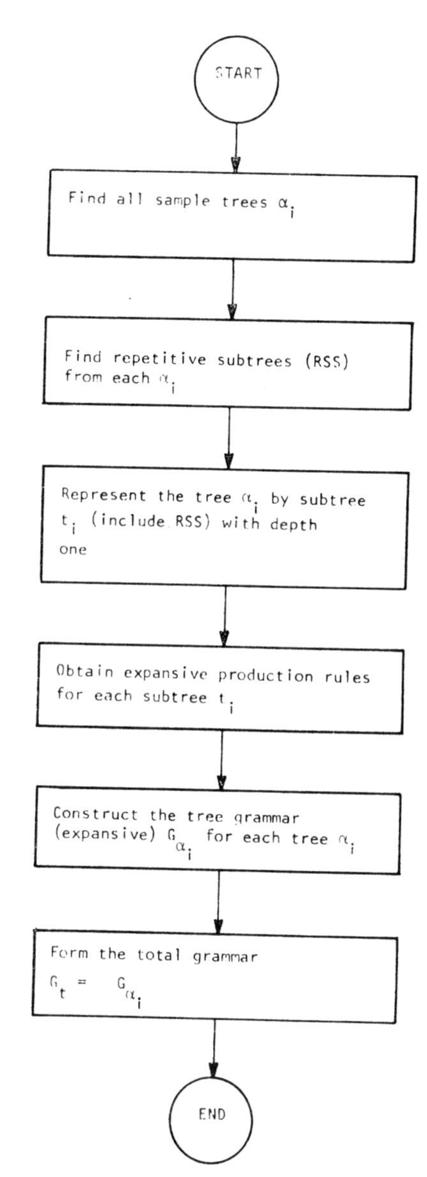

Figure 1. Flow Chart for Tree Grammar Inference

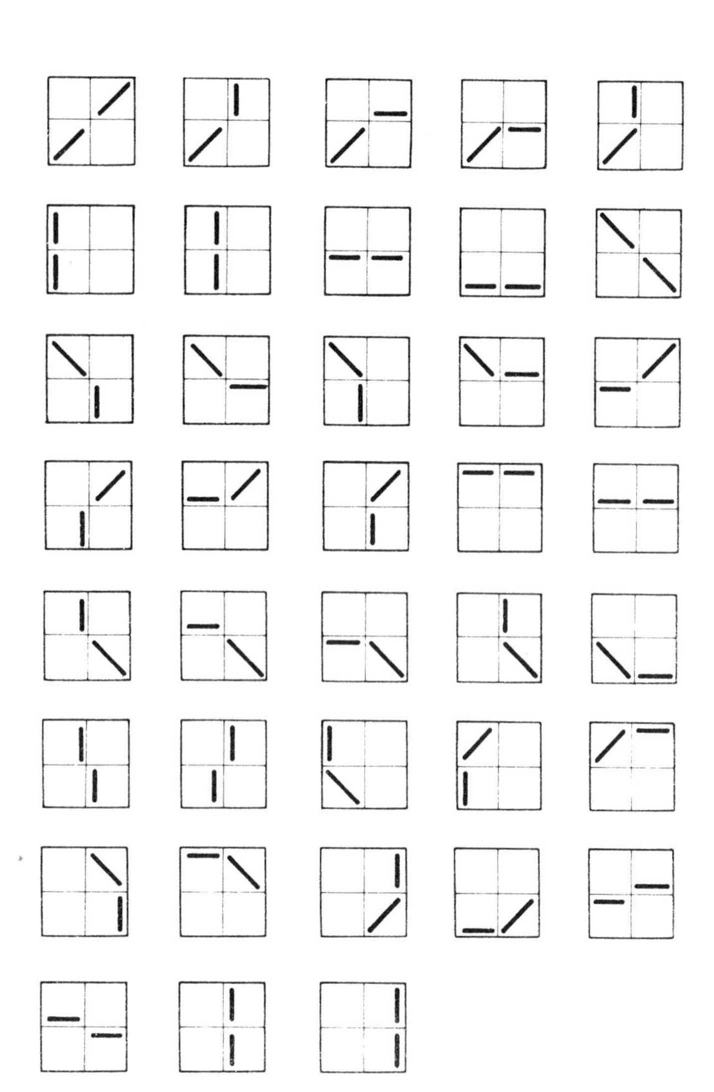

Figure 2 4-Tuple Patterns

Figure 3 Pointwise Classification of LANDSAT Data of the Lafayette Area

Figure 4 Syntactic Interpretation of Highway and River Patterns in the Lafayette Area

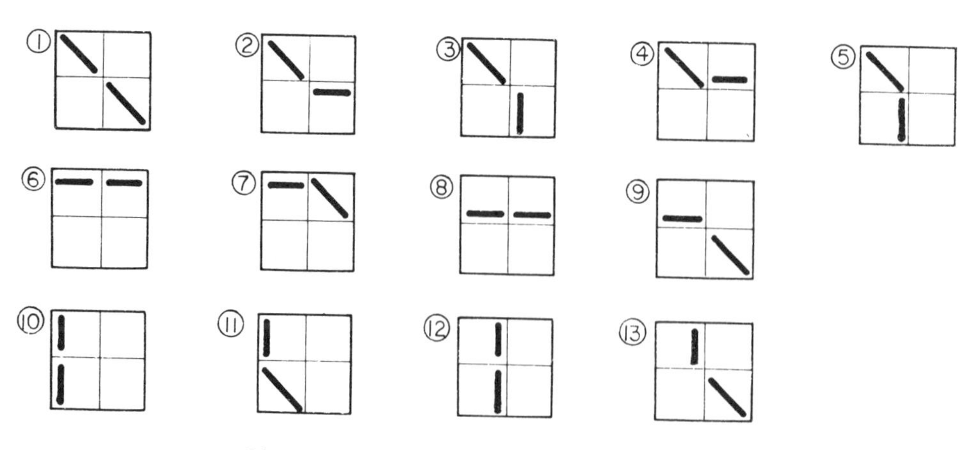

Figure 5 Basic (First Level) Highway Patterns

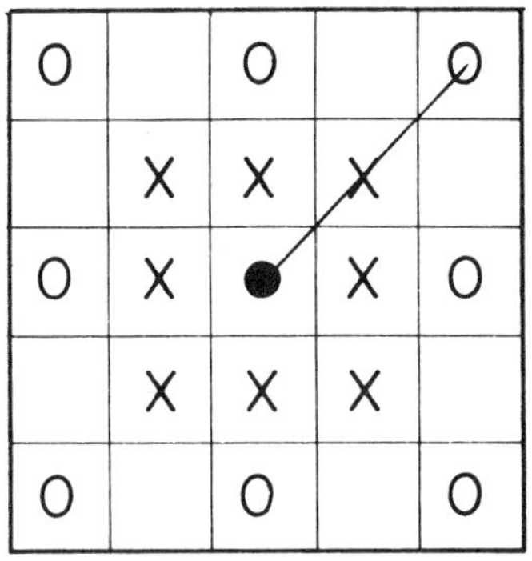

O		O		O
	X	X	X	
O	X	●	X	O
	X	X	X	
O		O		O

● point of reference pixel with sample a

O point at one pixel distance away with sample a

X points patched up by adding sample a to this pixel point if O and •'s relationship is established.

Figure 6 Preprocessing

Figure 7 Pointwise Classification of LANDSAT Data of a Suburban Area in Grand Rapids

Figure 9 Syntactic Interpretation of River and Highway Patterns in the Suburban Area of Grand Rapids

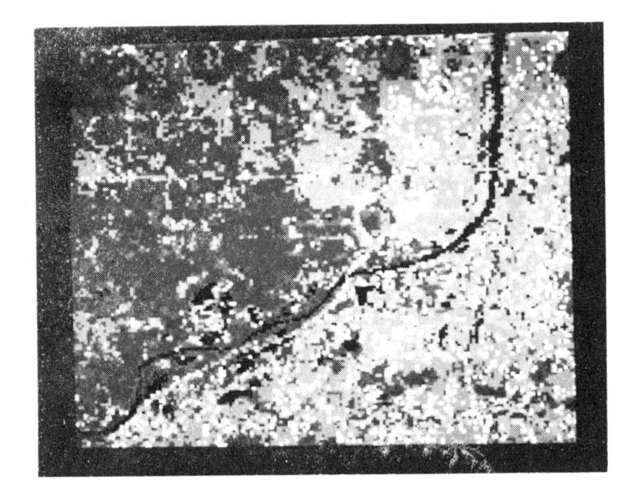

Figure 8 Pointwise Classification of LANDSAT Data of an Urban Area in Grand Rapids

Figure 10 Syntactic Interpretation of River Patterns in the Urban Area of Grand Rapids

Image Filtering—A Context Dependent Process

RUZENA BAJCSY AND MOHAMAD TAVAKOLI

***Abstract*—In this paper, we argue that in order to be able to build meaningful image filters, we have to have a world model, a description of the world that we are dealing with. This world model includes a model of the scene, a model of the eye (camera), and a model of the illumination. Secondly, using this description, we have to recognize objects in the scene which match this description before we can remove them, that is, filter them out. Thirdly, we show that the strategy of sequencing the procedures (deciding which object in the scene is recognized when) is guided by the world model and the visibility of objects on the scene. Since the filtering process involves recognition of the visual concepts of objects, we call it *conceptual filtering*. Similarly, since the guided strategy of the scene analysis, depends on partial recognition of the scene, we call it *conceptual focusing*. The preceding points are demonstrated by a concrete example of computer recognition of bridges, rivers, lakes, and islands from ERTS satellite pictures.**

I. Introduction

IT IS NOT the purpose of this paper to review the traditional filters in image processing. For that, there are many good articles and books available (see, for instance, [1] or [2]). Our paper will present a novel approach to image filtering from a computer scene analysis point of view.

The objective of scene analysis is to recognize and describe scenes via computers (see, for example, Binford and Tenenbaum [5]). The description is in terms of three-dimensional objects and their relationships in the scene. Having this goal in mind, we shall address ourselves in this paper to the problem of what is *noise*, what is *the relevant information* in a scene and/or picture, and *how can we separate* the two. Once we understand the problem of noise and the relevant information, we believe that we can design a filter for either of them.

The relevant information is that visual information in the scene and/or picture which is relevant to the observer at the given *context* and *time* in order to accomplish a certain *task*. The task could be image enhancement, or it could be target finding or it may be a complex scene description. Thus the relevant information is context, time, and task dependent. Now we can turn around this definition to a definition of noise. R. Kalman [11] pointed out that noise is everything in the signal which is not the relevant information and vice versa.

Consequently, in a scene and/or picture we regard everything as noise which is not of current interest to the viewer. This definition of noise is *dynamic*, that is, what is noise at one moment or in one context does not necessarily have to be noise at another moment or in a different context.

Actually, all classical filters have been designed with the assumption of knowing what the noise and the relevant information is in a certain context and time. Indeed, people have a very good mechanism for implementing dynamic filtering by focusing and defocusing attention on certain expected objects in the scene, or merely on the most visible objects in the scene.

The process of directed attention of the eye is guided by recognition of the scene as a whole and/or of objects in the scene (see Neisser [13]). Since this process is not just a simple focusing on some areas of the picture, but is rather a guided focusing on some objects by knowledge about them, we call it *conceptual focusing*. In other words, to be able to build filters which would decompose the scene into objects, we must first have a description of the world which is context and time dependent. Second, using this description, we have to recognize objects in the scene which match this description before we can remove them, that is, filter them out. Since this filtering process involves cognition, we call it a *conceptual filtering process*.

The importance of context for visual recognition has been recognized by all artificial intelligence researchers for the past ten years (see, for example, Roberts [15] and Minsky and Papert [12]). In this paper, we are just enhancing those ideas. Our approach clearly differs from the classical approaches of filtering known in electrical engineer-

Manuscript received April 1, 1974; revised September 30, 1974.
The authors are with the Moore School of Electrical Engineering, University of Pennsylvania, Philadelphia, Pa.

Reprinted from *IEEE Trans. Circuits and Syst.*, vol. CAS-22, pp. 463–474, May 1975.

ing. For instance, if a low-pass filter is applied to an image, the result is the filtering out of small spatial variations (high-frequency components). What are these small spatial variations? This depends on many factors, such as the spatial resolution of the picture, the distance between the observer and the scene from which the picture was taken, the gray level resolution, the relative expected size of objects on the scene, the viewing angle of the observer, etc. For the sake of discussion, let us assume that the images are pictures that one can see on a regular TV screen. In such a case, then, the small spatial variations are sometimes called the "snowing effect." A low-pass filter will remove the snowing effect, but it will also remove the sharp edges. This is certainly an undesirable side effect. If the snowing effect is noise, then the low-pass filter does serve a purpose, although in a limited way. Sometimes, however, the "snowing" may be the real part of the scene, and then it is clear that the low-pass filter is inappropriate. A similar case can be made for a high-pass filter which filters out large spatial variations, and, as a consequence, enhances the edges; however, it may also filter out some regions that contain essential data for the meaning of the picture.

The conceptual filtering and focusing process is a part of the overall scene analysis process that we have implemented. How and when different conceptual filters are used, and how the focusing process is directed are questions which are determined by the scene analysis strategy, also called the *higher level program*. In the next sections, we shall discuss in detail two different strategies that we use. The first is: *initial strategy* or how to decide the first object to be recognized in the scene. The second is: *continuous strategy* or after the first object is identified, how to go about deciding the next object to be recognized.

The underlying assumption in the previously mentioned process is that one has a suitable representation of the world to deal with. This representation, called the *World Model*, is composed of three submodels: the model of the scene, the model of the eye (the model of the camera), and the illumination model (the model of the light source, its position, etc.).

The model of the scene can have two different forms:

1) a network (graph) of expected objects in the scene and their relationships, similar to those introduced by Winston [19] (see Fig. 1); and
2) a network of expected objects in the scene and their relationships, which is, in addition, structured in such a way that the structure implies some analysis strategy, called *strategy net* (if, for instance, followed top-down, see Fig. 2). (While the strategy net may remind the vertical ordering (part-whole relationship) introduced in Preparata and Ray [14], in our case, it represents vertical ordering with respect to visibility of objects which includes the part-whole relationship.)

All of the ideas presented in this paper are demonstrated on a set of ERTS satellite pictures. The World Model for these pictures is shown in Fig. 2. Each object in the World Model has an associated list of attributes. An example of such a list is in Table I. The names of attributes correspond to procedures called *low-level operators* which extract particular features and desired descriptions from the picture data. In our case, the model of the camera is reduced to a simple statement, that is, ERTS satellite pictures are direct images of objects found on earth from a certain known distance.

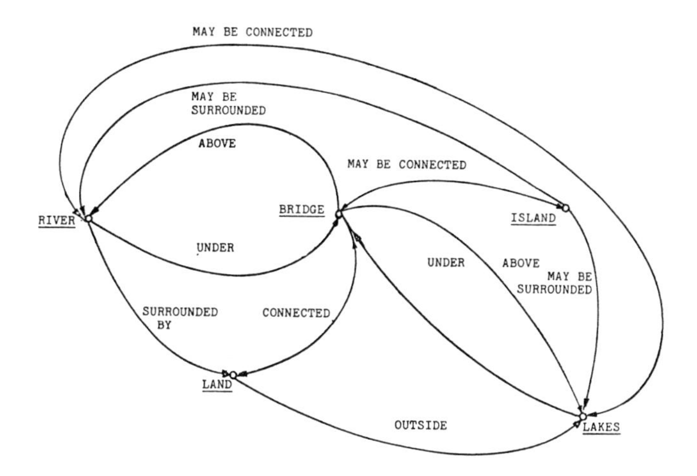

Fig. 1. Semantic net.

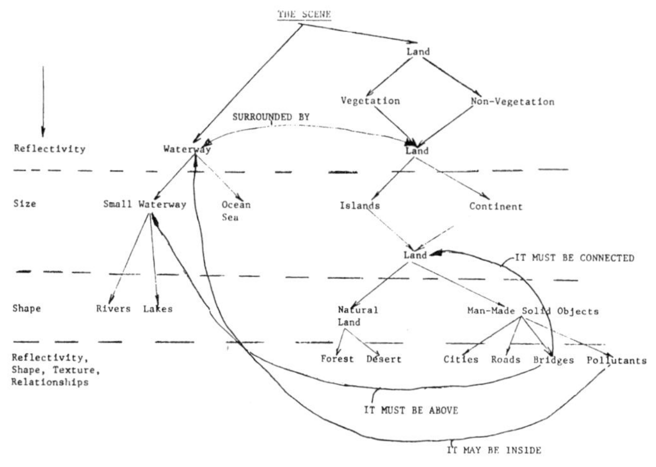

Fig. 2. Strategy net.

The illumination model in our application is again very simple. We assume diffuse daylight and no shadow effects.

Finally, we list the contributions of this research. We try to show via examples that the following are true.

1) In complex scenes, such as the real outdoor scenes, the context independent filters—the traditional filters—are not sufficient for scene analysis and that context dependent filters—conceptual filters—are necessary.

2) The context dependent focusing process—conceptual focusing—is necessary for going from the global structure of a picture into a detailed one in a meaningful way, our example of this case being the identification of pollutants in a waterway.

3) The higher level program must consider two different strategies during the scene analysis process.

TABLE I
ATTRIBUTES FOR SOME OBJECTS

Name	Waterway	Land	Lakes	River	Bridge
Brightness	Dark in Band 3	Light in Band 3	Same as the Waterway	Same as the Waterway	Same as the Nonvegetation
Contrast	Large with Land in Band 3	Large with Waterway in Band 3	Same as the Waterway	Same as the Waterway	Medium with Waterway
Texture	Homogeneous	Varies	Homogeneous	Homogeneous	Homogeneous
Size	Arbitrary	Not Important	Small	Small	Very Small
Shape	Arbitrary	Not Important	Arbitrary	Elongated	Thin Elongated
Boundary	Arbitrary	Not Important	Closed	Open	Open

a) One is the initial strategy, that is, how to recognize the first object on the scene, which is not as arbitrary as it may seem at first glance. We argue that the initial strategy follows either from some *a priori* knowledge about the scene or from the degree of visibility.

b) The other is the continuous strategy, also called recognition driven by World Model.

4) The recognition of three-dimensional spatial relationships implies partial occlusion which, in turn, enables one to complete the data of partially occluded objects and thus complete their recognition. An example of this case is recognition of a river after removal of recognized bridges.

We also introduce an operational definition of the visibility of objects.

II. Scene Analysis Strategy

In this section, we shall be concerned only with the organization of the scene analysis process, i.e., the sequence (or structure) of different operators extracting features for recognizing an object, for recognizing the first object in the scene, and for recognizing the next object and/or complex of objects after the first one.

The input for the higher level program consists of picture data, low-level operators, and World Model.

As we mentioned before, the world model represents knowledge that is available about the expected objects and their relationship in the scene, knowledge about the visual system of the observer, and finally, knowledge about the lighting conditions.

2.1 Initial Strategy in a Scene Analysis

We start with a question: What is the first approach in analysis and description of a picture or a scene? We propose two possible criteria that control this strategy.

One criterion is given by an *a priori knowledge* about the scene as well as about the *order* and *priority* of recognized objects. This *a priori* knowledge could be specified either *directly* or *indirectly*. By directly we mean that a detailed description is given of the first object to be recognized. An example of this case is a command to the machine: Find a red balloon with diameter of 3 in. Here a *simple matching* between the model of the object, its list of attributes and the picture data will do the recognition process. This reminds us very much of the matched filters known in image processing (for review, see Rosenfeld [16]), except that they were used as templates for matching with the data. Here, we match concepts (see Barrow *et al.* [4] and Fishler and Elschlager [7]), hence, the name *conceptual filter*.

By indirect knowledge, we mean that the description of the first desired object to be recognized is not given explicitly but rather implicitly; that is the explicit description is generated from the task of, say, a robot, or from the goal of the visual problem solver. As an example of such a case, the task, "to sit down," is translated to "find a place to sit, such as a chair, floor, bed, etc." Here the first object to be recognized will be a chair, floor, etc.

The other criterion for controlling the analysis strategy is given by the *degree of visibility* of different objects in the scene. Here we assume that the observer is not biased, at least in the beginning of the recognition process, by any preferences for one object over another.

Once we recognize the first object, however, the decision criterion in the choice of the next objects is a composition of the degree of visibility of the objects and the associations following from the world model.

The visibility of objects is a function of the following properties of the scene and/or objects in the scene:

1) the spatial and gray level resolution;
2) the contrast of objects with their background;
3) the illumination conditions of the scene;
4) the relative size of objects in comparison to other objects and/or to the whole scene;
5) the possible partial occlusion of objects by other objects; and
6) the view angle or the window size.

By presenting the two possible decision-making criteria in the initial approach in a scene analysis, we argue that these are ways that people use to sort the relevant information and noise.

In our implementation, the initial strategy follows from Fig. 2; that is, we start first by separating the waterway and land based on their differences in brightness.

2.2 The Continuous Strategy in a Scene Analysis

Once the first object in the scene is recognized, some existing structure among the objects (the previously identified object, spatially nearby, and related objects) will determine the next object to be recognized.

Let us go back to Fig. 2. As we stated before, the structure in Fig. 2 is a semantic net and, in addition, it is an ordered structure with respect to the visibility criteria. For example, it makes sense to classify large objects from small ones and then further classify them with respect to their shapes. It turns out that with more refined classification, we get more refined concepts. Consequently, this structure represents the relationships between general concepts and specific ones (waterway, ocean, rivers, and lakes)—"kind of" relationships as well as between the global objects and their details (waterway, island, and lake), and "part of" relationships. The "part of" relationships have natural translation into some spatial relationships, such as inside, surrounded, attached, next to, etc.

The structure in Fig. 2, if followed strictly top-down, satisfies only certain aspects of the order in which objects should be recognized. For instance, it is not obvious from the structure that if the bridges are above a river, you should first find them, then remove them so that you can find the shape of the river and thus distinguish the rivers from lakes. The bridges are smaller than rivers, and hence, from the simple visibility point of view, they should be recognized after the rivers or lakes. This seeming contradiction must be resolved by either an external rule or by finding a structure which would relate the concepts of relationship "above" and occluded objects.

2.3 Conceptual Filtering and Focusing as a Means for Scene Analysis

We mentioned at the end of the previous section the lack of such a structure in Fig. 2 for objects such as bridges, rivers, and lakes, from which the priority in recognition process for bridge before rivers and lakes would follow. The bridges, due to the 2-D projection, partially occlude the river or lake; in this case, the bridges visually cause a false partitioning of the waterway.

Occlusion or *partial occlusion* of objects occurs during our visual experiences since we see the three-dimensional world through the two-dimensional projection of it. It is true that we have at hand monocular and/or binocular depth cues which allow us to interpret the two-dimensional data in the three-dimensional world. The correctness and accuracy of this interpretation depends on the view of the non-occluded object, and on the amount of knowledge we have about the three-dimensional object, especially about its surface and form. Clearly the nonoccluded objects are more visible than the partially occluded objects. Then, following our analysis strategy, the nonoccluded objects will be recognized before the partially occluded ones.

This procedure is further justified by the fact that sometimes partially occluded objects cannot be interpreted properly for lack of sufficient data. An example of this case are the pieces of river divided by bridges, as shown in Fig. 6. The pieces of river could be interpreted as lakes (from the shape feature) unless we first recognize the bridges and their spatial relationship "above" the waterway. Once we recognize this, we can remove the bridges and fill in the missing data by water points. This process is shown in Fig. 8. It is evident that after removing—filtering out—the bridges, one can easily distinguish the rivers and lakes as shown in Figs. 16 and 12, respectively. Notice that we must first recognize the bridges as they are described in our world model, and then we can remove them. This process, as we said before, is called conceptual filtering. We could call it object filtering, but we prefer the first name since the visual concepts of objects are being matched with the input data rather than the physical objects.

Conceptual filtering decomposes the scene into single three-dimensional objects, preserving their mutual relationships. The important thing is indeed that the spatial relationships (as well as the objects) are interpreted in a three-dimensional world. While in the past, recognition of partially occluded objects was based on matching the pictorial information with the model (see for example, Roberts [15], Falk [6], and Grape [8]) we predict partial occlusion from an already recognized object (bridge) and its 3-D relationship (above) to another object (water).

Now we shall introduce *conceptual focusing*. In Section 2.2, we discussed the structure representing the relationships between the general and specific concepts. In addition to the generalization relationship, there is the part-whole relationship which determines a certain context for the parts from the whole.

Once we recognize the whole—the global structure of a picture—these results should guide us in the further recognition of possible parts. One can regard this process as the focusing by the world model. An example of this case is: in the waterway we look for pollutants, islands, etc.

The other aspect of conceptual focusing is that the same visual concepts may have different meaning in different contexts. It is as though the focusing process in memory is guided by the visual properties and the context. We look for the appropriate interpretation. An example of this can be the linear features—stripes are interpreted as roads in a satellite picture, but the same stripes in a biological image may be interpreted as veins under appropriate magnifications.

Rather than continuing our discussion on a general level, we shall demonstrate our ideas in a concrete example of recognition of rivers, lakes, islands, and bridges.

III. The Higher Level Program for Recognition of Rivers, Lakes, Islands, and Bridges

The higher level program which will be described is an implementation of the strategy shown in Fig. 2.

The picture data are obtained from ERTS satellite pictures. The spatial resolution on these pictures is constant, determined by the scanning system. Each point on the picture corresponds to 57×74 m^2. The illumination conditions are relatively constant, providing that the area covered by the analysis has the same daytime. All other characteristics of the visibility of objects, e.g., the contrast between the objects and their environment, partial occlusion, and the size of objects remain constant. This follows from

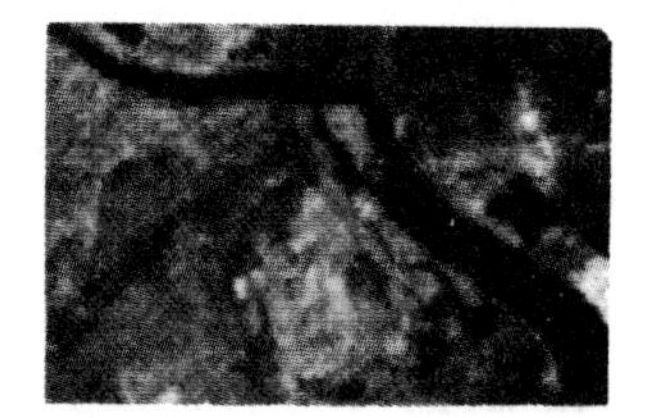

Fig. 3. Washington, D.C., area—gray scale image.

the fact that the objects of interest (rivers, roads, cities, etc.) do not change during the satellite turn-around. For this discussion, we assume that the window size of a picture will be constant during the analysis process. An example of a window from the Washington, D.C., area is shown in Fig. 3. In the algorithm that follows, we use several low-level operators, such as region finders, shape descriptions, connectivity operators, and others. We will not describe these operators here; for more details, see Tavakoli [18].

Fig. 4. Water after step 2).

3.1 The Algorithm

1) Find the watery areas based on the gray scale value estimates for water in Band 3. (Apply the *global region finder*.) If there is no watery area, then stop the program.

2) Apply the *local region expander* and find the missing points where the water could continue.

3) Apply the *skeleton operator* in order to find the shape of the land areas.

4) Find the thin elongated areas of the land; go to the world model and make an hypothesis about the object (in this case, a bridge or island). If no elongated areas can be found, then go to step 11).

5) Assume that the object is an hypothetical bridge; then, verify it by checking the connectivity between the hypothetical bridge and the land on its two sides. If this condition is satisfied, then go to step 7); otherwise, go to step 6).

6) Apply the connectivity operator in a given direction to check whether the previously missing connectivity to the land is just due to a threshold error, or if it is an indication that the object is not a bridge. If the hypothetical bridge and the land on its two sides are connected, then go to step 7); otherwise, go to step 11).

7) Verify the bridge by checking the relationship "surrounded by" the water on its two other sides.

8) If all the evidence, after steps 5), 6), and 7), shows that the hypothesis of a bridge has been verified, then mark all on the bridges on the picture and continue to step 9). Otherwise, reject the hypothesis about the bridges and form new ones, which, in this case, are that the objects are small islands or peninsulas. From here, continue to step 11).

9) Remove all the bridges from the picture and fill in the missing bridge points by water points.

10) Find a boundary between land and water.

11) Find the water regions which have closed boundaries; if there are no such regions, go to step 12). Check the world model for the possible hypothesis. In this case, the only possibility is a lake.

12) Find the land regions which have closed boundaries. If there are no such regions, go to step 13). Check the world model for the hypothesis. In this case, the only possible interpretation is an island.

13) If a watery region has two or more open boundaries, then it is hypothesized that this region is a river (or rivers). In the case that the region has one open boundary and the length of the boundary is larger than a certain threshold limit, hypothesize that as a river; otherwise, the region will be hypothesized as a possible candidate for part of a river, part of a lake, or part of an ocean (unidentified piece of water).

14) If the hypothesis is a river or rivers, then apply the skeleton operator in order to find the direction of flow.

15) Name each recognized region as islands, lakes, rivers or unidentified watery areas. Generate the points inside each region. Make a list of objects associated with the area measure, boundary measures and the coordinate of the center of gravity. In the case that a river has several branches, specify the number of branches. Record also the direction of water flow for each river.

16) If there is no bridge, go to STOP. Otherwise name all of the bridges as $(1,2,\cdots,)$. Make a list of all recognized bridges associated with their end coordinates, number of points, and their relationship with watery areas (e.g. bridge one is on lake 1 and river 2) for output purposes.

17) STOP.

3.2 The Description of Results

If we follow the higher level program applied on the data shown in Fig. 3, the following are the intermediate results. The first two steps separate the waterway from land, see Fig 4. The stars in the figure represent the water.

We lose a branch of the river due to the fixed value of the water threshold. This is not corrected even after we apply the local region grower, which uses a more relaxed water threshold, since the main river is visually separated from its branch by a bridge. Here, a more involved back tracking

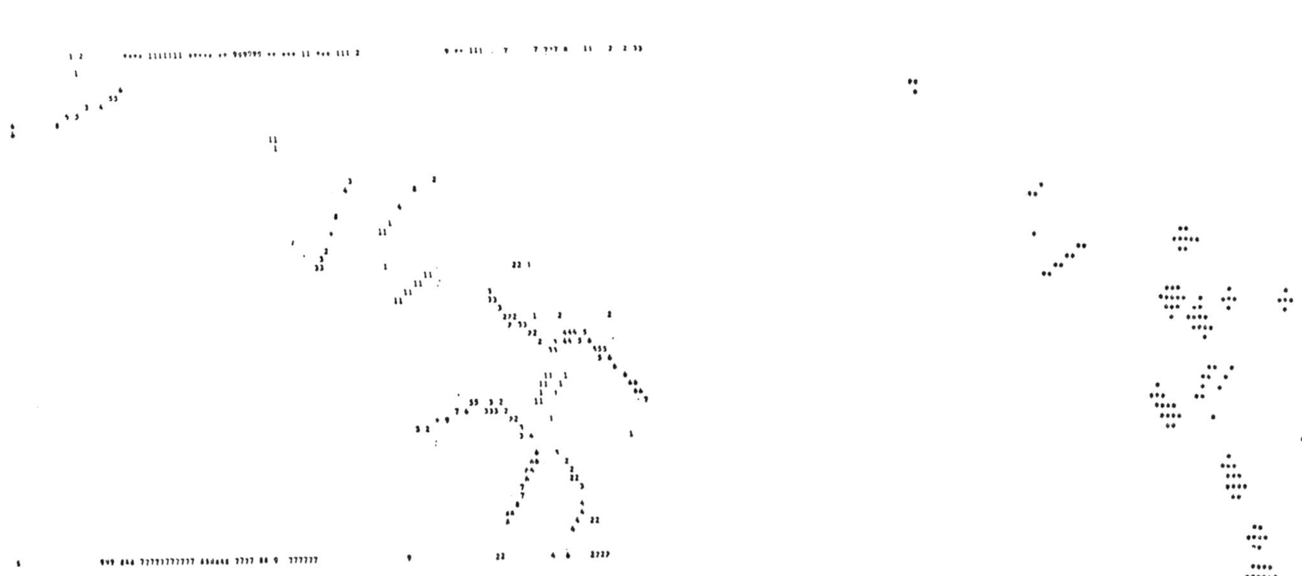

Fig. 5. Skeleton of land.

Fig. 6. Elongated areas of land.

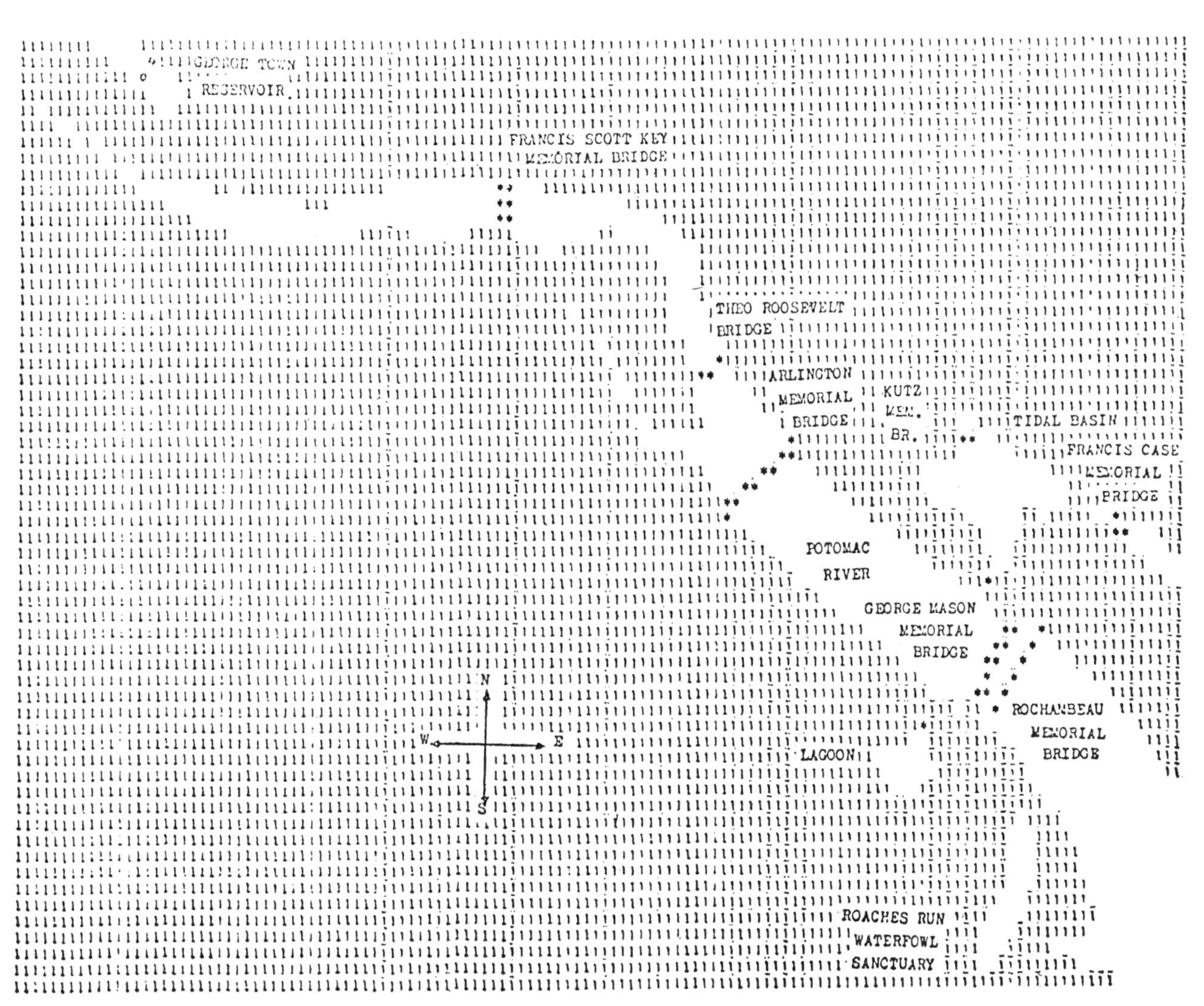

Fig. 7. Potomac River I—Complete Recognition.

procedure is needed. Fig. 5 displays the mod 10 representation of the skeleton of the land area. In Fig. 6, the stars show the main elongated areas of the land; this is the result after step 4). The output after steps 5), 6), 7), and 8) is in Fig. 7. The stars denote the recognized bridges. The number "1" denotes all points which belong to the land. The

Fig. 8. Potomac River and lakes without bridges.

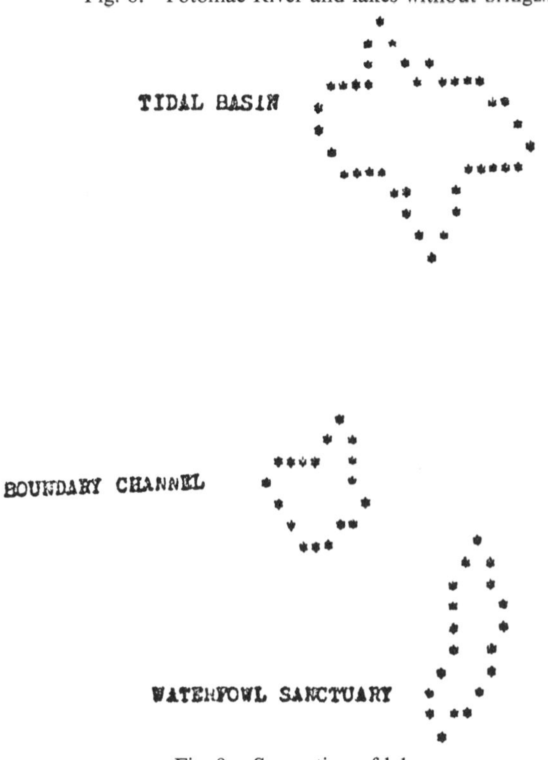

Fig. 9. Separation of lakes.

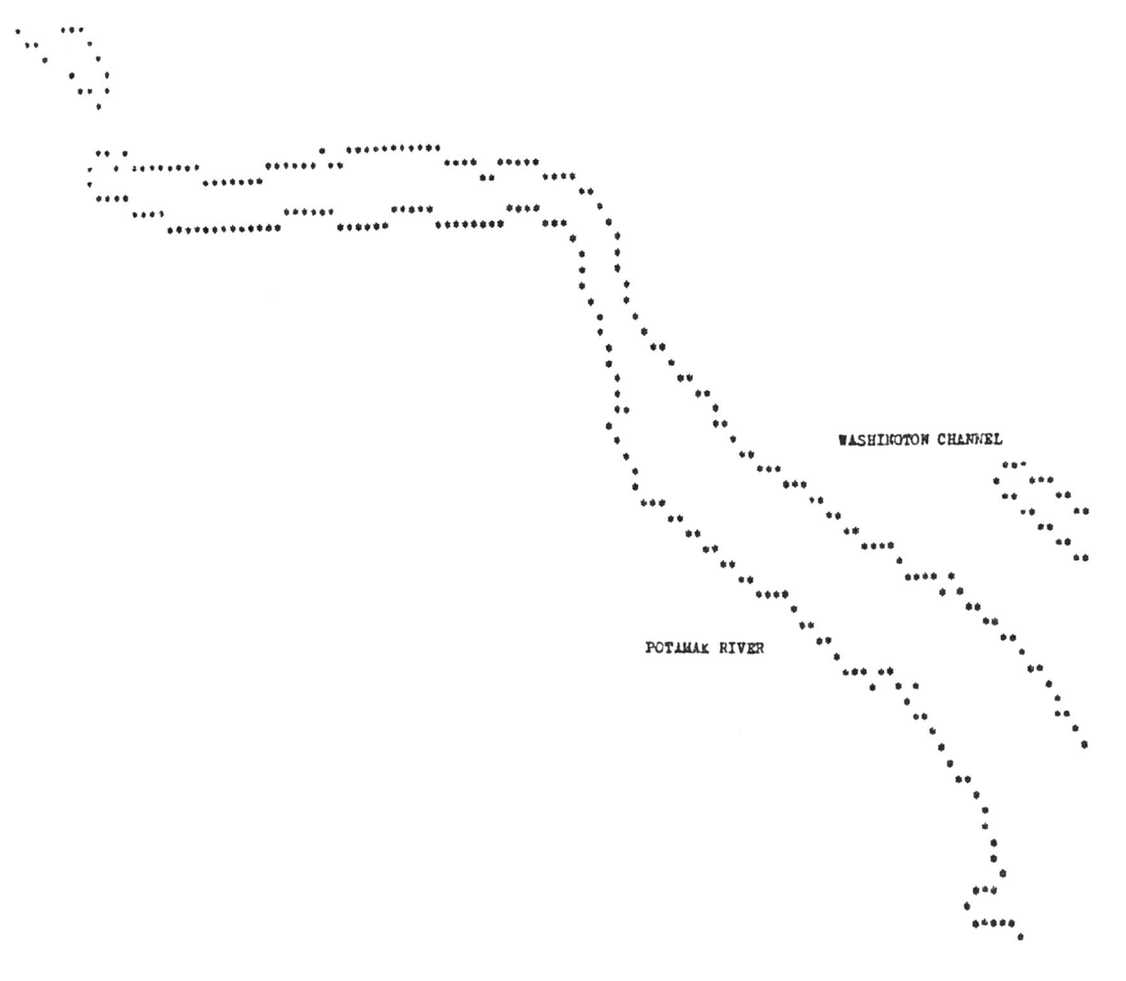

Fig. 10. Separation of rivers.

blank areas represent the water. The names of bridges and lakes were inserted manually. Once we recognize the bridges, we remove them and fill in the missing data by water points (see Fig. 8 which demonstrates the results after step 10)). The stars show the land boundary. Dots are representations of water and blanks are of land. The results of recognition of closed boundaries and open boundaries are demonstrated in Figs. 9 and 10. These two figures show the results after steps 11), 12), and 13). The program recognizes three lakes, one river, two unidentified pieces of water, and ten bridges.

This program has been implemented in FORTRAN on the Moore School Computer Facility (SPECTRA 70/40). The time for processing one window (128 × 60 points) is 200-s CPU time. We have tested the program on 3 windows (see Figs. 11 and 12).

IV. Evaluation of Results and Some Error Analysis

4.1 Kinds of Errors

In the process of evaluating the results, one can observe several kinds of errors. We shall discuss three of them:

1) errors due to inadequacy of the model (i.e., restricted definitions of objects);
2) errors caused by the weaknesses of some of the low-level operators; and
3) errors due to noise.

In what follows we further explain these errors in detail.

4.2 Errors Due to the Model

In the case of recognition of bridges, we will not recognize bridges which are not built over watery areas. On the other hand, we may say that a narrow piece of land which separates two pieces of water is a bridge. Clearly, this is a natural bridge as opposed to a man-made bridge.

4.3 Errors Due to Weaknesses of Some Low-Level Operators

During the bridge recognition process, we use threshold technique, shape operator, and operators which check the existence of two relationships: connectivity and "surrounded." Each operator applied by itself causes great error in recognition. However, all operators applied in an appropriate sequence yield 96-percent correct recognition.

An example of bridge points after only the threshold technique was applied is shown in Fig. 13. The threshold technique and the shape operator combined eliminates some false bridge points, but it still leaves a few incorrect ones (see Fig. 6). Finally, the correct recognition is shown in Fig. 7.

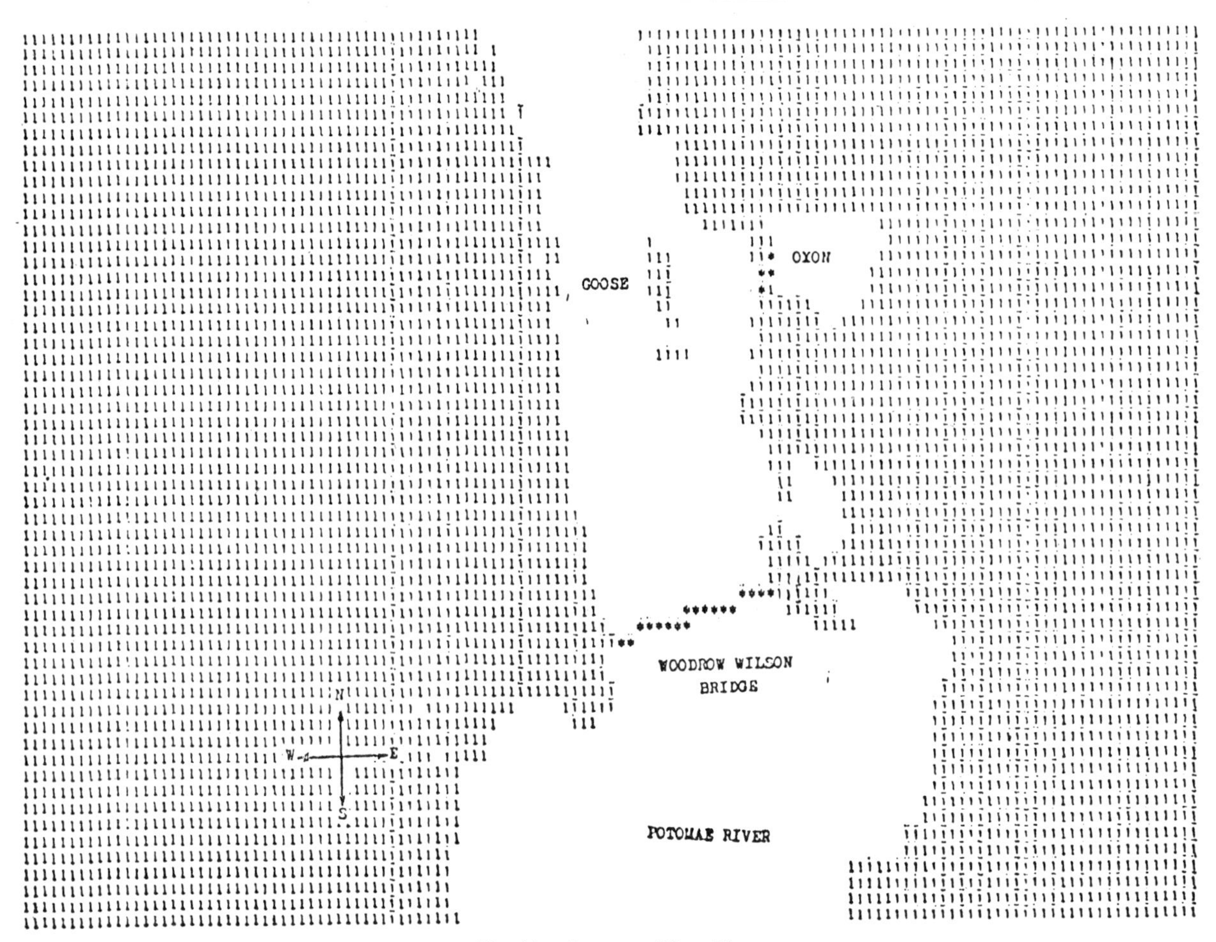

Fig. 11. Potomac River II.

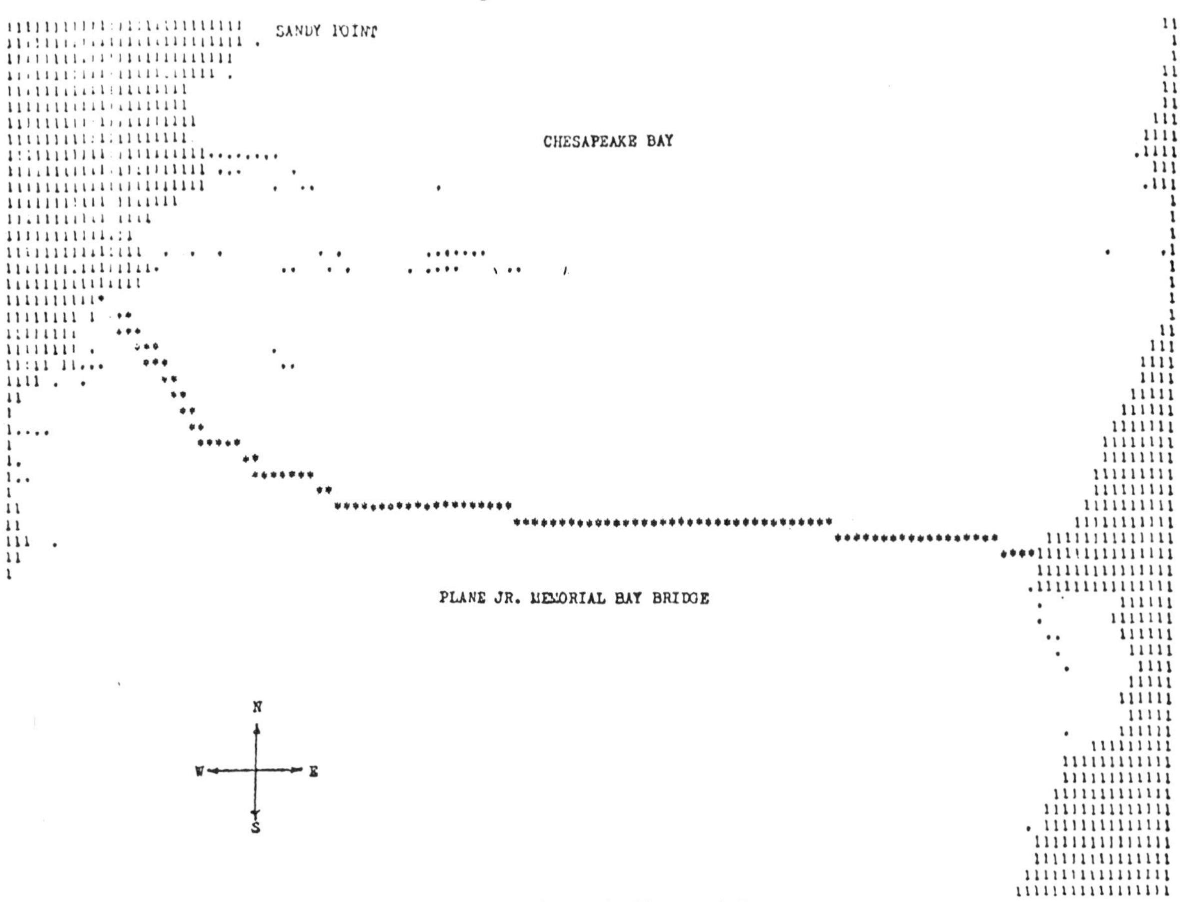

Fig. 12. A window in Chesapeak-Bay.

Fig. 13. Hypothetical bridges based only on gray value.

TABLE II
STATISTICAL ERROR EVALUATION—AMPLITUDE OF NOISE = ±1 UNIT

Spatial Distribution of Noise %	Amplitude of Noise	Number of Bridges Lost	Extra Bridges Recognized	Additional Islands	Additional Lakes
5	± 1	0	1	0	1
10	± 1	0	1	0	1
15	± 1	1	2	0	2
20	± 1	1	0	0	2
25	± 1	1	1	0	1
30	± 1	0	2	0	1
35	± 1	2	2	0	1
40	± 1	2	1	0	1
45	± 1	1	2	1	1
50	± 1	1	2	0	2
55	± 1	1	2	1	2
60	± 1	2	1	1	5
70	± 1	2	4	1	3
80	± 1	2	2	1	5
90	± 1	2	1	2	5
100	± 1	2	5	0	3

TABLE III
STATISTICAL ERROR EVALUATION—AMPLITUDE OF NOISE = ±2 UNITS

Spatial Distribution of Noise %	Amplitude of Noise	Number of Bridges Lost	Extra Bridges Recognized	Additional Islands	Additional Lakes
10	± 2	1	2	1	3
20	± 2	1	1	2	3
30	± 2	2	3	1	3
40	± 2	1	4	3	3
50	± 2	3	6	2	2
60	± 2	3	4	2	3
70	± 2	3	6	2	3
80	± 2	4	10	3	2
90	± 2	5	4	3	1
100	± 2	3	6	3	1

4.4 Recognition Performance Under Artificial Noise

To understand and to evaluate the performance of the system, noise is added to the data, and then we apply the program on the new data. The noise is produced by generating a random number between 0 and 1 for each point in the grid. The noise is added if the random number is greater than 0.5; otherwise, the noise is subtracted.

The percentage of points in the grid to which this noise is added or subtracted is controlled by another random number generator. As an example, suppose that we add or subtract noise to 10 percent of the points on the grid. The decision is governed by the random number; that is, if the random number is greater or equal to 0.9, then the point is chosen for adding or subtracting the noise; otherwise, the point is left unchanged.

The amount of noise added or subtracted should be a reasonable percentage of the average gray level of the object under study. To evaluate the program for recognition of bridges, islands, rivers, and lakes, we have done two experiments. In the first experiment, the value of the noise is chosen to be ± 1. Then, we start by adding this noise to 5 percent of the points and we increase that to 100 percent of the points. The average gray level for water is about 10 and for bridges 14 of 128 possible gray levels. If we add noise ± 1 gray level, then we have changed the gray value of water by ± 10 percent and the gray value of bridges by ± 7 percent. In the second experiment, the value of the noise is chosen as ± 2. This noise is two times larger than the first one. That is equivalent to adding or subtracting 20 percent of the average gray level for the water and ± 14 percent of the average gray level for the bridges. The result of these experiments are tabulated in Tables II and III.

Table II shows that as we increase the percentage of the points which receive noise, we recognize extra bridges and extra small lakes. The maximum number of recognized extra bridges is 5 with 100-percent noise, in comparison to 10 bridges. From the definition of the bridge we expect that the program recognizes extra bridges. The reasons are as follows.

1) The narrow land between the lake created by noise and other watery areas can be interpreted as a bridge.

2) When the river is narrow, one point or two points of noisy data can make a bridge across the river, and the program recognizes those points as a bridge.

Since the program is capable of recognizing small bridges, it is also sensitive to losing them. In other words, if the negative noise (subtractive) turns the bridge to be recognized into a watery area, the program recognizes that point as water and, therefore, we lose the bridge. Another reason why we lose bridges is due to additive noise. The added noise can produce a wider bridge which the program rejects as a bridge, since we do not allow bridges with widths of more than two picture points.

The experiment shows that the program is more sensitive to the recognition of extra bridges than to the loss of them. As we increase the percentage of points receiving noise, we also introduce more islands. This is also in agreement with Tables II and III. Table III shows the result of the same experiment under greater noise.

V. Conclusions

In this paper, we have argued that only in some special cases can one have an *a priori* knowledge of characteristics of noise in a picture. The noise and the relevant information are complementary, and, if one knows the nature of the noise, it implies that everything left in an image is relevant information. All the classical filters have been using some sort of knowledge-assumptions about noise in a picture.

These filters include all the preprocessing and feature extraction processors known in pattern recognition (see, for example, Holmes [10], Swoboda and Gerdes [17], Hawkins [9], and others) as well as many low-level operators known in artificial intelligence (for review see, for example, Rosenfeld [16]).

We know that in more complex situations, such as in the real outdoors, one does not have a full *a priori* knowledge about what is a noise. In such cases, the real problem is to find the relevant information, which is dependent upon context and time.

The basic difference between classical filters and conceptual filters is that while the first ones filter context independent features (for example edges, average density, etc.) from an image, the later ones filter objects according to their descriptions. These descriptions are partially composed of attributes which standing alone are context independent, but the structure is context dependent. The other part of the description is formed by the relationships of the object to other objects in the world model.

One can possibly argue that context independent filters are more general in their domain of application than the context dependent. By conceptual filtering and focusing we have tried to show that in real outdoor scenes and their images with imperfect-noisy data, one needs a context in order to separate the relevant information from noise and thus make correct recognition. We do not advocate custom-made procedures that are different for each object, like for a bridge, or river, etc. Rather, we propose a systematic way of using concepts, and their descriptions from a world model for recognition purposes; our examples from ERTS satellite pictures should demonstrate this attempt.

The process of recognizing a scene is essentially a scene decomposition and description of objects using conceptual filtering and focusing. This process can be done partially in parallel. Certain steps, however, must be done sequentially. The sequence of different procedures in this process is controlled by different strategies. We have discussed two such strategies: the initial and continuous ones. The strategies include decision criteria based on the visibility of objects, as well as on their expected associations with other objects (part-whole relationship). The three-dimensional interpretation of two-dimensional data plays a very important role in the overall recognition process. This is more evident in the case of recognition of partially occluded objects.

Acknowledgment

The authors are indebted to the referees of the earlier draft for their constructive criticism and valuable comments which helped them to improve the accuracy and the style of presentation.

References

[1] H. C. Andrews, *Computer Techniques in Image Processing.* New York: Academic Press, 1970.
[2] —— and L. H. Enloe, "Special issue on digital picture processing," *Proc. IEEE*, vol. 60, pp. 766–922, July 1972.

[3] R. Bajcsy and M. Tavakoli, "Computer recognition of roads from satellite pictures," Proc. 2nd Int. Pattern Recognition Conf., Copenhagen, 1974.
[4] H. B. Barrow, A. P. Ambler, and R. M. Burstall, "Some techniques for recognizing structures in pictures," in *Frontiers of Pattern Recognition*, S. Watanabe, Ed. New York: Academic Press, 1972.
[5] T. O. Binford and J. M. Tenenbaum, "Computer vision," *IEEE Trans. Computers,* vol. C–22 pp. 19–24, May 1973.
[6] G. Falk, "Computer interpretation of imperfect line data as a three-dimensional scene," *AIM 139*, Stanford University, Stanford, Calif., 1970.
[7] M. A. Fishler and R. A. Elschlager, "The representation and matching of pictorial structures," *IEEE Trans. Computers*, vol. C–22, pp. 67–82, May 1973.
[8] G. R. Grape, "Model based (Intermediate-Level) computer vision," AIM 201, Stanford University, Stanford, Calif. 1973.
[9] J. K. Hawkins, "Image processing: A review and projection," in *Automatic Interpretation and Classification of Images*. A. Graselli, Ed. New York: Academic Press, 1969, pp. 199–234.
[10] W. S. Holmes, "Automatic photointerpretation and target location," *Proc. IEEE*, vol. 54, pp. 1679–1686, Dec. 1966.
[11] R. Kalman, Lecture Notes in Filter Theory, Stanford University, Stanford, Calif., 1969.
[12] M. Minsky and S. Papert, "Research on intelligent automata," MIT Project MAC Progress, Rep. IV, pp. 7–15 July 1966–July 1967).
[13] U. Neisser, *Cognitive Psychology*. New York: Meredith, 1967.
[14] F. P. Preparata and S. R. Ray, "An approach to artificial nonsymbolic cognition," *Inf. Sci.*, vol. 4, pp. 65–86, 1972.
[15] L. G. Roberts, "Machine perception of three-dimensional solids," in *Optical and Electrooptical Information Processing*. J. T. Tippet, *et al.*, Eds. Cambridge, Mass.: MIT Press, 1965, pp. 159–197.
[16] A. Rosenfeld, "Picture processing 1972," *Computer Graphics and Image Processing*, vol. 1, pp. 394–416, 1973.
[17] W. Swoboda and J. W. Gerdes, "A system for demonstrating the effects of changing background on automatic target recognition," in *Pictorial Pattern Recognition*, G. C. Cheng *et al.*, Eds. Washington, D.C.: Thompson Book Company, 1968, pp. 33–43.
[18] M. Tavakoli, "Analysis of scenes as seen from the ERTS satellites," Ph.D. Dissertation, University of Pennsylvania, 1974.
[19] P. H. Winston, "Learning structural descriptions from examples," Rep. *AI TR-231*, Artificial Intelligence Laboratory, MIT, Cambridge, Mass., Sept. 1970.

Pattern Recognition and Image Processing

KING-SUN FU, FELLOW, IEEE, AND AZRIEL ROSENFELD, FELLOW, IEEE

Abstract—**Extensive research and development has taken place over the last 20 years in the areas of pattern recognition and image processing. Areas to which these disciplines have been applied include business (e.g., character recognition), medicine (diagnosis, abnormality detection), automation (robot vision), military intelligence, communications (data compression, speech recognition), and many others. This paper presents a very brief survey of recent developments in basic pattern recognition and image processing techniques.**

Index Terms—**Decision-theoretic recognition, image processing, image recognition, pattern recognition, syntactic recognition.**

I. INTRODUCTION

DURING the past twenty years, there has been a considerable growth of interest in problems of pattern recognition and image processing. This interest has created an increasing need for theoretical methods and experimental software and hardware for use in the design of pattern recognition and image processing systems. Over twenty books have been published in the area of pattern recognition [5], [8], [10], [11], [15], [16], [35], [41], [47], [79], [82], [86], [89], [110], [111], [118], [122], [123], [136], [137]. In addition, a number of edited books, conference proceedings, and journal special issues have also been published [40], [43], [45], [46], [57], [65], [67], [69], [77], [80], [96], [113], [121], [127], [128]. Cover [25] has given a comprehensive review of the five books published in 1972–1973 [5], [35], [47], [79], [89]. A specialized journal has existed for nearly ten years [73], and some special pattern recognition machines have been designed and built for practical use. Applications of pattern recognition and image processing include character recognition [37], [71], [123], target detection, medical diagnosis, analysis of biomedical signals and images [45], [57], [97], remote sensing [44], [57], identification of human faces and fingerprints [83], reliability [90], socio-economics [13], archaeology [12], speech recognition and understanding [43], [45], [98], and machine part recognition [3].

Many of the books and paper collections on pattern recognition contain material on image processing and recognition. In addition, there are four textbooks [4], [35], [99], [108] and several hardcover paper collections [21], [51], [60], [67], [74], [106], [132], [138] devoted especially to the subject, as of the end of 1976. There is a specialized journal in the field [107], and there have also been special issues of several other journals on the topic [1], [6], [7], [53]. For further references, the reader may consult a series of annual survey papers [100]–[105] which cover a significant fraction of the English language literature.

Although pattern recognition and image processing have developed as two separate disciplines, they are very closely related. The area of image processing consists not only of coding, filtering, enhancement, and restoration, but also analysis and recognition of images. On the other hand, the area of pattern recognition includes not only feature extraction and classification, but also preprocessing and description of patterns. It is true that image processing appears to consider only two-dimensional pictorial patterns and pattern recognition deals with one-dimensional, two-dimensional, and three-dimensional patterns in general. However, in many cases, information about one-dimensional and three-dimensional patterns is easily expressed as two-dimensional pictures, so that they are actually treated as pictorial patterns. Furthermore, many of the basic techniques used for pattern recognition and image processing are very similar in nature. Differences between the two disciplines do exist, but we also see an increasing overlap in interest and a sharing of methodologies between them in the future.

Within the length limitations of this paper, we provide a very brief survey of recent developments in pattern recognition and image processing.

II. PATTERN RECOGNITION

Pattern recognition is concerned primarily with the description and classification of measurements taken from physical or mental processes. Many definitions of pattern recognition have been proposed [112], [125], [127]. Our discussion is based on the above loose definition. In order to provide an effective and efficient description of patterns, preprocessing is often required to remove noise and redundancy in the measurements. Then a set of characteristic measurements, which could be numerical and/or nonnumerical, and relations among these measurements, are extracted for the representation of patterns. Classification and/or description of the patterns with respect to a specific goal is performed on the basis of the representation.

In order to determine a good set of characteristic measurements and their relations for the representation of patterns so good recognition performance can be expected, a careful analysis of the patterns under study is necessary. Knowledge about the statistical and structural charac-

Manuscript received April 16, 1976: revised June 14, 1976. This work was supported by the National Science Foundation under Grant ENG 74-17586 and by the National Science Foundation under Grant MCS-72-03610.

K.-S. Fu is with the School of Electrical Engineering, Purdue University, West Lafayette, IN 47907.

A Rosenfeld is with the Computer Science Center, University of Maryland, College Park, MD 20742.

Reprinted from *IEEE Trans. Comput.*, vol. C-25, pp. 1336–1346, Dec. 1976.

teristics of patterns should be fully utilized. From this point of view, the study of pattern recognition includes both the analysis of pattern characteristics and the design of recognition systems.

The many different mathematical techniques used to solve pattern recognition problems may be grouped into two general approaches. They are the decision-theoretic (or discriminant) approach and the syntactic (or structural) approach. In the decision-theoretic approach, a set of characteristic measurements, called features, are extracted from the patterns. Each pattern is represented by a feature vector, and the recognition of each pattern is usually made by partitioning the feature space. On the other hand, in the syntactic approach, each pattern is expressed as a composition of its components, called subpatterns or pattern primitives. This approach draws an analogy between the structure of patterns and the syntax of a language. The recognition of each pattern is usually made by parsing the pattern structure according to a given set of syntax rules. In some applications, both of these approaches may be used. For example, in a problem dealing with complex patterns, the decision-theoretic approach is usually effective in the recognition of pattern primitives, and the syntactic approach is then used for the recognition of subpatterns and of the pattern itself.

A. Decision-Theoretic Methods

A block diagram of a decision-theoretic pattern recognition system is shown in Fig. 1. The upper half of the diagram represents the recognition part, and the lower half the analysis part. The process of preprocessing is usually treated in the area of signal and image processing. Our discussions are limited to the feature extraction and selection, and the classification and learning. Several more extensive surveys on this subject have also appeared recently [18], [26], [32], [66], [120].

Feature extraction and selection: Recent developments in feature extraction and selection fall into the following two major approaches.

Feature space transformation: The purpose of this approach is to transform the original feature space into lower dimensional spaces for pattern representation and/or class discrimination. For pattern representation, least mean-square error and entropy criteria are often used as optimization criteria in determining the best transformation. For class discrimination, the maximization of interclass distances and/or the minimization of intraclass distances is often suggested as an optimization criterion. Both linear and nonlinear transformations have been suggested. Fourier, Walsh–Hadamard, and Haar transforms have been suggested for generating pattern features [5]. The Karhunen–Loeve expansion and the method of principal components [5], [39], [47] have been used quite often in practical applications for reducing the dimensionality of feature space.

In terms of the enhancement of class separability, nonlinear transformations are in general superior to linear transformations. A good class separation in feature space will certainly result in a simple classifier structure (e.g., a linear classifier). However, the implementation of nonlinear transformations usually requires complex computations compared with that of linear transformations. Results of transformations need to be updated when new pattern samples are taken into consideration. Iterative algorithms and/or interactive procedures are often suggested for implementing nonlinear transformations [45], [57].

In some cases, the results of transformations based on pattern representation and class discrimination respectively are in conflict. An optimization criterion for feature space transformation should be able to reflect the true performance of the recognition system. Some recent work appears to move in this direction [31].

Information and distance measures: The main goal of feature selection is to select a subset of l features from a given set of N features ($l < N$) without significantly degrading the performance of the recognition system, that is, the probability of misrecognition, or more generally, the risk of decision. Unfortunately, a direct calculation of the probability of misrecognition is often impossible or impractical partially due to the lack of general analytic expressions which are simple enough to be treated. One approach is to find indirect criteria to serve as a guide for feature selection.

The most common approach is to define an information or (statistical) distance measure, which is related to the upper and/or lower bounds on the probability of misrecognition, for feature selection [17], [19], [54], [66]. That is, the best feature subset is selected in the sense of maximizing a prespecified information or distance measure. Recently, Kanal [66] provided a fairly complete list of distance measures and their corresponding error bounds. Assuming that the most important characteristic of the distance measure is the resultant upper bound on the probability of misrecognition, the various measures can be arranged in increasing order of importance. For a two-class recognition problem, denoting the upper bound on the probability of misrecognition by P_e, for Bhattacharyya's distance by U_B, for Matusita distance by U_M, for equivocation by U_E, for Vajda's entropy by U_V, for Devijver's Bayesian distance by U_D, for Ito's measure (for $n = 0$) by U_I, for Kolmogorov's variational distance by U_K, and for the M_0-distance of Toussaint by U_T, the following point-wise relations hold [75]:

$$P_e = U_K \leq U_V = U_D = U_I = U_T \leq U_E \leq U_B = U_M.$$

The divergence and Kullback–Leibler numbers, which are simply related to each other, are excluded from the ordering because of the lack of a known upper bound except for the case of a normal distribution where its bound is larger than U_B. In terms of computational difficulty, however, the divergences and Bhattacharyya distance are easier to compute than the other distance measures.

It is interesting that the best bound on the probability of misrecognition (except U_K which is nothing but P_e itself) derived from the distance measures is equal to the

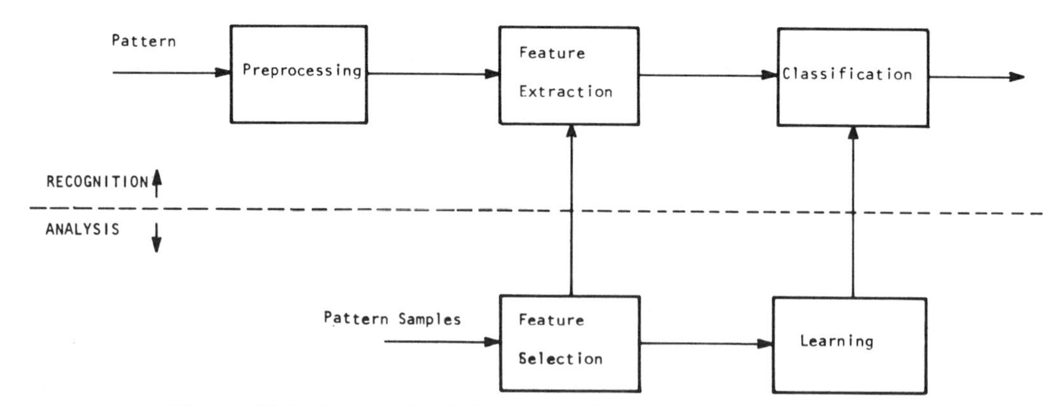

Fig. 1. Block diagram of a decision-theoretical pattern recognition system.

asymptotic error of the single nearest neighbor classifier. In addition to the information and distance measures mentioned above, a generalized Kolmogorov distance, called the J_α separability measure, was recently proposed as a feature selection criterion, and its upper and lower bounds on the probability of misrecognition derived [75]. When $\alpha = 1$, J_α is equivalent to the Kolmogorov distance. For $\alpha = 2$, the upper bound of the probability of misrecognition is equal to the asymptotic probability of error of the single nearest neighbor classifier.

Classification and learning: Most developments in pattern recognition involve classification and learning. When the conditional probability density functions of the feature vectors for each class (which we may call the class density functions) are known or can be accurately estimated, the Bayes classification rule that minimizes the average risk or the probability of misrecognition can be derived. When the class density functions are unknown, nonparametric classification schemes need to be used. In practice, when a large number of pattern samples is available, class density functions can be estimated or learned from the samples [24], [28], [126], and then an optimal classification rule can be obtained. If the parametric form of each class density function is known, only parameters need to be learned from the pattern samples. When the number of available pattern samples is small, the performance of density and parameter estimations is poor. Nonparametric classification schemes usually suggest a direct learning of the classification rule from pattern samples, for example, the learning of parameters of a decision boundary.

Depending upon whether or not the correct classification of the available pattern samples is known, the process of learning can be classified into supervised learning (or learning with a teacher) and nonsupervised learning (or learning without a teacher). Bayesian estimation and stochastic approximation and the potential function method have been suggested for the learning of class density functions or a decision boundary. When the learning is nonsupervised, a mixture density function can be formed from all the individual class density functions and *a priori* class probabilities. Nonsupervised learning of the parameters of each class density function can be treated as a supervised learning of parameters of the mixture density function from the unclassified pattern samples followed by a decomposition procedure. Under certain conditions, the decomposition can be accomplished and the estimates of the parameters of each class recovered. A related topic which has received an increasing amount of attention recently is learning with finite memory [26].

When the *a priori* information is sufficient, the classifier may be able to make decisions with good performance. In this case, the learning process could be carried out using the classifier's own decisions; that is, the unclassified pattern samples are now classified by the classifier itself. This type of nonsupervised learning is called decision-directed learning. When the classification of the pattern samples is incompletely known, learning with an imperfect teacher and learning with a probabilistic teacher have recently been proposed [27], [64]. An appropriate combination of supervised and nonsupervised modes of learning could result in a system of lower cost than those using a single learning mode [18], [23].

Classification based on clustering analysis has been regarded as a practically attractive approach, particularly in a nonsupervised situation with the number of classes not precisely known. Various similarity and (deterministic) distance measures have been suggested as criteria for clustering pattern samples in the feature space [33], [34]. Both hierarchical and nonhierarchical strategies are proposed for the clustering process. Often, some of the clustering parameters, such as the similarity measure and threshold, criteria for merging and/or splitting clusters, etc., need to be selected heuristically or through an interactive technique. It should be interesting to relate directly the distance measures for feature selection to those for clustering analysis [19], [135]. Recently, clustering algorithms using adaptive distance were proposed [34]. The similarity measure used in the clustering process varies according to the structure of the clusters already observed. Mode estimation, least mean-square optimization, graph theory and combinatorial optimization have been used as a possible theoretical basis for clustering analysis [5], [20],

[70], [80], [133]. Nevertheless, clustering analysis, at its present state-of-the-art, still appears to be an experiment-oriented "art."

Remarks: Most results obtained in feature selection and learning are based on the assumption that a large number of pattern samples is available, and, consequently, the required statistical information can be accurately estimated. The relationship between the dimensionality of feature space and the number of pattern samples required for learning has been an important subject of study. In many practical problems, a large number of pattern samples may not be available, and the results of small sample analysis could be quite misleading. The recognition system so designed will usually result in an unreliable performance. In such cases, the study of finite sample behavior of feature selection and learning is very important. The degradation of performance in feature selection, learning and error estimation [48], [119] due to the availability of only a small number of samples needs to be investigated.

In some practical applications, the number of features N and the number of pattern classes m are both very large. In such cases, it would be advantageous to use a multilevel recognition system based on a decision tree scheme. At the first level, m classes are classified into i groups using only N_1 features. Here, $i \ll m$ and $N_1 \ll N$, and the N_1 features selected are the best features to classify these i groups. In an extreme case, $i = 2$ so a two-class classifier can be used, or $N_1 = 1$ so a one-dimensional (thresholding) classifier can be used. The same procedure is then repeated at the second and third levels, etc., until each of the original m classes can be separately identified. Now, following each path in the decision tree, we should be able to recognize each of the m classes.

The idea of adaptively selecting a smaller number of features at different levels of classification appears to be very attractive in applications. An optimal design of such a tree scheme may be computationally quite complex. However, several heuristic design techniques have recently been suggested [44], [58], [72], [134].

B. Syntactic (or Structural) Methods

A block diagram of a syntactic pattern recognition system is shown in Fig. 2. Again, we divide the block diagram into the recognition part and the analysis part, where the recognition part consists of preprocessing, primitive extraction (including relations among primitives and subpatterns), and syntax (or structural) analysis, and the analysis part includes primitive selection and grammatical (or structural) inference.

In syntactic methods, a pattern is represented by a sentence in a language which is specified by a grammar. The language which provides the structural description of patterns, in terms of a set of pattern primitives and their composition relations, is sometimes called the "pattern description language." The rules governing the composition of primitives into patterns are specified by the so-called "pattern grammar." An alternative representation of the structural information of a pattern is to use a "relational graph," of which the nodes represent the subpatterns and the branches represent the relations between subpatterns.

Primitive extraction and selection: Since pattern primitives are the basic components of a pattern, presumably they are easy to recognize. Unfortunately, this is not necessarily the case in some practical applications. For example, strokes are considered good primitives for script handwriting, and so are phonemes for continuous speech; however, neither strokes nor phonemes can easily be extracted by machine. The segmentation problems for script handwriting and continuous speech, respectively, are still subjects of research. An approach to waveform segmentation through functional approximation has recently been reported [92]. Segmentation of pictorial patterns is discussed in Section III under Segmentation.

There is no general solution for the primitive selection problem at this time. For line patterns or patterns described by boundaries or skeletons, line segments are often suggested as primitives. A straight line segment could be characterized by the locations of its beginning (tail) and end (head), its length, and/or slope. Similarly, a curve segment might be described in terms of its head and tail and its curvature. The information characterizing the primitives can be considered as their associated semantic information or as features used for primitive recognition. Through the structural description and the semantic specification of a pattern, the semantic information associated with its subpatterns or the pattern itself can then be determined. For pattern description in terms of regions, half-planes have been proposed as primitives [91]. Shape and texture measurements are often used for the description of regions; see Section III under Properties.

Pattern grammars: After pattern primitives are selected, the next step is the construction of a grammar (or grammars) which will generate a language (or languages) to describe the patterns under study. It is known that increased descriptive power of a language is paid for in terms of increased complexity of the syntax analysis system (recognizer or acceptor). Finite-state automata are capable of recognizing finite-state languages although the descriptive power of finite-state languages is also known to be weaker than that of context-free and context-sensitive languages. On the other hand, nonfinite, nondeterministic procedures are required, in general, to recognize languages generated by context-free and context-sensitive grammars. The selection of a particular grammar for pattern description is affected by the primitives selected, and by the tradeoff between the grammar's descriptive power and analysis efficiency. Context-free programmed grammars, which maintain the simplicity of context-free grammars but can generate context-sensitive languages, have recently been suggested for pattern description [41].

A number of special languages have been proposed for the description of patterns such as English and Chinese characters, chromosome images, spark chamber pictures,

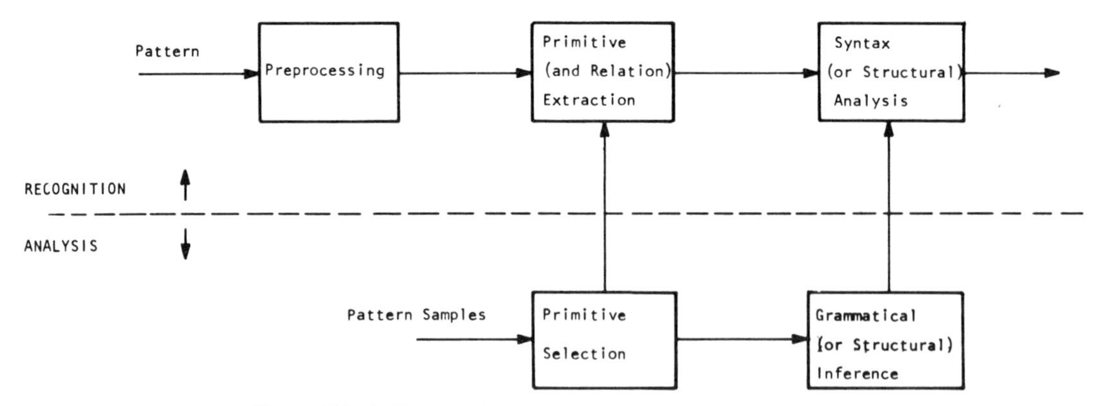

Fig. 2. Block diagram of a syntactic pattern recognition system.

two-dimensional mathematics, chemical structures, spoken words, and fingerprint patterns [41], [46]. For the purpose of effectively describing high dimensional patterns, high dimensional grammars such as web grammars, graph grammars, tree grammars, and shape grammars have been used for syntactic pattern recognition [41], [50], [88], [131]. Initial applications include fingerprint pattern recognition and the interpretation of Earth Resources Technology Satellite data [44], [46], [83].

Ideally speaking, it would be nice to have a grammatical (or structural) inference machine which would infer a grammar or structural description from a given set of patterns. Unfortunately, such a machine has not been available except for some very special cases [42]. In most cases so far, the designer constructs the grammar based on the *a priori* knowledge available and his experience.

In some practical applications, a certain amount of uncertainty exists in the process under study. For example, due to the presence of noise and variation in the pattern measurements, segmentation error and primitive extraction error may occur, causing ambiguities in the pattern description languages. In order to describe noisy and distorted patterns under ambiguous situations, the use of stochastic languages has been suggested [41], [52]. With probabilities associated with grammar rules, a stochastic grammar generates sentences with a probability distribution. The probability distribution of the sentences can be used to model the noisy situations. Other approaches for the description of noisy and distorted patterns using syntactic methods include the use of approximation and transformational grammars [43], [93]. The effectiveness of these approaches remains to be developed and tested.

Syntactic recognition: Conceptually, the simplest form of recognition is probably "template-matching." The sentence describing an input pattern is matched against sentences representing each prototype or reference pattern. Based on a selected "matching" or "similarity" criterion, the input pattern is classified in the same class as the prototype pattern which is the "best" to match the input. The structural information is not recovered. If a complete pattern description is required for recognition, a parsing or syntax analysis is necessary. In between the two extreme situations, there are a number of intermediate approaches. For example, a series of tests can be designed to test the occurrence or nonoccurrence of certain subpatterns (or primitives) or certain combinations of them. The result of the tests, through a table lookup, a decision tree, or a logical operation, is used for a classification decision.

A parsing procedure for recognition is, in general, nondeterministic and, hence, is regarded as computationally inefficient. Efficient parsing could be achieved by using special classes of languages such as finite-state and deterministic languages for pattern description. The tradeoff here between the descriptive power of the pattern grammar and its parsing efficiency is very much like that between the feature space selected and the classifier's discrimination power in a decision-theoretic recognition system. Special parsers using sequential procedures or other heuristic means for efficiency improvement in syntactic pattern recognition have recently been constructed [76], [94], [114].

Error-correcting parsers have been proposed for the recognition of noisy and distorted patterns [49], [117]. Different types of segmentation and primitive extraction errors (substitution, deletion and addition) are introduced into the pattern grammar. The recognition process is then based on the parsers designed according to the expanded pattern grammar. The error-correcting capability is achieved by using a minimum-distance criterion. Since the original grammar is expanded to include all possible error situations, the parser so designed is less efficient than that designed according to the original grammar. This tradeoff between error-correcting capability and parsing efficiency seems to be expected. Nevertheless, it could be a very serious drawback in practical applications.

When stochastic grammars are used for pattern description, the probability information is useful in resolving ambiguous situations. For example, if a sentence is found to be generated by two different pattern grammars, the ambiguity can be resolved by comparing the generation probabilities of the sentence in the two grammars. A maximum-likelihood or Bayes decision rule based on the two generation probabilities will yield the final recognition. Besides, the probability information can also be utilized to speed up the parsing process [41]. The use of a sequential decision procedure could result in further reducing the parsing time by slightly increasing the probability of misrecognition [76]. Of course, when a sequential procedure is used, the parsing procedure stops most of the time before a sentence is completely scanned, and, consequently, in these cases the complete structural information on the pattern cannot be recovered.

Remarks: Compared with decision-theoretic pattern recognition, syntactic pattern recognition is a newer area of research. When the patterns are complex and the number of pattern class is very large, it would be advantageous to describe each pattern in terms of its components and to consider description and classification of patterns rather than classification only. Of course, the practical utility of the syntactic approach depends upon our ability to recognize the simple pattern primitives and their relationships represented by the composition operations.

As research in both decision-theoretic and syntactic approaches is still in progress, heuristic methods are also being developed for specific purposes. New approaches proposed recently for pattern recognition include variable-value logic [81] and relation theory [55]. The effectiveness of these approaches still has to be tested.

III. Image Processing and Recognition

From its earliest beginnings, pattern recognition has dealt to a substantial extent with pictorial patterns. Section 3.2 of this paper reviews a few of the major themes in "image recognition." At the same time, extensive work has been done on aspects of image processing that are not directly related to pattern recognition—in particular, on image coding (for reduced-bandwidth transmission) and on image enhancement (for improving the appearance of images). Some aspects of this work are briefly discussed in Section 3.1.

For the purposes of this review, image processing refers to operations that transform images into other images, while *image recognition* is the mapping of images into (nonimage) descriptions. Different from both of these is *computer graphics,* which deals primarily with the computer synthesis and manipulation of images that are specified by descriptions; this subject is not reviewed here. We also do not cover optical (or other analog) methods of image processing, which has a large literature of its own.

A. Image Processing

Coding: In order to acceptably approximate a standard television image digitally, one normally needs an array of about 500 × 500 samples, each quantized to about 50 discrete gray levels—i.e., a total of about 6 bits for each of the 250 000 samples, or 1.5 million bits in all. The goal of image compression (or, as it is more commonly called, *image coding*) is to represent the image acceptably using a much smaller number of bits [61].

One basic approach to image coding is to apply an invertible transform to the given image, approximate the transform, and then construct the approximated image by inverting the transform. The transform can be designed so that it can be approximated more economically than the original image, and so that errors in this approximation become less noticeable when an image is reconstructed from its transform. For example, if we use the Fourier transform, we can achieve economical approximations because many of the Fourier coefficients (in the transform of a normal image) have negligible magnitudes, and so can be ignored or at least quantized very coarsely. Moreover, errors in approximating the Fourier coefficients are generally hard to notice when the image is reconstructed, because their effects are distributed over the entire image. Image compressions of as much as 10:1 can be achieved using this "transform coding" approach [7].

Many other approaches to image coding have been extensively investigated, but only a few of these can be mentioned here. One class of approaches takes differences between successive image samples; since these have a very nonuniform probability density (peaked at zero), they can be quantized acceptably using relatively few quantization levels. Note, however, that when the image is reconstructed from such differences by summing them, errors in the differences will tend to propagate, so that care is needed in using this type of approach. The differences used can be either spatial (intraframe) or temporal (interframe) [7].

The expected accuracy of an image coding system can be predicted theoretically if we assume a model for the class of images being encoded (usually, a homogeneous random field) and a specific error criterion (usually, mean squared error). Both of these assumptions are questionable. Images usually consist of distinct parts (objects or regions), so that a homogeneous random field model is inappropriate. On the other hand, the human visual system's sensitivity to errors is highly context-dependent, so that an integrated squared error criterion is inadequate. Work is needed on image coding techniques which segment the image into significant parts before attempting to approximate it; some image segmentation methods will be discussed here under Segmentation. At the same time, increased understanding of human visual capabilities is needed so that better error criteria for image coding systems can be developed.

Enhancement and restoration: There has been increasing interest in recent years in techniques for designing two-dimensional digital filters [60]. At the same time, much work is being done on digital methods of enhancing or restoring degraded images. Some enhancement techniques are conceptually very simple, and involve only pointwise modification of the image's grayscale. For example, one can analyze the gray levels in a neighborhood of each image point, determine a grayscale transformation that stretches these levels over the full displayable range, and apply this transformation to the given point (and the points in its immediate vicinity); this and similar techniques tend to give very good enhancement results [62].

A more sophisticated class of image enhancement techniques are designed to undo the effects of degradations on the image. It is customary to model these degradations as additive combinations of blurring and noise operations, where the blurring takes the form of a weighted sum or integral operation applied to the ideal image, and the noise is uncorrelated with the ideal image. A variety of methods have been developed for inverting the effects of the blurring operator; for example, pseudoinverse techniques can be used to define a deblurring operator which yields the best approximation to the ideal image in the expected least squares sense [7], [60]. Other classes of methods, e.g., based on Kalman filtering, have been devised to yield least-squares estimates of an ideal image corrupted by additive noise [7], [60]. As in the case of image coding, these approaches have usually been based on homogeneous random field models for the images (and noise), and on least-squares error criteria, both of which are questionable assumptions. Here too, image models based on segmentation of the image, and success criteria more closely related to human perceptual abilities, would be highly desirable.

A problem closely related to image restoration is that of reconstructing images; or three-dimensional objects, from sets of projections, e.g., from X-ray views taken from many angles. (The gray levels on a projection are linear combinations of the ideal gray levels, just like the gray levels on a blurred image.) Much work has been done in this area in the past few years, especially in connection with medical radiographic applications [22].

B. *Image Recognition*

The goal of image recognition is the classification or structural description of images. Image classification involves feature detection sr property measurement; image description involves, in addition, segmentation and relational structure extraction. Some significant ideas in each of these areas are reviewed in the following paragraphs. Historically, the techniques used have usually been developed on heuristic grounds, but there is increasing interest in deriving optimum techniques based on models for the classes of images to be analyzed.

Matching and feature detection: Detecting the presence of a specified pattern (such as an edge, a line, a particular shape, etc.) in an image requires matching the image with a "template," or standardized version of the pattern. This is a computationally costly process, but techniques have been developed for reducing its expected cost [9]. For example, one can match a subtemplate (or a reduced-resolution "coarse template") with the image at every point, and use the remainder of the template (or the full-resolution template) only at points where the initial match value is above some threshold. The subtemplate size, or the degree of coarseness, can be chosen to minimize the expected cost of this process [124]. In computing these matches, one should first check parts of the template that have large expected mismatch values (with a randomly chosen part of the picture), in order to minimize the expected amount of comparison that must be done before the possibility of a match at the given point is rejected [84]. Of course, the savings in computational cost must be weighed against the possible increased costs of false alarms or dismissals.

Template matching is often implemented as a linear operation in which the degree of match at a point is measured by a linear combination of image gray levels in a neighborhood of the point. However, the result of such a linear operation is generally ambiguous; for example, it may have the same value for a high-contrast partial match as it does for a lower contrast, but more complete, match. Such ambiguities can often be eliminated by breaking the template up into parts and requiring that specified match conditions be satisfied for each part, or for the most of the parts. This approach has been used to detect curves in noise [109], it needs to be extended to other types of image matching problems.

The use of template parts can also help overcome the sensitivity of template matching to geometrical distortion. Rather than matching the entire template with the image, one can match the parts, and then look for combinations of positions. Optimal combinations can be determined by mathematical programming techniques [38], or by simultaneous iterated reinforcement of the partial matches based on the presence of the other needed matches [63]. Research on these approaches is still at an exploratory stage.

Segmentation: Images are often composed of regions that have different ranges of gray levels, or of the values of some other local property. Such an image can be segmented by examining its gray level (or local property) histogram for the presence of peaks corresponding to the ranges, and using thresholds to single out individual peaks [30], [87], [116]. Detection of the peaks can be facilitated by hitogramming only a selected set of image points, e.g., points where the local property value is a local maximum [140], or points that lie on or near region boundaries (which can be identified by the presence of high values of a derivative operator) [130].

Parallel methods of region extraction based on thresholding are potentially less flexible than sequential methods, which can "learn as they go" about the geometrical, textural, and gray level properties of the region being extracted, and can compare them with any available information about the types of regions or objects that are supposed to be present in the image. Such information can be used to control merging and splitting processes with the aim of creating an acceptable partition of the image into regions [36], [59], [116].

An important special case of sequential region growing is tracking, which extracts regions (or region boundaries) in the form of thin curves. This technique can be regarded as a type of piecewise template matching, where the pieces are short line or curve segments, and a curve is any combination of these that smoothly continue one another; thus, here again, curves can be extracted by mathematical programming or iterated reinforcement techniques. The same is true for a wide variety of problems involving the selection of image parts that satisfy given sets of constraints [63].

Properties: Once regions have been extracted from an image, it becomes possible to describe the image in terms of properties of these regions. Much work has been done on defining and measuring basic geometrical properties of regions in a digitized image, such as connectedness, convexity, compactness, etc. Describing the shape of a region involves not only global properties such as those just listed, but also a hierarchically structured description in terms of "angles" and "sides" (i.e., polygonal approximation, of varying degrees of coarseness), symmetries, and so on [29].

Two "dual" methods of describing a region involve its boundary and its "skeleton." A region is determined by specifying the equations of its boundary curves; and it is also determined by specifying the centers and radii of the maximal disks that it contains [14]. (These disks define a sort of minimal piecewise approximation to the shape; such approximations can also be defined for grayscale images composed of regions that are approximately piecewise constant.) Skeleton descriptions can also be used in three dimensions, where a shape can be constructed out of "generalized cylinders," each of which is specified by a locus of centers and an associated radius function [2], [85].

Grayscale, as well as geometrical, properties of regions are of importance in image description. Of particular importance are textural properties (e.g., coarseness and directionality), which can be measured in terms of certain statistics of the second-order probability density of gray levels in the region [56]—or equivalently, statistics of the first-order probability density of gray level differences (or other local property values) [129]. Textures can be modeled as distorted periodic patterns [139], as two-dimensional "seasonal time series" [78], or in terms of random geometry; such models can be used to predict the values of the property measures for real-world textures.

Image and scene analysis: Image descriptions can usually be expressed in the form of relational structures which represent relationships among, and properties of, image parts. A major area of artificial intelligence research has been the study of how knowledge about the given class of scenes can be used to control the process of extracting such descriptions from an image. (See also the article on artificial intelligence in this issue.) In addition to the study of control structures for image analysis, there has also been recent interest in special data structures for image processing and description, e.g., "cone" or "pyramid" structures for variable-resolution image analysis [68], [115] and "fuzzy" structures for representing incompletely specified image parts.

The term "scene analysis" is generally used in connection with the description of images of three-dimensional objects seen from nearby, so that perspective and occlusion play major roles in the description [116]. (Note that the images being analyzed in applications such as document processing, photomicrography, radiology, and remote sensing are all basically two-dimensional.) Much work has been done on the extraction of three-dimensional depth information about scenes, using range sensors, stereo pairs of images, or single-image depth cues such as shading and texture gradients. These techniques are beginning to be applied to the analysis of various types of real-world indoor and outdoor scenes.

Another approach to image analysis involves the use of formal models derived from the theory of multidimensional formal languages. On this topic see the discussion of syntactic pattern analysis in Section II-B of this paper.

IV. Conclusions

It has been felt that in the past there was an unbalanced development between theory and practice in pattern recognition. Many theoretical results, especially in connection with the decision-theoretic approach, have been published. Practical applications have been gradually emphasized during the last five years, particularly in medical and remote sensing areas. Most of the practical results are considered inconclusive and require further refinement. Implementation of a practical system is often on a general purpose computer facility rather than on special purpose hardware. There is no doubt that, though heavily motivated by practical applications, pattern recognition is still very much an active research area.

In the decision-theoretic approach, we are still looking for effective and efficient feature extraction and selection techniques, particularly in nonparametric and small sample situations. The computational complexity of pattern recognition systems, in terms of time and memory, should be an interesting subject for investigation. In the syntactic approach, the problem of primitive extraction and selection certainly needs further attention. An appropriate selection of the pattern grammar directly affects the computational complexity or analysis efficiency of the resulting recognition system. Grammatical inference algorithms which are computationally feasible are still highly in demand.

In image processing, better models are needed for both the images and their user (the human visual system). Image models should also be used more extensively in the design of optimal image segmentation and feature extraction procedures. When the goal is not objectively specifiable, but rather involves general-purpose man-machine dialog about images, the computer will be also need to understand the visual capabilities and limitations of its human partner. Thus image and visual models need further development in both image processing and recognition.

References

[1] J. K. Aggarwal and R. O. Duda, Eds., "Special issue on digital filtering and image processing," *IEEE Trans. Circuits Syst.*, vol. CAS-2 pp. 161–304, 1975.

[2] G. J. Agin and T. O. Binford, "Computer description of curved objects," in *Proc. 3rd Int. Joint Conf. Artificial Intelligence*, 1973, pp. 629–640.

[3] G. J. Agin and R. O. Duda, "SRI vision research for advanced industrial automatic," in *Proc. 2nd USA-Japan Comput. Conf.*, Aug. 26–28, 1975, Tokyo, Japan.

[4] H. C. Andrews, *Computer Techniques in Image Processing.* New York: Academic, 1970.

[5] ——, *Introduction to Mathematical Techniques in Pattern*

Recognition, New York: Wiley, 1972.
[6] H. C. Andrews, Ed., "Special issue on digital picture processing," *Comput.*, vol. 7, pp. 17–87, May 1974.
[7] H. C. Andrews and L. H. Enloe, Eds., "Special issue on digital picture processing," *Proc. IEEE*, vol. 60, pp. 766–898, July 1972.
[8] A. G. Arkadev and E. M. Braverman, *Learning in Pattern Classification Machines*. Moscow: Nauka, 1971.
[9] D. I. Barnea and H. F. Silverman, "A class of algorithms for fast digital image registration," *IEEE Trans. Comput.*, vol. C-21, pp. 179–186, 1972.
[10] B. G. Batchelor, *Practical Approach to Pattern Classification*. New York: Plenum, 1974.
[11] P. W. Becker, An *Introduction to the Design of Pattern Recognition Devices*. New York: Springer, 1971.
[12] P. W. Becker, "Pattern recognition applications in work with ancient objects," in *Proc. NATO Advanced Study Institute on Pattern Recognition—Theory and Applications*, Sept. 8–17, 1975.
[13] J. M. Blin, *Patterns and Configurations in Economic Science*. Dordrect, The Netherlands: Reidel, 1973.
[14] H. Blum, "A transformation for extracting new descriptors of shape," in *Models for the Perception of Speech and Visual Form*, W. Wathen-Dunn, Ed. Cambridge, MA: M.I.T. Press, 1967, pp. 362–380.
[15] M. Bongard, *Pattern Recognition*. Moscow: Nauka, 1967 (English transl., Washington, DC: Spartan, 1970).
[16] C. H. Chen, *Statistical Pattern Recognition*. New York: Hayden, 1973.
[17] ——, "On a class of computationally efficient feature selection criteria," *Pattern Recognition*, vol. 7, no. 1/2, June 1975.
[18] ——, "Statistical pattern recognition—Review and outlook," Southeastern Massachusetts Univ., North Dartmouth, MA, Tech. Rep. TR-EE 75-4, June 25, 1975.
[19] ——, "On the use of distance and information measures in pattern recognition and applications," in *Proc. NATO Advanced Study Institute on Pattern Recognition—Theory and Applications*, Sept. 8–17, 1975.
[20] Z. Chen and K. S. Fu, "On the connectivity of clusters," *Inform. Sci.*, vol. 8, pp. 283–299, 1975.
[21] G. C. Cheng, D. K. Pollock, R. S. Ledley, and A. Rosenfeld, Eds., *Pictorial Pattern Recognition*. Washington, DC: Thompson, 1968.
[22] Z. H. Cho, Ed., *Special issue on physical and computational aspects of 3-dimensional image reconstruction, IEEE Trans. Nuclear Sci.*, vol. NS-21, June 1974.
[23] D. B. Cooper, "When should a learning machine ask for help," *IEEE Trans. Inform. Theory*, vol. IT-20, July 1974.
[24] T. Cover, "A hierarchy of probability density function estimates," in *Frontiers of Pattern Recognition*, S. Wantanabe, Ed. New York: Academic, 1972.
[25] ——, "Recent books on pattern recognition," *IEEE Trans. Inform. Theory*, vol. IT-19, Nov. 1973.
[26] T. M. Cover and T. J. Wagner, "Topics in statistical pattern recognition," in *Digital Pattern Recognition*, K. S. Fu, Ed. New York: Springer, 1976.
[27] B. V. Dasarathy and A. L. Lakshminarasimhan, "Sequential learning employing unfamiliar teacher hypothesis with concurrent estimation of both the parameters and teacher characteristics," *Int. J. Comput. Inform. Sci.*, vol. 5, Mar. 1976.
[28] H. I. Davies and E. J. Wagman, "Sequential nonparametric density estimation," *IEEE Trans. Inform. Theory*, vol. IT-21, Nov. 1975.
[29] L. S. Davis, "Understanding shape: Angles and sides," *IEEE Trans. Comput.*, to be published.
[30] L. S. Davis, A. Rosenfeld, and J. S. Weszka, "Region extraction by averaging and thresholding," *IEEE Trans. Syst., Man, Cybern.*, vol. SMC-5, pp. 383–388, 1975.
[31] R. J. P. de Figueiredo, "Optical linear and nonlinear feature extraction based on the minimization of the increased risk of misclassification," Rice Univ. Inst. Comput. Services Appl., Tech. Rep. 275-025-014, June 1974.
[32] P. A. Devijver, "Decision-theoretic and related approaches to pattern classification," in *Proc. NATO Advanced Study Institute on Pattern Recognition—Theory and Applications*, Sept. 8–17, 1975.
[33] E. Diday, "Recent progress in distance and similarity measures in pattern recognition," in *Proc. 2nd Int. Joint Conf. Pattern Recognition*, Aug. 13–15, 1974.
[34] E. Diday and J. C. Simon, "Clustering analysis," in *Digital Pattern Recognition*, K. S. Fu, Ed. New York: Springer, 1976.
[35] R. O. Duda and P. E. Hart, *Pattern Classification and Scene Analysis*. New York: Wiley, 1973.
[36] J. A. Feldman and Y. Yakimovsky, "Decision theory and artificial intelligence: I. A semantics-based region analyzer, "*Artificial Intelligence* vol. 5, pp. 349–471, 1975.
[37] G. L. Fischer, Jr., D. K. Pollock, B. Radack, and M. E. Stevens, Eds., *Optical Character Recognition*. Washington, DC: Spartan, 1962.
[38] M. A. Fischler and R. A. Elschlager, "The representation and matching of pictorial structures," *IEEE Trans. Comput.*, vol. C-22, pp. 67–92, 1973.
[39] K. S. Fu, *Sequential Methods in Pattern Recognition and Machine Learning*. New York: Academic, 1968.
[40] K. S. Fu, Ed., "Special issue on feature extraction and selection in pattern recognition," *IEEE Trans. Comput.*, vol. C-20, pp. 965–1120, Sept. 1971.
[41] ——, *Syntactic Methods in Pattern Recognition*. New York: Academic, 1974.
[42] K. S. Fu and T. L. Booth, "Grammatical inference—Introduction and survey," *IEEE Trans. Syst., Man, Cybern.*, vol. SMC-5, Jan. and July 1975.
[43] K. S. Fu, Ed., *Digital Pattern Recognition, Communication and Cybernetics*, vol. 10. New York: Springer, 1976.
[44] ——, "Pattern recognition in remote sensing of the earth's resources," *IEEE Trans. Geosci. Electron.*, vol. GE-14, Jan. 1976.
[45] K. S. Fu, Ed., *Special issue on pattern recognition, Computer* May 1976.
[46] K. S. Fu, Ed., *Applications of Syntactic Pattern Recognition*. New York: Springer, 1976.
[47] K. Fukunaga, *Introduction to Statistical Pattern Recognition*. New York: Academic, 1972.
[48] K. Fukunaga and L. D. Hostetler, "k-nearest-neighbor Bayes-risk estimation," *IEEE Trans. Inform. Theory*, vol. IT-21, May 1975.
[49] L. W. Fung and K. S. Fu, "Stochastic syntactic decoding for pattern classification," *IEEE Trans. Comput.*, vol. C-24, June 1975.
[50] J. Gips, *Shape Grammars and Their Uses*. Birkhaüser: Verlag, Basel and Stuttgart, 1975.
[51] A. Grasselli, Ed., *Automatic Interpretation and Classification of Images*. New York: Academic, 1969.
[52] U. Grenander, "Foundations of pattern analysis," *Quart. Appl. Math.*, vol. 27, pp. 1–55, 1969.
[53] E. L. Hall and C. F. George, Jr., Eds., "Special issue on two-dimensional digital signal processing," *IEEE Trans. Comput.*, vol. C-21, pp. 633–820, July 1972.
[54] K. Hanakata, "Feature selection and extraction for decision theoretic approach and structural approach," in *Proc. NATO Advanced Study Institute on Pattern Recognition—Theory and Applications*, Sept. 8–17, 1975.
[55] R. M. Haralick, "The pattern discrimination problem from the perspective of relation theory," *Pattern Recognition*, vol. 7, June 1975.
[56] R. M. Haralick, K. Shanmugam, and I. Dinstein, "Textural features for image classification," *IEEE Trans. Syst., Man, Cybern.*, vol. SMC-3, pp. 610–621, 1973.
[57] L. D. Harmon, Ed., "Special issue on digital pattern recognition," *Proc. IEEE*, vol. 60, pp. 1117–1233, Oct. 1972.
[58] H. Hauska and P. H. Swain, "The decision tree classifier: Design and potential," in *Proc. 2nd Symp. Machine Processing of Remotely Sensed Data*, June 3–5, 1975.
[59] S. L. Horowitz and T. Pavlidis, "Picture segmentation by a directed split-merge procedure," *Proc. 2nd Int. Joint Conf. Pattern Recognition*, 1974, pp. 424–433.
[60] T. S. Huang, Ed., *Picture Processing and Digital Filtering*. New York: Springer, 1975.
[61] T. S. Huang and O. J. Tretiak, *Picture Bandwidth Compression*. New York: Gordon and Breach, 1972.
[62] R. A. Hummel, "Histogram modification techniques," *Comput. Graphics Image Processing*, vol. 5, 1976.
[63] R. A. Hummel, S. W. Zucker, and A. Rosenfeld, "Scene labelling by relaxation operations," *IEEE Trans. Syst., Man, Cybern.*, to be published.
[64] T. Imai and M. Shimura, "Learning with probabilistic labelling," *Pattern Recognition*, vol. 8, Jan. 1976.
[65] L. Kanal, Ed., *Pattern Recognition*. Washington, DC: Thompson, 1968.
[66] ——, "Patterns in pattern recognition: 1968–1974, *IEEE Trans.*

Inform. Theory, vol. IT-20, Nov. 1974.

[67] S. Kaneff, Ed., *Picture Language Machines.* New York: Academic, 1971.

[68] A. Klinger and C. R. Dyer, "Experiments on picture representations using regular decomposition," *Comput. Graphics Image Processing,* vol. 5, pp. 68–105, 1976.

[69] P. A. Kolers and M. Eden, Eds., *Recognizing Patterns.* Cambridge, MA: M.I.T. Press, 1968.

[70] W. L. G. Koontz, P. M. Narendra, and K. Fukunaga, "A branch and bound clustering algorithm," *IEEE Trans. Comput.,* vol. C-24, Sept. 1975.

[71] V. A. Kovalevsky, *Character Readers and Pattern Recognition.* Washington, DC: Spartan, 1968.

[72] A. V. Kulkarni and L. N. Kanal, "An optimization approach to the design of decision trees," Dep. Comp. Sci., Univ. Maryland, College Park, MD, Tech. Rep. TR-396, Aug. 1975.

[73] R. S. Ledley, Ed., *Pattern Recognition.* New York: Pergamon, 1968.

[74] B. S. Lipkin and A. Rosenfeld, Eds., *Picture Processing and Psychopictorics.* New York: Academic, 1970.

[75] T. Lissack and K. S. Fu, "Error estimation in pattern recognition via L^α-distance between posterior density functions," *IEEE Trans. Inform. Theory,* vol. IT-22, Jan. 1976.

[76] S. Y. Lu and K. S. Fu, "Efficient error-correcting syntax analysis for recognition of noisy patterns," School Elec. Eng., Purdue Univ., West Lafayette, IN, Tech. Rep. TR-EE 76-9, Mar. 1976.

[77] *Machine Perception of Patterns and Pictures,* Conference Series 13, The Inst. Physics, London, England, 1972.

[78] B. H. McCormick and S. N. Jayaramamurthy, "Time series model for texture synthesis," *J. Comput. Inform. Sci.,* vol. 3, pp. 329–343, 1974.

——, "A decision theory model for the analysis of texture," *J. Comput. Inform. Sci.,* vol. 4, pp. 1–38, 1975.

[79] W. Meisel, *Computer-Oriented Approaches to Pattern Recognition.* New York: Academic, 1972.

[80] J. M. Mendel and K. S. Fu, Ed., *Adaptive, Learning and Pattern Recognition Systems: Theory and Applications.* New York: Academic, 1970.

[81] R. S. Michalski, "A variable-valued logic system as applied to picture description and recognition," in *Graphic Languages,* F. Nake and A. Rosenfeld, Ed. Amsterdam, The Netherlands: Holland, 1972.

[82] M. Minsky and S. Papert, *Perceptrons.* Cambridge, MA: M.I.T. Press, 1969.

[83] B. Moayer and K. S. Fu, "A tree system approach for fingerprint pattern recognition," *IEEE Trans. Comput.,* vol. C-25, Mar. 1976.

[84] R. N. Nagel and A. Rosenfeld, "Ordered search techniques in template matching," *Proc. IEEE,* vol. 60, pp. 242–244, 1972.

[85] R. Nevatia and T. O. Binford, "Structured descriptions of complex objects," in *Proc. 3rd Int. Joint Conf. Artificial Intelligence,* 1973, pp. 641–647.

[86] N. J. Nilsson, *Learning Machines—Foundations of Trainable Pattern-Classifying Systems.* New York: McGraw-Hill, 1965.

[87] R. B. Ohlander, "Analysis of natural scenes," Ph.D. dissertation, dep. Comput. Sci., Carnegie-Mellon Univ., Pittsburgh, PA, Apr. 1975.

[88] P. A. Ota, "Mosaic grammers," *Pattern Recognition,* vol. 7, June 1975.

[89] E. A. Patrick, *Fundamentals on Pattern Recognition.* Englewood Cliffs, NJ: Prentice-Hall, 1972.

[90] L. F. Pau, *Diagnostic Des Pannes Dans Les Systemas.* Cepadues-Edition, 1975.

[91] T. Pavlidis, "Analysis of set patterns," *Pattern Recognition,* vol. 1, Nov. 1968.

[92] ——, "Waveform segmentation through functional approximation," *IEEE Trans. Comput.,* vol. C-22, July 1973.

[93] T. Pavlidis and F. Ali, "Computer recognition of handwritten numerals by polygonal approximations," *IEEE Trans. Syst., Man, Cybern.,* vol. SMC-5, Nov. 1975.

[94] E. Persoon and K. S. Fu, "Sequential classification of strings generated by SCFG's," *Int. J. Comput. Inform. Sci.,* vol. 4, Sept. 1975.

[95] J. M. S. Prewitt and M. L. Mendelsohn, "The analysis of cell images," *Annals NY Academy of Sci.,* vol. 128, pp. 1035–1053, Jan. 1966.

[96] Proc. Int. Joint Conf. Pattern Recognition (First: Washington, DC, Oct. 1973; Second: Copenhagen, Denmark, Aug. 1974; Third: Coronado, CA, Nov. 1976).

[97] D. M. Ramsey, Ed., *Image Processing in Biological Sciences.* Berkeley, CA: Univ. California Press, 1969.

[98] D. R. Reddy, Ed., *Speech Recognition.* New York: Academic, 1975.

[99] A. Rosenfeld, *Picture Processing by Computer.* New York: Academic, 1969.

[100] ——, "Picture processing by computer," *Comput. Surveys,* vol. 1, pp. 147–176, 1969.

[101] ——, "Progress in picture processing: 1969–71," *Comput. Serveys,* vol. 5, pp. 81–108, 1973.

[102] ——, "Picture processing: 1972," *Comput. Graphics Image Processing,* vol. 1, pp. 394–416, 1972.

[103] ——, "Picture processing: 1973," *Comput. Graphics Image Processing,* vol. 3, pp. 178–194, 1974.

[104] ——, "Picture processing: 1974," *Comput. Graphics Image Processing,* vol. 4, pp. 133–155, 1975.

[105] ——, "Picture processing: 1975," *Comput. Graphics Image Processing,* vol. 5, 1976.

[106] A. Rosenfeld, Ed., *Digital Picture Analysis.* New York: Springer, 1976.

[107] A. Rosenfeld, H. Freeman, T. S. Huang, and A. van Dam, Eds., *Comput. Graphics Image Processing.* New York: Academic, 1972.

[108] A. Rosenfeld and A. C. Kak, *Digital Picture Processing.* New York: Academic, 1976.

[109] A. Rosenfeld and M. Thurston, "Edge and curve detection for visual scene analysis," *IEEE Trans. Comput.,* vol. C-20, pp. 562–569, 1971.

[110] K. M. Sayre, *Recognition: A Study in the Philosophy of Artificial Intelligence.* Notre Dame, IN: Univ. Notre Dame Press, 1965.

[111] G. S. Sebestyen, *Decision Processes in Pattern Recognition.* New York: Macmillan, 1962.

[112] J. C. Simon, "Recent progress to a formal approach of pattern recognition and scene analysis," in *Proc. 2nd Int. Joint Conf. Pattern Recognition,* Aug. 13–15, 1974.

[113] J. Sklansky, Ed., *Pattern Recognition: Introduction and Foundations.* Dowden, Hutchinson and Ross, 1973.

[114] G. Stockman, L. N. Kanal, and M. C. Kyle, "Structural pattern recognition of waveforms using a general waveform parsing system," Dep. Comput. Sci., Univ. Maryland, College Park, MD, Tech. Rep. TR-390, July 1975.

[115] S. Tanimoto and T. Pavlidis, "A hierarchical data structure for picture processing," *Comput. Graphics Image Processing,* vol. 4, pp. 104–119, 1975.

[116] J. M. Tenenbaum and H. G. Barrow, "Experiments in interpretation-guided segmentation," Standard Res. Inst., Menlo Park, CA, Tech. Note 123, Mar. 1976.

[117] R. A. Thompson, "Language correction using probabilistic grammars," *IEEE Trans. Comput.,* vol. C-25, Mar. 1976.

[118] J. T. Tou and R. C. Gonzalez, *Pattern Recognition Principles.* New York: Addison-Wesley, 1974.

[119] G. T. Toussaint, "Bibliography on estimation of misclassification," *IEEE Trans. Inform. Theory,* vol. IT-20, July 1974.

[120] ——, "Recent progress in statistical methods applied to pattern recognition," in *Proc. 2nd Int. Joint Conf. Pattern Recognition,* Aug. 13–15, 1974.

[121] L. Uhr, Ed., *Pattern Recognition.* New York: Wiley, 1966.

[122] ——, *Pattern Recognition, Learning and Thought.* Englewood Cliffs, NJ: Prentice-Hall, 1973.

[123] J. R. Ullmann, *Pattern Recognition Techniques.* Crane, Russak, & Co., 1973.

[124] G. J. VanderBrug and A. Rosenfeld, "Two-stage template matching," *IEEE Trans. Comput.,* to be published.

[125] C. J. D. M. Verhagen, "Some general remarks about pattern recognition; Its definition; Its relation with other disciplines; A literature survey," *Pattern Recognition,* vol. 7, Sept. 1975.

[126] T. J. Wagner, "Nonparametric estimates of probability densities," *IEEE Trans. Inform. Theory,* vol. IT-21, July 1975.

[127] S. Watanabe, Ed., *Methodologies of Pattern Recognition.* New York: Academic, 1969.

[128] ——, *Frontiers of Pattern Recognition.* New York: Academic, 1972.

[129] J. S. Weszka, C. R. Dyer, and A. Rosenfeld, "A comparative study of texture measures for terrain classification," *IEEE Trans. Syst., Man, Cybern.,* vol. SMC-6, pp. 269–285, 1976.

[130] J. S. Weszka, R. N. Nagel, and A. Rosenfeld, "A threshold selection technique," *IEEE Trans. Comput.,* vol. C-23, pp. 1322–1326, 1974.

[131] K. L. Williams, "A multidimensional approach to syntactic pattern

recognition," *Pattern Recognition,* vol. 7, Sept. 1975.
[132] P. H. Winston, Ed., *The Psychology of Computer Vision.* New York: McGraw-Hill, 1975.
[133] S. S. Yau and S. C. Chang, "A direct method for cluster analysis," *Pattern Recognition,* vol. 7, Dec. 1975.
[134] K. C. You and K. S. Fu, "An approach to the design of a linear binary tree classifier," in *Proc. 3rd Symp. Machine Processing of Remotely Sensed Data,* June 29–July 1, 1976.
[135] I. T. Young, "The prediction of performance in multi-class pattern classification," in *Proc. 2nd Int. Joint Conf. Pattern Recognition,* Aug. 13–15, 1974.
[136] T. Y. Young and T. W. Calvert, *Classification, Estimation, and Pattern Recognition.* New York: Elsevier, 1973.
[137] N. G. Zagoruyko, *Recognition Methods and Their Applications, Radio Sovetskoe.* Moscow, 1972.
[138] N. V. Zavalishin and I. B. Muchnik, *Models of Visual Perception and Algorithms for Image Analysis.* Moscow: Nauka, 1974.
[139] S. W. Zucker, "On the structure of texture," *Comput. Graphics Image Processing,* vol. 5, 1976.
[140] S. W. Zucker, A Rosenfeld, and L. S. Davis, "Picture segmentation by texture discrimination," *IEEE Trans. Comput.*, vol. C-24, pp. 1228–1233, 1975.

Patterns in Pattern Recognition: 1968–1974

Invited Paper

LAVEEN KANAL, FELLOW, IEEE

Abstract—**This paper selectively surveys contributions to major topics in pattern recognition since 1968. Representative books and surveys on pattern recognition published during this period are listed. Theoretical models for automatic pattern recognition are contrasted with practical design methodology. Research contributions to statistical and structural pattern recognition are selectively discussed, including contributions to error estimation and the experimental design of pattern classifiers. The survey concludes with a representative set of applications of pattern recognition technology.**

I. Introduction

WHAT IS a pattern that a machine may know it, and a machine that it may know a pattern? That is the fundamental mystery challenging research in automatic pattern recognition. This survey reviews the main paths followed since 1968 and examines some of the research performed in the quest for answers.

This paper complements two 1972 articles. The paper by Kanal and Chandrasekaran (1972) probed theoretical approaches based on alternate models for pattern recognition and assessed contributions to the problem of inferring grammars from samples. To make the present survey somewhat self-contained and accessible to readers not working in pattern recognition, a brief discussion of models is presented in Section III. However, work on interactive pattern analysis and classification systems is mentioned only in passing because the 1972 article in the PROCEEDINGS OF THE IEEE [Kanal (1972)] considered that topic at length.

The topics covered here are grouped under the following section headings:

II. Journals, Books, and Surveys
III. Models for Automatic Pattern Recognition
IV. Design Methodology For Automatic Pattern Recognition Systems
V. Statistical Feature Extraction, Evaluation, and Selection
VI. Dimensionality, Sample Size, and Error Estimation
VII. Statistical Classification
VIII. Structural Methods
IX. Applications
X. Prospects.

Section II gives a representative list of journals, books, and surveys for the period 1968–1974. Section III contrasts two models, the feature extraction statistical-classification model and the linguistic structural model, which have served as the basis for pattern recognition theory; it also briefly introduces a hybrid model. In Section IV, I describe how these theoretical models differ from the practical design methodology which has evolved during the last few years.

Prior to 1968 classification algorithms seemed to be the main output of theoretical research in statistical pattern recognition. Section V reflects the effort devoted since 1968 to theoretical approaches to problems of feature extraction, evaluation, and selection. In Section V, I examine recent approaches to defining pattern representation spaces and to deriving features that enhance class separability; theoretical and experimental investigations of distance measures and error bounds and their use in feature evaluation; and feature subset selection procedures.

Problems in the design and analysis of pattern classification experiments represent another area receiving increased attention since 1968. Section VI summarizes recent investigations and the resulting rules of thumb on the ratio that should be maintained between the number of design samples per class and the number of features. Insights gained from work on how best to use a fixed size sample in designing and testing a classifier are also summarized. In addition, Section VI presents results on the nonparametric estimation of the Bayes error and on the use of unlabeled samples in estimating the error rate.

Section VII is primarily concerned with nonparametric classification. It also briefly describes attempts to compare classification procedures.

Using examples in waveform segmentation and speech recognition in Section VIII, I comment on certain key concepts and differences that distinguish some recent problem-oriented contributions to segmentation, feature extraction, and structural analysis from other general numerical analysis and grammar based approaches. In addition, research on generalizing pattern grammars to overcome the limitations of string grammars is described.

Section IX considers the present status of applications, and Section X comments on how work in pattern recognition is likely to proceed in the near future. References follow Section X.

It is not feasible to cover the waterfront of pattern recognition in a journal article. The aim of the selective discussion of topics and contributions which I present here is to provide a perspective on how pattern recognition theories, techniques, and applications have evolved during the last few years.

Manuscript received July 5, 1974. This work was supported in part by the Air Force Office of Scientific Research under Grant AFOSR 71-1982, in part by the National Science Foundation under Grants GK39905 and GK41602, and in part by L.N.K. Corporation.
The author is with the Department of Computer Science, University of Maryland, College Park, Md. 20742, and with the L.N.K. Corporation, Silver Spring, Md. 20904.

Reprinted from *IEEE Trans. Inform. Theory*, vol. IT-20, pp. 697–722, Nov. 1974.

II. Journals, Books, and Surveys

Since 1968 more than five hundred journal articles on pattern recognition have appeared in the English language engineering literature alone. Within the family of IEEE journals, articles on pattern recognition have been regularly published in the Transactions on Computers, Transactions on Information Theory, Transactions on Systems, Man, and Cybernetics, and occasionally in the Transactions on Automatic Control, Transactions on Biomedical Engineering, and the Proceedings of the IEEE. Other journals regularly publishing papers in this area include *Pattern Recognition*, *Information Sciences*, and *Information and Control*. Statistical journals that frequently publish papers relevant to pattern classification include the *Journal of the American Statistical Association*, *Biometrics*, *Biometrika*, *Technometrics*, and the *Journal of the Royal Statistical Society*.

For years the closest item to a textbook in pattern recognition was a monograph entitled *Learning Machines* [Nilsson (1965)]. In 1974, one could choose from more than half a dozen textbooks of varying merit on statistical pattern recognition, e.g. [Andrews (1972), Young and Calvert (1974), Chen (1973), Duda and Hart (1973), Fukunaga (1972), Mendel and Fu (1970), Meisel (1972), and Patrick (1972)]; for a book review see [Cover (1973)]. In addition, one could turn to monographs devoted to, or including some discussion of, various aspects of pattern recognition [Anderberg (1973), Bongard (1970), Fu (1968), (1974), Lindsay and Norman (1972), Rosenfeld (1969), Tsypkin (1971), (1973), Uhr (1973), Ullmann (1973), and Watanabe (1969)] and numerous hardcover collections of papers and conference proceedings [e.g., Cacoullous (1973), Cheng *et al.* (1968), Fu (1970), Grasselli (1969), Kanal (1968), Kohlers and Eden (1968), Krishnaiah (1969), Tou (1970), (1971), and Watanabe (1969), (1972)].

The books by Duda and Hart, Meisel, and Ullmann provide broad coverage of the literature up to early 1972. Duda and Hart's bibliographic and historical remarks at the end of each chapter set a high standard of scholarship and give a "who did what, when, and where" picture of the pattern classification and scene analysis literature. Meisel's bibliography is also thorough. Ullmann starts with a description of a 1929 patent for a reading machine and gives the reader a guided tour of 451 references including several patents. Anderberg summarizes literature on clustering techniques, while Fukunaga examines the problems of error estimation in greater depth than the other textbooks. Young and Calvert include a chapter each on two specific applications, viz., electrocardiograms and optical character recognition (OCR).

In addition to these books, many survey articles, reviews, and bibliographies were also published in this period. The Proceedings of the IEEE devoted its October 1972 issue to papers extensively surveying applications of digital pattern recognition [Harmon (1972)]. Earlier survey papers that appeared in the Proceedings include [Ho and Agrawala (1968), Levine (1969), and Nagy (1968)]. A series of papers [Rosenfeld (1972), (1973), (1974)] covers developments in picture processing by computer during the period 1969 through 1973 and provides an extensive bibliography. For speech recognition, a survey [Hill (1971)], a study committee report [Newell *et al.* (1973)], a recent conference proceeding [Erman (1974)], and a forthcoming book [Reddy (1974)] provide adequate coverage. Additional survey articles on specific topics are cited in subsequent sections of this paper.

III. Models for Automatic Pattern Recognition

An early motivation for work on automatic pattern recognition was to model pattern recognition and intelligence as found in living systems; the Perceptron [Rosenblatt (1960)] and other 1960 vintage "learning" or "self-organizing" networks are examples of models that, at least initially, were biologically motivated. Although the excitement about them had been greatly dampened by 1968, such "bionic" models continued to attract a few circles interested in pattern recognition [Amari (1972)], adaptive control [*Business Week* (1974), Mucciardi (1972)], the implicit storage of a fixed set of patterns [Moore (1971)], modeling the cerebellum [Albus (1971), (1972)], and modeling the input–output relationships of other complex systems [Ivakhnenko (1971), Mucciardi (1974)]. The Proceedings of a 1974 Conference [Conf. on Biologically Motivated Automata (1974)] indicates a revival of interest in biologically motivated automata, neural models, and adaptive networks.

How well the "bionic" networks model the biological systems that served as their motivation is open to question. The point is moot if one accepts the view that recognition is an attainment or a goal rather than a process, method, or technique [Sayre (1965)]. Then machines can "recognize" certain patterns without necessarily having anything in common with the methods used by biological systems to recognize those same patterns [Kanal and Chandrasekaran (1968)]. Most of the theoretical work on machine recognition of patterns has not been biologically motivated but has adopted one or the other of two models, the feature extraction statistical-classification model or the linguistic model.

The period 1960–1968 witnessed extensive activity on decision-theoretic multivariate statistical procedures for the design of classifiers. However, the statistical decision theory approach was justly criticized for focusing entirely on statistical relationships among scalar features and ignoring other structural properties that seemed to characterize patterns. The general feature-extraction classification model, shown in Fig. 1, was also criticized for performing too severe data compression, since it provided only the class designation of a pattern rather than a description that would allow one to generate patterns belonging to a class.

These criticisms led to proposals for a linguistic model for pattern description whereby patterns are viewed as sentences in a language defined by a formal grammar. By 1968 these proposals together with the success of syntax-directed compilers had attracted many to research in pattern grammars. The linguistic or syntactic model for pattern recognition uses a "primitive extractor," which transforms the input data into a string of symbols or some general relational structure. The primitive extractor may itself be a feature extractor classifier. Then a structural pattern

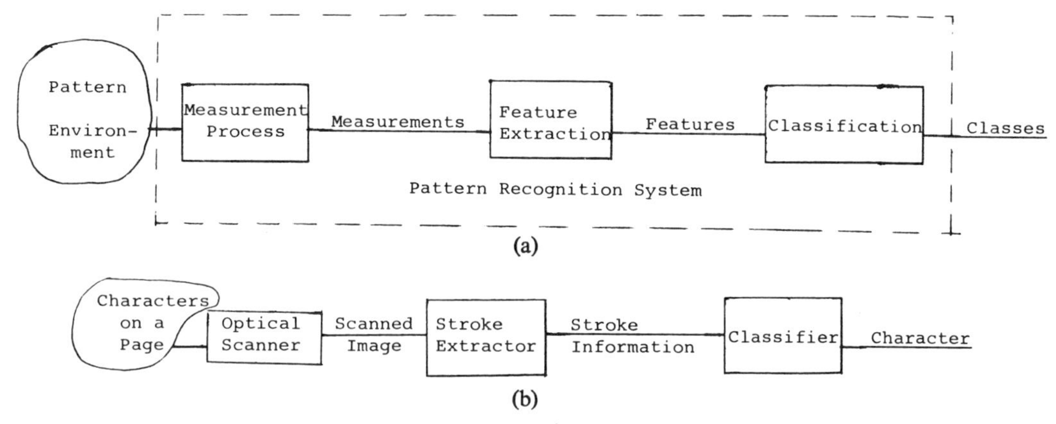

Fig. 1. (a) Operational system. (b) An example.

analyzer uses a formal grammar to parse the string and thus constructs a description of the pattern.

In the past, much has been made of the apparent difference between the two models. The stress on the distinction between the two models hides many similarities: in practice, in the syntactic model, the extraction of "primitives" can involve statistical classification procedures, and the association of patterns with generative grammars is equivalent to the classification of patterns into categories.

The definition of the formal linguistic model can be enlarged to include other familiar generative mechanisms, such as differential equations, functional equations, and finite-state Markov chains. Stochastic-syntactic models introduce probabilistic aspects into the linguistic model by specifying a discrete probability distribution over productions of a base grammar. For an N class problem, one could develop N stochastic grammars. Each parse provides a structure along with a probability that the structure represents the input pattern; the input is associated with the grammar giving the most probable parse [see, e.g., Fu and Swain (1971), Fu (1972)].

When a formal model is not explicitly present, the terms "ad-hoc" or "heuristic" are used. The phrase "structural pattern recognition" refers to all pattern recognition approaches based on defining primitives and identifying allowable structures in terms of relationships among primitives and substructures that combine primitives. This term represents less a specific set of procedures than an attitude, i.e., that pattern recognition algorithms should be based on the mechanisms that generate and deform patterns.

The structural pattern recognition model is reminiscent of the "analysis by synthesis" model proposed for speech recognition in this TRANSACTIONS [Halle and Stevens (1962)]. In the latter model, a synthetic pattern was generated and matched with the input pattern. The emphasis was on using a generative model that embodied the physical processes thought to govern speech pattern generation in humans, and driving this synthesizer with parameter values obtained from the input pattern. The set of parameter values that provided a match were then used to characterize and recognize the input pattern. The general flavor of the "analysis by synthesis" model and of the structural pattern recognition work now being done is similar, but the emphasis is no longer on identically matching the input pattern nor on matching the physical processes closely. More flexibility is obtained through "black-box" generative models that generate patterns "like" the input pattern without necessarily being closely related to the physical processes about which we may not have much information.

An outline of a formalism that attempts to combine the linguistic and statistical aspects of patterns has been presented in some thought-provoking papers by Grenander (1969), (1970). The major outlines of the proposal are fairly easy to follow but the details of the model are quite ambiguous and much interpretation must be provided by the reader. This model assumes we are given a set of primitive structural objects called signs, which together with known grammars or other known generative mechanisms, produce a set $\mathscr{I}$ of "pure images." Subsets of $\mathscr{I}$ satisfying certain similarity properties (which we do not define here) are called "pure patterns." The pure images are subjected to probabilistic deformations to give a set $\mathscr{I}^D$ of deformed images. Recognition algorithms would then have to define the inverse mappings from the set of deformed images to the pure patterns.

The formalism requires that there exist a method of analysis leading to a unique history of formation for any given image. In practice, in most interesting problems it is only the deformed patterns, further corrupted by noise, that are available, and the deformations and generative mechanism must be discovered from a limited set of samples. Because there will rarely be a unique definition of primitives and generative mechanisms, there will rarely be a unique analysis as required by the model.

In the period being considered, the fuzzy set model proposed by Zadeh (1965) has been applied to classification in a number of theoretical papers. Unlike classical sets, fuzzy sets are defined to have a membership function that can take on any real value between zero and one. This produces a nonexclusive assignment of a pattern to a class. It should be emphasized that this concept is different from a probabilistic assignment of patterns to classes even though a probability also takes on values between zero and one. In the latter case, for a two-class problem, for example, an individual pattern may be probabilistically assigned to one class or the other, but is not thought of as inherently belong-

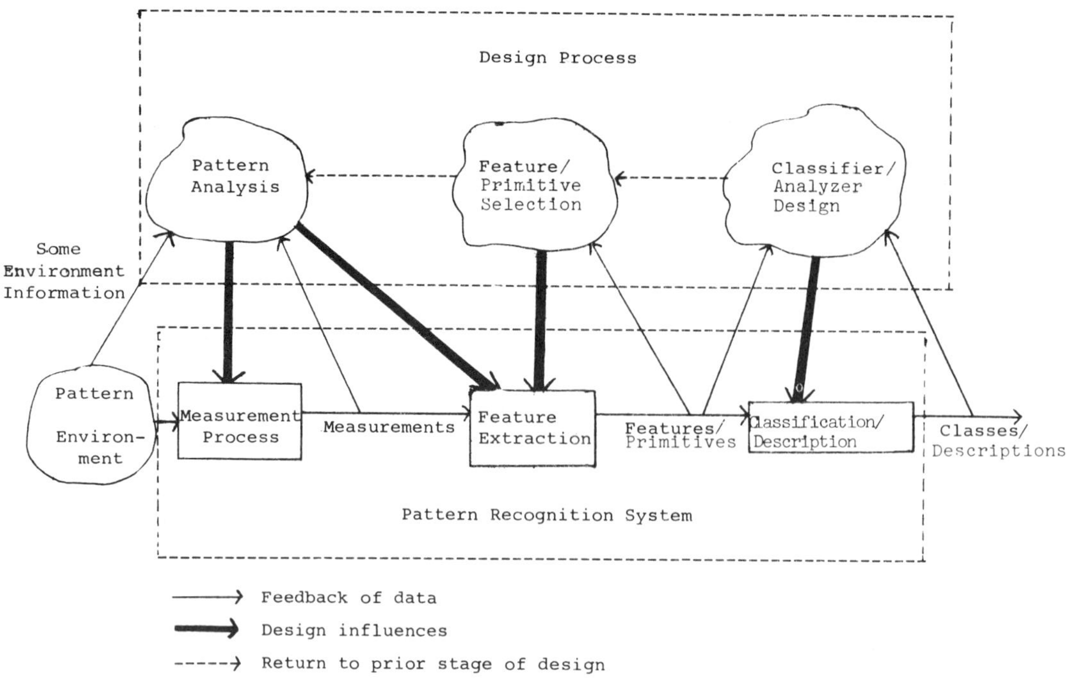

Fig. 2. Development of pattern recognition systems.

ing to both classes simultaneously, as is true in the fuzzy set model. An extensive bibliography on fuzzy sets is given in Kauffman (1973). Zadeh's papers remain the best source for understanding the idea and stimulating thinking about potential applications [Zadeh (1973)].

IV. Design Methodology for Automatic Pattern Recognition Systems

The term "pattern analysis" was not noticeably mentioned prior to 1968 and does not appear in the surveys on pattern recognition published that year in the Proceedings of the IEEE [Nagy (1968) and Ho and Agrawala (1968)]. Its widespread use in the literature seems to have followed the publication of Sammon's reports on the "On-Line Pattern Analysis and Recognition System" (OLPARS) [Sammon (1968)].

As it is understood today, pattern analysis consists of using whatever is known about the problem at hand to guide the gathering of data about the patterns and pattern classes which may exist in the environment being examined, and then subjecting the data to a variety of procedures for inferring deterministic and probabilistic structures that are present in the data. Statisticians call this exploratory data analysis. Histogram plots, scatter plots, cluster analysis routines, linear discriminant analysis, nonlinear mappings, analysis of variance, and regression analysis are examples of procedures used to detect and identify structures and substructures in the data. The purpose is to understand the regularities and peculiarities of a data base to enable better feature definition leading to simpler and better classification or description.

Pattern analysis is now considered an intrinsic and important part of the design process. In contrast, in the literature prior to 1968, automatic pattern recognition system design consisted primarily of designing the classifier. The available features and samples were not explored much but used directly, perhaps in a "learning machine" approach wherein parameters of a fixed structure are sequentially adjusted until correct classification is obtained for all "training" samples or until an error criterion is minimized; or the features were used in a fixed discriminant function the coefficients of which were statistically estimated from the available samples; or assuming parametric forms for the joint densities of feature vectors from each class, sequential and nonsequential statistical estimation procedures were proposed to estimate parameters of the densities for each class and derive classifier designs based on statistical decision theory.

Prior to 1968, it was acknowledged that the boundaries between feature definition, extraction, and classification were not sharp and that feedback between them was needed. However, this was not reflected in the work presented in the literature. At least all the theory-based papers assumed neat partitions between feature extraction and classification. The theoretical research published during the past few years on the syntactic approach to pattern recognition has, for the most part, continued on this path. For example, in the papers on the stochastic syntactic approach to pattern recognition [Fu (1972) and Lee and Fu (1972)], the extractor and analyzer functions are treated independently, which prevents the structural information available to the analyzer from influencing the primitive extraction. Without this feedback the representation provided by the extractor may not be well suited to the patterns being examined. Noting this limitation, models that incorporate feedback between the analyzer/classifier and the extractor have recently been proposed. These are described in Section VIII.

A major evolution that has occurred during the last few years is to view the design of a pattern recognition system as a highly iterative process. Fig. 2 illustrates the complex nature of this design process. The theoretical models, in which the flow of data and decisions is in only one direction, from the input pattern environment to the categorizer/structural analyzer, are indicative of the operational pattern recognition system one seeks as an end result.

The advantages of human interaction and intervention in all phases of the iterative design process and the important role of interactive computing and display technology in making this feasible have been elaborated upon in Kanal (1972), which also summarized the data analysis techniques based on clustering, statistical discriminant analysis nonlinear mapping, etc. New graphical representations to enable human understanding of multivariate data continue to be explored, e.g., see Chernoff (1973). While no *radically* new statistical approaches to data analysis have appeared since 1968, effort has been devoted to improving and comparing algorithms and interpreting their relevance to feature extraction, evaluation, and selection. The next section presents some of the results.

V. Feature Extraction, Evaluation, and Selection—"Statistical Methods"

Feature extraction and selection can have a variety of goals. We may be interested in: finding key features that permit the generation or reconstruction of the original patterns, selecting features that parsimoniously characterize patterns, finding features that are effective in discriminating between pattern classes, or some combination of these goals.

In this section, the word "feature" denotes an entity that is derived from some initial measurements; this implies that somehow we know what initial measurements should be made. Reducing the initial measurements can lead to economies in sensor hardware and data processing. A simple approach to discarding measurements might be to examine how closely a linear combination of the selected measurements represents a discarded measurement [Beale, Kendall and Mann (1967), and Allen (1971)]. Clearly, no reduction in measurement effort is achieved if the selection is from features that are combinations of all the initial measurements; this is true of many of the feature extraction selection procedures that have been proposed. However, in the resulting lower-dimensional space, the search for a classifier may be greatly simplified.

Much of the mathematical-statistical work on feature extraction and selection during the past few years has been on:

1) linear and nonlinear transformations to map patterns to lower-dimensional spaces for pattern representation or to enhance class separability;
2) feature evaluation criteria that bound the Bayes error probability and transformations that are optimum with respect to such criteria;
3) search procedures for suboptimal selection of a subset from a given set of measured or derived features.

Pattern Representation Spaces

Many transform techniques, such as Fourier, Walsh–Hadamard, and Haar, have been proposed for deriving feature domains [Andrews (1972)]. The method of principal components, which rank orders the eigenvalues of the pooled covariance matrix of all the classes according to the magnitude of their associated eigenvalues, has a long history in classical multivariate analysis. Some papers in this period have considered nonlinear principal component analysis where, given a class of possible nonlinear coordinates, one finds the coordinate along which the data variance is maximum, and then obtains another coordinate uncorrelated with the first, along which the variance is next largest, etc. [Gnanadesikan and Wilk (1969)].

Among linear transformations, the Karhunen–Loève (K–L) expansion in terms of the eigenvectors of the covariance matrix is in one sense the minimax or "most reliable" feature extractor [Young (1971)]. Watanabe and others have proposed feature domains based on eigenvectors of the pooled autocorrelation matrix and on the eigenvectors of the autocorrelation matrix of a given class [Watanabe (1969)]. A novel K–L type modification of the Fourier transform has been suggested recently for pictures [Fukunaga and Sherman (1973)].

Their "optimality" properties notwithstanding, for a given data set the preceding procedures may or may not provide effective representations. Other candidates to be tried include nonlinear mappings based on multidimensional scaling and intrinsic dimensionality algorithms. Multidimensional scaling and parametric mapping are techniques for finding a configuration of data points in the smallest dimensional space that, according to some defined error criterion, preserves the local structure of the points in the original n-dimensional space. It is possible that the data may tend to lie on a curve in the n-dimensional space; estimation of the parametric form of this curve would indicate the intrinsic dimensionality of the collection of data points.

These nonlinear mappings have been the subject of some investigation during the past few years [Bennett (1969), Calvert and Young (1969), Sammon (1969), Fukunaga and Olsen (1971), Trunk (1972), and Olsen and Fukunaga (1973)]. The basic ideas and references on these mappings were briefly summarized in Kanal (1972). Some recent contributions aimed at simplifying or improving nonlinear mapping algorithms are mentioned next.

Sammon's nonlinear mapping algorithm [Sammon (1969)] involves computing all the $K(K-1)/2$ interpoint distances in the lower-dimensional space. In Chang and Lee (1973) simultaneous adjustment of all the K points to minimize the error function is replaced by a heuristic relaxation procedure in which a pair of points is adjusted at a time.

Iterative algorithms for nonlinear mapping must be repeated for new data points. A noniterative nonlinear mapping is proposed in Koontz and Fukunaga (1972). Noniterative procedures, using K–L expansions for local regions, have also been proposed for estimating the intrinsic dimensionality of the nonlinear surface on which the data may lie [Fukunaga and Olsen (1971) and Olsen and Fukunaga (1973)]. The local dimensionality estimation could be affected by noisy data samples being distributed about their intrinsic dimensional surface, rather than falling exactly on it. In Fukunaga and Hostetler (1973), a method for density gradient estimation is presented, and it is suggested that the samples be moved according to the density gradient so as to condense them onto an intrinsic dimensional skeleton.

Iterative nonlinear mapping algorithms have often been useful for representing pattern data in a lower-dimensional space. Whether or not the noniterative procedures mentioned here truly improve existing implementations of nonlinear mapping and intrinsic dimensionality estimation remains to be seen.

Representations Enhancing Separability

Instead of the information preserving aspects of K–L representations, Fukunaga and Koontz (1970) emphasized the extraction of eigenvectors that enhance class separability. This was done by finding the linear transformation that when applied to the autocorrelation matrix of the mixture of the two classes gives an identity matrix. Then after rank ordering the eigenvalues for class 1, one has $1 \geq \lambda_1^{(1)} \geq \lambda_2^{(1)} \geq \cdots \geq \lambda_n^{(1)} \geq 0$ and for class 2, $\lambda_i^{(2)} = 1 - \lambda_i^{(1)}$. The recommendation that $|\lambda_i - 0.5|$ be the criterion to select the eigenvectors to be used as features has been disputed in Foley (1973) where a three-dimensional two-class counterexample is presented. In this example, there is complete overlap between the two classes along two of the dimensions with very little overlap in the remaining dimension, and it is shown that the Fukunaga–Koontz ranking procedure leads to the two features with zero discriminating power being given the same weight as the one feature that provides high discrimination. As an alternative, a generalization of an optimal discriminant plane [Sammon (1970)] is recommended.

Let

$$R(d) = \frac{d^t(\bar{X}^{(1)} - \bar{X}^{(2)})^2}{d^t A d}$$

where d is the (column) vector (of direction cosines) representing the direction on which the data is to be projected, $(\bar{X}^{(1)} - \bar{X}^{(2)})$ is the difference between sample mean vectors for the two classes, and $A = cW_1 + (1 - c)W_2$, where $0 \leq c \leq 1$ is a weight constant and W_i, $i = 1,2$, is the within-class scatter matrix for class i. Orthogonal discriminant directions are obtained by maximizing this "generalized Fisher ratio" and successively constraining each discriminant direction to be orthogonal to the previous set of discriminant directions. Foley (1973) presents examples in which this "discrim-vector" approach is superior to the Fukunaga–Koontz method.

For the multiclass case most of the work has either cast the M-class problem as $M(M - 1)/2$ two-class problems or employed multidimensional scatter ratios popular in classical statistical multiple discriminant analysis [Duda and Hart (1973)]. While based on linear operator theory, the attempt in Watanabe and Pakvasa (1973) to systemize the generation of orthonormal feature spaces for the multiclass problem, such that separability of the classes is emphasized, is different. Additional investigation is needed before its usefulness can be assessed.

Distance Measures and Error Bounds

If classification rather than description of patterns is the goal, the ultimate test of a set of features is their contribution to the Bayes error probability. The aim of feature selection is to reduce the number of features without adversely affecting error performance. Unfortunately, in most situations, even if the class conditional densities are known, a straightforward analytical relationship between the Bayes error probability and the features used is not available. Hence, various measures of information and distance have been proposed to measure the effectiveness of a given set of features. The major results relating such measures to the Bayes error probability are summarized in Table I.

The primary utility of the distance measures and corresponding bounds in Table I is for theoretical investigations. For example, the Bayesian distance and related bounds led Devijver (1973) to a theoretical justification for the least mean-squared error (LMSE) as a feature selection and ordering criterion [Wee (1968)] and to the relationship of the LMSE criterion to the nearest-neighbor rule. Of course, like the nearest-neighbor (NN) rules, the relationships derived from these bounds represent asymptotic results.

Despite the many papers published in this area, the net result of the extensive investigations on distance measure bounds for P_e seems to be that one should try to estimate the error probability itself in some direct manner.

Subset Selection and Heuristic Search

The feature selection problem can be viewed as a (combinatorial) optimization problem requiring a criterion function and a search procedure. All the literature on feature subset selection can be described in these terms. Some of the procedures were suggested years ago. I will cite here a few recent papers that illustrate the procedures.

In diFigueredo (1974) the probability of misclassification is the criterion functional to be minimized. Within a class of transformations that could include nonlinear transformations, the optimal transformation from the initial space to a feature space of prescribed lower dimension is determined, such that the increased misclassification error in the lower-dimensional space is minimized. The iterative algorithms presented are computationally much more complex than corresponding iterative algorithms for the Bhattacharyya bound [Decell and Quiren (1973)]. The k-NN bound was suggested as an evaluation criterion in Cover (1969) and used in Whitney (1971). Average information content $I(\Omega|X) = H(\Omega) - H(\Omega|X)$ was suggested for feature evaluation at least a decade ago. Experiments with this measure in different contexts continue to be reported [e.g., Simon *et al.* (1972) and Michael and Lin (1973)]. The divergence is another measure with a long history [Kailath

TABLE I
DISTANCE MEASURES AND ERROR BOUNDS

Name	Expression	Relationships
Bayes error probability	$P_e = 1 - \int_{S_x} \max_i \, [P_i p(X\|w_i)] \, dx$	
		m Class Bounds
1) Equivocation or Shannon entropy	$H(\Omega\|X) = E\left\{-\sum_{i=1}^{m} P_r(w_i\|x) \log P_r(w_i\|x)\right\}$	$\frac{1}{2}[1 - B(\Omega\|X)]$ $\leq [1 - \sqrt{B(\Omega\|X)}] \leq \frac{m-1}{m}\left[1 - \sqrt{\frac{mB(\Omega\|X) - 1}{m-1}}\right]$
2) Average conditional quadratic entropy [Vajda (1970)]	$h(\Omega\|X) = E\left\{\sum_{i=1}^{m} P_r(w_i\|x)[1 - P_r(w_i\|x)]\right\}$	$\leq P_e \leq [1 - B(\Omega\|X)]$
3) Bayesian distance [Devijver (1974)]	$B(\Omega\|X) = E\left\{\sum_{i=1}^{m} [P_r(w_i\|x)]^2\right\}$	$= h(\Omega\|X) = R_{NN} = \frac{m-1}{m} - M_0(\Omega\|X);$
4) Minkowski measures of nonuniformity [Toussaint (1973)]	$M_k(\Omega\|X) = E\left\{\sum_{i=1}^{m} \left\|P_r(w_i\|x) - \frac{1}{m}\right\|^{2(k+1)/2k+1}\right\}$	$P_e \leq \cdots \leq R_{KNN} \leq \cdots \leq R_{2NN} \leq R_{NN}$ [see Cover and Hart (1967) and Devijver (1974)]
		Two Class Bounds
5) Bhattacharyya bound [see Kailath (1967)]	$b(\Omega\|X) = E\{[P_r(w_1\|x) \cdot P_r(w_2\|x)]^{1/2}\}$	multicategory error: $P_e \leq \sum_{i=1}^{m} \sum_{j=i+1}^{m} P_e(w_i,w_j);$
6) Chernoff bound [see Kailath (1967)]	$C(\Omega\|X; s) = E\{[P_r(w_1\|x)^{1-s} \cdot P_r(w_2\|x)^s]\}$	$\frac{1}{2}\{1 - [J_\alpha(\Omega\|X)]^{1/\alpha}\} \leq P_e \leq \frac{1}{2}\{1 - J_\alpha(\Omega\|X)\},$ for $\alpha \geq 1;$
7) Kolmogorov variational distance [see Kailath (1967)]	$K(\Omega\|X) = \frac{1}{2}E\{\|P_r(w_1\|x) - P_r(w_2\|x)\|\}$	upper bound equals $[1 - B(\Omega\|X)]$, when $\alpha = 2;$ $Q_{n+1} \leq Q_n; \; Q_0 = 1 - B(\Omega\|X);$
8) Generalized Kolmogorov distance [Devijver (1974), Lissack and Fu (1973)]	$J_\alpha(\Omega\|X) = E\{\|P_r(w_1\|x) - P_r(w_2\|x)\|^\alpha\}, \quad 0 < \alpha < \infty$	
9) A family of approximating functions [Ito (1972)]	$Q_n(\Omega\|X) = \frac{1}{2} - \frac{1}{2}[E\{[P_r(w_1\|x) - P_r(w_2\|x)]^{2(n+1)/2n+1}\}]$	
10) The Matusita distance [see Kailath (1967)]	$\gamma = \left[\int_{S_x} \{p(x\|w_1) - p(x\|w_2)\}^2 \, dx\right]^{1/2}$	γ gives the same bound as $b(\Omega\|X)$; two-class bound relations: $P_e \leq Q_n(\Omega\|X) \leq Q_0(\Omega\|X) \leq \frac{1}{2}H(\Omega\|X) \leq b(\Omega\|X)$ [see Ito (1972) and Hellman and Raviv (1970)]

Notation: $\Omega = (w_i, i = 1,2 \cdots m; 2 \leq m < \infty)$—a set of pattern classes; P_i is an *a priori* probability of class w_i; X is a n dimensional vector random variable; S_x is a sample space of X; $p(X|w_i)$ is a conditional probability density function; $P_r(w_i|X)$ is a posterior probability of class w_i conditioned on X; $f(X) = \sum_{i=1}^{m} P_i p(x|w_i)$—the mixture distribution; E is an expectation over S_x with respect to $f(X)$; R_{NN} is an m class infinite sample nearest-neighbor risk; R_{KNN} is a k nearest-neighbor risk.

(1967)] on which experiments continue to be reported. Only for a few distributions is it possible to obtain analytical expressions for the distance measures of Table I and use them in feature selection. It is also generally necessary to estimate the distributions.

It is an annoying fact that the set of K individually best discriminating features is not necessarily the best discriminating feature set of size K, even for the case of (conditionally) independent features [Elashoff *et al.* (1967), Toussaint (1971), and Cover (1974)]. Unfortunately, the only way to ensure that the best subset of K features from a set of N is chosen is to explore all $\binom{N}{K}$ possible combinations. Since this is usually infeasible, various suboptimal search procedures are used.

A search procedure which has been used much in the past is the "forward sequential" selection procedure in which the best individual features are chosen on the first round, and then the best pair including the best individual feature are chosen, etc. An example of the use of this selection procedure is the experimental comparison of seven evaluation techniques in Mucciardi and Gose (1971). A much used counterpart to forward selection is the sequential rejection procedure in which one finds the best set of $(n - 1)$ features by discarding the worst one, then the best set of $(n - 2)$ among the preceding $(n - 1)$ selected features is chosen, etc. The dynamic programming formulations for feature selection presented in Fu (1968), Nelson and Levy (1968), and Chang (1973) translate problems of feature selection into the notation of dynamic programming.

Other systematic approaches to feature subset selection, which are likely to receive attention in the near future, are suggested by the possibility of posing many problems in pattern classification as graph searching problems. Branch and bound algorithms [Lawler and Wood (1966)] and heuristic search algorithms [Hart, Nilsson and Raphael (1968) and Nilsson (1972)] can be applied not only to clustering [Koontz *et al.* (1974)] but also to reducing the search involved in feature subset selection. Simple heuristic search procedures have been used with automatic feature generation procedures in Becker (1968) and Simon *et al.* (1972). The usefulness of the result, whether in feature generation or feature reduction, is, of course, dependent on the appropriateness of the evaluation function used in the search procedure.

Further Comments on Statistical Feature Extraction

The preceding approaches to feature extraction and evaluation start with the patterns as points in a multidimensional measurement space that has somehow been defined. The statistical procedures then act as if relationships such as joint probability distributions, interpoint distances, and scatter matrices were the only relationships that mattered in defining patterns and their class memberships. All the optimization, with respect to various criteria, glosses over the fact that the initial representation space (and the "semantic coordinate space") has nothing optimal about it but was arrived at arbitrarily by some accepted convention, or by a combination of intuition, problem knowledge, etc. There is no guarantee that with the representations chosen in a given situation the minimum achievable error will be acceptably low.

The initial representation space and the features selected must be iteratively refined in terms of one another and the classifier as described in Section IV; the proper role of the feature extraction, evaluation, and selection procedures described in this section is that of intermediate tools or subroutines in such a recursive interactive design procedure.

VI. Dimensionality, Sample Size, and Error Estimation

For feature selection and classifier assessment, estimates of the Bayes error probability are of interest, as are estimates of the probability of misclassification of any "suboptimal" classifier that is used. Very often, little is known about the underlying probability distributions, and performance must be estimated using whatever samples are available. In this context various questions arise concerning the relationships between the number of features, the limited size of the sample, the design of the classifier, and the estimation of its performance.

The questions and the answers available to them in 1968 were discussed in Kanal and Chandrasekaran (1968); see also Duda and Hart (1973). Here we summarize some recent results concerning:

1) quantitative estimation of the bias in the error estimate based on the design sample set;
2) whether statistical independence of measurements allows performance to be improved by using additional measurements;
3) how to best use a fixed size sample in designing and testing a classifier;
4) comparison of error estimation procedures based on counting misclassified samples with nonparametric estimation of the Bayes error probability using density estimation techniques;
5) use of unclassified test samples in error estimation.

"Testing on the training set" and "resubstitution" are names for the approach in which the entire set of available samples is first used to design the classifier, and future performance is predicted to be that obtained on the design set. The well-known optimistic bias of this approach was confirmed by various theoretical and experimental demonstrations [Hills (1966) and Lachenbrach and Mickey (1968)]. A classical alternative is the sample-partitioning or "hold-out" method, whereby some samples are used to design the classifier and the remaining to test it. Usually half the samples are held out. An attempt [Highleyman (1962)] at analytically determining the optimal partitioning in order to minimize the variance of the estimated error rate has been shown [Kanal and Chandrasekaran (1968)] to rest on shaky assumptions. Based on experimental comparisons reported in Lachenbruch and Mickey (1968) and elsewhere, the conclusion at the end of 1968 seemed to be that one should use the "leave-one-out" method. In this method, given N samples, a classifier is designed on $N - 1$ samples, tested on the remaining sample, and then the results of all such partitions of size $N - 1$ for the design set and one for the test set are averaged. Except in some special cases, this method takes N times the computation of the hold-out method.

In Glick (1972) it is shown that the resubstitution method is consistent for general nonparametric density estimation schemes, and Wagner (1973) proved that the leave-one-out method is consistent in the nonparametric case under certain mild conditions. As pointed out in Foley (1972), even if a sample partitioning scheme is used during the experimental phase of designing a classifier, the entire set of samples is likely to be used for the final design. Thus one would like to know the conditions under which the estimate using resubstitution is a good predictor of future performance, and the relationship between that and the optimal probability of error achievable by a Bayes classifier.

For two multivariate normal distributions with equal known covariance matrices and estimated mean vectors, Foley (1972) derived the amount of bias of the resubstitution estimate as a function of N/L, the ratio of the number of samples per class to the number of features. The practical qualitative recommendation that emerges from the analysis and simulations is that if N/L is greater than three, then (for the case considered) the expected error rate, using the resubstitution method, is reasonably close to one with an independent test set. An approximate upper bound of $1/8N$ for the variance of the design set error rate suggests that even if just a few features are used, there must be enough samples per class to get a good low-variance estimate of the error rate. Thus, for $N = 50$, regardless of the value of the expectation of the design set error rate, the variance is bounded above by 0.0025. In addition to this, the analysis in Foley (1972) reinforces a well-known result, viz., that by adding more and more features one can keep on decreasing the error rate on the design set and yet have the additional features provide no additional discrimination ability on independent test samples.

As mentioned in Kanal and Chandrasekaran (1968), the less that is known about the underlying probability structure, the larger is the ratio of sample size to dimensionality that is needed. This is borne out by the analyses and results in Mehrotra (1973), which extended the investigation of the N/L ratio in Foley (1972) to the case where the common covariance matrix of two multivariate normal distributions is no longer assumed known but has to be estimated from samples. The nature of the results is similar to those in Foley (1972) but now, even for a N/L ratio as large as five, the expected probability of error on the design set is shown to be considerably optimistically biased. The results of Foley and Mehrotra are based on certain expansions, approximations, and simulations and are meant to provide insight and rules of thumb for practice. They lead to the conclusion that the larger the ratio of training sample size to feature set dimensionality, the better is the error estimate obtained from the training set. Furthermore, a sufficiently large number of samples per class is required in order to have a low-variance error estimate.

What about the number of features? That is, for a given finite design sample size N, is there an optimal measurement complexity? Experimentally it has often been observed, given finite training sets, that as the number of measurements is increased, the performance of the classifier first improves, later reaches a peak, and finally falls off. The analyses in Hughes (1965) and Chandrasekaran and Harley (1969) convincingly demonstrate that, in general, there does exist an optimal measurement complexity at which the mean classification accuracy peaks and that the value of the optimal measurement complexity increases with increasing sample size. Later, the effect of constraining the measurements to be statistically independent was examined in Chandrasekaran (1971) and Chandrasekaran and Jain (1973). In the first paper, Chandrasekaran studied the optimal Bayesian decision function for the case of independent binary variables, with class conditional probabilities $\{p_i,(1 - p_i)\}$ and $\{q_i,(1 - q_i)\}$ under class one and class two, unknown and to be estimated from finite samples from the two classes, using uniform *a priori* distributions for p_i and q_i. His conclusion was that in this case the mean probability of correct classification monotonically increases with N, the number of measurements, giving perfect classification as $N \to \infty$. The resulting conjecture that independence of measurements guarantees an optimal measurement complexity of infinity was proved invalid in Chandrasekaran and Jain (1973). This second paper presents necessary and sufficient conditions to test whether or not the number of measurements in the statistically independent case should be arbitrarily increased. From this work and from the feature selection example in Elashoff *et al.* (1967), one learns that statistically independent variables can behave more strangely than one might suspect.

The qualitative practical conclusions to be drawn from the aforementioned investigations on dimensionality, sample size, and expected performance seem to be the following. Depending on the probability structure, our degree of knowledge about it, and the estimation procedure used: a) there exists a lower limit on the number N of design samples per class needed to achieve a low enough variance for the error estimate; b) the ratio of N to the dimensionality L must be "large enough" if we are to get a good estimate of the average probability of misclassification; c) for the given sample size there is an optimal value for L, i.e., an optimal measurement complexity consistent with N/L that satisfies b). These conclusions do not, of course, hold for the case of completely known statistics but the latter would be a fortunate situation enabling the use of simple statistical methods. It is apparent that only the surface has been scratched thus far, and the phenomena of dimensionality, sample size, and optimal measurement complexity need to be quantitatively investigated in a variety of contexts not hitherto examined.

Next consider how best to use a given fixed sample of size N in designing and testing a classifier. Toussaint (1974) gives an extensive bibliography on this and related topics in the estimation of misclassification. With the "rotation" or Π method recommended there is a compromise between the hold-out (H) method and the leave-one-out (U) method. It consists of partitioning the total set of N class tagged samples into a test set $\{X\}_i^{Ts} = \{X_1,\cdots,X_k\}$, where $1 \leq k \leq N/2$, N/k an integer, and a training set $\{X\}_i^{Tr} = \{X_{k+1},\cdots,X_N\}$; and then training the classifier on $\{X\}_i^{Tr}$ and testing it on $\{X\}_i^{Ts}$ to get an error estimate denoted by $\hat{P}_e[\Pi]_i$. The procedure is repeated with additional disjoint test sets $\{X\}_i^{Ts}$, $i = 2,\cdots,N/k$ and corresponding training sets, and the average over the various disjoint test sets results used for the expected error, i.e., $E[\hat{P}_e(\Pi)] = k/N \sum_{i=1}^{N/k} \hat{P}_e[\Pi]_i$. With $k = 1$ this is the leave-one-out method, and with $k = N/2$ this gives a version of the hold-out method well known in statistics as cross validation in both directions [Mosteller and Tukey (1968)]. The rotation method is also related to the "jackknifing" procedures described in Mosteller (1971).

The average resubstitution error rate $E\{\hat{P}_e(R)\}$ provides a lower bound on the true error probability while the other approaches yield upper bounds. In the graphs in Foley (1972) one finds that an average of the design set and test set results gives a good estimate of the true error probability. This leads Toussaint (1974) to recommend the estimate

$$\hat{P}_e{}^* = \alpha E\{\hat{P}_e(\Pi)\} + (1 - \alpha)\hat{P}_e(R)$$

where $0 \leq \alpha \leq 1$ is a constant depending on the sample size N, the feature size L, and the test set size k. In Toussaint and Sharpe (1973) it is reported that experimental work with $\alpha = 1/2$, $k/N = 1/10$, and $N = 300$, led to $\hat{P}_e{}^*$ essentially equal to $\hat{P}_e(U)$. To compute the leave-one-out estimate $\hat{P}_e(U)$ would take 300 training sessions, while to compute $\hat{P}_e{}^*$ takes only 11 training sessions, one for $\hat{P}_e(R)$ and ten for $E\{\hat{P}_e(\Pi)\}$.

Estimation of the Bayes error probability using classified, i.e., class-tagged design samples but unlabeled test samples has been investigated in a number of papers. These investigations use a result of Chow (1970) that, for optimal classification involving a reject option, a surprisingly simple fundamental relation exists between the error and reject rates. In a Bayes strategy, the conditional probability of error is

$$r(X) = 1 - \max_i \frac{P_i p_i(X)}{f(X)}$$

where $f(x)$ is the mixture density $\sum_i P_i p_i(x)$. With rejection allowed, the optimum strategy is to reject whenever $r(X) > t$, where t is the rejection threshold, and decide as before, otherwise. The reject rate $R(t) = \Pr[r(X) > t] = 1 - G(t)$, where $G(t)$ is the cumulative distribution function (cdf) of $r(X)$. The error rate is then given by

$$E(t) = \int_0^t y\, dG(y) = -\int_0^t y\, dR(y).$$

A plot of this relationship gives an error-reject tradeoff curve the slope of which at a given point is the rejection threshold. Chow (1970) noted that this simple integral relation allows the error rate and tradeoff curve to be determined from the empirically observed reject rate function $R(t)$ on unlabeled samples; it can also be used for model validation by comparing the empirical error-reject tradeoff curve with the theoretical one derived from the assumed P_i and $p_i(X)$.

This latter idea was applied in Fukunaga and Kessel (1972), which pointed out that the suggestion was equivalent to a goodness-of-fit test for the distributions $G(t)$ or $R(t)$. One of the methods examined is a test based on the expectation of the conditional probability of error $r(X)$; $E\{r(X)\}$ is just the Bayes error probability P_e, without the reject option. For the M-class case, the estimate

$$E = \frac{1}{N_t} \sum_{i=1}^{N_t} r(X_i)$$

based on N_t independent unlabeled samples from the mixture density $f(X)$, has a variance at least P_e/M less than the variance $P_e(1 - P_e)$ of the estimate based on counting misclassified labeled test samples. This paradoxical behavior, whereby one gets a better estimate by ignoring the class tags on test samples, is attributed to the fact that the error count estimate gives a binary quantization of the error on a test sample, while $r(X_i)$ assigns a real value.

The application in Fukunaga and Kessel (1972) of optimum error-reject rules to two-class multivariate normal problems, for equal and also unequal covariance matrices, provides some interesting comparisons with the work of Foley (1972) and Mehrotra (1973) described earlier, and the remarks made in Kanal and Chandrasekaran (1968) concerning the role of structure.

For the equal covariance case with sample means and sample covariance estimated from a total of $N_1 + N_2 = N_d$ design samples, the analysis in Fukunaga and Kessel (1972) suggests that N_d/L should be ten or greater in order for mean performance to reasonably approximate the optimum. In terms of number of samples per class, this suggestion is consistent with Mehrotra (1973), although the latter's result was obtained by a different approach not involving the reject option.

For the unequal covariance case with unequal as well as equal mean vectors, simulation experiments in Fukunaga and Kessel (1972) showed that the results still depended on the ratio of the number of samples per class to the feature dimensionality but that an even larger number of design samples is needed. Also, with estimated parameters the true error rate is greatly underestimated by the error rate calculated from the empirical reject function. The article concluded that "using the empirical reject rate to predict error rates can produce very inaccurate results if the model used in the classifier design is inaccurate."

As noted earlier, the asymptotic error rate of the nearest-neighbor classification rule provides a bound that is as close or closer to the Bayes error probability than any of the other bounds. Cover (1969) proposed that the number of misclassified samples when using a nearest-neighbor classifier be considered as an estimated bound for the Bayes error probability. As the total number of samples asymptotically increases, for increasing k, the k-NN rules do provide increasingly better asymptotic bounds on the Bayes error probability. Cover's suggestion was followed up in Fralick and Scott (1971) and Fukunaga and Kessel (1973), where nonparametric estimation of the Bayes error probability was investigated via a) error rates resulting when k nearest-neighbor classification was used, and b) error rates of approximate Bayes decision rules based on estimated density functions obtained by using multivariate extensions [Murthy (1965)] of Parzen estimators [Parzen (1962)].

Fukunaga and Kessel (1973) used labeled design samples and unlabeled test samples. For a test sample X_i from the test set of N_t unlabeled samples, consider its k nearest neighbors among the design set N_d. Of these k neighbors, let k_1 be from class w_1 and k_2 from class w_2, and let $r_k(X) = \min\{k_1/k, k_2/k\}$. Then the sample mean $\hat{E}_k = 1/N_t \sum_{i=1}^{N_t} r_k(X_i)$ has an expectation that is a lower bound on the Bayes error. An upper bound is obtained from an unbiased estimate of the conditional k nearest-neighbor error. For N_d very large, the average of the lower bound $r_k(X)$ and the upper bound, over the unlabeled samples, gives a good experimental estimate for the Bayes error. The use of unlabeled test samples results in a lower variance for this estimate than an error estimate based on labeled test samples.

The results in Fukunaga and Kessel (1973) and previous results in Fralick and Scott (1971) suggest that for a small number of design samples the approach using Parzen estimates performs better than the k nearest-neighbor procedures. Further comments are made in Section VII.

When designing a pattern classification device, it is expected that a large labeled design set will have to be gathered. These results suggest a way of estimating the minimum probability of error that is achievable with a given set of features, without having to also label a large set of test samples. The labeling of test samples is not only expensive but can often be an additional source of error, as has been found in some medical applications.

The investigations into dimensionality, sample size, and error estimation described in this section represent perhaps the most useful research in statistical pattern recognition during the period 1968–1974. Although incomplete, they do provide rules of thumb and guidance for designing pattern classification systems and analyzing their experimental performance.

VII. Statistical Classification

The basic assumption underlying statistical classification is that there exists a multivariate probability distribution for each class. Members of a pattern class are then treated as samples from a population, which are distributed in a n-dimensional feature space according to the distribution associated with that population. For two classes an observation x on the vector random variable X representing the features is treated as coming from one of two distributions F_1 or F_2.

This theoretical framework leads to subcategories ranging from complete statistical knowledge of the distributions to no knowledge except that which can be inferred from samples. The subcategories are

a) known distributions;
b) parametric families of distributions for which the functional forms are known, but some finite set of parameters need to be estimated;
c) the nonparametric case in which the distributions are not known.

Under b) and c) there are the possibilities that either some sample patterns of known classification are available, or unlabeled samples are available.

The subcategories a), b), and c) were discussed in Fix and Hodges (1951) and by various other authors in statistics. In Ho and Agrawala (1968) the basic categorization scheme was enlarged to include the additional aspect of unlabeled samples. The paper surveyed work on statistical classification algorithms presented in the engineering literature on pattern recognition through early 1968. Some topics under categories a) and b) that were considered in some detail are sequential and nonsequential statistical decision theoretic algorithms, recursive Bayesian procedures for "learning with a teacher" when labeled samples are available, and the Bayesian formulation of "learning without a teacher" when unlabeled samples are available. Under category c) the paper described: algorithms for learning the coefficients of linear decision functions based on iterative deterministic optimization procedures for solving linear inequalities under some criterion function; extensions of these procedures to deal with nonlinear inequalities or piecewise linear inequalities; algorithms based on stochastic approximation methods to find the coefficients of orthonormal series representations for the difference between the unknown *a posteriori* probability distributions for each class; and some clustering algorithms for unlabeled samples. Also mentioned was the result in Cover and Hart (1967) that for an infinite sample size, using the nearest-neighbor rule for classifying a sample leads to an error rate that is never worse than twice the Bayes error probability.

In the period since 1968, papers on classification under subcategories b) and c) for labeled and unlabeled samples have continued to appear; a survey of statistical classification similar to that in Ho and Agrawala (1968) could now easily be the sole topic of a very long journal article. However, the majority of recently published books in pattern recognition devote almost all their attention to statistical classification, estimation, and clustering procedures, and some of them, Duda and Hart (1973) in particular, provide very good surveys of the literature on these topics through early 1972. Thus I have limited the scope of this section to 1) some recent references and surveys for statistical classification procedures that derive from approaches covered in earlier surveys, and 2) brief descriptions and comments on some recent contributions. Under 2) I focus on topics in nonparametric classification. In recent years, this is the category of classification procedures that has been of greatest interest for work in pattern recognition.

Some Recent References

Prior to 1968 algorithms for the optimal solution of linear inequalities were often proposed in the pattern recognition literature. Papers on this topic continue to appear regularly. A recent example is Warmack and Gonzalez (1973) that claims to have the first direct algorithm, not based on gradient optimization techniques or linear programming, for the optimal solution of consistent and inconsistent strict linear inequalities. An accelerated relaxation-based procedure for finding piecewise linear discriminant functions is described in Chang (1973).

Many papers on decision-directed learning and on various other unsupervised learning schemes such as learning with a "probabilistic teacher" and learning with an "imperfect teacher" have appeared since 1968. In Agrawala (1973) schemes for learning with various types of teachers are reviewed, and simple block diagrams are presented to reveal their interrelationships.

For learning with various types of teachers and for many other problems in statistical pattern classification, e.g., automatic threshold adjustment, taking context into account, intersymbol interference, and distribution-free learning, at least conceptually, compound decision theory provides an integrated theoretical framework. A brief tutorial exposition of compound decision theory procedures appears in Kanal and Chandrasekaran (1969); see also the comments in Cover (1969) and the extended presentation in Abend (1968). In pattern recognition, examples of recent papers based on compound decision theory approaches are Welch and Salter (1971) and Hussain (1974). The optimal processing algorithms based on these approaches are generally unwieldy, and many approximations must be invoked.

Complementing the surveys of clustering presented in the pattern recognition literature is an excellent survey [Har-

TABLE II
NONPARAMETRIC PROBABILITY DENSITY FUNCTION ESTIMATORS [SEE COVER (1972)]

No.	Estimator	Formulation $f(X)$ density function; $X_1 \cdots X_n \cdots$ (independently identically distributed random variable)	Comments
1	Histogram	partition real line into sets $S_1, S_2, \cdots$, let $g_i(x)$ be indicator function for S_i, then $\hat{f}_n(x) = \sum_i^k g_i(x) \left\{ \frac{1}{n} \sum_{j=1}^n g_i(x_j) \right\}$	variance $\to 0$ as $\frac{1}{n}$, unbiased 1) selection of S_i's and their number is arbitrary 2) results in piecewise constant $\hat{f}_n(x)$
2	Orthogonal function	$\hat{f}_n(x) = \sum_{i=1}^k \hat{C}_i^{(n)} \psi_i(x)$ where $\hat{C}_i^{(n)} = \frac{1}{n} \sum_{j=1}^n \psi_i(x_j)$ which minimizes $J_n = \int (f(x) - \hat{f}_n(x))^2 \, dx$	1) reduces to histogram approach if $\psi_i = g_i$ 2) possibility of negative values for $\hat{f}_n(x)$ exists 3) scale of ψ_i must be selected before the data is observed [see also Crain (1973)]
3	Rosenblatt estimator	$\hat{f}_n(x) = \frac{[F_n(x + h) - F_n(x - h)]}{2h}$ where $F_n(x) = \frac{1}{n} \sum_{j=1}^n g'(x - x_j)$ $g'(x) = \begin{cases} 1, & x \geq 0 \\ 0, & x < 0 \end{cases}$	$E\hat{f}_n(x) = f(x) + \frac{h^2}{6} f''(x) + o(h^4)$ $E(\hat{f}_n(x) - f(x))^2 = \frac{f(x)}{2hn} + \frac{h^4}{36} \lvert f''(x) \rvert^2 + o\left(\frac{1}{nh} + h^4\right)$
4	Parzen estimator	$\hat{f}_n(x) = \frac{1}{h(n)} \sum_{i=1}^n K\left(\frac{x - x_i}{h(n)}\right)$ K is bounded, absolutely integrable Kernel function $\lvert xK(x) \rvert \to 0$ as $\lvert x \rvert \to \infty$ $\int K(y) \, dy = 1$	unbiased, var $(\hat{f}_n(x)) \leq \frac{1}{n} E\left(\frac{1}{h(n)} K\left(\frac{x - x_i}{hn}\right)\right)^2$ rate of convergence $\approx \frac{1}{n}$ (depends on continuity of f)
5	Loftsgaarden and Quesenberry	$\hat{f}_n(x) = \frac{k_n/n}{2d(n)}$ where k_n is an integer n, $d(n)$ is the distance to the k_nth closest sample point from x	if $k_n \to \infty$, $\frac{k_n}{n} \to 0$ $\Rightarrow$ consistent estimate

tigan (1974)], which provides many interesting recent references not cited in pattern recognition books and articles. The author aptly describes the present status of clustering theory as chaotic and says

> the probabilistic and statistical aspects of clustering are still immature, the principal body of knowledge being clustering algorithms which generate standard clustering structures such as trees or partitions from standard forms of input data such as a distance matrix or data matrix.

Dissatisfaction with heuristic approaches has led to some theoretical analyses of clustering. A recent reference is Wright (1973), who attempts an axiomatic formalization of clustering.

Some recent papers have considered the comparative evaluation of alternative discrimination procedures, a topic that is of direct interest to pattern recognition practice. In Moore (1973), five discriminant functions for binary variables are evaluated. These are the first- and second-order Bahadur approximations, linear and quadratic discriminant functions, and a full multinomial procedure based on estimating the class conditional probability distributions. Among the conclusions drawn is that for binary variables, the quadratic discriminant function rarely performs as well as the linear discriminant function. This confirms the experience reported by many persons working in pattern recognition [Kanal (1972)]. A comparative study of linear and quadratic discriminant functions for independent variables from three nonnormal continuous distributions is reported in Lachenbruck *et al.* (1973). Other comparative studies of some classification procedures are Gessaman and Gessaman (1972) and Odell and Duran (1974).

Many of the comparisons in these studies leave one less than satisfied as to the generality or objectivity of the conclusion. An objective approach suggested by decision theory is to develop admissibility criteria that would eliminate obviously bad algorithms. Admissibility of k-NN algorithms for classification is discussed in Cover and Hart (1967). In a recent study [Fisher and Van Ness (1973)], seven seemingly reasonable admissibility conditions were used in an attempt to compare eight classification procedures including k-NN rules, linear discriminant analysis, quadratic discriminant analysis, and Bayes procedures. Unfortunately, the approach did not provide much comparative information about the alternate algorithms.

Nonparametric Classification Procedures

Nonparametric approaches to classification include:

1) linear, nonlinear, and piecewise linear discriminant functions;
2) stochastic approximation and potential function methods for approximating the decision boundary;
3) clustering procedures;
4) density estimation methods for use in an optimal decision rule;
5) nearest-neighbor classification rules;
6) statistically equivalent blocks;
7) discrete variable methods when there is no inherent metric.

The basic concepts underlying the first five approaches are clearly presented in Duda and Hart (1973) and Fukunaga (1972). These books also briefly deal with discrete variables and present some series approximations for the joint probability function of binary variables. Extensive development and discussion of research contributions to the first five topics, up to 1972, are presented in Patrick (1972), which is the only pattern recognition book with any material on the topic of statistically equivalent blocks. In the following, I briefly describe some recent contributions to topics 4) to 6) that have not been previously surveyed and appear to merit comment.

Five commonly used nonparametric probability density function estimators are examined and compared in Cover (1972), which gives 87 references on nonparametric density estimation. Table II presents a summary of the five estimators. The application of B-splines to multivariate pdf estimation using Parzen estimators is the subject of a recent dissertation [Bennett (1974)]. B-splines are local rather than global approximating functions, so that each point in a set of data points being approximated with the B-spline basis functions has influence on a fixed fraction of the density estimate. For estimating a L dimensional pdf from n random L-vectors of data, Bennett presents an algorithm that uses a L dimensional density kernel estimator with a L-fold tensor product of B-splines as basis functions.

The k nearest-neighbor class of estimators of Loftsgaarden and Quesenberry derive from a method of nonparametric density estimation suggested by Fix and Hodges (1951). In this approach the volume of the region containing the k nearest neighbors of a point is used to estimate the density at that point. Thus the number of observations is fixed and the volume is random. This contrasts with the Parzen estimator approach in which the volume is fixed and the number of data points is random. This symmetry is suggestive and Fralick and Scott (1971) pointed out that the need to choose the kernel K and window (weighting) functions h in the Parzen estimator has its counterpart in the need to choose the number of nearest neighbors and the metric in the k nearest-neighbor approach. Using techniques similar to earlier work on the derivation of the optimum kernel function for Parzen estimators, Fukunaga and Hostetler (1973) obtained a functional form for the optimum k in terms of sample size, dimensionality, and the underlying probability distribution. The optimality is in the sense of minimizing an approximated mean-square error or integrated mean-square error criterion.

A number of papers, many of them published in this TRANSACTIONS, have been concerned with the asymptotic convergence of k-NN rules and certain variations thereof, [Cover (1968), Peterson (1970), Wilson (1972), Wagner (1971), (1973), and Wolverton (1973)]. Of course, the small sample behavior of any nonparametric decision rule is problematical. Cover (1969) has conjectured that

> The failure of the NN rule score to be near its limit is a good indication that every other decision rule based on the n samples will also be doomed to poor behavior. A small sample with respect to the NN rule is probably a smaller sample with respect to more complicated data processing rules.

The experiments of Fralick and Scott (1971) and Fukunaga and Hostetler (1973) would not seem to support that conjecture, as they appear to favor Parzen estimates. However, as Cover has pointed out, Parzen estimates involve a smoothing parameter that the experimenter can adjust after looking at the data. My own experimental comparisons, done in 1964, of nearest-neighbor rules with other competing classification procedures for a specific problem did not favor nearest-neighbor rules. However, these are isolated experiments, and theoretical analysis and systematic experimentation are needed to answer questions about the small sample performance of NN rules and, indeed, all competing classification procedures.

Other than the early work of Fix and Hodges (1952) for univariate and bivariate Gaussian distributions, the only published studies of the small sample performance of the NN rule seem to be Cover and Hart (1967) and the recent paper by Levine, Lustick, and Saltzberg (1973), for the case of samples from two uniform univariate distributions. Not surprisingly, for this case it is shown that the probability of misclassification is close to its asymptotic value even for extremely small samples. An unpublished result by W. Rogers and T. Wagner has been communicated to me by one of the reviewers of an early draft of this paper. For nearest-neighbor classifiers they find that with the leave-one-out method the variance in the risk estimate is less than $5/4n + 3/n^{3/2}$ independent of the underlying distribution. A similar result is claimed for any local classifier. This result is a nonparametric finite sample size result that should allow competing classification procedures to be compared using confidence intervals on the risk estimates.

During the period under consideration, a few papers on nonparametric classification using distribution-free tolerance regions have appeared. Unlike most of the pattern recognition literature, these papers take a non-Bayesian Neyman–Pearson approach to error performance and are thus of some interest.

In 1947, J. W. Tukey used the term "statistically equivalent block" for the multivariate analog of the interval between two adjacent order statistics; to extend the concept of order statistics to the multivariate case, it is necessary to introduce ordering functions. Given n observations from a continuous distribution, the sample space is divided by these observations into $n + 1$ blocks. For any block B_i the proportion of the population covered, referred to as the "coverage," is treated as the value of a random variable U_i. Subject to mild restrictions on the procedure used to divide the sample space, the random variables $U_1 \cdots U_{n+1}$ have a joint distribution that is independent of the distribution giving rise to the sample observations. The distribution is the Dirichlet distribution—a uniform distribution over a set prescribed by simple inequalities. This property of the coverages of the $(n + 1)$ blocks leads to the term statistically equivalent blocks.

Since the sum of any group of the U_i has a beta distribution, the marginal distribution of the proportion of the population that lies in a group of the blocks has the beta distribution. Enough blocks can be chosen to make a probability statement such as "in repeated sampling the probability is p that the region R contains at least α of the population." Thus a distribution-free tolerance region whose coverage has the beta distribution can be constructed in the multivariate case by defining ordering functions to generate statistically equivalent sample blocks.

In Quesenberry and Gessaman (1968), an optimal procedure, in the two-class case, is defined to be one that minimizes the probability of reserve judgment, i.e., rejection, while controlling the conditional error probabilities for each class within prescribed upper bounds. The paper presents a nonparametric classification procedure based on forming regions of reserve judgment from intersections of distribution-free tolerance regions. The choice of the ordering functions determines the usefulness of the procedures; for some families of distributions it is possible to select ordering functions that will make the nonparametric procedure consistent with the optimal procedure for the given family.

In the same context as the preceding paper, Anderson and Benning (1970) present a suboptimum nonparametric classification procedure for the two-class problem. In this paper, the set of ordering functions used to form tolerance regions for the first distribution are based on clusters of the sample drawn from the second distribution, and vice versa. Hyperspherical (Euclidean distance) and hyperelliptical ordering functions are suggested to order observations with respect to cluster means. Note that the general theory of distribution-free tolerance regions does not consider the case where the regions corresponding to a distribution depend on randomness from a source different than the observations on that distribution. Anderson and Benning (1970) prove that the theory does hold for this case.

An earlier paper in this area is Henrichon and Fu (1969). A recent paper is Beakley and Tuteur (1972), which presents three ordering procedures to develop nonparametric tolerance regions and uses them in automatic speaker verification. In Gessaman and Gessaman (1972) reserve judgment procedures based on nonparametric tolerance regions are compared with other standard procedures.

For discrete variables satisfying only a nominal scale (i.e., when there is no inherent metric), a switching theory-based approach is presented in Michalski (1972), (1973) and Stoffel (1972), (1974). In the past, various similarity and clustering metrics have been tried for nominal variables [Anderberg (1973), Goodall (1966), and Hills (1967)]. Sammon (1971) suggested procedures for transforming such discrete variables, termed Discrete Type II variables, into continuous features; the OLPARS Discrete Variable Subsystem commented upon in Kanal (1972) provides a number of such transformations. The contribution of Michalski (1972) and Stoffel (1972) is a feature generation and classification procedure that generates a small set of n-tuples for discrete nominal variables. These n-tuples, called "prime events" in Stoffel (1972), are claimed both to fit a specific class and to discriminate it from other classes.

The independent developments by Michalski and by Stoffel of essentially the same concepts and procedures are based on the idea of the "cover" of two events, and they are related to work on the synthesis of switching functions from incompletely specified input–output relations.

Michalski's work on a "covering theory" approach to switching and classification problems predates Stoffel's 1972 report, but his formulation, development, and exposition are imbedded in complex notation. Here I follow Stoffel's terminology.

An event is an n-tuple $(x_1,x_2,\cdots,x_n)$ in which a subset of the elements have specified values, and the unspecified elements are "don't care" variables. Event e_1 "covers" event e_2, if and only if every element of e_1 which has a specified value equals the value of the corresponding element in e_2. Thus event $e_1 = (2,\cdot,\cdot)$, where $\cdot$ denotes an unspecified element, covers event $e_2 = (2,1,0)$ and event $e_3 = (2,2,\cdot)$, but e_2 does not cover e_1 or e_3, and e_3 does not cover e_1 or e_2. A prime event is an event that covers only those measurement vectors assigned to one class by a Bayes classifier. Also a prime event is not covered by another prime event. An algorithm to generate a sufficient set of prime events that will cover the class is given; the resulting set may not be the smallest possible.

To account for vagaries in the sample, Hamming distance is used as a measure of similarity between events or between a measurement vector and an event. Classification is done by assigning a sample to that class the set of prime events of which covers the sample vector. If the distance from all prime events exceeds a threshold, then the sample is rejected.

The procedure is certainly a systematic approach to the generating of a small set of good templates. However, it generates prime events for each class versus the rest of the classes. It is easy to give examples where by grouping classes together and using a tree classification structure one can do as well with fewer prime events.

The last comment brings up the question as to whether the complexity of patterns and pattern representation schemes can be considered independently of the classification structure adopted.

In an essay on the complexity of patterns and pattern recognition systems [Kanal and Harley (1969)], it is argued that complexity of patterns is in the eye of the beholder and that one can consider evaluating the complexity of different patterns with respect to specific beholders, i.e., specific pattern recognition systems or recognition logics. This is the approach taken by Minsky and Papert (1968), who compute the complexity, or what they call the "order" with respect to single layer threshold logic of a number of interesting geometrical properties. Various aspects of the complexity of patterns and pattern recognition systems are also described heuristically in Kanal and Harley (1969). Cover (1973) represents the beginning of an attempt to obtain a measure of the intrinsic complexity of a pattern so that different beholders, i.e., computational systems, will arrive at the same complexity measure up to an additive constant. Until such time as this effort succeeds, it is likely that problem complexity and system complexity will be matched heuristically in the manner described in Harley *et al.* (1968).

Commentary on Statistical Classification

There has been some tendency to question, in the name of practicality and simplicity, the need for further theoretical studies in the area of statistical classification. No engineer will quarrel with the emphasis on practicality and simplicity but as demonstrated by the attempts to compare classification procedures, not having theoretical guidelines makes it difficult to select between competing techniques without extensive experimentation. A suitable criterion and analysis can be used to analytically decide on the best technique among a set of ad hoc methods. What recent attempts at comparing classification procedures show is that, if anything, more analytical studies are needed so that experimental comparisons can be more meaningfully conducted under theoretical guidelines. The study of classification procedures needs to be extended to cover the combination of hierarchical classification structures and discriminant procedures commonly employed in practice. Finally, the proper role of nonparametric procedures is in the early exploratory stage of a pattern recognition problem. There is no substitute for discovering the underlying structure and taking advantage of it. As an example, the use of a nonparametric tolerance region procedure in speaker verification is warranted in the early phase of an investigation but for such problems considerably more structural knowledge can be used, and approaches such as those considered in the next section seem more appropriate.

VIII. STRUCTURAL METHODS

The linguistic approach views patterns as complexes of primitive structural elements, called words or morphs, and relationships among the words are defined using syntactic or morphological rules. The primitive structural parts are perceptually and conceptually higher level objects than scalar measurements. For instance, the gray levels of individual points of a digitized picture would be too low-level to be meaningful units of that picture, nor would the individual amplitude levels of a digitized waveform be meaningful units for structural analysis.

In practice, the structural approach involves a set of interdependent processes: 1) identification and extraction of morphs—this is the segmentation problem; 2) identification of the relationships to be defined among the morphs; 3) recognition of allowable structures in terms of the morphs and the relationships among them. Two-dimensional line drawings, fingerprints, X-ray images, speech utterances, and other such patterns that exhibit strong deterministic structure and for which *a priori* information in the form of some model can be easily used, are natural candidates for the structural approach.

Ad hoc structural processing has a much longer history in pattern recognition practice than statistical methods, which are based on abstract relationships involving joint probability distributions and distances in multidimensional space among sets of scalar measurements. Commercial print readers involve ad hoc structural processing, and biomedical programs are usually of this kind.

Much of the literature on structural pattern recognition has been devoted to formal methods. Fu and Swain (1971) surveyed the literature prior to 1970 and the November 1971 and January 1972 issues of *Pattern Recognition* together constituted a special issue on syntactic pattern recognition. Much of the published work on structural methods for pattern recognition and scene analysis from 1969 to 1973 is mentioned in Rosenfeld (1972), (1973), (1974). A brief survey also appears in Klinger (1973). A book on *Syntactic Methods in Pattern Recognition* has recently been announced [Fu (1974)].

These general surveys allow us, in this section, to focus on a few key concepts and differences in approach that underlie current work on segmentation and structural analysis.

Segmentation and Structural Analysis

Pattern description has been viewed either as two distinct processes—segmentation followed by structural description, or as an integrated process—segmentation-structural description. The first approach often delegates segmentation to preprocessing and concentrates on formal models for structural description which assume that the patterns are already represented as a segmented structure [Evans (1968), Fu and Swain (1971), and Lee and Fu (1972)].

Piecewise functional approximation is one method for preprocessing waveform data [Pavlidis (1973) and Horowitz (1973)]. The data are fit according to an error criterion with line (or polynomial) segments. The output from the preprocessor is a string of triples $\{(x_i,y_i),A_i,B_i\}$, $i = 1,\cdots,S$, where S is the number of segments, $y = A_ix + B_i$ is the linear approximation to the ith data segment, and (x_i,y_i) is the right endpoint of the line segment. This string is translated into the terminal symbols (tokens) of a grammar under the control of parameters appropriate to the application. The structural analysis is accomplished by a left-to-right

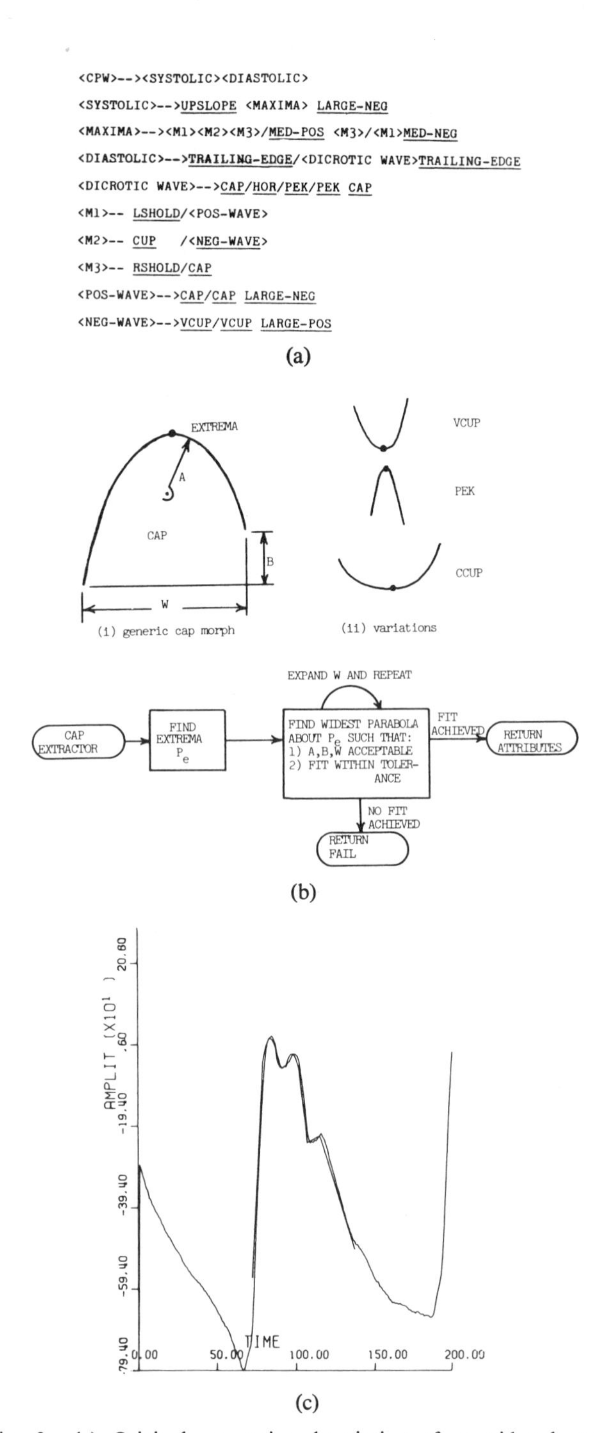

Fig. 3. (a) Original generative description of carotid pulsewaves through B.N.F. (b) Smooth extrema morphs used in pilot pulsewave analysis and rough extraction procedure. (c) One cycle of raw carotid pulsewave data juxtaposed with fitted representation obtained via parsing.

parser for a grammar that defines more complex relations among the terminal symbols, for example, several appropriate line segments could form a peak.

The advantages of this approach are preprocessing speed, generality, and the mathematical tractibility that approximation theory provides. However, the approach has more the flavor of numerical analysis than pattern analysis. This comment also applies to proposals to use truncated K–L series expansions in pattern analysis [McClure (1974) and Lavin (1972)]. Also such preprocessing usually results in arbitrary segmentations and requires excessive time for scanning and matching of all the data. Furthermore, the segments extracted may not be meaningful in the context of a specific application. In separating the analysis of structure from the extraction of morphs, each process is excluded from information available to the other. That is, the extraction of individual morphs must proceed in ignorance of the *a priori* combinatorial restrictions known by the structural component, and the structural component cannot profit from the intermediate work of extraction.

The alternative approach, integrated segmentation-structural description, is exemplified by the work of Stockman *et al.* (1973), (1974) for carotid artery pulse waveforms and Miller (1973) and Reddy *et al.* (1973) for speech recognition. The salient points of this approach are the following.

a) *Knowledge-based segmentation.* There is *a priori* knowledge of the possible segments in the data. For example, in a carotid pulse wave, it is known that an upslope, a trailing edge, a peak in between, and a dicrotic notch are likely to be found; see Fig. 3. Morphs can be defined that are believed to be of functional importance to the particular problem at hand and can vary in complexity from local extrema to exponential segments.

For scene analysis, Tennenbaum (1973), (1974), discusses the desirability of knowledge-based search for distinguishing features in preference to scanning the entire scene with low-level operators. The term "knowledge-based," currently popular in the artificial intelligence (AI) literature, generally refers to "nonstatistical" *a priori* information, although statistical information and Bayes' theorem are now also acceptable in AI [Yakimovsky and Feldman (1973)].

b) *A priori knowledge is represented by means of decision trees* [Narasimhan and Reddy (1971)], *graph models and decision graphs* [Harlow and Eisenbeis (1973)], *and grammars* [Stockman (1973)].

c) *Several levels of structural information are utilized.* For example, for speech recognition the following types of information have been used.

1) Acoustic-phonetic constraints that limit occurrences of given phonemes in segments of speech in the language under study and specify the phonemic content of vocabulary items.
2) Syntactic constraints in the form of a grammatical model that define what word sequences are allowable utterances of the language.
3) Semantic constraints that allow a hypothesized word sequence to be referred to a particular problem domain so that some measure of its reasonableness can be obtained.

d) *Parsing is bottom-up and top-down and non-left-to-right.* In other words, the more prominent morphs are sought first regardless of their location, and then the grammar is used to predict where other morphs are to be found. In the case of pulse waves, the most prominent morph might be the upslope. Having extracted the upslope, a grammar could be used to predict that the next morph to be scanned is the trailing edge. Having extracted the trailing edge, the next morphs to be scanned might be peaks and a dicrotic notch between the upslope and the trailing edge. In this fashion, the extracted morphs clue the system to the rest of the structures to be searched and also allow future search to be performed over restricted intervals.

In Miller (1973), parse trees are "seeded" by first scanning the entire input utterance for prominent vocabulary items (usually long words). Words in the local context of those found are then sought and partial parse trees (PPT) are assembled anywhere in the input. The PPT represents the grammatical structure of this part of the utterance, while the terminals of the tree are words in the vocabulary believed to exist in the input. PPT's are enlarged by using the grammar to guide the search for hypothesized words in the neighborhood of PPT's and to connect several PPT's into one. Analysis can terminate any time the complete utterance is "covered" by some PPT. The bottom-up non-left-right approach is attractive because of the ability to search first for reliable morphs regardless of location and to guide future analysis accordingly.

There is an analogous technique to this type of parsing, when decision trees rather than grammars are used to represent the possible structures. In the X-ray analysis in Harlow and Eisenbeis (1973), one searches for each lung lobe with respect to those lobes already found before it in the image. If processing leads to an impossible structural outcome, the analysis is backed up and extraction of morphs is taken up again. In Narisimhan and Reddy (1971), isolation of a given morph narrows the structural possibilities, while the present state of structural possibilities dictates the next morph extraction to be attempted.

The parsing approach can also be top-down and left-to-right [Walker (1974), Erman (1974)]. By going top-down only syntactically and semantically acceptable configurations are considered, and very expensive preprocessing can be minimized. By going left-to-right, classical parsing methods can be used and the state of analysis is easily recorded.

One can define a probability (or fit error) that a given morph matches a given segment of raw data. This probability can also be backed up a PPT to yield a figure of merit for a partial parse. The probability of a PPT can be changed by "semantic conditioning" [Erman (1974)]. By maintaining this figure of merit for all partial analyses only the most promising ones need to be extended, or in the case of complete analyses only the most likely cases need to be accepted.

Pattern Grammars

Much of the literature on structural pattern recognition deals with formal string grammars and their multidimensional generalizations. Recall that one of the reasons for introducing linguistic methods was the limited relationships handled in statistical pattern classification. However, phrase-structure string grammars are also severely limited in the relationships they model. Basically, they deal only with concatenation of primitives and immediate constituent structure.

According to a 1971 article [Uhr (1971)], past work on formal linguistic methods for pattern recognition has been either purely theoretical, or merely an incidental portion of some pattern recognition system, or incomplete in the sense of never actually resulting in a running program. As far as applications of purely formal syntactic methods are concerned, the situation remains essentially the same in 1974 as it was in 1971.

An incomplete effort of some interest (because it deals with a real problem) is Moayer and Fu (1973), which describes how a syntactic approach to fingerprint classification might be based on a context-free grammar model. This is one of the few formal grammar-based efforts aimed at obtaining a complete analysis from primitive extraction to classification. It contains: a) a careful study of the data environment yielding the choice of primitive syntactic elements; b) the hierarchical syntactic analysis that permits the one-dimensional concept of concatenation to readily apply to the two-dimensional representations used; and c) the "sequential recognition algorithm" that amalgamates syntactic recognition, primitive feature extraction, and sequential decision-making for computational and logical efficiency. However, no conclusive results have been reported yet on the performance of the technique.

Theoretical research in syntactic pattern recognition has been extensive during the past few years. Various generalizations and new formalisms have been proposed to overcome some of the limitations of string grammars. For example, stochastic finite state and context-free grammars obtained through the specification of a discrete probability distribution over each set of alternative productions have been used as a means of accounting for ambiguities of structure or generation [Fu and Swain (1971)].

We noted earlier the cleavage between the extractor and analyzer that occurs in formal models. In the work based on stochastic formal grammars [Fu and Swain (1971) and Lee and Fu (1972)], while the analyzer is designed to handle ambiguities of structure there seems to be no provision for ambiguities of representations handed over by the extractor. It seems clear that some data objects could be represented by alternative strings of primitives. Also, because the scheme is based on string grammars, it is forced to overemphasize the concatenation relation between primitives.

There have been several proposals for multidimensional generalizations of phrase structure string grammars. For example, in array grammars, instead of replacing one substring by another, rewrite rules are defined to replace a two-dimensional subarray by another subarray. Properties of array grammars and the relationship between array grammars and array automata have been investigated in Milgram and Rosenfeld (1971). In Siromoney *et al.* (1973), rectangular array models are generalized to *n* dimensions, and three-dimensional array models are used to describe the growth of crystals.

Plex grammars [Feder (1969), (1971)] involve primitive entities called napes. Each nape has a finite number of attaching points, each of which has an associated identifier. Napes are combined by bringing attaching points into coincidence. A picture description language [Shaw (1969, 1970)] can be used to describe pictorial patterns, the primitive elements of which have arbitrary shapes and distinguished heads and tails. The hierarchic structure of a picture is defined by using a "picture description grammar" to combine expressions in the picture description language. In coordinate grammars [Rosenfeld (1973) and Anderson (1968)], morphs have coordinates associated with them, and functions that compute the coordinates of the new morphs from the old morphs are associated with the rewriting rules. Coordinate grammars have been proposed for the description of mathematical notation, shapes, signs, textures, etc. [see, e.g., Chang (1970), (1971), Narasimhan and Reddy (1971), Carlucci (1972), Muchnik (1972), Nake (1971), and Simon and Checroun (1971)]. Graph grammars model various arbitrary relational structures among morphs. These include grammars that generate trees [Brainerd (1969) and Rounds (1969)] and labeled graphs, called webs [Pfaltz and Rosenfeld (1969)]. The notion of graph embedding has been generalized in Ehrig *et al.* (1973).

Bhargava and Fu (1973) present a scheme for representing line drawings in terms of trees. A tree grammar that generates trees is thus a formal description of the corresponding set of patterns. Among the examples given are tree grammars for the chemical structure of a natural rubber molecule and for two electrical circuit diagrams.

There are some problems in using trees and tree grammars in the manner of Bhargava and Fu (1973). First, since trees are acyclic graphs, a single tree cannot completely describe the connectivity of a closed figure. Second, trees introduce ambiguity into a pattern that may not itself be ambiguous. This ambiguity arises because the description of a figure by means of a tree requires a segmentation of the figure described, and an ordering of the segments. A different choice of segments and ordering would result in a different description of the pattern. Graphs and graph grammars are probably more appropriate structures for describing line drawings because cyclic graphs can completely describe the connectivity of closed figures, and a graph description need not order the parts of a figure.

It has often been suggested that transformational rules might be just as useful in pattern analysis as they have been in providing insights into the structure of natural languages. A transformational grammar is defined as $G_\phi = (G,\phi)$, where G is a reasonably simple "base" grammar such as a context-free grammar, and ϕ is a mapping that maps a structure in G, i.e., a tree, into a related tree. Joshi (1973) presents a detailed example of a transformational grammar derivation for a class of polygonal patterns. The example shows how a context-free base grammar and transformational rules for deletion of the interior lines of the generated patterns lead to much simpler derivation than a more direct approach involving a context-sensitive grammar.

Bhargava and Fu (1973) also discuss the application of transformational rules to trees generated by tree grammars. The paper considers transformations: 1) to duplicate patterns, i.e., to represent complex patterns as a periodic repetition of some simple pattern and 2) to relate two occurrences of the same pattern, one of which has undergone a shape-preserving transformation, e.g., rotation, linear translation, or reflection.

In Tauber and Rankin (1972), it is suggested that the syntactic structure of chemical structure diagrams be described by context-free web grammars or array grammars. However, there are some relations between chemical structure diagrams that cannot be described by context-free web grammars, for example, the relation between equivalent structural formulas of the same compound. Transformational rules can be used to transform equivalent diagrams into a canonical form, as well as to combine diagrams and to decompose diagrams into "kernel" diagrams. These are the ideas motivating the work of Underwood and Kanal (1973), which introduces the concept of a transformational web grammar. Because of the potential of graphs for describing patterns of practical interest, graph and transformational graph grammars are likely to receive increasing attention in syntactic pattern recognition.

IX. Applications

It is not difficult to formulate problems from a variety of fields in the terminology of pattern recognition. While that need not imply much about how amenable they are to solution, the terminology and methodology do provide a common framework for investigators in various fields as they become acquainted with work in pattern recognition. Some interesting recent examples are Pang *et al.* (1974), in which each set of measurements describing a state of a power system is treated as a pattern and classifiers, i.e., "security functions" are derived to indicate whether the system is in a secure or alert state; a dissertation in economics that shows how certain models proposed by economists for problems in voting theory and consumer demand can be interpreted and extended via the terminology and methodology of syntactic and decision-theoretic pattern recognition [Piccoli (1974)]; a classification of members of defined categories of stochastic nonlinear systems from input–output data vectors [Saridis and Hofstadter (1973)]; and a use of clustering and discriminant functions to classify characteristics of jobs received by a large digital computer in order to develop dynamic scheduling algorithms [Northouse and Fu (1973)].

Over the years certain investigators in chemistry have been quick to try whatever pattern classification algorithms happened to be popular, from Perceptron algorithms to k nearest-neighbor classifiers, on infrared spectrometric data, NMR spectra, and other chemical data [Kowalski *et al.* (1969, (1972) and Kowalski and Bender (1972), (1973)].

Most of these "applications" of pattern recognition in various fields have been academic demonstrations that suggest the relevance of the methodology. Similarly, many of the "applications" reported in the literature by persons developing pattern recognition techniques are merely incidental to the purpose of demonstrating that a new algorithm "works well," with the demonstration usually being performed on a limited data set. During the past two decades alphanumeric characters have been the favorite data for such "show and tell" experiments with new algorithms.

The focus of most of the published research in pattern recognition has been on techniques. This is very much in evidence in the books that have appeared in the last two years. They are almost entirely technique-oriented, giving a host of techniques with little insight into their comparative utility in different applications. This point is made rather colorfully in a review [Bremermann (1974) of Andrews (1972)]:

> While he should not be blamed for the unsatisfactory state of the art, he can be blamed for not making any attempt to convey to the reader a sense of the effectiveness and ineffectiveness of his methods. There are almost no applications (of 242 pages, only 6 are concerned with actual pattern recognition experiments). Thus a new sacred cow of mathematical machinery is created—its priesthood will probably make a good academic living regardless of whether the cow gives any milk.

Similar comments were also made until recently about information theory but one hears them less often now.

The blame for the state of affairs decried by Bremermann lies in part with the fact that the development of theoretical and heuristic insights, which are relevant to practical applications, requires the type of interaction between theoreticians and experimentalists that is very evident in physics. In pattern recognition, from time to time, such interaction has been fostered in industrial and government research laboratories by organizations interested in applications, such as optical character recognition, target recognition, electrocardiograms, and blood cells. For the most part, however, technique development has occurred without much feedback from experiments, since meaningful experimentation in pattern recognition often requires that significant resources be spent on collection, verification, and handling of large data bases. Some effort at sharing standard data bases is now present [IEEE Computer Society]; this should help. In many application areas, effective use of the data requires close interaction with persons knowledgeable about the processes that generate the data. Also required is a sustained effort devoted to the particular application.

Negative examples abound, wherein inadequate data bases were used and arbitrary operations were performed on data without knowledge of the field, thus leading to unconvincing results. A positive example is the work on the differential diagnosis of white blood cells. Much of the credit for the exploration of this application area goes to the pioneering and sustained work of Prewitt and her colleagues who, with adequate access to data gathering and experimental facilities, established many of the basic ideas for the development of this application [Prewitt and Mendelsohn (1966), Prewitt (1972)]. These ideas and other contributions [e.g., Bacus and Gose (1972), Ingram and Preston (1970), and Young (1969)] have led to the development of commercial products in this area. Whether or not these products of pattern recognition research become either a technical or a market success remains to be seen.

An annual review of progress in cell recognition and related areas occurs at the Engineering Foundation Conference on Automatic Cytology [Engineering Foundation, N.Y.C.]. Other examples of sustained efforts include various projects in electrocardiogram pattern recognition [see Caceres and Driefus (1970) and Zyweitz and Schneider (1973)]. One reason why this applications area is of considerable interest to theoreticians and experimentalists in pattern recognition is the extensive data base gathered by Pipberger [Pipberger *et al.* (1972)]. Although a few programs are supposed to work well, available pattern recognition products in electrocardiography are still undergoing evaluation [Cox *et al.* (1972) and Bailey (1974)]. Other areas of interest in medical pattern recognition include chromosomes [Castleman and Wall (1973)], X-rays [Chien and Fu (1974), Pratt (1973), Harlow and Eisenbeis (1973), and Ballard and Sklansky (1973), (1974)], and the process of diagnosis [Kulikowski (1970), Jacquez (1972), and Patrick *et al.* (1974)]. Only a few recent references have been cited here for these and other applications areas. Through these references, the interested reader can trace the literature in the area. A bibliography of articles on automatic quantitative microscopy is provided in Imanco (1973).

Much data are being gathered by remote sensing using the Earth Resources Technology Satellite (ERTS) and various other elevated platforms for pointing sensors back at the earth. It appears that this particular application grew more out of a need to justify continuation of the space program than out of experimental evidence that the data to be gathered would provide adequate information for discrimination of various phenomena. An excellent survey [Nagy (1972)] and subsequent symposia proceedings [ERTS Summaries, Remote Sensing Symposia] describe the extensive work being done in remote sensing. So far, only a small part of the work in this area is concerned with automatic classification but this is likely to change. Nagy has discussed, rather well, the prospects and pitfalls that await pattern classification studies on data gathered from satellites and aircraft. A study of the application of pattern recognition techniques to weather radar data is reported in Duda and Blackmer (1972) and Blackmer *et al.* (1973).

Table III presents a representative list of pattern recognition problems that have been attempted. It is interesting to note the recent activity in fingerprint and palmprint recognition and also in signature verification, which has been generated by the interest of law enforcement and military base security groups [Eleccion (1973), Nagel and Rosenfeld (1973), Proc. of the Electronic Crime Countermeasures Conf., and Sprouse *et al.* (1974)]. Pattern recognition methodology has also served as the basis for a study of the decision mechanisms used in palmistry [Oda *et al.* (1971)]. A more modern interpretation, viz., the genetic basis of dermatoglyphic patterns, underlies the preliminary pattern classification studies of palm and fingerprint patterns for their potential in diagnosing Down's syndrome, leukemia, and schizophrenia [Stowens and Sammon (1970), and Stowens, Sammon, and Proctor (1970)]; see also Penrose and Loesch (1971).

Applications of pattern recognition in industrial process control are being explored [see, e.g., *Business Week* (1974)].

TABLE III
SOME APPLICATIONS OF PATTERN RECOGNITION

Problem	Input to Pattern Recognition System	Output of Pattern Recognition System
Medical Applications		
Identification and counting of cells	slides of blood samples, microsections of tissue	types of cells
Detection and diagnosis of disease	electrocardiogram waveforms	types of cardiac conditions
	electroencephalogram waveforms	classes of brain conditions
	slides of blood samples	various types and proportions of normal and abnormal cells
Prosthetic control devices	mypotentials	categories of movements of limbs
X-ray diagnosis	X-ray photograph	presence or absence of specific conditions
Military Applications		
Interpretation of aerial reconnaissance imagery	visual, infra-red, radar, multispectral imagery	tanks, personnel carriers, weapons, missile launchers, airfields, campsites
Detection of enemy navy vessels	passive and active sonar waveforms	surface vessels, submarines, whales, fish
Detection of underground nuclear explosions	seismic waveforms	nuclear explosions, conventional explosions, earthquakes
Commercial and Government Applications		
Automatic detection of flaws— impurities in sheet glass, bottles, paper, textiles, printed circuit boards, integrated circuit masks	scanned image (visible on infrared, etc.)	acceptable vs. unacceptable, markings, bubbles, flaws, radiation patterns, etc.
Classification and identification of fingerprints	scanned image	fingerprint descriptions based on Henry system of classification
Traffic pattern study	aerial photographs of highways, intersections, bridges, road sensors	automobiles, trucks, motorcycles, etc., to determine the characteristics of the traffic flow
Natural resource identification	multispectral imagery	terrain forms, agricultural land, bodies of water, forests
Identification of crop diseases	multispectral imagery	normal and diseased crops
Economic prediction	time series of economic indicators	economic conditions
Speech recognition— remote manipulation of processes, parcel post sorting, management information systems, voice input to computers	speech waveform	spoken words, phonemes
Weather forecasting	weather data from various land-based, airborne, ocean, and satellite sensors	categories of weather
Object recognition— parts handling, inspection of parts, assembly	scanned image	object types
Character Recognition		
Bank checks	magnetic response waveform, optical scanned image	numeric characters, special symbols
Automatic processing of documents— utility bills, credit card charges, sale and inventory documents	optical scanned image	alphanumeric characters, special symbols
Journal tape reading	optical scanned image	numeric characters, special symbols
Page readers— automatic type setting, input to computers, reading for the blind	optical scanned image	alphanumeric characters, special symbols
Label readers	optical scanned image	alphanumeric characters, special symbols
Address readers	optical scanned image	letters and numerals combined into zip codes, city and state names, and street addresses
Other readers— licence plate readers, telephone traffic counter readers	optical scanned image	alphanumeric characters, special symbols

In some areas of automatic assembly, flaw detection using simple pattern classification techniques is feasible [Jensen (1973)], but the requirements on precision and low error rate may be quite severe. In other applications, e.g., stock market patterns, the problems are quite difficult but anything better than a 50-percent error rate may justify the effort. Another example is that of postal address readers, a number of which have been operating for several years. Even if they reject a substantial percentage of the letters they process, the sheer volume of mail they correctly sort can still make the installation worthwhile.

Character recognition is the only pattern recognition application area that has led to some commercially viable products. However, as a data entry device for converting written material into computer code, OCR equipment has so far not quite come up to earlier expectations in terms of revenues, customer acceptance, and technical achievement. Now there are new serious competitors to the data entry function performed by OCR; e.g., key-to-tape equipment, typewriter terminals using software editors, or on the near horizon, limited vocabulary isolated-word speech input to computers (also a pattern recognition technology). Also, the past few years have shown that the effectiveness of OCR installations depends upon complex interactions between computer systems, programming, forms design, imprinters, inks, systems and procedures, training, and so forth.

In summary, systems considerations have played a dominant role in the success of an installation, and this is likely to be true of many future pattern recognition products.

A survey of techniques for automatic recognition of print and script appears in Harmon (1972); for an elementary introduction to scanning techniques for OCR see Freedman (1974). Activity in this field continues, as evidenced by patents and papers [see, e.g., Sammon *et al.* (1973), Ullmann (1974)], and by the attendance at annual seminars of the Data Processing Supplies Association [DPSA (1971)] and other meetings. However, it is sobering to note that although many companies entered the OCR market over the last decade, not many have prospered in OCR activity. (Possible reasons why rosy predictions of rich markets did not quite materialize are offered in Kanal (1971).) Fortunately, the OCR experience has not stifled the entry of new companies into other pattern recognition product areas. In addition to blood cell and electrocardiogram recognition devices, isolated-word limited-vocabulary speech recognition devices are now available for trial in various applications. Perhaps these new areas will benefit from the lessons learned in OCR.

X. Prospects

In pattern recognition there have been repeated expressions of concern (similar to those familiar to information theorists) that theoretical research bears little relationship to practical applications. The research described in Sections VI and VIII suggests that this situation is changing. In the coming years, pattern recognition research is likely to intensify efforts to combine heuristic and formal methods and statistical and structural methods. Also likely to increase is the interplay between pattern recognition and various problem-solving techniques in artificial intelligence. The next few years should witness an increasing infiltration of pattern recognition techniques into various disciplines and an increase in serious collaborative investigations involving large data bases, especially in biomedical and remote sensing applications.

The interest in experimenting with real data bases should stimulate further theoretical and experimental studies on the design and analysis of pattern classification experiments. Because of their importance to practice, comparisons of various approaches to multiclass classification will no doubt be further investigated. In syntactic methods, the current interest in scenes, line figures, chemical structure, and electronic circuit diagrams is likely to stimulate further work on graphs and graph grammars.

Challenged by the question, "What is a pattern that a machine may know it?" perhaps someone will come up with a suitable definition of "pattern" in the way that Shannon gave precision to the colloquial word "information." If the resulting theory for such precisely defined patterns were relevant to a larger class of problems than existing theories—that would be a "Kendo" stroke of genius! However, the lack of such a theory does not prevent applied research from producing *now* results that would justify the promise held out by pattern recognition for the last fifteen years. Applications of pattern recognition technology to industrial automation, health care delivery, and other societal problems are being pursued and are likely to play a significant role in the near future.

Acknowledgment

I am grateful to A. K. Agrawala, B. Chandrasekaran, T. Cover, G. D. Forney, J. F. Lemmer, A. Rosenfeld, G. C. Stockman, W. E. Underwood, and two reviewers for their helpful comments on a rough draft of this paper and for their assistance. The construction of the introductory sentence of the paper parallels that used by Warren McCulloch in asking a similar question about numbers and man.

References

[1] K. Abend, "Compound decision procedures for unknown distributions and for dependent states of nature," in *Pattern Recognition*, L. Kanal Ed. Washington, D.C.: Thompson, 1968, pp. 207–249; also available from L.N.K. Corp. Silver Spring, Md.

[2] A. K. Agrawala, "Learning with various types of teachers," Comput. Sci. Center, Univ. Md., College Park, Tech. Rep. TR-276, Dec. 1973; see also *Proc. 1st Int. Joint Conf. Pattern Recognition*, IEEE Publ. no. 73 CHO 821-9c, pp. 453–461.

[3] A. K. Agrawala, "Learning with probabilistic teacher," *IEEE Trans. Inform. Theory*, vol. IT-16, pp. 373–379, July 1970.

[4] R. C. Ahlgren, H. F. Ryan, and C. W. Swonger, "A character recognition application of an iterative procedure for feature selection," *IEEE Trans. Comput.*, vol. C-20, pp. 1067–1075, Sept. 1971.

[5] J. S. Albus, "A theory of cerebellar function," *Math. Biosciences*, vol. 10, pp. 25–61, 1971.

[6] ——, "Theoretical and experimental aspects of a cerebellar model," Ph.d. dissertation, Univ. of Maryland, College Park, 1972.

[7] D. M. Allen, "Mean square error of prediction as a criterion for selecting variables," *Technometrics*, vol. 13, pp. 469–475, Aug. 1972.

[8] S. I. Amari, "Learning patterns and pattern sequences by self-organizing nets of threshold elements," *IEEE Trans. Comput.*, vol. C-21, pp. 1197–1206, Nov. 1972.

[9] M. R. Anderberg, *Cluster Analysis for Applications*. New York: Academic, 1973.

[10] R. H. Anderson, "Syntax-directed recognition of hand printed two dimensional mathematics," Ph.D. dissertation, Harvard Univ., Cambridge, Mass., Jan. 1968; also in *Interactive Systems For Experimental Applied Mathematics*. M. Kefrer and J. Reinfelds, Eds. New York: Academic, 1968, pp. 436–459.

[11] H. Andrews, *Introduction to Mathematical Techniques in Pattern Recognition*. New York: Wiley, 1972.

[12] J. W. Bacus, and E. E. Gose, "Leukocyte pattern recognition," *IEEE Trans. Syst., Man., Cybern.*, vol. SMC-2, Sept. 1972.

[13] J. J. Bailey, M. Horton, and S. Itscoitz, "A method for evaluating computer programs for electrocardiographic applications," Parts 1–3, *Circulation*, vol. XLIX, July 1974.

[14] D. Ballard and J. Sklansky, "Tumor detection in radiographs," *Comput. Biomedical Res.*, vol. 68, pp. 299–321.

[15] ——, "Hierarchic recognition of tumors in chest radiographs," Sch. of Eng., Univ. of Calif., Irvine, Mar. 1974.

[16] G. W. Beakley and F. B. Tuteur, "Distribution-free pattern verification using statistically equivalent blocks," *IEEE Trans. Comput.*, vol. C-21, pp. 1337–1347, Dec. 1972.

[17] E. M. Beale, M. G. Kendall, and D. W. Mann, "The discarding of variables in multivariate analysis," *Biometrica*, vol. 54, pp. 357–366, 1967.

[18] P. W. Becker, "Recognition of patterns using the frequencies of occurrence of binary words," Polyteknisk Forlag, Copenhagen, Denmark, 1968.

[19] J. O. Bennett, "Estimation of multivariate probability density functions using *B*-splines," Ph.D. dissertation, Rice Univ., Houston, Tex., May 1974.

[20] R. S. Bennett, "The intrinsic dimensionality of signal collections," *IEEE Trans. Inform. Theory.*, vol. IT-15, pp. 517–525, Sept. 1969.

[21] R. H. Blackmer, R. O. Duda, and R. Reboh, "Application of pattern recognition techniques to digitized weather radar data," Stanford Res. Inst., Menlo Park, Calif., Contract I-36072, SRI Project 1287, Apr. 1973.

[22] M. Bongard, *Pattern Recognition*, J. K. Hawkins, Ed. Transl. T. Cheron.

[23] W. S. Brainerd, "Tree generating regular systems," *Inform. Contr.*, vol. 14, pp. 217–231, 1969.

[24] H. J. Bremermann, Review of 'An introduction to mathematical techniques in pattern recognition,' by H. C. Andrews, *Amer. Scientist*, vol. 62, pp. 244–245, 1974.

[25] "Controls that learn to make their own decisions," *Business Week*, Apr. 6, 1974.

[26] C. A. Caceres and L. S. Dreifus, Eds., *Clinical Electrocardiography and Computers*. New York: Academic, 1970.
[27] T. Cacoullos, *Discriminant Analysis and Applications*. New, York: Academic, 1973.
[28] T. W. Calvert and T. Y. Young, "Randomly generated nonlinear transformations for pattern recognition," *IEEE Trans. Syst. Sci. Cybern.*, vol. SSC-5, pp. 266–273, Oct. 1969.
[29] L. Carlucci, "A formal system for texture language pattern recognition," vol. 4, no. 1, pp. 53–72, 1972.
[30] K. R. Castleman and R. J. Wall, "Automatic systems for chromosome identification," in Nobel Symp. 23: *Chromosome Identification*, T. Casperson and L. Zech, Eds. Nobel Foundation. New York: Academic, 1973.
[31] B. Chandrasekaran, "Independence of measurements and the mean recognition accuracy," *IEEE Trans. Inform. Theory*, vol. IT-17, pp. 452–456, July 1971. Correction, vol. IT-18, p. 217, Jan. 1972.
[32] B. Chandrasekaran and T. J. Harley, Comments "On the mean accuracy of statistical pattern recognizers," *IEEE Trans. Inform. Theory*, Corresp., vol. IT-15. pp. 421–423, May 1969.
[33] B. Chandrasekaran and A. K. Jain, "Distance functions for independent measurements and finite sample size," in *Proc. 2nd Int. Joint Conf. Pattern Recognition*, 1974; available from IEEE.
[34] ——, "Independence measurement complexity and classification performance," to appear in *IEEE Trans. Syst., Man, Cybern.*, 1974.
[35] B. Chandrasekaran, L. Kanal, and G. Nagy, Rep. on the 1968 IEEE Workshop Pattern Recognition, Delft, The Netherlands, IEEE Special Publ. 68 C 65-C, 1968.
[36] C. L. Chang, "Pattern recognition by piecewise linear discriminant functions," *IEEE Trans. Comput.*, vol. C-22, p. 859, Sept. 1973.
[37] C. L. Chang and R. C. Lee, "A heuristic relaxation method for nonlinear mapping in cluster analysis," *IEEE Trans. Syst., Man, Cybern.*, vol. SMC-3, pp. 197–200, Mar. 1973.
[38] C. Y. Chang, "Dynamic programming as applied to feature subject selection in a pattern recognition system," *IEEE Trans. Syst., Man, Cybern.*, vol. SMC-3, pp. 166–171, Mar. 1973.
[39] S. K. Chang, "A method for the structural analysis of two-dimensional mathematical expressions," *Inform. Sci.*, vol. 2, pp. 253–272, July 1970.
[40] ——, "Picture processing grammar and its applications," *Inform. Sci.*, vol. 3, pp. 121–148, Apr. 1971.
[41] C. H. Chen, *Statistical Pattern Recognition*. New York: Hayden, 1973.
[42] Cheng *et al.*, *Pictorial Pattern Recognition*. Washington, D.C.: Thompson, 1968.
[43] H. Chernoff, "Using faces to represent points in *K*-dimensional space graphically," *J. Amer. Statist. Ass.*, vol. 68, p. 361, June 1973.
[44] Y. T. Chien. "Bibliography on interactive pattern recognition and related topics," School of Eng., Univ. Conn., Storrs, Tech. Rep. CS-74-3, Feb., 1974.
[45] Y. T. Chien, and K. S. Fu, "Recognition of x-ray picture patterns," *IEEE Trans. Syst., Man, Cybern.*, vol. SMC-4, pp. 145–156, Mar. 1974.
[46] C. K. Chow, "On optimum recognition error and reject trade-off," *IEEE Trans. Inform. Theory*, vol. IT-16, pp. 41–46, Jan. 1970.
[47] T. M. Cover, "Rates of convergence of nearest neighbor decision procedures," in *1968 Hawaii Int. Conf. Systems Theory*, pp. 413–415.
[48] ——, "Learning in pattern recognition," in *Methodologies of Pattern Recognition*, S. Watanabe, Ed. New York: Academic 1969, pp. 111–132.
[49] ——, "A hierarchy of probability density function estimates," in *Frontiers of Pattern Recognition*, S. Watanabe, Ed. New York: Academic, 1972, pp. 83–98.
[50] ——, "Generalization of patterns using Kolmogorov complexity," in *Proc. 1st Joint Int. Conf. Pattern Recognition*, IEEE Publ. no. 73 CHO 821-9C, 1973.
[51] ——, "Recent books on pattern recognition," *IEEE Trans. Inform. Theory*, vol. IT-19, p. 827, Nov. 1973.
[52] ——, "The best two independent measurements are not the two best," *IEEE Trans. Syst., Man, Cybern.* (Corresp.), vol. SMC-4, pp. 116–117, Jan. 1974.
[53] T. M. Cover and P. F. Hart, "Nearest neighbor pattern classification," *IEEE Trans. Inform. Theory*, vol. IT-13, pp. 21–27, Jan. 1967.
[54] J. R. Cox, F. M. Noile, and R. M. Arthur, "Digital analysis of the electroencephalogram, the blood pressure wave, and the electrocardiogram," *Proc. IEEE*, vol. 60, pp. 1137–1164, Oct. 1972.
[55] B. R. Crain, "Density estimation using orthogonal expansions," *J. Amer. Statist. Ass.*, vol. 68, p. 964, Dec. 1973.
[56] R. J. deFigueiredo, "Optimal linear and nonlinear feature extraction based on the minimization of the increased risk of misclassification," in *Proc. 2nd Int. Joint Conf. Pattern Recognition*, 1974.
[57] H. P. Decell, Jr., and J. A. Quierein, "An iterative approach to the feature selection problem," in *Proc. Purdue Univ. Conf. Machine Processing of Remotely Sensed Data*, 1972, pp. 3B1–3B12.
[58] P. A. Devijver, "Relationships between statistical risks and the least-mean-square-error design criterion in pattern recognition," in *Proc. 1st Int. Joint Conf. Pattern Recognition*, IEEE Special Publ. CHO 821-9C, 1973, pp. 139–148.
[59] ——, "On a new class of bounds on Bayes risk in multihypothesis pattern recognition," *IEEE Trans. Comput.*, vol. C-23, pp. 70–80, Jan. 1974.
[60] H. L. Dreyfus, *What Computers Can't Do*. New York: Harper and Row, 1972.
[61] R. O. Duda and R. H. Blackmer, Jr., "Application of pattern recognition techniques to digitized weather radar data," Stanford Res. Inst., Menlo Park, Calif., Contract 1-36072, SRI Project 1287, Mar. 1972.
[62] R. O. Duda and P. E. Hart, *Pattern Classification and Scene Analysis*. New York: Wiley, 1973.
[63] A. R. Dyer, "Discrimination procedures for separate families of hypotheses," *J. Amer. Statist. Ass.*, vol. 68, p. 970, Dec. 1973.
[64] H. Ehrig, M. Pfender, and H. J. Schneider, "Graph grammars: An algebraic approach," in *Proc. 14th Annu. Conf. Switching and Automata Theory*, 1973.
[65] J. D. Elashoff, R. M. Elashoff, and G. E. Goldman, "On the choice of variables in classification problems with dichotomous variables," *Biometrika*, vol. 54, pp. 668–670, 1967.
[66] M. Eleccion, "Automatic fingerprint identification," *IEEE Spectrum*, pp. 36–45, Sept. 1973.
[67] L. Erman, Ed., *1974 IEEE Symp. Speech Recognition*, IEEE Cat. no. 74CHO878-9AE.
[68] T. G. Evans, "A grammar-controlled pattern analyzer," in *1968 Inform. Processing, Proc. IFIP Congr.*, vol. 2, A. J. Moreli, Ed., Amsterdam: North Holland, 1968, pp. 1592–1598.
[69] J. Feder, "Linguistic specification and analysis of classes of line patterns," Dep. Elec. Eng. New York Univ., N.Y.C., Tech. Rep. no. 403-2, Apr. 1969.
[70] ——, "Plex languages," *Inform. Sci.*, vol. 3, pp. 225–241, 1971.
[71] O. Firschein, *et al.*, "Forecasting and assessing the impact of artificial intelligence on society," in *Proc. Int. Joint Conf. Artificial Intelligence*, 1973, pp. 105–120.
[72] L. Fisher and J. W. Van Ness, 1973 "Admissible discriminant analysis," *J. Amer. Statist. Ass.*, vol. 68, pp. 603–607, Sept. 1973.
[73] E. Fix and J. L. Hodges, "Discriminatory analysis small sample performance," USAF Sch. of Aviation Medicine, Randolph Field, Tex., Project 21-49-004, Rep. 11, Contract AF-41-(128)-31, 1952.
[74] D. H. Foley, "Considerations of sample and feature size," *IEEE Trans. Inform. Theory*, vol. IT-18, pp. 618–626, Sept. 1972.
[75] ——, "Orthonormal expansion study for waveform processing system," Rome Air Develop. Center, AF Systems Command, Griffiss AFB, New York, Tech. Rep. RADC-TR-73-168, July 1973.
[76] S. C. Fralick and R. W. Scott, "Nonparametric Bayes-risk estimation," *IEEE Trans. Inform. Theory*, vol. IT-17, pp. 440–444, July 1971.
[77] M. D. Freedman, "Optical character recognition," *IEEE Spectrum*, vol. 11, pp. 44–52, Mar. 1974.
[78] K. S. Fu, *Sequential Methods in Pattern Recognition and Machine Learning*. New York: Academic, 1968.
[79] ——, "On syntactic pattern recognition and stochastic languages," in *Frontiers of Pattern Recognition*, S. Watanabe, Ed. New York: Academic, 1972, pp. 113–137.
[80] ——, *Syntactic Methods in Pattern Recognition*. New York: Academic, 1974.
[81] K. S. Fu and B. K. Bhargava, "Tree systems for syntactic pattern recognition," *IEEE Trans. Comput.*, vol. C-22, pp. 1087–1098, Dec. 1973.
[82] K. S. Fu and P. H. Swain, "On syntactic pattern recognition," in *Software Engineering*, vol. 2, J. T. Tou, Ed. New York: Academic, 1971.
[83] K. S. Fu, Ed. "Pattern recognition and machine learning," in *Proc. Japan–U.S. Seminar Learning Process in Control Systems*, New York: Plenum, 1970.
[84] K. S. Fu, P. J. Min, and T. J. Li, "Feature selection in pattern recognition," *IEEE Trans. Syst. Sci. Cybern.*, vol. SSC-6, p. 33–39, Jan. 1970.
[85] K. Fukunaga, *Introduction To Statistical Pattern Recognition*. New York: Academic, 1972.
[86] K. Fukunaga and L. D. Hostetler, "The estimation of the gradient of a density function and its applications in pattern recognition," presented at Int. Joint Conf. Pattern Recognition, Washington, D.C., Oct. 1973.

[87] ——, "Optimization of k-nearest neighbor density estimates," *IEEE Trans. Inform. Theory*, vol. IT-19, pp. 320–326, May 1973.
[88] K. Fukunaga and D. Kessell, "Application of optimum error-reject functions," *IEEE Trans. Inform. Theory*, (Corresp.), vol. IT-18, pp. 814–817, Nov. 1972.
[89] K. Fukunaga and D. L. Kessell, "Error evaluation and model validation in statistical pattern recognition," Sch. of Elec. Eng., Purdue Univ., Lafayette, Ind., Tech. Rep. TR-EE 72-23, Aug. 1972.
[90] ——, "Nonparametric Bayes error estimation using unclassified samples," *IEEE Trans. Inform. Theory*, vol. IT-19, pp. 434–440, July 1973.
[91] K. Fukunaga and W. Koontz, "Application of the Karhunen–Loève expansion to feature selection and ordering," *IEEE Trans. Comput.*, vol. C-19, pp. 311–318, Apr. 1970.
[92] K. Fukunaga and P. M. Narendra, "A branch and bound algorithm for computing k-nearest neighbors," available from IEEE Comput. Group depository.
[93] K. Fukunaga and D. R. Olsen, "An algorithm for finding intrinsic dimensionality of data," *IEEE Trans. Comput.*, vol. C-20, pp. 176–183, Feb. 1971.
[94] ——, Authors' reply to comments by G. V. Trunk on "An algorithm for finding intrinsic dimensionality of data," *IEEE Trans. Comput.* (Corresp.), vol. C-20, pp. 1615–1616, Dec. 1971.
[95] K. Fukunaga and G. V. Sherman, "Picture representation and classification by pseudo eigenvectors," presented at the IEEE Syst., Man, Cybern. Conf., Boston, Mass., Nov. 1973.
[96] K. Fukushima, "Visual feature extraction by a multilayered network of analog threshold elements," *IEEE Trans. Syst., Sci., Cybern.*, vol. SSC-5, pp. 322–333, Oct. 1969.
[97] ——, "A feature extractor for curvilinear patterns: a design suggested by the mammalian visual system," *Kybernetik*, vol., pp. 153–160, 1970.
[98] M. P. Gessaman and P. H. Gessaman, "Comparison of some multivariate discrimination procedures," *J. Amer. Statist. Ass.*, vol. 67, p. 468, June 1972.
[99] N. Glick, "Sample-based classification procedures derived from density estimators," *J. Amer. Statist. Ass.*, vol. 67, pp. 112–116, Mar. 1972.
[100] R. Gnanadesikan and M. B. Wilk, "Data Analytic Methods in Multivariate Statistical Analysis," in *Multivariate Analysis II*, P. R. Krishnaiah, Ed. New York: Academic, 1969, pp. 593–638.
[101] D. W. Goodall, "On a new similarity index based on probability," *Biometrics*, vol. 22, pp. 668–670, Dec. 1966.
[102] A. Grasselli, Ed. *Automatic Interpretation and Classification of Images*. New York: Academic, 1969.
[103] P. E Green and V. R. Rao, *Applied Multidimensional Scaling–A Comparison of Approaches and Algorithms*. New York: Holt, Rinehart and Winston, 1972.
[104] P. E. Green and F. J. Carmone, *Multidimensional Scaling and Related Techniques in Marketing Analysis*. Boston: Allyn and Bacon, 1970.
[105] U. Grenander, "Foundations of pattern analysis," *Quart. Appl. Math.*, vol. 27, pp. 1–55, Apr. 1969.
[106] U. Grenander, "A unified approach to pattern analysis," in *Advances in Computers*, vol. 10. New York: Academic, 1970.
[107] M. Halle and K. Stevens, "Speech recognition: A model and a program for research," *IRE Trans. Inform. Theory*, vol. IT-8, p. 155, Feb. 1962.
[108] R. M. Haralick, "Glossary and index to remotely sensed image pattern recognition concepts," *Pattern Recognition*, vol. 5, p. 391–403, 1973.
[109] T. J. Harley, L. Kanal, and N. C. Randall, "System considerations for automatic imagery screening," in *Pictorial Pattern Recognition*. Washington D.C.: Thompson, 1968, pp. 15–32.
[110] C. A. Harlow and S. A. Fisenbeis, "The analysis of radiographic images," *IEEE Trans. Comput.*, vol. C-22, pp. 678–689, July 1973.
[111] L. D. Harmon, "*Automatic Reading of Cursive Script Optical Character Recognition*, G. L. Fischer, Jr, *et al.*, Eds. New York: Spartan, 1962, pp. 151–152.
[112] *Special Issue on Digital Pattern Recognition, Proc. IEEE*, vol. 60, Oct. 1972.
[113] P. Hart, N. Nilsson, and B. Raphael, "A formal basis for the heuristic determination of minimal cost paths," *IEEE Trans. Syst. Sci. Cybern.*, vol. SSC-4, pp. 100–107, July 1968.
[114] J. A. Hartigan, "Clustering," *Rev. Biophysics and Bioengineering*, pp. 81–101, 1972.
[115] M. E. Hellman and J. Raviv, "Probability of error, equivocation, and the Chernoff bound," *IEEE Trans. Inform. Theory*, vol. IT-16, pp. 368–372, 1970.
[116] E. G. Henrichon, Jr., and K. S. Fu, "A nonparametric partitioning procedure for pattern classification," *IEEE Trans. Comput.*, vol. C-18, pp. 614–624, July 1969; comments by Beakley and Tuteur, *IEEE Trans. Comput.*, vol. C-19, pp. 362–363, Apr. 1970.
[117] W. H. Highleyman, "The design and analysis of pattern recognition experiments," *Bell Syst. Tech. J.*, vol. 41, pp. 723–744, 1962.
[118] D. R. Hill, "Man-Machine interaction using speech," in *Advances in Computers*, F. L. Alt *et al.*, Eds. New York: Academic, 1971, pp. 165–230.
[119] M. Hills, "Allocation rules and their error rates," *J. Roy. Stat. Soc.*, Ser. B, vol. 28, pp. 1–31, 1968.
[120] ——, "Discrimination and allocation with discrete data," *Appl. Statist.*, vol. 16, 1967.
[121] Y. C. Ho and A. K. Agrawala, "On pattern classification algorithms–Introduction and survey," *Proc. IEEE*, vol. 56, pp. 2101–2114, Dec. 1968.
[122] S. L. Horowitz, "A general peak detection algorithm with applications in the computer analysis of electrocardiograms," Comput. Sci. Lab., Dep. Elec. Eng., Princeton Univ., Princeton, N.J., Tech. Rep. 129, May 1973.
[123] G. F. Hughes, "On the mean accuracy of statistical pattern recognizers," *IEEE Trans. Inform. Theory*, vol. IT-14, pp. 55–63, Jan. 1968.
[124] A. B. Hussain, "Compound sequential probability ratio test for the classification of statistically dependent patterns," *IEEE Trans. Comput.*, vol. C-23, pp. 398–410, Apr. 1974.
[125] A. B. Hussain and R. W. Donaldson, "Suboptimal sequential decision schemes with on-line feature ordering," *IEEE Trans. Comput.*, vol. C-23, pp. 582–590, June 1974.
[126] M. Ichino and K. Hiramatsu, "Suboptimum linear feature selection in multiclass problem," *IEEE Syst., Man, Cybern.*, vol. SMC-4, pp. 28–33, Jan. 1974.
[127] Pattern Recognition Data Bases, available from IEEE Comput. Soc., referenced in Jan. 1974 issue of Computer.
[128] *Special Issue on Feature Extraction and Selection in Pattern Recognition, IEEE Trans. Comput.*, vol. C-20, Sept. 1971.
[129] IMANCO. Bibliography of articles on automatic quantitative microscopy, available from Image Analyzing Computers, Inc., Monsey, New York.
[130] M. Ingram and K. Preston, Jr., "Automatic analysis of blood cells," *Sci. Amer.*, vol. 223, pp. 72–83, Nov. 1970.
[131] T. Ito, "Approximate error bounds in pattern recognition," in *Machine Intelligence*, vol. 7, Edinburgh, Scotland: Edinburgh Univ. Press, Nov. 1972, pp. 369–376.
[132] A. G. Ivakhnenko, "Polynomial theory of complex systems," *IEEE Trans. Syst., Man, Cybern.*, vol. SMC-1, pp. 364–378, Oct. 1971.
[133] J. Jacquez, Ed. *Computer Diagnosis and Diagnostic Methods.* Springfield, Ill.: C. C. Thomas.
[134] N. Jensen, "Practical jobs for optical computers," *Machine Design*, pp. 94–100, Feb. 22, 1973.
[135] A. K. Joshi, "Remarks on some aspects of language structure and their relevance to pattern analysis," *Pattern Recognition*, vol. 5, pp. 365–381, Dec. 1973.
[136] T. Kailath, "The divergence and Bhattacharyya distance measures in signal selection," *IEEE Trans. Commun. Technol.*, vol. COM-15, pp. 52–60, Feb. 1967.
[137] L. Kanal, *Pattern Recognition.* Washington, D.C.: Thompson, also L.N.K., Inc., Silver Spring, Md.
[138] ——, "Generative, descriptive, formal and heuristic modeling in pattern analysis and classification," Rep. Aerosp. Med. Div. AF System Command, Wright-Patterson AFB, Ohio, Contract F33615-69-C-1571; also Univ. Md., College Park, CSC. Tech. Rep. TR-151, Mar. 1971.
[139] ——, "Applications and problems of optical character recognition," in *Proc. Data Processing Supplies Assoc. Annu. Input-Output Systems Meeting*, reprinted as Univ. Md., CSC Tech. Rep. TR-178 AFOSR 71-1982, Mar. 1972.
[140] ——, "Interactive pattern analysis and classification systems: A survey and commentary," *Proc. IEEE*, vol. 60, pp. 1200–1215, Oct. 1972.
[141] ——, Summary of Panel Discussion on Bridging the Gap between Theory and Implementation in Pattern Recognition, Rep. of Special Sessions held at *1st Joint Int. Conf. Pattern Recognition*, NSF, Washington, D.C., Oct. 1973.
[142] L. Kanal and B. Chandrasekaran, "On dimensionality and sample size in statistical pattern classification," in *Proc. Nat. Electronics Conf.*, 1968, pp. 2–7; also appears in *Pattern Recognition*, vol. 3, pp. 225–234, 1971.
[143] ——, "Recognition, machine 'recognition', and statistical approaches," in *Methodologies of Pattern Recognition.* M. S. Watanabe, Ed. New York: Academic, 1969, pp. 317–332.
[144] ——, "On linguistic, statistical and mixed models for pattern recognition," in *Frontiers of Pattern Recognition.* M. S. Watanabe, Ed. New York: Academic, 1972.
[145] L. Kanal and T. J. Harley, "The complexity of patterns and pattern recognition systems," Aerosp. Med. Res. Lab., Wright-Patterson AFB, Ohio, Tech. Rep. AMRI-TR-69-62, Nov. 1969.
[146] A. Kauffman, *Introduction A La Des Sous-Ensembles Flous*, vol. 1, Paris: Masson, 1973.

[147] A. Klinger, "Natural language, linguistic processing, and speech understanding: Recent research and future goals," *Rand Corp.*, R-1377-ARPA, ARPA order no. 189-1, Dec. 1973.
[148] P. A. Kolers and M. Eden, *Recognizing Patterns.* Cambridge, Mass.: M.I.T. Press, 1968.
[149] W. L. Koontz and K. Fukunaga, "A nonlinear feature extraction algorithm using distance transformation," *IEEE Trans. Comput.*, vol. C-21, pp. 56–63, 1972.
[150] W. L. Koontz, P. M. Narendra, and K. Fukunaga, "A branch and bound clustering algorithm," available from IEEE Comput. Group depository.
[151] B. R. Kowalski and C. F. Bender, "The *k*-nearest neighbor classification rule applied to NMR spectra," *Anal. Chem.*, vol. 44, p. 1405, 1972.
[152] B. R. Kowalski and C. F. Bender, "Pattern recognition II: linear and nonlinear methods for displaying chemical data," *J. Amer. Chem. Soc.*, vol. 95, p. 686, 1973.
[153] B. R. Kowalski, P. C. Jurs, and T. L. Isenhour, "Computerized learning machines applied to chemistry problems: Interpretation of infrared spectro metric data," *Anal. Chem.*, vol. 41, p. 1945, 1969.
[154] B. R. Kowalski, T. F. Schatzki, and F. H. Stross, "Classification of archeological artifacts by applying pattern recognition to trace element data," *Anal. Chem.*, vol. 44, p. 2176, 1972.
[155] P. R. Krishnaiah, Ed. *Multivariate Analysis*, vol. II. New York: Academic, 1969.
[156] C. A. Kulikowski, "Pattern recognition approach to medical diagnosis," *IEEE Trans. Syst. Sci. Cybern.*, vol. SSC-6, pp. 173–178, July 1970.
[157] M. C. Kyle and E. D. Freis, "Computer identification of systolic time intervals," *Comput. and Biomed. Res.*, vol. 3, pp. 637–651, 1971.
[158] P. A. Lachenbruch and R. M. Mickey, "Estimation of error rates in discriminant analysis," *Technometrics*, vol. 10, pp. 1–11, 1968.
[159] P. A. Lachenbruch, C. Sneeringer, and L. T. Revo, "Robustness of the linear and quadratic discriminant function to certain types of non-normality," *Commun. in Statist.*, vol. 1, p. 39–56, 1973.
[160] D. G. Lainiotis, "A class of upper bounds on probability of error for multihypotheses pattern recognition," *IEEE Trans. Inform. Theory.* (Corresp.), vol. IT-15, pp. 730–731, Nov. 1969.
[161] D. G. Lainiotis and S. K. Park, "Probability of error bounds," *IEEE Trans. Syst., Man, Cybern.*, vol. SMC-1, pp. 175–178, Apr. 1971.
[162] P. T. Lavin, "Stochastic feature selection," Ph.D. dissertation, Brown Univ., Providence, R.I., 1972; Rep. no. 3 on Pattern Analysis Center for Comp. and Inform. Sci., Div. of Appl. Math., Brown Univ., Providence, R.I.
[163] E. Lawler and D. Wood, "Branch and bound methods: A survey," *Oper. Res.*, vol. 14, pp. 699–719, July–Aug. 1966.
[164] H. C. Lee and K. S. Fu, "A stochastic syntax analysis procedure and its application to pattern classification," *IEEE Trans. Comput.*, vol. C-21, pp. 660–666, July 1972.
[165] A. Levine, L. Lustick, and B. Saltzberg, "The nearest neighbor rule for small samples drawn from uniform distributions," *IEEE Trans. Inform. Theory*, vol. IT-19, p. 697–699, Sept. 1973.
[166] M. D. Levine, "Feature extraction: A survey," *Proc. IEEE*, vol. 57, pp. 1391–1407, Aug. 1969.
[167] P. H. Lindsay and D. Norman, *Human Information Processing–An Introduction To Psychology.* New York: Academic, 1972.
[168] B. S. Lipkin and A. Rosenfeld, Eds. *Picture Processing and Psychopictorics.* New York: Academic, 1970.
[169] T. Lissack and K. S. Fu, "Error estimation and its application to feature extraction in pattern recognition," Purdue Univ. Lafayette, Ind., Rep. TR-EE73-25, Aug. 1973.
[170] H. Marko, "A biological approach to pattern recognition," *IEEE Syst., Man, Cybern.*, vol. SMC-4, pp. 34–39, Jan. 1974.
[171] D. E. McClure, "Problems and methods of nonlinear feature generation in pattern analysis," presented at the 1974 Princeton Conf. on Inform. Sciences and Systems.
[172] K. G. Mehrotra, "Some further considerations on probability of error in discriminant analysis," unpublished Rep. on RADC Contract no. F30602-72-C-0281, 1973.
[173] W. S. Meisel, *Computer-Oriented Approaches to Pattern Recognition.* New York: Academic, 1972.
[174] J. M. Mendel and K. S. Fu, *Adaptive Learning, and Pattern Recognition Systems: Theory and Applications.* New York: Academic, 1970.
[175] M. Michael and W. C. Lin, "Experimental study of information measure and inter-intra class distance ratios on feature selection and orderings," *IEEE Trans. Syst., Man, Cybern.*, vol. SMC-3, pp. 172–181, Mar. 1973.
[176] R. S. Michalski, "A variable-valued logic system as applied to picture description and recognition," in *Graphic Languages*, F. Nake and A. Rosenfeld, Eds. Amsterdam: North Holland, 1972, pp. 20–47.
[177] R. S. Michalski, "Aqval/1–Computer implementation of a variable-valued logic system VL and examples of its application to pattern recognition," in *Proc. 1st Int. Joint Conf. Pattern Recognition*, IEEE Publ. no. 73 CHO 821-90, 1973.
[178] D. Milgram and A. Rosenfeld, "Array automata and array grammars," *Proc. IFIPS Cong.* Amsterdam: North-Holland, Booklet TA-2, 1971, pp. 166–173.
[179] P. L. Miller, "A locally organized parser for spoken output," Lincoln Lab., M.I.T., Cambridge, Tech. Rep. 503, May 2, 1973.
[180] M. Minsky and S. Papert, *Perceptrons: An Introduction to Computational Geometry.* Cambridge, Mass.: M.I.T. Press, 1969.
[181] B. Moayer and K. S. Fu, "A syntactic approach to fingerprint pattern recognition," in *Proc. 1st Int. Joint Conf. Pattern Recognition*, 1973.
[182] W. S. Mohn, Jr, "Two statistical feature evaluation techniques applied to speaker identification," *IEEE Trans. Comput.*, vol. C-20, pp. 979–987, Sept. 1971.
[183] D. H. Moore, II, "Evaluation of five discrimination procedures for binary variables," *J. Amer. Statist. Ass.*, vol. 68, pp. 399–404, June 1973.
[184] F. Mosteller, "The jackknife," *Rev. Int. Stat. Inst.*, vol. 39, no. 3, pp. 363–368, 1971.
[185] F. Mosteller and J. W. Tukey, "Data analysis, including statistics," in *Handbook of Social Psychology*, G. Lindzey and E. Aronsen, Eds. Reading, Mass.: Addison-Wesley, 1968.
[186] A. N. Mucciardi, "Neuromime nets as the basis for the predictive component of robot brains," in *Cybernetics, Artificial Intelligence and Ecology*, E. Robinson and D. Knight, Eds. Washington, D.C.: Spartan, 1972.
[187] A. N. Mucciardi, "New developments in water shed pollution modeling using adaptive trainable networks," presented at 1974 WWEMA Industrial Water and Pollution Conf. and Exposition Sess. Technology Sources and Progress, Detroit, Michigan, Apr. 1–4, 1974.
[188] A. N. Mucciardi and E. E. Gose, "A comparison of seven techniques for choosing subsets of pattern recognition properties," *IEEE Trans. Comput.*, vol. C-20, pp. 1023–1031, Sept. 1971.
[189] I. Muchnik, "Simulation of process of forming the language for description and analysis of the forms of images," *Pattern Recognition*, vol. 4, pp. 101–140. 1972.
[190] V. K. Murthy, "Estimation of probability density," *Ann. Math. Stat.*, vol. 36, pp. 1027–1031, June 1965.
[191] R. N. Nagel and A. Rosenfeld, "Steps toward handwritten signature verification," in *Proc. 1st Int. Joint Conf. Pattern Recognition*, Publ. no. 73 CHO 821-9C, pp. 59–66, 1973.
[192] G. Nagy, "State of the art in pattern recognition," *Proc. IEEE*, vol. 56, pp. 836–862, May 1968.
[193] ——, "Digital image-processing activities in remote sensing for earth resources," *Proc. IEEE*, vol. 60, pp. 1177–1200, Oct. 1972.
[194] F. Nake, "A proposed language for the definition of arbitrary two-dimensional signs," in *Pattern Recognition in Biological and Technical Systems*, Grusser and Klinke, Eds. Berlin: Springer, 1971, pp. 396–402.
[195] R. Narasimhan and V. Reddy, "A syntax-aided recognition scheme for handprinted English letters," *Pattern Recognition*, vol. 3, pp. 345–361, Nov. 1971.
[196] G. D. Nelson and D. M. Levy, "A dynamic programming approach to the selection of pattern features," *IEEE Trans. Syst. Sci. Cybern.*, vol. SSC-4, pp. 145–151, July 1968.
[197] A. Newell, *et al.*, *Speech Understanding Systems.* New York: Elsevier, 1973.
[198] N. J. Nilsson, *Learning Machines.* New York: McGraw-Hill, 1965.
[199] N. J. Nilsson, *Problem-Solving Methods in Artificial Intelligence.* New York: McGraw-Hill, 1971.
[200] ——, "Artificial intelligence," in *Proc. IFIP Cong.*, 1974.
[201] R. A. Northouse and K. S. Fu, "Dynamic scheduling of large digital computer systems using adaptive control and clustering techniques," *IEEE Trans. Syst., Man, Cybern.*, vol. SMC-3, pp. 225–234, May 1973.
[202] M. Oda, B. F. Womack, and K. Tsubouchi, "A pattern recognizing study of palm reading," *IEEE Trans. Syst., Man, Cybern.*, vol. SMC-1, pp. 171–174, Apr. 1971.
[203] P. L. Odell and B. S. Duran, "Comparison of some classification techniques," *IEEE Trans. Comput.*, vol. C-23, pp. 591–596, June 1974.
[204] P. L. Odell, B. S. Duran, and W. A. Coberly, "On the table look-up in discriminate analysis," *J. Statist. Comput. Simul.*, vol. 2, pp. 171–188, 1973.
[205] M. Okamoto, "An asymptotic expansion for the distribution of the linear discriminant function," *Ann. Math. Stat.*, vol. 34, pp. 1286–1301, 1963.
[206] ——, "Correction to an asymptotic expansion for the linear discriminant function," *Ann. Math. Stat.*, vol. 39, pp. 135, 1968.

[207] D. R. Olsen and K. Fukunaga, "Representation of nonlinear data surfaces," *IEEE Trans. Comput.*, vol. C-22, pp. 915–922, Oct. 1973.
[208] P. A. Ota, "Mosaic grammars," Moore Sch. Elec. Eng., Univ. Penn., Philadelphia, Rep. no. 73-13, 1973.
[209] C. K. Pang, A. J. Koivo, and A. H. El-Abiad, "Application of pattern recognition to steady-state security evaluation in a power system," *IEEE Trans. Syst., Man, Cybern.*, vol. SMC-3, pp. 622–631, Nov. 1973.
[210] E. Parzen, "An estimation of a probability density function and mode," *Ann. Math. Stat.*, vol. 33, pp. 1065–1376, 1962.
[211] E. A. Patrick, *Fundamentals of Pattern Recognition.* Englewood Cliffs, N.J.: Prentice-Hall, 1972.
[212] E. A. Patrick and F. P. Fischer, II, "Nonparametric feature selection," *IEEE Trans. Inform. Theory.*, vol IT-15, pp. 577–584, Sept. 1969.
[213] E. A. Patrick, F. P. Stelmack, and L. Y. Shen, "Review of pattern recognition in medical diagnosis and consulting relative to a new system model," *IEEE Syst., Man, Cybern.*, vol. SMC-4, pp. 1–16, Jan. 1974.
[214] T. Pavlidis, "Linguistic analysis of waveforms," in *Software Engineering*, vol. 2, J. T. Tou, Ed. New York: Academic, 1971, pp. 203–225.
[215] T. Pavlidis, "Waveform segmentation through functional approximation," *IEEE Trans. Comput.*, vol. C-22, pp. 689–697, July 1973.
[216] L. S. Penrose and D. Loesch, "Dermatoglyphic patterns and clinical diagnosis by discriminant functions," *Ann. Hum. Genetics*, vol. 35, p. 51, 1971.
[217] L. S. Penrose and D. Loesch, "Diagnosis with dermatoglyphic discriminants," *J. Mental Def. Res.*, vol. 15, p. 185, 1971.
[218] D. W. Peterson, "Some convergence properties of a nearest neighbor decision rule," *IEEE Trans. Inform. Theory*, vol. IT-16, pp. 26–31, Jan. 1970.
[219] J. L. Pfaltz and A. Rosenfeld, "Web grammars," in *Proc. Joint Int. Conf. Artificial Intelligence*, 1969, pp. 609–619.
[220] H. Pipberger, R. Dunn, and J. Cornfield, "First and second generation computer programs for diagnostic ECG and VCG classification," in *Proc. XII Int. Colloquium Vectorcardiography*, Amsterdam: North Holland, 1972.
[221] M. L. Piccoli, "Pattern recognition and social choice theory," Ph.D. dissertation, Purdue Univ., Lafeyette, Ind., May 1974.
[222] J. Prewitt, "Parametric and nonparametric recognition by computer: An application to leukocyte image processing," in *Advances in Computers*, vol. 2. New York: Academic, 1972, pp. 285–414.
[223] J. Prewitt and M. Mendelsohn, "The analysis of cell images," *Ann. N.Y. Acad. Sci.*, vol. 128, pp. 1035–1053, 1966.
[224] C. P. Quesenberry and M. P. Gessaman, "Nonparametric discrimination using tolerance regions," *Ann. Math. Stat.*, vol. 39, no. 2, pp. 664–673, 1968.
[225] D. R. Reddy, *Invited Papers to the IEEE Symposium on Speech Recognition.* New York: Academic, 1974.
[226] D. R. Reddy, *et al.*, "The hearsay speech understanding system," in *Proc. 3rd Int. Joint Conf. Artificial Intelligence*, 1973, pp. 185–193.
[227] E. M. Riseman and A. R. Hanson, "A contextual postprocessing system for error correction using binary *n*-grams," *IEEE Trans. Comput.*, vol. C-23, pp. 480–493, May 1974.
[228] F. Rosenblatt, "Perceptron simulation experiments," *Proc. IRE*, vol. 48, pp. 301–309, Mar. 1960.
[229] A. Rosenfeld, *Picture Processing by Computer.* New York: Academic, 1969.
[230] A. Rosenfeld, "Picture processing: 1972," *Comput. Graphics and Image Processing*, vol. 1, pp. 394–416, 1972.
[231] ——, "Progress in picture processing: 1969–1971," *Computing Surveys*, vol. 5, pp. 81–108, 1973.
[232] ——, "Picture processing: 1973," Comput. Sci. Center, Univ. Md., Tech. Rep. TR-284, Jan. 1974; to appear in *Comput. Graphics and Image Processing.*
[233] P. Ross, "Computers in medical diagnosis," *CRC Critical Reviews in Radiological Science*, vol. 3, p. 197, 1972.
[234] W. C. Rounds, "Context-free grammars on trees," in *Proc. ACM Symp. Theory Computing*, 1969.
[235] J. W. Sammon, *et al.*, "Handprinted character recognition," U.S. Patent 3 755 780, Aug. 1973; also ASTIA document AD 755 936, Jan. 1973.
[236] J. W. Sammon, Jr., "On-line pattern analysis and recognition system (OLPARS)," Rome Air Develop. Center, Tech. Rep. TR-68-263, Aug. 1968.
[237] ——, "A nonlinear mapping for data structure analysis," *IEEE Trans. Comput.*, vol. C-18, pp. 401–409, May 1969.
[238] ——, "An optimal discriminant plane," *IEEE Trans. Comput.* (Short Notes), vol. C-19, pp. 826–829, Sept. 1970.
[239] ——, "Techniques for discrete type-II data processing on-line (AMOLPARS)," Tech. Rep. RADC-TR-71-243, 1971.
[240] G. N. Saridis and R. F. Hofstadter, "A pattern recognition approach to the classification of nonlinear systems from input-output data," presented at the IEEE Conf. Systems, Man, and Cybernetics, 1973.
[241] K. M. Sayre, *Recognition: A Study in the Philosophy of Artificial Intelligence.* Notre Dame, Ind.: Univ. of Notre Dame Press, 1965.
[242] H. J. Scudder, "Probability of error of some adaptive pattern-recognition machines," *IEEE Trans. Inform. Theory*, vol. IT-11, pp. 363–371, July 1965.
[243] A. C. Shaw, "Parsing of graph-representable pictures," *J. Ass. Comput. Mach.*, vol. 17, pp. 453–481, June 1970.
[244] J. C. Simon and A. Checroun, "Pattern linguistic analysis invariant for plane transformations," in *2nd Int. Joint Conf. A.I.*, British Comput. Soc., 1971, pp. 308–317.
[245] J. C. Simon, C. Roche, and G. Sabah, "On automatic generation of pattern recognition operators," in *Proc. 1972 Int. Conf. Cybernetics and Soc.*, pp. 232–238, IEEE Publ. no. 72 CHO647-8SMC.
[246] R. Siromoney, K. Krithivasan, and G. Siromoney, "*n*-dimensional array languages and description of crystal symmetry I and II," *Proc. Indian Acad. Sci.*, vol. 78, pp. 72–78 and 130–139, 1973.
[247] J. Sklansky, Ed. *Pattern Recognition: Introduction and Foundations.* Stroudsburg, Pa.: Dowden, Hutchinson, and Ross, 1973.
[248] L. V. Sprouse, D. L. Zuetle, and C. A. Harlow, "Automatic verification system for human signature images," in *Proc. 1974 Princton Conf. Information Sciences*, Dep. Elec. Eng., Princeton Univ., Princeton, N.J., 1974.
[249] G. Stockman, L. Kanal, and M. Kyle, "The design of a waveform parsing system," in *Proc. 1st Int. Joint Conf. Pattern Recognition*, 1973.
[250] ——, "An experimental waveform parsing system," in *Proc. 2nd Int. Joint Conf. Pattern Recognition*, 1974.
[251] J. C. Stoffel, "On discrete variables in pattern recognition," Ph.D. dissertation, Dep. Elec. Eng., Syracuse Univ., Syracuse, N.Y. Aug. 1972.
[252] ——, "A classifier design technique for discrete variable pattern recognition problems," *IEEE Trans. Comput.*, vol. C-23, pp. 428–441, Apr. 1974.
[253] D. Stowens and J. W. Sammon, Jr., "Dermatoglyphics and leukemia," *Lancet*, pp. 846, Apr. 18, 1970.
[254] D. Stowens, J. W. Sammon, and A. Proctor, "Dermatoglyphics in female schizophrenia," *Psychiatric Quart.*, vol. 44, pp. 516–532, July 1970.
[255] C. W. Swonger, "Property learning in pattern recognition systems using information content measurements," in *Pattern Recognition*, L. Kanal, Ed. Washington, D.C.: Thompson; L.N.K. Silver Spring, Md., 1966, pp. 329–347.
[256] S. J. Tauber and K. Rankin, "Valid structure diagrams and chemical gibberish," *J. Chem. Documentation*, vol. 12, no. 1, pp. 30–34, 1972.
[257] J. M. Tenenbaum, "On locating objects by their distinguishing features in multisensory images," A.I. Center, Stanford Res. Inst., Menlo Park, Calif., Tech. Note 84, SRI Project 1187 and 1530, Sept. 1973.
[258] J. M. Tenenbaum, *et al.*, "An interactive facility for scene analysis research," Artificial Intelligence Center, Stanford Res. Inst., Menlo Park, Calif., Tech. Note 87, SRI Project 1187, Jan. 1974.
[259] J. T. Tou, *Advances in Information Systems Science.* New York: Plenum, 1970.
[260] ——, *Software Engineering*, vol. II, New York: Academic, 1971.
[261] G. T. Toussaint, "Note on optimal selection of independent binary-valued features for pattern recognition," *IEEE Trans. Inform. Theory* (Corresp.), vol. IT-17, p. 618, Sept. 1971.
[262] ——, "Distance measures as measures of certainty and their application to statistical pattern recognition," in *Proc. Annu. Conf. Statistical Science Assoc. Canada*, 1973.
[263] ——, "Bibliography on estimation of misclassification," *IEEE Trans. Inform. Theory*, vol. IT-20, pp. 472–479, July 1974.
[264] G. T. Toussaint and R. W. Donaldson, "Some simple contextual decoding algorithms applied to recognition of handprinted text," in *Proc. Annu. Canadian Comput. Conf.*, 1972.
[265] G. T. Toussaint and P. M. Sharpe, "An efficient method for estimating the probability of misclassification applied to a problem in medical diagnosis," Sch. Comput. Sci., McGill Univ., Montreal, Canada, Nov. 1973.
[266] G. V. Trunk, "Parameter identification using intrinsic dimensionality," *IEEE Trans. Inform. Theory*, vol. IT-18, pp. 126–133, Jan. 1972.
[267] Ya. Z. Tsypkin, *Adaptation and Learning in Control Systems.* New York: Academic, 1971.
[268] ——, *Foundations of the Theory of Learning Systems.* New York: Academic, 1973.
[269] L. Uhr, "Flexible linguistic pattern recognition," in *Pattern*

Recognition, vol. 3, New York: Wiley, Nov. 1971, pp. 363–387.

[270] ——, *Pattern Recognition, Learning, and Thought.* Englewood Cliffs, N.J.: Prentice-Hall, 1973.

[271] J. R. Ullmann, *Pattern Recognition Techniques.* London: Butterworth, 1973.

[272] ——, "The use of continuity in character recognition," *IEEE Trans. Syst., Man, Cybern.*, vol. SMC-4, pp. 294–300, May 1974.

[273] W. E. Underwood and L. Kanal, "Structural descriptions, transformational rules, and pattern analysis," in *Proc. 1st Int. Joint Conf. Pattern Recognition*, Publ. no. 73 CHO 821-9C, 1973, pp. 434–444.

[274] I. Vajda, "Note on discrimination information and variation," *IEEE Trans. Inform. Theory*, vol. IT-16, pp. 771–773, Sept. 1970.

[275] C. J. Verhagen, "Some aspects of pattern recognition in the future," in *Proc. 1st Int. Joint Conf. Pattern Recognition*, IEEE Special Publ. 73 CHO 821-9C, 1973, pp. 541–545.

[276] T. R. Vilmansen, "Feature evaluation with measures of probability dependence," *IEEE Trans. Comput.*, vol. C-22, Apr. 1973.

[277] T. J. Wagner, "Convergence of the edited nearest neighbor," *IEEE Trans. Inform. Theory* (Corresp.), vol. IT-19, pp. 696–697, Sept. 1973.

[278] T. J. Wagner, "Convergence of the nearest neighbor rule," *IEEE Trans. Inform. Theory*, vol. IT-17, pp. 566–571, Sept. 1971.

[279] ——, "Deleted estimates of the Bayes risk," *Ann. Statist.*, vol. 1, pp. 359–362, Mar. 1973.

[280] D. E. Walker, "SRI speech understanding system," Contributed Papers to the *IEEE Symp. on Speech Recognition*, IEEE Catalog no. 74CHO878-9 AE, 1974, pp. 32–37.

[281] R. F. Warmack and R. C. Gonzalez, "An algorithm for the optimal solution of linear equalities and its application to pattern recognition," *IEEE Trans. Comput.*, vol. C-22, pp. 1065–1074, Dec. 1973.

[282] S. Watanabe, *Knowing and Guessing.* New York: Wiley, 1969.

[283] S. Watanabe, Ed., *Methodologies of Pattern Recognition.* New York: Academic, 1969; Review by Richard Duda in *IEEE Trans. Inform. Theory*, vol. IT-17, pp. 633–634, Sept. 1971.

[284] S. Watanabe, Ed., *Frontiers of Pattern Recognition.* New York: Academic, 1972.

[285] S. Watanabe and N. Pakvasa, "Subspace method in pattern recognition," in *Proc. 1st Joint Int. Conf. Pattern Recognition*, IEEE Special Publ. No. 73CHO821-9C, 1974.

[286] W. G. Wee, "Generalized inverse approach to adaptive multiclass pattern classification," *IEEE Trans. Comput.*, vol. C-17, pp. 1157–1164, Dec. 1968.

[287] A. Whitney, "A direct method of nonparametric measurement selection," *IEEE Trans. Comput.* (Short Notes), vol. C-20, pp. 1100–1103, Sept. 1971.

[288] D. L. Wilson, "Asymptotic properties of nearest neighbor rules using edited data," *IEEE Trans. Syst., Man, Cybern.*, vol. SMC-2, pp. 408–421, July 1972.

[289] C. T. Wolverton, "Strong consistency of an estimate of an asymptotic error probability of the nearest neighbor rule," *IEEE Trans. Inform. Theory*, vol. IT-19, p. 119, Jan. 1973.

[290] W. E. Wright, "A formalization of cluster analysis," *Pattern Recognition*, vol. 5, pp. 272–282, Sept. 1973.

[291] Y. Yakimovsky and J. A. Feldman, "A semantics-based decision region analyzer," in Advance Papers of *3rd Int. Joint Conf. Artificial Intelligence*, 1973, pp. 580–588; available from Stanford Res. Inst., Menlo Park, Calif.

[292] I. T. Young, "Automated leucocyte recognition," Ph.D. dissertation, Dep. Elec. Eng., M.I.T., Cambridge, Mass., June 1969.

[293] T. Young, "Reliability of linear feature extractors," *IEEE Trans. Comput.*, vol. C-20, pp. 967–971, Sept. 1971.

[294] T. Y. Young and T. W. Calvert, *Classification, Estimation and Pattern Recognition.* New York: Elsevier, 1974.

[295] L. A. Zadeh, "Fuzzy sets," *Inform. Contr.*, vol. 8, p. 338, 1965.

[296] ——, "Outline of a new approach to the analysis of complex systems and decision processes," *IEEE Trans. Syst., Man, Cybern.*, vol. SMC-3, pp. 28–44, Jan. 1973.

[297] C. Zywietz and B. Schneider, Eds., *Computer Application on ECG and VCG Analysis.* New York: Elsevier, 1973.

[298] Symp. Rec. IEEE Symp. Feature Extraction and Selection in Pattern Recognition, Argonne, Ill. Oct. 5–7, 1970.

[299] Data Processing Supplies Assoc. Ann. Input–Output Systems Meeting, DPSA, Stamford, Conn, 1971.

[300] *Special Issue of Pattern Recognition on Syntactic Pattern Recognition, Pattern Recognition*, vol. 3, 1971.

[301] *1st Int. Electron. Crime Countermeasures Conf.*, IEEE Special Pub. no. 73 CHO 813-6AES, 1973.

[302] *1st Int. Joint Conf. Pattern Recognition*, IEEE Publ. no. 73 CHO 821-9C, 1973.

[303] Symp. significant results obtained from Earth resources technology satellite-1, Goddard Space Flight Center, Greenbelt, Md., S. C. Freden, E. P. Mercanti, and D. E. Whitten, Eds., May 1973.

[304] *Proc. Conf. Biologically Motivated Automata Theory*, IEEE Publ. no. 74CHO889-6C, 1974.

Evaluation of Improved Digital-Processing Techniques of Landsat Data for Sulfide Mineral Prospecting

By R. G. Schmidt,
U.S. Geological Survey, Reston, Virginia
and Ralph Bernstein,
International Business Machines Corporation, Gaithersburg, Maryland

ABSTRACT

A relatively simple method of digital computer classification of multispectral scanner data was tested at an ideal porphyry copper deposit in a very arid part of Pakistan and was then successfully applied to mineral exploration in an adjacent region. The surface expressions of the already known porphyry copper deposit and the five new prospects discovered in this experiment are all characterized by abundant light-toned sulfate minerals and do not seem to contain much pigmentation by iron oxides.

Digital multispectral classification was performed at the IBM Digital Image Processing Facility by using reformatted computer-compatible tapes of one scene. A test area of 55 km^2 was selected in the Saindak, Pakistan, area, which included a well-mapped porphyry copper deposit. A "supervised" classification table was prepared of high and low limits of acceptable reflectance in each of the four spectral bands for each surface type by using control areas selected from a geologic map. The first classification table was revised several times until it was deemed capable of giving information useful in mineral exploration, even though we expected that many of the rock identifications would be incorrect. This revised table was applied to evaluation of 2,100 km^2 in the Chagai District, Pakistan. Of 50 sites classified as "mineralized" by the digital-processing program, 23 were selected as possibly similar to the control area and therefore deserving of inspection in the field. Nineteen of these were visited in October 1974; five of the sites, constituting a total area of 4.7 km^2, contain sulfide-bearing hydrothermally altered rock that is mostly quartz-feldspar porphyry. Parts of all five sites are believed to represent parts of porphyry copper systems. The presence of copper beyond trace amounts has been established at two points, and the prospects are under investigation by the Government of Pakistan.

It may be important when using remote-sensing methods in some geologic environments to seek altered parts of porphyry copper deposits that are not specifically enriched or pigmented by iron oxides.

Our experiment indicates that simple methods of digital classification of Landsat data can aid in the location of mineral deposits in desert terrain. These methods can complement conventional methods of mineral exploration.

A major problem in identifying the sulfide-bearing areas is the overlap of reflectance values of mineralized areas with the reflectance values from other high-albedo areas, such as parts of drywashes and, most commonly, areas of windblown sand. In an effort to solve this problem, new classification tables using several classes for each rock type have been prepared. Each class has relatively close high/low limits; several classes are therefore needed to cover the full range of albedo that can be expected from each major surface type, as for example, from similar rock occurring on slopes of different orientation to the Sun. The narrowed reflectance limits increase the possibility of discrimination between materials of generally similar reflectance values. Recent tests of these more sophisticated classification tables suggest that we can now achieve better discrimination of mineralized areas than we did in the evaluation study in 1974. Future satellite-borne sensors having higher sensitivity, wider

Reprinted from ***Proc. 1st Annu. William T. Pecora Memorial Symp.***, P. W. Woll and W. A. Fischer, Eds., U.S. Geol. Survey Prof. Paper 1015, Oct. 1975, pp. 201–212.

spectral range, narrower spectral bands, and greater resolution should help to alleviate the reflectance-overlap problem.

Our experiment used high-albedo areas associated with intense quartz-sericite alteration as guides to sulfide mineralization although some of the control areas do contain some hematitic and some jarositic pigmentation. Attempts to use strongly red stained areas in the outer pyrite zone as control areas resulted in no satisfactory delineation of the area of mineralization. All prophyry copper deposits may not include areas of light-toned sulfate and/or clay alteration; also, not all known deposits have "red-thumbs" or areas of strong red iron oxide pigmentation derived from decomposition of pyrite.

INTRODUCTION

The concept of using remote-sensing methods at the Saindak porphyry copper deposit predated Landsat–1 by almost 10 years, for it was discussed by the geologists during field mapping from 1961 to 1966, and some simple spectral tests were made at that time using ordinary color photography and colored filters (Schmidt, 1968, p. 59). Two investigations that have sought a simple method of identifying sulfide deposits using Landsat multispectral scanner data have used the Saindak deposit as a test site because the deposit is large, has well-developed alteration zones, and scant vegetation and is well exposed and undisturbed (fig. 1). The deposit has been mapped in detail at a scale of 1:6,000 (Khan, 1972), and the senior author was familiar with the deposit and region. Both investigations used image 1125–05545 (Nov. 25, 1972) and the first experiment also used image 1090–05595 (Oct. 21, 1972) and 1124–05491 (Nov. 24, 1972). In the first experiment, false-color composites were visually examined to locate possible favorable areas; in the second, favorable areas were

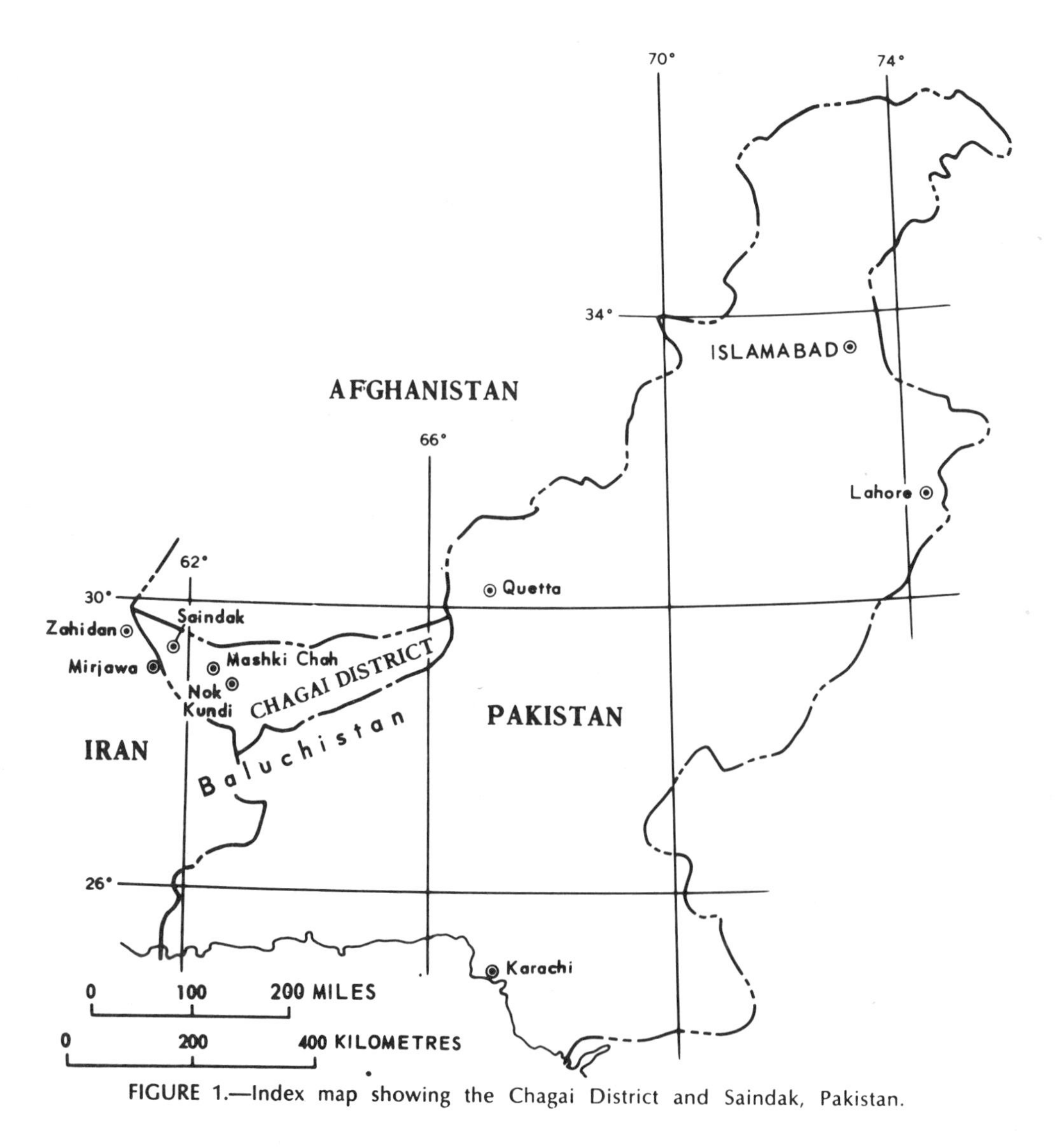

FIGURE 1.—Index map showing the Chagai District and Saindak, Pakistan.

DELLWIG AND MOORE

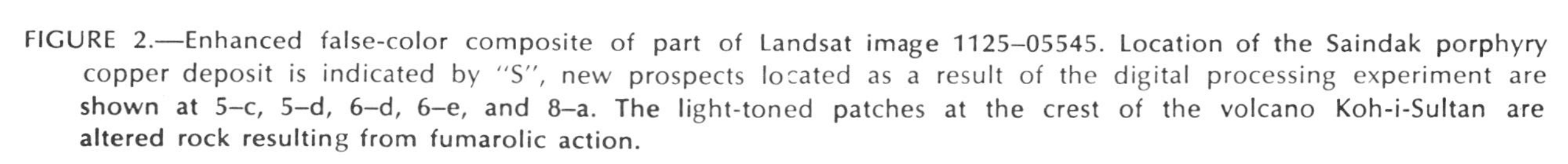

FIGURE 2.—Enhanced false-color composite of part of Landsat image 1125–05545. Location of the Saindak porphyry copper deposit is indicated by "S", new prospects located as a result of the digital processing experiment are shown at 5–c, 5–d, 6–d, 6–e, and 8–a. The light-toned patches at the crest of the volcano Koh-i-Sultan are altered rock resulting from fumarolic action.

SCHMIDT AND BERNSTEIN

selected by digital computer classification. The near-winter solstice date of these images and the resultant low Sun illumination angle was probably an advantage in the first experiment and a disadvantage in the second.

Both experiments were designed to test relatively simple methods of utilizing Landsat multispectral scanner (MSS) data in mineral exploration. Simplicity is considered important because we are looking for a method that may find wide usage in both government and private sector mineral exploration applications.

For the first experiment, false-color composites were made using NASA system corrected MSS images of three scenes over a very arid and vegetation-free part of the western Chagai District, Pakistan. Later, a color composite was also prepared from MSS bands 4, 5, and 7 by computer processing of the taped data to produce an image that was geometrically corrected for systematic errors and radiometrically adjusted (fig. 2, p. XXI). These composites were visually examined for tonal features resembling the known porphyry copper deposit at Saindak (Schmidt, 1974). Seven areas were selected for prospecting, and three of these were field checked. None of the three contained significant mineralization. Later analysis of the results indicated that too few areas were selected by visual examination and that this method must be considerably modified for greatest effectiveness in mineral exploration. Some further testing in which more light-toned areas are selected should be made before the results of this experiment can be considered final.

In the second experiment, machine processing was used. Digital multispectral classification was experimentally performed on the reformatted computer-compatible tapes of one scene. The IBM Digital Image Processing Facility was used for image correction, image enhancement, and multispectral classification (fig. 3). A 55-km ^{2} area in the Saindak vicinity was extracted and displayed in photographic form using the imagery printer, and a test area was selected which included known porphyry deposits. Areas representing specific rock types were identified, and their reflectance values were extracted for MSS bands 4, 5, 6, and 7. Provisional classification tables were prepared for given geological units; each table was used to classify the test area, and the tables were modified upon comparison of these results with the known geology. The cycle was repeated five times until a classification table was developed that could, it was hoped, provide information useful in mineral evaluation. High accuracy of identification was not considered to be necessary for success of the method, and was not expected.

Then the multispectral classification program, using the classification table already prepared, was used to evaluate 2,100 km ^{2} east of the test area, and a partial field check of the results was conducted. By evaluating the resultant digital classification map, 23 prospecting targets were selected as being similar to the Saindak altered rock area, and 19 of these were visited in the field. Of these 19, 5 localities constituting a total area of 4.7 km ^{2} contain hydrothermally altered rock, mostly quartz feldspar porphyry. A program for evaluation of these new prospects has been prepared by the Government of Pakistan.

Despite the apparent success of the 1974 application experiment, it used classification tables that were relatively simple. While the number of sites falsely classified as mineralized was not unreasonable when compared to other standard field methods of mineral prospecting, we had reason to believe that our method could be refined. This would result in more distinct delineation of mineralized areas, fewer areas falsely classified as mineralized, and perhaps improvement of quality of mapping of other surficial materials. This refinement has been the goal of our continuing experimentation.

Acknowledgments.—The original color composite experiment conducted by the U.S. Geological Survey was partly supported by the National Aeronautics and Space Administration. Financial support for part of the digital processing experiment and for the travel expenses of the field checking was provided by the EROS Program, and logistical support in the field was furnished by the Resource Development Corporation, a Pakistan Government corporation. The extraction of reflectance data for several sites in the experiment area by Jon D. Dykstra of Dartmouth College, using computer facilities and algorithm of the NASA Goddard Institute of Space Studies, New York, is gratefully acknowledged.

GEOLOGY OF THE EXPERIMENT AREA

The general geology of the western Chagai District is portrayed on 1:253,440 photogeologic maps (Hunting Survey Corporation, Ltd., 1960). The rocks of the region consist of marine and terrestrial sedimentary and volcanic rocks of Cretaceous to Holocene age, and a few shallow intrusive bodies. The influence of widespread volcanism is continuous in the geologic record from Cretaceous time to the present. The highest mountains in the area are the dormant volcanoes Koh-i-Sultan in Pakistan and Kuh-e-Taftan in nearby Iran.

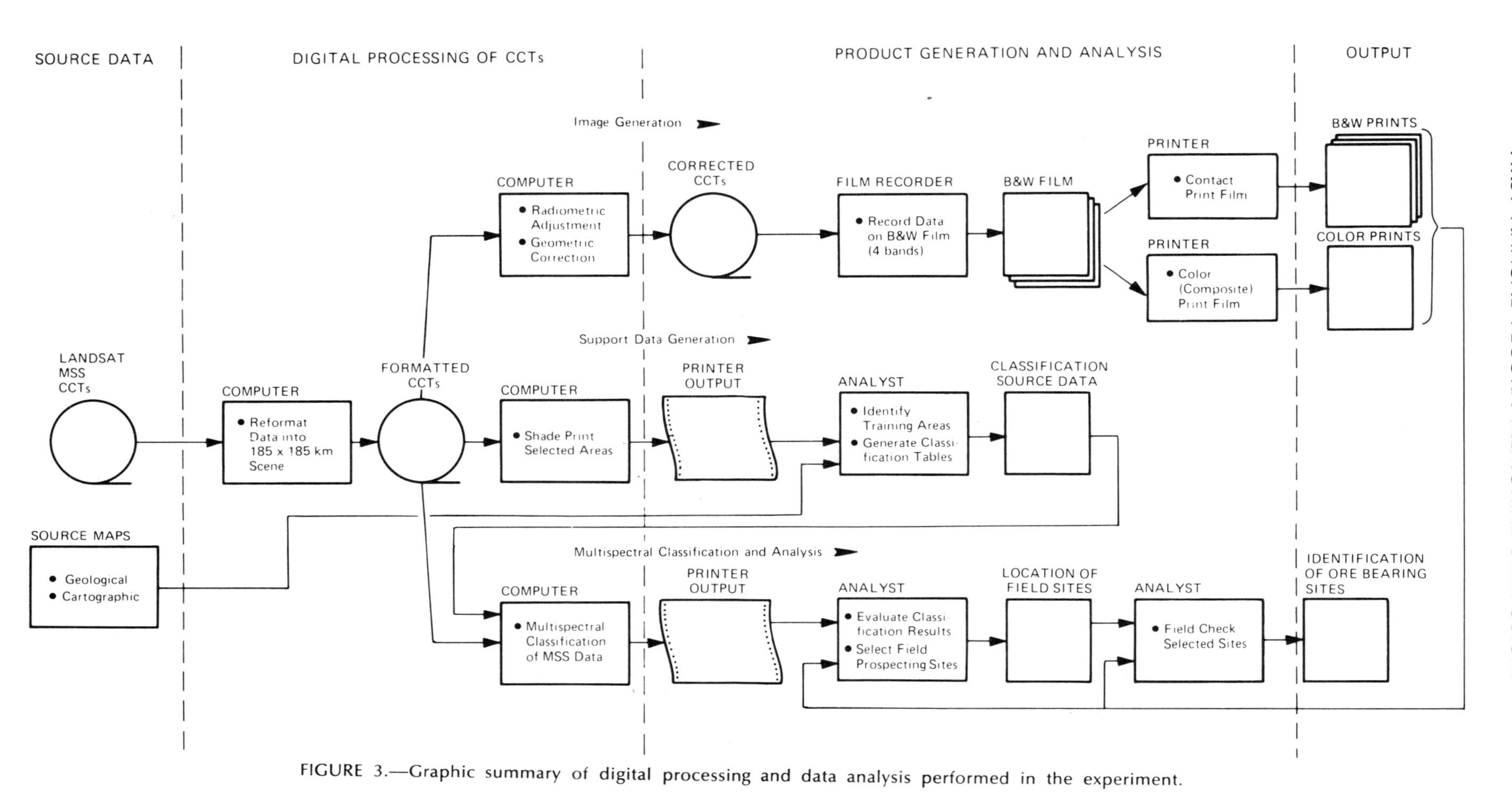

FIGURE 3.—Graphic summary of digital processing and data analysis performed in the experiment.

The Mirjawa range area, where the known porphyry copper deposit, Saindak, is located, has mostly folded and much faulted (along northwest trends) sedimentary and volcanic sedimentary strata. Except for a group of large and extensive pre-Tertiary sills, intrusive and extrusive volcanic rocks make up but little of the total outcrop area, but volcanic material is common in sedimentary strata. Cretaceous rocks represent a wide variety of marine and continental depositional environments; lower Tertiary rocks are mostly shallow marine deposits, and upper Tertiary and Quaternary strata are largely of continental origin.

The Mashki Chah region can be considered geologically as a western extension of the Chagai Hills, west of the volcano, Koh-i-Sultan. In this extension, gently folded Cretaceous and undeformed Tertiary volcanic rocks with low dips (probably initial) are common and widespread, and intrusive rocks, especially stocks, are more abundant than in the Mirjawa ranges. Faults are widespread. The folding lacks the strong linear pattern characteristic of the Mirjawa ranges, and the two areas are dissimilar structurally. Recently dried or still weakly flowing sinter-depositing saline springs, plus small weak fumaroles on Koh-i-Sultan, indicate that there is still a little hydrothermal activity.

ECONOMIC GEOLOGY

Several large areas of the Chagai District contain volcanic rocks of intermediate and felsic composition, within which small bodies of hypabyssal intrusive rocks may be considered to have a good potential for large sulfide deposits of the porphyry copper type. Mineral reconnaissance in the Chagai District has been spotty and mostly for high-grade deposits. Small copper occurrences are widespread in the region, but none had been known within several miles of the new prospects identified in this experiment.

At Saindak, a group of small copper-bearing porphyritic quartz diorite stocks cut northward across the folded lower Tertiary stratigraphic section (Ahmed, Khan, and Schmidt, 1972). The stocks may be cupolas on a single barely exposed granitic body 8 km (5 mi) long and as much as 1.5 km (1 mi) wide or separate but related intrusive bodies. The group of stocks is surrounded by zones of intense hydrothermal alteration very similar to those of the ideal model described by Lowell and Guilbert (1970). At Saindak, the innermost area seen at the surface is a zone of potassic alteration containing hydrothermal biotite (Shahid Noor Khan, oral communication). This is surrounded successively by zones of quartz-sericite alteration and finally a 1- to 3-km zone of propylitic alteration in which pyroclastic rocks in particular are altered to a hard, dark epidote-rich hornfelsic rock.

Copper mineralization is probably restricted to the potassic alteration zone and the inner part of the quartz-sericite zone, both mostly but not entirely located in the intrusive quartz diorite porphyry. The copper-bearing zone is in turn enclosed in a sulfide-rich zone that contains as much as 15 percent by volume pyrite (locally pyrrhotite), probably limited to the outer part of the quartz-sericite alteration zone. Mineralization in the propylitic zone is limited to lean lead-copper-silver veins (Ahmed, Kahn, and Schmidt, 1972, p. A19–A20) and local hematite and pyrrhotite skarns formed by replacement of limestone beds (Ahmed, Khan, and Schmidt, 1972, p. A19).

The pyrite-rich peripheral zone and the innermost quartz-sericite and potassic zones have been eroded to form a cross-structural valley in which the surficial materials are light toned and highly reflective. Soils related to certain parts of many porphyry copper deposits the world over have distinct red and orange anomalies, and this is true at part of the Saindak deposit as well, especially over the pyrite-rich zone in the valley north of the main areas of mineralization. The amount of red and yellow coloration in the most central part of the alteration zone at Saindak seems less than in some peripheral areas but, unfortunately, this has not been measured quantitatively. Alluvial and windblown materials cover part of the light-toned soil, and it is necessary to use much care in selecting specific control areas for any remote-sensing experiment.

COMPUTER-AIDED INFORMATION-EXTRACTION EXPERIMENTS

Computer-aided information-extraction experiments were conducted to identify potential sulfide-ore-bearing localities. The experimental approach is summarized in figure 3. Shown here are the source data used, the digital processing applied to the source data, the products generated, the analysis conducted, and the final products. By a combination of digital image processing and information extraction, and visual analysis and evaluation, three processing operations were performed: digital image generation, support data generation and analysis, and multispectral classification and analysis.

Digital image generation.—The uncorrected Landsat multispectral scanner (MSS) data were reformatted into band separated 185 km x 185 km areas, and each band was radiometrically (intensity) adjusted and systematically geometrically corrected (Bernstein,

1974). The resulting computer-compatible tape (CCT) data were then converted to an image on film from which black and white and color composite prints were made. These prints were used as aids in the selection of the field prospecting sites during the evaluation of the classification results and also during the field checking.

Support data generation and analysis.—The formatted but uncorrected CCT's were used for analysis, especially preparation of classification tables, prior to the multispectral classification operation. Shade prints (computer printouts providing the reflectance values in each spectral band) for selected areas were prepared and used as a geographic reference for precise location of individual data samples or picture elements (pixels) relative to known ground features and rock types. Numeric data for the four MSS bands were extracted for each pixel in the known areas, and maximum and minimum sensed-reflectance limits were chosen for each rock type. A known highly altered and copper sulfide-bearing part of the deposit at Saindak was the source of data used to prepare the classification tables. Five revisions were made and tested, and one alternate classification table was tried, resulting finally in classification table "B" shown on table 1. Table 1 shows three general types of materials: (1) *Mineralized rock*, including intensely hydrothermally altered (quartz-sericite zone) quartz diorite, and pyritic rock, (2) *Alluvial and eolian materials*, recently moved dry-wash alluvium and eolian sand being the materials most likely to be confused with the first group, and (3) *A loose category of dark surfaces* that includes both hornfels-type bedrock and many areas of desert-varnished lag gravels (especially in class 10). These tables of reflectance limits were then used on an interactive basis to classify a nearby region within the same Landsat scene in which copper sulfide-bearing areas were suspected but in which no deposits were known (application area).

Multispectral classification and analysis.—A spectral-intensity discrimination program was used for multispectral classification on the application area using the tables prepared for the Saindak deposit. The program tested the reflectance of each pixel within the application area against the maximum and minimum reflectance limits in the table and determined into which surface class (rock type) the pixel belonged. The symbol for that class was printed as part of a classification map. When the observed values fit more than one class (when classes were set up with overlapping limiting values), a pixel was placed in the class that was considered first in the search sequence.

The digital classification method using table "B" was used to evaluate an area of 2,100 km^2 near Saindak considered to have good potential for porphyry copper deposits (fig. 4). The results were printed out in 13 computer-generated vertical strip maps. These maps were examined for groups of pixels classified as one of the four mineralized classes (two classes each of mineralized quartz diorite and pyritic rock), and about 50 groups or concentrations were identified. Each was then evaluated for probability of correct classification, relationship to concentrations of other classes, and comparison with known rock types and known occurrences of hydrothermal mineralization. From this examination, 30 localities most deserving reconnaissance checking in the field were chosen. The locations of these targets were marked on an enlarged (1:253,440 scale) digitally enhanced image of MSS band 5 in order to simplify location on aerial photo-

TABLE 1.—*Digital classification table "B" used in 1974 mineral evaluation in the Chagai District, Pakistan*

Class number	Rock type	Symbol	Reflectance limits of multispectral scanner bands			
			4	5	6	7
1	Mineralized rock: Quartz diorite, High reliability	0	46–50	52–60	50–60	18–22
2	Mineralized rock: Quartz diorite, Low reliability	Ø	44–45	52–60	45–49	18–19
3	Mineralized rock: Pyritic rock, High reliability	¥	41–45	47–54	39–44	16–17
4	Mineralized rock: Pyritic rock, Low reliability	X	41–45	47–54	39–44	16–19
5	Dry-wash alluvium, High reliability	=	39–46	39–46	35–44	14–17
6	Dry-wash alluvium, Low reliability	-	41–45	46–51	42–49	18–19
7	Boulder fan	+	33–40	39–46	30–35	9–16
8	Eolian sand, High reliability	.	38–44	46–54	42–51	18–22
9	Eolian sand, Low reliability	,	45	46–127	42–127	18–63
10	Various dark surfaces including hornfelsic rock outcrops, desert-varnished lag gravels, and detrital black sand	I	33–36	28–38	29–35	11–15
11	Various dark surfaces including hornfelsic rock outcrops, desert-varnished lag gravels, and detrital black sand	#	24–35	19–27	20–32	8–12
12	Various dark surfaces including hornfelsic rock outcrops, desert-varnished lag gravels, and detrital black sand	II	29–36	28–38	20–28	9–14

TABLE 2.—*Revised classification table "SSD" prepared in April 1975*

Class Number	Rock type	Symbol	Reflectance limits of multispectral scanner bands 4	5	6	7
1	Detrital black sand	,	27–30	26–28	23–26	9
2		,	28–32	27–31	25–27	10
3	Dry wash alluvium	-	38–44	39–44	35–38	13–16
4		=	43–48	44–48	38–43	15–18
5	Eolian sand	.	49–54	59–66	54–62	23–27
6		.	43–48	50–58	46–55	19–24
7		.	37–43	41–50	38–47	15–20
8		.	33–38	35–42	32–39	12–16
9	Mineralized rock	0	39–42	41–45	33–38	14–16
10		θ	41–44	44–48	37–41	16–18
11		θ	43–46	47–51	41–45	17–20
12		0	44–46	50–53	45–48	18–20
13		0	46–49	53–56	48–51	19–21
14		0	49–52	56–59	51–54	20–22
15	Dark surfaces, except detrital black sand	I	33–36	28–38	29–35	11–15
16		#	24–35	19–27	20–32	8–12
17		H	29–36	28–38	20–28	9–14

TABLE 3.—*Revised classification table "S" prepared in September 1975*

Class number	Rock type	Symbol	Reflectance limits of multispectral scanner bands 4	5	6	7
1	Detrital black sand	,	27–30	26–28	23–26	9
2		,	28–32	27–31	25–27	10
3	Dark pebble desert	,,	35–36	34–38	31–34	12–13
4	Dry-wash alluvium	-	35–39	37–42	31–36	13–15
5		=	39–43	41–46	35–41	15–17
6		=	43–47	46–50	41–46	17–19
7	Eolian sand	.	37–43	41–50	38–47	15–20
8		.	43–48	50–58	46–55	19–24
9		.	49–55	59–68	54–63	23–27
10		.	54–60	67–76	62–70	27–30
11		.	60–66	76–85	70–78	30–33
12	Unmineralized felsic intrusive rock	F	48–52	55–60	50–54	21–24
13		F	44–48	51–56	47–51	19–21
14		F	40–44	46–51	43–47	17–19
15	Mineralized rock, pyritic	P	39–42	41–45	33–38	14–16
16		P	41–44	44–48	37–41	16–18
17		P	43–46	47–51	41–45	18–19
18	Mineralized rock, sericitic	Q	46–49	51–55	45–49	19–21
19		Q	48–51	54–58	49–53	21–22
20		Q	50–53	57–62	52–57	22–24
21	Dark surface, except detrital black sand	I	33–36	28–38	29–35	11–15
22		K	24–35	19–27	20–32	8–12
23		H	29–36	28–38	20–28	9–14

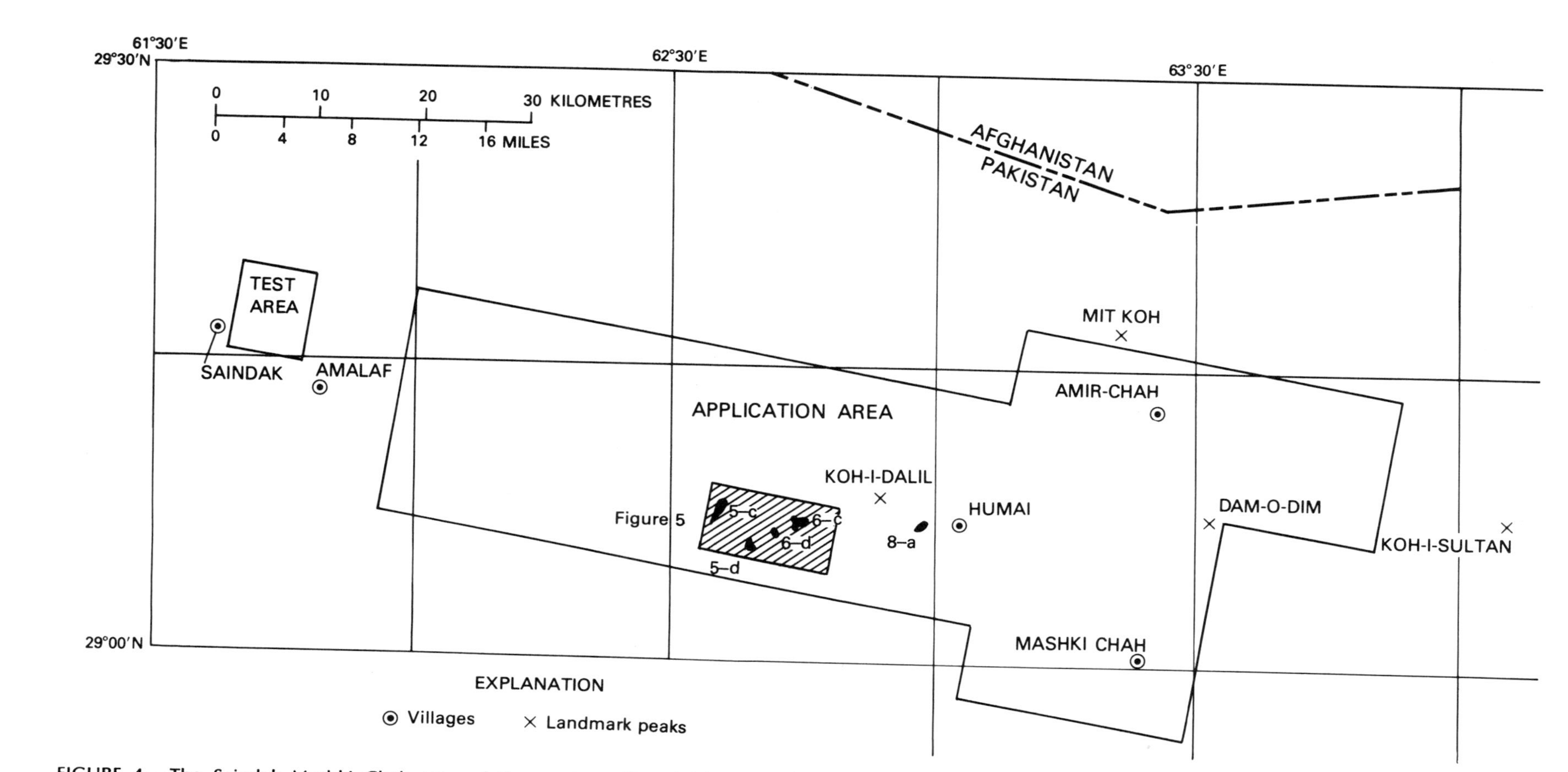

FIGURE 4.—The Saindak–Mashki Chah area of the western Chagai District, Pakistan, showing the control area near Saindak and the area where the digital-classification method was applied. The numbered locations are newly discovered mineralized prospect sites.

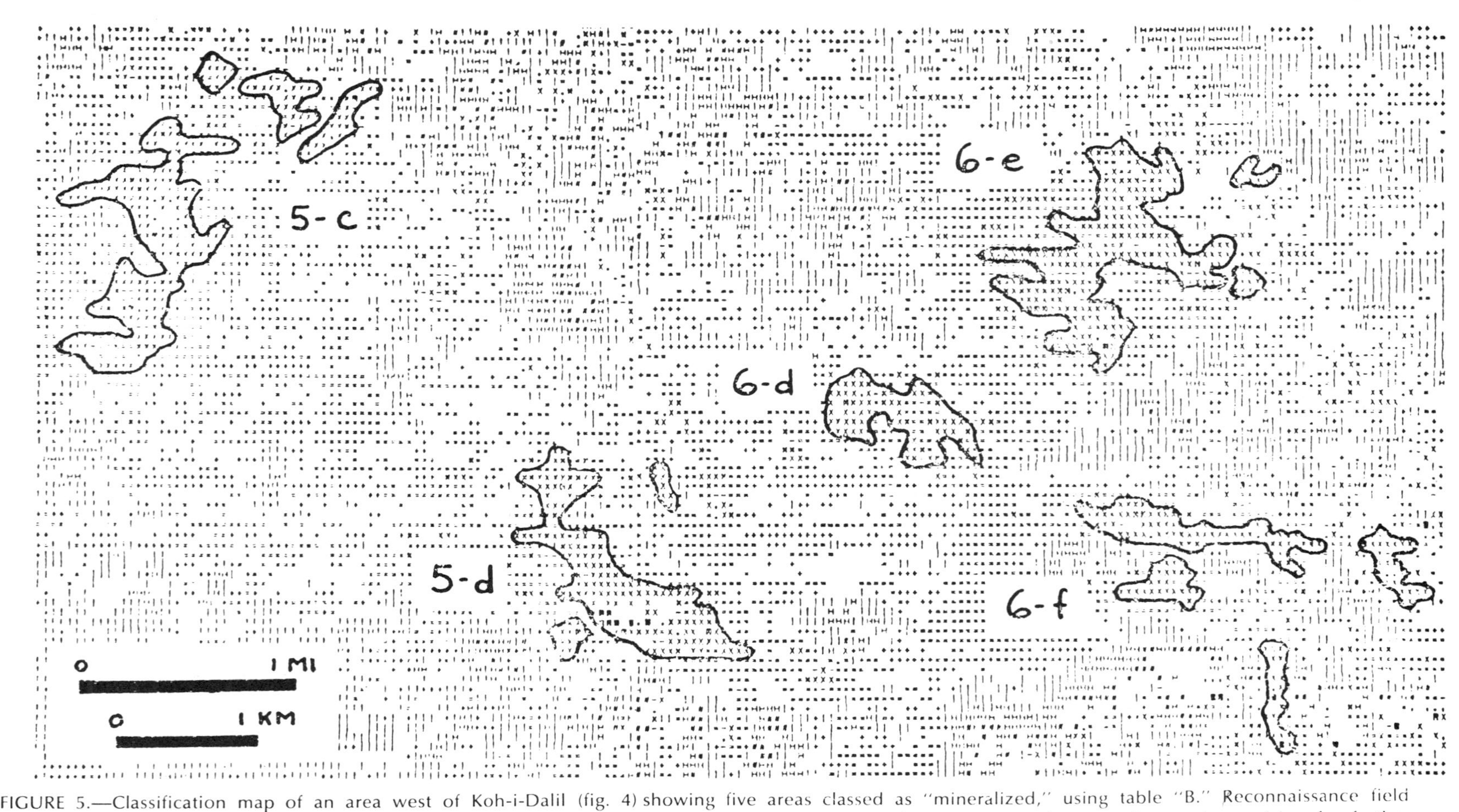

FIGURE 5.—Classification map of an area west of Koh-i-Dalil (fig. 4) showing five areas classed as "mineralized," using table "B." Reconnaissance field checks showed that some hydrothermal alteration and sulfide mineralization is present in each area. Area 6-d seems the most promising for further study. The northeast end of 5-c, the west end of 6-f, and the northeast end of 6-e were not adequately field checked. The east end of 6-f is mostly dune sand.

graphs and in the field. As part of the field check all anomalous areas were first examined on stereoscopic pairs of 1:40,000-scale aerial photographs. At this point, it was possible to reject seven more areas as probably related to windblown sand.

Nineteen sites were examined in the field, and four desirable sites were not reached in the field checks. Five sites were found to be extensive outcrops of hydrothermally altered sulfide-rich rock (figs. 4 and 5). Two additional sites contain altered rock with some sulfide but seem less attractive for prospecting at this time.

EVALUATION OF RESULTS

Classification using table "B".—In the evaluation using table "B," mineralized rock is believed to have been generally identified correctly, but much other light-toned surface was also classified as mineralized. Alluvium was mostly classified accurately, but some gravel fans derived from light-toned rocks were not. Perhaps only 25 percent of dune sand was correctly identified by means of table "B," most sand having been classified as mineralized rock. Classification of areas of dark rock outcrops and several types of lag gravels yielded different combinations of classes 10, 11, and 12 and made it possible to differentiate among some of these materials, and some of these differentiations were made among surfaces having surprisingly subtle differences.

In classification table "B," four of the main surface types are each represented by a high and low reliability class. The high reliability or more restrictive class has reflectance limits in at least one band that made it exclusive from all other high reliability classes; the secondary class has reflectance limits that overlap the limits of at least one other class. The classes of two reliability levels have not been used in later tables. The classification map made by using table "B" is compared with a geologic map (fig. 6).

In building table "B," as the acceptable reflectance limits of a class were narrowed, fewer matches were found. In classes 1 and 2 (mineralized quartz diorite, table 1), for example, if we chose limits that included most pixels in the training areas and also many pixels in all the areas of known mineralization, we also got false classifications in dry-wash material. The limits were subjectively adjusted wider or narrower so that enough points were classified correctly in the areas of known mineralization to call attention to those areas and the false classifications were reduced to an acceptable level. Because revisions to the tables are time-consuming and successive revisions generally achieve diminishing improvement, table "B" as used in this experiment was revised only a few times.

The three dark-surface classes were established on the basis of small samples of particular rock formations in training areas, but the classes were very unsatisfactory in discriminating among the three formations. The classes would have been abandoned except that various combinations of the three classes delineate areas of dark rock outcrops and lag gravels. The printout of class 11 (symbol #) in particular resulted in abundant pixel groups just inside the perimeter of the propylitic zone at Saindak (no similar propylitic zone in volcanic sedimentary rocks, as at Saindak, was found near the new prospects).

Classification using table "S".—The data collected in the field check made it possible to evaluate the accuracy of identification of all of the surface types, not the mineralized rock alone, and to revise the tables to improve accuracy. A new style of table evolved, having several relatively narrow classes spanning the expected albedo range for each rock type but no high and low reliability classes, resulting in classification table "SSD" (table 2) and finally classification table "S" (table 3). The limiting values for each class span relatively narrower ranges than those in table "B" and more classes are required. The use of several classes for a range in albedo from similar surface types is designed as a simple way to compensate for variations in reflectance angles, especially those caused by southeast- and northwest-facing slopes, and also albedo-related systematic changes in color balance of similar surfaces, regardless of their cause. Classification using table "SSD" was performed to reevaluate part of the application area. An area of 170 km^2 was tested that includes four of the new prospects. The results are very encouraging.

The area tested using table "SSD," included 13 of the groups of "mineralized"-class pixels originally evaluated using table "B." Of the 13 pixel groups, 7 had been selected as possibly mineralized rock and worth field checking, 5 were subsequently visited during the field check, and 4 were found to contain some mineralization. Evaluation of the same area using table "SSD" resulted in the delineation of only 5 areas as "mineralized;" 4 of these were known from earlier field checking of table "B" targets to have mineralization, 1 was identified in the earlier evaluation but not chosen as deserving a field check, and one was not delineated by the table "B" evaluation. (The latter two sites, not chosen for checking in the earlier study, have not had field examinations). One mineralized site identified using table "B" was omitted by the table "SSD" evaluation. Table "SSD"

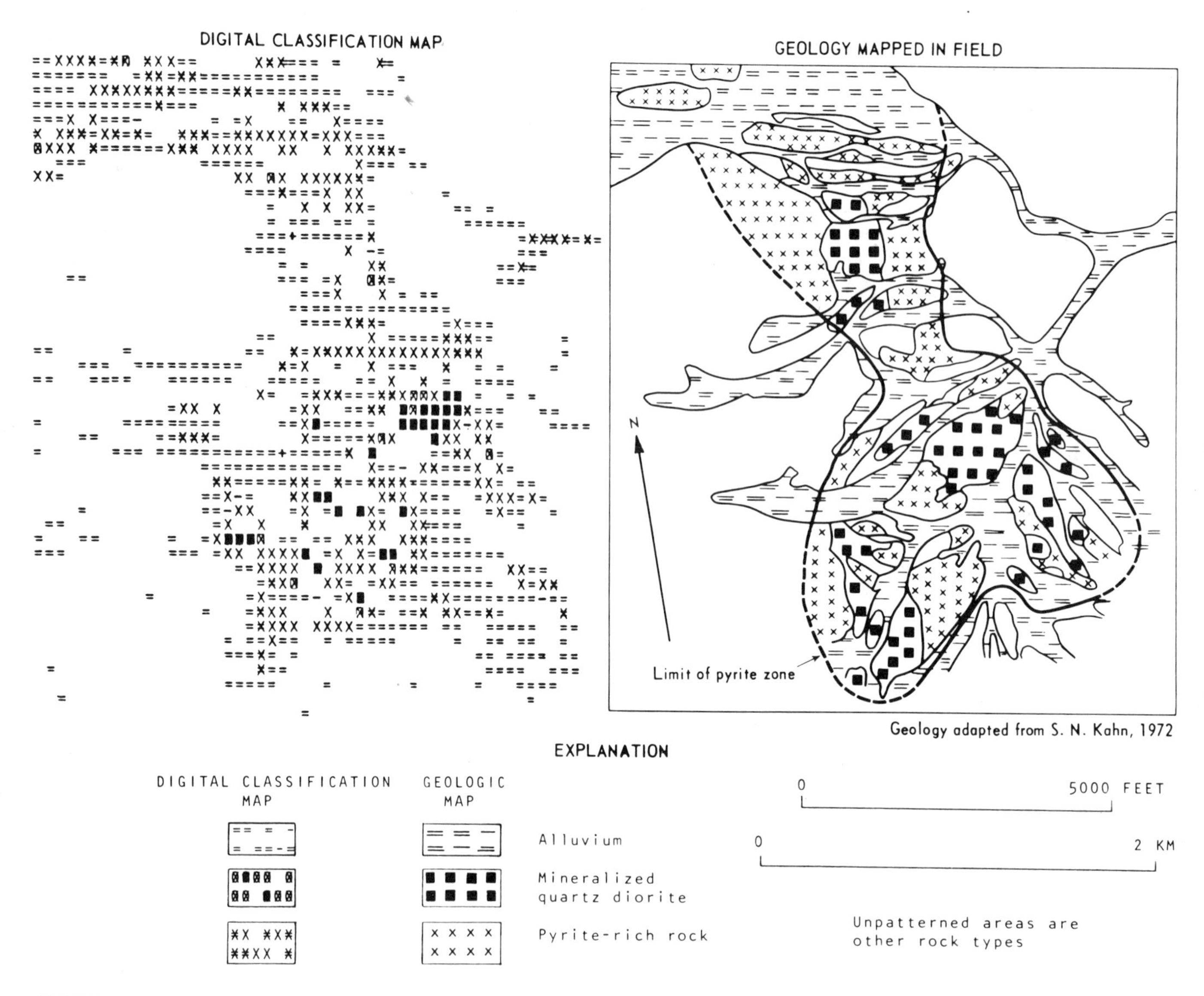

FIGURE 6.—Comparison of digital-classification map made by using table "B" and geology mapped in field, Saindak porphyry copper deposit, Pakistan.

does a much more effective job of separating mineralized areas from other light-toned materials, largely eliminating the need for human evaluation of a large number of "false" tonal anomalies. Omission of one of the best areas of mineralization points to the need for slight revision of the mineralized rock classes.

Revisions to table "SSD" have yielded table "S" which includes classes to cover sands of higher albedo and dry-wash material of lower albedo than were included in table "SSD," classes for dark pebble desert and unmineralized felsic rock (rock types not included in earlier tables), and modified reflectance limits for part of the mineralized rock classes. Only a few preliminary tests have been made of this new table.

SUMMARY

Visual evaluation of false color composites and a relatively simple method of digital classification were tested as aids to mineral exploration by using areas of known hydrothermal alteration and mineralized and altered intrusive stock as control areas. Both methods seem to be useful exploration tools, and greatest advantage is probably gained by using them together. The simple classification method was applied to evaluate 2,100 km^2 of area regarded as having a high potential for deposits of the porphyry copper type, leading to the discovery of five sizable areas of hydrothermally altered rock containing abundant sulfide, disseminated or in stockwork veins. Although the classification method used was relatively simple and unrefined, and 14 prospecting targets proved to be false leads, the number of sulfide-bearing areas identified was outstanding, and the falsely classified areas were not so many as to require an unreasonable amount of field checking. Further experimentation indicates that better discrimination of mineralized areas than we used in the field test can be achieved. Our study indicates that simple methods of digital classification of Landsat data can aid in exploration for mineral deposits in desert regions, and these methods can complement conventional exploration techniques.

REFERENCES CITED

Ahmed, Waheeduddin, Khan, S. N., and Schmidt, R. G., 1972, Geology and copper mineralization of the Saindak quadrangle, Chagai District, West Pakistan: U.S. Geol. Survey Prof. Paper 716–A, 21 p.

Bernstein, Ralph, 1974, Scene correction (precision processing) of ERTS sensor data using digital image processing techniques: Third Earth Resources Technology Satellite–1 Symposium, v. 1: Technical Presentations Section B, NASA SP–351, Dec. 10–14, 1973, p. 1909–1928.

Hunting Survey Corporation, Ltd., 1960, Reconnaissance geology of part of West Pakistan; a Colombo Plan Cooperative Project: Toronto, 550 p. (Report published by Government of Canada for the Government of Pakistan).

Khan, S. N., 1972, Interim report on copper deposit of Saindak, Chagai District (Baluchistan), Pakistan: Pakistan Geol. Survey, Saindak copper report no. 1.

Lowell, J. D., and Guilbert, J. M., 1970, Lateral and vertical alteration-mineralization zoning in porphyry ore deposits: Econ. Geology, v. 65, no. 4, p. 373–408.

Schmidt, R. G., 1968, Exploration possibilities in the western Chagai District, West Pakistan: Econ. Geology, v. 63, p. 51–60.

———1974, The use of ERTS–1 images in the search for large sulfide deposits in the Chagai District, Pakistan: U.S. Dept. of Commerce, Natl. Tech. Inf. Service, E74–10726, 38 p.

Part VI
Image Data Compression/Compaction

Organized by Azriel Rosenfeld, Paul E. Anuta, and Ruzena Bajcsy, Associate Editors

A Survey of Digital Picture Coding

Ali Habibi and Guner S. Robinson
University of Southern California

Introduction

The steady growth of modern communication requirements has resulted in a steady increase in the volume of pictorial data that must be transmitted from one location to another. In some cases, although image transmission to a remote location is not necessary, one does need to store the images for future retrieval and analysis.

In previous years, most of the image transmission has been accomplished through conventional analog techniques. Today, the trend in image transmission and storage is to use digital instead of analog techniques.[1] This may be attributed to the rapid growth in the use of digital computers as well as to the many inherent advantages of digital communication systems.

With the increasing emphasis on the use of digital communication systems, digital image compression techniques are drawing considerable interest as possible means for reducing the capacity requirements of the digital communication channels. An encoding scheme widely used for the transmission of digital signals is pulse code modulation (PCM), which has been refined considerably since its invention in 1939.[2] Such encoding generally involves the sampling of the analog input signal at a uniform rate and encoding the samples in a binary code. This provides an adequate number of quantizing levels, in order to maintain a required signal-to-noise ratio. Such techniques have been applied to voice, images and other signal waveforms such as telemetry and biomedical data.

Conventional PCM techniques require a high data rate (or equivalently large bandwidth) for the transmission of images.[3,4,5] Therefore, many new digital schemes have been explored in order to reduce the capacity requirements of the digital image communication systems, by making use of the statistical and psychovisual redundancies in images.[6]

Reprinted from *IEEE Computer*, vol. 7, pp. 22–34, May 1974.

Digitization and PCM Processing of Images

To achieve digital representation of an image, the image must be converted to a set of binary integers, which in turn can be operated upon to recover the picture with a minimum possible degradation. This requires first scanning an image to convert it to a set of one-dimensional waveforms, then sampling the analog waveforms.

In PCM transmission of an image, the continuous image is first sampled in the spatial domain to produce an $N \times N$ array of discrete samples which are then quantized in brightness by using 2^K levels. Then, B, the total number of bits per frame to be transmitted, is given by

$$B = N^2K$$

Subjective tests are made to determine the tradeoff between N and K which will result in the "best" picture.[7] The result necessarily depends on the type of material and the definition of the "best" picture. For commercial applications, however, the final decision is based on subjective tests. The values for U.S. commercial television systems are $N = 525$ and $K = 8$ bits per picture element (pixel).

In computer processing of images, considering pictures of the same quality as that of the television signal, one needs about $8(512)^2$ bits per frame. This results in $30 \times 8 \times (512)^2$ bits per second for 30 frames per second, the rate required for representing pictures that move at a normal speed.

The image degradation resulting when the quantization process uses a smaller number of bits per sample is commonly known as the contouring effect. This effect, the formation of discrete rather than gradual changes in brightness, becomes perceptible when M is reduced to six or fewer bits per pixel. For PCM picture transmission, some of the well-known techniques for removing the effect of the quantization noise are addition of pre-emphasis and de-emphasis networks to conventional PCM,[8] use of pseudo-random noise,[9] and nonuniform quantization.[10,11] Also, techniques have been proposed for shaping the quantizing noise spectrum by using feedback around the quantizer.[12,13] These techniques have found application in low resolution picture transmission systems.

The PCM system is capable of transmitting any picture, including ones that are composed of all uncorrelated samples. However, pictures with uncorrelated samples are viewed on the television screen only after the channel sign off, and they are of no particular interest.

The interesting and meaningful pictures are those that exhibit less sample-to-sample variation. That is, instead of all adjacent samples having different brightness levels, the pictures are composed of groups of samples having various brightness levels with samples within each group showing small variations. Such a picture is less random, has higher sample-to-sample correlation, shows more structure, or, in the language of information theory, has smaller entropy than the completely random picture. Indeed, statistical analysis of pictures has indicated that the entropy of typical pictures sampled for an appropriate viewing is about one bit per pixel; thus it should be possible to encode it using one bit per picture element, without any loss of information. Since samples corresponding to the background in an image remain stationary from one frame to the next, this number can be easily reduced by a factor of five if one makes use of the correlation or similarity that exists among the pixels of adjacent frames.

In the following sections some of the intra- and inter-frame image compression techniques will be briefly outlined, and some examples will be given using two images: the picture of a milk-drop (Figure 1) and a chest X-ray (Figure 2), both sampled as 256 x 256 pixels. Since it is impossible to cover all the important aspects and the results obtained by other researchers, the reader is referred to the references at the end of the paper.

Linear-Transform Processing of Images

The above discussion underscores the fact that even though a fine sampling and quantization of an image are essential for desirable subjective quality of a digital picture, from the viewpoint of a statistician the information in the picture can be conveyed quite adequately without all these variables. On the other hand, one cannot discard a part of these variates because of their equal statistical significance and the adverse effect that this would have on the subjective quality of the picture.

The next step is to explore the possibility of transforming these samples to a new set of variates that will have a varying degree of significance in contributing to both the information content and the subjective quality of the resulting picture. Then one can discard the less significant

Figure 1. Milkdrop, Original

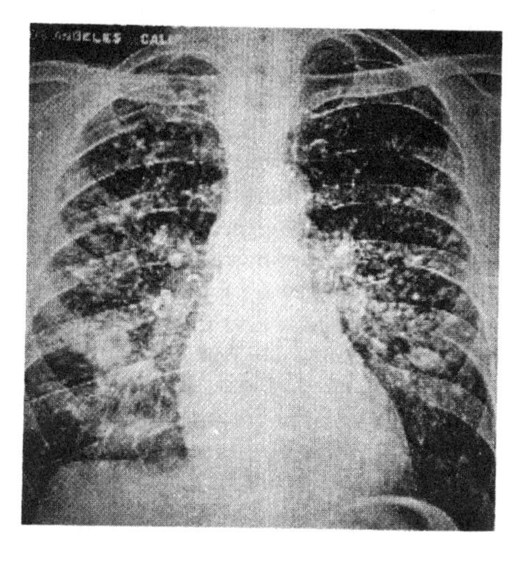

Figure 2. Enhanced X-ray, Original

of these variables without affecting the statistical information content of the picture or causing a severe degradation in the subjective quality of the resultant picture. The method of "principal components" is a coordinate transformation with the above properties.[18,19]

To illustrate the application of this coordinate transformation to pictorial data, let's consider two samples on a picture. Let X_1 and X_2 stand for the values that these two pixels could assume. From the fact that these variables are correlated, one easily concludes that the sample values often fall inside the region shown on Figure 3. Now if one rotates the coordinate system to positions indicated by y_1 and y_2, two new variables would be obtained – one having a larger variance than the other, though the sum total of

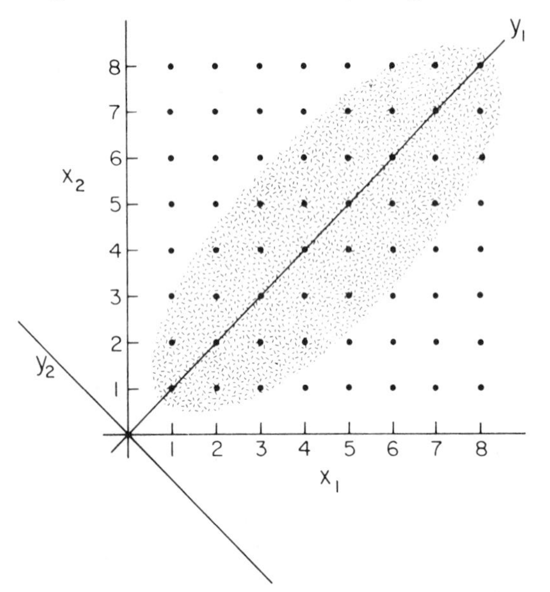

Figure 3. Coordinate System for the Gray Levels of Two Adjacent Pixels X_1 and X_2. The most likely values are the ones in shaded area. y_1 and y_2 are the new coordinate system. (This figure is from reference 16.)

the variances is invariant under the transformation. Performing the rotation of coordinates with all N^2 pixels, one obtains N^2 new variates that are uncorrelated and have monotonically decreasing variances. Indeed, the method of principal components, also known as the discrete Karhunen-Loève Transformation, results in variates with a maximum compaction of energy in the first M components that one desires to keep. Since the information content of the sampled picture is invariant under this linear transformation, and the variance of a variable is a measure of its information content, the coding strategy should be first to discard variates with low variances. Then, since quantizing each number corresponds to approximating its amplitude with the nearest rational number and thus corresponds to a loss of information, one should attempt to impose less of a distortion on those variates having larger variances. The above transformation corresponds to a matrix transformation of a vector. For a two-dimensional data (image) it is accomplished by ordering the sampled data in a vector form and performing the transformation. The block diagram of this transformation as well as quantization of the transformed data signal and the inverse transformation needed to recover a facsimile of the original image are shown in Figure 4. Denoting the data vector and the transform vectors by X and Y and the error introduced by quantization with vector Q, the vector X* reconstructed at the receiver is

$$X^* = X + A^{-1}Q \qquad (1)$$

where A corresponds to the operator in the method of principal components. Then the resulting mean squared error is

$$\epsilon^2 = \operatorname{tr} E\{A^{-1}QQ^T A^{-1^T}\} = \operatorname{tr}\{A^{-1} E\{QQ^T\} A^{-1^T}\} \qquad (2)$$

where tr denotes the trace of a matrix. Since AX is an uncorrelated vector, it follows that Q is also uncorrelated, thus $E\{QQ^T\}$ is diagonal with the ith element in its diagonal referring to the variance of the quantization error σ_q^2 in the ith component of Y vector. It is shown that the overall coding error is minimized if M_b binary digits are assigned to various components of the transformed vector Y so that the variance of the quantization error σ_q^2 for all components of Y are identical. Since the variance of the quantization error is shown to be directly proportional to the variance of the variable being quantized, an equal quantization noise for various components of Y implies assigning more bits to the components with large variances and fewer bits to those with small ones. Of course, assigning zero bits to a particular component of Y means that the particular component is ignored in the process of the transformation storage. Note that the above may not be possible since the number of binary digits assigned to each sample must be an integer. Thus, only an approximate solution is possible in practice.

Two-Dimensional Transformations

In the above discussion we indicated that a transformation of two-dimensional data is accomplished by first ordering the sampled image in a vector form, then proceeding to find the eigenvectors of the covariance of this vector and its subsequent transformation. This presents a problem, since even for small blocks of images one needs to calculate the eigenvector of a large matrix. This is avoided by defining two-dimensional Karhunen-Loève transformations. The forward and inverse two-dimensional Karhunen-Loève transformations are modeled as

$$u_{ij} = \sum_{x=1}^{N} \sum_{y=1}^{N} \phi_{ij}(x, y) u(x, y) \qquad (3)$$

$$u(x, y) = \sum_{i=1}^{N} \sum_{j=1}^{N} \phi_{ij}(x, y) u_{ij} \qquad (4)$$

It has been shown that the "eigenmatrices" of $\phi_{ij}(x,y)$ of the whole picture or sub-blocks of the picture that are composed of M by M pixels can be formed from outer products of eigenvectors of the covariance of the data in

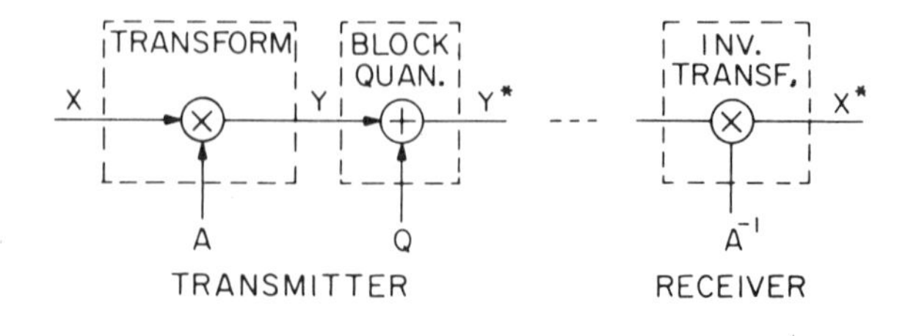

Figure 4. Block Diagram of the Transform Coding System.

the horizontal and the vertical directions if the covariance of the data is separable – that is, if

$$R(x,y,\hat{x},\hat{y}) = R_H(x,\hat{x})R_V(y,\hat{y}) \quad (5)$$

where R_H and R_V refer to the covariance matrices of the data in horizontal and vertical directions and $R(x,\hat{x},y,\hat{y})$ is the covariance "tensor" of the data. In the absence of this assumption the ordering of the two-dimensional data in a vector form is the only practical solution. Operations (3) and (4) correspond to operating on the rows of the image followed by operations on the columns of the horizontally transformed data to obtain the two-dimensional transformation. These individual operations in the horizontal and the vertical directions are the consequences of the separability of the covariance "tensor." The number of operations needed to perform (3) or (4) is N^4 multiplication/addition operations. Two-dimensional Karhunen-Loève transformation of images has been considered for data with a separable covariance

$$R(x,\hat{x},y,\hat{y}) = e^{-\alpha|x-\hat{x}|-\beta|y-\hat{y}|} \quad (6)$$

where α and β are estimated from the image. Some authors refer to these eigenmatrices as "basis pictures."[20]

The shortcomings of the method of principal components are the large number of operations required for forward and inverse transformation (3) and (4) and the difficulties of estimating the covariance of the data and calculating the eigenmatrices. To eliminate these shortcomings, a number of other transformations have been considered and are reviewed briefly below. (See references 20-27 for details.)

For discrete data, the two-dimensional Fourier transformation corresponds to choosing the basis matrices (images) of the form

$$\phi_{ij}(x,y) = \left\{\frac{1}{N} \exp - \frac{2\pi(-1)^{1/2}}{N}(ix+jy)\right\} \quad (7)$$

$$i,j,x,y = 1,2,\ldots,N$$

while the two-dimensional Hadamard transform[20,23] corresponds to choosing basis images as

$$\phi_{ij}(x,y) = \frac{1}{N}(-1)^{\sum_{h=0}^{\log_2 N-1} [b_h(x)b_h(y)+b_h(i)b_h(j)]} \quad (8)$$

where $b_h(\cdot)$ is the hth bit in the binary representation of $(\cdot)$ and N is a power of 2. Figure 5 shows the set of Hadamard "basis pictures" for an 8 by 8 transform. Both of the above transformations are members of a class of Kroneckered matrix transformations that have $2N^2 \log_2 N^2$ degrees of freedom and can therefore be implemented by $2N^2 \log_2 N^2$ computer operations as opposed to N^4 operations required by the Karhunen-Loève transformations. These transformations remedy the shortcomings of the Karhunen-Loève transformation by (1) eliminating the necessity of finding an operator matched to the covariance of the image and (2) significantly reducing the computational complexity.

In addition to these transformations a number of others possessing the above two properties have also been considered. For example, Cosine[26] and Slant transformations[27,28] have resulted in a better mean square error performance than either the Fourier or the Hadamard transformations. The performance of these transforms are still inferior to the performance of the Karhunen-Loève transformation, which is the only orthogonal transform that generates a set of uncorrelated signals. However, in most practical applications the computational simplicity and the ease of implementation of the Hadamard and other suboptimum transformations more than compensate for the suboptimal performance of these transforms. Figures 6 and 7 show the milk droplet and chest X-ray images coded using two-dimensional Hadamard transform for 2, 1, and 0.5 bits/pixel.

Predictive Coding and Delta Modulators

As pointed out earlier, the PCM system is inefficient in coding correlated data such as images since the encoder in the PCM assigns the same number of binary digits to each of a number of correlated variables. The transform coding system involved an operation on these variables to uncorrelate them prior to their quantization. Another approach to the problem of generating a set of uncorrelated variables uses the classical prediction theory. Consider a set of correlated variables $\{S_i\}$ with mean zero and variance σ^2. The set could represent a set of picture elements, as shown in Figure 8, where the mean value of the pixels is subtracted from each picture element. The linear predictor estimates the next sample value S_o by $\hat{S}_o$ based on the previous n sample values as

$$\hat{S}_o = \sum_{i=1}^{n} A_i S_i$$

The differential signal that corresponds to S_o is $e_o = S_o - S_o$. The best estimate in least mean-squared error sense is one where the weighting coefficients A_i are solutions of the n algebraic equations:

$$R_{0i} = \sum_{i=1}^{n} A_i R_{ij}, \quad i = 1,2,\ldots,n \quad (9)$$

where

$$R_{ij} = E(S_i S_j)$$

Then the variance of differential signal is

$$\sigma_e^2 = \sigma^2 - \sum_{i=1}^{n} A_i R_{0i} \quad (10)$$

As $n \to \infty$ the sequence of error samples can be made completely uncorrelated. However, if the sequence of samples $\{S_i\}$ is an nth order autogressive process, then using only n samples in forming estimates $\hat{S}_o$ will make the resulting sequence of error terms uncorrelated. In this case a further increase in the number of samples employed in forming the optimum predictor will not improve the quality of the estimate.

Experimental results with various images have indicated that sampled images can be modeled rather accurately[29] by a third-order Markov process. Thus using S_1, S_2, S_3 on Figure 8 will generate a set of almost uncorrelated variables. Equation (10) indicates that the variance of the differential signal is smaller than the variance of $\{S_i\}$ the reduction increases for more correlated data.

A DPCM encoder uses a linear prediction to generate a differential signal, then quantizes the signal with a quantizer that is designed for the probability density function of this signal. The block diagram of the transmitter and the receiver for an nth order DPCM system is shown in Figure 9. As the reader will note, the quantization of the data is performed in the feedback loop. This is essential to generate an uncorrelated coding error. This is particularly

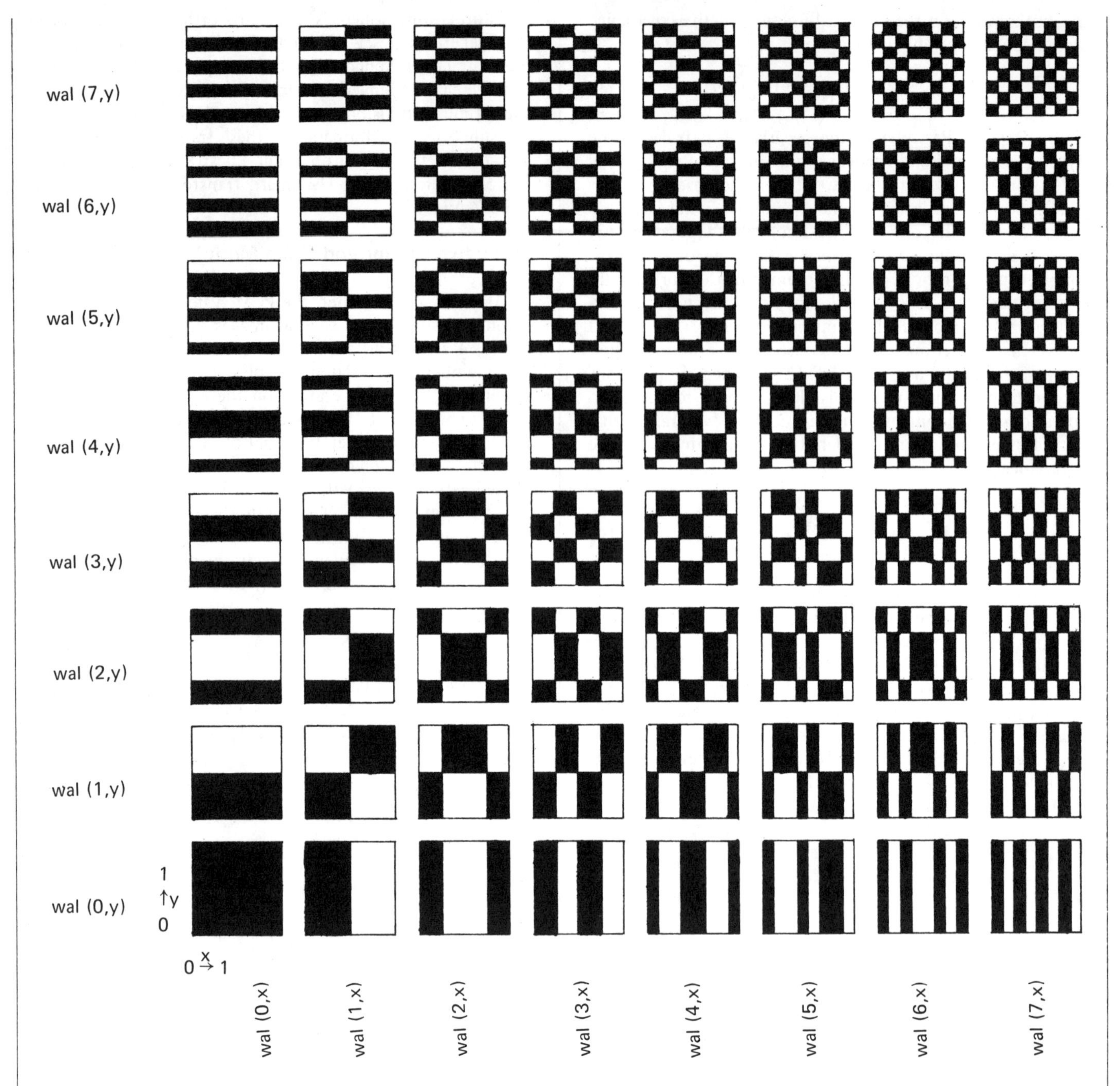

Figure 5. Two Dimensional Walsh-Hadamard Basis for N=8. Black areas represent +1/N and white areas represent -1/N.

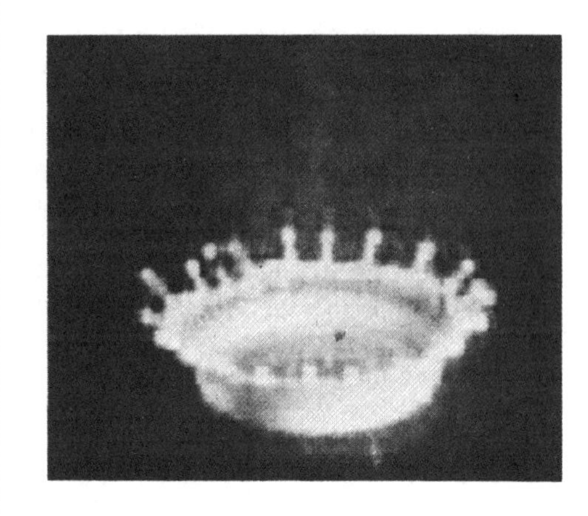

(a) 1/2 Bit/Pixel

(b) 1 Bit/Pixel

Figure 6. Milkdrop Coded Using Two-Dimensional Hadamard Transform (16 by 16) and Block Quantization.

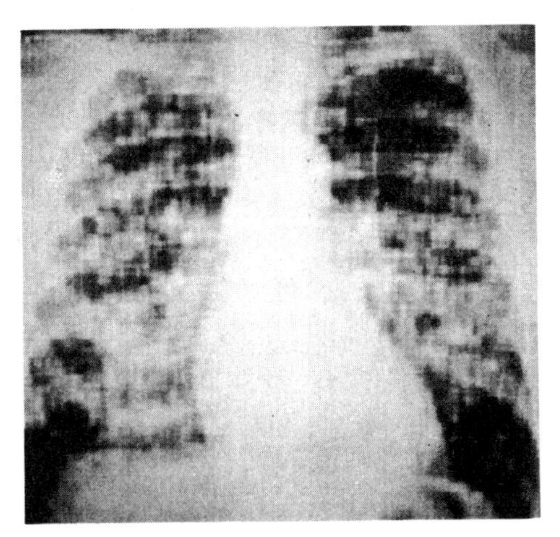

(a) 1/2 Bit/Pixel

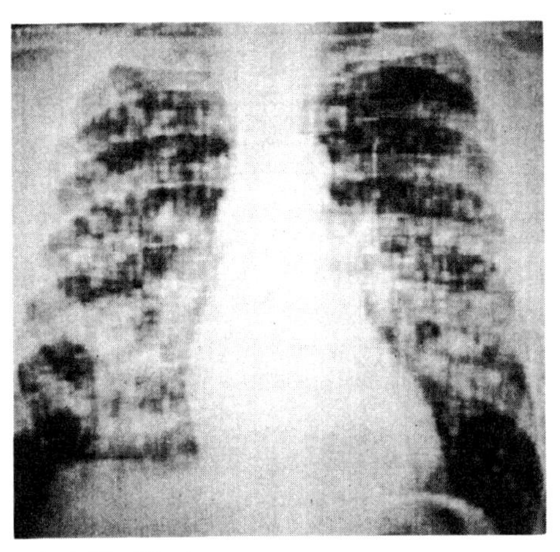

(b) 1 Bit/Pixel

Figure 7. X-Ray Coded Using Two-Dimensional Hadamard Transform (16 by 16) and Block Quantization.

important in coding pictures since human vision tolerates more degradation if the degradation is uncorrelated.

The DPCM encoder discussed above is a two-dimensional encoder, since $\{S_i\}$, as shown in Figure 8, refers to points adjacent to S_O in various spatial directions. Limiting one to points in only one direction, one could obtain a one-dimensional DPCM system that uncorrelates the data in only one spatial direction. The one-dimensional DPCM encoder gives inferior coding results; however, it is frequently used because of its simplicity[30-32].

The philosophy behind the DPCM encoder is the same as the one used to develop the transform coding system. Namely, in both cases one generates an uncorrelated signal which is then encoded by one or a set of memoryless quantizers. Indeed, it has been shown that if one uses a lower-triangular operator to uncorrelate the data vector X

Figure 8. Pixels Employed in Predicting S_0.

rather than an orthogonal operator employed in the method of principal components, one can obtain a generalized DPCM system that reduces to a simple DPCM encoder for Markov data.[33] This new approach to the theory of DPCM encoders unifies DPCM with the transform coding system – making one realize they both achieve the same end result using two different types of operators.

Analogous to the method of principal components, coding by DPCM system requires a knowledge of data statistics. That again forces one to estimate these statistics in addition to making the system signal-dependent. The suboptimum systems derived from the DPCM encoder are the element-differential quantizers and delta modulators. Element-differential quantizer is a suboptimal form of the one-dimensional DPCM system in which the predictor is replaced by a sample delay. This system is used with some success for coding both monochrome and color images.[31] The delta modulator uses a comparator instead of the quantizer and a gain circuit and an integrator instead of the predictor (see Figure 10). Note that this system can combine the operation of the sampler and the quantizer; thus it can operate directly on scanned lines of television signals. The delta modulator suffers from granular noise where the image has small variation, and overload noise at the edges of the image. Naturally the gain circuit in the feedback loop can be adjusted to reduce one of the above at the expense of increasing the other. This problem is overcome by using a logic in the feedback loop which would sense the type of data in the input and adjust the gain accordingly. One such system is shown in Figure 11 where at each instant i the polarity of three previous output bits, M_{i-2}, M_{i-1}, M_i, is examined. If these are of the same polarity,

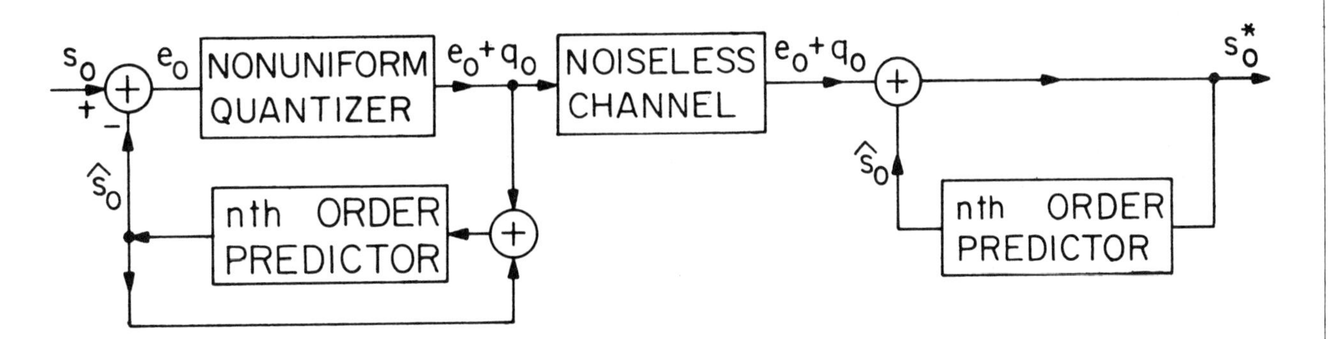

Figure 9. DPCM System.

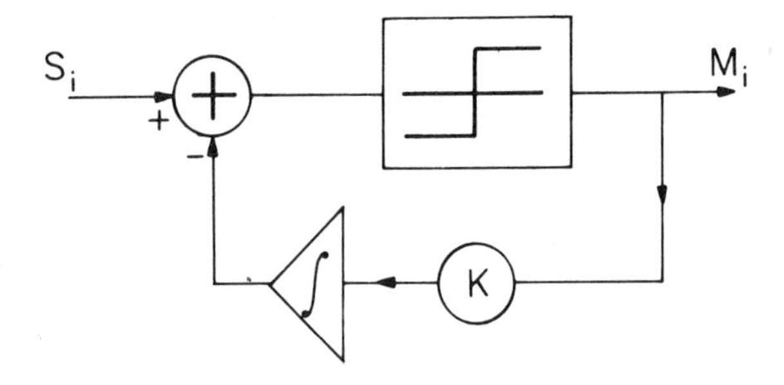

Figure 10. Simple Delta Modulation System.

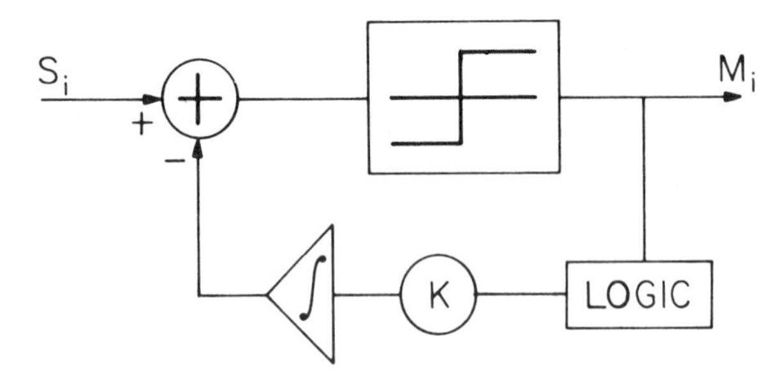

Figure 11. Adaptive Delta Modulation System.

the signal has a large slope and the feedback gain is increased. However, mixed polarity of output bits indicates slow signal variation, and this reduces the size of the feedback gain. The adaptive delta modulator has been used with some success in coding various types of images.[34-36] In addition to controlling the feedback gain, the logic in the feedback loop can also be used to control the sampling interval to further improve system performance. This system has a variable sampling interval which gives a variable bit rate, thus its operation requires buffering the binary digits prior to transmission.[37]

Hybrid Coding of Images

The common approach to the problem of transmitting pictorial data over digital communication channels entails processing the correlated data (images) to generate a set of uncorrelated (or as nearly uncorrelated as possible) set of signals which in turn are quantized using a memoryless quantizer. The quantized signal is then encoded using either fixed- or variable-length code words, and is transmitted over a digital channel. This is the general approach taken in designing DPCM and the systems that use unitary transformation and block quantization, as well as many other systems developed in recent literature. Both DPCM and transform coding techniques have been used with some success in coding pictorial data. A study of both these systems has indicated that each technique has some attractive characteristics and some limitations. The transform coding systems achieve superior coding performance at lower bit rates; they distribute the coding degradation in a manner less objectionable to a human viewer, show less sensitivity to data statistics (picture to picture variation), and are less vulnerable to channel noise. DPCM systems on the other hand, when designed to take advantage of spatial correlations of the data, achieve a better coding performance at a higher bit rate. The equipment complexity and the delay due to the coding operation are minimal. Perhaps the most desirable characteristic of this system is the ease of design and the speed of the operation that has made it possible for DPCM systems to be used in coding television signals in real time. The limitations of this system are the sensitivity of well-designed DPCM systems to picture statistics and the propagation of the channel error on the transmitted picture.

A hybrid coding system that combines the attractive features of both transform coding and the DPCM systems has also been studied. This system exploits the correlation of the data in the horizontal direction by taking a one-dimensional transform of each line of the picture, then operating on each column of the transformed data using a one-element predictor DPCM system. The unitary transformation involved is a one-dimensional transformation of individual lines of the pictorial data. Thus the equipment complexity and the number of computational operations are considerably less than that involved in a two-dimensional transformation. Theoretical and experimental results indicate that the system has good coding capability – one that surpasses both DPCM and the transform coding systems.

In the hybrid system the pictorial data is scanned to form N lines, then each line is sampled at a Nyquist rate. This sampled image is then divided into arrays of M by N picture elements u(x,y) where x and y index the rows and the columns in each individual array so that the number of samples in a line of image is an integer multiple of M. One-dimensional unitary transformation of the data and its inverse are modeled by the set of equations

$$u_i(y) = \sum_{x=1}^{M} u(x,y)\phi_i(x) \quad i = 1,2,\ldots,M \quad y = 1,2,\ldots,N \tag{11}$$

$$u(x,y) = \sum_{i=1}^{M} u_i(y)\phi_i(x) \tag{12}$$

where $\phi_i(x)$ is a set of M orthonormal basis vectors. The correlation of the transformed samples $u_i(y)$ and $u_i(y+\tau)$ is given by

$$C_i(\tau) = \sum_{x=1}^{M} \sum_{\hat{x}=1}^{M} R(x,\hat{x},y,y+\tau)\phi_i(x)\phi_j(\hat{x}) \tag{13}$$

where $R(x,\hat{x},y,\hat{y})$ is the spatial autocovariance of the data. Equation (13) indicates that the correlation of samples in various columns of the transformed array is different; thus a number of different DPCM systems should be used to encode each column of the transformed data. The block diagram of the hybrid encoder is shown on Figure 12. A replica of the original image $u^*(x,y)$ is formed at the receiver by inverse transforming the coded samples – i.e.,

$$u^*(x,y) = \sum_{i=1}^{n} v_i(y)\phi_i(x) \quad n \leq M$$

The hybrid system is used to encode pictures (see Figures 13 and 14) for N = 256 and M = 16. These encoded pictures use a cascade of Hadamard transform and DPCM encoders at various bit rates: 2, 1, and 0.5 bits/pixel.

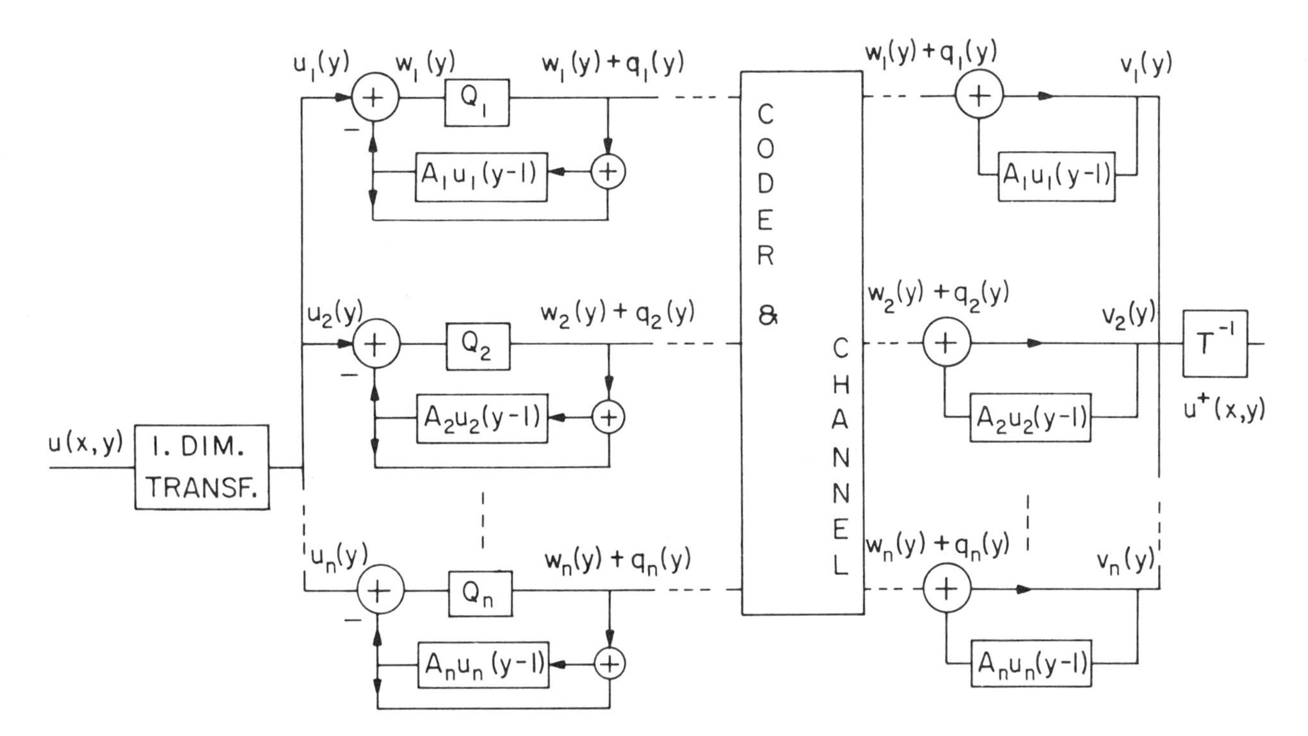

Figure 12. Block Diagram of the Hybrid Encoder using a One-Dimensional Unitary Transformation and a Block of DPCM Encoders.

(a) 1 Bit/Pixel

(b) 0.5 Bit/Pixel

Figure 13. Coded images Using a Cascade of One-Dimensional Hadamard Transform and a Black of DPCM Encoders.

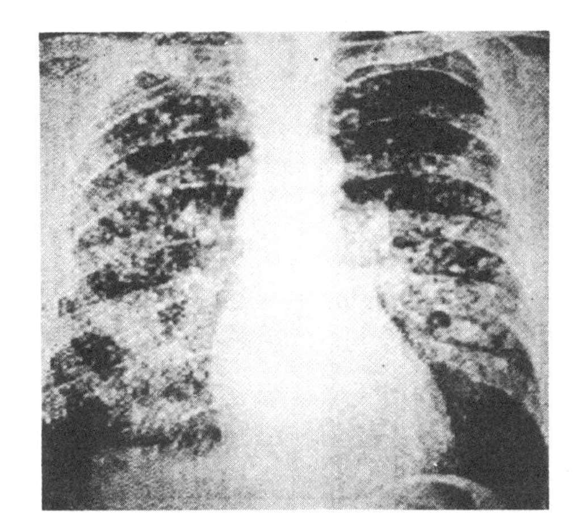

(a) 1 Bit/Pixel

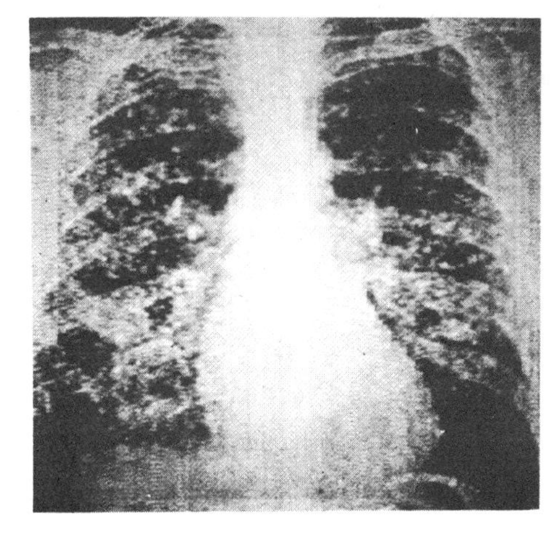

(b) 0.5 Bit/Pixel

Figure 14. Coded Images Using a Cascade of One-Dimensional Hadamard Transform and a Block of DPCM Encoders.

Image Coding by Tracing Contours of Constant Brightness

The concept of tracing contours of constant gray levels in a single frame is attractive since in a given digital picture many of the points in one region have the same gray level. Thus if a contour could be traced around these points, which are all the same gray level, then only the addressing information enabling the receiver to trace a similar contour along with the common level of elements in the contour is needed in the receiver. Naturally the scheme is more useful when the possible number of gray levels is small or when there is a high probability of having a large number of points at the same gray level. Figures 15a and 15b show the picture of a milkdrop quantized to 8 and 4 levels respectively. Figures 16a and 16b show the outer boundary of the contours present in 15a and 15b, respectively. Naturally 15a and b could be reconstructed from 16a and b accurately where 16a and 16b were coded using 0.71 and 0.28 bits/pixel, respectively. Wilkins and Wintz introduced a two-dimensional contour-tracing algorithm that locates and traces contours enclosing a maximum number of points of the same gray level.[39,40] The algorithm consists of two subalgorithms: one for locating the initial point of a new contour (IP algorithm), and the other for tracing the contours after they are located (T algorithm). The T algorithm traces the outer boundary of the largest connected set of elements having the same value as the initial point, and always terminates back at the initial point. The direction of travel on the T contour can be limited to 4 or 8 spatial directions to limit the directional information to 2 or 3 bits respectively. All elements enclosed by the contour and having the same value as the contour are neglected, but they can be reconstructed at the receiver. Wilkins and Wintz also developed an algorithm for reconstructing the original data from the system output, and they coded a number of still pictures using the contour-tracing algorithm. If some contours, say contours that consist of only single points. are deleted and are replaced by the gray level of the neighboring contours, the algorithm will result in some degradation of the encoded data. But at the same time this reduces the number of binary digits essential to reconstruct the data at the receiver. In transmitting the addressing information, various types of coding could be used for a further reduction in the bit rate.

Interframe Coding of Television Signals

In a television signal a large fraction of picture elements correspond to the background material and do not change from one frame to the next. On the other hand a relatively small number of picture elements change from one frame

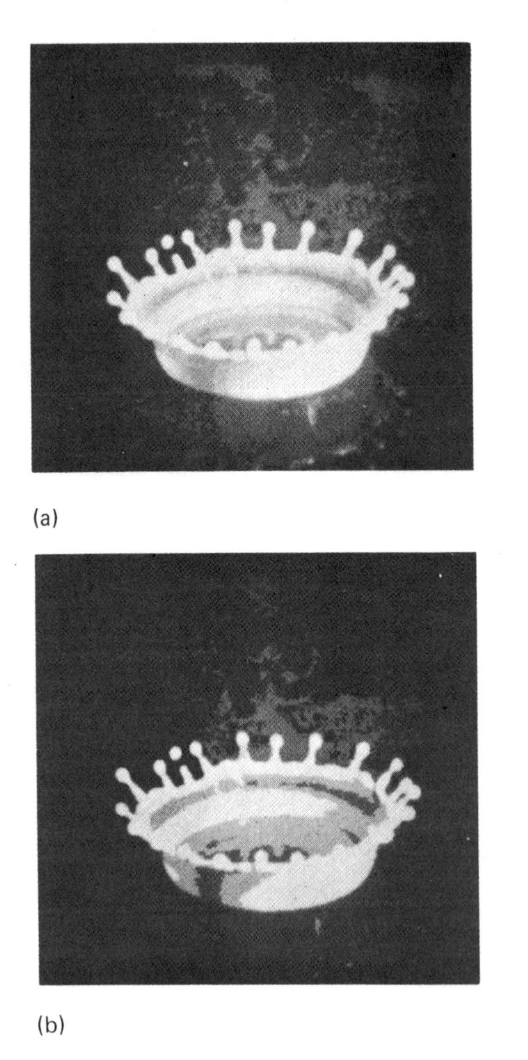

Figure 15. The Picture of Milkdrop Quantized to (a) 8 Levels (b) 4 Levels

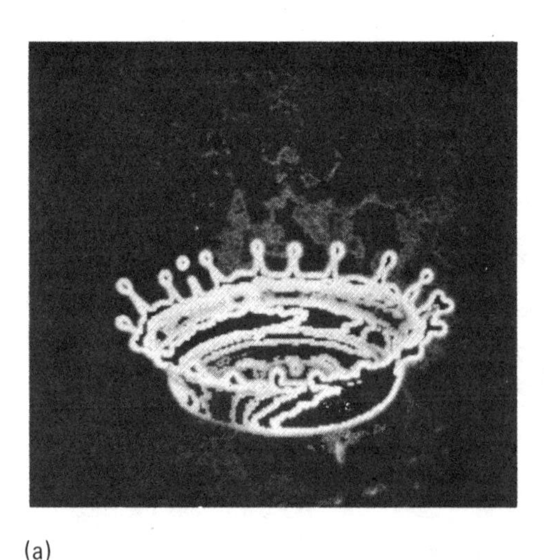

(a)

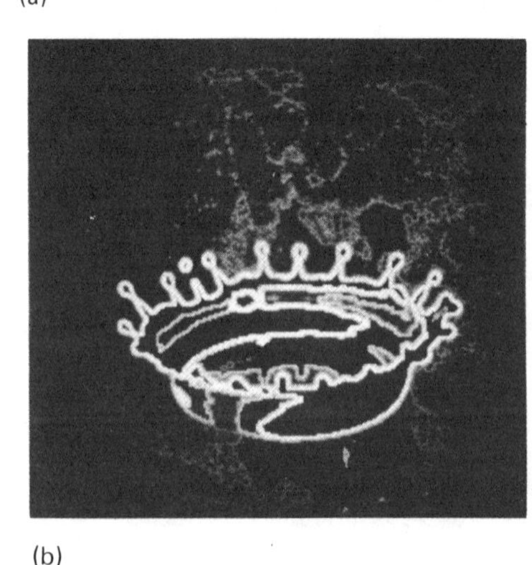

(b)

Figure 16. Exterior Contours of (a) Figure 15a (b) Figure 15b

to the next to convey the new information that is generated by a relative movement of camera with the object in each frame. From a statistical viewpoint, the similarity of pixels from one frame to the next corresponds to a high level of interframe correlation. Thus the statistical coding techniques exploiting the spatial correlation of the data that was considered for coding single frames of data could be extended to take advantage of the frame-to-frame correlation, further reducing the bit rate required to transmit the data. Indeed some research in the area of three-dimensional Fourier and Hadamard transformations has indicated that the bit rates could be reduced at least by a factor of five by incorporating the correlation in the temporal direction.[41] However, the three-dimensional transform encoding systems suffer from the serious shortcomings of computational complexity and require large amounts of storage. For this reason most practical researchers have avoided extending transform coding systems to a third dimension; instead they have suggested suboptimum coding systems that do not require extensive amounts of memory or computations. The design of these systems is based on the fact that only a small percentage of picture points in a television signal change in successive frames. Thus a typical frame differential picture will consist of a large gray area (after a shift in gray scale to eliminate the negative components), some bright and dark spots along the moving edges, and also a number of scattered points in the background that are caused by granular noise. A typical interframe difference picture is shown in Figure 17.

Figure 17. Frame-Differential Picture.

An efficient technique of interframe coding of video data is simply to transmit the gray levels of the elements that have changed in successive frames with proper addressing information. The receiver then generates the successive frames by replenishing the previous frames with the transmitted information. Naturally less information needs to be transmitted if only the elements that change beyond a particular threshold level are updated. Using an appropriate value for the threshold eliminates the granular noise in the background without noticeably degrading the encoded video data. Experiments with picturephone signals using the conditional frame replenishment technique have indicated good coding results for an average of one bit per picture element.[42] The only shortcoming of the system is that the data is generated at an uneven rate. This is caused by the change in percentage of those pixels that change beyond a threshold in each frame. This percentage is in turn dependent upon the types of the motion involved. To transmit this data over a fixed channel requires buffering the data prior to its transmission. The size of the buffer in addition to the bit rate limits the volume of the motion in the data for which this system could be employed.

Since the gray levels of most of the picture elements that change in successive frames vary by a small amount and only a very small percentage of the pixels change significantly from one frame to the next, the above coding scheme could be made more efficient by transmitting the change in the gray level of the picture elements whose values have changed beyond a given threshold. This improvement is similar to the improvement that one obtains using a DPCM over a PCM system. This and various other coding techniques have been considered for a further reduction of the bit rate and reducing the buffer needed in the conditional frame replenishment of video data.[43-45]

Coding Color Images

Conventional color television and facsimile systems represent colors at each image point proportional to the tristimulus values of a color at that point where the tristimulus values of a color specify the red, green, and blue content of the color. Thus a color video signal is composed of three video signals, each corresponding to one of the tristimulus components.

In coding color video data, essentially three different signals should be coded. An efficient scheme of doing this is to transform the National Television Standard Committee tristimulus color coordinates (R, G, and B) to generate a set of uncorrelated color coordinates, code these signals, then reconstruct the original color components by inverse transforming of the coded signals. This transformation is one that generates the Karhunen-Loève tristimulus color coordinates. However, NTSC transmission color components Y, I, and Q are obtained similarly and are almost as uncorrelated as the Karhunen-Loève tristimulus color coordinates with a similar distribution of energy in the coordinates.[46] NTSC transmission color coordinates have the advantage that Y represents the luminance of the color signal. Coding Y, I, and Q components separately using the element-differential quantization technique, Limb et al.[34] have obtained good results.

Other techniques of coding color signals are (1) to consider the three frames corresponding to the tristimulus components of a color video signal as three correlated frames of data, and (2) to encode the color signals using the techniques discussed in interframe coding of video data. One such system uses a three-dimensional unitary transformation with an efficient bit assignment procedure in the transformed domain to encode one frame of a color video signal. Figures 18 and 19 show the originals as well as the coded images using a three-dimensional Fourier transform with an efficient bit assignment in the transformed domain.[41]* These figures show that reasonably good quality color pictures can be obtained using a bit rate of about 1.2 bits/pixel.

*The color video signals were supplied by A. G. Tescher and relate to reference 41.

Acknowledgement

This research was supported by the Naval Undersea Center, San Diego, under contract N00123-73-C-1507 and by the Advanced Research Projects Agency of the Department of Defense. It was monitored by the Air Force Eastern Test Range under contract F08606-72-C-0008.

References

1. T. S. Huang, W. F. Schreiber, and O. J. Tretiak, "Image Processing," *IEEE Proceedings,* Vol. 59, No. 11 pp. 1586-1609 (November 1971).

2. E. M. Deloraine and A. H. Reeves, "The 25th Anniversary of Pulse Code Modulation," *Spectrum,* Vol. 2, No. 5, pp. 56-63 (May 1965).

3. W. M. Goodall, "Television by Pulse Code Modulation," *Bell System Technical Journal,* Vol. 1, pp. 33-49 (January 1951).

4. R. L. Carbrey, "Video Transmission over Telephone Cable Pairs by Pulse Code Modulation," *IRE Proceedings,* Vol. 48, pp. 1546-1561 (September 1960).

5. T. S. Huang, "PCM Picture Transmission," *Spectrum,* Vol. 2, No. 12, pp. 57-60 (December 1965).

6. W. K. Pratt, "Bibliography on Digital Image Processing and Related Topics," Univ. of Southern California, USCEE Report 453, September 1, 1973.

7. F. W. Scoville, and T. S. Huang, "The Subjective Effect of Spatial and Brightness Quantization in PCM Picture Transmission," *NEREM Proceedings,* pp. 234-235 (November 1965).

8. R. A. Bruce, "Optimum Pre-emphasis and De-emphasis Networks for Transmission of Television by PCM," *IEEE Transactions on Communication Technology,* Vol. COM-12, pp. 91-96 (September 1964).

9. L. G. Roberts, "Picture Coding using Pseudo-Random Noise," *IRE Transactions on Information Theory,* Vol. IT-8, pp. 145-154 (February 1962).

10. P. F. Panter, and W. Dite, "Quantization Distortion in Pulse Count Modulation with Nonuniform Spacing of Levels," *IRE Proceedings,* Vol. 39, pp. 44-48 (January 1951).

11. T. Max, "Quantizing for Minimum Distortion," *IRE Transactions on Information Theory,* Vol. IT-16, pp. 7-12 (March 1960).

12. H. A. Spang, and P. R. Schultheiss, "Reduction of Quantizing Noise by Use of Feedback," *IRE Transactions on Communication Systems,* pp. 373-380 (December 1962).

13. E. G. Kimme, and F. F. Kuo, "Synthesis of Optimal Filters for a Feedback Quantization System," *IEEE Transactions on Circuit Theory,* Vol. CT-10, pp. 405-413 (September 1963).

14. *IEEE Proceedings,* Special Issue on Redundancy Reduction, Vol. 55, No. 3 (March 1967).

15. *IEEE Transactions on Communication Technology,* Special Issue on Signal Processing for Digital Communications, Vol. COM-19, No. 6, Part I (December 1971).

16. *IEEE Proceedings,* Special Issue on Digital Picture Processing, Vol. 60, No. 7 (July 1972).

17. *IEEE Transactions on Computers,* Special Issue on Two-Dimensional Signal Processing, Vol. C-21, No. 7 (July 1972).

18. H. Hotelling, "Analysis of Complex of Statistical Variables into Principal Components," *Journal of Educational Psychology,* Vol. 24, pp. 417-441, 498-520 (1933).

19. T. T. Y. Huang, and P. M. Schultheiss, "Block Quantization of Correlated Gaussian Random Variables," *IRE Transactions on Communication Systems,* Vol. CS-11, No. 3, pp. 289-296 (September 1963).

20. P. A. Wintz, "Transform Picture Coding," *IEEE Proceedings,* Vol. 60, No. 7, pp. 809-820 (July 1972).

21. H. C. Andrews, *Computer Techniques in Image Processing,* Academic Press, New York, 1970.

22. A. Habibi, and P. A. Wintz, "Image Coding by Linear Transformation and Block Quantization," *IEEE Transactions on Communication Technology,* Vol. COM-19, No. 1, pp. 50-63 (February 1971).

23. G. S. Robinson, "Orthogonal Transform Feasibility Study," *NASA Final Report* NASA-CR-115314, N72-13143 (176 pp.) (submitted by COMSAT Labs. to NASA Manned Spacecraft Center, Houston, Texas) November 1971.

24. W. K. Pratt, J. Kane, and H. C. Andrews, "Hadamard Transform Image Coding," *IEEE Proceedings,* Vol. 57, No. 1, pp. 58-68 (January 1969).

25. G. S. Robinson, "Quantization Noise Considerations in Walsh Transform Image Processing," *Proceedings of the 1972 Symposium on Applications of Walsh Functions (AD-744 650),* pp. 240-247.

26. N. Ahmed, T. Natarajan, and K. R. Rao, "Discrete Cosine Transform," *IEEE Transactions on Computers* (to appear).

27. W. K. Pratt, L. R. Welch, and W. Chen, "Slant Transforms for Image Coding," *Proceedings of the 1972 Symposium on Applications of Walsh Functions, (AD-744 650),* pp. 229-234.

28. W. H. Chen, and W. K. Pratt, "Color Image Coding with the Slant Transform," *Proceedings of the 1973 Symposium on Applications of Walsh Functions*

29. A. Habibi, "Comparison of nth-order DPCM Encoder with Linear Transformations and Block Quantization Techniques," *IEEE Transactions on Communication Technology,* Vol. COM-19, No. 6, Part I, pp. 948-956 (December 1971).

30. J. B. O'Neal, Jr., "Predictive Quantizing Systems (Differential Pulse Code Modulation) for the Transmission of Television Signals," *Bell System Technical Journal,* Vol. 45, May-June 1966.

(a) Original

(b) Bit rate 0.55

(c) Bit rate 1.19

Figure 18. Original and Coded Color Signals Using a Three-Dimensional Fourier Transformation.

(a) Original

(b) Bit rate 0.55

(c) Bit rate 1.19

Figure 19. Original and Coded Color Signals Using a Three-Dimensional Fourier Transformation

(*Note:* The reader is referred to the original publication for color illustrations.)

31. T. O. Limb, C. B. Rubinstein, and K. A. Walsh, "Digital Coding of Color Picturephone Signals by Element-Differential Quantization," *IEEE Transactions on Communication Technology,* Vol. COM-19, No. 6, Part I, pp. 992-1006 (December 1971).

32. R. P. Abbott, "A Differential Pulse-Code-Modulation Code for Video Telephoning using Four Bits per Sample," *Bell System Technical Journal,* Vol. 45, pp. 907-912 (May-June 1966).

33. A. Habibi, "Application of Lower-Triangular Transformations in Coding and Restoration of Two-Dimensional Sources," *Proceedings of the 1973 National Telecommunications Conference,* November 26-28, 1973, Atlanta, Georgia.

34. A. Habibi, "Delta Modulation and DPCM Coding of Color Signals," *Proceedings of the International Telemetry Conference,* Vol. 8, October 10-12, 1972, Los Angeles, pp. 333-343.

35. N. S. Jayant, "Adaptive Delta Modulation with a One-Bit Memory," *Bell System Technical Journal,* Vol. 49, No. 3, pp. 321-342 (March 1970).

36. C. L. Song, T. Garodnick, and D. L. Schilling, "An Adaptive Delta Modulator for Speech and Video Processing," *Proceedings of the IEEE International Conference on Communications (ICC-72),* June 19-21, 1972, Philadelphia, Pennsylvania, pp. 21-30, 21-31.

37. T. A. Hawkes, and P. A. Simonpieri, "Signal Coding using Asynchronous Delta Modulation," *IEEE Transactions on Communication Systems* (to appear).

38. A. Habibi, "Hybrid Coding of Pictorial Data," *IEEE Transactions on Communications* (to appear).

39. L. C. Wilkins, and P. A. Wintz, "A Contour Tracing Algorithm for Data Compression for Two-Dimensional Data," Technical Report No. TR-EE69-3, Purdue University School of Engineering, Lafayette, Indiana (January 1969).

40. P. A. Wintz, and L. C. Wilkins, "Studies on Data Compression, Part I: Picture Coding by Contours," Technical Report No. TR-EE70-17, Purdue University School of Engineering, Lafayette, Indiana (September 1970).

41. A. G. Tescher, *The Role of Phase in Adaptive Image Coding,* Ph.D Thesis, University of Southern California, Electrical Engineering Department, January 1974.

42. F. W. Mounts, "A Video Encoding System using Conditional Picture-Element Replenishment," *Bell System Technical Journal,* Vol. 48, No. 7, pp. 2545-2555 (September 1969).

43. T. O. Limb, "Buffering of Data Generated by the Coding of Moving Images," *Bell System Technical Journal,* Vol. 51, No. 1, pp. 239-261 (January 1972).

44. B. G. Haskell, F. W. Mounts, and T. C. Candy, "Interface Coding of Video Telephone Pictures," *IEEE Proceedings,* Vol. 60, No. 7, pp. 792-800 (July 1972).

45. B. G. Haskell, "Buffer and Channel Shaping by Several Interframe Picturephone Coders," *Bell System Technical Journal,* Vol. 51, No. 1, pp. 261-291 (January 1972).

46. W. K. Pratt, "Spatial Transform Coding of Color Images," *IEEE Transactions on Communication Technology,* Vol. COM-19, No. 6, pp. 980-992 (December 1971).

Intraframe Coding for Picture Transmission

DENIS J. CONNOR, RALPH C. BRAINARD, MEMBER, IEEE,
AND JOHN O. LIMB

Abstract—A great variety of intraframe coding techniques have been proposed and demonstrated in recent years. Most of them exploit the properties of both the image and the human viewer. These properties are discussed briefly so as to provide a framework for describing the various coding techniques.

The development of intraframe coders has been an evolutionary process. Techniques such as delta modulation (DM) and differential pulse code modulation (DPCM) which were proposed in the early 1950's are still being actively studied and improved today. A survey of this early work is given as a preliminary to more detailed discussion of the work from 1967 to the present. Three main approaches are evident in these studies. First, by making linearizing assumptions, frequency domain analysis can be applied to the problem of minimizing the visibility of granular coding noise. Second, in order to exploit the nonlinear properties of the human visual system, a time domain approach to the development of coding algorithms is found most useful. Third, by exploiting only the statistics of the source, efficient reversible encoding operations can be developed. These three approaches are dealt with in some detail.

Possible avenues for further research are pointed out.

Manuscript received November 15, 1971; revised March 10, 1972.
The authors are with Bell Telephone Laboratories, Inc., Holmdel, N. J. 07733.

I. Introduction

GIVEN a computer and digitized pictures it is a simple matter to measure many of the statistics of a particular source: a number of coding techniques have been developed to trade on the redundancy revealed by these measurements. However, when it comes to the receiver, fewer relevant measurements have been made and approaches to incorporating the properties of the viewer are more intuitive. In particular, the nonlinear and picture-dependent properties are poorly understood, but it is precisely from these properties that we feel the greatest coding efficiency will be achieved for pictures that are to be viewed by humans.

The problem is significantly changed if human viewing is not the primary objective. In coding pictures from planetary probes or from resource survey satellites, the ultimate objective in many cases is the recovery of scientific data. Consequently, after the initial linear quantization of the picture, only reversible or information-preserving coding operations are permissible; these can only be based on the picture statistics. For human viewing the initial quantization may

Reprinted from *Proc. IEEE*, vol. 60, pp. 779–791, July 1972.

include substantial irreversible operations provided the result is a picture of acceptable quality.

The actual implementation of a coding algorithm must take into account other factors as well: the level of complexity that is permissible, the channel requirements, etc. Thus, if a constant-rate channel is being used, either the coder must produce a constant rate or the data must be buffered. Much recent work has accepted the limitation of constant-rate coding, but as the cost of storage is reduced, variable rate schemes become more attractive.

To facilitate our discussion of in-frame[1] picture coding some background knowledge of the properties of pictures and of the human viewer is given in Section II. (Detailed coverage of these properties is given elsewhere in this issue.) Section III describes some early attempts at utilizing these properties in the design of in-frame coders. The next three sections, IV–VI, are devoted to a discussion of more recent work, mainly from 1967 to the present.

For the analysis of many coding systems it has been convenient to make linearizing assumptions. This linearization has permitted the application of frequency domain analysis to the problem of minimizing the visibility of granular coding noise. In Section IV we discuss a group of papers based mainly on this approach; by and large, linearizing assumptions are made and in most cases frequency domain analysis is used.

In chaotic, contrasty, or highly detailed areas of a picture the nonlinear properties of the human visual system are dominant, and a time domain approach is found most useful. Subjective testing is usually employed at some point in the development of a time domain algorithm. Thus the intuitive or 'gut' approach starts with a basic coder design and selects the coder parameters on the basis of subjective tests. The more fundamental approach is to measure specific properties of the human observer via subjective testing, and to incorporate these results in a coder design. The design is then checked by further subjective evaluation. In Section V we discuss a number of coding algorithms which have adopted one or the other of these essentially time domain approaches.

In Section VI we depart from our concern with coding for human viewing and describe two reversible coding algorithms. Finally, in Section VII we summarize the paper and pcint out some areas for possible research in the future.

In writing this paper the major problem has been the volume of material that had to be covered. (The picture coding bibliography of Wilkins and Wintz [1] lists some 600 papers.) We have been forced to omit some important topics, e.g., analog bandwidth compression [2], run length coding [3], edge coding [4], coding of color images [5]–[12], and coding of two-tone images [13]. Even in the areas covered, by no means have we been exhaustive. Indeed, only representative articles are discussed for purposes of illustration. These are mainly chosen from work with which we are intimately familiar, and consequently tend to overemphasize work that has been done at Bell Telephone Laboratories.

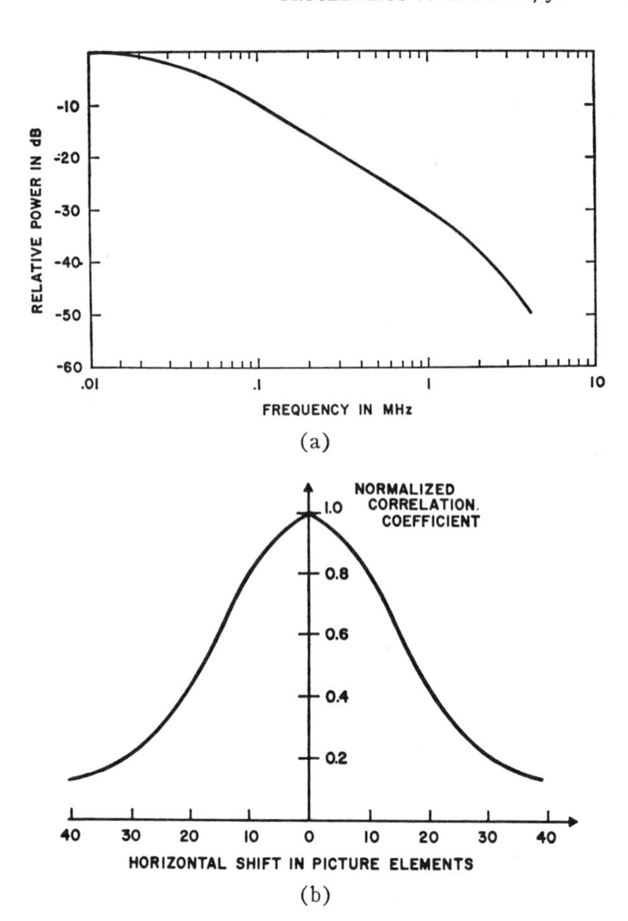

Fig. 1. (a) Power spectrum of video signal [14]. (b) Horizontal autocorrelation function of picture [16].

II. The Properties of the Source and Receiver

The earliest statement about the global properties of a video signal was its power spectrum. A typical spectrum is shown in Fig. 1(a) [14]. The envelope of the spectrum is relatively flat out to about twice the line rate (about 30 kHz for broadcast rates) where it begins to drop at about 6 dB/octave. This is for a signal with a nominal bandwidth of 5 MHz! Evidently the overwhelming portion of the power in a video signal is in the low frequencies. These provide the information about large area brightness levels. Although the power at high frequencies is small, the high frequencies are essential if the processed image is to have clean sharp edges.

The autocorrelation function of a signal is related to its power spectral density by the Fourier transform. Kretzmer [15] measured the autocorrelation function for a number of pictures and found, as one would expect from the power spectral density, a high degree of local correlation for all except the most detailed of pictures. A "typical" autocorrelation function in the horizontal direction is shown in Fig. 1(b) [16]. The vertical correlation function is similar. Such correlations are indicative of redundant information in the signal. They can be removed by the use of linear prediction filters [17].

Kretzmer [15] also measured certain probability density functions of the video signal. He found that while the amplitude density for a single picture would be quite nonuniform, the density for an ensemble of pictures was nearly uniform. The horizontal element-difference density in both cases was highly peaked at zero. Although this latter result is to be expected for a signal that has its predominant power at low frequencies, it did provide the first quantitative results that could be used to estimate the redundancy of a video signal. Kretzmer found that the entropy of a 64-level element-difference signal which would result from a 5-bit original was about 2.6 bit. In other words, linear prediction using only the previous element permits the elimination of 2.4 bit of redundancy. Kretzmer's figures may be compared with later measurements of Schreiber [18] on a 6-bit original in which

[1] Contraction of the word intraframe.

he obtains an entropy of 1.85 bit/pel (picture element) for a simple picture and 3.36 bit/pel for a more complex scheme. More importantly, Schreiber concludes that most of the second-order redundancy is removed by coding the difference signal, thus reducing the coding to a one-dimensional problem.

Concurrent with Kretzmer, Harrison [17] exploited the correlation in the video signal in order to reduce the average power required to transmit it. This was done using a predictive or preemphasis filter at the transmitter and the inverse filter at the receiver. He found that whereas a substantial reduction in the average power could be obtained using just the previous element, little further reduction was possible using more of the adjacent elements on the same and preceding lines.

The third-order statistics measured by Schreiber are in complete accord with these results. He found that a coding system based on the statistics of three horizontally adjacent elements would not yield a substantial reduction in entropy over that obtained using just the two adjacent elements. Further supporting evidence comes from the study of Powers *et al.* [19], [20]. Under the assumption that the correlation between pels decays exponentially with distance, they showed that the greatest part of the linear redundancy[2] can be removed by using either the horizontally adjacent or the vertically adjacent element.

Thus we have seen that the signal is characterized in the frequency domain by a power spectrum that is highly peaked at low frequencies. The signal power can be greatly reduced by predictive or preemphasis filters. In the time domain the signal is characterized by its autocorrelation function. By using simple predictive techniques and taking advantage of the highly peaked nature of the resulting distribution, the redundancy in the signal can be greatly reduced.

Let us now turn to the properties of the receiver, the human visual system. One of the most important aspects of human perception to picture coder design is the differential sensitivity of the eye. Any variation of visual sensitivity can be exploited by adapting the encoding accuracy to the variation.

Teer [21] considers three aspects of differential sensitivity which can be grouped into two categories.

1) Picture-independent sensitivity variations: the noise-detection threshold increases with increasing noise frequency (this will be referred to as the frequency dependence).

2) Picture-dependent sensitivity variations: a) the noise-detection threshold increases with increasing picture detail (detail dependence); b) the noise detection threshold increases with increasing luminance (luminance dependence).

The first aspect (which, by the way, does not hold at very low frequencies [22]) is best illustrated by pictures which duplicate an effect noted by Goodall [23] in his pioneering studies on PCM-encoded pictures. In Fig. 2(a) we show a 5-bit PCM-encoded picture from a camera having a 60-dB S/N ratio. Note the contouring effect in the sky. This is low-frequency quantizing error or noise caused by the coarse quantizer settings. In Fig. 2(b) high-frequency noise is added to the signal from the camera before quantization. This noise causes the quantizer to oscillate between levels in the neighborhood of the contours seen in Fig. 2(a). Thus the quantizing noise is still there (actually the rms value is increased) but it is concentrated at a higher frequency. The result is that the contours are no longer visible. In effect the eye acts like a low-pass filter. This property of the visual system can be modeled in the frequency domain by the frequency response shown in Fig. 3(a) [24]–[27]. In the time domain it is described by the impulse response shown in Fig. 3(b) [28], [29]. (This impulse response is referred to in the psychophysical literature as the point spread function of the eye [30].)

[2] Linear redundancy is that part of the total redundancy that can be eliminated using linear prediction.

(a)

(b)

Fig. 2. (a) Contouring is visible in this low-noise 5-bit PCM picture. (b) Increasing noise level reduces visibility of contours.

The other two aspects of the differential sensitivity of the eye make the visibility of the noise dependent on the local properties of the picture. They are essentially nonlinear and their use in a coding scheme is hampered by the lack of quantitative descriptions. In practice it is found that they are more easily handled in the time domain than in the frequency domain.

The detail dependence of noise visibility [31], [32] is obvious in Fig. 2(a) if one looks for contouring in the detailed areas. None is visible. The brightness dependence [33], [34] is evident if one compares the visibility of contours in the darker areas (highly visible) to that in the light areas (vir-

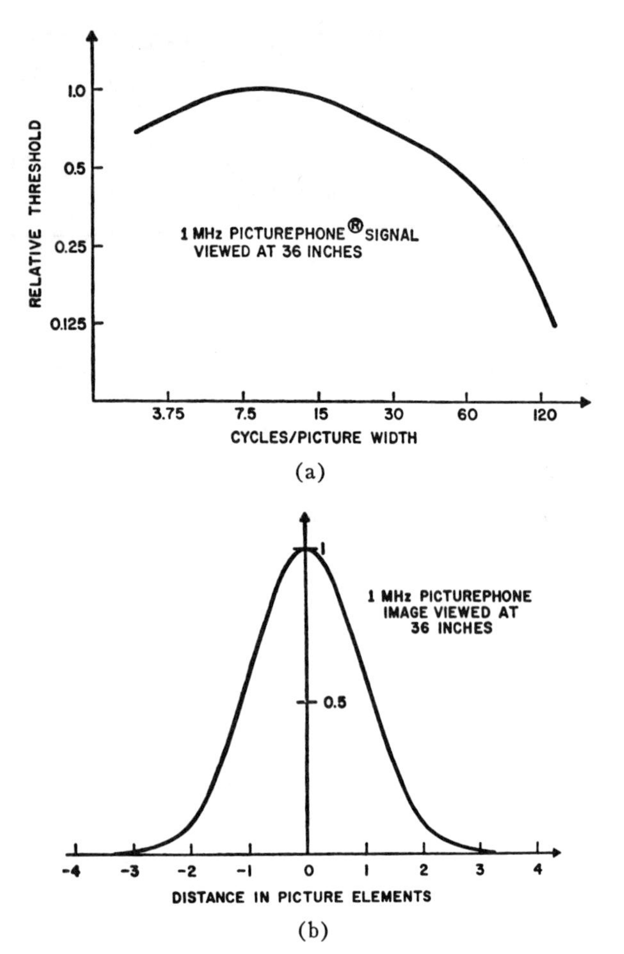

Fig. 3. (a) Sine-wave response of the human visual system [22]. (b) Normalized impulse response of the human visual system [28], [29.]

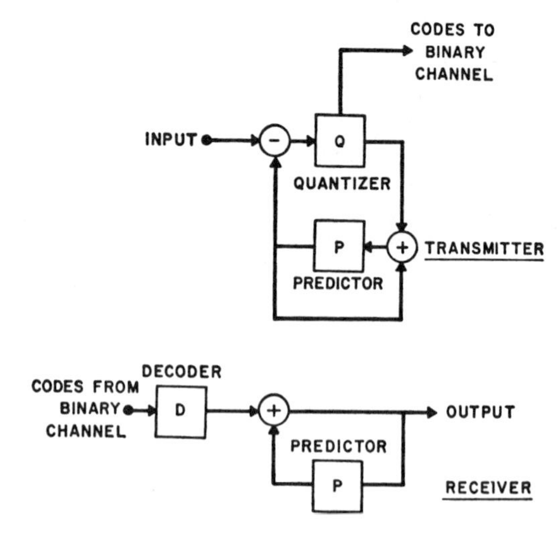

Fig. 4. General form of DPCM transmitter and receiver.

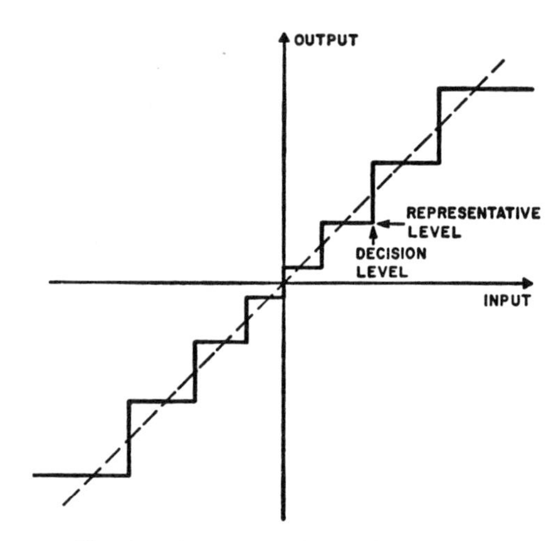

Fig. 5. 8-level tapered quantization scale.

tually invisible). This brightness dependence is easily exploited in a video transmission system by compressing the signal before transmission (logarithmic amplification, black stretching[3]), and reversing the operation at the receiver [35]. Consequently, it need not concern us further.

With this brief characterization of the properties of the source and the receiver in both the frequency and time domains, let us examine how various workers have exploited these properties in the design of efficient in-frame coders.

III. Source–Receiver Encoding: Initial Approaches

The first approaches to the problem of efficiently encoding a video signal were mainly intuitive. Kretzmer's reduced alphabet technique [36] sought to exploit the detail sensitivity of the eye by splitting the signal into a 0–0.5-MHz band and a 0.5–4-MHz band. The low-frequency band needed to be sampled at only 1 MHz, but had to be quantized finely because of the sensitivity of the eye to low-frequency noise. The high-frequency band had to be sampled more frequently, but could be quantized coarsely because of the decreased sensitivity in high-frequency or detailed regions. Moreover, a tapered 5-level quantizer was used because, ". . . it seems plausible that small changes should be rendered with greater absolute accuracy than large changes, both because the observer is likely to be more error sensitive where small changes are concerned and because small changes are much more frequent than large ones" [36].

The reproduction of edges in the encoded picture was impaired by echo-like noise and edge busyness. The author cautions about drawing too many conclusions from the experiment due to the exploratory nature of the implementation. A similar technique, demonstrated by Schreiber, was more successful [37].

The most practical encoding scheme that is compatible with Kretzmer's insight is differential quantization or differential pulse code modulation (DPCM) [38]. It is illustrated in its most general form in Fig. 4. A prediction for the value of the next sample is made based on the past history of the encoded signal, not the video signal itself. The difference between the prediction and the signal is quantized into one of n levels and a code is transmitted to indicate which of the n levels occurs. (Delta modulation (DM) is a particular form of DPCM in which $n=2$ [39], [40].)

R. E. Graham recognized that the use of Kretzmer's tapered quantization scale in a DPCM coder would provide a "truly perceptual coding technique" [41]. The 8-level tapered scale Graham proposed is shown in Fig. 5. Its use assures that larger differences are quantized more coarsely than fine ones. Thus the DPCM configuration permits the simultaneous exploitation of the receiver properties by appropriate selection of the quantizing scale, and of the source

[3] This usually takes place to a certain extent since the signal from the television camera is nonlinearly processed (gamma corrected) to compensate for the nonlinear characteristic of cathode-ray tubes.

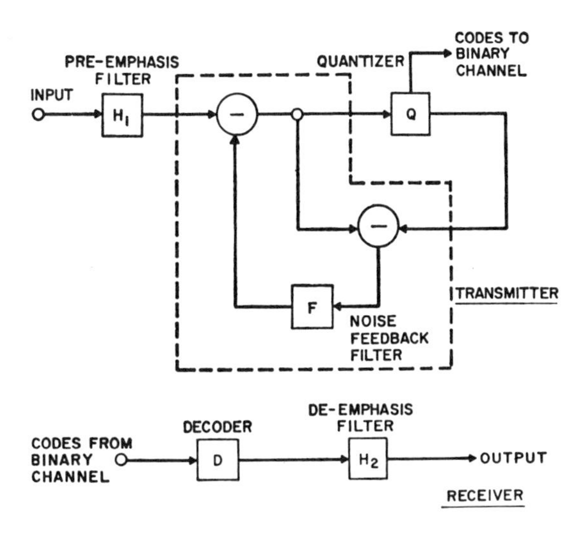

Fig. 6. Noise-feedback coder configuration.

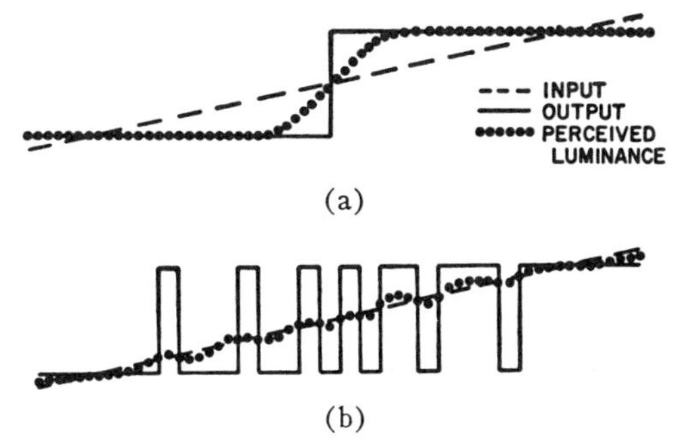

Fig. 7. (a) Quantizer output and perceived luminance without addition of dither. (b) Addition of dither causes perceived luminance to follow actual input more closely.

statistics by appropriate linear or nonlinear prediction algorithms.

The novel feature of the DPCM coder is the feedback path around the quantizer. Its use is not indicated from considerations of source and/or receiver properties alone. The choice of a digital channel and the desire to transmit a difference signal (because of source properties) combine in dictating the use of feedback. Geddes [42] shows what happens when one tries to quantize and transmit the time derivative directly. The quantizing errors are accumulated in the integrating filter at the receiver producing gross streaking in the received picture.

After the pioneering work of Kretzmer and Graham people apparently lost sight of the fact that there were two properties of the receiver that could be exploited [27], [43]–[45]. They concentrated more on the frequency dependence of noise visibility and ignored the detail dependence. To some extent this was dictated by analytic considerations since the frequency dependence could be treated using linear techniques. In addition, theoretical studies of nonuniform quantizers had led to Max's criteria for minimizing the rms distortion caused by quantization [46]. They are 1) the decision levels must lie midway between the output levels, and 2) the output levels must lie at the centroid of that part of the probability density function of the signal occurring between the adjacent decision levels. (In effect these criteria legitimize the second half of Kretzmer's intuitive reasons for tapering the quantizer scale.)

When quantizing a signal with a highly peaked probability density function, application of the Max criteria will result in a companded quantizer scale. Thus for example, Kimme and Kuo [43] suggest using a companded quantizer, not to exploit the lack of sensitivity to noise in detailed regions, but simply to minimize the rms distortion of the quantized difference signal.[4]

The noise feedback coder [85], [71] investigated by Kimme and Kuo is illustrated in Fig. 6 [43]. The predictive filter H_1 is taken out of the feedback loop and is put in front of the coder to preemphasize the signal, i.e., to form a differential signal. The filter H_2 performs the inverse operation. The noise caused by the quantizer is detected by subtracting the input to the quantizer from its output. The noise is filtered by F to emphasize the high-frequency noise which is then added to the input signal. This noise causes the quantizer to oscillate between levels. Thus the quantization contours are obscured, while the average output is maintained equal to the input (similar to the action of dither, see below).

For analytic reasons they replace the quantizer with an independent noise source. This approach ignores the correlation between signal and noise that occurs with a companded quantizer. They obtain designs for H_1, H_2, and F which minimize the noise visibility given the power spectrum of the original signal, and the visibility curve for additive noise. The frequency domain approach is obviously powerful in minimizing the uncorrelated or granular noise.

Another method for reducing the granularity is the addition of dither to the signal before quantizing. Roberts [44] explored the effect, first observed by Goodall [23], that the addition of noise to the original improved the overall appearance of a coarsely quantized picture. The objectionable contours occurring in flat areas of the picture are eliminated at the cost of a small increase in the overall noise in the picture. Roberts showed that the probability density function of the noise (dither signal) should be rectangular and have an amplitude equal to the step size of the quantizer.

The action of the dither signal is depicted in Fig. 7. In Fig. 7(a) the input waveform, a slowly changing ramp, goes from one quantizing level to the next. With no dither the quantized signal switches in one step from one level to the next. Also shown is a dotted waveform, which represents the appearance of the display as seen by an observer at normal viewing distance. In Fig. 7(b) the quantizer is "modulated" by the added noise. The perceived waveform now lies closer to the input waveform because the visual filtering attenuates the error which now is at higher spatial frequencies. The retinal size of the pel will determine the efficacy of such techniques since the size of a pel, and hence the visibility of the high-frequency error, will decrease as the square of the viewing distance. For this reason dither is more helpful at broadcast television rates than it is at present videotelephone rates (where the size of a pel is about four times larger than it is for broadcast television).

Although the average value of output is maintained equal to the input by the addition of a random dither component, the rms noise after quantization is double what it would be without dither. If, however, the dither component is made pseudorandom, it can be subtracted at the receiving end, since it is a simple matter to synchronize two identical pseudo-

[4] It is probably not coincidental that by working with the difference signal (i.e., minimizing the error of the difference signal) one exploits the lack of sensitivity in detailed regions, since the visual system is apparently more concerned with the differential of luminance (e.g., edges) than with luminance itself.

random noise generators over the channel. Subtraction at the receiving end has the effect of halving the rms noise relative to the noise that occurs without subtraction [44].

All of the above techniques take advantage of the decreasing sensitivity of the eye to noise of increasing frequency. One technique demonstrated in 1964 did exploit, as well, the detail dependence of the visual system in a fashion reminiscent of the dual mode schemes of Kretzmer and Schreiber. Fukushima and Ando [47] proposed that every fourth point on a line be transmitted as 6-bit PCM. Intervening values were obtained by quantizing the difference between the signal and an interpolative prediction. Note that the high-frequency information is contained in the differences. The quantization scale used was a function of the magnitude of the difference in brightness of the pels used to obtain the interpolative prediction.

IV. Frequency Domain Studies in Recent Years

In this section we examine the present status of the frequency domain approach in designing video encoders. Valuable theoretical results on rms noise in digital encoders have been obtained by O'Neal [48]–[50]. Although a number of simplifying assumptions were made, these results still serve as a basis for heuristic projections. In the actual design of encoders two main techniques have been explored, both of which depend on the addition of high-frequency noise to a signal before quantization. The closed-loop technique utilizes the noise feedback idea, whereas the open-loop approach uses an additive dither signal.

Closed-Loop Encoders

The closed-loop feedback coder is dealt with in two papers. In the first paper, by Brainard, the coder illustrated in Fig. 4 was the subject of an extensive subjective testing program [51]. As indicated in the previous section, Kimme and Kuo [43] had specified the three filters for an optimal system. Although these specific filters were found to be inadequate, the general design approach was useful in that for the simple element-difference filter pair specified below:

$$H_1(\omega, \alpha) = 1 - \alpha\Delta$$

$$H_2(\omega, \alpha) = \frac{1}{1 - \alpha\Delta}$$

where Δ represents the delay operator, $e^{-i\omega T}$. A value of $\alpha = 0.9$ was found to give the maximum subjective S/N ratio. This value most nearly approximates the optimal frequency domain characteristic from which the optimal filters were derived. In a comparison test betwen the noise feedback system with $\alpha = 0.9$ and a DPCM system using a previous element predictor, there was no detectable difference in picture quality. In both cases the 8-level quantizer characteristic was obtained by application of the Max criteria. No tests were carried out to determine if this setting was subjectively optimal.

The second paper by Brainard and Candy deals with a simplification of the noise feedback coder which they have called a direct feedback coder [52]. They observed that the lefthand loop inside the dashed lines of Fig. 6 functions as an integrating amplifier. Thus it can be reconfigured as shown in Fig. 8. By making the linearizing assumption that the quantizer adds uncorrelated noise to the signal, they obtain expressions for the filters H and H^{-1} in terms of the signal spectrum, the frequency weighting function for the visibility

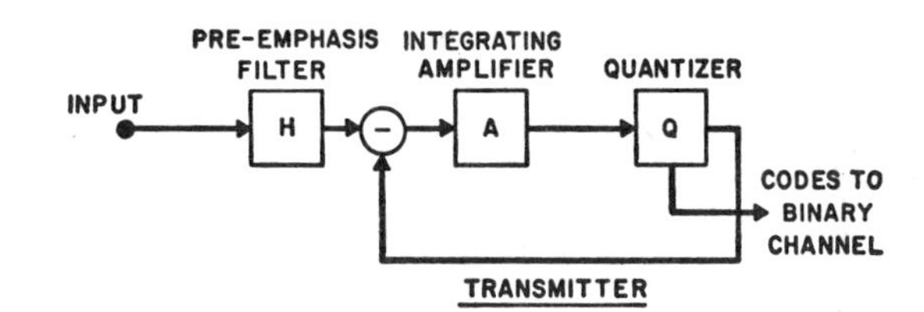

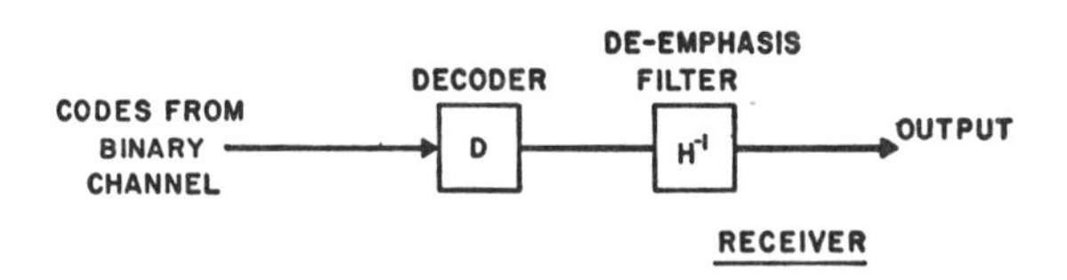

Fig. 8. Direct feedback coder configuration, transmitter, and receiver [52].

of uncorrelated noise, and the transfer function of the amplifier A. This last function is a compromise between loop stability considerations and the desire to have a large gain at low frequencies in order that the feedback minimize the low-frequency noise. They conclude that for video signals, A should be a simple integration with a long time constant. Then the filter functions H and H^{-1} follow, with H having a differentiating characteristic and H^{-1} being a leaky integrator. The final form for these filters is obtained empirically using the derived formulas as guides.

1-, 2-, and 3-bit coders were constructed to encode 1-MHz video signals for transmission at 6 Mbit/s. Using a uniform quantizing scale, a predicted 2-dB difference in performance between the three coders was not detected. Companding the quantizer scales using the Max criteria led to a 5-dB improvement in subjective noise performance for the 3-bit coder but gave no improvement with the 2-bit coder. With the companded scale there was no detectable difference between the direct feedback coders and the equivalent differential coders. Evidently the greater flexibility achieved by separating the preemphasis and the noise shaping filters in the direct feedback coder does not lead to markedly improved coder operation. We conclude that for video signals the DPCM coder is to be preferred because it is simpler.

Open-Loop Encoders

Open-loop encoders for video signals are of interest because of their simplicity. They generally consist of a 3- or 4-bit PCM encoder operating on a signal to which noise has been added. The addition of noise, usually pseudorandom noise or dither, serves to break up the objectionable contours caused by coarse quantization. Instead of seeing the high-frequency oscillations that are actually present in the picture, the eye forms a weighted average of the noisy signal over a region determined by its point spread function (i.e., impulse response). It is this average that the viewer perceives.

Thompson and Sparkes constructed a 2-bit dithered quantizer [53]. They used a fixed temporal dither pattern with a two-frame period to which they added a pseudorandom spatial dither in order to disperse flickering contours. The result was ". . . an image subjectively equivalent to the original degraded with white Gaussian noise to approximately 30-dB S/N ratio."

Limb evaluated the visibility of granular noise in coarsely quantized dithered signals using a low-pass filter model for threshold vision [54]. Visibility was determined as a function of the signal position within the quantizing interval and the amount of correlation in the random dither signal. The noise is most visible when the original signal coincides with a quantizer decision level. The noise visibility decreases rapidly as the correlation changes from $+1$ to -1.

In a second paper, Limb designs one- and two-dimensional deterministic dither patterns having low visibility [28]. Deterministic dither patterns both with and without additional random noise are evaluated using the visibility model mentioned above. Extending the results to include temporal dither patterns would require a model which included temporal effects, such as the sensitivity to flicker. The effect of using these deterministic patterns to reduce the visibility of contours in a DPCM-encoded picture is described in a companion paper [55].

In a recently published paper, Thompson reports on a study in which the video signal is preemphazised in that part of its spectrum where the visibility of the dither noise peaks [56]. The dither signal, with its carefully chosen spectrum, is then added to the signal and the result is quantized. At the receiver the dither is subtracted and the signal plus quantization noise is deemphasized.

The subjective S/N ratio falls 3.2 dB short of the theoretical prediction of 43.0 dB for a full amplitude signal. Interactive effects between camera noise (target grain in particular) and the dither signal apparently prevent the achievement of predicted S/N ratios. These ratios are achieved, however, using noise-free input signals such as ramps.

Thompson compared his dithered system with an equivalent bit-rate feedback coder. He found that the latter gives a superior picture, the difference being most obvious at edges where clipping of the emphasized input in the dithered system causes smearing. It would appear that the major advantage of direct-feedback or DPCM coders lies in the use of a companded quantizer which permits more perceptually acceptable coding of edges. Perhaps this advantage would disappear if a way could be found to use nonlinear quantization in the open-loop dithered system. Since significant low-frequency components of the signal are transmitted, the problem of quantization errors accumulating in the deemphasis or integrating filter at the receiver should be less than in open-loop systems that attempt to send a true differential signal.

V. Time Domain Encoder Design

The time domain approach to encoder design tacitly assumes that the video image is composed of a large array of separate entities called pels. These are usually obtained by sampling the video waveform at or above the Nyquist rate. A coding algorithm is evaluated by examining its detailed pel-by-pel operation in flat areas and at edges. In flat regions the amplitude should be accurately encoded. At edges more amplitude distortion is permitted provided the apparent edge position is reproduced accurately. Given these requirements most people attempting to encode video signals have settled on a DPCM encoder (Fig. 4) with a companded quantizing scale as their basic system. Within this context, however, there have been a great variety of approaches which, for simplicity of exposition, can be broken down into three classes: fixed-format encoders, edge adaptive encoders, and area adaptive encoders.

Fixed-Format Encoders

A good example of recent advances in the development of DPCM encoders is described by Limb and Mounts [55] and by Estournet [57]. The design philosophy is the same as that proposed by Graham [41]: linear prediction to take advantage of the statistical redundancy of the source, plus nonuniform quantization of the prediction error so as to exploit the changing sensitivity of vision. Implementation of this philosophy is greatly simplified by extensive use of digital circuitry. In particular, the problem of tracking (i.e., decoding exactly the same analog signal at the transmitter and at any one of many receivers) is simplified by digital implementation of the weighter (level assignment unit) and accumulator at both the transmitter and receiver. An "infinite" time constant is used: resetting both accumulators at the end of every line serves to limit the effect of transmission errors.

Recent interest in the encoding of video signals has been spurred by a desire to provide videotelephone service. Millard and Maunsell describe an early version of a DPCM encoder for Picturephone® service [58]. They include an extensive discussion of the many aspects involved in the design of a practical encoder. They describe and illustrate the various subjective impairments caused by DPCM encoding of a video signal—slope overload, granular noise, contouring, and edge busyness.

An improved version of this coder is described by Abbott [59]. Instead of the 3-bit or 8-level quantizer used by Millard and Maunsell, this improved version uses a 4-bit 16-level quantizer. Since the coder output must still be transmitted at the same bit rate as previously (6.3 Mbit/s), the increase in the number of bits per sample is offset by two factors: 1) a small buffer is included so that the horizontal sync period can be used for picture data transmission; 2) the signal is sampled at about 93 percent of the nominal Nyquist rate. This latter technique may introduce further aliasing in regions with fine vertical detail, but this additional impairment is more than offset by the subjective improvement due to using the 16-level quantizer. This system also has extensive digital implementation.

Theoretical studies on S/N ratio predict only a small improvement (up to 3 dB) in picture quality by going to two-dimensional predictive algorithms [49]. However, the S/N ratio can be a misleading measure of picture quality. Subjective evaluation of some two-dimensional linear and nonlinear predictive algorithms by Connor, Pease, and Scholes indicates that for 3-bit DPCM the improved rendition of vertical edges is dramatic [60]. Although some of the impairment is transferred to horizontal edges, it is much less noticeable because, first, it is masked by the line structure of the video image, and second, it does not manifest itself as an attention-getting "busyness."

This study also revealed an unexpected property of two-dimensional DPCM encoders—the effects of transmission errors for some configurations are much less visible than for previous element DPCM. The decoded image and the transmission error patterns for an element-difference DPCM en-encoder are shown in Fig. 9(a) and (b). The same transmission errors are present in Fig. 9(c), which is the decoded image from a DPCM encoder that uses as a prediction the average of the previous element and the element above and to the right of the element being encoded. The error patterns shown in Fig. 9(d) can be used to locate the errors in Fig. 9(c). The difference in visibility of the errors in the two pictures is striking.

® Registered service mark of the American Telephone and Telegraph Company.

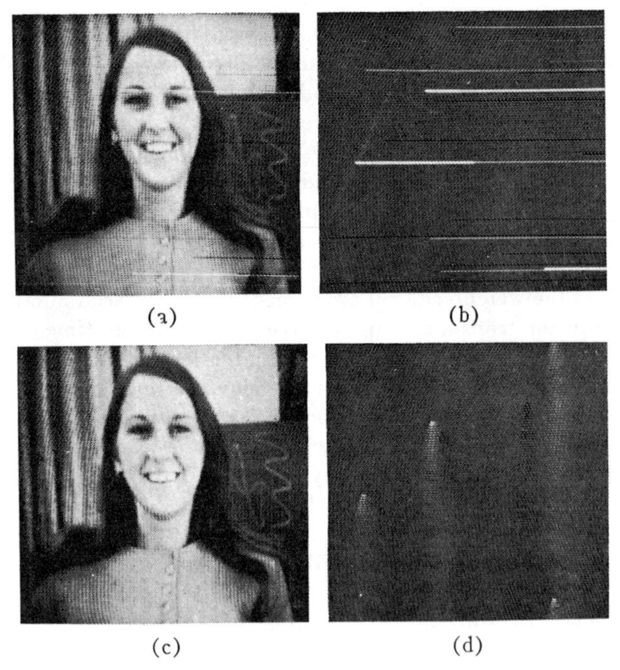

Fig. 9. Transmission errors in DPCM encoders [60]. (a) Previous element prediction. (b) Error patterns for previous element prediction. (c) Spatial average predicton. (d) Error patterns for spatial average prediction.

Edge Adaptive Encoders

In this and the following section a number of signal-dependent coding algorithms are described. The algorithms discussed in this section use "instantaneous" adaptation to improve their subjective rendition of rapidly changing signals, i.e., at edges in the video image. The algorithms of the next section employ "syllabic" adaptation in order to select the strategy most suited to the signal properties.

Edge Adaptive Delta Modulation: Instantaneous adaptation has been applied most successfully in improving the response of DM encoders. DM is a special case of differential coding in which the output at each sampling instant is restricted to one of two codes representing a positive or a negative step respectively [39], [40]. J. B. O'Neal has shown analytically that for good quality pictures multibit differential coders have larger S/N ratios than equivalent bit-rate delta modulators [50]. Thus to obtain equivalent picture quality a linear delta modulator requires a higher bit rate. However, in situations where the cost of implementation is more important than the bit rate, the simple circuitry of the delta modulator makes it an attractive alternative to multibit differential encoding.

A tradeoff between sampling rate (or bit rate) and coder complexity is possible by going to adaptive, or companded, delta modulation in which the step size is made dependent on the local properties of the signal. Two approaches are possible: syllabic companding, in which the step size is dependent on some local average of the signal, e.g., loudness or pitch in telephony [61], [62]; the alternative is instantaneous companding, in which the step size is dependent on instantaneous signal values, e.g., the slope of a video waveform.

The problem being dealt with and the general method of approach is best illustrated by considering one of the algorithms in some detail. In Fig. 10 a general block diagram for a companded codec (coder–decoder combination) is given. Assume for the moment that the companding is inoperative

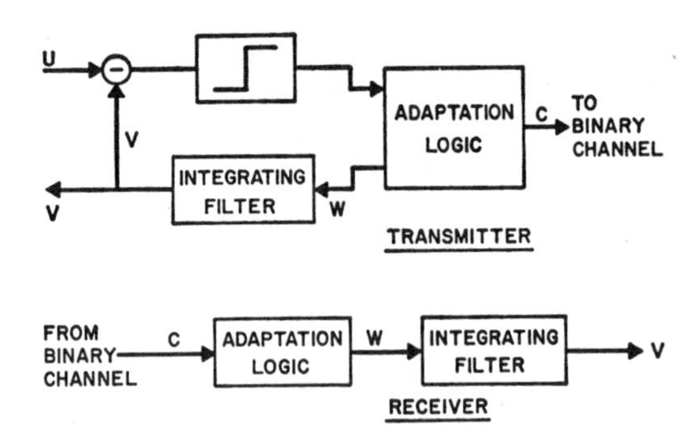

Fig. 10. Block diagram of companded delta modulator. U: input waveform; V: processed output waveform; W: weighting sequence; C: code sequence.

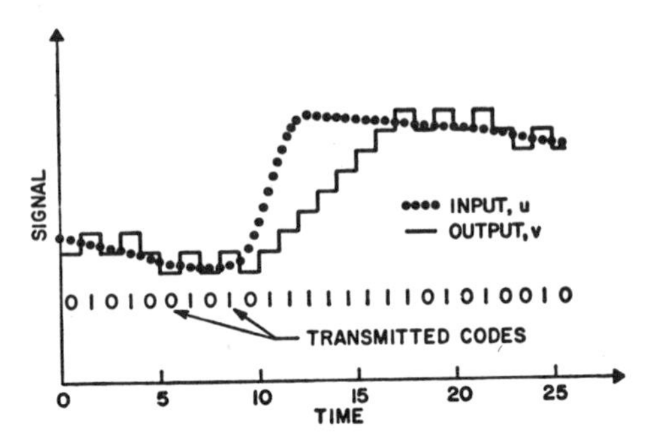

Fig. 11. Input and processed waveform plus output code from a linear delta modulator.

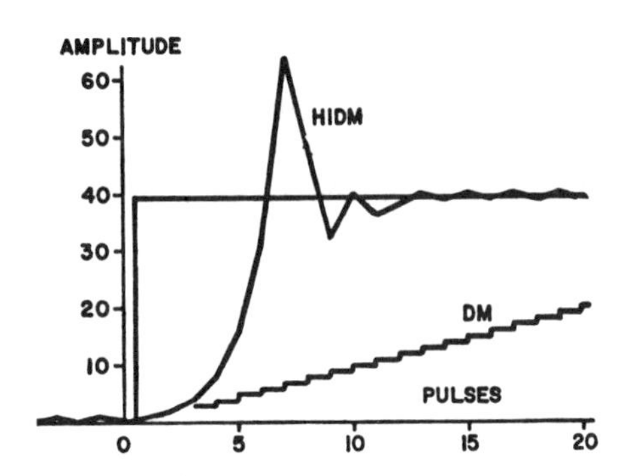

Fig. 12. High information delta modulation (HIDM) response and linear delta-modulator response at an edge [63].

and that the codec is operating as a linear delta modulator whose step size has been optimized according to some criterion. Because it is impossible to compand a two-step quantizing scale, a tradeoff has been made between the fidelity of reproduction in flat areas and that in detailed areas. Thus the level of quantization noise in flat areas has been balanced against the slope overload noise on sharp vertical edges. Given the input waveform u in Fig. 11, the linear delta modulator will produce the output waveform v in Fig. 11. The digital code for the waveform is shown below it. From the code it is evident that in flat areas the digits alternate whereas in areas where slope overload occurs the digits are the same.

Winkler pointed out that the output code could be used to detect the presence of an edge [63], [64]. Consequently, instead of using a fixed step size in the delta modulator, he made it depdndent on the previous history of ones and zeros. If three successive pulses have the same polarity, the step size associated with the third pulse is double that of the second pulse. If two successive pulses differ in polarity, the step size

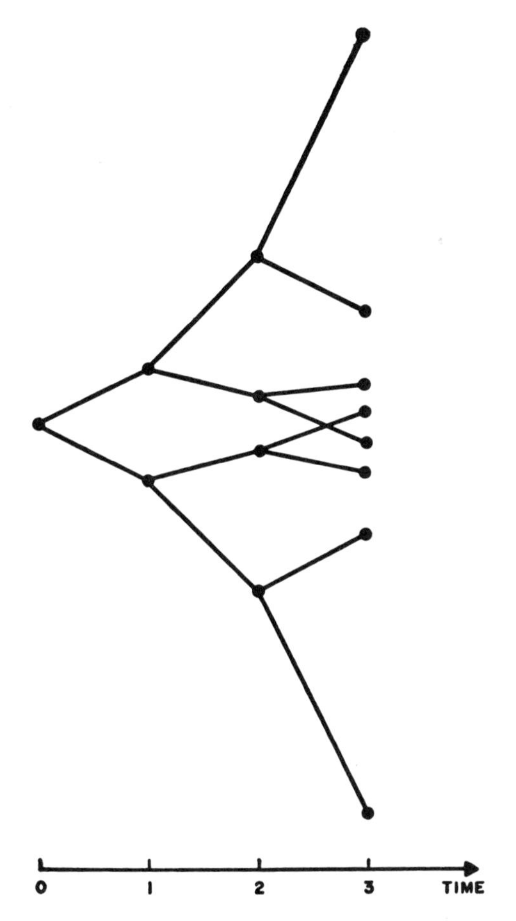

Fig. 13. Encoding tree for companded delta modulator with delayed encoding.

for the second pulse is half that for the first. A comparison of Winkler's high information delta modulation (HIDM) response and the DM response at an edge is shown in Fig. 12. Jayant gives a theoretical treatment of the problem of selecting factors P and Q for increasing and for decreasing the step size respectively [65]. He concludes that $PQ=1$ for reasons of stability, and that $P \cong 1.5$ for optimum performance in encoding both telephone and video signals.

Abate generalized the adaptive delta modulator so that the step size chosen from a library was dependent on the particular code sequence being produced [62]. Candy adopted this approach and carried out an extensive program of subjective testing in order to obtain the subjectively optimum weighting sequence for encoding 1-MHz video signals [66], [67]. He found that the sequence 1, 1, 2, 3, 5, 5 · · · was best, with the sequence restarting at every change of code polarity. Note two features of this weighting sequence. The two ones at the beginning of the sequence permit flat areas with nonzero slope to be coded using the smallest weight. About 90 percent of a picture can thus be coded using the smallest step size. The second feature is the immediate return to the smallest step size once a change in code polarity indicates a flattening of the waveform. This avoids using the larger steps for coding flat areas of the picture where the eye is sensitive to noise.

Aughenbaugh, Irwin, and O'Neal [68] suggested a technique called "delayed encoding" for DPCM, which has since been more successfully applied to delta modulators. To illustrate (Fig. 13), if the operation of a delta modulator is likened to an encoding tree, then from the last point that was encoded there are two possible encodings for the present point. Each of these paths branch to give four options for the next sample, eight for two samples into the future, etc. Thus, for the present sample and n future samples there are 2^{n+1} possible paths corresponding to signals that can be transmitted. Each of these possible signals is compared sample by sample with the input signal and a weighted difference function is formed. The encoding option for the *present* point that lies on the path for which this function is a minimum is selected and transmitted. The window $n+1$ samples wide, is moved ahead one sample and the process is repeated.[5]

Cutler has applied delayed encoding as part of a complex delta modulator for video signals [69]. He employs double integration in the predictor and Jayant's adpative step size algorithm to obtain a rapid change in step size at an edge. Such a configuration is normally unstable in the sense that the output oscillates about the input level. However, the addition of delayed encoding causes the unstable conditions to be anticipated and hence avoided.

The use of the present sample and the first future sample in a variety of error weighting functions was studied by Cutler. He found that the best S/N ratio was obtained by selecting the code for the present sample that caused the error on the upcoming sample to be a minimum. The S/N ratio was improved from 41.6 dB with an adaptive algorithm but no delayed encoding (double integration was of no help in this situation) to 44.8 dB with delayed encoding and full double integration in the feedback path. Pictures coded with the algorithm have sharper smoother edges than the adaptive coder.

The original appeal of DM was its simplicity of implementation. The more complex the adaptation algorithm, the less this is true. Implementing Cutler's algorithm, for example, would appear to yield a device approaching or surpassing the complexity of a DPCM encoder.

Other Edge Adaptive Encoders: Since edges usually occur rarely in a picture signal they can be transmitted efficiently by using a larger number of quantizer levels, and a variable length code (see Section VI). If the resultant signal is to be sent down a constant rate channel, a buffer must be provided to smooth the data flow. If the data generation rate varies widely, quite a large buffer may be required. Because of the resulting increase in complexity and cost, fixed-rate edge adaptive coders are of interest.

Kitsopoulos and Kretzmer tested element differences in order to choose between two modes of operation [70]. If the element-to-element difference exceeds about 6 percent of the peak signal amplitude, a 4-bit sample is transmitted. Otherwise a 7-bit sample is transmitted every alternate sample. The resulting signal can be transmitted in 4 bit/pel although presumably, certain PCM values are deleted to make room for the code change information. While these deletions were not simulated they should affect picture quality very little if chosen carefully.

Bisignani, Richards, and Whelan, in their coarse–fine scheme transmit in rapidly changing areas only the three most significant bits of a 6-bit PCM signal [71]. The occurrence of two pels with the same PCM value causes the coder

[5] A close relative of delayed encoding is the "threshold control" technique proposed by Candy [67]. Gross overshoots in the response of a companded delta modulator can be prevented by selecting the code foı the present sample that minimizes the magnitude of the quantizing error.

to transmit the three least significant bits instead (fine information). The mode change information is fitted in at no extra cost largely by inhibiting the transmission of fine data just after transition to the fine mode. There is only a small reduction in picture quality relative to 6-bit PCM.

Kaminski and Brown start with a 10-level differential signal [72]. This is converted to an 8-level signal by switching between an 8-level coarse mode and a 10-level fine mode. Only the transition from the fine mode to the coarse mode (which is the more challenging) will be described. The six inner levels of the fine mode are transmitted with a 3-bit code. The remaining two words of the 3-bit code specify the amplitude but not the sign of the remaining four levels (the two outer pairs). When one of the four outer levels are used the coder switches to the coarse mode and the last bit of the previous word is used to indicate the sign of the outer levels. This restricts the accuracy with which the previous sample can be coded but since it is adjacent to a large change in luminance one can get away with it. The final result is that the 10-level signal can be transmitted in 3 bit with virtually no additional degradation.

Finally, we briefly mention some studies in which the final bit rate is not constant [41], [73], [74]. In the reservoir scheme of Graham he transmits a 20-level differentially quantized signal with either a 3-bit or a 6-bit code, a 3-bit code for the inner levels and a 6-bit code for the outer levels [41]. Because the outer levels are used so rarely the additional bit rate can generally be incorporated in the horizontal blanking interval. This scheme is implemented in a way that requires little additional equipment over an 8-level coder. However, the penalty paid is that the position of the levels is far from optimum and the performance of the coder is probably closer to a 12- or 14-level coder than to a 20-level coder. Another, more recent, implementation has been presented by Brown [73].

Area Adaptive Encoders

In general, pictures are not homogeneous. Areas of soft texture can lie adjacent to flat areas or contrasty areas. As we have seen, perceptual performance will change markedly between such areas. In view of these facts, greater coding efficiency may be achieved by an algorithm that selects the coding operation best suited to the local characteristics of the image. Choosing the best operating mode presumes a method for segmenting the picture into a number of distinct classes. Although a continuously adaptive algorithm is feasible the only report of such an encoder is the preliminary work of Hayes [75]. Anderson and Huang report on a transform coding algorithm in which a piecewise segmentation of the picture is implicit [76]. Explicit segmenting algorithms are reported by Limb [77] and by Tasto and Wintz [78], [79].

Limb examined each pel sequentially by applying a simple algorithm on the eight adjacent pels in the same line [77]. The amount of detail in the surrounding points was determined from the usage of the quantizing levels in a differential quantizer. The algorithm which divided the picture into high- and low-detail areas was purposely made hysteretic so as to yield large contiguous areas. Hence, less information is required to specify the category of the segment. The algorithm is successful to the extent that the sgementation corresponded pretty much with ones subjective segmentation into high-

(a)

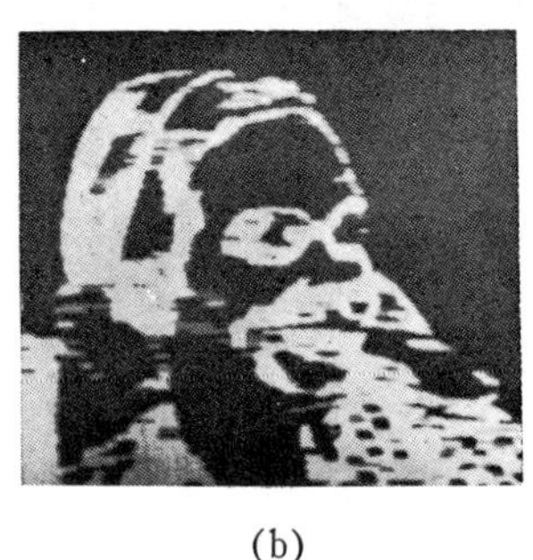

(b)

Fig. 14. (a) Original picture with high- and low-detailed areas. (b) Segmented picture with high-detailed area shown in white [77].

and low-detailed areas. An example is shown in Fig. 14. The white area in Fig. 14(b) indicates the high-detail areas of Fig. 14(a.)

The coding process consisted of an irreversible stage followed by a reversible stage. The high-detailed and low-detailed segments were coded differently in the irreversible stage of processing. The coding efficiency was evaluated by calculating the entropy of each category separately (corresponding to different treatment at the reversible processing stage also). The coding for both categories was based on the differential quantizer. In the low-detail category, a level-dependent sampling scheme was used in which sharp edges were reproduced with full spatial accuracy while small amplitude changes were reproduced with reduced spatial resolution. In the high-detail category the quantization accuracy was reduced using a number of techniques. The bit rate for the low-detail class varied between 1.0 and 1.5 bit/pel while for the high-detail class it was about 2.2 bit/pel. Thus the final bit rate for the picture depended very much on the percentage of the picture that was classified in the high-detail category. A conclusion was that the main gain from the adaptive processing came not from the different statistics of the two areas, but rather from the different irreversible processing that was possible due to different receiver perception in the two categories.

Tasto and Wintz have used a sophisticated scheme for segmenting the picture into categories which, in theory, involves both the source statistics and the receiver properties to achieve a minimum entropy segmentation [78], [79]. The algorithm is an iterative one which converges quickly and requires that one, in advance, specify the number of categories required and the probability that each category is used. The sampled picture array is divided into 6×6 square subarrays of pels (a 6×1 horizontal array is used in another example) with the pel amplitudes defining a 36-component vector. Thus for any group of vectors one can calculate a mean and a covariance matrix for the group. An initial segmentation is obtained by first assuming a mean and covariance matrix for each category. Each vector of the sample population used to determine the categories is then assigned to the category "closest" to it, where the "distance" depends on the covariance matrices of the categories (and in the more general case, on the distortion criteria of the receiver). This segmentation is used to calculate *actual* category covariances and means from those vectors falling within each category from which a new set of categories can be calculated, and so on. Three or four iterative steps were sufficient. The success of the final segmentation depends very much on the initial probability assumed for the high-detail category and the best value varies considerably from picture to picture.

Three categories were always used and they had the appearances of a high-detailed area, a dark low-detailed area, and a light low-detailed area. One should be careful not to attach too much significance to this division as far as the receiver is concerned because the rate-distortion calculations on which the segmentation depends use only a mean-square error criterion. The division into the latter two categories could as well be due to some quirk of the optical–electrical converter (such as a variation in background noise with signal level) as of the original scheme itself. Nevertheless, the segmentation technique is a powerful one and as receiver properties are incorporated into the distortion criteria the technique will be even more useful. Subsequent encoding of categories by Tasto and Wintz involved the use of transform techniques and as such will be discussed elsewhere in this issue [84].

The Tasto–Wintz segmentation algorithm will probably not be widely used due to computational complexity. A task of the future will be to develop and test simple algorithms like Limb's to see now closely they compare with rate-distortion based algorithms [77]. In all probability the simpler techniques will be at least as good until the rate-distortion approach can handle more realistic distortion criteria.

VI. Reversible Encoding Operations

The task of encoding a video signal can be decomposed into two stages: irreversible encoding and reversible encoding. For example, generation of a difference signal is a reversible operation. If the difference signal were transmitted exactly, no loss of information would occur, and the picture at the receiver would be identical to that at the input to the transmitter. If a digital signal is to be transmitted, however, quantization must occur at some point before transmission. Quantization is an irreversible process in that signals lying between decision levels are all assigned to the same representative level. Once quantization has occurred, a reversible encoding operation can reduce the number of bits that must be transmitted provided the symbols from the quantizer (i.e., the sequence of representative levels) have a nonuniform probability distribution, or are statistically dependent (i.e., redundancy is present). In the former case variable word length coding (e.g., a Huffman code) can reduce the number of bits required to transmit the symbol sequence [80]. In the latter case, if there is linear dependence, we have seen that linear prediction can be used to reduce the bit rate. In the two papers we discuss below reversible encoding operations are described that take advantage of both possibilities.

Rice and Plaunt start with a picture that has been linearly quantized in an 8-bit PCM encoder, and restrict themselves to coding operations that are completely reversible [81]. Consequently, the coding technique trades only on the statistics of the PCM signal. The data are broken into blocks 21 elements long and each complete block is coded in the same way. First, the 9-bit element-to-element differences are formed and coded with the variable length code shown in Table I. (Note the long sequences of zeros in the codes for the larger element differences.) The sequence of code words for a block is formed, and this sequence is itself coded by a second variable length code if the first variable length encoding is not very efficient. Efficiency is measured by counting the number of 0's in the code sequence from the first stage. (The number of 1's is always equal to the number of elements, i.e., 21.) If the number of 0's is sufficiently different from the number of 1's, the second coding stage is used with the provision that if there are more 1's the data are complemented before being applied to the second encoding. A code word is sent every block to indicate the mode.

TABLE I
Variable Length Code for Differences Used by Rice and Plaunt

9-Bit Difference	Variable Length Code
0	1
+1	01
−1	001
+2	0001
−2	00001
.	.
.	.
.	.

The algorithm was first tested by selecting sections of pictures that were uniformly detailed, computing the entropies of the element differences in each section, and then processing the sections using the algorithm. Although the picture sections ranged from flat areas with entropies approaching zero to active or detailed areas having entropies up to 4.5 bit/pel, the encoded data rate in all cases came to within 0.2 bit/pel of the first-order entropy.

Since the particular encoding used for each block of data is selected according to an efficiency measure, the algorithm can adapt to changes in the amount of detail from one part of a picture to another. To test this adaptive feature, two nonuniform pictures were processed. The encoded data rate in both instances was slightly below the element-difference entropy for the complete picture. This result indicates that the rather novel idea of consecutive variable word length encodings of an element-difference signal permits the efficient encoding of a variety of pictures.

Chow discusses the more restricted problem of nonadaptive variable-word-length encoding of the output of a 24-level DPCM encoder [82]. Because both high-contrast printed material and continuous tone images were to be encoded, a rather coarse quantizing scale was used. Two different models for the encoder as a symbol source were considered; a memoryless process which assumes the element differences are independent, and a first-order Markov process which assumes some statistical dependence between consecutive element-difference values.

The Shannon–Fano encoding algorithm was used to design optimum variable length codes for both source models based on statistics derived from ten pictures with widely varying amounts of detail. Given the memoryless model, 24 different code words are used to encode the quantized difference signals. For the first-order Markov model, 24 different code dictionaries of 24 words each are needed, one for each previous state.

Experimental results for four pictures indicates that the average source entropy for the Markov model is about 0.5 bit/pel less than that for the memoryless model. This result is in agreement with Schreiber's measurements on higher order conditional entropies [18]. Markov variable length coding results in an average code length across the four pictures that is 0.3 bit/pel greater than the corresponding entropy of the Markov model source. The variable length code for the memoryless source model is relatively more efficient in that the average code length is only 0.1 bit/pel greater than the respective source entropy. Hence, the use of the more complex Markov source model with its much more elaborate code structure results in a saving of 0.3 bit/pel in average code length relative to the memoryless case. Use of a much more easily implemented "cyclic difference code" [82] for the Markov source coder increases the average code length to the point that it is only marginally less than that for the memoryless coder except for the most highly detailed pictures.

The major advantage of the Markov source coder is that it produces data at a much more uniform rate. Consequently, the buffer required to smooth the data flow for transmission down a constant rate channel can be much smaller. For example, the average buffer size for the four pictures used in this study was 4750 bit for the memoryless coder, but only 1420 bit for the Markov coder.

VII. Perspective

On the basis of the papers discussed above, we can give some rough estimates of the number of bits per resolvable picture element that are required to transmit a reasonable quality picture. Under the assumption that a good predictive algorithm is being used (e.g., previous element, etc.), a companded quantizer with between 8 and 16 levels is required to quantize the difference between the prediction and the actual intensity. If a fixed-length code is being used, between 3 and 4 bit/pel are required to transmit the quantizer output. If a variable length code is used, this rate can be reduced to 2 to 3 bit/pel. Application of adaptive techniques leads to a reduction to between 1.2 and 2.2 bit/pel. Using transform techniques (tested only on single frames) a reduction to between 1 and 2 bit/pel is apparently feasible [84]. The prospects for a dramatic decrease in the rates mentioned above are few. The most likely possibility involves changing our philosophy of picture transmission from one of reproduction to one of representation. This is the approach being adopted by those working on the segmentation and analysis of pictures, although their main interest is in pattern recognition [83].

What then of the work we have discussed? It is typified by a heuristic approach due to the lack of a good distortion measure, i.e., due to the lack of a good model for the properties of the human visual system. Such a model will be hard won and will ultimately be based on an understanding of the neurophysiology and the psychophysics of the visual system. The most obvious deficiency in our knowledge in terms of picture coding is a quantitative description of the visibility of noise as a function of a complex surround. For example, we need to know the way in which the noise threshold changes in the neighborhood of edges, stripes, soft texture, etc. Given these better descriptions for the properties of the viewer, they can be incorporated in the encoder design via an extension of the existing techniques.

References

[1] L. C. Wilkins and P. A. Wintz, "Bibliography on data compression, picture properties, and picture coding," School of Elec. Eng., Purdue Univ., Lafayette, Ind., Tech. Rep. TR-EE69-10, July 1969.

[2] W. K. E. Geddes and G. F. Newell, "Tests of three systems of bandwidth compression of TV signals," *Proc. Inst. Elec. Eng.*, vol. 109, pt. 5, pp. 311–324, July 1962.

[3] A. H. Robinson and C. Cherry, "Results of a prototype television bandwidth compression scheme," *Proc. IEEE (Special Issue on Redundancy Reduction)*, vol. 55, pp. 356–364, Mar. 1967.

[4] D. N. Graham, "Image transmission by two-dimensional contour coding," *Proc. IEEE (Special Issue on Redundancy Reduction)*, vol. 55, pp. 336–346, Mar. 1967.

[5] C. F. Teacher and R. W. Yutz, "Secure color video techniques," Defense Documentation Center, ASTIA Doc. AD462528, Contract AF30(602)-3254, Contractor: Philco Corp., Blue Bell, Pa., Feb. 1965.

[6] U. F. Gronemann, "Coding color pictures," Res. Lab. of Electron., Mass. Inst. Technol., Cambridge, Tech. Rep. 422, June 1964.

[7] A. K. Bhushan, "Efficient transmission and coding of color pictures," thesis, Mass. Inst. Technol., Cambridge, June 1967.

[8] J. O. Limb, C. B. Rubinstein, and K. A. Walsh, "Digital coding of color Picturephone signals by element-differential quantization," *IEEE Trans. Commun. Technol. (Special Issue on Signal Processing for Digital Communications)*, vol. COM-19, pp. 992–1006, Dec. 1971.

[9] W. K. Pratt, "Spatial transform coding of color images," *IEEE Trans. Commun. Technol. (Special Issue on Signal Processing for Digital Communications)*, vol. COM-19, pp. 980–992, Dec. 1971.

[10] H. Enomota and K. Shibata, "Orthogonal transform coding system for television signals," presented at Transform Coding Symp., Washington, D.C., Apr. 1971.

[11] L. S. Golding and R. Garlow, "Frequency interleaved sampling of a color television signal," *IEEE Trans. Commun. Technol. (Special Issue on Signal Processing for Digital Communications)*, vol. COM-19, pp. 972–979, Dec. 1971.

[12] R. Schaphorst, "Frame-to-frame coding of NTSC color TV," presented at Symp. Picture Bandwidth Compression, Mass. Inst. Technol., Cambridge, Mass., Apr. 1969.

[13] S. B. Gray, "Local properties of binary images in two dimensions," *IEEE Trans. Comput.*, vol. C-20, pp. 551–561, May 1971.

[14] D. G. Fink, *Television Engineering Handbook.* New York: McGraw-Hill 1957, sect. 10.7.

[15] E. R. Kretzmer, "Statistics of television signals," *Bell Syst. Tech. J.*, vol. 31, pp. 751–763, July 1952.

[16] J. Capon, "Bounds to the entropy of television signals," Res. Lab. of Electron., Mass. Inst. Technol., Cambridge, Tech. Rep. 296, May 1955.

[17] C. W. Harrison, "Experiments with linear prediction in television," *Bell Syst. Tech. J.*, vol. 31, pp. 764–783, July 1952.

[18] W. F. Schreiber, "The measurement of third order probability distributions of television signals," *IRE Trans. Inform. Theory*, vol. IT-2, pp. 94–105, Sept. 1956.

[19] K. H. Powers and H. Staras, "Some relations between television picture redundancy and bandwidth requirements," *Trans. AIEE (Commun. Electron.)*, vol. 52, pp. 492–496, Sept. 1957.

[20] K. H. Powers, H. Staras, and G. L. Fredendall, "Some relations between television picture redundancy and bandwidth requirements," *Acta Electron.*, vol. 2, pp. 378–383, 1957–1958.

[21] K. Teer, "Some investigations on redundancy and possible bandwidth compression in television transmission," thesis, Technische Hogeschool Delft, Delft, The Netherlands, Sept. 1959.

[22] R. C. Brainard, "Low resolution TV: Subjective effects of noise added to a signal," *Bell Syst. Tech. J.*, vol. 46, pp. 233–260, Jan. 1967.

[23] W. M. Goodall, "Television by pulse code modulation," *Bell Syst. Tech. J.*, vol. 30, pp. 33–49, Jan. 1951.

[24] J. M. Barstow and H. N. Christopher, "Measurement of random video interference to monochrome and color TV," *AIEE Trans. (Commun. Electron.)*, pp. 313–320, Nov. 1962.

[25] J. Muller and G. Wengenroth, "Perception of interferences from random noise in color television pictures of the NTSC system," *Nachrichtentech. Z.*, vol. 15, pp. 438–441, Sept., 1962.

[26] R. D. Prosser, J. W. Allnatt, and N. W. Lewis, "Quality grading of impaired television pictures," *Proc. Inst. Elec. Eng.*, vol. 111, pp. 491–502, Mar. 1964.

[27] R. C. Brainard, F. W. Kammerer, and E. G. Kimme, "Estimation of the subjective effects of noise in low-resolution television systems," *IRE Trans. Inform. Theory*, vol. IT-8, pp. 99–106, Feb. 1962.

[28] J. O. Limb, "Design of dither waveforms for quantized visual signals," *Bell Syst. Tech. J.*, vol. 48, pp. 2555–2582, Sept. 1969.

[29] Z. L. Budrikis, "Visual thresholds and the visibility of random noise in TV," *Proc. IRE* (Australia), vol. 22, pp. 751–759, Dec. 1961.

[30] H. R. Blackwell, "Neural theories of simple visual discriminations," *J. Opt. Soc. Amer.*, vol. 53, pp. 129–160, Jan. 1963.

[31] S. Novak and G. Sperling, "Visual thresholds near a continuously visible or briefly presented light-dark boundary," *Opt. Acta*, vol. 10, pp. 87–91, Apr. 1963.

[32] J. O. Limb, "Source-receiver encoding of television signals," *Proc. IEEE (Special Issue on Redundancy Reduction)*, vol. 55, pp. 364–379, Mar. 1967.

[33] P. Moon and D. E. Spencer, "The visual effect of nonuniform surrounds," *J. Opt. Soc. Amer.*, vol. 35, pp. 233–248, Mar. 1945.

[34] G. F. Newell and W. K. E. Geddes, "The visibility of small luminance perturbations in television displays," BBC Eng. Div., Res. Dept., Tech. Rep. T106, 1963.

[35] A. V. Oppenheim, R. W. Schafer, and T. G. Stockham, Jr., "Nonlinear filtering of multiplied and convolved signals," *Proc. IEEE*, vol. 56, pp. 1264–1291, Aug. 1968.

[36] E. R. Kretzmer, "Reduced-alphabet representation of television

signals," in *IRE Nat. Conv. Rec.*, vol. 4, pt. 4, pp. 140–147, 1956.
[37] W. F. Schreiber, "The mathematical foundation of the synthetic highs system," Res. Lab. of Electron., Mass. Inst. Technol., Cambridge, Quart. Prog. Rep. 68, p. 140, Jan. 1963.
[38] C. C. Cutler, "Differential quantization of communication signals," U. S. Patent 2 605 361, July 29, 1952.
[39] L. J. Libois, "A novel method of code modulation—delta modulation," *Onde Elec.*, vol. 32, pp. 26–31, Jan. 1952.
[40] F. DeJager, "Delta modulation: A method of PCM transmission using a one-unit code," *Phillips Res. Rep.*, vol. 7, pp. 442–466, 1952.
[41] R. E. Graham, "Predictive quantizing of television signals," in *1958 IRE WESCON Conv. Rec.*, pt. 4, pp. 147–157.
[42] W. K. E. Geddes, "Picture processing by quantization of the time derivative," BBC Eng. Div., Tech. Rep. T-114, 1963.
[43] E. G. Kimme and F. F. Kuo, "Synthesis of optimal filters for a feedback quantization system," *IEEE Trans. Circuit Theory*, vol. CT-10, pp. 405–413, Sept. 1963.
[44] L. G. Roberts, "Picture coding using pseudo-random noise," *IRE Trans. Inform. Theory*, vol. IT-8, pp. 145–154, Feb. 1962.
[45] R. A. Bruce, "Optimum pre-emphasis and de-emphasis networks for transmission of television by PCM," *IEEE Trans. Commun. Technol.*, vol. COM-12, pp. 91–96, Sept. 1964.
[46] J. Max, "Quantizing for minimum distortion," *IRE Trans. Inform. Theory*, vol. IT-6, pp. 7–12, Mar. 1960.
[47] K. Fukushima and H. Ando, "Television band compression by multimode interpolation," *J. Inst. Electron. Commun. Eng. Jap.*, vol. 47, pp. 55–64, 1964.
[48] J. B. O'Neal, "Delta modulation quantizing noise analytical and computer simulation results for Gaussian and TV input signals," *Bell Syst. Tech. J.*, vol. 45, pp. 117–142, Jan. 1966.
[49] ——, "Predictive quantizing system (differential pulse code modulation) for the transmission of television signals," *Bell Syst. Tech. J.*, vol. 45, pp. 689–721, May–June, 1966.
[50] ——, "A bound on signal-to-quantizing noise ratios for digital encoding systems," *Proc. IEEE (Special Issue on Redundancy Reduction)*, vol. 55, pp. 287–292, Mar. 1967.
[51] R. C. Brainard, "Subjective evaluation of PCM noise-feedback coder for television," *Proc. IEEE (Special Issue on Redundancy Reduction)*, vol. 55, pp. 346–353, Mar. 1967.
[52] R. C. Brainard and J. C. Candy, "Direct-feedback coders: Design and performance with television signals," *Proc. IEEE*, vol. 57, pp. 776–786, May 1969.
[53] J. E. Thompson and J. J. Sparkes, "A pseudo-random quantizer for television signals," *Proc. IEEE (Special Issue on Redundancy Reduction)*, vol. 55, pp. 353–355, Mar. 1967.
[54] J. O. Limb, "Coarse quantization of visual signals," *Aust. Telecommun. Res.*, vol. 1, pp. 32–42, Nov. 1967.
[55] J. O. Limb and F. W. Mounts, "Digital differential quantizer for television," *Bell Syst. Tech. J.*, vol. 48, pp. 2583–2599, Sept. 1969.
[56] J. E. Thompson, "A 36 Mbit/s television codec employing pseudorandom quantization," *IEEE Trans. Commun. Technol. (Special Issue on Signal Processing for Digital Communications)*, vol. COM-19, pp. 872–879, Dec. 1971.
[57] D. Estournet, "Compression d'information de signaux d'images par les systemes differential codes," *Onde Elec.*, vol. 49, pp. 858–867, Sept. 1969.
[58] J. B. Millard and H. I. Maunsell, "Digital encoding of the video signal," *Bell Syst. Tech. J.*, vol. 50, pp. 459–497, Feb. 1971.
[59] R. P. Abbott, "A differential pulse-code-modulation codec for videotelephony using four bits per sample," *IEEE Trans. Commun. Technol. (Special Issue on Signal Processing for Digital Communications)*, vol. COM-19, pp. 907–912, Dec. 1971.
[60] D. J. Connor, R. F. W. Pease, and W. G. Scholes, "Television coding using two-dimensional spatial prediction," *Bell Syst. Tech. J.*, vol. 50, pp. 1049–1061, Mar. 1971.
[61] S. J. Brolin and J. M. Brown, "Companded delta modulation for telephony," *IEEE Trans. Commun. Technol.*, vol. COM-16, pp. 157–162, Feb. 1968.
[62] J. E. Abate, "Linear and adaptive delta modulation," *Proc. IEEE (Special Issue on Redundancy Reduction)*, vol. 55, pp. 298–308, Mar. 1967.
[63] M. R. Winkler, "High information delta modulation," in *IEEE Int. Conv. Rec.*, pt. 8, pp. 260–265, 1963.
[64] ——, "Pictorial transmission with HIDM," in *IEEE Int. Conv. Rec.*, pt. 1, pp. 285–290, 1965.
[65] N. S. Jayant, "Adaptive delta modulation with one bit memory," *Bell Syst. Tech. J.*, vol. 49, pp. 321–342, Mar. 1970.
[66] R. H. Bosworth and J. C. Candy, "A companded one-bit coder for television transmission," *Bell Syst. Tech. J.*, vol. 48, pp. 1459–1479, May–June 1969.
[67] J. C. Candy, "Refinement of a delta modulator," presented at Symp. Picture Bandwidth Compression, Mass. Inst. Technol., Cambridge, Apr. 1969.
[68] G. W. Aughenbaugh, J. D. Irwin, and J. B. O'Neal, "Delayed differential pulse code modulation," in *Proc. 2nd Ann. Princeton Conf.*, pp. 125–130, Oct. 1970.
[69] C. C. Cutler, "Delayed encoding, stabilizer for adaptive coders," *IEEE Trans. Commun. Technol. (Special Issue on Signal Processing for Digital Communications)*, vol. COM-19, pp. 898–907, Dec. 1971.
[70] S. C. Kitsopoulos and E. R. Kretzmer, "Computer simulation of a television coding scheme," *Proc. IRE* (Corresp.), vol. 49, pp. 1076–1077, June 1961.
[71] W. T. Bisignani, G. P. Richards, and J. W. Whelan, "The improved gray scale and the coarse-fine PCM systems, two new digital TV bandwidth reduction techniques," *Proc. IEEE*, vol. 54, pp. 376–390, Mar. 1966.
[72] W. Kaminski and E. F. Brown, "An edge adaptive three-bit, ten-level differential PCM coder for television," *IEEE Trans. Commun. Technol. (Special Issue on Signal Processing for Digital Communications)*, vol. COM-19, pp. 944–947, Dec. 1971.
[73] E. F. Brown, "A sliding scale direct-feedback PCM coder for television," *Bell Syst. Tech. J.*, vol. 48, pp. 1537–1553, May–June 1969.
[74] A. H. Frei, H. R. Schindler, and P. Vettiger, "An adaptive dual-mode coder/decoder for television signals," *IEEE Trans. Commun. Technol. (Special Issue on Signal Processing for Digital Communications)*, vol. COM-19, pp. 933–944, Dec. 1971.
[75] J. F. Hayes, "Experimental results on picture bandwidth compression," in *Proc. UMR-Mervin J. Kelly Commun. Conf.*, Univ. Missouri at Rolla, Rolla, Mo., Oct. 1970.
[76] G. B. Anderson and T. S. Huang, "Piecewise Fourier transformation for picture bandwidth compression," *IEEE Trans. Commun. Technol.*, vol. COM-19, pp. 133–140, Apr. 1971.
[77] J. O. Limb, "Adaptive encoding of picture signals," presented at Symp. Picture Bandwidth Compression, Mass. Inst. Technol., Cambridge, Apr. 1969.
[78] M. Tasto and P. A. Wintz, "Picture bandwidth compression by adaptive block quantization," School Elec. Eng., Purdue Univ., Lafayette, Ind., Tech. Rep. TR-EE 70-14. July 1970.
[79] ——, "Image coding by adaptive block quantization," *IEEE Trans. Commun. Technol. (Special Issue on Signal Processing for Digital Communications)*, vol. COM-19, pp. 957–972, Dec. 1971.
[80] R. G. Gallagher, *Information Theory and Reliable Communication.* New York: Wiley, 1968.
[81] R. F. Rice and J. R. Plaunt, "Adaptive variable-length coding for efficient compression of spacecraft television data," *IEEE Trans. Commun. Technol. (Special Issue on Signal Processing for Digital Communications)*, vol. COM-19, pp. 889–897, Dec. 1971.
[82] M. C. Chow, "Variable-length redundancy removal coders for differentially coded video telephone signals," *IEEE Trans. Commun. Technol. (Special Issue on Signal Processing for Digital Communications)*, vol. COM-19, pp. 923–926, Dec. 1971.
[83] A. Rosenfeld, *Picture Processing by Computer.* New York: Academic Press, 1969.
[84] P. A. Wintz, "Transform picture coding," this issue, pp. 809–820.
[85] C. C. Cutler, "Transmission systems employing quantization," U. S. Patent 2 927 962, Mar. 8, 1960.

Transform Picture Coding

PAUL A. WINTZ, MEMBER, IEEE

***Abstract*—Picture coding by first dividing the picture into subpictures and then performing a linear transformation on each subpicture and quantizing and coding the resulting coefficients is introduced from a heuristic point of view. Various transformation, quantization, and coding strategies are discussed. A survey of all known applications of these techniques to monochromatic image coding is presented along with a summary of the dependence of performance on the basic system parameters and some conclusions.**

I. Introduction

Bits and Pieces

CONSIDER a digitized image consisting of an $N \times N$ array of pels (picture elements) x_{ij} $i, j=1, 2, \cdots, N$ each of which is quantized to one of 2^K gray levels $1, 2, \cdots, 2^K$. Any such array can be encoded by a sequence of N^2 K-bit code words that specify the gray levels of the N^2 pels, i.e., PCM encoding [1]. PCM requires K bit/pel or KN^2 bit/picture to code any of the 2^{KN^2} possible pictures. For $K=8$ and $N=256$ we have $N^2=65\,536$ pels, $KN^2=524\,288$ bit/picture, 8 bit/pel, and $2^{524\,288} \approx 10^{158\,000}$ possible pictures. The cameraman picture presented in Fig. 1 was reconstructed from a 256×256 array of 8-bit pels.

Fig. 1. The cameraman was reconstructed from a 256×256 array of 8-bit pels.

We can conceive of a more efficient encoder that has stored a unique code word for each of the $2^{524\,288}$ pictures. If the Huffman code is used to assign short code words to the more likely pictures and long code words to the least likely pictures the average number of bits required per picture would be close to the entropy

$$-\sum_{i=1}^{2^{524\,288}} p_i \log p_i$$

where p_i is the probability of the ith picture. Two practical considerations prohibit this approach: We do not know the probabilities p_i $i=1, 2, \cdots, 2^{524\,288}$ and the storage and speed requirements for implementing the code book look-up procedure are beyond present-day technology.

One possibility that comes to mind is to block code the picture by first partitioning the picture into a number of $n \times n$ arrays of subpictures with $n \ll N$ as illustrated in Fig. 2, and then use the code book look-up procedure for each subpicture. However, a few quick calculations show that the code book is still too large. For example, the number of 4×4 arrays of pels with each pel quantized to 2^8 gray levels is $2^{128} \approx 10^{40}$.

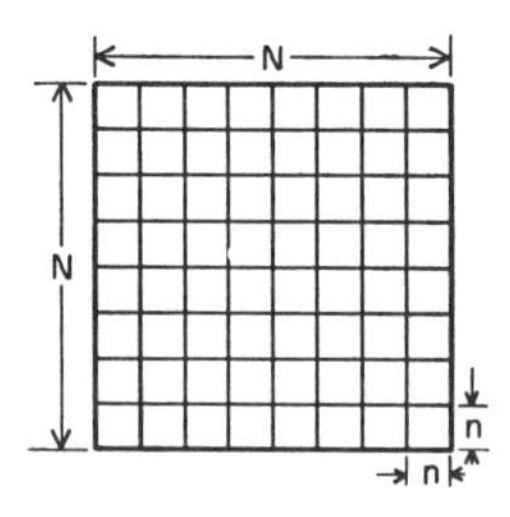

Fig. 2. Partitioning of the $N \times N$ picture into $(Nn)^2$ $n \times n$ subpictures.

Picture Structure

Pictorial data contain significant structure. Structure is a departure from randomness. Efficient source encoding can be accomplished by first determining the structure of the data and then developing encoding algorithms that are efficient for data having that structure. The more structure inherent in the data the greater the efficiency that can be achieved by matching the encoder to the data structure. On the other hand, if an encoder designed for a particular data structure is used to encode data having a different structure, the performance is degraded proportional to the amount of mismatch between the design structure and the actual structure.

Pictorial data are not homogenous—different regions of a picture contain different structures. Nonadaptive encoders are matched to the average data structure. Adaptive encoders are matched to the local data structure by first determining the local structure and then processing the local data with an algorithm that is efficient for that structure.

The structure inherent in pictorial data is not well understood. Pictures of natural objects have some structure due to the structure inherent in the universe, e.g., the shape of the earth, the direction of gravitational forces, etc. Man-made objects tend to have more structure. A few simple kinds of structure in pictures are more or less obvious. Pictures consist of a number of areas of nearly constant brightness. Statistics on the numbers of areas, their brightness, sizes, etc., have been collected by Nishikawa, Massa, and Mott-Smith [2], Gattis and Wintz [3], and others. See [3] for a literature survey.[1] A definite structure also exists in the boundaries between the areas of nearly constant brightness. These boundaries are

Manuscript received September 20, 1971; revised February 25, 1972.
The author is with the School of Electrical Engineering, Purdue University, Lafayette, Ind. 47907.

[1] For a more general survey see L. C. Wilkins and P. A. Wintz, "Bibliography on data compression, pictures properties, and picture coding," *IEEE Trans. Inform. Theory*, vol. IT-17, pp. 180–197, Mar. 1971.

Reprinted from *Proc. IEEE*, vol. 60, pp. 809–820, July 1972.

usually smooth lines as illustrated in [4, figs. 6(a) and 7(a)]. Coding strategies that take these structures into account include the contour coding techniques developed at MIT [4] and Purdue [3], [5].

Statistical Coding

One manifestation of picture structure is in the picture statistics. Encoding techniques that match to the data statistics are referred to as *statistical coding* [6], [7]. Although Schreiber [8] measured a few third-order statistics, only first- and second-order moments can be thoroughly measured. Furthermore, techniques for matching to moments of higher order than first and second have not been developed. Unfortunately, means, covariances, and first-order probability density functions are very gross measures of picture structure. This can be demonstrated by measuring them for a set of pictures and then programming a computer to generate sample pictures having these same statistics [9]. The resulting pictures do not contain birds and bees, etc., but look like random noise. We conclude that pictures contain significantly more structure than can be accounted for by first- and second-order moments.

Psychovisual Coding

The sensitivity of the human visual system to errors in the reconstructed picture depends on the frequency spectrum of the error, the gray level, and amount of detail in the picture in the vicinity of the error, etc. Hence it is possible to increase the efficiency of the coder by allowing distortions that do not degrade subjective quality (how the picture looks to a human observer). Picture coders that take the characteristics of the human visual system into account are called *psychovisual coders* [6] or *psychophysical* coders [7]. Schreiber [7] gives a good summary of the properties of human vision and presents some coding techniques that take them into account.

Transform Coding

Transform coding is a method for accomplishing some aspects of both statistical and psychovisual coding. Transform coders perform a sequence of two operations, the first of which is based on statistical considerations and the second on psychovisual considerations. The first operation is a linear transformation that transforms the set of statistically dependent pels into a set of "more independent" coefficients. The second operation is to individually quantize and code each of the coefficients. The number of bits required to code the coefficients depends on the number of quantizer levels which is dictated by the sensitivity of the human vision to the subjective effect of the quantization error.

II. Linear Transformations

Coordinate System for Pictures

Let us interpret an $n \times n$ subpicture $n \leq N$ as a point in an n^2-dimensional coordinate system where each of the n^2 coordinates corresponds to one of the n^2 pels; the value of each coordinate is the gray level of the corresponding pel.

Consider a simplified example. $n=1$ is too simple while $n=2$ requires a four-dimensional space that is difficult to visualize. Hence we consider a 1×2 array (a subpicture consisting of two adjacent pels). Let $K=3$ so that each of the two pels can take on any of $2^3=8$ gray levels. Then each of the $2^{2 \cdot 3}=64$ possible 1×2 arrays can be represented by one of the 64 points in the two-dimensional space illustrated in Fig. 3(a). Since adjacent pels are likely to have nearly the same gray level the most likely subpictures are the ones in the vicinity of $x_1=x_2$, i.e., those in the shaded area.

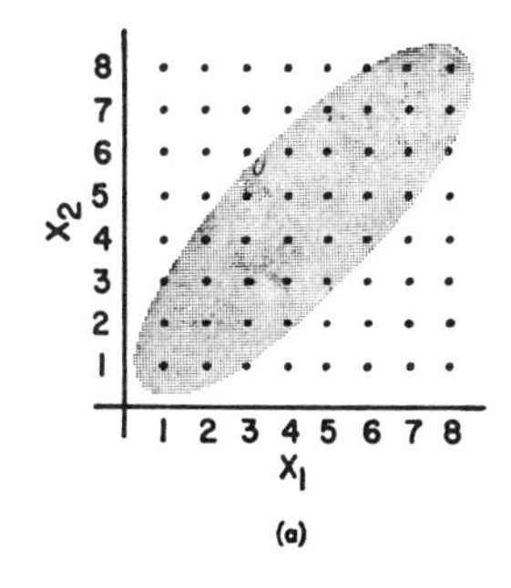

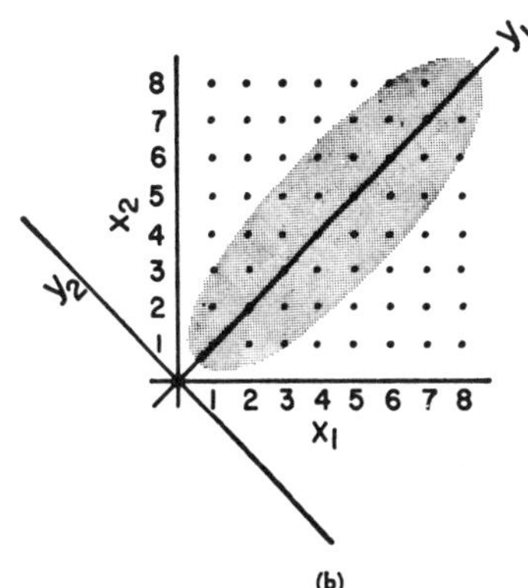

Fig. 3. (a) Coordinate system for the gray levels of two adjacent pels x_1 and x_2. The most likely subpictures are the ones in the shaded area. (b) New coordinate system.

Now rotate the coordinate system as illustrated in Fig. 3(b). In the new coordinate system the more likely subpictures are not in the vicinity of $y_1=y_2$ but are lined up with the y_1 axis. Hence the variables y_1 and y_2 are "more independent" than were x_1 and x_2, e.g., y_2 is likely to be small independent of the value of y_1. Rotating the coordinate system also rearranged the variances. The total is the same, $\sigma_{y_1}^2+\sigma_{y_2}^2=\sigma_{x_1}^2+\sigma_{x_2}^2$, but whereas both pels had the same variance, $\sigma_{x_1}^2=\sigma_{x_2}^2$, more of the variance is now in the first coefficient, $\sigma_{y_1}^2>\sigma_{y_2}^2$. Finally, note that given the coefficients y_1 and y_2 we can perform the inverse rotation to obtain the pels x_1 and x_2.

The same procedure can be used for an $n \times n$ array of pels each of which is quantized to one of 2^K gray levels. In this case an n^2-dimensional coordinate system is required with each coordinate labeled with the values $1, 2, \cdots, 2^K$. Each of the 2^{n^2K} points corresponds to one of the 2^{n^2K} possible pictures.

One-Dimensional Transformations

Rotating the n^2-dimensional coordinate system of the n^2 pels corresponds to arranging the n^2 pels into the n^2 vector $\boldsymbol{x}=[x_1, x_2, \cdots, x_{n^2}]^t$ and performing the linear transformation

$$\boldsymbol{y} = \boldsymbol{a}\boldsymbol{x} \tag{1}$$

where $\boldsymbol{a}$ is an $n^2 \times n^2$ unitary matrix with elements a_{ki} $k, i=1, 2, \cdots, n^2$ that determines the rotation and $\boldsymbol{y}=[y_1, y_2, \cdots, y_{n^2}]^t$ is an n^2 vector whose elements give the values of the coefficients on the rotated coordinates. The inverse rotation is accomplished by the inverse transformation

$$\boldsymbol{x} = \boldsymbol{a}^t\boldsymbol{y} \tag{2}$$

where $\boldsymbol{a}^t$ is the transpose of $\boldsymbol{a}$. (For unitary matrices $\boldsymbol{a}^{-1}=\boldsymbol{a}^t$.)

According to (1) each coefficient y_k is a linear combination of all of the pels

$$y_k = \sum_{i=1}^{n^2} a_{ki}x_i, \qquad k = 1, 2, \cdots, n^2 \tag{3}$$

and similarly (2) gives each pel as a linear combination of all the coefficients

$$x_i = \sum_{k=1}^{n^2} a_{ik}x_k, \qquad i = 1, 2, \cdots, n^2. \tag{4}$$

The best reversible transformation, linear or otherwise, would be one that results in independent variables y. This transformation cannot be determined for two reasons. First, it evidently depends on very detailed statistics—the joint probably density function of the n^2 pels—which have not been deduced from basic physical laws and cannot be measured. Second, even if the joint density function of the n^2 pels was known, the problem of determining a reversible transformation that results in independent coefficients is unsolved. The closest we can get with linear transformations to a transformation that produces independent coefficients is the one that produces uncorrelated coefficients. The resulting coefficients are uncorrelated but not necessarily independent.

The transformation matrix $\boldsymbol{a}$ that produces uncorrelated coefficients can be computed from the covariance matrix of the pels

$$\boldsymbol{C}_x = E\{(\boldsymbol{x} - E\{\boldsymbol{x}\})(\boldsymbol{x} - E\{\boldsymbol{x}\})^t\}. \tag{5}$$

The columns of $\boldsymbol{a}$ are the normalized eigenvectors of $\boldsymbol{C}_x$, i.e., the n^2 column vector solutions $\boldsymbol{\phi}$ to the matrix equation

$$\boldsymbol{C}_x\boldsymbol{\phi} = \lambda\boldsymbol{\phi}. \tag{6}$$

The covariance matrix of the coefficients is then given by

$$\boldsymbol{C}_y = \begin{bmatrix} \lambda_1 & 0 & \cdot & \cdot & \cdot & 0 \\ 0 & \lambda_2 & \cdot & \cdot & \cdot & 0 \\ \cdot & \cdot & \cdot & & & \cdot \\ \cdot & \cdot & & \cdot & & \cdot \\ \cdot & \cdot & & & \cdot & \cdot \\ 0 & 0 & \cdot & \cdot & \cdot & \lambda_{n^2} \end{bmatrix} \tag{7}$$

where $\lambda_1, \lambda_2, \cdots, \lambda_{n^2}$ are the eigenvalues of (6). This transformation is known as the *eigenvector transformation* or the *Hotelling transformation* [10].

Two-Dimensional Transformations

It is also possible to arrange the $n \times n$ array of pels into an $n \times n$ matrix $\boldsymbol{X}$ with elements x_{ij} $i, j = 1, 2, \cdots, n$

$$\boldsymbol{X} = \begin{bmatrix} x_{11} & x_{12} & \cdot & \cdot & \cdot & x_{1n} \\ x_{21} & x_{22} & \cdot & \cdot & \cdot & x_{2n} \\ \cdot & \cdot & \cdot & & & \cdot \\ \cdot & \cdot & & \cdot & & \cdot \\ \cdot & \cdot & & & \cdot & \cdot \\ x_{n1} & x_{n2} & \cdot & \cdot & \cdot & x_{nn} \end{bmatrix} \tag{8}$$

and then transform this $n \times n$ matrix into another $n \times n$ matrix $\boldsymbol{Y}$ with elements y_{kl} $k, l = 1, 2, \cdots, n$

$$\boldsymbol{Y} = \begin{bmatrix} y_{11} & y_{12} & \cdot & \cdot & \cdot & y_{1n} \\ y_{21} & y_{22} & \cdot & \cdot & \cdot & y_{2n} \\ \cdot & \cdot & \cdot & & & \cdot \\ \cdot & \cdot & & \cdot & & \cdot \\ \cdot & \cdot & & & \cdot & \cdot \\ y_{n1} & y_{n2} & \cdot & \cdot & \cdot & y_{nn} \end{bmatrix} \tag{9}$$

by premultiplying $\boldsymbol{X}$ by the fourth-order point tensor $\boldsymbol{A}$ with elements a_{klij} $k, l, i, j = 1, 2 \cdots, n$. Each coefficient is now given by

$$y_{kl} = \sum_{i=1}^{n}\sum_{j=1}^{n} a_{klij}x_{ij}, \qquad k, l = 1, 2, \cdots, k. \tag{10}$$

Similarly, the inverse transformation gives each pel as a linear combination of the coefficients

$$x_{ij} = \sum_{k=1}^{n}\sum_{l=1}^{n} a_{ijkl}y_{kl}, \qquad i, j = 1, 2, \cdots, n. \tag{11}$$

Clearly, both the one-dimensional transformation (3) and the two-dimensional transformation (10) result in a set of n^2 coefficients each of which is a linear combination of the n^2 pels. Hence the difference in these two transformations is simply a matter of notation.

A number of two-dimensional linear transformations other than the eigenvector transformation have been proposed for picture coding. These include the Fourier transformation

$$a_{klij} = \frac{1}{n} \exp\left[-2\pi\sqrt{-1}\,(ki + lj)/n\right] \tag{12}$$

and the Hadamard transformation

$$a_{klij} = \frac{1}{n}(-1)^{b(k,l,i,j)} \tag{13}$$

where

$$b(k, l, i, j) = \sum_{h=0}^{\log_2 n - 1} \left[b_h(k)b_h(l) + b_h(i)b_h(j)\right]$$

$b_h(\cdot)$ is the hth bit in the binary representation of $(\cdot)$, and n is a power of 2. Both are members of a class of Kroneckered matrix transformations [11] that have $2n^2 \log_2 n^2$ degrees of freedom and can, therefore, be implemented with $2n^2 \log_2 n^2$ computer operations as opposed to the n^4 operations required by the Hotelling transformation. The Fourier transformation requires $2n^2 \log_2 n^2$ multiplications and a like number of additions whereas the Hadamard transformation requires only $2n^2 \log_2 n^2$ additions since all of the entries (13) are $+1$ or -1 except for the normalizing constant n. The coefficients produced by both of these transformations are usually more dependent than those produced by the Hotelling transformation. Andrews [12] and Pearl [13] have investigated "distances" between the Hotelling, Fourier, Hadamard, and Haar transformations.

Basis Pictures

Another interpretation of the two-dimensional transformation of the $n \times n$ array of pels $\boldsymbol{X}$ into the $n \times n$ array of coefficients $\boldsymbol{Y}$ is possible. Let us write (11) in the form

$$X = \sum_{k=1}^{n} \sum_{l=1}^{n} y_{kl} \boldsymbol{a}_{kl} \tag{14}$$

and interpret this as a series expansion of the $n \times n$ picture X onto the n^2 $n \times n$ basis pictures

$$\boldsymbol{a}_{kl} = \begin{bmatrix} a_{kl11} & a_{kl12} & \cdots & a_{kl1n} \\ a_{kl21} & a_{kl22} & \cdots & a_{kl2n} \\ \cdot & \cdot & & \cdot \\ \cdot & \cdot & & \cdot \\ \cdot & \cdot & & \cdot \\ a_{kln1} & a_{kln2} & \cdots & a_{klnn} \end{bmatrix}, \quad k, l, = 1, 2, \cdots, n \tag{15}$$

with the y_{kl} $k, l = 1, 2, \cdots, n$ the coefficients of the expansion. Hence (14) gives the picture X as a weighted sum of the basis pictures $\boldsymbol{a}_{kl}$. The weights y_{kl} are given by (10) which can be written in the form

$$y_{kl} = \boldsymbol{a}_{kl} X. \tag{16}$$

The weight y_{kl} given to basis picture $\boldsymbol{a}_{kl}$ in the sum (14) can be interpreted as the amount of correlation between the picture X and the basis picture $\boldsymbol{a}_{kl}$.

The 256 16×16 basis pictures for the Hotelling and Hadamard transformations for $n = 16$ are presented in Fig. 4. The Hadamard basis pictures are not picture dependent whereas the Hotelling basis pictures are the "eigenpictures" for the cameraman.[2]

We have already noted that if the set of basis pictures is chosen such that the coefficients are "more independent" than the pels, then the variances of the coefficients are unequal. Therefore, we can index the basis pictures such that the terms in the sum (14) are ordered according to the variances of the coefficients so that successive terms contribute proportionally less and less, on the average, to the total. Indeed, for some choices of basis pictures the coefficients become insignificant after the first, say η, terms so that an approximate representation for the picture can be obtained by truncating the series after the first η terms, i.e.,

$$X = \sum_{k=1}^{n} \sum_{l=1}^{n} y_{kl} \boldsymbol{a}_{kl} \approx \sum_{k=1}^{\eta}\sum_{l=1} y_{kl} \boldsymbol{a}_{kl} = X'. \tag{17}$$

The mean-square approximation error between the original picture X and the approximate picture X' is given by

$$\begin{aligned} \epsilon_a^2 &= E\{\|X - X'\|^2\} \\ &= E\left\{\left\| \sum_{k=1}^{n}\sum_{l=1}^{n} y_{kl}\boldsymbol{a}_{kl} - \sum_{k=1}^{\eta}\sum_{l=1} y_{kl}\boldsymbol{a}_{kl}\right\|^2\right\} \\ &= E\left\{\left\| \sum_{\eta+1}^{k=n}\sum^{l=n} y_{kl}\boldsymbol{a}_{kl}\right\|^2\right\} \\ &= \sum_{\eta+1}^{k=n}\sum^{l=n} \sigma_{y_{kl}}^2. \end{aligned} \tag{18}$$

The last step follows because the basis pictures are orthonormal. Equation (18) states that the mean-square approximation error is given by the sum of variances of the discarded coefficients. Equation (16) states that the y_{kl} and, therefore, their variances depend on the basis pictures.

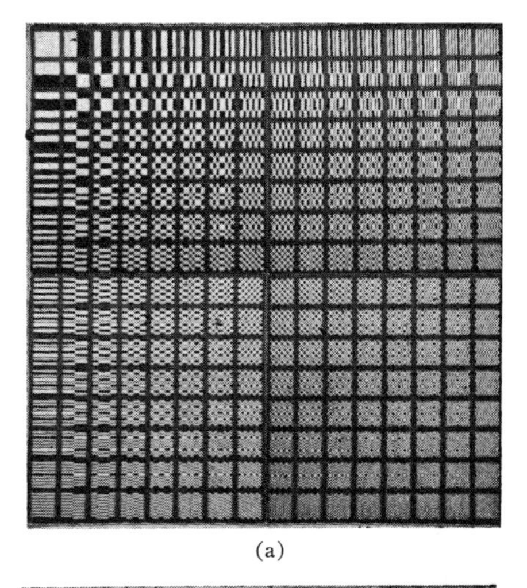

(a)

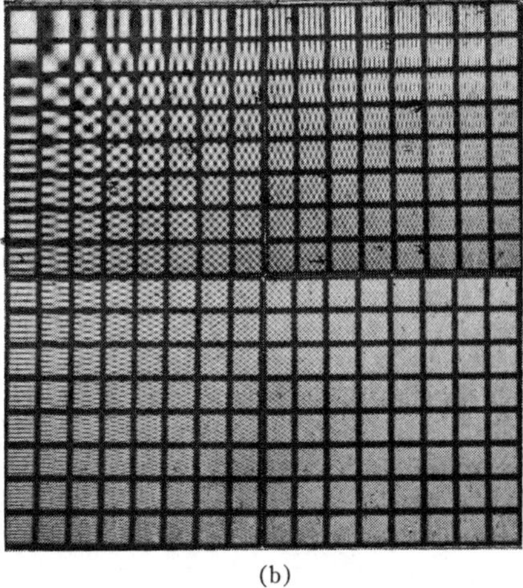

(b)

Fig. 4. (a) The 256 16×16 sequency ordered basis pictures for the Hadamard transformation. (b) The 256 16×16 Hotelling basis pictures for the cameraman picture of Fig. 1.

We now pose the following problem: What set of basis pictures minimizes the mean-square error (18) by packing the most variance into the first η coefficients? The solution to this problem is the same as the solution to the seemingly unrelated problem of determining the set of basis pictures that produce uncorrelated coefficients. The Hotelling transformation: a) produces uncorrelated coefficients; b) minimizes the mean-square approximation error; and c) packs the maximum amount of variance into the first η coordinates (for any η).

The cameraman picture was divided into 16×16 subpictures and each subpicture expanded in a Hotelling, Fourier, and Hadamard series expansion. The sample variances of the coefficients, presented in Fig. 5(a), can be interpreted as generalized power spectra of the pictures. The Fourier variances give the usual power spectrum. Note from Fig. 5(b) that all three transformations are approximately equally efficient in packing the variances into lower order coefficients.

Retaining the η coefficients with the largest variances corresponds to dividing the domain of the coefficients (9) into two zones one of which contains the retained coefficients and the other the discarded coefficients. Hence we can view the series truncation process (17) as a "zonal filtering" or "masking" in the transform domain. This procedure is sometimes referred to as "zonal sampling." Zonal sampling is a nonadap-

[2] Computing the actual Hotelling basis pictures would have required inverting a 256×256 matrix. Therefore, the basis pictures presented in Fig. 4(b) and used in the sequel were obtained by approximating the elements in the covariance matrix $C_x(\Delta_h, \Delta_v)$ with the elements exp $(-0.125|\Delta_h| - 0.249|\Delta_v|)$ which is a good approximation for the cameraman [63], [68].

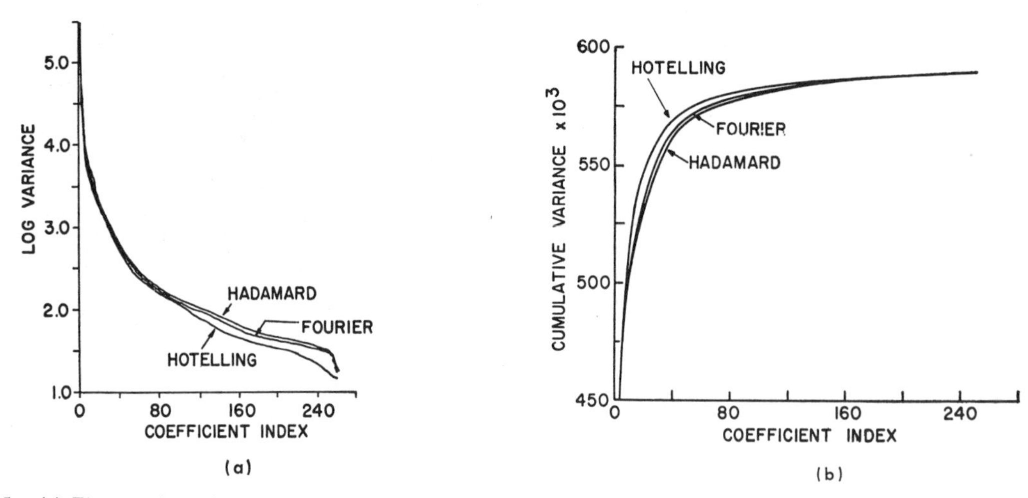

Fig. 5. (a) The sample variances σ_{kl}^2 $k, l=1, 2, \cdots, 16$ of the 256 coefficients y_{kl} averaged over the 256 subpictures of the cameraman and ordered according to their magnitudes. (b) Cumulative sums of the variances.

(a) (b) (c)

Fig. 6. Reconstructed pictures X' obtained by retaining the first $\eta=128$ of the $n^2=256$ terms in the sum (17) for each of the 256 16×16 subpictures of the 256×256 cameraman. (a) Hotelling transformation, $\epsilon_a^2=0.34$ percent: (b) Fourier transformation, $\epsilon_a^2=0.45$ percent. (c) Hadamard transformation, $\epsilon_a^2=0.49$ percent.

(a) (b) (c)

Fig. 7. Reconstructed pictures X' obtained by retaining the first $\eta=64$ of the $n^2=256$ terms in the sum (17) for each of the 256 16×16 subpictures of the 256×256 cameraman. (a) Hotelling transformation, $\epsilon_a^2=0.86$ percent. (b) Fourier transformation, $\epsilon_a^2=1.06$ percent. (c) Hadamard transformation, $\epsilon_a^2=1.14$ percent.

tive technique for retaining those terms in the sum (17) that, on the average, have the largest energy. Another alternative would be to first evaluate all n^2 coefficients (16) and then retain only those coefficients that exceed a preset threshold. This is an adaptive technique that retains only those coefficients that are large for the particular picture being processed and is sometimes referred to as "threshold sampling." When threshold sampling is used certain "bookkeeping information" that specifies which coefficients have been retained must also be coded.

Some approximations to the cameraman picture are presented in Figs. 6 and 7 along with the mean-square errors

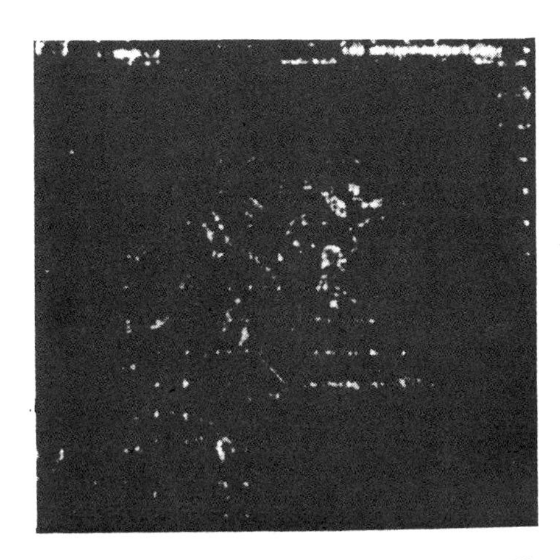

Fig. 8. The magnitude error between the original picture X of Fig. 1 and the picture X' of Fig. 6(b) scaled by eight, i.e., $8|X-X'|$. Black corresponds to zero error and white to an error of $256/8=32$ gray levels. The error was scaled by 8 to make it more visible.

between them and the original picture. These pictures were obtained by dividing the picture into 16×16 subpictures, representing each subpicture with its expansion onto the 16×16 basis pictures, and truncating the expansion after η terms (zonal sampling). Approximately one half of the terms ($\eta=128$) can be discarded with no visible degradation in picture quality although some mean-square error (18) is incurred. Truncating the expansion prior to the 128th term results in a loss of resolution due to the low-pass filtering effect. That is, from the basis pictures presented in Fig. 4 we note that increasing index corresponds to increasing spatial frequency content. Hence truncating the expansion (17) at η terms corresponds to discarding all picture energy at frequencies higher than those corresponding to the first η basis pictures. Also note that for small η the subpicture edges are visible. Fig. 8 shows the error between the reconstructed picture of Fig. 6(b) and the original picture of Fig. 1. Note that the largest errors occur in the high detail (high-frequency) parts of the picture and at the block edges.

Finally, the reader is referred to Landau and Slepian [44] who describe a somewhat different picture coding philosophy based on the basis picture concept. Whereas we seek basis pictures that result in independent coefficients (basis pictures matched to the picture statistics) Landau and Slepian seek a set of basis pictures matched to the area properties of pictures. They argue that the Hadamard basis are such a set.

Historical Notes

Hotelling [10] was the first to derive and publish the transformation that transforms discrete variables into uncorrelated coefficients. He referred to this technique as *the method of principal components*. His paper gives considerable insight into the method and is worth reading. Hotelling's transformation was rediscovered by Kramer and Mathews [14] and Huang and Schultheiss [15].

The analogous transformation for transforming continuous data into a set of uncorrelated coefficients was discovered by Karhunen [16] and Loéve [17] and is called the Karhunen–Loéve expansion. (See Selin [18] for an excellent discussion.) The result that the Karhunen–Loéve expansion minimizes the mean-square truncation error was first published by Koschman [19] and rediscovered by Brown [20]. (See also Totty [21].)

Transform picture coding evolved as a natural extension to pictorial data of the basic transform coding techniques developed for one-dimensional data such as speech, electrocardiograms, etc. Since pictorial data are sometimes modeled as a Markov process we list a few references concerned with the transform coding of one-dimensional Markov data [12]–[15], [22]–[35].

III. Quantizing the Coefficients

After the $n\times n$ array of pels has been transformed into $\eta\leq n^2$ coefficients each coefficient must be quantized and coded.

Since the variances of the coefficients vary widely as illustrated in Fig. 5(a) it would be inefficient to use the same quantizer for all coefficients. That is, if the quantizer output levels are adjusted to span the range of the coefficient with the largest variance, then the coefficients with much smaller variances would fall in a much smaller range with the result that most of the quantizer levels are not used. This effect can be negated by first scaling each coefficient by the inverse of its standard deviation to form the normalized coefficients $\gamma_{kl}=y_{kl}/\sigma_{kl}$ all of which have unit variance and can be efficiently quantized with the same quantizer. The decoder must, of course, scale each γ_{kl} by σ_{kl} to obtain the y_{kl}.

Using the same quantizer for each normalized coefficient γ_{kl} and the natural or gray code to assign equal length code words to all quantizer output levels results in each coefficient requiring the same number of bits. Since the coefficients with the larger variances contribute significantly more to the reconstructed picture (17), on the average, than the coefficients with the smaller variances, it appears that the total distortion due to quantizing the coefficients could be lessened by alloting more quantization levels and/or bits to the coefficients with the larger variances and proportionally fewer to the coefficients with the smaller variances.

Furthermore, recall that each coefficient corresponds to a particular spatial frequency band. Since the sensitivity of the human visual system to distortion is dependent on the frequency of the distortions, it appears that better subjective quality could be obtained by alloting more quantization levels and/or bits to those coefficients corresponding to the frequencies to which the eye is most sensitive.

Quantizing for Minimum Mean-Square Error

The total mean-square quantization error is usually defined as

$$\epsilon_q^2 = E\left\{ \sum_{k=1}^{n} \sum_{l=1}^{n} (y_{kl} - \hat{y}_{kl})^2 \right\} \tag{19}$$

where $\hat{y}_{kl}$ is the quantized value of y_{kl}. Equation (19) depends on the joint probability density function of y_{kl} and $\hat{y}_{kl}$. Since each coefficient is a linear combination of n^2 pels the central limit theorem indicates that the distributions of the y_{kl}'s tend toward Gaussian since some of the pels are more or less independent. Indeed, histograms for the coefficients for various transformations have been constructed and found to be roughly "bell-shaped." This effect becomes more pronounced with increasing index.

Panter and Dite [36] and Max [37] investigated quantization strategies that minimize the mean-square error $E\{(y_{kl}-\hat{y}_{kl})^2\}$ of a single coefficient. They found that if the probability density function of y_{kl} is uniform a uniform quantizer (uniformly space output levels) is optimum. For other distributions the mean-square error can be decreased by using a nonuniform quantizer with the spacing between output levels decreased in regions of high probability and increased in regions of low probability. For the Gaussian distribution the nonuniform quantizer can be 20–30 percent more efficient than the uniform quantizer [37].

Nonuniform quantization is usually accomplished by companding [38]. This involves passing y_{kl} through a nonlinearity called a compressor and then into a uniform quantizer. The combined effect is identical to that of a nonuniform quantizer. The decoder must then pass $\hat{y}_{kl}$ through the inverse nonlinearity called an expandor.

Gish and Pierce [39] and Wood [40] showed that the uniform quantizer minimizes the entropy of the quantizer output for a large class of distortion measures independent of the probability distribution of the input provided the quantization is done sufficiently fine. This indicates that uniform quantizers are inherently more efficient than nonuniform quantizers for a single coefficient. Sophisticated coding schemes, e.g., the Huffman code, are usually required to achieve this efficiency.

Quantization strategies for minimizing the total mean-square error (19) have been investigated by Huang and Schultheiss [15] and Hayes and Bobilin [41]. Huang and Schultheiss determined the optimum allocation of a total of M bits to the n^2 coefficients. They found that the number of bits m_{kl} used to code coefficient y_{kl} should be proportional to $\log \sigma_{kl}^2$. They give an algorithm for computing the m_{kl} $k, l=1, 2, \cdots, n$ such that (19) is minimized for a given $M=\sum_{k=1}^{n}\sum_{l=1}^{n} m_{kl}$ and set of variances σ_{kl}^2 $k, l=1, 2, \cdots, n$. Ready and Wintz [74] give an algorithm for computing the m_{kl} $k, l=1, 2, \cdots, n$ and M such that (19) is minimized for a given n^2 and set of variances σ_{kl}^2 $k, l=1, 2, \cdots, n$. This technique is called *block quantization*. It is significantly more efficient than using the same number of bits $m_{kl}=M/n^2$ for all coefficients. Its disadvantage lies in the problems inherent in handling binary words of unequal lengths.

Hayes and Bobilin suggested a different approach. They use the same uniform quantizer for all of the coefficients. The number of quantizer output levels and the spacings are determined to achieve adequate quantization of the coefficient with the largest variance. The coefficients with smaller variances will then tend to fall near the center of the quantizer range. Hence the probabilities associated with the quantizer outputs, averaged over all coefficients, are large near the center of the range and small at the ends. This set of probabilities is used to generate a Huffman code. This quantization strategy is significantly more efficient than block quantization. Its disadvantage is the complexity required to implement the Huffman code.

Quantizing for Best Subjective Quality

Experiments have been performed to determine the bit assignments m_{kl} for the block quantizer that yield reconstructed pictures having the best subjective quality. In general, the results indicate bit assignments quite close to those given by the mean-square error criterion although in some cases the subjective quality was slightly improved by assigning more bits to the coefficients with the larger variances and fewer to the coefficients with the smaller variances than suggested by the rule $m_{kl}\sim\log \sigma_{kl}^2$. This is probably due to the fact that the sensitivity of the human visual system to distortion as well as the coefficient variances is inversely proportional to spatial frequency.

Experiments have also been performed to optimize Hayes and Bobilin's method relative to subjective quality by finding the set of numbers u_{kl} $k, l=1, 2, \cdots, n$ such that if coefficient y_{kl} is first scaled by u_{kl} and then quantized with the same uniform quantizer, an approximately equal amount of subjective distortion is generated by each coefficient [66].

Effect of Quantization Error

The effect of quantizing the coefficients on the reconstructed picture is illustrated in Figs. 9 and 10. The cameraman was divided into 16×16 subpictures and each subpicture expanded onto the Fourier basis pictures. The expansions were truncated at $\eta=128$ terms and the 128 retained coefficients quantized by each of the three methods discussed in the preceding paragraphs. Fig. 11 shows the error between the pictures of Figs. 9(b) and 6(b).

IV. Discussion

Transform Coding Parameters

Transform coding performance depends on a number of parameters: 1) the transformation; 2) the quantization strategy; 3) the subpicture size; 4) the subpicture shape.

Transformation: The best transformation from both a mean-square error and subjective quality viewpoint is the Hotelling transformation, but it is closely followed by the Fourier transformation which is closely followed by the Hadamard transformation. Each is seprated by 0.1 or 0.2 bit/pel for $n=8$ or 16. For $n=4$ the performances are essentially the same.

Quantization Strategy: Both mean-square error and subjective quality are quite sensitive to the efficiency with which bits are used to code the coefficients. The most simple strategy is to form the normalized coefficients $\gamma_{kl}=y_{kl}/\sigma_{kl}$ and use the same quantizer (number of bits) for each coefficient. If η coefficients are retained and m bits used to code each coefficient then a total of $m\eta/n^2$ bit/pel are required. For good quality reproductions half the coefficients must be retained and at least 7 bit per coefficient must be used. Hence $m\eta/n^2\approx m(n^2/2)/n^2=m/2=3.5$ bit/pel are required. If the normalized coefficients are quantized with the same 7- or 8-bit quantizer, but only the $m_{kl}\sim\log \sigma_{y_{kl}}$ most significant bits retained (block quantization) the same quality pictures can be obtained with a savings of about 1 bit/pel. Sometimes a further 0.1 bit/pel can be saved by choosing the bit assignments to give the best subjective quality. A further saving of 0.2 to 0.3 bit/pel can be achieved by quantizing the coefficients with the same 9-bit quantizer and using a Huffman code to assign code words of unequal lengths to the quantizer output levels.

Subpicture Size: Mean-square error performance should improve with increasing n since the number of correlations taken into account increases with n. However, most pictures contain significant correlations between pels for only about 20 adjacent pels although this number is strongly dependent on the amount of detail in the picture. (See [68, fig. 2].) Hence a point of diminishing returns is reached and $n>16$ is not warranted. Even smaller n, say $n=8$, does not significantly increase the error.

This argument does not appear to apply when subjective quality is the criterion of goodness. The subjective quality appears to be essentially independent of n for $n\geq 4$. Since the

Fig. 9. Reconstructed pictures obtained by quantizing the $\eta=128$ coefficients of the picture in Fig. 6(b) using an average of 2 bit/pel (4 bit per retained coefficient). (a) All 128 normalized coefficients y_{kl}/σ_{kl} quantized to 16 levels, $\epsilon_q^2=2.09$ percent. (b) The 128 coefficients block quantized, i.e., the number of quantization levels for coefficient $y_{kl}\sim\sigma_{kl}$, $\epsilon_q^2=0.78$ percent. (c) The 128 coefficients quantized with the same 746-level quantizer with a Huffman code, $\epsilon_q^2=0.02$ percent.

Fig. 10. Reconstructed pictures obtained by quantizing the $\eta=128$ coefficients of the picture of Fig. 6(b) using an average of 1 bit/pel (2 bit per retained coefficient). (a) All 128 normalized coefficients y_{kl}/σ_{kl} quantized to 4 levels, $\epsilon_q^2=8.68$ percent. (b) The 128 coefficients block quantized, i.e., the number of quantization levels for coefficient $y_{kl}\sim\sigma_{kl}$, $\epsilon_q^2=2.21$ percent. (c) The 128 coefficients quantized with the same 158 level quantizer with a Huffman code, $\epsilon_q^2=0.26$ percent.

number of computations per pel is proportional to n, 4×4 is a reasonable choice for subpicture size.

An optimum subpicture size is expected for the adaptive techniques since too large a subpicture size, say $n=N$, does not allow adaptation to the local picture characteristics while too small a size, say $n=1$, does not allow pel correlations to be taken into account. It appears that the best size is between $n=4$ and $n=8$ with $n=4$, 6, 8 yielding essentially the same performance relative to both mean-square error and subjective picture quality.

Subpicture Shape: Transforming two-dimensional arrays ($n\times n$ arrays) of pels yields better performance than one-dimensional ($1\times n$) arrays of pels but the gain is surprisingly small—about 0.2 bit/pel [42], [66], [71]. However, a larger n is required for one-dimensional arrays. Whereas two-dimensional arrays of size 4×4 are reasonable, one-dimensional arrays of 1×16 arrays are reasonable.

Adaptive Techniques

Transform coders can be made to adapt to local picture structure by allowing a number of modes of operation and, for each subpicture, choosing the mode that is most efficient for that subpicture. Bookkeeping information that indicates which mode was used must be coded along with the subpicture. In general, increasing the number of modes decreases the number of bits required to code the subpicture pels, but increases the number of bits required to code the bookkeeping information.

Since the Hotelling transformation is matched to the subpicture statistics one is tempted to use different transformations for subpictures with different statistics. However, if we segregate the cameraman subpictures with different statistics into different groups and compute the eigenpictures for each group we would find them strikingly similar. Hence using different transformations for different subpictures is not generally warranted. On the other hand, once the transformation is chosen (whether it be Hotelling, Fourier, Hadamard, or another) significant efficiencies can be achieved by adapting to the coefficients generated for each subpicture. A number of schemes for accomplishing this have been reported, but most are variations of the following:

Method 1: Compute all n^2 coefficients (ordered according to their variances) and determine the smallest η for which

$$\sum_{k=1}^{\eta}\sum_{l=1} |y_{kl}|^2 \Big/ \sum_{k=1}^{n}\sum_{l=1}^{n} |y_{kl}|^2$$

Fig. 11. The magnitude error between the pictures of Fig. 6(b) and Fig. 9(b) scaled by eight. Black corresponds to zero error and white to an error of $256/8=32$ gray levels. The error was scaled by 8 to make it more visible.

(a)

(b)

Fig. 12. Reconstructed pictures obtained after dividing the picture into 16×16 subpictures, expanding each subpicture into a Fourier series expansion and retaining the first $\eta=128$ terms, block quantizing the coefficients using 2 bit/pel, and making random bit error in the coefficients. (a) Bit error rate$=10^{-3}$. (b) Bit error rate$=10^{-2}$.

exceeds a predetermined threshold (like 0.99). Code the first η coefficients and code the number η. Making the threshold dependent on the average subpicture brightness y_{11} improves the subjective quality because of the Weber–Fechner law.

Coding method	bits/pel
PCM	8
DPCM	3
DPCM, 2-dimensional, adaptive	2.3
Transform, 1-dimensional	2.3
Transform, 2-dimensional	2
Transform, 2-dimensional, adaptive	1

Fig. 13. Comparison of coding methods on the basis of the number of bits per pel required to achieve approximately the same quality as 7-bit PCM for moderate detail pictures such as the cameraman.

Method 2: Compute all n^2 coefficients (ordered according to their variances) and retain all that exceed a predetermined threshold. Code the retained coefficients and code which subset of coefficients was retained. As in method 1 the threshold can depend on the average brightness.

Note that method 1 wastes bits by coding all the small coefficients contained within the first η coefficients, but requires only a few bits to code the bookkeeping information. Method 2 codes only the large coefficients, but requires many more bits to code the bookkeeping information. Surprisingly, both of these schemes, as well as most variations of them, achieve almost identical performances. They all require about 1 bit/pel less than the nonadaptive techniques to achieve the same mean-square error or subjective quality.

Picture Complexity

The numbers given in the preceding subsections are for pictures containing a moderate amount of detail such as the camerman. For pictures with less detail, such as head and shoulder portraits, the same mean-square error and subjective quality can be achieved with about 0.5 fewer bit/pel. More detailed pictures such as aerial photographs require about 0.5 more bit/pel for the same quality.

Effect of Bit Errors in the Coefficients

Since the decoder reconstructs the pels from linear combinations of the coefficients, an error in a coefficient leads to errors in all the pels reconstructed from it. If the complete picture is transformed as a unit, i.e., $n=N$, then one or more errors in the coefficients result in some error in all the reconstructed pels [92], [93]. If the $N\times N$ picture is first divided into $n\times n$ subpictures and each subpicture coded independently, then only the subpictures with errors are affected.

We obtained the pictures presented in Fig. 12 by starting with the picture of Fig. 9(b) and making random bit errors in the coefficients at error rates of 10^{-2} and 10^{-3}, respectively. The structure of the blocks containing errors can be explained by recalling that each subpicture is a weighted sum of basis pictures as in (17). Hence changing the value of a coefficient results in the corresponding basis picture receiving the wrong weight in the reconstruction process.

V. Conclusions

Fig. 13 lists a number of picture coding techniques and their approximate performances. Since adaptive two-dimensional DPCM is capable of approximately the same performance as nonadaptive transform coding while being easier to implement it appears that nonadaptive transform coding has no general utility. Adaptive transform coding appears to be the only interframe technique capable of rates approaching 1 bit/pel. Habibi [43] contains a more detailed comparison of DPCM and transform techniques relative to performance, complexity, and picture-to-picture variations.

The adaptive transform techniques appear to do a good job of matching to first- and second-order picture statistics. Significant improvements in performance by developing strategies that achieve a better match to first- and second-order moments are not likely.

Pictures contain significant amounts of structure that cannot be accounted for with means, covariances, and first-order probability density functions. Significant advances in picture coding performance are possible, but must be based on a better understanding of picture structure.

Appendix

Survey of Contributions to Transform Picture Coding

Bell Laboratories [44]

Landau and Slepian used the Hadamard transform with 4×4 subpictures and block quantization. The number of bits used to code the coefficients were chosen on the basis of subjective tests. Good quality pictures were obtained at 2 bit/pel. The paper contains an excellent philosophical discussion.

Boeing [45]

Claire, Farber, and Green used the Hadamard transform with 16×16, 64×64, and 256×256 subpictures of a 256×256 aerial photograph. All coefficients for all subpictures were computed first, and then threshold sampling applied. (Different numbers of coefficients were retained for different subpictures.) All retained coefficients were quantized with the same 8-bit quantized.

Centre National D'Etudes des Telecommunications [46]–[50]

Schwartz, Marano, Poncin, and Fino used the Hadamard, Fourier, and Haar transformations with block quantization of the coefficients. Both zonal and threshold sampling were investigated. Hadamard and Haar coding were shown to be subjectively equivalent.

COMSAT [51]

Robinson compared the Hotelling, Fourier, and Hadamard transformations on 16×16 subpictures of a 512×512 picture (moonscape). Zonal sampling and block quantization was used with the bit assignment chosen on the basis of the coefficient variances. The picture was coded with 7, 4, and 2 bit/pel and compared to PCM coding of the same picture with respect to both subjective quality and the improvement obtained in the signal-to-quantizing-noise ratio. The paper includes an extensive list of references on various orthogonal transforms and their application to speech and image processing.

McDonnell Douglas [52], [53]

Kennedy, Clark, and Parkyn describe some experiments with the Hadamard transformation and threshold sampling. The effects of quantization and the bookkeeping bits required to specify which coefficients are retained were neglected. They also expanded the picture $\boldsymbol{X}$ in terms of the eigenvalues and eigenvectors of $\boldsymbol{X}\boldsymbol{X}^t$ and $\boldsymbol{X}^t\boldsymbol{X}$.

MIT [54]–[58]

Huang and Woods divided 256×256 pictures into $n\times n$ subpictures with $n=4$, 8, 16, 256 and used the Hadamard transform, zonal sampling, and block quantization. They also found the structure required of a covariance matrix in order that it can be diagonalized by the Hadamard transformation. For small block sizes ($n=4$) the covariance matrices of pictures are approximately of this form, but for $n=8$, 16, etc., they diverge considerably.

Anderson and Huang developed an adaptive technique based on the Fourier transform and threshold sampling. $n\times n$ blocks with $n=8$, 16, 32, and 256 were investigated. They also used a windowing technique to combat the edge effect between the subpictures. The average picture energy over the $n\times n$ subpicture was first computed and this number used to determine a set of three thresholds which are then used to determine whether the magnitudes and phases of the coefficients are quantized to 2, 3, 4, or 5 bit. 7 bit were used to code the average gray level y_{11} and 10 bit to specify the largest coefficient aside from y_{11}. A run length (Huffman) code was used to code the location of the retained coefficients. The same technique was used with one-dimensional blocks each of which consists of one row of the 256×256 array of pels. 1.25 bit/pel are required to achieve negligible visible distortion with the two-dimensional technique.

Purdue University [59]–[74]

Habibi and Wintz divided 256×256 pictures into 16×16 arrays and used the Hadamard, Fourier, and Hotelling transformations and zonal sampling. Block quantization was used with the bit assignments optimized relative to both the mean-square error and subjective quality criterion. The performances of the four transformations are compared at 0.5, 1.0, and 2.0 bit/pel.

Tasto and Wintz developed an adaptive technique that takes into account the sensitivity of the human visual system to both gray level and spatial frequency. The 256×256 pictures were divided into 6×6 arrays, transformed with the Hotelling transformation, and classified as one of three types. The number of coefficients retained and the quantization strategy used to code them depends on the classification. Both block quantization and the Hayes and Bobilin method were optimized relative to mean-square error and subjective quality. Some relationships between mean-square error and subjective quality are presented. 0.75–1.5 bit/pel depending on the picture gray level and spatial frequency content are required for negligible distortion.

Proctor and Wintz divided the picture into 6×6 arrays of subpictures and used the Hotelling and Hadamard transformations. The system was made adaptive by truncating the expansion after η terms with η determined by computing the fraction of retained energy

$$\sum_{k=1}^{\eta}\sum_{l=1} y_{kl}^2 \Big/ \sum_{k=1}^{n}\sum_{l=1}^{n} y_{kl}^2$$

and comparing it to a threshold that depends on the average gray level of the subpicture. Three quantization strategies were tried. The effect of bit errors was demonstrated.

Ready and Wintz used these techniques to code aircraft and spacecraft multispectral scanner data.

Teledyne Brown Engineering [75], [76]

Bowyer used the Hadamard transformation and threshold sampling on a 32 gray-level test pattern. All retained coefficients were quantized with a 2^7 level quantizer and transmitted over a noisy channel. A BCH code was used for error control.

Tokyo Institute of Technology and KDD Research Laboratory [77]–[83]

Enomoto and Shibata performed experiments with 1×8 and 2×4 arrays of pels arranged into 8 vectors and transformed by the one-dimensional Hadamard transformation and some variations of it. Block quantization was used. They also constructed hardware for coding TV 2.5 to 3 bit/pel resulted in picture quality comparable to that obtained with 6-bit PCM coding of the video signal. They also suggested that color TV could be coded with an additional 0.75 bit/pel by increasing the number of bits assigned to the sixth and seventh coefficients since these correspond to the frequencies containing the color signal in the composite color TV signal.

TRW [84]

Agarwal and Stephens implemented a hardware Hadamard transform coder using 16×16 subpictures and threshold sampling. All coefficients exceeding the threshold were quantized with a 6-bit quantizer. A run length code was used to specify which coefficients exceeded the threshold.

University of Southern California [85]–[99]

Pratt, Andrews *et al.* used the Hotelling, Hadamard, and Fourier transformations and both zonal and threshold sampling with 256×256 pictures and $n = 16$ and 256. An equal number of bits used to code all retained coefficients which were scaled before being quantized. "Bookkeeping information" that indicates which coefficients have been retained was included by using a run length code to indicate the sequence of retained coefficients. They also investigated the effect of bit errors in the coefficients on the quality of the reconstructed pictures and performed experiments using BCH codes for error control. They also applied these techniques to color pictures.

Acknowledgment

The author wishes to thank H. Andrews, D. Slepian, and R. Totty for their critical comments on the original manuscript.

References

[1] T. S. Huang, "PCM picture transmission," *IEEE Spectrum*, vol. 2, pp. 57–63, Dec. 1965.
[2] A. Nishikawa, R. J. Massa, and J. C. Mott-Smith, "Area properties of television pictures," *IEEE Trans. Inform. Theory*, vol. IT-11, pp. 348–352, July 1965.
[3] J. L. Gattis and P. A. Wintz, "Automated techniques for data analysis and transmission," School of Electrical Engineering, Purdue University, Lafayette, Ind., Tech. Rep. TR-EE 71-37, Aug. 1971.
[4] D. N. Graham, "Image transmission by two-dimensional contour coding," *Proc. IEEE* (*Special Issue on Redundancy Reduction*), vol. 55, pp. 336–346, Mar. 1967.
[5] L. C. Wilkins and P. A. Wintz, "Studies on data compression, Part I: Picture coding by contours, Part II: Error analysis of run-length codes," School of Electrical Engineering, Purdue University, Lafayette, Ind., Tech. Rep. TR-EE 70-17, Sept. 1970.
[6] T. S. Huang, "Digital picture coding," in *Proc. Nat. Electron. Conf.*, pp. 793–797, 1966.
[7] W. F. Schreiber, "Picture coding," *Proc. IEEE* (*Special Issue on Redundancy Reduction*), vol. 55, pp. 320–330, Mar. 1967.
[8] ——, "The measurement of third order probability distributions of television signals," *IRE Trans. Inform. Theory*, vol. IT-2, pp. 94–105, Sept. 1956.
[9] N. A. Broste, "Digital generation of random sequences with specified autocorrelation and probability density functions," U. S. Army Missile Command, Redstone Arsenal, Ala., Rep. RE-TR-70-5, Mar. 1970.
[10] H. Hotelling, "Analysis of a complex of statistical variables into principal components," *J. Educ. Psychology*, vol. 24, pp. 417–441, 498–520, 1933.
[11] H. C. Andrews and K. L. Caspari, "Degrees of freedom and modular structure in matrix multiplication," *IEEE Trans. Comput.*, vol. C-20, pp. 133–141, Feb. 1971.
[12] H. C. Andrews, "Some unitary transformations in pattern recognition and image processing," in *IFIP Congress '71* (Yugoslavia, Aug. 1971).
[13] J. Pearl, "Basis-restricted transformations and performance measures for spectral representations," in *Proc. 1971 Hawaiian Conf.*; also, *IEEE Trans. Inform. Theory* (Corresp.), vol. IT-17, pp. 751–752, Nov. 1971.
[14] H. P. Kramer and M. V. Mathews, "A linear coding for transmitting a set of correlated signals," *IRE Trans. Inform. Theory*, vol. IT-2, pp. 41–46, Sept. 1956.
[15] J. J. Y. Huang and P. M. Schultheiss, "Block quantization of correlated Gaussian random variables," *IEEE Trans. Commun. Syst.*, vol. CS-11, pp. 289–296, Sept. 1963.
[16] H. Karhunen, "Über lineare Methoden in der Wahrscheinlich-Keitsrechnung," *Ann. Acad. Sci. Fenn.*, Ser. A.I. 37, Helsinki, 1947. (An English translation is available as "On linear methods in probability theory" (I. Selin transl.), The RAND Corp., Doc. T-131, Aug. 11, 1960.)
[17] M. Loéve, "Fonctions aléatoires de seconde ordre," in P. Lévy, *Processus Stochastiques et Mouvement Brownien.* Paris, France: Hermann, 1948.
[18] I. Selin, *Detection Theory.* Princeton, N. J.: Princeton Univ. Press, 1965.
[19] A. Koschman, "On the filtering of nonstationary time series," in *Proc. 1954 Nat. Electron. Conf.*, p. 126.
[20] J. L. Brown, Jr., "Mean-square truncation error in series expansions of random functions," *J. SIAM*, vol. 8, pp. 18–32, Mar. 1960.
[21] R. E. Totty and J. C. Hancock, "On optimum finite-dimensional signal representation," in *1963 Proc. 1st Ann. Allerton Conf. on Circuits and System Theory.*
[22] J. Pearl, "Walsh processing of random signals," University of California at Los Angeles, Los Angeles, Calif.
[23] W. R. Crawther and C. M. Rader, "Efficient coding of vocoder channel signals using linear transformation," *Proc. IEEE* (Lett.), vol. 54, pp. 1594–1595, Nov. 1966.
[24] L. M. Goodman, "A binary linear transformation for redundancy reduction," *Proc. IEEE* (Lett.) (*Special Issue on Redundancy Reduction*), vol. 55, pp. 467–468, Mar. 1967.
[25] P. A. Wintz and A. J. Kurtenbach, "Waveform error control in PCM telemetry," *IEEE Trans. Inform. Theory*, vol. IT-14, pp. 650–661, Sept. 1968.
[26] C. J. Palermo, R. V. Palermo, and H. Horwitz, "The use of data omission for redundancy removal," in *1965 Rec. Int. Space Electron. and Telemetry Symp.*, pp. (11)D1–(11)D16.
[27] A. J. Kurtenbach and P. A. Wintz, "Data compression for second order processes," in *1968 Proc. Nat. Telemetering Conf.*
[28] C. E. Shannon, "Coding theorems for a discrete source with a fidelity criterion," in *1959 IRE Nat. Conv. Rec.*, pt. 4, pp. 142–163.
[29] T. J. Goblick, Jr., and J. L. Hoslinger, "Analog source digitation: A comparison of theory and practice," *IEEE Trans. Inform. Theory* (Corresp.), vol IT-13, pp. 323–326, Ar. 1967.
[30] L. M. Goodman and P. R. Drouihet, Jr., "Asymptotically optimum pre-emphasis and de-emphasis networks for sampling and quantizing," *Proc. IEEE* (Lett.), vol. 54, pp. 795–796, May 1966.
[31] S. Watanabe, "The Loéve-Karhunen expansion as a means of information compression for classification of continuous signals," IBM Watson Res. Cen., Yorktown Heights, N. Y., Tech. Rep. AMRL-TR-65-114.
[32] T. Berger, *Rate Distortion Theory.* Englewood Cliffs, N. J.: Prentice-Hall, 1971.
[33] E. Donchin, "A multivariate approach to the analysis of average evoked potentials," *IEEE Trans. Biomed. Eng.*, vol. BME-13, pp. 131–139, July 1966.
[34] C. K. Stidd, "The use of eigenvectors for climatic estimates," *J. Appl. Meteorology*, vol. 6, pp. 255–264, Apr. 1967.
[35] J. Pearl, "On the distance between representations," presented at Symp. Picture Coding, North Carolina State Univ., Raleigh, N. C., Sept. 1970.
[36] P. F. Panter and W. Dite, "Quantization distortion in pulse-count modulation with nonuniform spacing of levels," *Proc. IRE*, vol. 39, pp. 44–48, Jan. 1951.
[37] J. Max, "Quantizing for minimum distortion," *IRE Trans. Inform. Theory*, vol. IT-6, pp. 7–12, Mar. 1960.
[38] B. Smith, "Instantaneous companding of quantized signals," *Bell Syst. Tech. J.*, vol. 36, pp. 653–709, May 1957.
[39] H. Gish and J. N. Pierce, "Asymptotically efficient quantizing," *IEEE Trans. Inform. Theory*, vol. IT-14, pp. 676–683, Sept. 1968.
[40] R. C. Wood, "On optimum quantization," *IEEE Trans. Inform. Theory*, vol. IT-15, pp. 248–252, Mar. 1969.
[41] J. F. Hayes and R. Bobilin, "Efficient waveform encoding," School of Electrical Engineering, Purdue University, Lafayette, Ind., Tech. Rep. TR-EE 69-4, Feb. 1969.
[42] D. J. Sakrison and V. R. Algazi, "Comparison of line-by-line and two-dimensional encoding of random images," *IEEE Trans. Inform. Theory*, vol. IT-17, pp. 386–398, July 1971.
[43] A. Habibi, "Comparison of *n*th order DPCM encoder with linear transformations and block quantization techhiques," *IEEE Trans.*

Commun. Technol., vol. COM-19 (pt. 1), pp. 948–956, Dec. 1971.

[44] H. J. Landau and D. Slepian, "Some computer experiments in picture processing for bandwidth reduction," *Bell Syst. Tech. J.*, vol. 50, pp. 1525–1540, May–June 1971.

[45] E. J. Claire, S. M. Farber, and R. R. Green, "Practical techniques for transform data compression/image coding," in *Proc. 1971 Applications of Walsh Function Symp.*, pp. 2–6 (Washington, D. C., Apr. 1971).

[46] P. Marano and P. Y. Schwartz, "Compression d'information sur la transformee de Fourier d'une image," *Onde Elec.*, vol. 50, pp. 908–919, Dec. 1970.

[47] P. Y. Schwartz, "Analysis of Fourier compression on images" (note technique interne est.TDS/423), presented at Symp. Picture Coding, Raleigh, N. C., Sept. 1970.

[48] ——, "Analyse de la compression d'information sur la transformee de Fourier d'une image," *Ann. Telecommun.*, Mar.–Apr. 1971.

[49] J. Poncin, "Utilisation de la transformation de Hadamard pour le codage et la compression de sigaux d'images," *Ann. Telecommun.*, July–Aug. 1971.

[50] B. Fino, "Traitement d'images nar transformations orthogonales," Note Tech. ITD/TTI/52.

[51] G. S. Ribonson, "Orthogonal transform feasibility study," prepared for NASA/MSC by COMSAT, Contract no. NAS 9-11240, Suppl. Reps. 1 (Aug. 1971), 2 (Sept. 1971), 3 (Oct. 1971), and Final Rep. (Oct. 1971).

[52] J. D. Kennedy, S. J. Clark, and W. A. Parkyn, Jr., "Digital imagery data compression techniques," McDonnell Douglas Corp., Huntington Beach, Calif., Rep. MDC G0402, Jan. 1970.

[53] J. D. Kennedy, "Walsh function imagery analysis," in *Proc. 1971 Applications of Walsh Functions Symp.* (Washington, D. C., Apr. 1971), pp. 7–10.

[54] J. W. Woods, "Video bandwidth compression by linear transformation," MIT Research Laboratory of Electronics, Quarterly Progress Rep. 91, Oct. 15, 1968, pp. 219–224.

[55] T. S. Huang and J. W. Woods, "Picture bandwidth compression by block quantization," presented at the 1969 Int. Symp. Inform. Theory, Ellenville, N. Y.

[56] J. W. Woods and T. S. Huang, "Picture bandwidth compression by linear transformation and block quantization," presented at the 1969 Symp. Picture Bandwidth Compression, MIT, Cambridge, Mass.

[57] G. B. Anderson and T. S. Huang, "Picture bandwidth compression by pieceiwse Fourier transformation," in *Proc. Purdue Centennial Year Symp. Inform. Processing*, 1969.

[58] ——, "Piecewise Fourier transformation for picture bandwidth compression," *IEEE Trans. Commun. Technol.*, vol. COM-19, pp. 133–140, Apr. 1971.

[59] J. E. Essman and P. A. Wintz, "Redundancy reduction of pictorial information," presented at the Midwestern Simulation Council Meeting on Advanced Mathematical Methods in Simulation, Oakland University, 1968.

[60] G. G. Apple and P. A. Wintz, "Experimental PCM system employing Karhunen–Loéve sampling," presented at the 1969 Int. Symp. Inform. Theory, Ellenville, N. Y., Jan. 1969.

[61] A. Habibi and P. A. Wintz, "Optimum linear transformations for encoding 2-dimensional data," presented at the Symp. Picture Bandwidth Compression, MIT, Apr. 1969.

[62] ——, "Optimum linear transformations for encoding 2-dimensional data," School of Electrical Engineering, Purdue University, Lafayette, Ind., Tech. Rep. TR-EE 69-15, May 1969.

[63] ——, "Linear transformations for encoding 2-dimensional sources," School of Electrical Engineering, Purdue University, Lafayette, Ind., Tech. Rep. TR-EE 70-2, Jan. 1970.

[64] M. Tasto and P. A. Wintz, "An adaptive picture bandwidth compression system using pattern recognition techniques," in *Proc. 3rd Hawaii Int. Conf. on System Sciences*, pp. 513–515, Jan. 1970.

[65] A. Habibi and P. A. Wintz, "Picture coding by linear transformation," presented at the 1970 IEEE Int. Symp. Inform. Theory, Noordwijk, The Netherlands, June 1970.

[66] M. Tasto and P. A. Wintz, "Picture bandwidth compression by adaptive block quantization," School of Electrical Engineering, Purdue University, Lafayette, Ind., Tech. Rep. TR-EE 70-14, June 1970.

[67] M. Tasto, A. Habibi, and P. A. Wintz, "Adaptive and nonadaptive image coding by linear transformation and block quantization," presented at the Symp. Picture Coding, North Carolina State Univ., Raleigh, N. C., Sept. 10, 1970.

[68] A. Habibi and P. A. Wintz, "Image coding by linear transformations and block quantization," *IEEE Trans. Commun. Technol.*, vol. COM-19, pp. 50–60, Feb. 1971.

[69] M. Tasto and P. A. Wintz, "Image coding by adaptive block quantization," *IEEE Trans. on Commun. Technol.*, vol. COM-19, pp. 957–971, Dec. 1971.

[70] C. W. Proctor and P. A. Wintz, "Picture bandwidth reduction for noisy channels," School of Electrical Engineering, Purdue Univ., Lafayette, Ind., Tech. Rep. TR-EE 71-30, Aug. 1971.

[71] M. Tasto and P. A. Wintz, "A bound on the rate-distortion function and application to images," *IEEE Trans. Inform. Theory*, vol. IT-18, pp. 150–159, Jan. 1972.

[72] P. J. Ready, P. A. Wintz, and D. A. Landgrebe, "A linear transformation for data compression and feature selection in multispectral imagery," Lab. for Applications of Remote Sensing, Purdue Univ., Lafayette, Ind., Info. Note 072071, July 1971.

[73] P. J. Ready, P. A. Wintz, S. J. Whitsitt, and D. A. Landgrebe, "Effects of compression and random noise on multispectral data," in *Proc. 7th Int. Symp. on Remote Sensing of Environment* (Univ. of Michigan, May 1971).

[74] P. J. Ready and P. A. Wintz, "Multispectral data compression thru transform coding and block quantization," School of Electrical Engineering, Purdue Univ., Lafayette, Ind., Tech. Rep. TR-EE 72-2, Jan. 1972.

[75] D. E. Bowyer, "Rate calculations for compressed Hadamard-transformed data," Teledyne Brown Engineering, Huntsville, Ala., Summary Rep. MSD-INT-1338, June 1971.

[76] ——, "Walsh functions, Hadamard matrices and data compression," presented at 1971 Walsh Function Symp.

[77] H. Enomoto and K. Shibata, "Features of Hadamard transformed television signal," presented at 1965 Nat. Conf. IECE Jap., Paper 881.

[78] ——, "Television signal coding method by orthogonal transformations," presented at 1966 Joint Conv. IECE Jap., Paper 1430.

[79] ——, "Television signal coding method by orthogonal transformation," presented to 6th Research Group on Television Signal Transmission. The Institute of Television Engineers of Japan (May 1968).

[80] S. Inoue and K. Shibata, "Quantizing noise of orthogonal transform TV PCM system," presented at 1968 Nat. Conf. IECE Jap., Paper 1310.

[81] H. Enomoto, K. Shibata *et al.*, "Experiments on television PCM system by orthogonal transformation," presented at 1969 Joint Conv. IECE Jap., Paper 2219.

[82] K. Shibata, T. Ohira *et al.*, "PCM terminal equipment for bandwidth compression of television signal," presented at 1969 Joint Conv. IECE Jap., Paper 2619.

[83] H. Enomoto and K. Shibata, "Orthogonal transform coding system for television signals," *Television, J. Inst. TV Eng. of Japan*, vol. 24, pp. 99–108, Feb. 1970; also in *Proc. 1971 Applications of Walsh Functions Symp.* (Washington, D. C., April 1971), pp. 11–17.

[84] V. K. Agarwal and T. J. Stephens, Jr., "On-board data processor (picture bandwidth compression)," Data Systems Lab., IRAD Rep. R 733.3-32 Feb. 1970.

[85] W. K. Pratt and H. C. Andrews, "Fourier transform coding of images," in *Proc. Hawaii Int. Conf. on System Sciences*, Jan. 1968.

[86] ——, "Television bandwidth reduction by Fourier image coding," presented at 103rd Society of Motion Picture and Television Eng. Conf., 1968.

[87] ——, "Television bandwidth reduction by encoding spatial frequencies," *J. Motion Picture and Television Eng.*, pp. 1279–1281, Dec. 1968.

[88] ——, "Two dimensional transform coding for images," presented at 1969 IEEE Int. Symp. on Inform. Theory, Jan. 1969.

[89] ——, "Transformation coding for noise immunity and bandwidth reduction," in *Proc. 2nd Hawaii Int. Conf. on System.Sci.*, Jan. 1969.

[90] W. K. Pratt, J. Kane, and H. C. Andrews, "Hadamard transform image coding," *Proc. IEEE*, vol. 57, pp. 58–68, Jan. 1969.

[91] W. K. Pratt and H. C. Andrews, "Application of Fourier-Hadamard transformation to bandwidth compression," presented at MIT Symp. on Picture Bandwidth, Apr. 1969.

[92] H. C. Andrews and W. K. Pratt, "Transform image coding," presented at Polytech. Inst. of Brooklyn Int. Symp. on Computer Processing in Communications, Apr. 1969.

[93] W. K. Pratt and H. C. Andrews, "Transform image coding," University of Southern California, Los Angeles, Calif. Final Rep. USCEE 387, Mar. 1970.

[94] H. C. Andrews and W. K. Pratt, "Digital image transform processing," presented at Symp. Applications of Walsh Functions, Apr. 1970.

[95] W. K. Pratt, "Karhunen-Loéve transform coding of images," presented at 1970 IEEE Int. Symp. Inform. Theory, June 1970.

[96] ——, "Application of transform coding to color images," presented at Symp. Picture Coding, North Carolina State Univ., Raleigh, N. C., Sept. 1970.

[97] ——, "A comparison of digital image transforms," presented at University of Missouri at Rolla—Mervin J. Kelly Communications Conf., Rolla, Mo., Sept. 1970.

[98] H. C. Andrews, "Fourier and Hadamard image transform channel error tolerance," in *Proc. UMR—Mervin J. Kelly Communications Conf.*, Rolla, Mo., pp. 10-4-1–10-4-6, Oct. 1970.

[99] W. K. Pratt, "Spatial transform coding of color images," *IEEE Trans. Commun. Technol.*, vol. COM-19, pp. 980–991, Dec. 1971.

Glossary

This glossary includes terms, abbreviations, and acronyms commonly employed in remote sensing and image processing. Some of these terms were abstracted from the *Manual of Remote Sensing*, published by the American Society of Photogrammetry in 1975, *Remote Sensing, Principles and Interpretation*, by F. F. Sabins, Jr., W. H. Freeman and Company, 1978, and a glossary prepared by R. Haralick, University of Kansas.

absorption band Wavelength interval at which electromagnetic radiation is absorbed by the atmosphere or by other substances. For example, there is an atmospheric absorption band at 5 to 8 μm, caused by water vapor, that absorbs thermal IR radiation of those wavelengths.

absorptivity The capacity of a material to absorb incident radiant energy.

active remote sensing Remote-sensing methods that provide their own source of electromagnetic radiation. Radar is one example.

airborne scanner A scanner designed for use on aircraft or spacecraft in which the forward motion of the vehicle provides coverage normal to the scan direction.

albedo The ratio of the amount of electromagnetic energy reflected by a surface to the amount of energy incident upon it. The symbol is *A*.

altitude The distance of an aircraft or spacecraft above the earth or other planet.

amplitude For waves this term represents the vertical distance from crest to trough.

analog A form of data display in which values are shown in graphic form, such as curves. Also a form of computing in which values are represented by directly measurable quantities, such as voltages or resistances. Analog computing methods contrast with digital methods in which values are treated numerically.

angle of incidence In SLAR systems this is the angle between the vertical and a line connecting the antenna and a target.

angular field of view The angle subtended by lines from a remote-sensing system to the outer margins of the strip of terrain that is viewed by the system.

antenna The device that transmits and receives microwave and radio energy in SLAR systems.

aperture The opening in a remote-sensing system that admits electromagnetic radiation to the film or detector.

Apollo The United States lunar program of satellites with crews of three.

aspect angle Same as angle of incidence, which is generally the preferred term.

atmosphere The layer of gases that surround some planets.

atmospheric windows Wavelength intervals at which the atmosphere transmits most electromagnetic radiation.

attitude The angular orientation of a remote-sensing system with respect to a geographic reference system.

azimuth The geographic orientation of a line given as an angle measured clockwise from north.

background The area on an image or the terrain that surrounds an area of interest, or target.

band A wavelength interval in the electromagnetic spectrum. For example, in Landsat the bands designate specific wavelength intervals at which images are acquired.

band-separated format The type of image data organization which consists of one record per scan line with all scan lines from the same band. One MSS scene requires a data set for each spectral band.

base-height ratio Air base divided by aircraft height. This ratio determines vertical exaggeration on stereo models.

batch processing The method of data processing in which data and programs are entered into a computer, which then carries out the entire processing operation with no further instructions.

beam A focused pulse of energy.

bilinear interpolation A resampling algorithm using a four-point neighborhood of pixels vs. input, and linear intensity interpolation between these points for the value of the output pixel.

binary A numerical system using the base 2.

bit In digital computer terminology, this is a binary digit that is an exponent of the base 2.

boresight To align the axis of one remote-sensing system with the axis of another system.

brightness Magnitude of the response produced in the eye by light.

byte A group of eight bits of digital data.

calibration The process of comparing measurements made by an instrument with a standard.

cathode-ray tube A vacuum tube with a phosphorescent screen upon which images are displayed by an electron beam. The abbreviation is CRT.

CCT Computer compatible tape; the magnetic tape upon which the digital data for Landsat MSS images are recorded.

change detection The process by which two images may be compared, resolution cell by resolution cell, and an output generated whenever corresponding resolution cells have different enough grey tones.

change-detection images Images prepared by digitally comparing two original images acquired at different times. The gray tones of each pixel on a change-detection image portray the amount of difference between the original images.

classification The process of assigning individual pixels of a multispectral image to categories, generally on the basis of spectral-reflectance characteristics.

color-composite image A color image prepared by projecting individual black-and-white multispectral images in color.

color-ratio composite image A color-composite image prepared by combining, in different colors, individual spectral band ratio images of a scene.

complementary colors Two colors of light that produce white light when added together, such as red and cyan.

compositor viewer Device in which black-and-white multispectral images are registered and projected with colored lights to produce a color-composite image.

congruencing The process by which two images of a multi-image set are transformed so that the size and shape of any object on one image is the same as the size and shape of that object on the other image. In other words, when two images are congruenced, their geometries are the same.

contrast ratio The ratio between the reflectance of the brightest and darkest parts of the image. Commonly referred to as contrast.

contrast stretching Improving the contrast of images by digital processing. The original range of digital values is expanded to utilize the full contrast range of the recording film or display device.

cubic-convolution resampling A cubic spline approximation to the theoretically perfect sinc(x) resampling function. It uses a 16-point (4 X 4) neighborhood of pixels as input to compute the value of the output pixel.

cycle A single oscillation of a wave.

data collection system On Landsat the system that acquires in-situ information from seismometers, flood gauges, and other measuring devices. These data are relayed to an earth receiving station. The abbreviation is DCS.

density, of images A measure of the opacity, or darkness, of an image.

density slicing The process of converting the continuous gray tone of an image into a series of density intervals, or slices, each corresponding to a specific digital range.

detectability A measure of the smallest object that can be discerned on an image.

detector The component of a remote-sensing system that converts electromagnetic radiation into a signal that is recorded.

development The chemical processing of exposed photographic material to produce an image.

digital image A digital image or digitized image or digital picture function of a photograph is obtained by partitioning the area of the photograph into a finite two-dimensional array of small uniformly shaped mutually exclusive regions, called resolution cells, and assigning a "representative" grey tone to each such spatial region. A digital image may be abstractly thought of as a function whose domain is the finite two-dimensional set of resolution cells and whose range is the set of grey tones.

digital image processing Computer manipulation of the digital values or picture elements of an image.

digitization The process of converting an image recorded originally on photographic material into numerical format.

digitizer A device for scanning an image and converting it into numerical picture elements.

distortion On an image, this refers to changes in shape and position of objects with respect to their true shape and position.

diurnal Daily.

DN Digital number. The value of reflectance recorded for each pixel on Landsat CCT's.

EDC EROS Data Center. The U.S. Geological Survey facility at Sioux Falls, South Dakota. Here aircraft and satellite images are archived and made available for purchase. This facility also provides an indexing service for photographs and images acquired by NASA and the U.S. Geological Survey. (See also EROS.)

Ektachrome A Kodak color-positive film.

electromagnetic radiation Energy propagated in the form of an advancing interaction between electric and magnetic fields.

emission The process by which a body emits electromagnetic radiation, usually as a consequence of its kinetic temperature.

emissivity The ratio of radiant flux from a body to that from a black body at the same kinetic temperature. The symbol is ϵ.

emulsion A suspension of photosensitive silver halide grains in gelatin that constitutes the image-forming layer on photographic materials.

enhancement The process of altering the appearance of an image so that the interpreter can extract more information. Enhancement may be done by digital or photographic methods.

EREP Earth Resources Experiment Package on Skylab, which consisted of multispectral cameras, earth terrain camera, multispectral scanner, and nonimaging systems.

EROS Earth Resource Observation System, administered by U.S. Geological Survey.

ERTS Earth Resource Technology Satellite, now called Landsat.

filter, digital A mathematical procedure for removing unwanted values from numerical data or selectively enhancing values.

filter, optical A material that, by absorption or reflection, selectively modifies the radiation transmitted through an optical system.

focal length In cameras the distance measured along the optical axis from the optical center of the lens to the plane at which the image of a very distant object is brought into focus.

focus To adjust a remote-sensing system to produce a sharp, distinct image.

format The size and scale of an image.

frequency The number of wave oscillations per unit time or the number of wavelengths that pass a point per unit time. The symbol is ν.

GCP Ground control point. A geographic feature of known location that is recognizable on images and can be used to determine geometric corrections.

Gemini The United States program of two-man earth-orbiting satellites in 1965 and 1966.

geometric correction The removal of sensor, platform, or scene induced geometric errors such that the data conform to a desired projection. This involves the creation of a new digital image by resampling the input digital image.

geothermal Refers to heat from sources within the earth.

GMT Greenwich mean time. This international 24-hour system is used to designate the time at which Landsat images are acquired.

gray scale A calibrated sequence of gray tones ranging from black to white.

grid-point correspondence A table of corresponding pixel locations in two images that is used to define a geometric transformation.

ground receiving station A facility that records image data transmitted by Landsat.

ground resolution cell The area on the terrain that is covered by the instantaneous field of view of a detector. Size of the ground resolution cell is determined by the altitude of the remote-sensing system and the instantaneous field of view of the detector.

GSFC Goddard Space Flight Center. The NASA facility at Greenbelt, Maryland that is a Landsat earth receiving station. Landsat data from all United States receiving stations are processed and converted into digital and photographic images at GSFC.

highlights Areas of bright tone on an image.

histogram A graphical method of representing a frequency of occurrence distribution of image pixel intensities.

hue The attribute of a color that differentiates it from gray of the same brilliance and that allows it to be classed as blue, green, red, or intermediate shades of these colors.

image An image may be abstractly thought of as a continuous function I of two variables defined on some bounded region of a plane. When the image is a photograph, the range of the function I is the set of grey tones G usually considered to be normalized to the interval [0, 1]. The grey tone located at spatial coordinate (x, y) is proportional to the radiant energy in the electromagnetic band the sensor is sensitive and is denoted by $I(x, y)$. When the image is a map, the range of the function I is a set of symbols or colors and the symbol or color located at spatial coordinate (x, y) is denoted by $I(x, y)$.

image compression An operation which preserves all or most of the information in the image and which reduces the amount of memory needed to store an image or the time needed to transmit an image.

image restoration A process by which a degraded image is restored to its original condition. Image restoration is possible only to the extent that the degradation transform is mathematically invertible.

image enhancement Any one of a group of operations which improve the detectability of the targets of interest. These operations include, but are not limited to, contrast improvement, edge enhancement, spatial filtering, and noise suppression, image smoothing, and image sharpening.

image processing Encompasses all the various operations which can be applied to photographic or image format data. These include, but are not limited to, image compression, image restoration, image enhancement, preprocessing, quantization, spatial filtering, and other image pattern recognition techniques.

incident energy Electromagnetic radiation impinging on a surface.

index of refraction The ratio of the wavelength or velocity of electromagnetic radiation in a vacuum to that in a substance.

instantaneous field of view The solid angle through which a detector is sensitive to radiation. In a scanning system this refers to the solid angle subtended by the detector when the scanning motion is stopped. Instantaneous field of view is commonly expressed in milliradians.

interactive processing The method of data processing in which the operator views preliminary results and can alter the instructions to the computer to achieve optimum results.

interpretation The extraction of information from an image.

interpretation key A characteristic or combination of characteristics that enable an object or material to be identified on an image.

IR The infrared region of the electromagnetic spectrum that includes wavelengths from 0.7 μm to 1 mm.

IR color film A color film consisting of three layers in which the red-imaging layer responds to photographic IR radiation ranging in wavelength from 0.7 to 0.9 μm. The green-imaging layer responds to red light and the blue-imaging layer responds to green light.

JPL Jet Propulsion Laboratory. A NASA facility at Pasadena, California operated by the California Institute of Technology.

Lambert conic conformal projection A map projection on which all geographic meridians are represented by straight lines that meet at a common point outside the limits of the map, and the geographic parallels are represented by a series of arcs having this common point as a center. Meridians and parallels intersect at right angles and angles on the earth are correctly represented on the projection.

Landsat An unmanned earth-orbiting NASA satellite that detects and transmits multispectral images in the 0.5 to 10.6 μm region to earth receiving stations (formerly called ERTS).

light Visible radiation from 0.4 to 0.7 μm in wavelength that is detectable by the eye.

line-interleaved format A type of data organization for the four (or five) spectral bands of MSS digital data. It consists of four (or five) data fields per record with each data field containing a scan line from a different band. Each of the four (or five) data fields represents the same ground area.

line-length-adjustment pixels Extra pixels inserted by NASA or EDC to make all scan lines in an input MSS tape have the same number of pixels.

line-pair A pair of light and dark bars of equal sizes. The number of such line-pairs that can be distinguished per unit distance is used to express resolving power of imaging systems.

lineament A linear topographic or tonal feature on the terrain and on images and maps that may represent a zone of structural weakness.

mapping function The output image-space to input image-space function that defines the geometric transformation used in a geometric correction process.

Mercury The United States program of one-man, earth-orbiting satellites in 1962 and 1963.

microwave The region of the electromagnetic spectrum in the wavelength range from 1 mm to beyond 1 m.

modulate To vary the frequency, phase, or amplitude of electromagnetic waves.

modulation transfer function The spatial frequency response of a system or phenomenon.

mosaic An image or photograph made by piecing together individual images or photographs covering adjacent areas.

MSS Multispectral scanner system of Landsat that acquires images at four or five wavelength bands in the visible, reflected, and thermal IR regions.

multi-image A set of images or photographs, each acquired of the same subject at different times, or from different positions, or with different sensors, or at different electromagnetic frequencies, or with different polarizations. Although there is a high degree of information redundancy between images in a multi-image set, each image usually has information not available in any one of or combinations of the other images in the set.

multispectral camera A system that simultaneously acquires photographs at different wavelengths of the same scene. Also called multiband camera.

multispectral scanner A scanner system that simultaneously acquires images in various wavelength regions of the same scene.

nadir The point on the ground vertically beneath the center of a remote-sensing system.

NASA National Aeronautics and Space Administration.

nearest neighbor resampling A resampling algorithm using a one-point neighborhood of pixels as input. The pixel nearest the resampling location is used as the value of the output pixel.

noise Random or repetitive events that obscure or interfere with the desired information.

nonsystematic distortion Geometric irregularities on images that are not constant and cannot be predicted from the characteristics of the imaging system.

oblique photograph A photograph acquired with the camera axis intentionally directed between the horizontal and vertical orientations.

orbit The path of a satellite around a body under the influence of gravity.

overlap The extent to which adjacent images or photographs cover the same terrain, expressed in percent.

oversampling The intentional partial overlapping of picture elements in a sensor by digitizing the detector output more frequently than that corresponding to the detector instantaneous field of view. This is sometimes used in the along-scan direction to improve the frequency response of the sensor.

pass In digital filters this refers to the spatial frequency of data transmitted by the filter. High-pass filters transmit high-frequency data; low-pass filters transmit low-frequency data.

passive remote sensing Remote sensing of energy naturally reflected or radiated from the terrain.

pattern The regular repetition of tonal variations on an image or photograph.

pattern recognition Concerned with, but not limited to, problems of: pattern discrimination, pattern classification, feature selection, the pattern identification, cluster identification, feature extraction, filtering, enhancement, and pattern segmentation.

pattern segmentation In pattern recognition problems such as target discrimination, where the category of interest is some specific formation of resolution cells with characteristic shape or tone-texture composition, the problem of pattern segmentation occurs. Pattern segmentation is the problem of determining which regions or areas in the image constitute the patterns of interest, i.e., which resolution cells should be included and which excluded from the pattern measurements.

photodetector A device for measuring energy in the photographic band.

photograph A representation of targets formed by the action of light on silver halide grains of an emulsion.

photographic IR The short wavelength portion of the IR band from 0.7 to 0.9 μm wavelength that is detectable by IR color film or IR black-and-white film.

photon The elementary quantity of radiant energy.

picture element A pair whose first member is a resolution cell and whose second member is the grey tone assigned by the digital image to that resolution cell. Sometimes pixel is used in a sense which is synonomous with resolution cell and other times synonomous with the grey tone.

pitch Rotation of an aircraft about the horizontal axis normal to its longitudinal axis that causes a nose-up or nose-down attitude.

pixel A contraction of picture element.

point-shift algorithm A fast computer algorithm for performing nearest neighbor resampling of a digital image. The point-shift algorithm achieves its speed by moving line segments instead of individual pixels to create the output image array.

preprocessing Commonly used to describe corrections and processing done to image data before information extraction. Includes geometric and radiometric correction, mosaicking, resampling, and formatting.

principal point The center of an aerial photograph.

printout Display of computer data in alphanumeric format.

quantizing The process by which each grey tone in a digital image or photograph is assigned a new value from a finite set of new grey tone values. There are three often used methods of quantizing:

1) in *equal interval quantizing*, the range of grey tones from maximum grey tone to minimum grey tone is divided into contiguous intervals, each of equal length, and grey tone is assigned to the quantized class which corresponds to the interval it lies within;

2) in *equal probability* quantizing, the range of grey tones is divided into contiguous intervals such that after the grey tones are assigned to their quantized class there is an equal frequency of occurrence for each grey tone in the quantized digital image or photograph;

3) in *minimum variance* quantizing, the range of grey tones is divided into contiguous intervals such that the weighted sum of the variance of the grey tone intervals is minimized. The weights are usually chosen to be the grey tone interval probabilities which are computed as the proportional area on the photograph or digital image which have grey tones in the given interval.

radar Acronym for radio detection and ranging, an active form of remote sensing that operates at wavelengths from 1 mm to 1m.

radian The angle subtended by an arc of a circle equal in length to the radius of the circle; 57.3°.

radiance level A measure of the energy radiated by the object. In general, radiance is a function of viewing angle and spectral wavelength and is expressed as energy per solid angle.

radiant temperature Concentration of the radiant flux from a material. Radiant temperature is the product of the kinetic temperature multiplied by the emissivity to the one-fourth power.

radiation The propagation of energy in the form of electromagnetic waves.

radiometric correction Processing of sensor data to calibrate and correct the radiation data provided by the sensor detectors.

random-line dropout A defect in scanner images caused by the loss of data from individual scan lines in a nonsystematic fashion.

raster lines The individual lines swept by an electron beam across the face of a CRT that constitute the image display.

ratio image An image prepared by processing digital multispectral data. For each pixel the value for one band is divided by that of another. The resulting digital values are displayed as an image.

RBV Return-beam vidicon.

real time To make images or data available for inspection simultaneously with their acquisition.

rectification A congruencing process by which the geometry of an image area is made planimetric. For example, if the image is taken of an equally spaced rectangular grid pattern, then the rectified image will be an image of an equally spaced rectangular grid pattern. Rectification does not remove relief distortion or perspective distortion.

rectilinear Refers to images with no geometric distortion in which the scales in the X and Y directions are the same.

reference source In thermal IR scanners and radiometers these are electrically heated cavities maintained at a known radiant temperature and used for calibrating the radiant temperature detected from the target.

reflectance The ratio of the radiant energy reflected by a body to that incident upon it.

reflectance, spectral Reflectance measured at a specific wavelength interval.

reflected IR Wavelengths from 0.7 to about 3 μm that are primarily reflected solar radiation.

refraction The bending of electromagnetic rays as they pass from one medium into another.

registration The process of superimposing two or more images or photographs so that equivalent geographic points coincide. Registration may be done digitally or photographically.

relief The vertical irregularities of a surface.

relief displacement The geometric distortion on vertical aerial photographs. The tops of objects are located on the photograph radially outward from the base.

remote sensing The collection of information about an object without being in physical contact with the object. Remote sensing is restricted to methods that record the electromagnetic radiation reflected or radiated from an object, which excludes magnetic and gravity surveys that record force fields.

resampling The computation of the intensity of a picture element between input or known picture element intensities.

resolution The ability to distinguish closely spaced objects on an image or photograph. Commonly expressed as the spacing, in line-pairs per unit distance, of the most closely spaced lines that can be distinguished.

resolution cell The smallest, most elementary areal constituent of grey tones under consideration in an image. A resolution cell is referenced by its spatial coordinates.

resolution target Regularly spaced pairs of light and dark bars that are used to evaluate the resolution of images or photographs.

resolving power A measure of the ability of individual components, and of remote-sensing systems, to define closely spaced targets.

return-beam vidicon An imaging system on Landsat that consists of three cameras for Landsat 1 and 2, operating in the green, red, and photographic IR spectral regions. Landsat 3 uses two panchromatic sensors. Instead of using film, the images are formed on the photosensitive surface of a vacuum tube. The image is scanned with an electron beam and transmitted to earth receiving stations.

roll Rotation of an aircraft about the longitudinal axis to cause a wing-up or wing-down attitude.

satellite An object in orbit around a celestial body.

scale The ratio of distance on an image to the equivalent distance on the ground.

scan line One row of a digital image, generally in a cross-track direction.

scan skew Distortion of scanner images caused by forward motion of the aircraft or satellite during the time required to complete a scan.

scanner An optical-mechanical imaging system in which a rotating or oscillating mirror sweeps the instantaneous field of view of the detector across the terrain. The two basic types of scanners are airborne and stationary.

scanner distortion The geometric distortion that is characteristic of scanner images. The scale of the image is constant in the direction parallel with the aircraft or spacecraft flight direction. At right angles to this direction, however, the image scale becomes progressively smaller from the nadir line outward toward either margin of the image. Also includes scanner-induced distortions due to nonlinear sensor effects such as the mirror scan nonlinearities.

scattering Multiple reflection of electromagnetic waves by gases or particles in the atmosphere.

scene The area on the ground that is covered by an image or photograph.

sensitivity The degree to which a detector responds to electromagnetic energy incident upon it.

sensor A device that receives electromagnetic radiation and converts it into a signal that can be recorded and displayed as numerical data or as an image.

shade print A gray scale representation of an image produced on a line printer or other similar device.

sidelap The extent of lateral overlap between images acquired on adjacent flight lines.

signature A characteristic, or combination of characteristics, by which a material or an object may be identified on an image or photograph.

silver halide Silver salts that are sensitive to electromagnetic radiation and convert to metallic silver when developed.

sixth-line banding A defect on Landsat MSS images in which every sixth scan line is brighter or darker than the others. Caused by the sensitivity of one detector being higher or lower than the others.

sixth-line dropout An occasional defect on Landsat MSS images in which no data are recorded for every sixth scan line, which is black on the image.

Skylab The United States' earth-orbiting workshop that housed three crews of three men in 1973 and 1974.

SLAR Acronym of side-looking airborne radar.

software The programs that control computer operations.

spatial filter A kind of image transformation which can be applied to photographs or digital images and which usually maintains the one-to-one correspondence between the spatial (x, y) coordinates of the domain image and the range image and is usually used to lessen noise or enhance certain characteristics of the image.

spectral band A range of electromagnetic wavelengths that are sensed by an imaging detector. Landsat 1 and 2 MSS has four spectral bands which operate in the solar reflected spectral region from 0.5 to 1.1 μm. Landsat 3 has an additional thermal infrared band in the 10.4 to 12.6 μm region.

spectral reflectance The reflectance of electromagnetic energy at specified wavelength intervals.

spectral sensitivity The response, or sensitivity, of a film or detector to radiation in different spectral regions.

spectrum A continuous sequence of energy arranged according to wavelength or frequency.

stereo pair Two overlapping images or photographs that may be viewed stereoscopically.

sun synchronous An earth satellite orbit in which the orbit plane is near polar and the altitude such that the satellite passes over all places on earth having the same latitude twice daily at the same local sun time.

synthetic-aperture radar SLAR system in which high resolution in the azimuth direction is achieved by utilizing the Doppler principle to give the effect of a very long antenna.

synthetic stereo images A stereo model made by digital processing of a single image. Topographic data are used in calculating the geometric distortion.

systematic distortion Geometric irregularities on images that are caused by the characteristics of the imaging system and are predictable.

target An object on the terrain of specific interest in a remote-sensing investigation.

template matching An operation which can be used to find out how well two photographs or images match one another. The degree of matching is often determined by cross-correlating the two images or by evaluating the sum of the squared image differences. Template matching can also be used to best match a measurement pattern with a prototype pattern.

terrain The surface of the earth.

texture Concerned with the spatial distribution of the grey tones and discrete tonal features. When a small area of the image has little variation of discrete tonal features, the dominant property of that area is tone. When a small area has wide variation of discrete tonal features, the dominant property of that area is texture. There are three things crucial in this distinction: 1) the size of the small area, 2) the relative sizes of the discrete tonal features, and 3) the number of distinguishable discrete tonal features.

thermal IR The portion of the IR region from approximately 3 to 14 μm that corresponds to heat radiation. This spectral region spans the radiant power peak of the earth.

thermal IR image An image acquired by a scanner that records radiation within the electromagnetic band that ranges from approximately 3 to 14 μm in wavelength.

tone Each distinguishable shade of gray from white to black on an image.

transmissivity The property of a material that determines the amount of energy that can pass through the material.

transparency A positive or negative image on a transparent photographic material. The capability of a material to transmit light.

two-pixel-interleaved format The type of data organization used by EDC for the four spectral bands of MSS digital data. It consists of 8-byte words, each containing four

2-byte elements from each band. The elements in a given word all represent the same ground area.

UTM Universal Transform Mercator, a common cylinderical map projection.

UV The ultraviolet region consisting of wavelengths from 0.01 to 0.4 μm.

vidicon An imaging tube having a photosensitive surface. It is a device used to convert image data from photographic format to electronic video signal format. An electron beam is scanned like a TV raster across the photosensitive surface and it generates a signal whose amplitude corresponds to the light intensity focused on the surface at each point. This electron beam signal is then amplified to a usable video signal.

visible radiation Energy at wavelengths from 0.4 to 0.7 μm that is detectable by the eye.

watt Unit of electrical power equal to rate of work done by one ampere under a pressure of one volt. The symbol is W.

wavelength The distance between successive wave crests, or other equivalent points, in a harmonic wave. The symbol is λ.

***X* band** Radar wavelength region from 2.4 to 3.8 cm.

yaw Rotation of an aircraft about its vertical axis to cause the longitudinal axis to deviate from the flight line.

zenith The point in the celestial sphere that is exactly overhead.

Bibliography

Remote Sensing References

J. Lintz, Jr., and D. S. Simonett, Eds., *Remote Sensing of Environment.* Reading, MA: Addison-Wesley, 1976.

F. F. Sabins, Jr., *Remote Sensing—Principles and Interpretation.* San Francisco, CA: Freeman, 1978.

E. C. Barrett and L. F. Curtis, *Introduction to Environmental Remote Sensing.* New York: Halsted Press, Wiley, 1976.

R. G. Reeves, A. Anson, and D. Landen, Eds., *Manual of Remote Sensing, Vol. 1—Theory, Instruments, and Techniques; Vol. 2—Interpretation and Applications.* Falls Church, VA: American Society of Photogrammetry, 1975.

Remote Sensing with Special Reference to Agriculture and Forestry, National Academy of Sciences, Washington, DC, 1970.

N. M. Short, P. D. Lowman, S. C. Freden, and W. A. Finch, *Mission to Earth: Landsat Views the Earth*, NASA SP-360, 1976.

Resource Sensing from Space, Prospects for Developing Countries, National Academy of Sciences, 1977.

Proceedings of the NASA Earth Resources Survey Symposium, 1975, Houston, TX, NASA TM X-58168, JSC-09930.

F. Shahrokhi, Ed., *Remote Sensing of Earth Resources, vol. I-VI, Selected Papers from 1972, 1973, 1974, 1975, 1976, 1977 Remote Sensing of Earth Resources Conferences*, Univ. Tennessee Space Inst., Tullahoma, TN 37388, 1978.

Vision and Perception

H. B. Barlow, R. Narasimhan, and A. Rosenfeld, "Visual pattern analysis in machines and animals," *Science*, vol. 177, pp. 567-575, Aug. 1972.

M. M. Benarie, "Optimal encoding of the visual image," *J. Opt. Soc. Amer.*, vol. 51, pp. 371-372, Mar. 1961.

I. Biederman, "Perceiving real-world scenes," *Science*, vol. 177, pp. 77-79, July 1972.

R. M. Boynton, "Progress in physiological optics," *Appl. Opt.*, vol. 6, pp. 1283-1293, Aug. 1967.

S. Coren and J. S. Girgus, "Visual spatial illusions: Many explanations," *Science*, vol. 179, pp. 503-504, Feb. 1973.

R. H. Day, "Visual spatial illusions: A general explanation," *Science*, vol. 175, pp. 1335-1340, Mar. 1972.

L. D. Harmon and B. Julesz, "Masking in visual recognition: Effects of two-dimensional filtered noise," *Science*, vol. 180, pp. 1194-1197, June 1973.

J. M. Heyning, "The human observer," from *Proceedings of The Human in the Photo-Optical System Seminar, SPIE*, NYC, Apr. 1966, pp. III-1-III-50.

A. J. Hill, "A mathematical and experimental foundation for steroscopic photography," *J. SMPTE*, vol. 61, pp. 461-486, Oct. 1953.

J. Hochberg, "Perception and depiction," *Science*, vol. 172, pp. 685-686, May 1971.

J. Krauskopf, "Light distribution in human retinal images," *J. Opt. Soc. Amer.*, vol. 52, pp. 1046-1050, Sept. 1962.

B. S. Lipkin, "Psychopictorics and pattern recognition," *SPIE J.*, vol. 8, pp. 126-138, May 1970.

E. M. Lowry and J. J. DePalma, "Sine-wave response of the visual system, I. The mach phenomenon," *J. Opt. Soc. Amer.*, vol. 51, pp. 740-746, July 1961; "II. Sine-wave and square-wave contrast sensitivity," *J. Opt. Soc. Amer.*, vol. 52, pp. 328-335, Mar. 1962.

D. Marr and T. Poggio, "Cooperative computation of stereo disperity," *Science*, vol. 194, pp. 283-287, Oct. 1976.

D. L. MacAdam, "Stereoscopic perceptions of size, shape, distance and direction," *J. SMPTE*, vol. 62, pp. 271-293, Apr. 1954.

C. McCamy, H. Marcus, and J. G. Davidson, "A color-rendition chart," *J. Appl. Photo. Eng.*, vol. 2, pp. 95-99, Spring 1976.

O. H. Schade, Sr., "Optical and photoelectric analog of the eye," *J. Opt. Soc. Amer.*, vol. 46, pp. 721-739, Sept. 1956.

R. N. Shepard and S. A. Judd, "Perceptual illusion of rotation of three-dimensional objects," *Science*, vol. 191, pp. 952-954, 1976.

J. J. Sheppard, Jr., *Human Color Perception.* New York: American Elsevier, 1968.

C. R. R. Snyder, "Selection, inspection, and naming in visual search," *J. Experimental Psychol.*, vol. 92, no. 3, pp. 428-431, 1972.

B. W. Tansley and R. M. Boynton, "A line, not a space, represents visual distinctness of borders formed by different colors," *Science*, vol. 191, pp. 954-957, 1967.

R. N. Wolfe, "Width of the human visual spread function as determined psychometrically," *J. Opt. Soc. Amer.*, vol. 52, pp. 460-469, Apr. 1962.

Resolution and Image Quality

J. H. Altman, "Image quality criteria for data recording and storage," *J. SMPTE*, vol. 76, pp. 629-634, July 1967.

T. W. Barnard, "Image evaluation by means of target recognition," *Photo. Sci. Eng.*, vol. 16, pp. 144–150, Mar.–Apr. 1972.

J. Bhushan, A. K. Dimri, and V. P. Kathura, "Overall performance of reconaissance photographic camera," *J. Appl. Photo. Eng.*, vol. 2, pp. 28–30, Winter 1976.

F. C. Billingsley, "Noise considerations in digital image processing hardware," in *Topics in Applied Physics*, T. Huang, Ed. New York: Springer-Verlag, 1975.

M. Born and E. Wolf, *Principles of Optics-Electromagnetic Theory of Propagation, Interference and Diffraction of Light*. New York: Pergamon, 1959.

W. N. Charman, *Spatial Frequently Spectra and Other Properties of Conventional Resolution Targets*, PS&E 8, 1964.

J. W. Coltman, "The specification of imaging properties by response to a sine wave input," *J. Opt. Soc. Amer.*, vol. 44, June 1954.

H. D. Friedman, "On the expected error in the probability of misclassification," *Proc. IEEE*, p. 658, June 1965.

G. L. Fultz, "The effect of source noise on quantization accuracy and on PE statistics," JPL Tech. Memo. 3341-65-5, Jan. 1965.

G. A. Fry, "Relation of blur functions to resolving power," *J. Opt. Soc. Amer.*, vol. 51, pp. 560–563, May 1961.

J. W. Goodman, *Introduction to Fourier Optics.* New York: McGraw-Hill, 1968.

W. E. Grabau, "Pixel problems," U.S. Army Eng. Waterways Exp. Sta. Misc., Paper M-76-9, Vicksburg, MS, 1976.

K. Hacking, "The relative visibility of random noise over the grey scale," *J. Brit. IRE*, pp. 307–310, Apr. 1962.

H. B. Hammill and T. M. Holladay, "The effects of certain approximations in image quality evaluation from edge traces," *SPIE J.*, Sept. 1970.

J. L. Harris, "Diffraction and resolving power," *J. Opt. Soc. Amer.*, vol. 54, July 1964.

H. Horwitz, R. F. Nalepka, and J. Morgenstern, "Estimating the proportions within a single resolution element of a multispectral scanner," in *Proc. 7th Int. Symp. on Remote Sensing of Environment*, Ann Arbor, MI, May 1971.

T. S. Huang, "The subjective effect of two-dimensional pictorial noise," *IEEE Trans. Inform. Theory*, pp. 43–53, Jan. 1965.

J. R. Jenness, Jr., W. A. Eliot, and J. A. Ake, "Intensity ripple in a raster generated by a Gaussian scanning spot," *J. SMPTE*, p. 76, June 1967.

N. Jensen, *Optical and Photographic Reconnaissance Systems.* New York: Wiley, 1968.

Jet Propulsion Lab., Tech. Brief 67-10005 and 67-10630, "Technical support package on image processing system," JPL, California Inst. Technol., Pasadena, 1970.

C. B. Johnson, "Modulation transfer functions: A simplified approach," *Electro-Optical Systems Design*, Nov. 1972.

D. M. Johnston, "Target recognition on TV as a function of horizontal resolution and shades of gray," *Human Factors*, vol. 10, June 1968.

R. A. Jones, "An automated technique for deriving MTF's from edge traces," *Photo. Sci. Eng.*, vol. 11, Mar.–Apr. 1967.

R. A. Jones, "Image evaluation by edge gradient analysis," presented at the Society of Photo-Optical Instrumentation Engineers 11th Annu. Tech. Symp., Aug. 22–26, 1965.

R. C. Jones, "On the point and line spread functions of photographic images," *J. Opt. Soc. Amer.*, vol. 48, Dec. 1958.

S. J. Katzberg, F. O. Huck, and S. D. Wall, "Photosensor aperture shaping to reduce aliasing in optical-mechanical line-scan imaging systems," *Appl. Opt.*, vol. 12, May 1973.

D. H. Kelly, "Spatial frequency, bandwidth, and resolution," *Appl. Opt.*, vol. 4, Apr. 1965.

V. Kratky, "Cartographic accuracy of ERTS," *Photogram. Eng.*, vol. XL, no. 2, pp. 203–212, 1974.

D. R. Knudson, S. N. Teicher, J. F. Reintjes, and U. F. Gronemann, "Experimental evaluation of the resolution capabilities of image-transmission systems," *Inform. Display*, Sept.–Oct. 1968.

R. L. Lamberts, "Relationship between the sine-wave response and the distribution of energy in the optical image of a line," *J. Opt. Soc. Amer.*, vol. 48, July 1958.

R. L. Lamberts, "Application of sine-wave techniques to image-forming systems," *J. Opt. Soc. Motion Picture and Television Eng.*, vol. 71, Sept. 1962.

N. W. Lewis and T. V. Hauser, "Microcontrast and blur in imaging systems," *J. Photo. Science*, vol. 10, 1962.

E. W. Marchand, "Derivation of the point spread function from the line spread function," *J. Opt. Soc. Amer.*, vol. 54, July 1964.

E. W. Marchand, "From line to point spread function: The general case," *J. Opt. Soc. Amer.*, vol. 55, Apr. 1965.

E. L. O'Neil, *Introduction to Statistical Optics.* Reading, MA: Addison-Wesley, 1963.

A. V. Oppenheim, R. W. Schafer, and T. G. Stockham, "Nonlinear filtering of multiplied and convolved signals," *Proc. IEEE*, vol. 56, pp. 1264–1291, Aug. 1968.

Perkin-Elmer Corp., "The practical application of modulation transfer functions," a symp. held at Perkin-Elmer, Mar. 6, 1963.

F. H. Perrin, "Methods of appraising photographic systems, Part II—Manipulation and significance of the sine-wave response function," *J. SMPTE*, vol. 69, Apr. 1960.

E. F. Puccinelli, "Ground location of satellite scanner data," *Photogram. Eng.*, vol. 42, pp. 537-543, 1976.

M. D. Rosenau, Jr., "Image evaluation by edge gradient analysis," Society of Photo-Optical Instrumentation Engineers, 11th Annu. Tech. Symp., Aug. 22-26, 1966.

J. W. Schoonmaker, Jr., "Geometric evaluation of MSS images from ERTS-1," in *Proc. 40th Annu. Meeting ASP*, 1974, pp. 582-587.

M. Schwartz, *Information Transmission, Modulation and Noise.* New York: McGraw-Hill.

R. Scott, R. M. Scott, and R. V. Shack, "The use of edge gradients in determining modulation-transfer functions," *Photo. Sci. Eng.*, vol. 7, no. 6, 1963.

C. F. Shelton, H. H. Herd, and J. J. Leybourne, "Grey-level resolution of flying spot scanner systems," presented at the SPIE Photo Optical Systems Seminar, Rochester, NY, May 1967.

C. R. Shelton, "Spatial frequency response of flying-spot-scanner systems," a paper from the 11th SPIE Tech. Symp., Aug. 22, 1966.

B. Titian, "Method for obtaining the transfer function from the edge response function," *J. Opt. Soc. Amer.*, vol. 55, no. 8, 1965.

C. E. Thomas, A. Kahn, G. Tisdale, and T. Spink, "Detection of TV imagery by humans vs. machines," *Appl. Opt.*, vol. 11, no. 5, 19-7-1056, 1972.

A. Van der Ziel, *Noise.* Englewood Cliffs, NJ: Prentice-Hall.

A. Van der Ziel, *Information Transmission, Modulation and Noise.* New York: McGraw-Hill.

C. R. Walli, "Quantizing and sampling errors in hybrid computation," in *Proc. Fall Joint Comput. Conf.*, 1964, pp. 545-558.

R. Welch, "Modulation transfer functions," *Photogram. Eng.*, 1971.

J. Zimmerman, "A practical approach toward the determination of resolution requirements," in *Proc. Society of Photo-Optical Instrumental Engineers, 10th Tech. Symp.*, Aug. 1965, pp. 16-20.

Sampling

H. A. Andrews and C. L. Patterson, "Digital interpolation of discrete images," *IEEE Trans. Comput.*, vol. C-25, Feb. 1976.

A. V. Balakrishnan, "On the problem of time jitter in sampling," *IRE Trans. Inform. Theory*, pp. 226-236, 1962.

T. Berger, "On detection of sparsely sampled data," *Proc. IEEE*, pp. 700-701, May 1967.

R. Bernstein, "All-digital precision processing of ERTS (Landsat) images," IBM Corp., Final Rep., Contract NAS5-21716 for NASA Goddard Space Flight Center, Greenbelt, MD, Apr. 1975.

W. M. Brown and C. J. Palermo, "Effects of phase errors on the ambiguity function," in *IRE Conv. Rec.*, vol. 4, 1963, p. 118.

W. M. Brown and C. J. Palermo, "Effects of phase errors on resolution," *IEEE Trans. Mil. Electron.*, pp. 4-9, Jan. 1965.

G. L. Cariolaro, "Dependence on sampling frequency of amplitude statistical parameters of signals," *Alta Frequenza*, vol. 27, Nov. 1968.

S. C. Chao, "Flutter and time errors in magnetic data recorders," *IEEE Trans. Aerosp. Electron. Syst.*, p. 214, Mar. 1966.

N. Chu, C. McGilliam, and P. Anuta, "Effects of interpolation and enhancement on classification of Landsat MSS data," Lab. for Application of Remote Sensing, Tech. Rep., in publication.

C. P. Colby, S. G. Wheeler, and W. Miller, "Thematic mapper design parameter investigation," paper to be published in *Proc. Summer Computer Simulation Conf.*, 1978.

"Recovery of randomly sampled time sequences," *IRE Trans. Commun. Syst.*, pp. 214-216, June 1962.

R. E. Crochiere and L. R. Rabiner, "Further considerations in the design of decimators and interpolators," *IEEE Trans. Acoust., Speech, Signal Processing*, vol. ASSP-24, pp. 296-311, 1976.

B. Dunbridge, "Analysis of analog-to-digital converter time aperture error," presented at the Nat. Telemetering Conf., Boston, MA; *Telemetry J.*, pp. 32-36, Aug./Sept. 1966.

D. G. Ferneyhough, Jr. and C. W. Niblack, Resampling Study: IBM Final Rep., NAS5-21865, NASA/GSFC, 1977.

N. T. Gaarder, "A note on the multidimensional sampling theorem," *Proc. IEEE*, pp. 247-248, Feb. 1972.

L. W. Gardenhire, "The use of digital data systems," presented at Nat. Telemetering Conf., Albuquerque, NM, May 1963.

J. V. Gaven, Jr. and J. Tavitian, "The informative value of sampled images as a function of the number of gray levels used in encoding the images," *Photo. Sci. Eng.*, vol. 14, pp. 16-20, 1970.

A. J. Gibbs, "Simulation of a low-pass transmission system for sampled data," *Proc. IEEE*, pp. 443-444, 1966.

A. F. H. Goetz and F. C. Billingsley, "Digital image enhancement techniques used in some ERTS application problems," in *Proc. 3rd ERTS-1 Symp.*, vol. 1-B, 1971-1992, Dec. 1973.

L. M. Goodman, "Optimum sampling and quantizing rates," *Proc. IEEE*, pp. 90-92, Jan. 1966.

P. A. Hollanda and F. Scott, "The informative value of sampled images as a function of the number of scans per scene object and the signal-to-noise ratio," *Photo. Sci. Eng.*, vol. 14, no. 6, pp. 407–412, 1970.

H. S. Hou, "Least squares image restoration using spline interpolation," Univ. Southern California, Rep. USCIPI-650, Feb. 1976.

H. S. Hou and H. C. Andrews, "Fundamental limits and degrees of freedom of imaging systems," in *Proc. Optical Soc. Amer. Topical Meeting*, Asilomar, CA, Feb. 1976.

R. R. Jayroe, "Nearest neighbor, bilinear bicubic interpolation, geographic correction effects on Landsat imagery," NASA TMX-73348, NASA/MSFC, 1976.

J. G. Kennedy and A. N. Williamson, "A technique for achieving geometric accordance of Landsat digital data," U.S. Army Eng. Waterways Exp. Sta., Misc. Paper M-76-16, Vicksburg, MS, 1976.

P. M. Liebman, "Determination of modulated pulse train spectrums using woodward methods," *Proc. IEEE*, pp. 401–402, Mar. 1966.

B. Liu and T. P. Stanley, "Error bounds for jittered sampling," *IEEE Trans. Automat. Contr.*, vol. AC-10, Oct. 1965.

A. W. Lohmann and D. P. Paris, "Influence of longitudinal vibrations on image quality," *Appl. Opt.*, vol. 4, pp. 393–397, Apr. 1965.

R. Myszko, "Effect of vibration on the photographic image," *J. Opt. Soc. Amer.*, vol. 53, pp. 935–940, Aug. 1963.

G. Oetken, T. W. Parks, and H. W. Schussler, "New results in the design of digital interpolators," *IEEE Trans. Acoust., Speech, Signal Processing*, vol. ASSP-23, June 1975.

A. V. Oppenheim and D. H. Johnson, "Discrete representation of signals," *Proc. IEEE*, vol. 60, pp. 681–691, June 1972.

D. P. Paris, "Influence of image motion on the resolution of a photographic system," *Photo. Sci. Eng.*, vol. 6, no. 1, pp. 55--59, 1962.

D. P. Paris, "Influence of image motion on the resolution of a photographic system, II," *Photo. Sci. Eng.*, vol. 7, no. 4, pp. 233–236, 1963.

Perkin-Elmer Corp., "The practical application of modulation transfer functions," Mar. 6, 1963.

Perkin-Elmer Corp., a symposium on sampled images, IS10763, 1971.

M. J. Peyrovian, "Image restoration by spline functions," Univ. Southern California, Rep. USCIPI-680, Aug. 1976.

L. R. Rabiner and R. E. Crochier, "On the design of all-pass signals with peak amplitude constraints," *Bell Syst. Tech. J.*, vol. 55, pp. 395–407, Apr. 1976.

A. G. Ratz, "The effect of tape transport flutter on spectrum and correlation analysis," *IEEE Trans. Space Electron. Telemetry*, Dec. 1964.

S. S. Rifman and D. M. McKinnon, "Evaluation of digital correction techniques for ERTS images," TRW Corp., Final Rep. 20634-6003-TU-00 for NASA Goddard Space Flight Center, Greenbelt, MD, Mar. 1974. Also, related Rep. 26232-6004-TU-01.

M. D. Rosenau, Jr., "Parabolic image-motion," *Photogram. Eng.*, pp. 421–427, 1962.

F. Scott and P. A. Hollanda, "The informative value of sampled images as a function of the number of scans per scene object," *Photo. Sci. Eng.*, vol. 14, no. 1, pp. 21–27, 1970.

R. M. Scott, "Contrast rendition as a design tool," *Photo. Sci. Eng.*, vol. 3, no. 4, pp. 201–209, 1959.

S. Twomey, "Information content in remote sensing," *Appl. Opt.*, vol. 13, no. 4, pp. 942–945, 1974.

E. Van Vlecks, a TRW study on resampling for data compression and its effects on classification accuracy.

W. N. Waggener, "Digital TV and the sampling theorem," *Electron. Communicator*, Mar./Apr. 1967.

P. R. Wallace, "Real-time measurement of element differences in television programs," *Proc. IEEE*, pp. 1576–1577, Nov. 1966.

Digital Image Processing

H. C. Andrews, "Outer product expansions and their uses in digital image processing," *IEEE Trans. Comput.*, vol. C-25, Feb. 1976.

C. Arcelli and S. Levialdi, "Picture processing and overlapping blobs," *IEEE Trans. Comput.*, vol. C-20, pp. 1111–1115, Sept. 1971.

Bendix Corp., "Image data processing," *Bendix Tech.*, vol. 5, Spring 1972.

R. Bernstein and D. G. Ferneyhough, Jr., "Digital image processing," *Photogram. Eng.*, vol. 41, pp. 1465–1476, Dec. 1975.

R. Bernstein, "Scene correction (precision processing) of ERTS sensor data using digital image processing techniques," in *Proc. 3rd ERTS-1 Symp.*, vol. I, sect. B, 1973, pp. 1909–1928.

R. Bernstein, "Digital image processing of earth observation sensor data," *IBM J. Res. Develop.*, vol. 20, pp. 40–57, Jan. 1976.

R. Bernstein, "Digital image processing—Past, present, future," presented at the Int. Symp. on Image Processing, Interactions with Photogrammetry and Remote Sensing, Technical Univ., Graz, Austria, 1977.

R. Bernstein and L. P. Schoene, "Advances in digital image processing of earth observation sensor data," presented at the 13th Space Congr., Cocoa Beach, FL, 1976.

R. Bernstein and G. C. Stierhoff, "Precision processing of earth image data," *Amer. Sci.*, vol. 64, pp. 500–508, 1976.

F. C. Billingsley, "Computer-generated color image display of lunar spectral reflectance ratios," *Photo. Sci. Eng.*, vol. 16, p. 51, 1972.

F. C. Billingsley, "Digital image processing for information extraction," *Int. J. Man-Machine Studies*, vol. 5, pp. 203–214, 1973.

F. C. Billingsley, "Some digital techniques for enhancing ERTS imagery," in *Proc. ASP Remote Sensing Symp.*, Sioux Falls, SD, Oct. 1973.

F. C. Billingsley, A. F. H. Goetz, and J. N. Lindsley, "Color differentiation by computer image processing," *Photo. Sci. Eng.*, vol. 14, p. 28, 1970.

F. C. Billingsley and A. F. H. Goetz, "Computer techniques used for some enhancements of ERTS images," presented at the Symp. on Significant Results Obtained from ERTS-1, NASA SP-327, 1973.

F. C. Billingsley, "Computer-generated color image display of lunar spectral reflectance ratios," *Photo. Sci. Eng.*, vol. 16, 1972.

R. J. Blackwell, "Fingerprint image enhancement by computer methods," presented at the 1970 Carnahan Conf. on Electronic Crime Countermeasures, Lexington, KY, Apr. 17, 1970.

F. Brestenreiner, U. Greis, J. Helmberger, and K. Stadler, "Visibility and correction of periodic interference structures in line-by-line recorded images," *J. Appl. Photo. Eng.*, vol. 2, pp. 86-92, Spring 1976.

R. E. Cummings, "Application and evaluation of an operational unsupervised multispectral classification technique for earth resources data," presented at the Symp. on Management and Utilization of Remote Sensing Data, Sioux Falls, SD, 1973.

J. L. Dragg and N. W. Naugle, "Photometric reduction of lunar orbiter video tapes," *SPIE J.*, vol. 9, pp. 159–165, July 1971.

J. A. Dunn, W. D. Stromberg, R. M. Ruiz, S. A. Collins, and T. E. Thorpe, "Maximum discriminability versions of the near encounter marine pictures," *J. Geophys. Res.*, vol. 76, p. 438, 1971.

Earth Resources Observation Systems Data Center, "Handling and processing systems for LANDSAT data," LANDSAT Data User Notes #1, 2–4, USGS/EROS Data Center, Sioux Falls, SD, 1973.

D. S. Eigen and R. A. Northhouse, "Feature selection based on discrete clustering techniques," in *Proc. Symp. on Applications of Minicomputers to Control and Robotics Systems*, Univ. Wisconsin-Milwaukee, 1973.

M. P. Ekstrom, "On the numerical feasibility of digital image restoration," *Proc. IEEE*, pp. 1155–1156, Aug. 1973.

L. A. Gambino and M. A. Crombie, "Digital mapping and digital image processing," *Photogram. Eng.*, vol. XL, no. 11, pp. 1295–1302, 1974.

L. A. Gambino and B. L. Schrock, "An experimental digital interactive facility," *IEEE Computer*, pp. 22–28, Aug. 1977.

J. V. Gaven, Jr., J. Tavitian, and P. A. Hollanda, "Relative effects of Gaussian and Poisson noise on subjective image quality," *Appl. Opt.*, vol. 10, pp. 2171–2178, Sept. 1971.

A. R. Gillespie and J. M. Soha, "An orthographic photomap of the south pole of Mars," *Icarus*, vol. 16, p. 522, 1972.

A. F. H. Goetz, F. C. Billingsley, E. Yost, and T. B. McCord, "Apollo 12 multispectral photography experiment," in *Proc. 2nd Lunar Science Conf.*, Suppl. 2, *Geochim. Cosmochim. Acta.*, vol. 3, MIT Press, 2301, 1971.

R. C. Gonzalez and P. Wintz, *Digital Image Processing*. Reading, MA: Addison-Wesley, 1978.

W. B. Green *et al.*, "Removal of instrument signatures from Mariner 9 television images of Mars," *Appl. Opt.*, vol. 14, p. 105, Jan. 1975.

T. J. Janssen, G. C. Kozlowski, and A. Luther, "Real time digital subtraction and enhancement of video pictures," *SPIE J.*, vol. 6, pp. 120–124, Apr./May 1968.

R. R. Jayroe, "Nearest neighbors, bilinear interpolation, and bicubic interpolation geographic correction effects on Landsat imagery," NASA TM X-73348, Marshall Space Flight Center, 1976.

B. Julesz, "Computers, patterns, and depth perception," *Bell Lab. Rec.*, pp. 261–266, Sept. 1966.

T. S. Huang, W. F. Schreiber, and O. J. Tretiak, "Image processing," *Proc. IEEE*, vol. 59, pp. 1586-1609, Nov. 1971.

B. R. Hunt, "Digital image processing (survey)," *Proc. IEEE*, vol. 63, pp. 693–708, 1975.

N. S. Kapany and J. J. Burke, "Various image assessment parameters," *J. Opt. Soc. Amer.*, vol. 52, pp. 1351--1361, 1962.

R. J. Kohler and H. K. Howell, "Photographic image enhancement by super-imposition of multiple images," *Photo. Sci. Eng.*, vol. 7, no. 4, pp. 241–245, 1963.

J. V. Lamar and P. M. Merrifield, "Pseudocolor transformation of ERTS imagery," in *Proc. Symp. on Significant Results Obtained from ERTS-1*, NASA SP-327, 1187, 1973.

D. A. Landgrebe, "Systems approach to the use of remote sensing," LARS Information Note 041571, Lab. for the Application of Remote Sensing, Purdue Univ., Lafayette, IN, 1971.

M. D. Levine, D. A. O'Handley, and G. M. Yagi, "Determination of depth maps," *Computer Graphics and Image Processing*, vol. 2, no. 2, 1973.

B. S. Lipkin and A. Rosenfeld, *Picture Processing and Psychopictorics.* New York: Academic.

Machine Processing of Remotely Sensed Data:
IEEE Catalog Number 77CH 1218-7 MPRSD 1977

IEEE Catalog Number 76CH 1103-1 MPRSD 1976
IEEE Catalog Number 75CH 1009-O-C 1975
IEEE Catalog Number 73CH 0834-2 GE 1973

D. F. Marble and D. J. Peugeut, "Computer software for spatial data handling: Current status and future developmental needs," Geographic Information Systems Lab. Research Rep. R-77/8, SUNY, Buffalo, 1977.

H. Markarian, R. Bernstein, D. G. Ferneyhough, L. E. Gregg, and F. S. Sharp, "Correction for high resolution images," *Photogram. Eng.*, pp. 1311–1320, 1973.

C. W. McMillan, F. C. Billingsley, and R. E. Frazer, "Fast scan EM with digital image processing as applied to dynamic experiments," *Wood Science*, vol. 6, no. 3, pp. 272–277, 1974.

S. W. Murphrey, R. D. Depew, and R. Bernstein, "Digital processing of conical scanner data," *Photogram. Eng. and Remote Sensing*, vol. 43, pp. 155–167, Feb. 1977.

H. H. Nagel, "Formation of an object concept by analysis of systematic time variations in an optically perceptible environment," Rep. Ifl-HH-B-27/76, Institut fur Informatik, D 2000 Hamburg 13, Schluterstrasse 70, Germany.

NASA LANDSAT Data Users Handbook, NASA/GSFC Document No. 76SDS4258, 1976.

R. Nathan, *Image Processing for Electron Microscopy: I. Enhancement Procedures*, *Advances in Optical and Elec. Microscopy*, vol. 4. New York: Academic, pp. 85–125.

A. V. Oppenheim, R. W. Schafer, and T. G. Stockham, Jr., "Nonlinear filtering of multiplied and convolved signals," *Proc. IEEE*, vol. 56, Aug. 1968.

Perkin-Elmer Corp., symposium on sampled images, Perkin-Elmer Corp., Optical Group, Norwalk, CT, 1971 (completed issue).

D. J. Peuguet, "Raster data handling in geographic information systems," Geographic Information Systems Lab. Research Rep. R-77/7, SUNY, Buffalo, 1977.

K. Preston, Jr. and P. E. Norgren, "Interactive image processor speeds pattern recognition by computer," *Electronics*, pp. 89–98, Oct. 23, 1972.

"Digital picture processing," *Proc. IEEE*, vol. 60, July 1972.

S. F. Rifman *et al.*, "Experimental study of digital image processing techniques for Landsat data: Final report," NAS5-20085, NASA/GSFC.

T. C. Rindfleisch, "Photometric method for lunar topography," *Photogram. Eng.*, vol. 32, pp. 262–276, Mar. 1966.

T. C. Rindfleisch, J. A. Dunne, H. J. Frieden, W. D. Stromberg, and R. M. Ruiz, "Digital processing of the Mariner 6 and 7 pictures," *J. Geophys. Res.*, vol. 76, p. 394, 1971.

A. Rosenfeld, "Progress in picture processing: 1969–71 (A bibliography)," *Computing Surveys*, vol. 5, no. 2, pp. 81–108, 1973.

A. Rosenfeld, "Picture processing (surveys)"
1972 Comp. Graphics and Image Proc.
1973 Comp. Graphics and Image Proc.
1974 Comp. Graphics and Image Proc., vol. 4, 1975, pp. 133–155
1975 Comp. Graphics and Image Proc., vol. 5, 1976, pp. 215–237

A. Rosenfeld and A. C. Kak, *Digital Picture Processing.* New York: Academic, 1976.

S. Ruben, R. Faiss, J. Lyons, and M. Quinn, "Applications of a parallel processing computer in LACIE," presented at the *Int. Conf. on Parallel Processing*, Waldenwoods, MI, Aug. 1976.

J. W. Schoonmaker, Jr., "Geometric evaluation of MSS images from ERTS-1," in *Proc. ASP 40th Annu. Meeting*, St. Louis, MO, 1974, pp. 582–587.

J. B. Seidman, "Some practical applications of digital filtering in image processing," in *Proc. Symp. on Computer Image Processing and Recognition*, Univ. Missouri-Columbia, Aug. 24–26, 1972.

R. H. Selzer, "The use of computers to improve biomedical image quality," in *Proc. Fall Joint Comput. Conf.*, 1968, pp. 817–834.

M. Svedlow, C. D. McGillem, and P. E. Anuta, "Experimental examination of similarity measures and preprocessing methods used for image registration," presented at the *Symp. on Machine Processing of Remote Sensing Data*, LARS, Lafayette, IN, June 1976.

J. E. Tabor, "Evaluation of digitally corrected ERTS images," in *Proc. 3rd ERTS-1 Symp.*, vol. I, sect. B, 1973, pp. 1837–1842.

V. L. Thomas, "Generation and physical characteristics of the Landsat-1 and 2 MSS computer compatible tapes," NASA/GSFC X-563-75-233, 1975.

TRW, "EROS digital image processing system: An overview," TRW Defense and Space Systems Group, Redondo Beach, CA.

J. D. Turinetti and O. W. Mintzer, "Computer analysis of imagery," *Photogram. Eng.*, pp. 501–505, 1973.

A. N. Williamson, "Corrected Landsat images using a small digital computer," *Photogram. Eng.*, vol. 43, no. 9, pp. 1153–1159, 1977.

K. W. Wong, "Geometric and cartographic accuracy of ERTS-1 imagery," *Photogram. Eng.*, vol. 41, no. 5, pp. 621–635, 1975.

Y. Yakimovsky, "On the problem of embedding picture elements in regions," Tech. Memo. 33-774, Jet Propulsion Lab., Pasadena, CA, June 1976.

Y. Yakimovsky, "Boundary and object detection in real world images," Tech. Memo. 33-709, Jet Propulsion Lab., Pasadena, CA, Nov. 1974.

Y. Yakimovsky and R. Cunningham, "A system for extracting 3-dimensional measurements from a stereo pair of TV cameras," *Computer Graphics and Image Processing*, 1977.

Multispectral Sensing and Image Analysis

J. D. Addington, "A hybrid classifier using the parallelepiped and Bayesian techniques," in *Proc. Amer. Soc. Photogrammetry 41st Annu. Meeting*, Mar. 1975, p. 727.

H. C. Andrews, *Introduction to Mathematical Techniques in Pattern Recognition.* New York: Wiley-Interscience, 1972.

J. N. P. Beers, "Collection and processing of multispectral data," Netherlands Interdepartmental Working Community for the Application and Research of Remote Sensing Techniques (NIWARS), Publication No. 4, 3 Kanaalweg, Delft, The Netherlands, 1972.

R. J. Blackwell and D. H. Boland, "The trophic classification of lakes using ERTS multispectral scanner data," in *Proc. Amer. Soc. Photogrammetry 41st Annu. Meeting*, Mar. 1975, p. 393.

C. A. Cato, *Basic Principles of Earth Resources Sensing, Remote Sensing of Earth Resources*, vol. 1, Univ. Tennessee, 1972, p. 64.

Cheng *et al.*, *Pictorial Pattern Recognition.* Washington, DC: Thompson, 1968.

A. P. Colvocoresses, "ERTS-A satellite imagery," *Photogram. Eng.*, vol. 36, no. 6, 1970.

A. P. Colvocoresses, "Proposed parameters for an operational Landsat," *Photogram. Eng.*, vol. 43, no. 9, pp. 1139–1145, 1977.

N. Congard, *Pattern Recognition.* New York: Spartan-Macmillan, 1970.

F. J. Doyle, "The next decade of satellite remote sensing," *Photogram. Eng.*, vol. 44, no. 2, pp. 155–164, 1978.

R. Duda and P. Hart, *Pattern Classification Scene Analysis.* New York: Wiley-Interscience, 1972.

Earth Resources Survey Symposium, NASA TM X-58168, 1975.

D. D. Egberg and F. T. Ulaby, "Effect of angles on reflectivity," *Photogram. Eng.*, pp. 556–564, 1972.

K. S. Fu, *Sequential Methods in Pattern Recognition and Machine Learning.* New York: Academic, 1968.

K. Fukunaga, *Introduction to Statistical Pattern Recognition.* New York: Academic, 1972.

A. F. H. Goetz, F. C. Billingsley, A. R. Gillespie, M. J. Abrams, R. L. Squires, E. M. Schoemaker, I. Lucchitta, and D. P. Elston, NASA TR-32-1597, "Application of ERTS images and image processing to regional geologic problems and geologic mapping in Northern Arizona," Jet Propulsion Lab., May 15, 1975.

I. L. Goldberg, "Design considerations for a multispectral scanner for ERTS," in *Proc. Symp. of Inform. Processing*, Purdue Univ., West Lafayette, IN, 1969.

M. Goldberg and S. Shlein, "A clustering scheme for multispectral images," *IEEE Trans Syst., Man, Cybern.*, vol. SMC-8, Feb. 1978.

G. J. Grebowsky, *Characteristics of Digital Multispectral Scanner Data*, NASA/GSFC X-563-75-169, 1975.

R. A. Holmes, "Data requirements and data processing earth resources surveys," *SPIE J.*, vol. 9, pp. 52–56, 1971.

M. R. Holter and F. C. Polcyn, "Comparative multispectral sensing," Infrared and Optics Lab., Univ. Michigan, Rep. 2900-484-5, 1964.

P. L. Johnson, *Remote Sensing in Ecology.* Atlanta: Univ. Georgia Press.

L. N. Kanal, *Pattern Recognition.* Washington, DC: Thompson, 1968.

R. J. Kauth and G. S. Thomas, "The tasselled cap—A graphic description of the spectral-temporal development of agricultural crops as seen by Landsat," in *Proc. Symp. on Machine Processing of Remotely Sensed Data*, Purdue Univ., West Lafayette, IN, 1976, pp. 4B-41–4B-51.

R. H. Kidd and R. H. Wolfe, "Performance modeling of earth resources remote sensors," *IBM J. Res. Develop.*, pp. 29–39, 1976.

E. B. Knipling, "Physical and physiological basis for the reflectance of visible and near-infrared radiation from vegetation," *Remote Sensing of Environment*, pp. 155–159, 1970.

Laboratory for Agricultural Remote Sensing: Remote Multispectral Sensing in Agriculture:
- vol. 1, Annual Report, Agricultural Experiment Station Research Bulletin No. 831, 1967
- vol. 2, Bulletin No. 832, 1967
- vol. 3, Bulletin No. 844, 1968
- vol. 4, Bulletin No. 873, 1970

P. A. Kolers and M. Eden, *Recognizing Patterns, Studies in Living and Automatic Systems.* Cambridge, MA: MIT Press.

J. Lintz, Jr. and D. S. Simonett, Eds., *Remote Sensing of Environment.* Reading, MA: Addison-Wesley, 1976.

D. S. Lowe, J. G. Braithwaite, and V. L. Larrowe, "An investigative study of a spectrum-matching imaging system," Willow Run Lab., Univ. Michigan, Rep. 8201-1-F, 1966.

T. J. Lynch, *Data Compression Requirements for the Landsat Follow-On Mission*, NASA/GSFC X-930-76-55, 1976a.

T. J. Lynch, *Ground Data Handling for Landsat-D,* NASA/GSFC X-930-76-264.

R. J. P. Lyon, "The multiband approach to geological mapping from orbiting satellites: Is it redundant or vital?," *Remote Sensing of Environment*, vol. 1, pp. 155–159, 1970.

R. B. MacDonald, F. G. Hall, R. B. Erb, P. J. Waite, R. I. Dideriksen, and J. F. Murphy, "The large area crop inventory experiment (LACIE)," in *Proc. NASA Earth Resources Survey Symp.*, NASA TM X-58168, 1975, pp. 43-74.

T. H. Maugh, II, "ERTS: Surveying Earth's resources from space," *Science*, vol. 180, p. 49, Apr. 1973.

W. S. Meisel, *Computer-Oriented Approaches to Pattern Recognition.* New York: Academic, 1972.

J. M. Mendel and K. S. Fu, *Adaptive, Learning and Pattern Recognition Systems: Theory and Applications.* New York: Academic, 1970.

M. Minsky and S. Papers, *Perceptrons.* Cambridge, MA: MIT Press, 1968.

G. Nagy, "State of the art in pattern recognition," *Proc. IEEE*, vol. 56, pp. 836-862, 1968.

Proc. NASA Earth Resources Symp., Houston, TX, NASA Publication TM X-58168, June 1975.

NASA-Landsat Data Users Handbook, NASA Document 76SDS4258, Goddard Space Flight Center. Replaces earlier *ERTS Data Users Handbook*, Doc. 715D4249, 1976.

V. T. Norwood, "Optimilization of a multispectral scanner for ERTS," in *Proc. 6th Int. Symp. on Remote Sensing*, Univ. Michigan, Ann Arbor, 1969.

J. Otterman and R. S. Fraser, "Earth-atmosphere system and surface reflectivities in and regions from Landsat MSS data," *Remote Sensing of Environment*, vol. 5, pp. 247-266, 1976.

J. J. Palgen, "International bibliography of pictoral pattern recognition," Allied Research Associates, Inc., Concord, MA, Apr. 1970.

E. Patrick, *Statistical Pattern Recognition.* Englewood Cliffs, NJ: Prentice-Hall, 1972.

Proc. 4th Int. Symp. on Remote Sensing of Environment, Inst. of Science and Technology, Univ. Michigan, Ann Arbor, 1966.

Proc. 5th Int. Symp. on Remote Sensing of Environment, Inst. of Science and Technology, Univ. Michigan, Ann Arbor, 1968.

Proc. 6th Int. Symp. on Remote Sensing of Environment, Inst. of Science and Technology, Univ. Michigan, Ann Arbor, 1969.

Proc. 7th Int. Symp. on Remote Sensing of Environment, Inst. of Science and Technology, Univ. Michigan, Ann Arbor, 1971.

Proc. 8th Int. Symp. on Remote Sensing of Environment, Inst. of Science and Technology, Univ. Michigan, Ann Arbor, 1972.

Proc. 9th Int. Symp. on Remote Sensing of Environment, Inst. of Science and Technology, Univ. Michigan, Ann Arbor, 1974.

Proc. 38th Annu. Meeting of the Amer. Society of Photogrammetry, Washington, DC, 1972.

L. C. Rowan, "Application of satellites to geologic exploration," *Amer. Sci.*, vol. 63, pp. 393-403, 1975.

L. C. Rowan, P. H. Wetlanfer, A. F. H. Goetz, F. C. Billingsley, and J. H. Stewart, "Discrimination of rock types and detection of hydrothermally altered areas in South Central Nevada by the use of computer—Enhanced ERTS images," U.S. Geological Society Professional Paper 883, U.S. Government Printing Office, 1974.

G. S. Sebestyen, *Decision Making Processes in Pattern Recognition.* New York: Macmillan.

Proc. 2nd Int. Joint Conf. on Pattern Recognition, IEEE Catalog No. 74CH0885-46, 1974.

F. Shahrocki, *Remote Sensing of Earth Resources*, vol. 1, Univ. Tennessee, Tullahoma, 1972. (Also subsequent symposia.)

B. S. Siegal and M. J. Abrams, "Comparison and analysis of some classification techniques applied to geologic mapping using Landsat data," *Photogram. Eng.*, vol. XLII, pp. 325-337, Mar. 1976.

D. S. Simonett, T. R. Smith, W. Tobler, D. G. Marks, J. E. Frew, and J. C. Dozier, "Geobase information system impacts on space image formats," Univ. California, Santa Barbara, SBRSU Tech. Rep. 3, NASA Contract NASW-3118, Apr. 1978.

P. H. Swain, "Land use classification and mapping by machine assisted analysis of Landsat multispectral scanner data," LARS Information Note 111276, Purdue Univ., West Lafayette, IN, 1976.

J. T. Tou, *Pattern Recognition Principles.* Reading, MA: Addison-Wesley, 1978.

Proc. Symp. on Significant Results Obtained from the Earth Resources Technology Satellite-1, NASA SP-327, 1973.

L. Uhr, *Pattern Recognition.* New York: Wiley, 1966.

R. K. Vincent and F. S. Thomson, "Rock type discrimination from ratioed infrared scanner images of pigsah crater," *California Sci.*, vol. 175, pp. 986-988, 1972.

S. Watanabe, *Methodologies of Pattern Recognition.* New York: Academic, 1969.

S. G. Wheeler and D. S. Ingram, "Approximations for the probability of misclassification," *Pattern Recognition*, vol. 8, pp. 121-128, 1976.

S. G. Wheeler and P. N. Misra, "Linear dimensionality of LANDSAT agricultural data with implications for classification," in *Proc. Annu. Purdue Conf. on Remote Sensing*, 1976, pp. A1-A9.

W. G. Woo and E. G. Donelan, "A multispectral point scan camera for earth resources technology satellite," AIAA Paper No. 70-319.

E. M. Zaitzeff, C. L. Wilson, and D. H. Ebert, "MSDS—An experimental 24-channel multispectral scanner system," *Bendix Tech. J.*

Sources of Earth Observation Data[1]

Landsat Satellite Data

Landsat film and digital data are available through the Department of Interior EROS (Earth Resources Observations System) Data Center (EDC), Sioux Falls, SD. EDC is operated by the U.S. Geological Survey and is an integral part of the National Cartographic Information Center (NCIC). Orders for data should be directed to EDC:

User Services
EROS Data Center
Sioux Falls, SD 57198
Phone: (605) 594-6411 X151
FTS 784-7511

or by computer terminal link from the following regional field offices:

Topographic Office
U.S. Geological Survey
900 Pine Street
Rolla, MO 65401
Phone: (314) 364-3680
Hours: 7:45–4:15

NCIC Information Unit
National Center—Stop 507
12201 Sunrise Valley Drive
Reston, VA 22092
Phone: (703) 860-6045
Hours: 7:45–4:15

Air Photo Sales
U.S. Geological Survey
Federal Center, Building #25
Denver, CO 80225
Phone: (303) 234-2326
Hours: 7:45–4:15

Map and Air Photo Sales
U.S. Geological Survey
345 Middlefield Road
Menlo Park, CA 94025
Phone: (415) 323-2157
Hours: 7:45–8:15

EROS Applications Assistance Facilities maintain microfilm copies of data archived at the Center and provide computer terminal inquiry and order capability to the EROS Data Center. The facility addresses are:

EROS Applications Branch
U.S. Geological Survey
Room 202, Building 3
345 Middlefield Road
Menlo Park, CA 94025
Phone: (415) 323-8111

EROS Applications Assistance Facility
EROS Data Center
U.S. Geological Survey
Sioux Falls, SD 57198
Phone: (605) 594-6511

U.S. Geological Survey
Room B-207-A, Building 1100
National Space Technology Laboratories
Bay St. Louis, MS 39520
Phone: (601) 688-3541

EROS Applications Assistance Facility
University of Alaska
Geophysical Institute
College, Fairbanks, AK 99701
Phone: (907) 470-7558
Hours: 8:00–5:00

EROS Applications Assistance Facility
HQ Inter American Geodetic Survey
Headquarters Building
Drawer 934
Fort Clayton, Canal Zone
Phone: 83-3897

EROS Applications Assistance Facility
U.S. Geological Survey
Suite 1880
Valley Bank Center
Phoenix, AZ 85073
Phone: (602) 261-3188

EROS Applications Assistance Facility
U.S. Geological Survey
1925 Newton Square East
Reston, VA 22090
Phone: (703) 860-7871
FTS: (900) 860-7871

EROS Data Reference Files have been established throughout the United States which maintain microfilm copies of data available from the EROS Data Center. The addresses are as follows:

EROS Data Reference File
Public Inquiries Office
U.S. Geological Survey
108 Skyline Building
508 Second Avenue
Anchorage, AK 99501
Phone: (907) 277-0577

EROS Data Reference File
Public Inquiries Office
U.S. Geological Survey
Room 7638, Federal Building

[1] Extracted from D. S. Simonett *et al.*, "Geobase information system impacts on space image formats," Santa Barbara Remote Sensing Unit, Univ. California, SBRSU Tech. Rep. 3, NASA Contract NASW-3118, 1978.

300 North Los Angeles Street
Los Angeles, CA 90012
Phone: (213) 688-2850

EROS Data Reference File
Water Resources Division
U.S. Geological Survey
Room 343, Post Office and Court
House Building
Albany, NY 12201
Phone: (518) 474-3107 or 6042

EROS Data Reference File
Bureau of Land Management
729 NE Oregon Street
Portland, OR 97208
Phone: (503) 234-3361 X400

EROS Data Reference File
Topographic Office
U.S. Geological Survey
900 Pine Street
Rolla, MO 65401
Phone: (314) 364-3680

EROS Data Reference File
Water Resources Division
U.S. Geological Survey
975 West Third Avenue
Columbus, OH 43212
Phone: (614) 469-5553

EROS Data Reference File
University of Hawaii
Department of Geography
Room 313C, Physical Science
Building
Honolulu, HI 96825
Phone: (808) 9-4-8643

EROS Data Reference File
Maps and Surveys Branch
Tennessee Valley Authority
20 Haney Building
311 Broad Street
Chattanooga, TN 37401
Phone: (615) 755-2149

In addition to the regular Landsat Facilities, the NOAA Satellite Data Services Branch (SDSB) has 16 mm browse files of Landsat data. There are 21 SDSB browse files:

University of Alaska
Arctic Environmental Information
and Data Center
142 East Third Avenue
Anchorage, AK 99501
Phone: (907) 279-4523

Lake Survey Center—CLx13
630 Federal Building & U.S.
Courthouse
Detroit, MI 48226
Phone: (313) 226-6126

Inter-American Tropical Tuna
Commission
Scripps Institute of Oceanography
Post Office Box 109
LaJolla, CA 92037
Phone: (714) 453-2820

National Geophysical and Solar
Terrestrial Data Center
Solid Earth Data Service Branch
Boulder, CO 80302
Phone: (303) 499-1000 X 6915

National Oceanographic Data Center
Environmental Data Service
2001 Wisconsin Avenue
Washington, DC 20235
Phone: (202) 634-7510

Atlantic Oceanographic and
Meteorological Laboratories
15 Rickenbacker Causeway,
Virginia Key
Miami, FL 33149
Phone: (305) 361-3361

National Weather Service,
Pacific Region
Bethel-Pauaha Building, WFP 3
1149 Bethel Street
Honolulu, HI 96811
Phone: (808) 841-5028

National Ocean Survey—C3415
Building #1, Room 526
6001 Executive Boulevard
Rockville, MD 20852
Phone: (301) 496-8601

Atmospheric Sciences Library—D821
Gramax Building, Room 526
8060 13th Street
Silver Spring, MD 20910
Phone: (301) 427-7800

National Environmental Satellite
Service
Environmental Sciences Group
Suitland, MD 20233
Phone: (301) 673-5981

National Weather Service,
Central Region
601 East 12th Street
Kansas City, MO 64106
Phone: (816) 374-5672

National Weather Service,
Eastern Region
585 Stewart Avenue
Garden City, NY 11530
Phone: (516) 248-2105

National Climatic Center
Federal Building
Asheville, NC 28801
Phone: (704) 258-2850 X620

National Severe Storms Lab
1313 Halley Circle
Normal, OK 73069
Phone: (405) 329-0388

Remote Sensing Center
Texas A & M University
College Station, TX 77843
Phone: (713) 845-5422

National Weather Service,
Southern Region
819 Taylor Street
Fort Worth, TX 76102
Phone: (817) 334-2671

National Weather Service,
Western Region
125 South State Street
Salt Lake City, UT 84111
Phone: (801) 524-5131

Atlantic Marine Center—CAMO2
439 West York Street
Norfolk, VA 23510
Phone: (804) 441-6201

Northeast Fisheries Center
Post Office Box 6
Woods Hole, MA 02543
Phone: (617) 548-5123

University of Wisconsin
Office of Sea Grant
610 North Walnut Street
Madison, WI 53705
Phone: (608) 263-4836

Northwest Marine Fisheries Center
2725 Montlake Boulevard East
Seattle, WA 98112
Phone: (206) 442-4760

NOAA Satellite Data

NOAA data are used primarily for meteorological studies and are of a much coarser spatial resolution, although higher radiometric resolution (8-bit versus 6- or 7-bit quantization), and are of interest to earth resource monitors. Real time data products may be ordered and sent by mail or user's communication links by contacting:

National Environmental Satellite Service (NESS)
Director of Operations, FOB-4
Washington, DC 20233

Archived data products may be obtained from:

National Climatic Center (NCC)
Satellite Data Services Branch
Rm. 606, World Weather Bldg.
Washington, DC 20233

Author Index

Subject Index

R

S

T

V

W

Editor's Biography

Ralph Bernstein (S'55-M'57-SM'68) received the B.S. degree in electrical engineering from the University of Connecticut, Storrs, and the M.S. degree, also in electrical engineering, from Syracuse University, Syracuse, NY.

He is a Senior Engineer and Manager of the IBM Federal Systems Division Advanced Image Processing Analysis and Development Department, Gaithersburg, MD. He is responsible for the design, development, and implementation of image processing technology at FSD. He was a Principal Investigator on the NASA Earth Resources Technology Satellite (ERTS), now LANDSAT Program, within which he demonstrated the feasibility of digitally processing Landsat sensor data and correcting and registering the data to subpicture element accuracies. He has led the requirements definition, design, and implementation of various image processing systems, and has directed efforts dealing with image processing research and development. He has also performed a number of digital information extraction experiments on Landsat data. His experience also includes the analysis, computer simulation, design, and development of various systems. He has written a number of papers in the areas of geophysical data processing, automatic digital control systems, radio navigation systems, aircraft simulation and control, and digital processing of aerospace imagery. He is a contributor to the book, *Pollution—Engineering and Scientific Solutions*; to the book, *Geoscience Electronics*; and to the American Society of Photogrammetry's *Manual of Remote Sensing*. He has several patents granted and published. He is also a Lecturer at the George Washington University School of Engineering and Applied Science, Continuing Engineering Education Program.

Mr. Bernstein is a member of the American Society of Photogrammetry and a past member of the Administrative Committee of the IEEE Geoscience Electronics Group. He served as Chairman of the 1971 IEEE International Geoscience Electronics Symposium, and is Past Chairman of the Washington, DC Chapter of the IEEE Geoscience Electronics Group. He is a member of the Space Science Board of the National Research Council of the National Academy of Science. He received the NASA Medal for Exceptional Scientific Achievement and the IBM Outstanding Contribution Award, and is listed in *American Men and Women of Science.*